FPGA-Based Embedded System Developer's Guide

FPGA-Based Embedded System Developer's Guide

A. Arockia Bazil Raj

CRC Press is an imprint of the
Taylor & Francis Group, an **informa** business

MATLAB®, Simulink®, and Xilinx® are trademarks of the Math Works, Inc. and are used with permission. The Mathworks does not warrant the accuracy of the text or exercises in this book. This book's use or discussion of MATLAB®, Simulink®, and Xilinx® software or related products does not constitute endorsement or sponsorship by the Math Works of a particular pedagogical approach or particular use of the MATLAB®, Simulink®, and Xilinx® software.

CRC Press
Taylor & Francis Group
6000 Broken Sound Parkway NW, Suite 300
Boca Raton, FL 33487-2742

© 2018 by Taylor & Francis Group, LLC
CRC Press is an imprint of Taylor & Francis Group, an Informa business

No claim to original U.S. Government works

Printed on acid-free paper

International Standard Book Number-13: 978-1-4987-9675-0 (Hardback)

This book contains information obtained from authentic and highly regarded sources. Reasonable efforts have been made to publish reliable data and information, but the author and publisher cannot assume responsibility for the validity of all materials or the consequences of their use. The authors and publishers have attempted to trace the copyright holders of all material reproduced in this publication and apologize to copyright holders if permission to publish in this form has not been obtained. If any copyright material has not been acknowledged please write and let us know so we may rectify in any future reprint.

Except as permitted under U.S. Copyright Law, no part of this book may be reprinted, reproduced, transmitted, or utilized in any form by any electronic, mechanical, or other means, now known or hereafter invented, including photocopying, microfilming, and recording, or in any information storage or retrieval system, without written permission from the publishers.

For permission to photocopy or use material electronically from this work, please access www.copyright.com (http://www.copyright.com/) or contact the Copyright Clearance Center, Inc. (CCC), 222 Rosewood Drive, Danvers, MA 01923, 978-750-8400. CCC is a not-for-profit organization that provides licenses and registration for a variety of users. For organizations that have been granted a photocopy license by the CCC, a separate system of payment has been arranged.

Trademark Notice: Product or corporate names may be trademarks or registered trademarks, and are used only for identification and explanation without intent to infringe.

Library of Congress Cataloging-in-Publication Data

Names: Raj, A. Arockia Bazil, author.
Title: FPGA-Based Embedded System Developer's Guide / A. Arockia Bazil Raj.
Description: Boca Raton : Taylor & Francis, CRC Press, 2018. | Includes bibliographical references and index.
Identifiers: LCCN 2017045457| ISBN 9781498796750 (hardback : alk. paper) | ISBN 9781315156200 (ebook)
Subjects: LCSH: Embedded computer systems--Design and construction. | Field programmable gate arrays--Programming. | VHDL (Computer hardware description language)
Classification: LCC TK7895.E42 R347 2018 | DDC 006.2/2--dc23
LC record available at https://lccn.loc.gov/2017045457

Visit the Taylor & Francis Web site at
http://www.taylorandfrancis.com

and the CRC Press Web site at
http://www.crcpress.com

To my mother, A. Jothy Mary, and my father, A. Anthoni Samy;

To my wife, A. Johnsi Rosita, and my daughter, A. Feona Hydee Mitchell; and

To Mr. J. Niranjan Samuel, who has admirably and unreservedly extended his helping hand and support throughout this book.

Contents

List of Figures xiii
List of Tables xxv
List of Abbreviations xxix
Preface xxxv
Author xxxix

Part I Basic System Modeling and Programming Techniques

1. Very-Large-Scale Integration Technology: History and Features 3
1.1 Introduction and Preview 3
1.2 A Review of Microelectronics 4
1.3 Complementary Metal Oxide Semiconductor Technology and Gate Configuration 7
1.4 CMOS Fabrication and Layout 10
1.5 VLSI Design Flow 12
1.6 Combinational and Sequential Circuit Design 14
1.6.1 Combinational Logic Circuits 14
1.6.2 Sequential Logic Circuits 15
1.7 Subsystem Design and Layout 15
1.8 Types of Application-Specific Integrated Circuits and Their Design Flow 17
1.9 VHDL Requirements and Features 19
Laboratory Exercises 21

2. Digital Circuit Design with Very-High-Speed Integrated Circuit Hardware Description Language 23
2.1 Introduction and Preview 23
2.2 Code Design Structures 24
2.3 Data Types and Their Conversions 34
2.4 Operators and Attributes 45
2.5 Concurrent Code 50
2.6 Sequential Code 55
2.7 Flip-Flops and Their Conversions 65
2.7.1 Flip-Flops 65
2.7.2 Flip-Flop Conversions 70
2.8 Data Shift Registers 77
2.9 Multifrequency Generator 84
Laboratory Exercises 89

3. Simple System Design Techniques 91
3.1 Introduction 91
3.2 Half and Full Adder 91
3.3 Half and Full Subtractor 96
3.4 Signed Magnitude Comparator 98

3.5 Seven-Segment Display Interfacing 104
3.6 Counter Design and Interfacing 107
3.7 Digital Clock Design and Interfacing 115
3.8 Pulse Width Modulation Signal Generation 123
3.9 Special System Design Techniques 127
3.9.1 Packages and Libraries 127
3.9.2 Functions and Procedures 131
Laboratory Exercises 134

4. Arithmetic and Logical Programming 135
4.1 Introduction and Preview 135
4.2 Arithmetic Operations: Adders and Subtractors 136
4.2.1 Serial Adder 137
4.2.2 Parallel and Pipelined Adders 141
4.2.3 Subtractors 146
4.3 Arithmetic Operations: Multipliers 152
4.4 Arithmetic Operations: Dividers 163
4.5 Trigonometric Computations Using the COordinate Rotation DIgital Computer (CORDIC) Algorithm 170
4.6 Multiply-Accumulation Circuit 185
4.7 Arithmetic and Logical Unit 188
4.8 Read-Only Memory Design and Logic Implementations 194
4.9 Random Access Memory Design 200
Laboratory Exercises 203

Part II Custom Input/Output Peripheral Interfacing

5. Input/Output Bank Programming and Interfacing 207
5.1 Introduction and Preview 207
5.2 Optical Display Interfacing 208
5.2.1 Light-Emitting Diode Displays 208
5.2.2 Multisegment Display 210
5.3 Buzzer Control 215
5.4 Liquid Crystal Display Interfacing and Programming 217
5.4.1 Liquid Crystal Display 217
5.4.2 Graphical Liquid Crystal Display 223
5.5 General-Purpose Switch Interfacing 228
5.5.1 Dual Inline Package Switch 228
5.5.2 Bidirectional Port/Switch Design 231
5.5.3 Matrix Keypad Interfacing 233
5.6 Dual-Tone Multifrequency Decoder 235
5.7 Optical Sensor Interfacing 238
5.7.1 Infrared Sensors 238
5.7.2 Proximity Sensor 242
5.8 Special Sensor Interfacing 244
5.8.1 Passive Infrared Sensor 245
5.8.2 Metal Detector 248
5.8.3 Light-Dependent Resistor 251

5.9 Wind-Speed Sensor Interfacing 255
Laboratory Exercises 259

6. System Design with Finite and Algorithmic State Machine Approaches 261
6.1 Introduction and Preview 261
6.2 Finite State Machine Design: Moore and Mealy Models 261
6.2.1 Moore Finite State Machine Design 269
6.2.2 Mealy Finite State Machine Design 271
6.2.3 Finite State Machine Model Conversion 276
6.3 Code Classifier and Binary to Binary-Coded Decimal Converters 277
6.3.1 Input Code Classifier 278
6.3.2 Binary-to-Binary-Coded-Decimal Converter and Its Arithmetic 282
6.4 Binary Sequence Recognizer 286
6.5 Vending Machine Controller 293
6.6 Traffic Light Controller 301
6.7 Escalator, Dice Game and Model Train Controller Designs 309
6.7.1 Escalator Controller Design 309
6.7.2 Dice Game Controller Design 314
6.7.3 Electronic Model Train Controller Design 322
6.8 Algorithmic State Machine Charts 328
6.9 Algorithmic State Machine-Based Digital System Design 332
Laboratory Exercises 336

7. Interfacing Digital Logic to the Real World: Sensors, Analog to Digital, and Digital to Analog 339
7.1 Introduction and Preview 339
7.2 Basics of Signal Conditioning for Sensor Interfacing 339
7.2.1 Analog to Digital Conversion 340
7.2.2 Digital to Analog Conversion 341
7.3 Principles of Sensor Interfacing and Measurement Techniques 344
7.3.1 Optical Power Measurement 344
7.3.2 Temperature Measurement 346
7.3.3 Strain Measurement 349
7.3.4 Magnitude Comparator 350
7.3.5 ADC0804 Interfacing 351
7.4 Universal Asynchronous Receiver-Transmitter Design 356
7.4.1 Serial Communication: Data Reading/Writing Using MATLAB® 358
7.4.2 UART: Transmitter Design 365
7.4.3 Universal Asynchronous Receiver-Transmitter Receiver Design 371
7.5 Multichannel Data Logging 378
7.5.1 ADC0808/ADC0809 Interfacing 379
7.5.2 ADC0848 Interfacing 384
7.5.3 Analog to Digital Converter MAX1112 Interfacing and Serial Data Fetching 396
7.6 Bipolar Signal Conditioning and Data Logging 399
7.6.1 Bidirectional Analog to Digital Converter Interfacing 399
7.6.2 Opto-Electronic Position Detector Interfacing 403
7.7 Encoder/Decoder Interfacing for Remote Control Applications 411
7.7.1 Optical Tx/Rx Wireless Control 411

7.7.2 Radio Frequency Transmitter/Receiver Wireless Control 412
7.8 Pseudorandom Binary Sequence Generator and Time-Division Multiple Access 415
7.8.1 Serial Pseudorandom Binary Sequence Generator 415
7.8.2 Parallel Pseudorandom Binary Sequence Generator 418
7.8.3 Kasami Sequence Generator 422
7.8.4 Analog Time Division Multiplexing and M-Array Pulse Amplitude Modulation 424
7.9 Signal Generator Design and Interfacing 426
7.9.1 Low Voltage Digital to Analog Conversion Using DAC0808 426
7.9.2 High-Voltage Digital to Analog Conversion Using DAC7728 434
Laboratory Exercises 438

Part III Hardware Accelerated Designs

8. Real-Time Clock and Interface Protocol Programming 441
8.1 Introduction and Preview 441
8.2 Real-Time Clock (DS12887) Interface Programming 442
8.3 Inter-Integrated Circuit Interface Programming 449
8.4 Two-Wire Interface (SHT11 Sensor) Programming 456
8.5 Serial Peripheral Interface (SCP1000D) Programming 461
8.6 Global System for Mobile Communications Interface Programming 467
8.7 Global Positioning System Interface Programming 478
8.8 Personal System/2 Interface Programming 480
8.9 Video Graphics Array Interface Programming 484
Laboratory Exercises 492

9. Real-World Control Device Interfacing 493
9.1 Introduction and Preview 493
9.2 Relay, Solenoid Valve, Opto-Isolator, and Direct Current Motor Interfacing and Control 493
9.2.1 Relay Control 494
9.2.2 Solenoid Valve Control 497
9.2.3 Opto-Isolator Interfacing 501
9.2.4 Direct Current Motor Control 504
9.3 Servo and BLDC Motor Interfacing and Control 508
9.3.1 Servo Motor Control 508
9.3.2 Brushless Direct Current Motor Control 511
9.4 Stepper Motor Control 515
9.5 Liquid/Fuel Level Control 518
9.6 Voltage and Current Measurement 522
9.7 Power Electronic Device Interfacing and Control 526
9.8 Power Electronics Bidirectional Switch Interfacing and Control 532
9.8.1 Triac Control 532
9.8.2 Diac Controller 536
9.9 Real-Time Process Controller Design 538
Laboratory Exercises 544

10. Floating-Point Computations with Very-High-Speed Integrated Circuit Hardware Description Language and Xilinx System Generator (SysGen) Tools ... 547
10.1 Introduction and Preview ... 547
10.2 Representation of Fixed and Floating-Point Binary Numbers ... 549
10.2.1 Fixed-Point Number System ... 549
10.2.2 IEEE754 Single-Precision Floating-Point Number System ... 553
10.2.3 IEEE754 Double-Precision Floating Point Number System ... 556
10.2.4 Customized Floating-Point Number System ... 558
10.3 Floating-Point Arithmetic ... 561
10.3.1 Floating-Point Addition ... 562
10.3.2 Floating-Point Subtraction ... 565
10.3.3 Floating-Point Multiplication ... 568
10.3.4 Floating-Point Division ... 573
10.4 Xilinx System Generator (SysGen) Tools ... 575
10.4.1 Use and Interfacing Methods of Some Blocksets ... 577
10.4.2 System Design and Implementation Using SysGen Tools ... 583
10.5 Fractional-Point Computation Using SysGen Tools ... 594
10.6 System Engine Model Using Xilinx Simulink Block Sets ... 603
10.7 MATLAB® Code Interfacing with SysGen Tools ... 610
10.8 Very-High-Speed Integrated Circuit Hardware Description Language Code Interfacing with SysGen Tools ... 614
10.9 Real-Time Verification and Reconfigurable Architecture Design ... 620
10.9.1 Design Flow for Hardware Co-Simulation ... 620
10.9.2 Reconfigurable Architecture Design ... 622
Laboratory Exercises ... 624

Part IV Miscellaneous Design and Applications

11. Digital Signal Processing with Field-Programmable Gate Array ... 627
11.1 Introduction and Preview ... 627
11.2 Discrete Fourier Transform ... 628
11.3 Digital Finite Impulse Response Filter Design ... 641
11.4 Digital Infinite Impulse Response Filter Design ... 644
11.5 Multirate Signal Processing ... 650
11.6 Modulo Adder and Residual Number Arithmetic Systems ... 663
11.6.1 Modulo 2^n and 2^n-1 Adder Design ... 663
11.6.2 Residue Number System ... 665
11.7 Distributed Arithmetic-Based Computations ... 672
11.8 Booth Multiplication Algorithm and Design ... 684
11.9 Adaptive Filter/Equalizer Design ... 690
Laboratory Exercises ... 698

12. Advanced SysGen-Based System Designs ... 701
12.1 Introduction and Preview ... 701
12.2 Fast Fourier Transform Computation Using SysGen Design ... 701
12.3 Finite Impulse Response Digital Filter Design ... 704
12.4 Infinite Impulse Response Digital Filter Design ... 710

12.5 Multiply Accumulation Finite Impulse Response Filter Using SysGen Design 712
12.6 Cascaded Integrator Comb Filter Design 715
12.7 COordinate Rotation DIgital Computer Design Using SysGen Tools 719
12.8 Image Processing Using Discrete Wavelet Transform 722
12.9 Very-High-Speed Integrated Circuit Hardware Description Language Design Debugging Techniques 727
12.9.1 ChipScope Pro Analyzer 728
12.9.2 Very-High-Speed Integrated Circuit Hardware Description Language Test-Bench Design 729
12.9.3 Data/Text File Reading/Writing 732
Laboratory Exercises 739

13. Contemporary Design and Applications 741
13.1 Introduction and Preview 741
13.2 Differential Pulse Code Modulation System Design 741
13.3 Data Encryption System 744
13.4 Soft Computing Algorithms 750
13.4.1 Artificial Neural Network 751
13.4.2 Fuzzy Logic Controller 753
13.5 Bit Error Rate Tester Design 755
13.6 Optical Up/Down Data Link 761
13.7 Channel Coding Techniques 766
13.7.1 Linear Block Code 767
13.7.2 Convolutional Code 769
13.8 Pick-and-Place Robot Controller 773
13.9 Audio Codec (AC97) Interfacing 775
Laboratory Exercises 781

Appendix A 783

References 791

Index 797

List of Figures

Figure 1.1 Switch models of MOS transistors: nMOS switch (a) and pMOS switch (b) 7

Figure 1.2 Symbol, circuit structure, and truth table of a CMOS inverter 8

Figure 1.3 CMOS configuration of OR (a) and AND (b) gates 9

Figure 1.4 CMOS configuration of NAND (a) and NOR (b) gates with their truth tables ...9

Figure 1.5 Illustrations of CMOS fabrication process 11

Figure 1.6 VLSI design flow 12

Figure 1.7 Configuration of combinational (a) and sequential (b) circuit 14

Figure 1.8 Circuit of a full adder 16

Figure 1.9 Low-level implementation of a full adder circuit for sum (left) and for carry (right) 16

Figure 1.10 Classification of ASICs....17

Figure 1.11 Typical ASIC design flow 18

Figure 1.12 VHDL design environment 20

Figure 1.13 Configuration of parallelism using FPGA: x(n) is I/P samples, Z^{-1} is the unit delay element, $\otimes$ is the binary multiplier, $\oplus$ is the binary adder, and C_0, C_1, ... C_n are filter coefficients. 21

Figure 2.1 Symbolic representation of FFs: SR-FF (a), JK-FF (b), D-FF (c), and T-FF (d)....66

Figure 2.2 General design of SR-FF to JK-FF conversion....70

Figure 2.3 K-map simplification and circuit diagram for SR-FF to JK-FF conversion 71

Figure 2.4 K-map simplification and circuit diagram for JK-FF to SR-FF conversion 72

Figure 2.5 K-map simplification and circuit diagram for SR-FF to D-FF conversion....73

Figure 2.6 K-map simplification and circuit diagram for D-FF to SR-FF conversion 73

Figure 2.7 K-map simplification and circuit diagram for JK-FF to T-FF conversion 74

Figure 2.8 K-map simplification and circuit diagram for JK-FF to D-FF conversion....75

Figure 2.9 K-map simplification and circuit diagram for D-FF to JK-FF conversion 76

Figure 2.10 Design of a SISO shift register (a), SIPO shift register (b), PISO shift register (c), and PIPO shift register (d) 77

Figure 2.11 Design of a bidirectional shift register....79

Figure 2.12 Circuit diagram of a 4-bit bidirectional universal shift register 80

Figure 2.13 Clock divider circuit with D-FF (a) and clock divider output (b) 85

Figure 2.14 Counter-based clock divider circuit 86
Figure 3.1 Combinational circuit of half (a) and full (b) adder 92
Figure 3.2 Construction of a full adder using two half adder circuits (a) and picture of a CLA adder IC74LS83 (b) 93
Figure 3.3 Combinational circuit of a half (a) and full (b) subtractor 96
Figure 3.4 Construction of a full subtractor using two half subtractor circuits 98
Figure 3.5 Construction of a single-bit magnitude comparator 99
Figure 3.6 Circuit for a 4-bit magnitude comparator 100
Figure 3.7 Seven-segment display array (a) and different configuration of display modules (b) 105
Figure 3.8 Picture of a seven-segment display decoder IC (a) and a decoder circuit configuration (b) 106
Figure 3.9 Binary 4-bit synchronous up counter 108
Figure 3.10 Schematic design of digital real-time clock 115
Figure 3.11 Waveform of an in-phase PWM signal 124
Figure 4.1 Schematic diagram of a serial adder 138
Figure 4.2 State transition diagram of a serial adder 138
Figure 4.3 General n-bit parallel adder with example addition 141
Figure 4.4 Digital circuit: standard (a) and pipelined (b), input processing in pipelined digital circuit (c), and schematic diagram of a 31-bit pipelined parallel adder (d) 144
Figure 4.5 Half subtractor (a), full subtractor using two half subtractors (b), and a typical full subtractor (c) 148
Figure 4.6 Design of a 4-bit parallel subtractor 148
Figure 4.7 Design of a 4-bit parallel adder/subtractor 149
Figure 4.8 A 4-bit unsigned binary multiplication (a) and its implementation using half and full adders (b) 153
Figure 4.9 Design of an add and shift method-based multiplier 154
Figure 4.10 Fast array multiplier 162
Figure 4.11 Digital architecture for division operation 166
Figure 4.12 Illustration of coordinate rotations 171
Figure 4.13 Results of coordinate rotation with different values of x and y 173
Figure 4.14 Illustration of coordinate rotations 175
Figure 4.15 Profile of the CORDIC scaling factor (K_i) 177
Figure 4.16 Digital design for implementation of CORDIC 178

Figure 4.17 A typical MAC unit with pipelined register 186

Figure 4.18 Simulation result of the CORDIC algorithm 186

Figure 4.19 Entity of a basic ALU (a), single-bit ALU circuit module (b), and pin details of a standard ALU: IC74181 (c) 189

Figure 4.20 Entity of a ROM unit (a), and internal logic of (k–1) x (n–1) ROM unit (b) 194

Figure 4.21 Programming the ROM according to Table 4.17 196

Figure 4.22 Function implementation using PLA (a) and using PAL (b) 197

Figure 4.23 ROM implementation of Design Example 4.23 198

Figure 4.24 Entity of a RAM module (a) and internal logic of a RAM cell (b) 201

Figure 5.1 Appearance and typical wiring of LED (a), color LEDs switching circuit (b), and LED wiring with reverse bias protection diode (c) 208

Figure 5.2 A multisegment display chip and arrangement of LEDs for a 7 × 5 dot-matrix display 211

Figure 5.3 State diagram to display "A" and illustration of character "A" in a 7 × 5 dot-matrix display 211

Figure 5.4 Schematic diagram of a scrolling text display system 212

Figure 5.5 General circuit diagram for a scrolling display system 213

Figure 5.6 Buzzer driver circuits using DC source and digital pulses 215

Figure 5.7 Picture and basic wiring of 16 × 2 LCD 218

Figure 5.8 Picture and pin wiring of a GLCD 224

Figure 5.9 Page and byte mapping of a 128 × 4 GLCD 225

Figure 5.10 Picture of a DIP switch (a), application circuit (b), and reed switch (c) 229

Figure 5.11 Bidirectional (a) and tristate switches (b) 231

Figure 5.12 Construction of a matrix keypad 233

Figure 5.13 Appearance and application circuit of a DTMF decoder 236

Figure 5.14 Appearance of IR Tx/Rx (a), IR Rx application circuits (b), photodiode with a comparator (c), appearance of a phototransistor (d), and application circuit of a phototransistor (e) 239

Figure 5.15 Circuit for rotating disc speed measurement 239

Figure 5.16 IR proximity or obstruction sensor 242

Figure 5.17 Appearance and working principles of a PIR sensor 245

Figure 5.18 Appearance and application circuit of a metal detector 249

Figure 5.19 Appearance (a) and typical nonlinear characteristic response of an LDR (b) 251

Figure 5.20 LDR-based light/dark detector circuit (a) and LDR with an op-amp comparator (b) ... 252

Figure 5.21 Cup anemometer assembly and wind-direction-finding vane ... 256

Figure 5.22 Circuit and architecture for cup anemometer interfacing ... 257

Figure 6.1 State diagram for sensing a button and producing a logic high pulse for only one cycle ... 262

Figure 6.2 K-map and logical simplification with D-FF ... 264

Figure 6.3 FSM realization of Equation 6.1 ... 265

Figure 6.4 K-map and logical simplification with JK FF ... 266

Figure 6.5 FSM realization of Equation 6.2 ... 266

Figure 6.6 State diagram for NRZ to Manchester code conversion ... 267

Figure 6.7 Standard Moore FSM model ... 269

Figure 6.8 State diagram for a simple Moore FSM model ... 270

Figure 6.9 Structure of a Mealy FSM model ... 272

Figure 6.10 Mealy model state diagram for Table 6.6 ... 273

Figure 6.11 Mealy model state diagram for BCD to excess-3 converter ... 273

Figure 6.12 Sequence detector-high level block diagram (a), state diagram without overlapping (b), and state diagram with overlapping (c) ... 287

Figure 6.13 Mealy FSM model for recognizing a pattern "1011" with overlapping ... 290

Figure 6.14 Moore FSM model for recognizing a pattern "10010" without overlapping ... 292

Figure 6.15 FSM design for a vending machine control – top-level illustration (a) and control flow state diagram (b) ... 296

Figure 6.16 Photograph of a traffic light (a), circuit for simple traffic light wiring (b), and illustration of traffic control sequences (c) ... 302

Figure 6.17 Master unit in wireless traffic light system (a), geometrical installation of slave units (b), direction and seven-segment (down counter) display at the slave unit (c), configuration of light displays with slave unit (d), and state diagram of control engine (e) ... 304

Figure 6.18 Photograph and design layout of a single escalator unit ... 310

Figure 6.19 State diagram of an escalator controller ... 311

Figure 6.20 Photographs of different dice ... 315

Figure 6.21 Top-level schematic diagram of a dice game controller ... 315

Figure 6.22 Illustrations of an electronic model train and its routes ... 322

Figure 6.23 Track layout for an electronic model training control ... 323

Figure 6.24 Top-level partitioning of a digital system (a) and simple complex system, that is, combination of Moore and Mealy models (b) 329

Figure 6.25 ASM blocks: state box (a), decision box (b), and conditional output box (c) 329

Figure 6.26 A simple ASM chart with one entry and four exit paths (a), a complex state diagram (b), and corresponding ASM chart (c) 330

Figure 6.27 ASM chart for Design Example 6.22 331

Figure 6.28 ASM-based digital system design entity 332

Figure 6.29 ASM chart for Design Example 6.23 333

Figure 6.30 ASM chart for Design Example 6.24 335

Figure 7.1 Basic illustration of sensor interfacing and data processing 340

Figure 7.2 Illustration of basic process of A/D conversion 341

Figure 7.3 Working principles of D/A converter using binary weighted resistors 342

Figure 7.4 D/A converter using R/2R ladder (a) and illustration of Thevenin simplification (b) 343

Figure 7.5 Optical power measurement-transimpedance amplifier 345

Figure 7.6 Appearance and application circuit of an LM35 sensor 346

Figure 7.7 Appearance of a thermistor and its characteristic response 347

Figure 7.8 Application circuits of a thermistor 348

Figure 7.9 Appearance and construction of temperature measurement system using a thermocouple 348

Figure 7.10 Appearance and application circuit of a strain gauge 349

Figure 7.11 Basic operation of a voltage comparator: negative input (a), positive input (b), and a 4-bit application circuit (c) 350

Figure 7.12 Two-bit binary A/D converter using LM324 351

Figure 7.13 Application circuit and interfacing timing diagram of ADC0804 352

Figure 7.14 Data frame format of a serial communication: RS232 voltage level (a), data bit packing for serial transmission (b), and voltage level swing for transferring an ASCII character "A" (c) 357

Figure 7.15 Electrical and mechanical interfacing of a DB9 connector: RS232 cable (a), three wire connections at the serial port (b), circuit schematic of MAX233 line driver (c), and RS232-to-USB converter cable (d) 358

Figure 7.16 Digital architecture of UART-Tx and data transmission frame format 365

Figure 7.17 Data receiving process flow in an UART-Rx 372

Figure 7.18 A/D converter application circuit and timing diagram of control signals 379

Figure 7.19 Application circuit of ADC0848 and timing diagram of control signals 385

Figure 7.20 A diurnal period profile of weather data 386

Figure 7.21 A diurnal period profile of C_n^2 387

Figure 7.22 Application circuit of MAX1112 A/D converter 397

Figure 7.23 Timing diagram of single conversion process of MAX1112 A/D converter 397

Figure 7.24 Photograph of a signal conditioning board using ADC1674 core and its operational timing diagram 400

Figure 7.25 Appearance of an OPD (a), centre beam spot on an OPD surface (b), displaced beam spot on OPD (c), and MPAC (d) 403

Figure 7.26 Optical transmitter (IR transmitter) (a), optical receiver (TSOP1738) (b), HT12E application circuit (c), and HT12D application circuit (d) 412

Figure 7.27 RF transmitter module (a), RF receiver module (b), encoder application circuit (c), and decoder application circuit (d) 413

Figure 7.28 Simple PRBS generator circuits using serially connected LFSRs 416

Figure 7.29 Illustration of application of a parallel PRBS generator 418

Figure 7.30 General and different designs of parallel PRBS generator 419

Figure 7.31 Schematic of a Kasami code generator 422

Figure 7.32 Schematic of a simplex TDM access 424

Figure 7.33 Circuit diagram and simulation result of a 8-level PAM 425

Figure 7.34 Circuit diagram of configuration of DAC0808 with FPGA 427

Figure 7.35 Sine wave with circuit diagram of configuration of DAC0808 with FPGA 428

Figure 7.36 Functional block diagram of DAC7728 435

Figure 7.37 Timing diagram of write operation-1 436

Figure 8.1 Pin details and application circuit of DS12887 442

Figure 8.2 DS12887 address map 443

Figure 8.3 Configuration of master-slave over I^2C bus (a) and pattern of start and stop sequences of I^2C protocol (b) 449

Figure 8.4 I^2C bus address/data: write to slave device's register (a) and read from slave's register (b) 450

Figure 8.5 Picture of an SHT11 sensor (a), circuit configuration of SHT11 sensor (b), and digital architecture for temperature and relative humidity measurement (c) 457

Figure 8.6 FSM control engine state transition flow for temperature and relative humidity measurement ... 458

Figure 8.7 Picture (a), application circuit (b), and interface digital architecture of SCP1000-D01 sensor (c) ... 462

Figure 8.8 FSM control engine state transition flow for pressure measurement in triggered mode ... 463

Figure 8.9 Picture and interface circuit of GSM module ... 468

Figure 8.10 GPS module picture (a) and interface circuit with FPGA (b) ... 479

Figure 8.11 Picture of PS/2 connector (a) and its male pin out details (b) ... 480

Figure 8.12 PS/2 interface circuit (a) and PS/2 keyboard transmission timing diagram (b) ... 481

Figure 8.13 PS/2 keyboard interface logic architecture ... 482

Figure 8.14 Picture and pin details of a VGA DB15 connector ... 485

Figure 8.15 VGA DB15 port interfacing circuits: direct configuration (a), 3-bit configuration (b), video DAC–based configuration (c), and picture of a video driver (ADV7123) board (d) ... 486

Figure 8.16 Mechanism of displaying image on the 640 × 480 screen ... 486

Figure 8.17 Schematic diagram of VGA interface ... 488

Figure 9.1 Photograph of four SPDT relays (a), photograph of DPDT relay (b), and schematic of SPDT relay (c) ... 494

Figure 9.2 Circuit of DPDT relay (a), typical relay control circuit (b), relay control circuit with Darlington pair of transistors (c), and 8-channel Darlington current amplifier ICULN2803A (d) ... 495

Figure 9.3 Solenoid outflow control valve (a), three-port solenoid valve, switching the outflow between two outlet ports (b), and solenoid plunger (c) ... 496

Figure 9.4 Solenoid valve for precise liquid flow control (a) and solenoid control circuit (b) ... 498

Figure 9.5 IC package of an opto-isolator (a), opto-isolator interfacing circuit (b), and application circuit of an opto-isolator (c) ... 502

Figure 9.6 Picture of a DC motor (a) and constructional layout of a DC motor (b) ... 504

Figure 9.7 PWM control signals (a) and DC motor speed control circuit (b) ... 505

Figure 9.8 Relay-based DC motor driver circuit: clockwise direction (a) and anti-clockwise direction (b), picture of an L293D DC motor driver IC (c), and application circuit of L239D (d) ... 506

Figure 9.9 Picture of a servo motor with label for its important parts ... 509

Figure 9.10 Duty cycles and corresponding positions of the shaft ... 509

Figure 9.11 Picture and coil winding of a BLDC motor (a–d) ... 512

Figure 9.12 Patterns of BLDC motor switching sequence (a–d)....................................513
Figure 9.13 One of the BLDC motor switching circuits....................................513
Figure 9.14 Picture of a stepper motor (a) and coil winding and shaft rotation of a stepper motor (b)....................................516
Figure 9.15 An application circuit for stepper motor interfacing with the FPGA.........517
Figure 9.16 Liquid level monitoring: using float sensor (a) and capacitive sensor (b)....................................519
Figure 9.17 Liquid level monitoring: using radar sensor....................................520
Figure 9.18 Pin details of LM1830 fluid lever sensor (a) and a liquid level control application circuit (b)....................................520
Figure 9.19 Simple illustration of step-up and step-down transformers....................522
Figure 9.20 Simple voltage measurement circuit....................................523
Figure 9.21 Illustration of current transformers....................................524
Figure 9.22 Simple current measurement circuit....................................525
Figure 9.23 Picture of an AC motor (a) and simple design circuit of an AC motor (b)....................................527
Figure 9.24 AC motor speed control in closed-loop control configuration....................527
Figure 9.25 Electrical symbol, picture, and simple application circuit of a SCR...........528
Figure 9.26 Simple SCR switching circuit: by DC source (a) and by AC source (b). The loads in both cases are DC and AC lamps respectively....................529
Figure 9.27 SCR phase control circuit....................................529
Figure 9.28 SCR phase control and its application circuits....................................530
Figure 9.29 Symbol, picture, and equivalent circuit of a Triac....................................532
Figure 9.30 Application circuits of a Triac....................................533
Figure 9.31 Triac opto-coupling relay....................................533
Figure 9.32 Symbol and general configuration of a Diac....................................536
Figure 9.33 Application circuits with a Diac....................................537
Figure 9.34 Schematic diagram of a closed-loop control system....................................539
Figure 9.35 Illustration of: a simple control system (a) and its response over time (b)....................................541
Figure 10.1 FPGA-based signal-processing system....................................548
Figure 10.2 General representation of a fixed-point number....................................550
Figure 10.3 Length of double-precision floating-point format....................................556
Figure 10.4 Customized floating-point format of size (1,6,10)....................................559
Figure 10.5 MATLAB-Xilinx simulation environment window....................................576

Figure 10.6 Some of the Xilinx System generator blocks 577

Figure 10.7 A simple Xilinx model for realizing Equation 10.4 584

Figure 10.8 Input pin configuration for implementation 587

Figure 10.9 Configurations of target device details 588

Figure 10.10 SysGen VHDL code simulation window (time versus data) 588

Figure 10.11 A MAC unit design 590

Figure 10.12 Performance simulation of a MAC unit (magnitude versus time) 591

Figure 10.13 A design to add and display two sine waves 591

Figure 10.14 Simulation result of a design shown in Figure 10.13 (magnitude versus time) 592

Figure 10.15 A design to sample a sine wave with a PRBS sequence 593

Figure 10.16 Simulation result of Figure 10.15 (magnitude versus time) 594

Figure 10.17 Fractional-point calculations with stagewise-result display 595

Figure 10.18 Xilinx SysGen-based system as per design Example 10.17 597

Figure 10.19 Simulation result of second example of Figure 10.18 599

Figure 10.20 Design resource estimation 600

Figure 10.21 Time-domain input-output response of a FIR filter 602

Figure 10.22 A design for FIR filter $y[n] = x[n] + x[n - 1]$ 603

Figure 10.23 Xilinx SysGen simulation result of Figure 10.22 (time versus amplitude) 604

Figure 10.24 System modeling using expression block 606

Figure 10.25 Simulation result of Figure 10.24 (time versus amplitude) 607

Figure 10.26 Subsystem design with DSP48 macro 2.0 IP core 608

Figure 10.27 DSP system design with subsystems 609

Figure 10.28 Simulation result of Figure 10.27 (time versus amplitude) 609

Figure 10.29 Design for Xilinx SysGen tool and MATLAB code interfacing: Data types are displayed at all the stages 611

Figure 10.30 Mealy model sequence detector using SysGen tools and MATLAB function 613

Figure 10.31 Simulation result of Figure 10.29 (time versus amplitude) 614

Figure 10.32 VHDL code interfacing with SysGen tools 616

Figure 10.33 Simulation result of Figure 10.32 (time versus amplitude) 617

Figure 10.34 IP core ADC interfacing with Black Box 618

Figure 10.35 Simulation result of Figure 10.34 (time versus amplitude) 619

Figure 10.36 System generator token – window ... 621

Figure 10.37 Hardware co-simulation window ... 621

Figure 10.38 Illustration of temporal and spatial computation architecture ... 623

Figure 11.1 Illustration of time wave and frequency spectrum ... 628

Figure 11.2 Illustration of signal correlation measurement. Cosine and sine basis functions with 1 cycle over N samples ... 630

Figure 11.3 Random (sensor) signal with 350 samples ... 631

Figure 11.4 Basis function correlation with k = 0 to 3 over 200 samples of `x[n]`. In Figure 11.4, black, red, green, and blue represent the `x[n]`, DC, sine and cosine functions, respectively. The correlation results are shown below the plots corresponding to k ... 634

Figure 11.5 Same as Figure 11.4 with k = 4 to 7 ... 636

Figure 11.6 Same as Figure 11.4 with k = 8 to 10 and real and imaginary magnitude spectrum ... 638

Figure 11.7 Power, magnitude, and phase spectrum over 200 samples ... 639

Figure 11.8 Standard structure of a FIR filter ... 641

Figure 11.9 General lattice structure of an IIR filter ... 644

Figure 11.10 Direct form-I of an IIR filter ... 645

Figure 11.11 Representation and illustration of decimation and interpolations ... 651

Figure 11.12 Structure of a simple multirate filter ... 651

Figure 11.13 Down/up-sampling result for the input values of N = 50 and M = 2 ... 653

Figure 11.14 Spectral characteristics of up-sampled and interpolated samples for the input values of N = 100 and M = 2 ... 655

Figure 11.15 Schematic diagram of 2^n and 2^n–1 modulo adders ... 664

Figure 11.16 Simple structure of an adaptive filter ... 690

Figure 11.17 Signal flow graph of an adaptive channel equalizer ... 691

Figure 11.18 Actual and noise signal with error profile for the input values of 50 and 0.3 ... 694

Figure 11.19 Profile of weight variations, correction voltage and recovered signal for the input values of 50 and 0.3 ... 695

Figure 11.20 Desired and noise signal profile along with filter output and error function for the input values of 1000 and 0.005 ... 696

Figure 11.21 Weights (W0 and W1) and error-square profile for the input values of 1000 and 0.005 ... 697

Figure 12.1 Representation of N- (a) and N/2- (b) point DFT ... 702

Figure 12.2 Representation of calculation of IFFT ... 703

Figure 12.3 Main design module with two FFT blocks 703

Figure 12.4 First subsystem design with one FFT 8.0 core 704

Figure 12.5 Second subsystem design with one FFT 8.0 core 704

Figure 12.6 Filter design using FDA core 706

Figure 12.7 Filter design using FDA core 706

Figure 12.8 Time-domain result of the lowpass filter 707

Figure 12.9 Filter design using FDA core 707

Figure 12.10 Filter design using FDA core 708

Figure 12.11 Time-domain simulation result of the designed bandpass FIR filter 709

Figure 12.12 Direct Form-II representation of a biquad IIR filter 711

Figure 12.13 Top-level design of an IIR filter 712

Figure 12.14 Design of a MAC FIR filter 713

Figure 12.15 Top-level design of a MAC-FIR filter 714

Figure 12.16 Structure of moving average (a), recursive running sum (b), and CIC (c) filters 715

Figure 12.17 Time-domain impulse responses of single-stage CIC filter: comb (a), integrator (b), and CIC (c) 717

Figure 12.18 Structure of single-stage CIC filters with decimation (a) and interpolation (b) 718

Figure 12.19 Structure of a simple CIC filter 718

Figure 12.20 Representation of polar and rectangular coordinate variables 719

Figure 12.21 CORDIC-based SysGen design for rectangular-to-polar conversion 720

Figure 12.22 Design of display subsystem 721

Figure 12.23 Design of phase error measurement subsystem 721

Figure 12.24 Representation of subband coding scheme 724

Figure 12.25 Wavelet-transformed images 726

Figure 12.26 Design of image-processing system using DWT 727

Figure 12.27 DWT-applied subband coded images 727

Figure 13.1 MATLAB simulation results of fuzzy logic controller. (a) Behavior of unknown system, (b) training data and testing data, (c) original o/p and NN o/p without training, and (d) original o/p and trained NN o/p 753

Figure 13.2 MATLAB simulation correlation errors 754

Figure 13.3 MATLAB simulation results of fuzzy logic controller 756

List of Tables

Table 1.1 VHDL Milestones 20

Table 2.1 Std Library Package 36

Table 2.2 List of Main Operators in VHDL 46

Table 2.3 List of Attributes in VHDL 49

Table 2.4 Characteristic and State Change Control of SR-FF 66

Table 2.5 Characteristic and State Change Control of JK-FF 67

Table 2.6 Characteristic and State Change Control of D-FF 68

Table 2.7 Characteristic and State Change Control of T-FF 68

Table 2.8 SR-FF to JK-FF Conversion Table 71

Table 2.9 JK-FF to SR-FF Conversion Table 72

Table 2.10 Conversion of SR-FF to D-FF 73

Table 2.11 Conversion of D-FF to SR-FF 74

Table 2.12 Conversion of JK-FF to T-FF 75

Table 2.13 Conversion of JK-FF to D-FF 75

Table 2.14 Conversion of D-FF to JK-FF 76

Table 3.1 Function Table of Half and Full Adders 93

Table 3.2 Function Table of Half and Full Subtractors 97

Table 3.3 Truth Table of a 1-Bit Magnitude Comparator 99

Table 3.4 Truth Table of Two 4-Bit Magnitude Comparators 101

Table 4.1 Arithmetic Operators 137

Table 4.2 Carry Function Table of a Full Adder 139

Table 4.3 Truth Table of Half and Full Subtractor 147

Table 4.4 Function of a Parallel Adder/Subtractor Circuit 150

Table 4.5 Add and Shift-Based Multiplication: `Md` $= (13)_{10}$ and `Mr` $= (11)_{10}$ 154

Table 4.6 Look-Up Table of an 8-Bit Fast Array Multiplier 161

Table 4.7 Classical Binary Division for 135/13 163

Table 4.8a Initialization of Division Algorithm 164

Table 4.8b Division Algorithm 164

Table 4.8c Division Algorithm 165

Table 4.8d Division Algorithm 165
Table 4.8e Division Algorithm 165
Table 4.9 Division Algorithm for 8-Bit/4-Bit Data 167
Table 4.10 Division Algorithm for 8-Bit/4-Bit Data 168
Table 4.11 Order of Rotation Angle in the CORDIC Algorithm 174
Table 4.12 Recursive Equations of the CORDIC Algorithm 176
Table 4.13 CORDIC Iteration for $\theta = 5.0385°$ 179
Table 4.14 CORDIC Iteration for $\theta = 28.027°$ 180
Table 4.15 Design of a Single-Bit ALU Module 190
Table 4.16 Function Table of ALU: IC74181 192
Table 4.17 ROM Truth Table 195
Table 4.18 ROM Truth Table for Design Example 4.23 198
Table 5.1 Pin Details of a 16×2 LCD 218
Table 5.2 16×2 LCD Module Command Data and the Corresponding Functions 219
Table 5.3 GLCD Pin Details and Their Functions 224
Table 5.4 Display Control Instructions 225
Table 6.1 State or Transition Table 263
Table 6.2 Excitation Table with D-FFs 264
Table 6.3 Excitation Table with JK-FFs 265
Table 6.4 Excitation Table with SR-FFs 267
Table 6.5 General State Transition Table—Moore Model 270
Table 6.6 State Transition Table for Mealy Model 272
Table 6.7 Excitation Table for BCD to Excess-3 Converter 274
Table 6.8 State Transition Table for Moore FSM Conversion 276
Table 6.9 State Transition Table for Mealy FSM Conversion 277
Table 6.10 Binary to BCD Conversion 283
Table 6.11 Master to Slave Control and Display Data Pattern 305
Table 7.1 Reference Input Voltages and Corresponding Conversion Step Size 352
Table 7.2 Data Logging in a Serial Port Buffer 359
Table 7.3 Screen Snapshot of the Results of Design Example 7.5 362
Table 7.4 Analog Channel Selection Control 380
Table 7.5 Multiplexed Channel Selection Controls 385

Table 7.6 Structure of the Control Byte Register 396

Table 7.7 ADC1674 Enable and Channel Selection Details 401

Table 7.8 Function Table of a PRBS Generator 417

Table 7.9 A Portion of Calculated Values to be Sent to DAC0808 to Generate Sine Wave 429

Table 8.1 Address Location for Time, Calendar, and Alarm IN DS12887 444

Table 8.2 Register A and Its Contents 445

Table 8.3 Register B and Its Contents 445

Table 8.4 A Portion of Keyboard Scan Code 482

Table 8.5 Control Signal Generation Details for Several Pixel Resolutions 487

Table 9.1 Switching and Corresponding Action of a L293D 507

Table 9.2 BLDC Motor Switching Sequence 514

Table 9.3 Data Sequence to Energise Stepper Motor Coils 516

Table 10.1 Signed/Unsigned Representation of 4-Bit Binary Word 548

Table 10.2 Binary Weighing of a Fixed-Point Number 550

Table 10.3 Fixed-Point Fractional Part Conversion 551

Table 10.4 Conversion of Decimal to Fixed-Point Binary Value 551

Table 10.5 Feature Comparison of Single- and Double-Precision Floating-Point Numbers 559

Table 11.1 Operation of a Multirate Signal Processor 652

Table 11.2 RNS-Based Addition and Multiplication 666

Table 11.3 RNA-Based Subtraction 666

Table 11.4 DA-LUT for 3-Bit Samples 676

Table 11.5 Filter Response Computation Using DA Technique 676

Table 11.6 DA_LUT for Signed Samples and Coefficients 677

Table 11.7 Filter Response Computation for Signed Samples/Coefficients Using DA Techniques 678

Table 11.8 Initialization Process of Booth's Multiplication 684

Table 11.9 Booth's Algorithm – Multiplication Example 685

Table 11.10 Multiplication of 14 × –5 Using Booth's Algorithm 686

Table 12.1 Filter Coefficients 706

List of Abbreviations

Acronym	Expansion
1D	One dimension
2DA	Two dimensional array
A/D	Analog to digital
AC	Alternate current
ACK	Acknowledgement
ADC	Analog to digital convertor
AGND	Analog ground
ALE	Address latch enable
ALU	Arithmetic logic unit
ANN	Artificial neural network
AS	Address strobe
ASCII	American standard code for information interchange
ASIC	Application specific integrated circuit
ASK	Amplitude shift keying
ASM	Algorithmic state machine
ASR	Addressable shift register
AT	Attention
ATA	Attention answer
ATD	Attention dial
ATH	Attention hang-up
ATM	Automatic teller machine
ATO	Attention online
AU	Arithmetic unit
AWG	Additive white gaussian
AWGN	Additive white gaussian noise
BC	Binary cell
BCD	Binary coded decimal
BCH	Bose-Chaudhuri-Hocquenghem
BER	Bit error rate
BGA	Ball grid array
BLDC	Brush less direct current
BPSK	Binary phase shift keying
BRAM	Block random access memory
BRC	Baud rate counter
BSR	Bidirectional shift register
CAD	Computer aided design
CE	Clock enable
CHN	Channel N
CIC	Cascaded integrated comb
CISC	Complex instruction set computer
CLAA	Carry look ahead adder
CMOS	Complementary metal oxide semiconductor
CNC	Computer numerical control

CORDIC	Coordinate rotation digital computer
CPU	Central processing unit
CRC	Cyclic redundancy check
CRT	Chinese remainder theorem
CS	Chip select
CVD	Chemical vapor deposition
D/A	Digital to analog
DA	Distributed arithmetic
DAC	Digital to analog convertor
DAFIR	Distributive arithmetic finite impulse response
DC	Direct current
DCM	Digital clock manager
DCT	Discrete cosine transform
DDR	Double data rate
DDS	Direct digital synthesizer
D-FF	Delay flip flop
DFT	Discrete Fourier transform
DGPS	Differential global positioning system
DIAC	Diode for alternative current
DIP	Dual inline package
DLC	Delay line canceller
DOD	Department of defence
DPCM	Differential pulse code modulation
DRAM	Dynamic random access memory
DRDY	Data ready
DSD	Data sequence detector
DSO	Digital storage oscilloscope
DSP	Digital signal processing
DSR	Data storage register
DSSS	Direct sequence spread spectrum
DTMF	Dual tone multi frequency
DUT	Device under test
DWT	Discrete wavelet transform
EDA	Electronic design automation
EDK	Embedded development kit
EDO	Extended data out
EEPROM	Electronically erasable programmable read only memory
EMF	Electro motive force
EMI	Electromagnetic interference
EMIF	External memory interface
ENIAC	Electronics numerical integrator and computer
EOC	End of conversion
EPROM	Electronically programmable read only memory
FA	Full adder
FAM	Fast array multiplication
FEC	Forward error correction
FET	Field effect transistor
FF	Flip flop
FFT	Fast Fourier transform

FIFO	First in first out
FIR	Finite impulse response
FLC	Fuzzy logic controller
FM	Frequency modulation
FMCW	Frequency modulated continuous wave
FPGA	Field programmable gate array
FPM	Frames per minute
FSK	Frequency shift keying
FSL	Fast simplex link
FSM	Finite state machine
FSO	Free space optics
GDS	Geometric data stream
GLCD	Graphical liquid crystal display
GNSS	Global navigation satellite system
GPIO	General purpose input output
GPRMC	Global positioning remote machine code
GPRS	General packet radio service
GPS	Global positioning system
GSI	Giant scale integration
GSM	Global system for mobile
GUI	Graphical user interface
HA	Half adder
HCDTA	Hi current Darlington transistor array
HDL	Hardware description language
HDOP	Horizontal dilution of precision
I/O	Input/output
I^2C	Inter-integrated circuit
IBM	International business machine
IC	Integrated circuit
IDFT	Inverse discrete Fourier transform
IET	Institute of Engineering and Technology
IFF	Identify Friend or Foe
IFT	Inverse Fourier transform
IIR	Infinite impulse response
ILA	Integrated logic analyzer
INTR	Interrupt
IP	Intellectual property
IRQ	Interrupt request
ISE	Integral square error
ISTE	Indian Society for Technical Education
IVR	Interactive voice response
JK-FF	Jack-Kilby flip flop
JPEG	Joint photographic expert group
JTAG	Joint text action group
LASER	Light amplification by stimulated emission of radiation
LC	Load current
LCD	Liquid crystal display
LCL	Laser communication laboratory
LDR	Light dependent resistor

LED	Light emitting diode
LFSR	Linear feedback shift register
LMS	Least mean square
LOS	Line of sight
LSB	Least significant bit
LSI	Large scale integration
LU	Logic unit
LUT	Look up table
LV	Load voltage
M/m	Measurement
MAC	Multiply and accumulator
MAS	Micro architecture specification
MCM	Multichip modules
M-FSK	M array frequency shift keying
MilStd	Military standard
MIMO	Multiple input multiple output
MISO	Master in slave out
MOS	Metal oxide semiconductor
MOSFET	Metal oxide semiconductor field effect transistor
MOSI	Master out slave in
MPAC	Mono pulse arithmetic circuit
MPEG	Motion pictures expert group
M-PSK	M array phase shift keying
MSB	Most significant bit
MSE	Mean square error
MSI	Medium scale integration
MUX	Multiplexer
NaN	Not a number
NMEA	National marine electronics association
NMOS	Negative channel metal oxide semiconductor
NPN	Negative-positive-negative
NRZ	Non return to zero
NRZ-L	Non return to zero level
NRZ-M	Non return to zero mark
OE	Output enable
OPD	Optical position detector
PAL	Programmable array logic
PAM	Pulse amplitude modulation
PCB	Printed circuit board
PCM	Pulse code modulation
PD	Photo diode
PGA	Pin grid array
PID	Proportional integral derivative
PIPO	Parallel in parallel out
PIR	Pyrometric infrared
PISO	Parallel in serial out
PISOSR	Parallel in serial out shift register
PLA	Programmable logic array
PLC	Programmable logic circuits

PLD	Programmable logic device
PLL	Phase locked loop
PM	Permanent magnet
PMOS	Positive channel metal oxide semiconductor
PNP	Positive-negative-positive
PNPN	Positive-negative-positive-negative
PNRZ	Polar non return to zero
POS	Product of sum
POT	Potentiometer
PPI	Pulse position indicator
PPS	Pulse per second
PRBS	Pseudo random binary sequence
PROM	Programmable read only memory
PRZ	Polar return to zero
PSK	Phase shift keying
PT	Photo transistor
PWM	Pulse width modulation
QFP	Quad flat package
QPSK	Quadrature phase shift keying
R/W	Read/write
RADAR	Radio detection and ranging
RAM	Random access memory
RBR	Receive buffer register
RCA	Ripple carry adder
RD	Read
RF	Radio frequency
RGB	Red green blue
RISC	Reduced instruction set computer
RMS	Root mean square
RNA	Residual number arithmetic
ROL	Left circular rotate
ROM	Read only memory
ROR	Right circular rotate
RPM	Revolution per minute
RS	Register select
RS	Reed Solomon
RSA	Right shift arithmetic
RSC	Right shift circulate
RST	Reset
RTC	Real time clock
RTL	Register transfer logic
RZ-AMI	Return to zero alternate mark inversion
S/H	Sample and hold
SBSD	Start bit sequence detector
SC	Start conversion
SCLK	Serial clock
SCR	Silicon control rectifier
SDA	Serial data
SDR	Software defined radio

SDRAM	Synchronous dynamic random access memory
SHDL	Shut down
SIPO	Serial in parallel out
SISO	Serial in serial out
SLA	Shift left Arithmetic
SLL	Shift left logic
SMS	Short message service
SNR	Signal to noise ratio
SOC	System on chip
SOP	Sum of product
SPI	Serial peripheral interface
SPIE	Society of Photographic Instrumentation Engineers
SPST	Single pole single throw
SQRT	Square root
SQW	Square wave
SQWE	Square wave enable
SR	Shift register
SR/FF	Set reset / flip flop
SRA	Shift right arithmetic
SRL	Shift right logic
SSD	Solid state drive
SSI	Small scale integration
STFT	Short term Fourier transform
STRB	Serial strobe
SysGen	System generator
TCP/IP	Transmission control protocol/ internet protocol
TDM	Time division multiplexing
TDMA	Time division multiple access
TE	Transmit enable
T-FF	Toggle flip flop
TI	Texas instrument
TTL	Transistor transistor logic
TV	Television
TWI	Two wire interface
UART	Universal asynchronous receive-transmit
ULSI	Ultra large scale integration
UPRZ	Uni polar return to zero
USB	Universal serial bus
USR	Universal shift register
VCO	Voltage controlled oscillator
VDC	Voltage divider circuit
VGA	Video graphics array
VHDCI	Very high density cable interconnect
VHDL	Very high speed integrated circuit (VHSIC) hardware description language
VHPI	VHDL procedural interface
VHSIC	Very high speed integrated circuit
VLSI	Very large scale integration
XPS	Xilinx platform studio

Preface

Very-high-speed hardware description language (VHDL) has been at the heart of electronic design productivity since its initial ratification by the Institute of Electrical and Electronics Engineers (IEEE) in 1987. For almost 15 years, the electronic design automation industry has expanded the use of VHDL from the initial concept of design documentation to design implementation and functional verification. It can be said that VHDL fuelled modern synthesis technology and enabled the development of application-specific integrated circuit (ASIC) semiconductor industries. The use of VHDL has evolved and its importance increased as semiconductor device dimensions have shrunk. A major revolution in digital design has taken place over the past decade. Field-programmable gate arrays (FPGAs) can now contain over millions of millions of equivalent logic gates and tens of thousands of flip-flops. This means that it is not necessary to use traditional methods of logic design involving the drawing of logic diagrams when the digital circuit may contain thousands of gates. The reality is that today, digital systems are designed by writing software in the form of HDLs, which are in widespread use. When using VHDL, the designer typically describes the behavior of the logic circuit rather than writing traditional Boolean logic equations. Computer-aided design tools are used to both simulate the VHDL design and synthesize the design to actual hardware. With the maturity and availability of VHDL and synthesis software, using it to design custom digital hardware has become a mainstream practice. FPGA technology minimizes the wiring and engineering design complexities.

Unique Features of the Book: This book is a digital system design and hardware interfacing text. VHDL synthesis and simulation software are used as tools to realize the intended designs. Several unique features that distinguish the book are:

- A new approach is used to explain the very-large-scale integration (VLSI) technology, VHDL-based real-world applications and system generator–assisted top-level designs.
- The suggested coding style shows a clear relationship between VHDL constructs and hardware components.
- Easy-to-understand conceptual diagrams, rather than cell-level netlists, explain the realization of VHDL codes.
- The book emphasizes the re-use aspect of code throughout.
- More than 300 design examples (VHDL and MATLAB® code) with computation flow tables, explanations about the code design, function procedures, simulation (timing) results, data analysis, peripheral interfacing, hardware implementation and so on make the design "technology neutral" so that the developed VHDL code can be synthesized using demo-version synthesis software provided by FPGA vendors.
- The book contains a large number of nontrivial, practical examples to illustrate and reinforce the design concepts, procedures and techniques.
- Chapters consist of (i) illustrations corresponding to the digital architectures built inside the FPGA for various applications, (ii) explanations of the architectural

design and external component wiring and (iii) scientific specifications of all the peripherals/devices required for the applications covered in the book.

- Many advanced real-time applications such as specialized sensor interfacing, bidirectional signal conditioner designs, audio codec interfacing, VGA interfacing, data communication protocol design and so on are detailed with the necessary circuits, RTL schematics and color photographs (for the fast and reliable diagnosis of component assembly) of the experimental test beds.

Book Organization: A systematic, step-by-step approach is used to cover various aspects of VHDL programming and FPGA interfacing. Many examples and instances of sample code are given to clarify the concepts and provide readers with an opportunity to learn by doing. Exercise designs are given at the end of every chapter to reinforce the relevant additional applications of that section. The book is basically divided into four major parts. The first part, Chapters 1 to 4, provides a comprehensive overview of VLSI technology, digital circuit design with VHDL, programming with packages, components, functions and procedures and arithmetic designs. The second part, Chapters 5 to 7, covers the core of external I/O programming, algorithmic state machine (ASM)–based system design and real-world interfacing examples. The third part, Chapters 8 to 10, describes the programming of data communication protocols, real-time industrial controls and floating point computations and system design with system generator tools. The fourth part, Chapters 11 to 13, covers designs for digital signal processing, IP-based system design examples and contemporary design applications. More detailed descriptions of the chapters follow.

Part I – Basic System Modeling and Programming Techniques

Chapter 1 presents the history of VLSI technology with the features and architecture of FPGA. Reviews of microelectronics, device technologies, complementary metal oxide semiconductor (CMOS) layout design, subsystem development, ASIC design flow and the requirements of VHDL are also presented.

Chapter 2 provides digital system design, system representation, development flow, software tools, and usage and capability of a hardware description language (HDL). A series of simple codes is used to introduce the basic modeling concepts of VHDL, data type conversions (signed, unsigned, integer, std_logic_vector, numbers, bit vector), operators and attributes and concurrent and sequential codes. Flip-flops, parallel to serial converters and multifrequency and signal generators are used as examples to explain simple application circuit design with VHDL.

Chapter 3 describes system design based on packages, components, functions and procedures. Advantages of digital circuit design using these standards are explained and their significance is highlighted with systems developed for a signal generator, seven-segment display, half/full adder/subtractor, N-bit signed magnitude comparator, digital clock design, counter designs and pulse width modulation (PWM) signal generation.

Chapter 4 covers arithmetic, logical and special function programming. Arithmetic and logical operations, trigonometric function approximation, serial/parallel adders/subtractors, multipliers, divider multiply-accumulate units, arithmetic-logical units, read-only memory (ROM), programmable logic arrays, programmable array logic and programmable logic devices are the design examples in this chapter.

Part II – Custom Input/Output Peripheral Interfacing

Chapter 5 presents external input/output (I/O) device interfacing and programming techniques. This chapter explains the system construction methods for interfacing light-emitting diodes and multisegment displays, buzzer controls, liquid crystal and graphical liquid crystal displays, dip switch and matrix keypads, dual-tone multifrequency encoders, infrared and proximity sensors, pyrometric sensors, metal detectors, light-dependent resisters and cup anemometers (wind speed measurement).

Chapter 6 gives an in-depth overview of formulation of the ASM and realization of relevant real-time systems. Finite-state machine design based on the Moore and Mealy techniques is also detailed. Further, the designs of input code classifiers, sequence detectors, code converters, vending machine controllers, traffic light controllers, escalator controllers, dice games and electronic model train controllers are discussed in detail to explore more about system design with ASM.

Chapter 7 covers the construction of more sophisticated combinational and sequential real-world interfacing circuits. Interfacing digital logic includes analog to digital converter (A/D), optical and temperature sensors, universal asynchronous receiver transmitters, multichannel data logging, bidirectional A/D, optical beam tracking, aligning and positioning, radio frequency (RF) Tx/Rx, pseudorandom binary sequence generators, time division multiple access and low/high voltage digital to analog converters (D/A). These examples show how to transform conceptual ideas into real-time working systems/hardware.

Part III – Hardware Accelerated Designs

Chapter 8 deals with the principles, construction and design methodology of real-time clock and data communication protocol programming. This chapter explains digital circuit design for interfacing many specialized sensors that follow a two-wire interface, serial peripheral interface, inter-integrated circuit (I^2C) interface, global system for mobile (GSM) Wr/Rd interface, global positioning system (GPS) interface, video graphics array (VGA) interface and Ps/2 interface for data transmission, initialization, resetting, writing/reading the measurement data, monitoring the battery level and so on.

Chapter 9 is devoted to describing various motor controls and switching of high-voltage control devices. The design examples include relay and direct current (DC) motor control, alternating current (AC) motor and brush less direct current (BLDC) motor control, stepper motor control systems, automatic fuel level (solenoid valve) control, voltage and current measurement, power electronics such as thyristors/silicon-controlled rectifiers (SCRs), Triacs and Diacs and real-time process control.

Chapter 10 is devoted to floating-point number representations, their arithmetic computations and top-level designs using the system generator tools. Varieties of design examples are given to illustrate how the system generator tools are integrated into the MATLAB®/VHDL environment. Design procedures and methodology that can be used for different types of designs are explained and the relevant issues are highlighted. Real number representation, fixed and floating point arithmetic operations, hardware co-simulations and system generator implementations are the main examples in this chapter.

Part IV – Miscellaneous Design and Applications

Chapter 11 covers some of the important digital signal processing applications. Architectural design and hardware implementation corresponding to all the applications covered in this chapter are detailed. Real-time modules for Z-transforms, finite impulse response (FIR) filters, infinite impulse response (IIR) filters, discrete Fourier transform (DFT), residual number arithmetic systems, distributed arithmetic systems, booth multipliers and adaptive equalizers are designed, and the results associated with the implementation are explained.

Chapter 12 gives different advanced IP-based design examples. This chapter is designed with system generator–based fast Fourier transform (FFT), DFT design, coordinate rotation digital computer (CORDIC) algorithm design, FIR and IIR filter design, multiply and accumulator-based bandpass filter design, image processing and data-text reading/writing examples.

Chapter 13 describes differential pulse code modulation (DPCM), RF data encryption/decryption, optical up/down link data coding, fuzzy logic controllers, artificial neural network controllers, bit error rate tester design, error control codes (linear and convolutional codes), pick and place robotic controls and audio-codec interfaces.

Audience: This book is framed based on industrial evolution to provide in-depth knowledge/coverage of system development and the synthesis of efficient, portable and scalable programming using VHDL. This book is intended for use in university/college-level bachelor's/master's degree courses teaching advanced digital system design and real-world hardware interfaces. It is an ideal source for gaining knowledge very rapidly and starting project design straightaway. Readers should have taken an introductory digital course. Knowledge of VHDL would be helpful but is not necessary, since this book emphasizes hardware-interfacing methodology rather than language constructs. Therefore, this book is a more informative and handy guide to building real-world systems for academicians, research scholars, postdoctoral students, engineers, scientists, small/medium-level system developers, project designers, practicing technicians, hardware engineers, electronics scientists, hobbyists and so on. It not only establishes a foundation of VHDL, but also provides a comprehensive treatment of FPGA interfacing for various engineering design requirements that make this book useful for career interviews and competitive/comprehensive examinations. From this background, the design and interfacing of FPGA-based real-time digital systems can be explored, and readers can sharpen their design skills and learn the effective use of today's synthesis software and tools.

Dr. A. Arockia Bazil Raj

MATLAB®, Simulink®, and Xilinx® are registered trademark of The MathWorks, Inc. For product information, please contact:

The MathWorks, Inc.
3 Apple Hill Drive
Natick, MA 01760-2098 USA Tel: 508 647 7000
Fax: 508-647-7001
E-mail: info@mathworks.com
Web: www.mathworks.com

Author

Dr. A. Arockia Bazil Raj received his BE in Electronics and Communication Engineering from Bharathidasan University, Tiruchirappalli, India, and his ME in Communication Systems and his PhD in Information and Communication Technology from Anna University, Chennai, India. His PhD research on free-space optical communication was fully funded by the Defence Research and Development Organization (DRDO), New Delhi, India, under the Extramural Research and Intellectual Property Right (ER&IPR) project. He has delivered several invited talks and headed national/international conference sessions. He has authored several papers published in reputed international journals and conferences. He has published four books that are designed based on his teaching and research experience. He is a reviewer for various journals of IEEE, Springer, OSA, Wiley, Taylor & Francis, SAGE, Elsevier and so on, and he is a member of Indian Society for Technical Education (ISTE), IEEE, Institute of Engineering and Technology (IET) and Society of Photographic Instrumentation Engineers (SPIE). He was an assistant professor in the Kings College of Engineering, Thanjavur, India, from 2002 to 2006. He has been an associate professor in the Research, Development and Establishment (RDE) section of the Laser Communication Laboratory (LCL) facility at the same institute from 2007 to October 2015. Presently, he has been working in the field of radar system design and radar signal processing at the Defence Institute of Advanced Technology, Pune, Maharashtra, India, from November 2015 onwards. He has successfully completed/investigated various research projects sponsored by government and nongovernment sectors/laboratories. His current research interests cover free-space optical communication, radar system design, photonics radar design, MIMO radar design and radar signal processing.

Part I

Basic System Modeling and Programming Techniques

1

Very-Large-Scale Integration Technology: History and Features

1.1 Introduction and Preview

Before we learn about field programmable gate arrays (FPGAs) and building the digital systems inside them, let us have a look at what very large scale integration (VLSI) technology is; how it reduces the size of electronic circuits, making their function faster and more reliable; and its uses. The invention of the transistor was the driving factor of growth in VLSI technology. Electronics deals with circuits which involve various active and passive components. These circuits are used in various electronic devices and are called electronic systems. Originally, the components used in electronic systems, like diodes, were made up of vacuum tubes and were called discrete components. Later, when the solid state device (SSD) was invented, the components were made up of semiconductors. Vacuum tubes had the disadvantage of size, power requirements and reliability [1]. An integrated circuit (IC) is the circuit in which all the passive and active components are fabricated onto a single chip. Initially, the integrated chip could accommodate only a few components, but over time, the devices became more complex and required more circuits, which made the devices look bulky. Instead of accommodating more circuits in the system, integration technology was developed to increase the number of components that are to be placed on a single chip.

This technology not only helped reduce the size of the devices but also improved their speed. Depending upon the number of components, that is, transistors, to be integrated, they were categorized as small-scale integration (SSI), medium-scale integration (MSI), large-scale integration (LSI), VLSI, ultra-large-scale integration (ULSI) and giant-scale integration (GSI). The transistor-packing density of all these generations are: in SSI, 1–100 transistors were fabricated on a single chip, for example, gates and flip-flops; in MSI, 100–1000 transistors could be integrated on a single chip, for example, bit microprocessors; in LSI, 1000–10,000 transistors could be integrated on a single chip, for example, 8-bit microprocessors, random access memory (RAM), and read only memory (ROM); in VLSI, 10,000–1 million transistors could be accommodated, for example, 16–32 bit microprocessors; in ULSI, 1–10 million transistors could be accommodated, for example, special purpose registers; and in GSI, more than 10 million transistors could be accommodated, for example, in embedded systems. As we have seen, VLSI and further generations are the technology by which millions of transistors can be fabricated on a single chip.

Now, what is the necessity of fabricating that many transistors on a single chip? During the vacuum-tube era, electronic devices were huge, required more power, dissipated more heat, and were not as reliable. Thus, there was certainly a need to reduce the size of these devices and their heat dissipation. After the invention of SSDs, the size and heat produced by devices were drastically reduced, but over time, the requirement of additional features in electronic devices increased, which again made the devices look bulky and complex. This gave birth to the invention of technology that can fabricate more components onto a single chip. As the need for additional features in electronic devices grew, the growth of VLSI technology improved. In 1965, Gordon Moore, an industry pioneer, predicted that the number of transistors on a chip would double every 18 to 24 months. He also predicted that semiconductor technology would double its effectiveness every 18 months and many other factors would grow exponentially. VLSI technology has many advantages, such as reducing the size of the circuits, reducing the effective cost of the devices, increasing the operating speed of the circuits, requiring less power than discrete components, providing higher reliability, and occupying a relatively smaller area. In today's scientific world, VLSI chips are widely used in various branches of engineering applications, such as voice/data communication, networks, digital signal processing (DSP), computers, commercial electronics, automobiles, medical electronics, defense applications, and many more [1].

In the later sections of this chapter, some topics relevant to microelectronics, complementary metal oxide semiconductor (CMOS) technology, VLSI design flow, subsystem design, and HDL requirements are summarized. Several books on these topics are available through which the reader can gain more information and knowledge.

1.2 A Review of Microelectronics

Microelectronics has become more useful in embedded systems thanks to the development of semiconductor technology. The goal of this section is to briefly review the main features of electronics and microelectronics. The most significant step in modern electronics was the development of the transistor by Bell Laboratories in 1948. The development of vacuum tubes soon led to the simple radio; then came more complex circuits such as communication systems. Modern electronic systems now allow us to communicate with other parts of the world via satellite. Data are now collected from space without the presence of man because of microelectronic technology. Sophisticated control systems allow us to operate equipment by remote control in hazardous situations. We can remotely pilot aircraft from takeoff to landing. We can also make course corrections to spacecraft millions of miles from earth. Space flight, computers, and even video games would not be possible without microelectronics.

The solid-state diode and transistor opened the door to microelectronics, which is defined as "an area of technology associated to the realization of electronic systems made of extremely small electronic parts or elements". The term *microelectronics* is normally associated with IC; however, many other types of circuits also fall into the microelectronics category. During the Second World War, the need to reduce the size, weight, and power of military electronic systems became important because of the increased use of these systems. As systems became more complex, their size, weight and power requirements rapidly increased. The increases finally reached a point that was unacceptable, especially in aircraft and for infantry personnel who carried equipment in combat. These unacceptable factors were the driving force in the development of smaller, lighter, and more efficient electronic circuit components. Such requirements continue to be important factors in the development of new systems, both for military and commercial markets.

The earliest electronic circuits were fairly simple. They were composed of a few tubes, transformers, resistors, capacitors, wiring, and so on. As more was learned by designers, they began to increase both the size and complexity of circuits. Component limitations were soon identified as this technology developed. Vacuum tubes were found to have several built-in problems. Although the tubes were lightweight, associated components and chassis were quite heavy. It was not uncommon for such a chassis to weigh 20 to 60 kilograms. In addition, the tubes generated a lot of heat, required a warm-up time from 1 to 2 minutes, and required power supply voltages of 300 or more DC volts. No two tubes of the same type were exactly alike in output characteristics. Therefore, designers were required to produce circuits that could work with any tube of a particular type. This meant that additional components were often required to tune the circuit to the output characteristics required for the tube used. The actual size of the transformer is approximately $4 \times 4 \times 3$ inches and the capacitors are approximately 1×3 inches. The components used in the circuits are very large when compared to modern microelectronics. A circuit could be designed either as a complete system or as a functional part of a larger system. In complex systems, such as radar, many separate circuits were needed to accomplish the desired tasks. Multiple function tubes, such as dual diodes, dual triodes, tetrodes and others helped considerably to reduce the size of circuits. However, weight, heat, and power consumption continued to be problems that plagued designers. Another major problem with vacuum-tube circuits was the method of wiring components referred to as point-to-point wiring. Not only does this wiring look like a rat's nest, but it also often causes unwanted interactions between components.

Vacuum-tube circuits proved to be reliable under many conditions. Still, the drawbacks of large size, heavy weight, and significant power consumption made them undesirable in most situations; for example, computer systems using tubes were extremely large and difficult to maintain. The electronics numerical integrator and computer (ENIAC) built the electronic computer in 1945, which contained 18,000 tubes and required a full day just to locate and replace the faulty tubes. One concept that eased the technician's job was that of modular packaging. Instead of building a system on one large chassis, it was built of modules or blocks. Each module performed a necessary function of the system. Modules could easily be removed and replaced during troubleshooting and repair. For instance, a faulty power supply could be exchanged with a good one to keep the system operational. The faulty unit could then be repaired while out of the system. This is an example of how the module concept improved the efficiency of electronic systems. Even with these advantages, vacuum tube modules still had faults. Tubes and point-to-point wiring increased the size, weight, and power consumption, which is emphasized in the field of microelectronics.

The transition from vacuum tubes to SSDs took place rapidly. As new types of transistors and diodes were created, they were adapted to the circuits. The reductions in size, weight, and power were impressive. Circuits that earlier weighed as much as 25 kilograms were reduced to just a few grams by replacing bulky components with the much lighter SSDs. The earliest solid-state circuits still relied on point-to-point wiring which caused many of the disadvantages mentioned earlier. A metal chassis, similar to the type used with tubes, was required to provide physical support for the components. The solid-state chassis was still considerably smaller and lighter than the older tube chassis. One of the most significant developments in circuit packaging has been the printed circuit board (PCB). The PCB is usually an epoxy board on which the circuit leads have been added by the photo-etching process. This process leaves the copper strips that are used to connect the components. In general, PCBs eliminate both the heavy metal chassis and the point-to-point wiring. Although PCBs represent a major improvement over tube technology, they are not without fault. For example, the number of components on each board is limited by the sizes and shapes of components. Also, while vacuum tubes are easily removed for testing or replacement, PCB components are soldered into place and are not as easily removed. Normally, each PCB contains a single circuit or a subassembly of a system. All PCBs within the system are routinely interconnected through cabling harnesses.

Another mounting form that has been used to increase the number of components in a given space is the cordwood module, where the components are placed perpendicular to the end plates. The components are packed very closely together, appearing like cordwood for a fireplace. Cordwood modules may or may not be encapsulated, but in either case, they are difficult to repair. Many advertisements for electronic equipment refer to ICs or solid-state technology. The accepted definition for an IC is that "it consists of elements inseparably associated and formed on or within a single substrate [2]". In other words, the circuit components and all interconnections are formed as an unit. Connections were made with three types of ICs namely, monolithic, film, and hybrid. Monolithic integrated circuits are those that are formed completely within a semiconductor substrate.

These integrated circuits are commonly referred to as silicon chips. Film-integrated circuits are broken down into two categories, thin film and thick film. Film components are made of either conductive or nonconductive material that is deposited in desired patterns on a ceramic or glass substrate. Film can be used only as passive circuit components, such as resistors and capacitors. Transistors and/or diodes are added to the substrate to complete the circuit. Hybrid integrated circuits combine two or more IC types and discrete components. Microelectronic technology today includes thin film, thick film, hybrid and IC, and combinations of these. Such circuits are applied in digital switching and linear circuits.

LSI and VLSI are the results of improvements in microelectronics production technology. The entire substrate wafer is used instead of one that has been separated into individual circuits. In LSI and VLSI, a variety of circuits can be implanted on a wafer, resulting in size and weight reduction. ICs in modern computers may contain the entire memory and processing circuits on a single substrate. LSI is generally applied to ICs consisting of 1000 to 2000 logic gates or from 1000 to 64,000 bits of memory. VLSI is used in ICs containing over 2000 logic gates or greater than 64,000 bits of memory. Development of a microelectronic device begins with a demand from industry or as the result of research. A device that is needed by industry may be a simple diode network or a complex circuit consisting of thousands of components. No matter how complex the device, the basic steps of production are similar. Each type of device requires circuit design, component arrangement, preparation of a substrate and the depositing of proper materials on the substrate. The first consideration in the development of a new device is to determine what the device is to accomplish. Once this has been decided, engineers can design the device. During the design phase, the engineers will determine the numbers and types of components and the interconnections needed to complete the planned circuit. Planning the component arrangement for a microelectronic device is a critical phase of production. Care must be taken to ensure the most efficient use of space available.

A computer is used to prepare the layout for complex devices. The computer is able to store the characteristics of thousands of components and can provide a printout of the most efficient component placement. Component placement is then transferred to extremely large drawings. During this step, care is taken to maintain the patterns as they will appear on the substrate. A wafer mask is used to deposit materials on a substrate. By changing the pattern of the mask, we can change the component arrangement of the circuit. Several different masks may be used to produce a simple microelectronic device. When used in proper sequence, conductor, semiconductor, or insulator materials may be applied to the substrate to form transistors, resistors, capacitors, and interconnecting leads [2].

1.3 Complementary Metal Oxide Semiconductor Technology and Gate Configuration

Over the past two decades, CMOS technology has played a very important role in the global IC industries. Although the basic principle of the MOS field effect transistor (MOSFET) was explained by J. Lilienfeld in 1925, commercial success of MOS devices could be ensured only during the 1960s with the invention of the silicon planar process. Then, MOS devices fabricated by n-channel MOS (nMOS)-silicon-gate technology came into use in the early of 1970s, prior to which only single-polarity p-channel transistors were in use. At the same time, P.K. Weimer and F. Wanlass demonstrated the possibility of using both polarity devices on the same substrate. With the implementation of the CMOS inverter, NOR gate and NAND gate, the CMOS concept took root and showed low power-dissipation devices. Initially, the requirements of more complex processing technology and larger silicon area compared to single polarity transistors led to limited application of CMOS transistors in general system designs. The CMOS technology rapidly improved to support large chip sizes; however, the issue of power consumption became more and more critical. Silicon is predominantly used in the fabrication of semiconductor devices and microcircuits. A MOS transistor structure is built by stacking several layers of conducting, insulating and semiconductor materials. The fabrication process is carried out on a single crystal of silicon available as thin circular wafers of diameter about 30 cm. CMOS technology makes way for two kinds of transistors, namely nMOS transistors and p-channel MOS (pMOS) transistors built by diffusing n-type impurities (rich in electrons) and positively doped silicon (rich in holes), respectively [3].

The structure of the n-channel transistor is made of a p-type silicon substrate accommodating two diffused islands of n-type silicon. Selected areas of the p-substrate are altered by diffusion or implantation of n-type impurities. On top of the area, separating the n-type provides a thin insulating layer of silicon dioxide (SiO_2) above which there is a conducting layer called the gate (G), as shown in Figure 1.1a. A p-channel transistor, on the other hand, is made of an n-substrate separating two diffused p-type islands as shown in Figure 1.1b. Apart from the gate electrode, an nMOS transistor has two more terminals known as the source (S) and the drain (D) which connect the two n-diffused regions (p-diffused regions

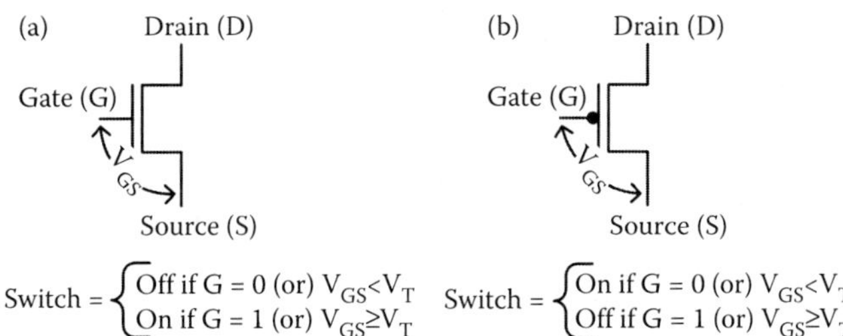

FIGURE 1.1
Switch models of MOS transistors: nMOS switch (a) and pMOS switch (b).

in a pMOS transistor) with the external environment of the device. The gate acts as a control input, regulating the current flow between the source and the drain. Although the source and the drain are physically equivalent, the name source is reserved for the terminal by which the current carriers enter the device, whereas the drain refers to the terminal by which the carriers leave the device. The substrate, also called the body, happens to be the fourth terminal of a MOS transistor. The voltage applied at the gate terminal regulates the flow of current between the source and the drain. In this way, a MOS transistor may be viewed as a simple on/off switch. Assume that logic 1, power on or V_{DD} denotes a high voltage normally between 1.5 to 15 volts, and logic 0, ground or V_{SS} stands for a low voltage normally set to zero volts.

The switch models of both nMOS and pMOS transistors are shown in Figure 1.1; as shown there, an nMOS switch is deemed closed or on if the gate voltage is at logic 1 or, more precisely, if the potential between the G and the S terminals (V_{GS}) happens to be greater than the threshold voltage (V_T). A closed nMOS switch implies the existence of a continuous channel between the S and D terminals. On the other hand, an off or open nMOS switch indicates the absence of a connecting channel between the S and the D. Similarly, a pMOS switch is considered on or closed if the potential V_{GS} is smaller or more negative than the V_T.

Now, we will discuss the configuration of CMOS logic elements such as inverters, OR, AND, NAND, and NOR gates. Any combinational, sequential, and memory circuits can be designed using these basic logic elements. The inverter is universally accepted as the most basic logic gate doing a Boolean operation on a single input variable. Figure 1.2 shows the symbol, truth table, and a general structure of a CMOS inverter. As shown there, the simple structure consists of a combination of a pMOS transistor at the top and an nMOS transistor at the bottom.

Combinational logic circuits perform Boolean operations on multiple input variables and determine the outputs as Boolean functions of the inputs. The basic two-input OR functions can be realized by series and parallel combinations of nMOS and pMOS transistors, as shown in Figure 1.3a. The configuration of two nMOS switches in parallel and two pMOS switches in series produce the OR function. On the other hand, a two-input AND function is realized by placing three pMOS transistors and three nMOS transistors as shown in Figure 1.3b. The logical symbols and function tables corresponding to the OR and AND logic gates are also shown in Figure 1.3. For nMOS transistors, if the input is logic 1, the switch is on; otherwise, it is off. On the other hand, for the pMOS, if the input

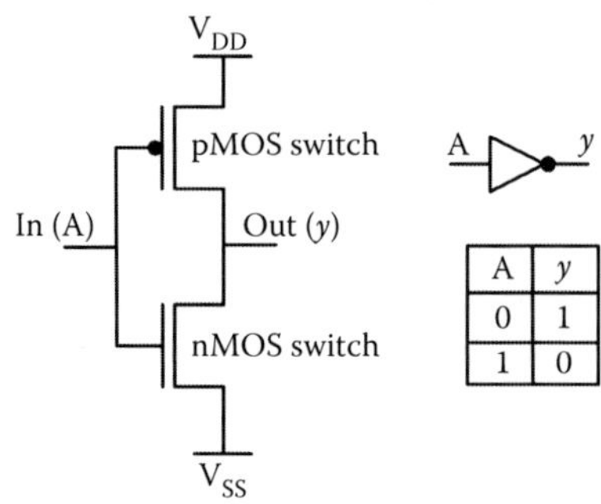

A	y
0	1
1	0

FIGURE 1.2
Symbol, circuit structure, and truth table of a CMOS inverter.

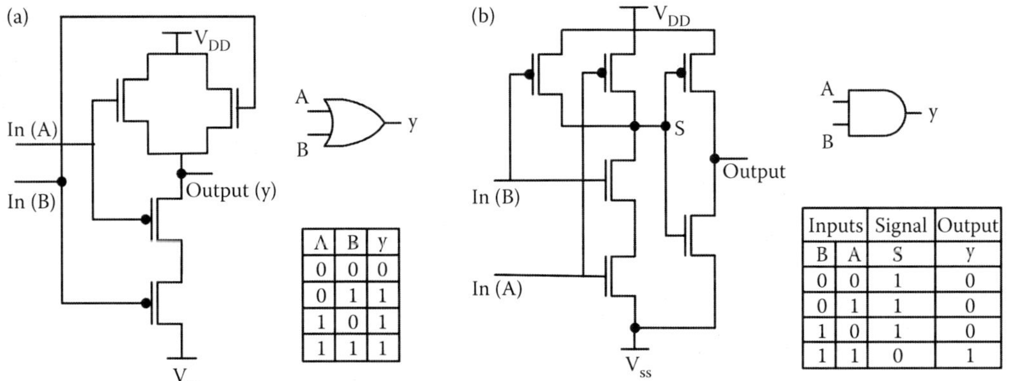

A	B	y
0	0	0
0	1	1
1	0	1
1	1	1

Inputs		Signal	Output
B	A	S	y
0	0	1	0
0	1	1	0
1	0	1	0
1	1	0	1

FIGURE 1.3
CMOS configuration of OR (a) and AND (b) gates.

is logic 0, the transistor is on; otherwise, the transistor is off. When a circuit contains both nMOS and pMOS transistors, we say it is implemented in CMOS [3].

Figure 1.4 shows the CMOS implementation of a two-input NAND gate and two-input NOR gate. In the NAND gate, the pull-down subcircuit is made of a series combination of two nMOS transistors. These are responsible for conducting logic 0 to the output node when both of the gates are at logic 1. The pull-up path, on the other hand, consists of a parallel combination of two pMOS transistors. If either is at logic 0, the output node gets the logic value 1. For a two-input NAND gate, the Boolean expression is $(AB)' = A' + B'$.

In a two-input NOR gate, the pull-up subcircuit is made of a series combination of two pMOS transistors. These are responsible for conducting logic 1 to the output node when both of the gates are at logic 0. The pull-down path, on the other hand, consists of a parallel combination of two nMOS transistors. If either of the inputs is at logic 1, the output node gets the logic value 0. For a two-input NOR gate, the Boolean expression is $(A+B)' = A'B'$.

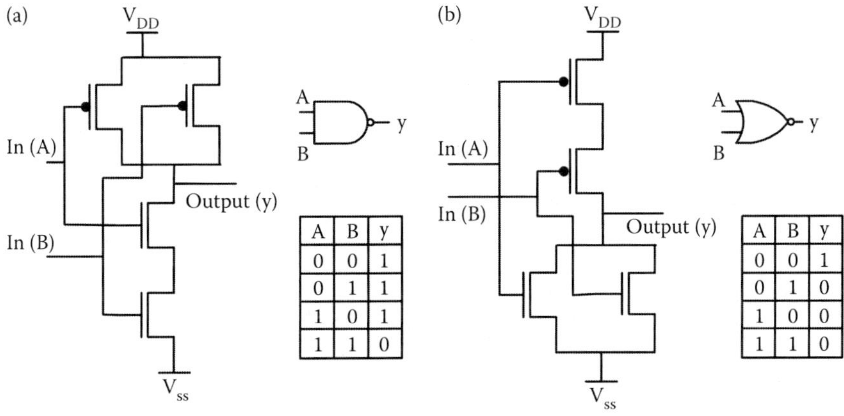

A	B	y
0	0	1
0	1	1
1	0	1
1	1	0

A	B	y
0	0	1
0	1	0
1	0	0
1	1	0

FIGURE 1.4
CMOS configuration of NAND (a) and NOR (b) gates with their truth tables.

1.4 CMOS Fabrication and Layout

In the early 1960s, the semiconductor manufacturing process was initiated in Texas, and in 1963, CMOS was patented by Frank Wanlass. ICs are manufactured by utilizing the semiconductor device fabrication process. These ICs are major components of every electronic device we use in our daily life. Many simple and complex electronic circuits are designed on a wafer made of semiconductor compounds, mostly silicon, by using different fabrication steps/processes. This technology is used in developing microprocessors, microcontrollers, digital logic circuits and many other ICs. It facilitates low power dissipation and high packing density with much less noise margin. The CMOS can be fabricated using different processes such as the N-well process, P-well process and twin tub process. The fabrication of a CMOS can be done by following 20 steps, through which the CMOS can be obtained by integrating both the nMOS and pMOS transistors on the same chip substrate. The integration of the nMOS and pMOS devices on a chip is done via a special region called a well or tub at which semiconductor type and substrate type are opposite each other. A P-well has to be created on a N-substrate and an N-well has to be created on a P-substrate. In this section, the fabrication of the CMOS is described using the P-substrate, in which the nMOS transistor is fabricated on a P-type substrate.

The fabrication processes are briefly listed below and the corresponding illustrations are shown in Figure 1.5. The steps are as follows:

1. Take a P-substrate.
2. The oxidation process is done by using high-purity oxygen and hydrogen that are exposed at approximately 1000°C.
3. A light-sensitive polymer is applied as a photoresist layer.
4. The photoresist is exposed to UV rays through the N-well mask.
5. A part of the photoresist layer is removed by treating the wafer with a basic or acidic solution.
6. The SiO_2 oxidation layer is removed through the open area made by the removal of the photoresist using hydrofluoric acid.
7. The entire photoresist layer is stripped off, as shown in Figure 1.5.
8. By using ion implantation or a diffusion process, the N-well is formed.
9. Using hydrofluoric acid, the remaining SiO_2 is removed.
10. The chemical vapor deposition (CVD) process is used to deposit a very thin layer of gate oxide.
11. Except the two small regions required for forming the gates of nMOS and pMOS, the remaining layer is stripped off.
12. Next, an oxidation layer is formed on this layer with two small regions for the formation of the gate terminals of nMOS and pMOS.
13. By using the masking process, small gaps are made for the purpose of N-diffusion. The n-type (n+) dopants are diffused or ion implanted, and the three n+ are formed for the formation of the terminals of the nMOS.
14. The remaining oxidation layer is stripped off.
15. Similar to the above N-diffusion process, the P-diffusion regions are diffused to form the terminals of the pMOS.

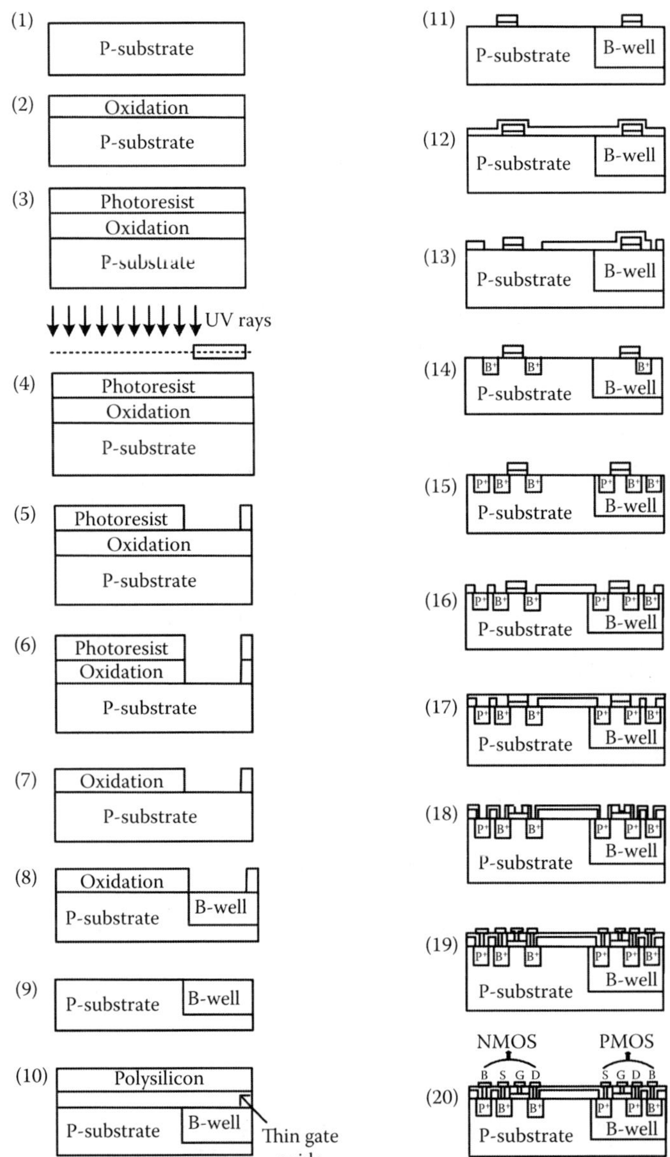

FIGURE 1.5
Illustrations of CMOS fabrication process.

16. A thick-field oxide is formed in all regions except the terminals of the pMOS and nMOS.
17. Aluminum is sputtered on the whole wafer.
18. The excess metal is removed from the wafer layer.
19. The terminals of the pMOS and nMOS are made from the respective gaps.
20. The names of the terminals of the NMOS and pMOS are assigned.

More details about the each and every step can be found in Ref. 4.

1.5 VLSI Design Flow

The VLSI design cycle starts with a formal specification of a VLSI chip, as listed in the previous section. It follows series of steps and eventually produces a packaged chip. A typical design cycle at the top level may be represented by the flow chart shown in Figure 1.6. In this section, our emphasis is on the physical design step of the VLSI design cycle. To gain a global perspective, we briefly outline all the steps of the VLSI design cycle below.

System Specification: The first step of any design process is to lay down the specifications of the system. System specification is a high-level representation of the system. The factors to be considered in this process include performance, functionality and physical dimensions, that is, the size of the die (chip). The fabrication technology and design techniques are also considered. The specification of a system is a compromise between market requirements, technology, and economical viability. The end results are specifications for the size, speed, power, and functionality of the VLSI system.

Architectural Design: The basic architecture of the system is designed in this step. This step includes reduced instruction set computer (RISC), complex instruction set computer (CISC), a number of arithmetic logic units (ALUs), floating-point units, the number and structure of pipelines, and the size of the caches, among others. The outcome of architectural design is a micro-architectural specification (MAS). While MAS is a textual description, architects can accurately predict the performance, power, and die size of the design based on such a description.

Functional Design: In this step, main functional units of the system are identified. This also identifies the interconnection requirements between the units. The area, power, and other parameters of each unit are estimated. The behavioral aspects of the system are considered without implementing the specific information. For example, it may specify that a multiplication is required, but exactly in which mode such multiplication may be executed is not specified. We may use a variety of multiplication hardware depending on the speed and word size requirements. The key idea is to specify behavior in terms of input, output, and timing of each unit, without specifying its internal structure. This information leads to improvement of the overall design process and reduction of the complexity of subsequent phases.

Logic Design: In this step, the control flow, word widths, register allocation, arithmetic operations, and logic operations of the design that represent the functional

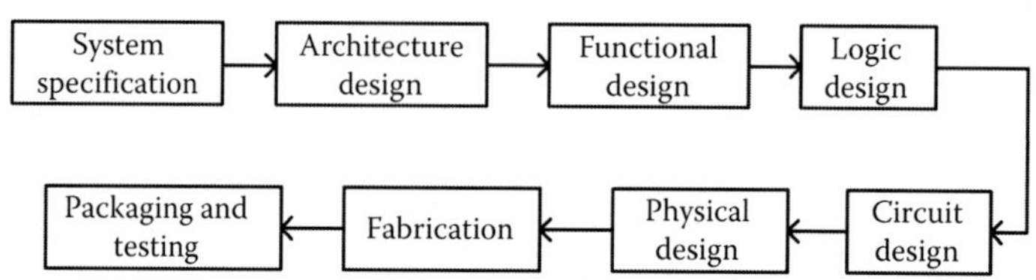

FIGURE 1.6
VLSI design flow.

design are derived and tested. This description is called the register transfer level (RTL) description. RTL is expressed in an HDL such as VHDL or Verilog HDL. This description can be used in simulation and verification. This description consists of Boolean expressions and timing information. The Boolean expressions are minimized to achieve the smallest logic design which conforms to the functional design. This logic design of the system is simulated and tested to verify its correctness. In some special cases, logic design can be automated using high-level synthesis tools. These tools produce an RTL description from a behavioral description of the design.

Circuit Design: The purpose of circuit design is to develop a circuit representation based on the logical design. The Boolean expressions are converted into a circuit representation by taking into consideration the speed and power requirements of the original design. Circuit simulation is used to verify the correctness and timing of each component. The circuit design is usually expressed in a detailed circuit diagram. This diagram shows the circuit elements, such as cells, macros, gates, transistors, and so on, and the interconnection between these elements. This representation is also called a netlist. Tools used to manually enter such a description are called schematic capture tools. In many cases, a netlist can be created automatically from logic (RTL) description by using logic synthesis tools.

Physical Design: In this step, the circuit representation or netlist is converted into a geometric representation. As stated earlier, this geometric representation of a circuit is called a layout and is created by converting each logic component into a geometric representation, that is, specific shapes in multiple layers which perform the intended logic function of the corresponding component. Connections between different components are also expressed as geometric patterns, typically lines in multiple layers. The exact details of the layout also depend on design rules, which are guidelines based on the limitations of the fabrication process and the electrical properties of the fabrication materials. Physical design is a very complex process and therefore it is usually broken down into various substeps.

Fabrication: After layout and verification, the design is ready for fabrication. Since layout data are typically sent to fabrication on a tape, the event of release of data is called tape out. Layout data is converted (or fractured) into photo-lithographic masks, one for each layer. Masks identify spaces on the wafer where certain materials need to be deposited, diffused, or even removed. Silicon crystals are grown and sliced to produce wafers. Extremely small dimensions of VLSI devices require that the wafers be polished to near perfection. The fabrication process consists of several steps involving deposition and diffusion of various materials on the wafer. During each step, one mask is used. Several dozen masks may be used to complete the fabrication process. A large wafer is 20 cm (8 inches) in diameter and can be used to produce hundreds of chips, depending on the size of the chip.

Packaging, Testing, and Debugging: Finally, the wafer is fabricated and diced into individual chips in a fabrication facility. Each chip is then packaged and tested to ensure that it meets all the design specifications and that it functions properly. Chips used in PCBs are packed in a dual in-line package (DIP), pin grid array (PGA), ball grid array (BGA), and quad flat package (QFP). Chips used in multichip modules (MCMs) are not packed, since they use bare or naked chips [1,5].

1.6 Combinational and Sequential Circuit Design

The computer's central processing unit (CPU) utilizes the hardware circuitry that includes electronic kits, microchips, ICs, and software (programming) technology. However, we must know how this hardware and software interact to perform all these required operations. We know that computers perform all their operations using binary digit format. For ease of operation, they use binary instead of decimal values. These binary (logical) operations are carried out by the circuits formed using various types of digital logic gates. Primarily, we must know what digital logic circuits are. Digital logic gates are the physical building blocks of ICs used for the execution of logical operations or tasks by utilizing Boolean logic. These logic gates are primarily implemented using electronic semiconductor switches such as diodes or transistors. However, later, pneumatic logic, molecules, optics, fluidic logic, electromagnetic relay logic, and mechanical elements are used to implement logic gates, though, practically, logic gates are built using CMOS technology, field effect transistors (FETs), and MOSFETs. Different types of logic gates such as the AND, OR, NOT, NAND, NOR, XOR, and XNOR gates are used to form different digital logic circuits. These basic logic gate and digital logic circuits utilize Boolean functions to execute output. Hence, these digital logic gates are also called binary logic or Boolean logic gates. These digital logic circuits can be classified into two categories, combinational logic circuits and sequential logic circuits. Before studying the difference between combinational and sequential logic circuits, we must primarily know what a combinational logic circuit is and what a sequential logic circuit is.

1.6.1 Combinational Logic Circuits

The simple time-independent logic circuits that are implemented using Boolean circuits whose output logic value depends only on the input logic values can be called combinational logic circuits. Figure 1.7a shows three major components of a combinational logic circuit, the logic diagram, truth table, and Boolean expression. The digital logic circuits

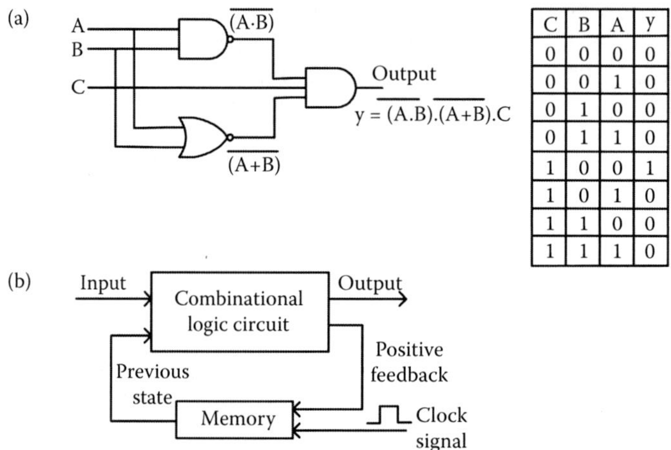

C	B	A	y
0	0	0	0
0	0	1	0
0	1	0	0
0	1	1	0
1	0	0	1
1	0	1	0
1	1	0	0
1	1	1	0

FIGURE 1.7
Configuration of combinational (a) and sequential (b) circuit.

whose outputs can be determined using the logic function of the current state input are combinational logic circuits; hence, these are also called time-independent logic circuits. These combinational digital logic circuits do not have the capability to store a state inside them. Hence, the combinational logic circuits do not contain any memory elements. The arithmetic operations performed on the data stored in the computers are done using combinational logic circuits. The combinational digital logic circuits are fundamentally implemented using different types of devices such as multiplexers, demultiplexers, encoders, decoders, half adders, full adders, and so on. The components of the arithmetic and logic units of the computers are generally composed of combinational digital logic circuits. The independent working states of the combinational logic circuits are represented with Boolean algebra and, after simplification, the circuit can be implemented by using logical gates. The combinational digital circuits do not require any feedback and are independent of the clock. As there are no clocks used in these digital logic circuits, they do not need any triggering. The combinational logic circuit's behavior can be defined by using the set of output functions. In general, the sum of product (SOP) or products of sum (POS) method is used for the construction of combinational logic [6].

1.6.2 Sequential Logic Circuits

The simple logic circuit whose output logic value depends on the input logic values and also on the stored information is called a sequential logic circuit. Figure 1.7b represents the block diagram of the sequential logic circuit. The difference between combinational logic circuits and sequential logic circuits can be easily understood by knowing about each circuit in detail. Digital logic circuits whose outputs can be determined using the logic function of current and past state inputs are called sequential logic circuits. These sequential digital logic circuits are able to retain the earlier state of the system based on the current inputs and earlier state. Hence, unlike combinational logic circuits, sequential digital logic circuits are capable of storing the data in a digital circuit. Sequential logic circuits contain memory elements. The latch is considered the simplest element used to retain the earlier memory or state in the sequential digital logic. If we consider the true structural form, it can be viewed as a combinational circuit with one or more outputs fed back as inputs. These sequential digital logic circuits are used in memory elements and also in finite state machines (FSMs), which are digital circuit models with finite possible states. The maximum number of sequential logic circuits uses a clock for triggering flip-flop operation. If the flip-flop in the digital logic circuit is triggered, then the circuit is called a synchronous sequential circuit and the other circuits (which are simultaneously not triggered) are called asynchronous sequential circuits. The sequential digital logic circuits utilize the feedback from outputs to inputs. The sequential logic circuit's behavior can be defined by using the sets of output functions and next state or memory functions. In practical digital logic circuits, combinational digital logic circuits and sequential digital logic circuits are used separately and/or in combination [6,7].

1.7 Subsystem Design and Layout

Any digital system can be implemented by using the necessary subsystems or components. For example, consider a full adder circuit that can be designed using few basic gates.

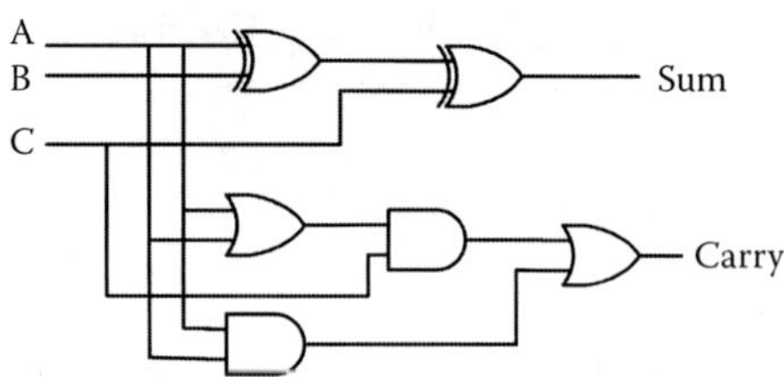

FIGURE 1.8
Circuit of a full adder.

In Figure 1.8, the full adder is designed with two XOR gates, two AND gates and two OR gates. All these basic gates have to be implemented while preparing the hardware using the nMOS and pMOS transistors, which are actually low-level implementations of the intended digital circuit [1,3–5].

As we have seen in Section 1.3, any combinational and sequential circuits can be implemented using CMOS transistors. The low-level design layout for a full adder is shown in Figure 1.9. The layout for the sum is shown on the left and that for carry is shown on the right in Figure 1.9 with their respective truth tables. Based on the configuration of the combinational circuits, the layout with the nMOS and pMOS transistors is chosen and placed in the design. Inputs in both the circuits are applied directly, that is, A, B, C and so on, as well as their complemented form, that is, A′, B′, C′ and so on.

Now, we analyze the function of these two circuits for some set of inputs. Consider the input ABC is "111"; then, the first three pMOS transistors whose inputs are C′, A′, and B′ contribute to generate the output logic 1. In the same way, the input "111" generates the carrying of logic 1 with the contribution of two pMOS transistors. The suitable transistors are driven using the inputs A, B, and C in accordance with the truth tables to implement a full adder.

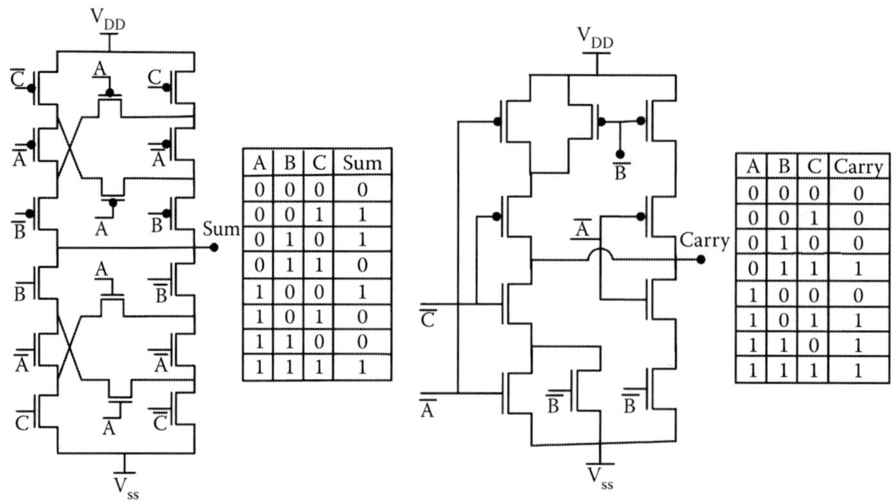

A	B	C	Sum
0	0	0	0
0	0	1	1
0	1	0	1
0	1	1	0
1	0	0	1
1	0	1	0
1	1	0	0
1	1	1	1

A	B	C	Carry
0	0	0	0
0	0	1	0
0	1	0	0
0	1	1	1
1	0	0	0
1	0	1	1
1	1	0	1
1	1	1	1

FIGURE 1.9
Low-level implementation of a full adder circuit for sum (left) and for carry (right).

1.8 Types of Application-Specific Integrated Circuits and Their Design Flow

Application specific integrated circuit (ASIC) is not a standard IC since they are designed for a specific use or application. Here, a complete system or product is integrated into a chip and virtually no other components are required. The cost of designing an ASIC is very high and therefore they tend to be reserved for high-volume products. However, ASICs can be very cost effective for many applications where they are produced on a large scale. ASICs are classified as shown in Figure 1.10. The main types briefed below are (i) full-custom ASICs, (ii) standard-cell–based ASICs, and (iii) gate-array–based ASICs. In the full-custom ASIC type, the complete IC design and developments are customized by the designer. Full custom offers the highest performance and lowest part cost. The advantages of full-custom design are reduced area and thus recurring component cost, better performance and increased reliability. The disadvantages of this type of ASIC are increased design time, complexity, increased nonrecurring engineering costs, design expense, requirement of a much more highly skilled design team and the highest risk. Examples of full-custom ASICs are mobile processors, sensor ICs, actuator ICs and so on.

In standard-cell–based ASICs, the ASIC manufacturer creates functional blocks with known electrical characteristics, such as propagation delay, capacitance and inductance. Utilizing these functional blocks, a standard cell of very high gate density and good electrical performance is designed. This gives a high degree of flexibility, which provides the standard functions that are able to meet the requirements. The significant advantage in standard-cell–based ASICs is that this uses the manufacturer's cell libraries that have been used in potentially hundreds of other design implementations. Thus, there is a very low risk associated with this compared to full-custom design. Standard cells produce a design density that is cost effective; the manufacturing time is also shorter, and they can also integrate IP cores and SRAM effectively.

In gate-array–based ASICs, a large number of standard functions, as shown in Figure 1.11, are required and can be connected in a particular manner to meet the given requirements. The manufacturing lead time of gate-array-based ASICs is between two days and two weeks. The design flow is a sequence of steps to design an ASIC. Following are the steps of ASIC design:

1. Design entry: Using an HDL or schematic entry, the design is entered into an ASIC design system.
2. Logic synthesis: Using an HDL (VHDL or Verilog) and a logic synthesis tool, a netlist is produced. A netlist is a description of the logic cells and their connections.
3. System partitioning: In this step, a large system is divided into ASIC-sized pieces.

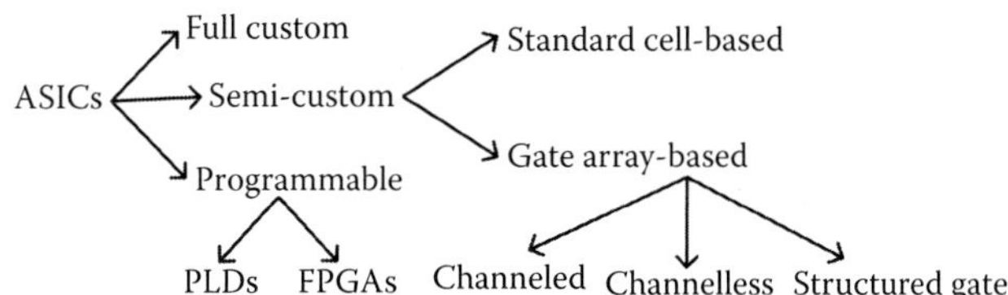

FIGURE 1.10
Classification of ASICs.

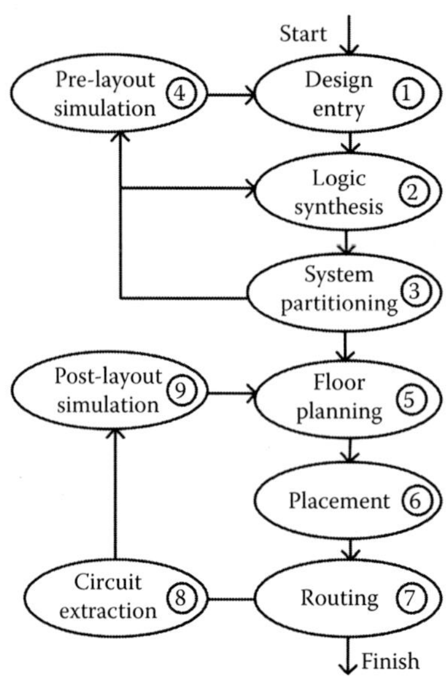

FIGURE 1.11
Typical ASIC design flow.

4. Prelayout simulation: Functioning of the design is checked to see if it is working correctly.
5. Floor planning: The blocks of the netlist are arranged on the chip such a way that they take up the optimum space. In floor planning, distribution of connections and intereffects of electrical parameters are also considered.
6. Placement: The locations of the cells in a block are decided.
7. Routing: The connections between cells and blocks are made.
8. Extraction: The resistance and capacitance of the interconnects are determined.
9. Postlayout simulation: After the interconnections have been made, the working of the design is tested.

Today, ASIC design flow is a very solid and mature process. The overall ASIC design flow and the various steps within it have proven to be both practical and robust in millions of previous ASIC designs. Each and every step of the ASIC design flow has a dedicated electronics design automation (EDA) tool that covers all the aspects related to the specific task perfectly. Most importantly, all the EDA tools can import and export the different file types to help make a flexible ASIC design flow that uses multiple tools from different vendors. ASIC design flow is not exactly a push-button process. To succeed in the ASIC design flow process, one must have a robust and silicon-proven flow, a good understanding of the chip specifications and constraints, and an absolute domination over the required EDA tools and their reports. The next section covers the main ASIC design flow steps at a very high level, and more in-depth details can be found in References 1, 5, 8, and 9.

ASIC System Design: Assuming the ASIC specifications are completed and approved by the different parties, it is time to start thinking about the architectural design. In the ASIC system design phase, the entire chip functionality is broken down into small pieces with a clear understanding of the block implementation. For example, for an encryption block, should we use a CPU or a state machine? Some other large blocks need to be divided into subsystems, and the relationship between the various blocks has to be defined. In this phase, the working environment is documentation.

RTL: For digital ASICs or digital blocks within a mixed-signal chip, this phase is basically the detailed logic implementation of the entire ASIC. This is where the detailed system specifications are converted into VHDL or Verilog HDL language. In addition to the digital implementation, a functional verification is performed to ensure the RTL design that is done according to the specifications. When all the blocks are implemented and verified, the RTL is then converted into a gate-level netlist.

Synthesis: In this phase, the hardware description (RTL) is converted to a gate-level netlist. This process is performed by a synthesis tool that takes a standard cell library, constraints and the RTL code and produces a gate-level netlist. Synthesis tools are run different implementations to provide best gate-level netlist that meets the constraints. It takes power, speed and size into account; therefore, the results can vary significantly from each other. Verification of whether the synthesis tool has correctly generated the gate-level netlist becomes significant.

Layout: In this stage, the gate-level netlist is converted to a complete physical geometric representation. The first step is floor planning, which is a process of placing the various blocks and the I/O pads across the chip area based on the design constraints. Then, placement of physical elements within each block and integration of analog blocks or external IP cores are performed. When all the elements are placed, a global and detailed routing is run to connect all the elements together. Also, after this phase, a complete simulation is required to ensure the layout phase has been properly done. The file produced at the output of the layout is the GDS2 file, which is the file used by the foundry to fabricate the silicon. The layout should be done according to the silicon foundry design rules [8,9].

1.9 VHDL Requirements and Features

The development of VHDL was initiated in 1981 by the United States Department of Defense (DoD) to address the hardware lifecycle crisis. The cost of reproducing electronic hardware became obsolete and was reaching crisis point because the function of the parts was not adequately documented and the various components making up a system were individually verified using a wide range of different and incompatible simulation languages and tools. The requirement was for a language with a wide range of descriptive capability that would work on any simulator and was independent of technology or design methodology. The standardization process for VHDL was unique in that participation and feedback from the industry were sought at an early stage. A baseline language (version 7.2) was published two years before the standard so that tool development could begin in earnest in advance of the standard. All rights to the language definition were given away by the Department of Defense to the IEEE in order to encourage industry acceptance and investment. DoD Mil Std 454 mandates the supply of a comprehensive VHDL description with every ASIC delivered to the DoD.

The best way to provide the required level of description is to use VHDL throughout the design process. As an IEEE standard, VHDL must undergo a review process every five

TABLE 1.1
VHDL Milestones

Year	VHDL Realization
1981	Initiated by US DoD to address hardware lifecycle crisis
1983–85	Development of baseline language by Intermetrics, IBM, and TI
1986	All rights transferred to IEEE
1987	Publication of IEEE standard
1987	Mil Std 454 requires comprehensive VHDL descriptions to be delivered with ASICs
1994	Revised standard (named VHDL 1076–1993)
2000	Revised standard (named VHDL 1076 2000 Edition)
2002	Revised standard (named VHDL 1076–2002)
2007	VHDL Procedural Language Application Interface standard (VHDL 1076c–2007)
2009	Revised standard (named VHDL 1076–2008)

Source: https://www.doulos.com/knowhow/vhdl_designers_guide/a_brief_history_of_vhdl/; https://www.xilinx.com/products/technology/dsp.html

years (or more often) to ensure its ongoing relevance to the industry. A few of them are given in Table 1.1. The first such revision was completed in September 1993, and this is still the most widely supported version of VHDL. One of the features introduced in VHDL-1993 was shared variables. Unfortunately, it wasn't possible to use these in any meaningful way. A working group eventually resolved this by proposing the addition of protected types to VHDL. The VHDL 2000 Edition is simply VHDL-1993 with protected types. VHDL-2002 is a minor revision of VHDL 2000 Edition. There is one significant change, though; the rules on using buffer ports are relaxed, which makes these much more useful than before. In 2007, an amendment to VHDL 2002 was created. This introduces the VHDL procedural interface (VHPI) and also makes a few minor changes to the text of VHDL 2002. Apart from the VHPI itself, no new features were added to VHDL. The VHPI allows tools programmable access to a VHDL model before and during simulation. In other words, you can write programs in a language such as C that interact with a VHDL simulator. The next revision of VHDL was released in January 2009 and is referred to as VHDL-2008.

VHDL is an industry-standard language used to describe hardware from the abstract to the concrete level, as shown in Figure 1.12. The language not only defines the syntax but also defines very clear simulation semantics for each language construct. It is a strongly typed language and is often verbose to write. It provides an extensive range of modeling capabilities, and it is possible to quickly assimilate a core subset of the language that is

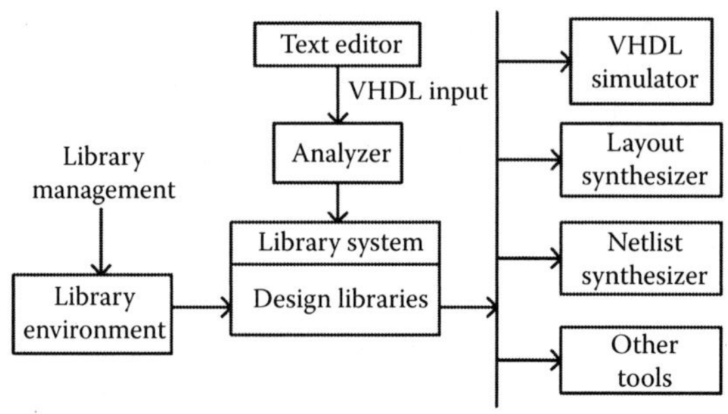

FIGURE 1.12
VHDL design environment.

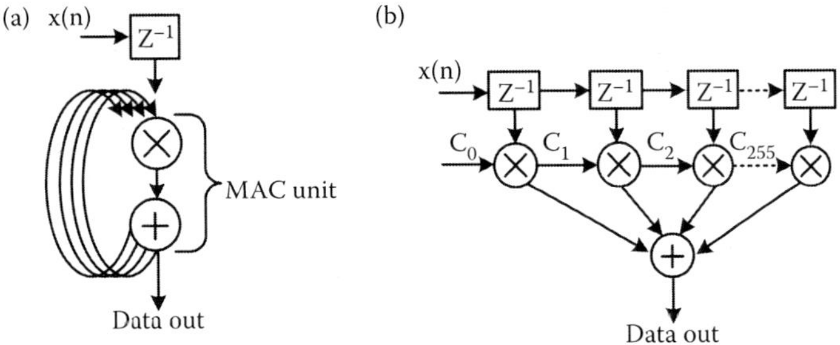

FIGURE 1.13
Configuration of parallelism using FPGA: x(n) is I/P samples, Z^{-1} is the unit delay element, $\otimes$ is the binary multiplier, $\oplus$ is the binary adder, and C_0, C_1, ... C_n are filter coefficients.

easy to understand without learning the more complex features. Why use VHDL? A quick time-to-market allows designers to quickly develop designs requiring tens of thousands of logic gates, provides powerful high-level constructs for describing complex logic, and supports modular design methodology and multiple levels of hierarchy [10,11].

VHDL is one language for the design, simulation and creation of device-independent designs that are portable to multiple vendors. With their inherent flexibility, Xilinx FPGAs and all programmable systems on chips (SoCs) are ideal for high-performance or multi-channel digital signal processing (DSP) applications that can take advantage of hardware parallelism. Xilinx FPGAs and SoCs combine this processing bandwidth with comprehensive solutions, including easy-to-use design tools for hardware designers, software developers, and system architects. A standard Von Neumann DSP architecture requires 256 cycles to complete a 256-tap FIR filter, while Xilinx FPGAs can achieve the same result in a single clock cycle, as shown in Figure 1.13. More details on parallel architecture design in FPGA can be found in References 10 and 11.

LABORATORY EXERCISES

1. Study and understand the techniques used for realizing Boolean operations before developing SSI technology.
2. Study and understand the size, power requirements, cooling setup, startup time, weight, thermal radiation, and so on of the older technology used for realizing electronics circuits before developing SSI technology.
3. Visit a semiconductor laboratory and understand the crystal growth process.
4. Design and draw a bit serial and bit parallel adders using a CMOS circuit.
5. Design and draw an 8-bit conditional-sum adder using a CMOS circuit.
6. Design and draw a binary subtractor using a CMOS circuit.
7. Design and draw a binary multiplier using a CMOS circuit.
8. Study and understand the architecture of different FPGAs.
9. Collect user manuals and understand the pin details for at least five standard FPGA development boards.
10. Prepare a list that consists of details of peripherals connected to the FPGA, their pin details, interfacing protocols, and general purpose input output (GPIO) port details of any standard FPGA development boards.

2

Digital Circuit Design with Very-High-Speed Integrated Circuit Hardware Description Language

2.1 Introduction and Preview

VHDL is a hardware description language for modeling digital circuits that can range from simple gate designs to complex system designs. VHDL is an acronym for very-high-speed integrated circuit hardware description language. This chapter gives a brief overview of the basic VHDL elements and its syntax with design applications. In 1980, the US Department of Defense and the IEEE sponsored the development of a hardware description language with the goal of developing very-high-speed integrated circuits, which has now become one of the industry's standard languages used to describe digital systems [12–14]. VHDL languages look similar to conventional programming languages, but there are some important differences. A hardware description language is inherently parallel; that is, commands, which correspond to logic gates, are executed (computed) in parallel as soon as a new input arrives. A VHDL program mimics the behavior of a physical, usually digital, system. It also allows incorporation of timing specifications (gate delays) as well as description of a system as an interconnection of different components [12,13]. A digital system can be represented at different levels of abstraction, which keeps the description and design of complex digital systems manageable. The levels are (i) data flow model, (ii) behavioral model, (iii) structural model, and (iv) algorithmic model. The data flow model describes how data move through the system. This is typically done in terms of data flow between registers, that is, RTL. The data flow model makes use of concurrent statements that are executed in parallel as soon as data arrives at the input. The behavioral model describes a system in terms of what it needs to do rather than in terms of its components and interconnections. A behavioral description specifies the relationship between the input and output signals. This could be a Boolean expression or a more abstract description such as the RTL. The structural model describes a system as a collection of gates and components that are interconnected to perform a desired function. A structural description could be compared to a schematic of interconnected logic gates. It is a representation that is usually closer to the physical realization of a system. The algorithmic model describes the digital systems by sequential statements which are executed in the sequence they are specified. The model allows both concurrent and sequential signal assignments that will determine the manner in which they have to be executed.

2.2 Code Design Structures

A digital system in VHDL consists of a design entity that can contain other entities that are then considered components of the top-level entity. Each entity is modeled by an entity declaration and an architecture body. One can consider the entity declaration as the interface to the outside world that defines the input and output signals, while the architecture body contains the description of the entity and is composed of interconnected entities, processes, and components, all operating concurrently. In a typical design, there will be many such entities connected together to perform the desired function. VHDL uses reserved keywords that cannot be used as signal names or identifiers. Keywords and user-defined identifiers are case insensitive. Lines with comments start with two adjacent hyphens, "- -", and will be ignored by the compiler. VHDL also ignores line breaks and extra spaces. VHDL is a strongly typed language, which implies that one always has to declare the type of every object that can have a value, such as signals, constants, and variables. The entity declaration defines the name of the entity and lists the input and output ports. The general form is as follows:

```
entity <name-of-entity> is [ generic generic_declarations);]
port (signal_names: mode type;
signal_names: mode type;
            :
signal_names: mode type);
end <name-of-entity>;
```

An entity always starts with the keyword `entity`, followed by its name and the keyword. Next are the port declarations using the keyword `port`. An entity declaration always ends with the keyword `end`. `<name_of_entity>` is a user-selected identifier. Signal names consist of a comma-separated list of one or more user-selected identifiers that specify external interface signals. `mode` is one of the reserved words to indicate the signal direction as (i) `in` indicates the signal is an input, (ii) `out` indicates the signal is an output of the entity whose value can be read only by other entities that use it, (iii) `buffer` indicates the signal is an output of the entity whose value can be read inside the entity's architecture, and (iv) `inout` indicates the signal can be an input or an output. For example, the types may be `bit`, `bit_vector`, `Boolean`, `character`, `std_logic` and `std_ulogic`. `bit` can have the value 0 or 1, `bit_vector` is a vector of bit values (e.g., `bit_vector [0 to 7]`), and `std_logic`, `std_ulogic`, `std_logic_vector`, and `std_ulogic_vector` can have nine values to indicate the value and strength of a signal. `std_ulogic` and `std_logic` are preferred over the `bit` or `bit_vector` types. `Boolean` can have the value true or false, `integer` can have a range of integer values, `real` can have a range of real values, `character` represents any printing character, and `time` indicates real time; generic declarations are optional and determine the local constants used for timing and sizing (e.g., bus widths) the entity. A generic can have a default value. The syntax for a generic is as follows:

```
generic (constant_name: type [:=value];
constant_name: type [:=value];
                        :
constant_name: type [:=value]);
```

Since VHDL is a strongly typed language, each port has a defined type. In this case, we specified the `std_logic` type. This is the preferred type for digital signals. In contrast to the bit type that can only have the values 1 and 0, the `std_logic` and `std_ulogic` types can have nine values. This is important to describe a digital system accurately including the binary values 0 and 1, as well as the unknown value X, the uninitialized value U, "-" for do not care, Z for high impedance, and several symbols to indicate the signal strength, for example, L for weak 0, H for weak 1, and W for weak unknown. The `std_logic` type is defined in the `std_logic_1164` package of the IEEE library. The type defines the set of values an object can have. This has the advantage of helping with the creation of models and helps reduce errors. For instance, if one tries to assign an illegal value to an object, the compiler will flag the error. An example with these delectations for 4:1 mux is as follows:

```
entity mux4_to_1 is
port (I0,I1,I2,I3: in std_logic_vector(7 downto 0):="00000000";
SEL: in std_logic_vector (1 downto 0):="00";
OUT1: out std_logic_vector(7 downto 0):="00000000");
end mux4_to_1;
```

The architecture body specifies how the circuit operates and how it is implemented. As discussed earlier, an entity or circuit can be specified in a variety of ways, such as data flow, behavioral, structural, and algorithmic or a combination of the above. The main body of the architecture starts with the keyword `begins` and gives the Boolean expression of the function. The <= symbol represents an assignment operator that assigns the value of the expression on the right to the signal on the left. The architecture body ends with an `end` keyword followed by the architecture name. The behavioral description of a two-input AND gate is shown below.

```
entity XNOR1 is
port (A, B: in std_logic;
Z: out std_logic);
end XNOR1;
architecture behavioral_xnor of XNOR1 is
signal X, Y: std_logic;
begin
X <= A and B;
Y <= (not A) and (not B);
Z <= X or Y;
End behavioral_xnor
```

The statements in the body of the architecture make use of logic operators. Logic operators that are allowed are `and`, `or`, `nand`, `nor`, `xor`, `xnor`, and `not`. In addition, other types of operators, including relational, shift, and arithmetic, are also allowed. The signals are executed when one or more of the variables on the right-hand side change their value; that is, an event occurs on one of the signals. For instance, when the input `A` changes, the internal signals `X` and `Y` change values, which in turn causes the last statement to update the output `Z`. There may obviously be a propagation delay associated with this change. Digital systems are basically data driven and an event which occurs on one signal will lead to an event on another signal and so on. The execution of the statements is determined by the flow of signal values. As a result, the order in which these statements are given does not matter; that is, moving the statement for the output `Z` ahead of that for `X` and `Y` does not change the outcome. This is in contrast to conventional software programs that execute the statements in a sequential or procedural manner. The digital circuit can also be described using a structural model that specifies what gates are used and how they are interconnected as given below

```
architecture structural of BUZZER is
component AND1
port (in1, in2: in std_logic;
out1: out std_logic);
end component;
component OR1
port (in1, in2: in std_logic;
out1: out std_logic);
end component;
component NOT1
port (in1: in std_logic;
out1: out std_logic);
end component;
signal DOOR_NOT, SBELT_NOT, B1, B2: std_logic;
begin
U0: NOT1 port map (DOOR, DOOR_NOT);
U1: NOT1 port map (SBELT, SBELT_NOT);
U2: AND2 port map (IGNITION, DOOR_NOT, B1);
U3: AND2 port map (IGNITION, SBELT_NOT, B2);
U4: OR2 port map (B1, B2, WARNING);
end structural;
```

Following the header is the declarative part that gives the components (gates) that are going to be used in the description of the circuits. In the above example, we use a two-input AND gate, two-input OR gate, and NOT (inverter) gate. These gates have to be defined first; that is, they will need an entity declaration and architecture body. These can be stored in one of the packages referred to in the header of the file. The declarations for the components give the inputs (`in1`, `in2`) and output (`out1`). Next, one has to define internal nets (signal names). In this example, the signals are called `DOOR_NOT`, `SBELT_NOT`, `B1`, and `B2`. Note that one always has to declare the type of the signal. The statements after the `begin` keyword give the instantiations of the components and describe how these are interconnected. A component instantiation statement creates a new level of hierarchy. Each line starts with an instance name (e.g., `U0`) followed by a colon, a component name, and the keyword `port map`. This keyword defines how the components are connected. In this example, this is done through positional association: `DOOR` corresponds to the input `in1` of the `NOT1` gate and `DOOR_NOT` to the output. Similarly, for the `AND2` gate, the first two signals (`IGNITION` and `DOOR_NOT`) correspond to the inputs `in1` and `in2`, respectively, and the signal `B1` to the output `out1`. An alternative method is to use explicit association between the ports as given below

```
label: component-name port map (port1=>signal1, port2=> signal2,…
port3=>signaln);
```

For example:

```
U0: NOT1 port map (in1 => DOOR, out1 => DOOR_NOT);
U1: NOT1 port map (in1 => SBELT, out1 => SBELT_NOT);
U2: AND2 port map (in1 => IGNITION, in2 => DOOR_NOT, out1 => B1);
U3: AND2 port map (in1 => IGNITION, in2 => SBELT_NOT, B2);
U4: OR2 port map (in1 => B1, in2 => B2, out1 => WARNING);
```

Note that the order in which these statements are written has no bearing on the execution since these statements are concurrent and therefore executed in parallel. Indeed, the schematic that is described by these statements is the same independent of the order of the statements. Structural modeling of design lends itself to hierarchical design in which one can define components of units that are used over and over again. Once these components are defined, they can be used as blocks, cells, or macros in a higher-level entity. This can significantly reduce the complexity of large designs. Hierarchical design approaches are always preferred over flat designs. A full adder can be described by the Boolean expressions for the `sum` and `carry` signals as `sum` $= (A \oplus B) \oplus C$ and `carry` $= AB + C(A \oplus B)$. In the below VHDL file, we defined a component for the full adder first; then, we used several instantiations of the full adder to build the structure of a 4-bit adder.

```
library ieee;
use ieee.std_logic_1164.all;
entity FULLADDER is
port (a, b, c: in std_logic;
sum, carry: out std_logic);
end FULLADDER;
architecture fulladder_behav of FULLADDER is
begin
sum <= (a xor b) xor c;
carry <= (a and b) or (c and (a xor b));end fulladder_behav;
-------------------------------------------------------------------------
library ieee;
use ieee.std_logic_1164.all;
entity FOURBITADD is
port (a, b: in std_logic_vector(3 downto 0);
Cin : in std_logic;
sum: out std_logic_vector (3 downto 0);
Cout, V: out std_logic);
end FOURBITADD;
architecture fouradder_structure of FOURBITADD is
signal c: std_logic_vector (4 downto 0);
component FULLADDER
port(a, b, c: in std_logic;
sum, carry: out std_logic);
end component;
begin
FA0: FULLADDER
port map (a(0), b(0), Cin, sum(0), c(1));
FA1: FULLADDER
port map (a(1), b(1), C(1), sum(1), c(2));
FA2: FULLADDER
port map (a(2), b(2), C(2), sum(2), c(3));
FA3: FULLADDER
port map (a(3), b(3), C(3), sum(3), c(4));
V <= c(3) xor c(4);
Cout <= c(4);
end fouradder_structure;
```

Note that the same input names, `a` and `b`, were used for the ports of the full adder and the 4-bit adder. This does not pose a problem in VHDL since they refer to different levels. We needed to define the internal signals `c[4:0]` to indicate the nets that connect the output carry to the input carry of the next full adder. For the first input, we used the input signal Cin. For the last carry, we defined c[4] as an internal signal since the last carry is needed as the input to the xor gate. A VHDL library can be considered a place where the compiler stores information about a design project. A VHDL package is a file or module that contains declarations of commonly used objects, data types, component declarations, signals, procedures and functions that can be shared among different VHDL models. In order to use std_logic, one needs to specify the library and package. This is done at the beginning of the VHDL file using the library and use keywords as follows:

```
library ieee;
use ieee.std_logic_1164.all;
```

std_logic_1164 package defines the standard data types. std_logic_arith package provides arithmetic, conversion, and comparison functions for the signed, unsigned, integer, std_ulogic, std_logic, and std_logic_vector types. std_logic_unsigned and std_logic_misc packages define supplemental types, subtypes, constants, and functions for the std_logic_1164 package. The .all extension indicates all of the ieee.std_logic_1164 packages that are to be used. The Xilinx Foundation Express comes with several packages. To use any of these, one must include the library and use clauses:

```
library ieee;
use ieee.std_logic_1164.all;
use ieee.std_logic_arith.all;
use ieee.std_logic_unsigned.all;
```

In addition, the synopsis library has the attributes package:

```
library SYNOPSYS;
use SYNOPSYS.attributes.all;
```

One can add other libraries and packages. The syntax to declare a package is as follows:

```
-- Package declaration
package name_of_package is
package declarations
end package name_of_package;
-- Package body declarations
package body name_of_package is
package body declarations
end package body name_of_package;
```

For instance, the basic functions of the AND2, OR2, NAND2, NOR2, XOR2, and other components need to be defined before one can use them. This can be done in a package, for example, basic_func, for each of these components, as follows:

```
------------------------- Package declaration
library ieee;
use ieee.std_logic_1164.all;
package basic_func is
-- AND2 declaration
```

```
component AND2
generic (DELAY: time :=5ns);
port (in1, in2: in std_logic; out1: out std_logic);
end component;
-- OR2 declaration
component OR2
generic (DELAY: time :=5ns);
port (in1, in2: in std_logic; out1: out std_logic);
end component;
end package basic_func;
------------------------ Package body declarations
library ieee;
use ieee.std_logic_1164.all;
package body basic_func is
-- 2 input AND gate
entity AND2 is
generic (DELAY: time);
port (in1, in2: in std_logic; out1: out std_logic);
end AND2;
architecture model_conc of AND2 is
begin
out1 <= in1 and in2 after DELAY;
end model_conc;
-- 2 input OR gateentity OR2 is
generic (DELAY: time);
port (in1, in2: in std_logic; out1: out std_logic);
end OR2;
architecture model_conc2 of AND2 is
begin
out1 <= in1 or in2 after DELAY;
end model_conc2;
end package body basic_func;
```

Note that we included a delay of 5 ns; however, it should be noted that delay specifications are ignored by the foundation synthesis tool. We made use of the predefined type `std_logic` that is declared in the package `std_logic_1164`. We have included the `library` and `use` clauses for this package. This package needs to be compiled and placed in a library. Let us call this library `my_func`. To use the components of this package, one has to declare it using the `library` and `use` clauses:

```
library ieee, my_func;
use ieee.std_logic_1164.all;
my_func.basic_func.all;
```

One can concatenate a series of names separated by periods to select a package. The `library` and `use` statements are connected to the subsequent entity statement. The `library` and `use` statements have to be repeated for each entity declaration. Identifiers are user-defined words used to name objects in VHDL models. The first character must be a letter and the last one cannot be an underscore. An identifier cannot include two consecutive underscores. An identifier is case insensitive; for example, `And2`, `AND2` and `and2` refer to the same object. An identifier can be of any length, for example `X10`, `x_10`, `My_gate1` and so on. Some invalid identifiers are `_X10`, `my gate@input`, `gate-input`, and so on. Certain identifiers are used by the system as keywords for special use, such as specific constructs. These keywords cannot be used as identifiers for signals or objects we define. We have seen some of these reserved words already, such as `in`, `out`, `or`, `and`, `port`, `map`, `inout`, `end`, and so on. The keywords often appear in boldface. The default number representation is the decimal system. VHDL allows integer literals and real literals. Integer literals consist of whole numbers without a decimal point, while real literals always include a decimal point. Exponential notation is allowed using the letter "E" or "e". For integer literals, the exponent must always be positive, for example, integer literals: `12`, `10`, `256E3`, and `12e+6` and real literals `v1.2`, `256.24`, and `3.14E-2`. The number –12 is a combination of a negation operator and an integer literal. To express a number in a base different from base 10, use the following convention: `base#number#`, for example, Base 2: `2#11101#` (representing the decimal number 29), Base 16: `16#1D#` and Base 8: `8#35#`. To make the readability of large numbers easier, one can insert underscores in the numbers as long as the underscore is not used at the beginning or the end, for example:

```
2#1001_1101_1100_0010#215_123.
```

To use a character literal in VHDL code, put it in single quotation marks, such as 'a,' 'B,' and ';'. On the other hand, a string of characters are placed in double quotation marks, for example, "This is a string," "To use a double quotation mark inside a string, use two double quotation marks," and "This is a "String"." Any printing character can be included inside a string. A bit-string represents a sequence of bit values. In order to indicate a bit string, place a "B" in front of the string: `B"1001"`. One can also use strings in the hexagonal or octal base by using the `X` or `O` specifiers, respectively, for example, binary: `B"1100_1001"`, `b"1001011"`; hexagonal: `X"C9"`, `X"4b"` and octal: `O"311"`, `o"113"`. Notice that in the hexadecimal system, each digit represents exactly 4 bits. As a result, the number `b"1001011"` is not the same as `X"4b"` since the former has only 7 bits while the latter represents a sequence of 8 bits. For the same reason, `O"113"` (representing 9 bits) is not the same sequence as `X"4b"` (representing 8 bits) [12–14].

DESIGN EXAMPLE 2.1

Design and develop a digital circuit to realize a binary-to-gray-code converter using the behavioral model.

Solution

```
library ieee;
use ieee.std_logic_1164.all;
use ieee.std_logic_unsigned.all;
use ieee.numeric_std.all;
entity bin_gra_behav is
port (i:in std_logic_vector(3 downto 0):="0000";
y:out std_logic_vector(3 downto 0):="0000");
end bin_gra_behav;
architecture behavioral of bin_gra_behav is
begin
y<= "0000" when i="0000" else
"0001" when i="0001" else
"0011" when i="0010" else
"0010" when i="0011" else
"0110" when i="0100" else
"0111" when i="0101" else
"0101" when i="0110" else
```

```
"0100" when i="0111" else
"1100" when i="1000" else
"1101" when i="1001" else
"1111" when i="1010" else
"1110" when i="1011" else
"1010" when i="1100" else
"1011" when i="1101" else
"1001" when i="1110" else
"1000" when i="1111";
end behavioral;
```

DESIGN EXAMPLE 2.2

Develop a digital system to realize an 8:2 multiplexer using the data flow model.

Solution

```
library ieee;
use ieee.std_logic_1164.all;
use ieee.std_logic_unsigned.all;
use ieee.numeric_std.all;
entity mux_data_flow is
port (i0,i1,i2,i3,i4,i5,i6,i7:in std_logic_vector(1 downto 0):="00";
s0,s1,s2:in std_logic:='0';
y:out std_logic_vector(1 downto 0):="00");
end mux_data_flow;
architecture behavioral of mux_data_flow is
begin
y(0)<=((i0(0) and (not (s0)) and (not(s1)) and (not (s2))) or
(i1(0) and (s0) and (not(s1)) and (not (s2))) or
(i2(0) and (not (s0)) and (s1) and (not (s2))) or
(i3(0) and (s0) and (s1) and (not (s2))) or
(i4(0) and (not (s0)) and (not(s1)) and (s2)) or
(i5(0) and (s0) and (not(s1)) and (s2)) or
(i6(0) and (not (s0)) and (s1) and (s2)) or
(i7(0) and (s0) and (s1) and (s2)));
y(1)<=((i0(1) and (not (s0)) and (not(s1)) and (not (s2))) or
(i1(1) and (s0) and (not(s1)) and (not (s2))) or
(i2(1) and (not (s0)) and (s1) and (not (s2))) or
(i3(1) and (s0) and (s1) and (not (s2))) or
(i4(1) and (not (s0)) and (not(s1)) and (s2)) or
(i5(1) and (s0) and (not(s1)) and (s2)) or
(i6(1) and (not (s0)) and (s1) and (s2)) or
(i7(1) and (s0) and (s1) and (s2)));
end behavioral;
```

DESIGN EXAMPLE 2.3

Develop a digital system to realize a priority encoder using the structural model.

Solution

The logical expression for an 8-bit priority encoder is $y(0)=I_6'(I_4'I_2'I_1+I_4'I_3+I_5)+I_7$, $y(1)=I_5'I_4'(I_2+I_3)+I_6+I_7$ and $I_4+I_5+I_6+I_7$, where I_n represents the priority encoder's binary input at the nth position and y represents the priority encoder's output. As we know, for an 8-bit priority encoder, we require 3-bit output, which is denoted in this design as `y[2:0]`, i.e., `y(0)`, `y(1)`, and `y(0)`. We implement this logic by using just two components: one 3-bit `AND` gate and one 2-bit `OR` gate, as given below. Since we plan to use one 3-bit `AND` gate and one 2-bit `OR` gate, we need to spilt the above logical expressions suitably to subject them to those components.

Component (Three Input AND Gate)

```
library IEEE;
use IEEE.STD_LOGIC_1164.ALL;
entity and_gate is
Port (a,b,c : in STD_LOGIC;
y0 : out STD_LOGIC);
end and_gate;
architecture Behavioral of and_gate is
begin
y0<= (a and b and c);
end Behavioral;
```

Component (Two Input OR Gate)

```
library IEEE;
use IEEE.STD_LOGIC_1164.ALL;
entity or_gate is
Port (d,e : in STD_LOGIC;
y1 : out STD_LOGIC);
end or_gate;
architecture Behavioral of or_gate is
begin
y1<= (d or e);
end Behavioral;
```

Main Code

```
library ieee;
use ieee.std_logic_1164.all;
use ieee.std_logic_unsigned.all;
use ieee.numeric_std.all;
entity pert_encod_stru is
port (i0,i1,i2,i3,i4,i5,i6,i7:in std_logic:='0';
y:out std_logic_vector(2 downto 0):="000");
end pert_encod_stru;
architecture behavioral of pert_encod_stru is
component and_gate
port(a : in std_logic;
b : in std_logic;
c : in std_logic;
y0 : out std_logic);
end component;
component or_gate
port(d : in std_logic;
e : in std_logic;
y1 : out std_logic);
end component;
signal temp0,temp1,temp2,temp3,s0,s1,s2,s3,s4,s5,s6,s7,s8,s9:
std_logic:='0';
begin
temp0<=not (i4);
temp1<=not (i2);
temp2<=not (i6);
temp3<=not (i5);
inst_and_gate0: and_gate port map(
a => temp0,
b => temp1,
c => i1,
y0 =>s0);
```

```
inst_and_gate1: and_gate port map(
a =>temp0,
b => i3,
c => '1',
y0 => s1);
inst_or_gate0: or_gate port map(
d =>s0,
e =>s1,
y1 =>s2);
inst_or_gate1: or_gate port map(
d =>s2,
e =>i5,
y1 =>s3);
inst_and_gate2: and_gate port map(
a =>temp2,
b => s3,
c => '1',
y0 => s4);
inst_or_gate2: or_gate port map(
d =>s4,
e =>i7,
y1 =>y(0));
----------------------
inst_or_gate3: or_gate port map(
d =>i2,
e =>i3,
y1 =>s5);
inst_and_gate3: and_gate port map(
a =>s5,
b => temp0,
c => temp3,
y0 => s6);
inst_or_gate4: or_gate port map(
d =>s6,
e =>i6,
y1 =>s7);
inst_or_gate5: or_gate port map(
d =>s7,
e =>i7,
y1 =>y(1));
---------------------------
inst_or_gate6: or_gate port map(
d =>i4,
e =>i5,
y1 =>s8);
inst_or_gate7: or_gate port map(
d =>s8,
e =>i6,
y1 =>s9);
inst_or_gate8: or_gate port map(
d =>s9,
e =>i7,
y1 =>y(2));
end behavioral;
```

Digital circuit design using the algorithmic approach is discussed appropriately in Section 2.6.

2.3 Data Types and Their Conversions

A data object is created by an object declaration and has a value and type associated with it. An object can be a constant, variable, signal, or file. We have seen signals that were used as input or output ports or internal nets. Signals can be considered wires in a schematic that can have current and future values and are a function of the signal assignment statements. On the other hand, variables and constants are used to model the behavior of a circuit and are used in processes, procedures, and functions, similarly to how they would be in a programming language. A constant can have a single value of a given type and cannot be changed during the simulation. A constant is declared as follows: `constant list_of_name_of_constant: type [:= initial value]`, where the initial value is optional. Constants can be declared at the start of an architecture and can then be used anywhere within the architecture. Constants declared within a process can be used only inside that specific process, for example, `constant RISE_FALL_TME: time: =2ns` and `constant DATA_BUS: integer:=16`. A variable can have a single value, as with a constant, but a variable can be updated using a variable assignment statement. The variable is updated without any delay as soon as the statement is executed. Variables must be declared inside a process as `variable list_of_variable_names: type [:= initial value]`, for example, `variable CNTR_BIT: bit :=0`, `variable VAR1: boolean :=FALSE`, `variable SUM: integer range 0 to 256 :=16` and `variable STS_BIT: bit_vector (7 downto 0)`. The variable `SUM` is an integer that has a range from 0 to 256 with an initial value of 16 at the start of the simulation. The fourth example defines a bit vector of eight elements: `STS_BIT(7)`, `STS_BIT(6)`,... `STS_BIT(0)`. A variable can be updated using a variable assignment statement such as `Variable_name := expression`. As soon as the expression is executed, the variable is updated without any delay. The signals are declared outside the process as `signal list_of_signal_names: type [ := initial value]`, for example, `signal SUM, CARRY: std_logic`, `signal CLOCK: bit`, `signal TRIGGER: integer :=0`, `signal DATA_BUS: bit_vector (0 to 7)`, `signal VALUE: integer range 0 to 100` and so on. Signals are updated when their signal assignment statement is executed after a given or delta delay, for example, `SUM <= (A xor B)` after 2 ns, where the signal (`SUM`) will be updated after a 2-ns delay; in the case where no delay is specified, the `SUM` is updated after a delta delay. One can also specify multiple waveforms using multiple events, such as `signal wavefrm : std_logic`, `wavefrm <= '0', '1' after 5 ns, '0' after 10 ns` and `'1' after 20 ns`. It is important to understand the difference between variables and signals, particularly how it relates to when their value changes. A variable changes instantaneously when the variable assignment is executed. On the other hand, a signal changes on a delay after the assignment expression is evaluated. If no delay is specified, the signal will change after a delta delay. This has important consequences for the updated values of variables and signals [12,13]. Let us compare two example designs in which a process is used to calculate the `signal RESULT [7]` as given below:

Example with variable:

```
library ieee;
use ieee.std_logic_1164.all;
use ieee.std_logic_unsigned.all;
use ieee.numeric_std.all;
entity var is
port (a,clk:in std_logic:='0';
```

```
b:out std_logic:='0');
end var;
architecture var1 of var is
signal result: integer := 0;
begin
b<= not a;
p0:process
variable variable1: integer :=1;
variable variable2: integer :=2;
variable variable3: integer :=3;
begin
wait on clk;
variable1 := variable2;
variable2 := variable1 + variable3;
variable3 := variable2;
result <= variable1 + variable2 + variable3;
end process;
end var1;
```

Example with signal:

```
library ieee;
use ieee.std_logic_1164.all;
use ieee.std_logic_unsigned.all;
use ieee.numeric_std.all;
entity sig is
port (a,clk:in std_logic:='0';
b:out std_logic:='0');
end sig;
architecture sig1 of sig is
signal result : integer :=0;
signal signal1: integer :=1;
signal signal2: integer :=2;
signal signal3: integer :=3;
begin
process
begin
wait on clk;
signal1 <= signal2;
signal2 <= signal1 + signal3;
signal3 <= signal2;
result <= signal1 + signal2 + signal3;
end process;
end sig1;
```

Note that the above two designs are not synthesizable; however, they can be simulated to understand the execution of variables and signals. In the first design, the variables `variable1`, `variable2`, and `variable3` are computed sequentially and their values updated instantaneously after the `clock (clk)` signal arrives. Next, the `RESULT`, which is a signal, is computed using the new values of the variables and updated a time delta after `clk` arrives. This results in the following values (after a time `clk`): `variable1 = 2, variable2 = 5 (=2+3), variable3= 5`). Since `RESULT` is a signal, it will be computed at the time `clk` and updated at the time `clk + Delta`. Its value will be `RESULT=12`. On the other hand, in the second design, the signals will be computed at the time `clk`. All of these signals are computed at the same time using the old values of `signal1`, `2`, and `3`. All the signals will be updated at Delta time after the `clk` has arrived. Thus, the signals will have these values: `signal1= 2`, `signal2= 4 (=1+3)`, `signal3=2`, and `RESULT=6`. Each data object has a type associated with it. The type defines the set of values that the object can have and the set of operations that are allowed on it. There are four classes of data types: scalar, composite, access and file. The scalar types represent a single value and are ordered so that relational operations can be performed on them. The scalar type includes integer, real and enumerated types of Booleans and characters. The VHDL has several predefined types in the standard package as listed in Table 2.1. To use this package, one has to include the clauses `library std, work` and `use std.standard.all`.

TABLE 2.1

Std Library Package

Type	Range of Values	Example
`bit`	"0", "1"	`signal A: bit :='1';`
`bit_vector`	An array with each element of type bit	`signal INBUS: bit_vector(7 downto 0);`
`boolean`	FALSE, TRUE	`variable TEST: Boolean :=FALSE'`
`character`	Any legal VHDL character printable characters must be placed between single quotes (e.g., '#')	`variable VAL: character :='$';`
`file_open_kind`	`read_mode`, `write_mode`, `append_mode`	–
`file_open_status`	`open_ok`, `status_error`, `name_error`, `mode_error`	–
`integer`	Range is implementation-dependent but includes at least $-(2^{31}-1)$ to $+(2^{31}-1)$	`constant CONST1: integer :=129;`
`natural`	Integer starting with 0 up to the max specified in the implementation	`variable VAR1: natural :=2;`
`positive`	Integer starting from 1 up to the max specified in the implementation	`variable VAR2: positive :=2;`
`real`	Floating-point number in the range of -1.0×10^{38} to $+1.0x\ 10^{38}$ (can be implementation-dependent. Not supported by the Foundation synthesis program)	`variable VAR3: real :=+64.2E12;`
`severity_level`	Note, warning, error, failure	–
`string`	Array of which each element is of the type `character`	`variable VAR4: string(1 to 12):= "@$#ABC*()_%Z";`
`time`	An integer number for which the range is implementation-defined; units can be expressed in sec, ms, us, ns, ps, fs, min and hr. Not supported by the Foundation synthesis program	`variable DELAY: time :=5 ns;`

One can introduce new types by using the type declaration, which names the type and specifies its value range. The syntax is type identifier is type_definition, for example, type small_int is range 0 to 1024, type my_word_length is range 31 downto 0, subtype int_small is integer range −1024 to +1024 and subtype data_word is my_word_length range 7 downto 0. A subtype is a subset of a previously defined type. It defines a type called data_word that is a subtype of my_word_length whose range is restricted from 7 to 0. The floating-point types can be declared as type cmos_level is range 0.0 to 3.3, type pmos_level is range −5.0 to 0.0, type probability is range 0.0 to 1.0 and subtype cmos_low_V is cmos_level range 0.0 to +1.8. The physical type definition includes a units identifier as follows: type conductance is range 0 to 2E−9, units: mho; mmho = 1E−3 mho; umho = 1E−6 mho; nmho = 1E−9 mho; pmho = 1E−12 mho; end. The variables and conductance can be declared in VHDL as variable BUS_WIDTH: small_int :=24, signal DATA_BUS: my_word_length, variable VAR1: cmos_level range 0.0 to 2.5 and constant LINE_COND: conductance:= 125 umho. Notice that a space must be left before the unit name. The physical data types are not supported by the Xilinx Foundation Express synthesis program. In order to use our own types, we need to either include the type definition inside an architecture body or to declare the type in a package. An enumerated type consists of lists of character literals or identifiers. The enumerated type can be very handy when writing models at an abstract level. The syntax for an enumerated type is type type_name is (identifier list or character literal), for example, type my_3values is ('0', '1', 'Z'), type PC_OPER is (load, store, add, sub, div, mult, shiftl, shiftr), type hex_digit is ('0', '1', '2', '3', '4', '5', '6', '7', 8', '9', 'A', 'B', 'C', 'D', 'E', 'F') and type state_type is (S0, S1, S2, S3). If one does not initialize the signal, the default initialization is the leftmost element of the list. Enumerated types have to be defined in the architecture body or inside a package. An example of an enumerated type that has been defined in the std_logic_1164 package is the std_ulogic type:

```
type STD_ULOGIC is (

'U',    -- uninitialized
'X',    -- forcing unknown
'0',    -- forcing 0
'1',    -- forcing 1
'Z',    -- high impedance
'W',    -- weak unknown
'L',    -- weak 0
'H'.    -- weak 1
'-');   -- don't care
```

In order to use this type, one has to include library ieee; use ieee.std_logic_1164.all; before the entity declaration. It is possible that multiple drivers are driving a signal. In that case, there could be a conflict and the output signal would be undetermined. For instance, the outputs of an AND gate and NOT gate are connected into the output net OUT1. In order to resolve the value of the output, one can call up a resolution function. These are usually user-written functions that will resolve the signal. If the signal is of the type std_ulogic and has multiple drivers, one needs to use a resolution function. The std_logic_1164 package has such a resolution function, called RESOLVED, predefined. One can then use the signal OUT1: resolved: std_ulogic; declaration for the signal

OUT1. If there is contention, the RESOLVED function will be used to intermediate the conflict and determine the value of the signal. Composite data objects consist of a collection of related data elements in the form of an array or record. Before we can use such objects, one has to declare the composite type first as type array_name is array (indexing scheme) of element_type, for example, type MY_WORD is array (15 downto 0) of std_logic, type YOUR_WORD is array (0 to 15) of std_logic, type VAR is array (0 to 7) of integer and type STD_LOGIC_1D is array (std_ulogic) of std_logic. In the first two examples above, we have defined a 1D array of elements of the type std_logic indexed from 15 down to 0 and 0 up to 15, respectively. The last example defines a 1D type std_logic elements that uses the type std_ulogic to define the index constraint. Thus, this array looks like: Index: 'U' 'X' '0' '1' 'Z' 'W' 'L' 'H' '-'. We can now declare objects of these data types as signal MEM_ADDR: MY_WORD, signal DATA_WORD: YOUR_WORD := B"1101100101010110" and constant SETTING: VAR := (2,4,6,8,10,12,14,16). In the first example, the signal MEM_ADDR is an array of 16 bits, initialized to all zeros. To access individual elements of an array, we specify the index. For example, MEM_ACCR(15) accesses the leftmost bit of the array, while DATA_WORD(15) accesses the rightmost bit of the array with the value "0". To access a subrange, one specifies the index range, MEM_ADDR(15 downto 8) or DATA_WORD(0 to 7). Multidimensional arrays can be declared as well by using a similar syntax as above, for example, type MY_MATRIX3X2 is array (1 to 3, 1 to 2) of natural, type YOUR_MATRIX4X2 is array (1 to 4, 1 to 2) of integer, type STD_LOGIC_2D is array (std_ulogic, std_ulogic) of std_logic and variable DATA_ARR: MY_MATRIX :=((0,2), (1,3), (4,6), (5,7)). The variable array DATA_ARR will then be initialized to 0 2, 1 3, 4 6, 5 7. To access an element, one specifies the index; e.g., DATA_ARR(3,1) returns the value 4. The last example defines a 9×9 array or table with an index of the elements of the std_ulogic type. Sometimes it is more convenient not to specify the dimension of the array when the array type is declared. This is called an unconstrained array type. The syntax for the array declaration is type array_name is array (type range <>) of element_type, for example, type MATRIX is array (integer range <>) of integer, type VECTOR_INT is array (natural range <>) of integer and type VECTOR2 is array (natural range <>, natural range <>) of std_logic. The range is now specified when one declares the array object as variable MATRIX8: MATRIX (2 downto -8) := (3, 5, 1, 4, 7, 9, 12, 14, 20, 18) and variable ARRAY3×2: VECTOR2 (1 to 4, 1 to 3)) := (('1','0'), ('0','-'), (1, 'Z')). A second composite type is the records type, which consists of multiple elements that may be of different types. The syntax for a record type is type name is record identifier :subtype_indication; ...identifier :subtype_indication; end record, for example:

```
type MY_MODULE is

record
RISE_TIME     : time;
FALL_TIME     : time;
SIZE          : integer range 0 to 200;
DATA          : bit_vector (15 downto 0);
end record;
signal A, B: MY_MODULE;
```

To access values or assign values to records, one can use one of the following methods:

```
A.RISE_TIME <= 5 ns;
```

```
A.SIZE <= 120;
B <= A;
```

Any given VHDL FPGA design may have multiple VHDL types being used. Since VHDL is a strongly typed language, one cannot assign a value of one data type to a signal of a different data type. To allow assigning data between objects of different types, one needs to convert one type to the other. Fortunately, there are functions available in several packages in the ieee library, such as the std_logic_1164 and the std_logic_arith packages. The IEEE std_logic_unsigned and std_logic_arith packages allow additional conversions such as from an integer to std_logic_vector and vice versa. The most common VHDL types used in synthesizable VHDL code are std_logic, std_logic_vector, signed, unsigned, and integer. Because VHDL is a strongly typed language, most often differing types cannot be used in the same expression. In cases where you can directly combine two types into one expression, you are really leaving it up to the compiler or synthesis tool to determine how the expression should behave, which is a dangerous thing to do. An easy way to remember when to use a function or typecast is to remember that both the std_logic_vector and signed/unsigned types are defined with a specific bit width, while integers do not define a bit width. A typecast between std_logic_vector and signed/unsigned can be used as long as the origin and destination signals have the same bit width. Integers do not have a set bit width, which is why the conversion function from integer to signed/unsigned includes a specification of the intended bit width. Notice that there is no direct conversion path between the std_logic_vector type and the integer type. Integer types do not have a set width, unlike signed, unsigned and std_logic_vector types. To convert between integer and std_logic_vector types, we must first convert to signed or unsigned. If we do not restrict the range when defining an integer, the compiler will assume a 32-bit width. Depending on your synthesis tool and its settings, the default bit width of 32 may or may not be optimized out to the appropriate bit width [12,15,16]. The most common conversions in VHDL are discussed below one by one.

Integer to Signed Using numeric_std: The below example uses the to_signed conversion, which requires two input parameters. The first is the signal that you want to convert, and the second is the length of the resulting vector, for example:

```
signal input_3 : integer;
signal output_3 : signed(3 downto 0);
output_3 <= to_signed(input_3, output_3'length);
```

Integer to std_logic_vector Using numeric_std: First, we need to think about the range of values stored in the integer. Can our integer be positive and negative? If so, we need to use the to_signed() conversion. If our integer is only positive, we need to use the to_unsigned() conversion. Both of these conversion functions require two input parameters. The first is the signal that we want to convert, and the second is the length of the resulting vector, for example:

```
signal input_1 : integer;
signal output_1a : std_logic_vector(3 downto 0);
signal output_1b : std_logic_vector(3 downto 0);
-- This line demonstrates how to convert positive integers
output_1a <= std_logic_vector(to_unsigned(input_1, output_1a'length));
-- This line demonstrates how to convert positive or negative integers
output_1b <= std_logic_vector(to_signed(input_1, output_1b'length));
```

Integer to Unsigned Using numeric_std: The below example uses the `to_unsigned` conversion, which requires two input parameters. The first is the signal that we want to convert, and the second is the length of the resulting vector.

```
signal input_2 : integer;
signal output_2 : unsigned(3 downto 0);
output_2 <= to_unsigned(input_2, output_2'length);
```

std_logic_vector to Integer Using numeric_std: First, we need to think about the data that are represented by our `std_logic_vector`. Is it signed or unsigned data? Signed data mean that our `std_logic_vector` can be a positive or negative number. Unsigned data mean that our `std_logic_vector` is only a positive number. The example below uses the `unsigned()` typecast, but if our data can be negative, we need to use the `signed()` typecast. Once we cast our input `std_logic_vector` as unsigned or signed, then we can convert it to an integer as given below [15,16].

```
signal input_4 : std_logic_vector(3 downto 0);
signal output_4a : integer;
signal output_4b : integer;
-- This line demonstrates the unsigned case
output_4a <= to_integer(unsigned(input_4));
-- This line demonstrates the signed case
output_4b <= to_integer(signed(input_4));
```

std_logic_vector to Signed Using numeric_std: This is an easy conversion; all we need to do is cast the `std_logic_vector` as signed as given below.

```
signal input_6 : std_logic_vector(3 downto 0);
signal output_6 : signed(3 downto 0);
output_6 <= signed(input_6);
```

std_logic_vector to Unsigned Using numeric_std: This is an easy conversion; all we need to do is cast the `std_logic_vector` as unsigned as shown below.

```
signal input_5 : std_logic_vector(3 downto 0);
signal output_5 : unsigned(3 downto 0);
output_5 <= unsigned(input_5);
```

Signed to Integer Using numeric_std: This is an easy conversion; we need to use the `to_integer` function call from `numeric_std` as given below.

```
signal input_10 : signed(3 downto 0);
signal output_10 : integer;
output_10 <= to_integer(input_10);
```

Signed to std_logic_vector Using numeric_std: This is an easy conversion; we need to use the `std_logic_vector` cast as given below.

```
signal input_11 : signed(3 downto 0);
signal output_11 : std_logic_vector(3 downto 0);
output_11 <= std_logic_vector(input_11);
```

Signed to Unsigned Using numeric_std: This is an easy conversion; we need to use the `unsigned` cast as given below.

```
signal input_12 : signed(3 downto 0);
signal output_12 : unsigned(3 downto 0);
output_12 <= unsigned(input_12);
```

Unsigned to Integer Using numeric_std: This is an easy conversion; we need to use the `to_integer` function call from `numeric_std` as given below.

```
signal input_7 : unsigned(3 downto 0);
signal output_7 : integer;
output_7 <= to_integer(input_7);
```

Unsigned to Signed Using numeric_std: This is an easy conversion; we need to use the `signed` cast as given below.

```
signal input_9 : unsigned(3 downto 0);
signal output_9 : signed(3 downto 0);
output_9 <= signed(input_9);
```

Unsigned to std_logic_vector Using numeric_std: This is an easy conversion; we need to use the `std_logic_vector` cast as given below.

```
signal input_8 : unsigned(3 downto 0);
signal output_8 : std_logic_vector(3 downto 0);
output_8 <= std_logic_vector(input_8);
```

Integer to Signed Using std_logic_arith: The below example uses the `conv_signed` conversion, which requires two input parameters. The first is the signal that we want to convert, and the second is the length of the resulting vector.

```
signal input_3 : integer;
signal output_3 : signed(3 downto 0);
output_3 <= conv_signed(input_3, output_3'length);
```

Integer to std_logic_vector Using std_logic_arith: The below example uses the `conv_std_logic_vector` conversion, which requires two input parameters. The first is the signal that we want to convert, and the second is the length of the resulting vector. One thing to note here is that if we input a negative number into this conversion, then our output `std_logic_vector` will be represented in 2's complement signed notation.

```
signal input_1 : integer;
signal output_1 : std_logic_vector(3 downto 0);
output_1 <= conv_std_logic_vector(input_1, output_1'length);
```

Integer to Unsigned Using std_logic_arith: The below example uses the `conv_unsigned` conversion, which requires two input parameters. The first is the signal that we want to convert, and the second is the length of the resulting vector.

```
signal input_2 : integer;
signal output_2 : unsigned(3 downto 0);
output_2 <= conv_unsigned(input_2, output_2'length);
```

std_logic_vector to Integer Using std_logic_arith: First, we need to think about the data that are represented by our `std_logic_vector`. Is it signed or unsigned data? Signed data mean that our `std_logic_vector` can be a positive or negative number. Unsigned data mean that our `std_logic_vector` is only a positive number. The example below uses the `unsigned()` typecast, but if our data can be negative, we need to use the `signed()` typecast. Once our input `std_logic_vector` is unsigned or signed, then we can convert it to an integer as given below [15,16].

```
signal input_4 : std_logic_vector(3 downto 0);
signal output_4a : integer;
signal output_4b : integer;
-- This line demonstrates the unsigned case
output_4a <= conv_integer(unsigned(input_4));
-- This line demonstrates the signed case
output_4b <= conv_integer(signed(input_4));
```

std_logic_vector to Signed Using std_logic_arith: This is an easy conversion; all we need to do is cast the `std_logic_vector` as signed as given below.

```
signal input_6 : std_logic_vector(3 downto 0);
signal output_6 : signed(3 downto 0);
output_6 <= signed(input_6);
```

std_logic_vector to Unsigned Using std_logic_arith: This is an easy conversion; all we need to do is cast the `std_logic_vector` as unsigned as given below.

```
signal input_5 : std_logic_vector(3 downto 0);
signal output_5 : unsigned(3 downto 0);
output_5 <= unsigned(input_5);
```

Signed to Integer Using std_logic_arith: This is an easy conversion; we need to use the `conv_integer` function call from `std_logic_arith` as given below.

```
signal input_10 : signed(3 downto 0);
signal output_10 : integer;
output_10 <= conv_integer(input_10);
```

Signed to std_logic_vector Using std_logic_arith: This is an easy conversion; we need to use the `std_logic_vector` cast as given below.

```
signal input_11 : signed(3 downto 0);
signal output_11 : std_logic_vector(3 downto 0);
output_11 <= std_logic_vector(input_11);
```

Signed to Unsigned Using std_logic_arith: This is an easy conversion; we need to use the `unsigned` cast as given below.

```
signal input_12 : signed(3 downto 0);
signal output_12 : unsigned(3 downto 0);
output_12 <= unsigned(input_12);
```

Unsigned to Integer Using std_logic_arith: This is an easy conversion; we need to use the `conv_integer` function call from `std_logic_arith` as given below.

```
signal input_7 : unsigned(3 downto 0);
signal output_7 : integer;
output_7 <= conv_integer(input_7);
```

Unsigned to Signed Using std_logic_arith: This is an easy conversion; we need to use the `signed` cast as given below.

```
signal input_9 : unsigned(3 downto 0);
signal output_9 : signed(3 downto 0);
output_9 <= signed(input_9);
```

Unsigned to std_logic_vector Using std_logic_arith: This is an easy conversion; we need to use the `std_logic_vector` typecast as given below.

```
signal input_8 : unsigned(3 downto 0);
signal output_8 : std_logic_vector(3 downto 0);
output_8 <= std_logic_vector(input_8);
```

We see some design examples that cover the concepts, data types, and type conversions we discussed so far below.

DESIGN EXAMPLE 2.4

Design and develop a digital system to realize a 16-bit binary-to-integer convertor.

Solution

```
library ieee;
use ieee.std_logic_1164.all;
use ieee.std_logic_arith.all;
use ieee.std_logic_unsigned.all;
use ieee.numeric_bit.all;
use ieee.numeric_std.all;
entity bin_to_int is
port (data_in: in std_logic_vector(15 downto 0):="1111111111111111";
data_out: out std_logic_vector(15 downto 0));
end bin_to_int;
architecture behavioral of bin_to_int is
signal t1:integer;
begin
t1 <= conv_integer(data_in);
data_out<=data_in;
end behavioral;
```

DESIGN EXAMPLE 2.5

Design and develop a digital system to identify and output that the given integer is either an even or odd number. Select the integer "40" if the selection input is "0"; otherwise, select "41".

Solution

```
library ieee;
use ieee.std_logic_1164.all;
use ieee.std_logic_arith.all;
use ieee.std_logic_unsigned.all;
use ieee.numeric_std.all;
entity int_to_bin is
```

```
Port (sel : in std_logic:='0';
odd,even: out std_logic:='0');
end int_to_bin;
architecture Behavioral of int_to_bin is
signal a:std_logic_vector (5 downto 0):="000000";
begin
a<= std_logic_vector(to_unsigned(40,6)) when sel='0' else
std_logic_vector(to_unsigned(41,6));
p0:process(a)
variable i,j,k:std_logic:='0';
begin
j:=a(5) xor a(4);
for i in 3 downto 0 loop
k:=j xor a(i);
j:=k;
end loop;
if (j='1') then
odd<='1';
even<='0';
else
odd<='0';
even<='1';
end if;
end process;
end Behavioral;
```

DESIGN EXAMPLE 2.6

Design and develop a digital system to realize a 32-bit error counter. Assume that there is a system which represents the error by logic 1 at the appropriate position in a 255-bit binary error pattern. Add the number of 1s present the 255-bit error pattern and give the result in a 32-bit binary data form. All these computations have to happen when `enable = "0"`; otherwise, the system has to reset with "0".

Solution

```
library IEEE;
use IEEE.STD_LOGIC_1164.ALL;
use IEEE.STD_LOGIC_ARITH.ALL;
use IEEE.STD_LOGIC_UNSIGNED.ALL;
use IEEE.NUMERIC_STD.ALL;
entity error_count is
Port (data : in std_logic_vector(255 downto 0):="00000000000000000000000
000000000000000000000000000000000000000000000000000000000000000000000000
000000000000000000000000000000000000000000000000000000000000000000000000
000000000000000000000000000000000000000000000000000000000000000000000000
00000000000000000";
enable:in std_logic:='0';
outp:out std_logic_vector(31 downto 0):="0000000000000000000000000000000
0");
end error_count;
architecture Behavioral of error_count is
begin
p1:process(data,enable)
variable error_cnt:integer:=0;
begin
if (enable='1') then
error_cnt:=0;
outp<=conv_std_logic_vector(error_cnt,32);
else
for i in 0 to 255 loop
```

```
if (data(i)='1') then
error_cnt:=error_cnt+ 1;
else
error_cnt:=error_cnt;
end if;
end loop;
outp<=conv_std_logic_vector(error_cnt,32);
end if;
end process;
end Behavioral;
```

DESIGN EXAMPLE 2.7

Design and develop a digital system to realize a 1D binary array, 1D integer array, and multidimensional array in VHDL. The system should also have provisions to read the array element based on the user selection input.

Solution

```
library ieee;
use ieee.std_logic_1164.all;
use ieee.std_logic_arith.all;
use ieee.std_logic_unsigned.all;
use ieee.numeric_std.all;
entity type_dec is
port (sel:in std_logic:='0';
data_temp1,data_temp2,data_temp3:out std_logic_vector(7 downto
0):="00000000");
end type_dec;
architecture behavioral of type_dec is
type dis_data is array (0 to 3) of std_logic_vector(7 downto 0);
constant data_d : dis_data :=("11111111","00000001",
"00000011","10001111");
type com_data is array (7 downto 0) of integer range 0 to 255;
constant data_c : com_data:= (20,60,40,8,0,0,1,239);
type lutable is array (0 to 4, 0 to 2) of integer range 0 to 255;
--(0 to 4) is row;
signal sample_array: lutable:=((10, 20, 30), ---(0 to 2) is column
                                (40, 30, 20),
                                (100, 200, 100),
                                (1,2,3),
                                (5,6,7));
begin
data_temp1<=conv_std_logic_vector(sample_array(1,1),8) when sel='0'
else
conv_std_logic_vector(sample_array(3,2),8);
data_temp2<=conv_std_logic_vector(data_c(1),8) when sel='0' else
conv_std_logic_vector(data_c(3),8);
data_temp3<=data_d(2) when sel='0' else
data_d(3);
end behavioral;
```

2.4 Operators and Attributes

VHDL supports different classes of operators and attributes that act on signals, variables, and constants. A few of them have already been seen in the preceding sections, and the

TABLE 2.2

List of Main Operators in VHDL

S.No	Actions	Operators							
1	Logical operations	and	or	nand	nor	xor	xnor	not	
2	Relational operations	=	/=	<	<=	>	>=	:=	:
3	Shift operations	sll	srl	sla	sra	rol	ror		
4	Multiplying operations	*	/	mod	rem				
5	Addition operations	+	=	&					
6	Unary operations	+	-						
7	Miscellaneous operations	**	abs						

remaining classes of operators and attributes are discussed in this section. Almost all the main operators listed in Table 2.2 have equal precedence for building digital systems in FPGA. Based on the brackets we insert, the operators will be executed during the operation of the system. Operators of the same class have the same precedence and are applied from left to right in an expression. We discuss these operators below one by one. As we have seen in previous sections, logical operators are used to perform various logical or Boolean functions/expressions bit by bit or array by array. As we have seen, these operators can be applied to signals, variables, and constants [12–16].

Relational operators test the relative values of two scalar types and give as a result a Boolean output of "TRUE" or "FALSE". Notice that symbol of the operator <= (less than or equal to) is the same as the assignment operator used to assign a value to a signal or variable. The operators are equality (=), inequality (/=), less than (<), greater than (>), greater than or equal to (>=), less than or equal to (<=), integer assignment (:=), and constant value assignment (:). Some examples of relational operations are

```
variable STS  : Boolean;
constant A    : integer :=24;
constant B_COUNT: integer :=32;
constant C    : integer :=14;
STS <= (A < B_COUNT);
STS <= ((A >= B_COUNT) or (A > C));
STS <= (std_logic ('1', '0', '1') < std_logic('0', '1','1'));
type new_std_logic is ('0', '1', 'Z', '-');
variable A1: new_std_logic :='1';
variable A2: new_std_logic :='Z';
STS <= (A1 < A2);
```

The shift operators perform a bitwise shift or rotate operation on a one-dimensional array of elements of the type `bit`, `std_logic` or `Boolean`. The operators are shift left logical (`sll`), filling right vacated bits by 0; shift right logical (`srl`), filling left vacated bits by 0; shift left arithmetic (`sla`), filling right vacated bits by rightmost bit; shift right arithmetic (`sra`), filling left vacated bits by leftmost bit; left circular rotation (`rol`), and right circular rotation (`ror`). The operand is on the left of the operator and the number (integer) of shifts is on the right side of the operator; for example, `variable NUM1:bit_vector := "10010110"; NUM1 srl 2;` will result in the number "00100101". When a negative integer is given, the opposite action occurs; that is, a shift to the left will be a shift to the right. For example, `NUM1 srl -2` would be equivalent to `NUM1 sll 2` and give the result "01011000". Some other examples for shift and rotation operation on a `bit_vector` `A="101001"` are `(A sll 2)`, resulting in "100100"; `(A srl 2)`, resulting in "001010"; `(A sla 2)`, resulting in "100111"; `(A sra 2)`, resulting in "111010"; `(A rol 2)`, resulting in "100110"; and `(A ror 2)`, resulting in "011010".

The multiplying operators are used to perform mathematical functions on numeric types (integer or floating point). The main operators are multiplication (`*`), division (/), modulus (`mod`), and remainder (`rem`). The multiplication operator is also defined when one of the operands is a physical type and the other an integer or real type. The remainder and modulus are defined as `(A rem B)`, which is equal to `(A -(A/B)*B)`, where A and B are integers and `(A mod B)` is equal to `(A-B * N)`, and `N` is also an integer. The result of the `rem` operator has the sign of its first operand, while the result of the `mod` operators has the sign of the second operand. Some examples of these operators are `(11 rem 4)`, resulting in 3; `((−11) rem 4)`, resulting in –3; `(9 mod 4)`, resulting in 1; `(7 mod (-4))`, resulting in –1 and `(7-4*2)`, resulting in –1.

The addition operators are used to perform arithmetic operation (addition and subtraction) on operands of any numeric type. The operators are addition (+), subtraction (–) and concatenation (`&`). The `&` operator is used to concatenate two vectors together to make a longer one. For example, `signal MYBUS:std_logic_vector (15 downto 0);signal STATUS:std_logic_vector (2 downto 0);signal RW, CS1, CS2:std_logic; signal MDATA:std_logic_vector (0 to 9);MYBUS <= STATUS & RW & CS1 & SC2 & MDATA; MYARRAY (15 downto 0) <= "1111_1111" & MDATA (2 to 9)` and `NEWWORD <= "VHDL" & "93"`. `MYBUS` results in filling the first 8 leftmost bits of `MYARRAY` with 1s and the rest with the 8 rightmost bits of `MDATA`. `NEWWORD` results in an array of characters, "VHDL93". In order to use these operators, one has to specify the IEEE library `ieee.std_logic_unsigned.all` or `std_logic_arith` package in addition to the `ieee.std_logic_1164` package [12–14].

The unary operators + and – are used to specify the sign of a numeric type. The operators are identity (+) and negation (-).

The miscellaneous operators are the absolute value (`abs`) and exponentiation (`**`) operators that can be applied to numeric types. The exponential operation can be performed on integers as well as floating point types.

DESIGN EXAMPLE 2.8

Design and develop a digital system to verify the functions of some of the operators of VHDL.

Solution

```
library IEEE;
use IEEE.STD_LOGIC_1164.ALL;
use IEEE.STD_LOGIC_ARITH.ALL;
use IEEE.STD_LOGIC_UNSIGNED.ALL;
use IEEE.NUMERIC_STD.ALL;
entity operators_exp is
port(a,b: in integer:=0;
a1,b1: in std_logic_vector (2 downto 0):="000";
r1,r2,r3,r4,r5,r6:out integer:=0;
r7: out std_logic_vector (5 downto 0):="000000");
end operators_exp;
architecture Behavioral of operators_exp is
begin
r1<=a+b;
r2<=a-b;
```

```
r3<=a*b;
r4<=a/2;
r5<=a mod 2;
r6<=a rem 2;
r7<=(a1 & b1);
end Behavioral;
```

DESIGN EXAMPLE 2.9

Design and develop a digital system to verify the functions of some of the shift operators of VHDL.

Solution

```
library IEEE;
use IEEE.STD_LOGIC_1164.ALL;
use IEEE.STD_LOGIC_ARITH.ALL;
use IEEE.STD_LOGIC_UNSIGNED.ALL;
use IEEE.NUMERIC_STD.ALL;
entity shift_ope is
port(m_clk: in std_logic:='0';
a: in bit_vector (3 downto 0):="0101";
out1, out2,out3,out4,out5,out6: inout bit_vector(3 downto 0):="0000");
end shift_ope;
architecture shif_oper of shift_ope is
signal a1,a2,a3,a4,a5,a6: bit_vector(3 downto 0):="0000";
begin
p0: process(m_clk)
variable cnt: integer:=1;
begin
if rising_edge(m_clk) then
if (cnt=1) then
a1<= a; a2<=a; a3<=a; a4<=a; a5<=a; a6<=a;
cnt:=cnt+1;
elsif (cnt=2) then
out1<= (a1 sll 1); out2<= (a1 srl 1);
out3<= (a1 sla 1);out4<= (a1 sra 1);
out5<= (a1 rol 1);out6<= (a1 ror 1);
cnt:=cnt+1;
elsif (cnt=3 or cnt=5 or cnt=7 or cnt=9) then
a1<= out1; a2<= out2; a3<= out3;
a4<= out4; a5<= out5; a6<= out6;
cnt:=cnt+1;
elsif (cnt=4) then
out1<= (a1 sll 1); out2<= (a1 srl 1);
out3<= (a1 sla 1);out4<= (a1 sra 1);
out5<= (a1 rol 1);out6<= (a1 ror 1);
cnt:=cnt+1;
elsif (cnt=6) then
out1<= (a1 sll 1); out2<= (a1 srl 1);
out3<= (a1 sla 1);out4<= (a1 sra 1);
out5<= (a1 rol 1);out6<= (a1 ror 1);
cnt:=cnt+1;
elsif (cnt=8) then
out1<= (a1 sll 1); out2<= (a1 srl 1);
out3<= (a1 sla 1);out4<= (a1 sra 1);
out5<= (a1 rol 1);out6<= (a1 ror 1);
cnt:=cnt+1;
elsif (cnt=10) then
out1<= (a1 sll 1); out2<= (a1 srl 1);
out3<= (a1 sla 1);out4<= (a1 sra 1);
```

```
out5<= (a1 rol 1);out6<= (a1 ror 1);
cnt:=1;
end if;
end if;
end process;
end shif_oper;
```

Now we discuss attributes. VHDL supports five types of attributes. Predefined attributes are always applied to a prefix such as a signal name, variable name, or type. Attributes are used to return various types of information about a signal, variable, or type. Attributes consist of a quote mark (') followed by the name of the attribute. Table 2.3 gives several signal attributes.

An example of an attribute is `if (CLOCK'event and CLOCK='1');` this expression checks for the arrival of a positive clock edge. To find out how much time has passed since the last clock edge, one can use the attribute `CLOCK'last_event`. Several attributes of a scalar type are also supported in VHDL as given in Table 2.3. A few examples for the scalar types are:

TABLE 2.3

List of Attributes in VHDL

S. No.	Attribute	Function
1	`signal_name'event`	Returns the Boolean value `True` if an event on the signal occurred; otherwise, gives a `False`
2	`signal_name'active`	Returns the Boolean value `True` if there has been a transaction (assignment) on the signal; otherwise, gives a `False`
3	`signal_name'transaction`	Returns a signal of the type `bit` that toggles (0 to 1 or 1 to 0) every time there is a transaction on the signal
4	`signal_name'last_event`	Returns the time interval since the last event on the signal
5	`signal_name'last_active`	Returns the time interval since the last transaction on the signal
6	`signal_name'last_value`	Gives the value of the signal before the last event occurred on the signal
7	`signal_name'delayed(T)`	Gives a signal that is the delayed version (by time `T`) of the original one (`T` is optional, default `T` = 0)
8	`signal_name'stable(T)`	Returns a Boolean value, `True`, if no event has occurred on the signal during the interval T; otherwise, returns a `False` (`T` is optional, default `T` = 0)
9	`signal_name'quiet(T)`	Returns a Boolean value, `True`, if no transaction has occurred on the signal during the interval `T`; otherwise, returns a `False` (`T` is optional, default `T` = 0)
10	`scalar_type'left`	Returns the first or leftmost value of scalar type in its defined range
11	`scalar_type'right`	Returns the last or rightmost value of scalar type in its defined range
12	`scalar_type'low`	Returns the lowest value of scalar type in its defined range
13	`scalar_type'high`	Returns the greatest value of scalar type in its defined range
14	`scalar_type'ascending`	`True` if `T` is an ascending range, otherwise `False`
15	`scalar_type'value(s)`	Returns the value in `T` that is represented by `s` (`s` stands for string value)

```
type conductance is range 1E-6 to 1E3
units mho;
end units conductance;
type my_index is range 3 to 15;
type my_levels is (low, high, dontcare, highZ);
conductance'right            returns:      1E3
conductance'high             1E3
conductance'low                     1E-6
my_index'left                3
my_index'value(5)            "5"
my_levels'left                      low
my_levels'low                low
my_levels'high                      highZ
my_levels'value(dontcare)   "dontcare"
```

Another type of attribute is array attributes, which return an index value corresponding to the array range. VHDL supports the following attributes:

```
MATRIX'left(N) returns leftmost element index
MATRIX'right(N) returns rightmost index
MATRIX'high(N) returns upper bound
MATRIX'low(N) returns lower bound
MATRIX'length(N) returns the number of elements
MATRIX'range(N) returns range
MATRIX'reverse_range(N) returns reverse range
MATRIX'ascending(N) returns a Boolean value TRUE if the index is an
ascending range, and otherwise FALSE. The number N between brackets
refers to the dimension. For a one-dimensional array, one can omit the
number N as given below
type MYARR8×4 is an array (8 downto 1, 0 to 3) of Booleans
type MYARR1 is array (-2 to 4) of integers
MYARR1'left                  returns: -2
MYARR1'right                        4
MYARR1'high                         4
MYARR1'reverse_range              4 downto to −2
MYARR8×4'left(1)                  8
MYARR8×4'left(2)                  0
MYARR8×4'right(2)                 3
MYARR8×4'high(1)                  8
MYARR8×4'low(1)                   1
MYARR8×4'ascending(1)        False
```

We will see some design examples in the forthcoming chapters.

2.5 Concurrent Code

Behavioral modeling can be done with sequential statements using the process construct or with concurrent statements. The first method was described in the previous section and is useful to describe complex digital systems. In this section, we will use concurrent statements to describe the behavior of the digital system. As we know, this method is usually called data-flow modeling, which describes a circuit in terms of its function and the flow of data through the circuit. This is different from the structural modeling that describes a circuit in terms of the interconnection of components. Concurrent signal assignments are event triggered and executed as soon as an event on one of the signals occurs. In the remainder of this section, we will describe several concurrent constructs for use in data-flow modeling. We have discussed several concurrent examples earlier in this chapter. In this section, we will review the different types of concurrent signal assignments. A simple concurrent signal assignment is given in the following examples,

```
Sum <= (A xor B) xor Cin;
Carry <= (A and B);
Z <= (not X) or Y after 2 ns;
```

The actual syntax is `Target_signal <= expression;` in which the value of the expression is transferred to the `target_signal`. As soon as an event occurs on one of the signals, the expression will be evaluated. The type of the `target_signal` has to be the same as the type of the value of the expression. Note that we need to include library `IEEE.std_logic_unsigned;` in order to use the + operator. The syntax for the conditional signal assignment is `Target_signal <= expression when Boolean_condition else expression when Boolean_condition … else expression;`. The target signal will receive the value of the first expression whose Boolean condition is `TRUE`. If no condition is found to be `TRUE`, the target signal will receive the value of the final expression. If more than one condition is true, the value of the first condition that is `TRUE` will be assigned [12–16]. For example, a 4:1 multiplexer using conditional signal assignments is given below.

```
library IEEE;
use IEEE.STD_LOGIC_1164.ALL;
use IEEE.NUMERIC_STD.ALL;
entity mux_4_1_conc is
port (s1, s0, a, b, c, d: in std_logic:='0';
z: out std_logic:='0');
end mux_4_1_conc;
architecture concurr_mux41 of mux_4_1_conc is
begin
z <=  a when s1='0' and s0='0' else
b when s1='0' and s0='1' else
c when s1='1' and s0='0' else
d;
end concurr_mux41;
```

The conditional signal assignment will be re-evaluated as soon as any of the signals in the conditions or expression change. The `when-else` construct is useful to express logic functions in the form of a truth table. For example, the same multiplexer as above is given below in a more compact form.

```
library ieee;
use ieee.std_logic_1164.all;
use ieee.numeric_std.all;
entity mux_4_1_functab is
port (a, b, c, d: in std_logic;
sel: in std_logic_vector (1 downto 0):="00";
z: out std_logic:='0');
end mux_4_1_functab;
```

```
architecture concurr_mux41 of mux_4_1_functab is
begin
z <=  a when sel = "00" else
b when sel = "01" else
c when sel = "10" else
d;
end concurr_mux41;
```

Note that this construct is simpler than the `if-then-else` construct using the `process` statement or the `case` statement. An alternative way to define the multiplexer is the `case` construct inside a `process` statement, as discussed earlier. The selected signal assignment is similar to the conditional one as described above, whose syntax is:

```
with choice_expression select
target_name <= expression when choices,
target_name <= expression when choices,
:
target_name <= expression when choices;
```

The target is a signal that will receive the value of an expression whose choice includes the value of the `choice_expression`. The expression selected is the first with a matching choice. The choice can be a static expression or a range expression. The rules that must be followed for the choices are (i) no two choices can overlap and (ii) all possible values of `choice_expression` must be covered by the set of choices, unless another choice is present. An example of a 4:1 multiplexer is

```
library ieee;
use ieee.std_logic_1164.all;
entity mux_4_1_conc2 is
port (a, b, c, d: in std_logic:='0';
sel: in std_logic_vector(1 downto 0):="00";
z: out std_logic:='0');
end mux_4_1_conc2;
architecture concurr_mux41b of mux_4_1_conc2 is
begin
with sel select
z <= a when "00",
b when "01",
c when "10",
d when "11",
a when others;
end concurr_mux41b;
```

The equivalent `process` statement would make use of the `case` construct. Similar to the `when-else` construct, the selected signal assignment is useful to express a function as a truth table. The choices can express a single value, a range or combined choices as:

```
target <= value1 when "000",
value2 when "001" | "011" | "101",
value3 when others;
```

In this example, all eight choices are covered only once. The other choice must be the last one used. Note that the Xilinx Foundation Express does not allow a vector as choice_expression, such as std_logic_vector' (A,B,C). For example, let us consider a full adder with inputs A, B, and C and outputs sum and cout, then:

```
library ieee;
use ieee.std_logic_1164.all;
entity FullAdd_Conc is
port (A, B, C: in std_logic;
sum, cout: out std_logic);
end FullAdd_Conc;
architecture FullAdd_Conc of FullAdd_Conc is
signal INS: std_logic_vector (2 downto 0);
begin
INS(2) <= A;
INS(1) <= B;
INS(0) <= C;
with INS select
(sum, cout) <= std_logic_vector'("00") when "000",
std_logic_vector'("10") when "001",
std_logic_vector'("10") when "010",
std_logic_vector'("01") when "011",
std_logic_vector'("10") when "100",
std_logic_vector'("01") when "101",
std_logic_vector'("01") when "110",
std_logic_vector'("11") when "111",
std_logic_vector'("11") when others;
end FullAdd_Conc;
```

In the above example, we had to define an internal vector INS (A,B,C) of the input signals to use as part of the with-select-when statement. This was done because the Xilinx Foundation Express does not support the construct std_logic_vector' (A,B,C). A structural way of modeling describes a circuit in terms of components and its interconnection. Each component is supposed to be defined earlier and can be described as structural, behavioral, or data flow model. At the lowest hierarchy, each component is described as a behavioral model using the basic logic operators defined in VHDL. In general, structural modeling is very good for describing complex digital systems through a set of components in a hierarchical fashion. A structural description can best be compared to a schematic block diagram that can be described by the components and the interconnections. VHDL provides a formal way to do this by (i) declaring a list of components being used and (ii) declaring the signals which define the nets that interconnect components. Label multiple instances of the same component so that each instance is uniquely defined. The components and signals are declared within the architecture body as:

```
architecture architecture_name of NAME_OF_ENTITY is
-- Declarations
component declarations
signal declarations
begin
-- Statements
component instantiation and connections
:
end architecture_name;
```

Before instantiating the components, they need to be declared in the architecture declaration or package declaration section. The component declaration consists of the component name and the interface ports. The syntax is as follows:

```
component component_name [is]
[port (port_signal_names: mode type;
port_signal_names: mode type;
:
port_signal_names: mode type);]
end component [component_name];
```

The component name refers to either the name of an entity defined in a library or an entity explicitly defined in the VHDL file. The list of interface ports gives the name, mode, and type of each port, similar to what is done in the entity declaration. A few examples of component declarations are:

```
component OR2
port (in1, in2: in std_logic:='0';
out1: out std_logic);
end component;
component PROC
port (CLK, RST, RW, STP: in std_logic;
ADDRBUS: out std_logic_vector (31 downto 0);
DATA: inout integer range 0 to 1024);
component FULLADDER
port(a, b, c: in std_logic;
sum, carry: out std_logic);
end component;
```

As mentioned earlier, the component declaration has to be done either in the architecture body or the package declaration. If the component is declared in a package, one does not have to declare it again in the architecture body as long as one uses the `library` and `use` clauses. The component instantiation statement references a component that can be previously defined at the current level of the hierarchy or defined in a technology library (vendor's library). The syntax for component instantiation is as follows: `instance_name: component name port map (port1=>signal1, port2=> signal2,... port3=>signaln);`. The instance name or label can be any legal identifier and is the name of this particular instance. The component name is the name of the component declared earlier using the component declaration statement. The port name is the name of the port, and the signal is the name of the signal to which the specific port is connected. The above port map associates the ports to the signals through named association. An alternative method is positional association as `port map (signal1, signal2,... signaln);` in which the first port in the component declaration corresponds to the first signal, the second port to the second signal and so on. The signal position must be in the same order as the declared component's ports. One can mix named and positional associations as long as one puts all positional associations before the named ones [12–14].

DESIGN EXAMPLE 2.10

Design VHDL code to realize the serial and concordant execution of a digital system.

Solution

```
library ieee;
use ieee.std_logic_1164.all;
```

```
use ieee.std_logic_arith.all;
use ieee.std_logic_unsigned.all;
use IEEE.numeric_bit.all;
entity ex1 is
port(m_clk,a,b,c: in std_logic:='0';
d: out std_logic:='0');
end ex1;
architecture se_ADC of ex1 is
signal e:std_logic:='0';
begin
p0: process(m_clk)
begin
if rising_edge(m_clk) then
e<=(a and b);
d<=(c or not(e));
end if;
end process;
end se_ADC;
```

DESIGN EXAMPLE 2.11

Design VHDL code to realize concurrent data flow using a digital multiplexer system.

Solution

```
library IEEE;
use IEEE.STD_LOGIC_1164.ALL;
use IEEE.STD_LOGIC_arith.ALL;
use IEEE.STD_LOGIC_unsigned.ALL;
entity concurrent_flow_mux is
port(i0,i1,i2,i3:in std_logic_vector(7 downto 0):="00000000";
s:in std_logic_vector(1 downto 0):="00";
z:out std_logic_vector(7 downto 0):="00000000");
end concurrent_flow_mux;
architecture Behavioral of concurrent_flow_mux is
begin
z<=i0 when s="00" else
i1 when s="01" else
i2 when s="10" else
i3;
end Behavioral;
```

2.6 Sequential Code

As discussed in earlier sections, VHDL provides the means to represent digital circuits at different levels of representation of abstraction, such as behavioral and structural modeling. In this section, we will discuss different constructs for describing the behavior of components and circuits in terms of sequential statements, which is also called algorithmic design/modeling. The basis for sequential modeling is the process construct. As you will see, the process construct allows us to model complex digital systems, in particular sequential circuits. A process statement is the main construct in behavioral modeling that allows us to use sequential statements to describe the behavior of a system over time. The syntax for a process statement is

```
[process_label:] process [ (sensitivity_list) ] [is]
[ process_declarations]
```

```
begin
list of sequential statements such as:
signal assignments
variable assignments
case statement
exit statement
if statement
loop statement
next statement
null statement
procedure call
wait statement
end process [process_label];
```

For example, a positive edge-triggered D-flip-flop (FF) with asynchronous clear input is:

```
library ieee;
use ieee.std_logic_1164.all;
entity dff_clear is
port (clk, clear, d : in std_logic:='0';
q : out std_logic:='0');
end dff_clear;
architecture behav_dff of dff_clear is
begin
dff_process: process (clk, clear)
begin
if (clear = '1') then
q <= '0';
elsif (clk'event and clk = '1') then
q <= d;
end if;
end process;
end behav_dff;
```

A process is declared within the architecture and is a concurrent statement; however, the statements inside a process are executed sequentially. Like other concurrent statements, a process reads and writes signals and values of the interface (input and output) ports to communicate with the rest of the architecture. One can thus make assignments to signals that are defined externally to the process, such as the Q output of the FF in the above example. The expression `CLK'event and CLK = '1';` or `rising_edge (CLK) then;` checks for a positive clock edge (`clock event AND clock high`). The sensitivity list is a set of signals to which the process is sensitive. Any change in the value of the signals in the sensitivity list will cause immediate execution of the process. If the sensitivity list is not specified, one has to include a `wait` statement to make sure that the process will halt. Notice that one cannot include both a sensitivity list and a `wait` statement. Variables and constants that are used inside a process have to be defined in the `process_declarations` part before the keywords begin. The keywords begin signals that are the start of the computational part of the process. The statements are sequentially executed, similarly to a conventional software program. It should be noted that variable assignments inside a process are executed immediately and denoted by the := operator. This is in contrast to signal assignments denoted by <=, in which changes occur after a delay. As a result, changes made to variables will be available immediately to all subsequent statements within the same process. The previous example of the D-FF illustrates how to describe a sequential circuit with the process statement. Although the process is mainly used to describe sequential circuits, one can also describe combinational circuits with the process construct [12,13]. The following example illustrates this for a full adder composed of two half adders. This example also illustrates how one process can generate signals that will trigger other processes when events on the signals in its sensitivity list occur. We can write the Boolean expression of a half adder and full adder as:

```
Sum and carry of the half adder is
S_ha = (A⊕B); and
C_ha = AB;
Sum and carry of a full adder is
Sum = (A⊕B)⊕Cin; = S_ha ⊕Cin;
Cout = (A⊕B)Cin + AB; = S_ha.Cin + C_ha;
```

In this example, adders are modeled with two processes, P1 and P2, as:

```
library ieee;
use ieee.std_logic_1164.all;
entity full_adder is
port (a, b, cin : in std_logic:='0';
sum, cout : out std_logic:='0');
end full_adder;
architecture behav_fa of full_adder is
signal int1, int2, int3: std_logic:='0';
begin
p1: process (a, b)
begin
int1<= a xor b;
int2<= a and b;
end process;
p2: process (int1, int2, cin)
begin
sum <= int1 xor cin;
int3 <= int1 and cin;
cout <= int2 or int3;
end process;
end behav_fa;
```

Of course, one could simplify the behavioral model significantly by using a single process. The `if` statement executes a sequence of statements whose sequence depends on one or more conditions. The syntax is:

```
if condition then
sequential statements
[elsif condition then
sequential statements ]
[else
sequential statements ]
end if;
```

Each condition is a Boolean expression. The `if` statement is performed by checking each condition in the order they are presented until a "true" is found. Nesting of `if` statements is allowed. An example of an `if` statement was given earlier for a D-FF with asynchronous clear input. The `if` statement can be used to describe combinational circuits as well. The following example illustrates this for a 4:1 multiplexer with inputs `A`, `B`, `C`, and `D` and select signals S0 and S1. This statement must be inside a process construct. We will see that other constructs, such as the conditional signal assignment `when else` or the `select` construct, may be more convenient for these types of combinational circuits.

```
library ieee;
use ieee.std_logic_1164.all;
entity MUX_4_1a is
port (S1,S0,A,B,C,D: in std_logic:='0';
Z: out std_logic:='0');
end MUX_4_1a;
architecture behav_MUX41a of MUX_4_1a is
begin
P1: process (S1,S0,A,B,C,D)
begin
if ((not S1 and not S0)='1') then
Z <= A;
elsif ((not S1 and S0) = '1') then
Z<=B;
elsif ((S1 and not S0) ='1') then
Z <=C;
else
Z<=D;
end if;
end process P1;
end behav_MUX41a;
```

A slightly different way of modeling the same multiplexer is shown below:

```
library ieee;
use ieee.std_logic_1164.all;
entity MUX_4_1a is
port (S1,S0,A,B,C,D: in std_logic:='0';
Z: out std_logic:='0');
end MUX_4_1a;
architecture behav_MUX41a of MUX_4_1a is
begin
P1: process (S1,S0,A,B,C,D)
begin
if ((not S1 and not S0)='1') then
if S1='0' and S0='0' then
Z <= A;
elsif S1='0' and S0='1' then
Z <= B;
elsif S1='1' and S0='0' then
Z <= C;
elsif S1='1' and S0='1' then
Z <= D;
end if;
end if;
end process P1;
end behav_MUX41a;
```

The `case` statement executes one of several sequences of statements based on the value of a single expression. The syntax is:

```
case expression is
when choices =>
sequential statements
when choices =>
sequential statements
-- branches are allowed
[ when others => sequential statements ]
end case;
```

The expression must evaluate to an integer, an enumerated type of a 1D array such as a `bit_vector`. The `case` statement evaluates the expression and compares the value to each of the choices. The `when` clause corresponding to the matching choice will have its statements executed. The following rules must be adhered to: no two choices can overlap; that is, each choice can be covered only once. If the `when others` choice is not present, all possible values of the expression must be covered by the set of choices. An example of a `case` statement using an enumerated type is given below, where an output `D = 1` when the signal grades have a value between 51 and 60, `C = 1` for grades between 61 and 70 and `when others` covers all the other grades and results in an `F = 1`.

```
library ieee;
use ieee.std_logic_1164.all;
entity grd_201 is
port(value: in integer range 0 to 100:=10;
a, b, c, d,f: out bit:='0');
end grd_201;
architecture behav_grd of grd_201 is
begin
process (value)
begin
case value is
when 51 to 60 =>
d <= '1';
when 61 to 70 | 71 to 75 =>
c <= '1';
when 76 to 85 =>
b <= '1';
when 86 to 100 =>
a <= '1';
when others =>
f <= '1';
end case;
end process;
end behav_grd;
```

We used the vertical bar (|), which is equivalent to the `or` operator, to illustrate how to express a range of values. This is a useful operator to indicate ranges that are not adjacent, for example, `0 to 4 | 6 to 10`. It should be noted that the combinational circuits can also be expressed in other ways, using concurrent statements such as the `with-select` construct. Since the `case` statement is a sequential statement, one can have nested `case` statements as well. A loop statement is used to repeatedly execute a sequence of sequential statements. The syntax for a loop is:

```
[ loop_label :]iteration_scheme loop
sequential statements
[next [label] [when condition];
[exit [label] [when condition];
end loop [loop_label];
```

Labels are optional but are useful when writing nested loops. The next and exit statements are sequential statements that can only be used inside a loop. The next statement terminates the rest of the current loop iteration, and execution will proceed to the next loop iteration. The exit statement skips the rest of the statements, terminating the loop entirely, and continues with the next statement after the exited loop. There are three types of iteration schemes: basic loop, while... loop, and for... loop. We will discuss these one by one below.

This loop has no iteration scheme. It will be executed continuously until it encounters an exit or next statement. The syntax of this loop statement is:

```
[ loop_label :] loop
sequential statements
[next [label] [when condition];
[exit [label] [when condition];
end loop [ loop_label];
```

We defined a variable intern_value inside the process because output ports cannot be read inside the process.

The while... loop evaluates a Boolean iteration condition. When the condition is TRUE, the loop repeats; otherwise, the loop is skipped and the execution will halt. The syntax for the while... loop is:

```
[ loop_label :] while condition loop
sequential statements
[next [label] [when condition];
[exit [label] [when condition];
end loop[ loop_label ];
```

The condition of the loop is tested before each iteration, including the first iteration. If it is false, the loop is terminated.

The for... loop uses an integer iteration scheme that determines the number of iterations. The syntax is as follows:

```
[ loop_label :] for identifier in range loop
sequential statements
[next [label] [when condition];
[exit [label] [when condition];
end loop[ loop_label ];
```

Below is an example of the for... loop statement:

```
library ieee;
use ieee.std_logic_1164.all;
```

```
entity parity is
port (input: in std_logic_vector (7 downto 0);
output_even: out std_logic;
output_odd : out std_logic);
end parity;
architecture behavioral of parity is
function parity_odd(input:std_logic_vector (7 downto 0)) return std_logic
is
variable temp: std_logic:='0';
begin
for i in 0 to 7 loop
temp:=temp xor input(i);
end loop;
return (not temp);
end parity_odd;
function parity_even(input:std_logic_vector (7 downto 0)) return std_
logic is
variable temp: std_logic:='0';
begin
for i in input'range loop
temp:=temp xor input(i);
end loop;
return temp;
end parity_even;
begin
output_even <=parity_even(input);
output_odd<=parity_odd(input);
end behavioral;
```

The identifier (`index`) is automatically declared by the loop itself; thus, one does not need to declare it separately. The value of the identifier can only be read inside the loop and is not available outside its loop. One cannot assign or change the value of the index. This is in contrast to the `while`... loop, whose condition can involve variables that are modified inside the loop. The range must be a computable integer range in one of the following forms, in which `integer_expression` must evaluate to an integer: `integer_expression to integer_expression` and `integer_expression downto integer_expression`. The `next` statement skips execution to the next iteration of a loop statement and proceeds with the next iteration. The syntax is:

```
next [label] [when condition];
```

The `when` keyword is optional and will execute the next statement when its condition evaluates to the Boolean value `TRUE`. The `exit` statement skips the rest of the statements, terminating the loop entirely, and continues with the next statement after the exited loop. The syntax is as follows:

```
exit [label] [when condition];
```

The `when` keyword is optional and will execute the `next` statement when its condition evaluates to the Boolean value `true`. Notice that the difference between the `next` and `exit` statement is that the `exit` statement terminates the loop. The `wait` statement will halt a process until an event occurs. There are several forms of the `wait` statement, such as:

```
wait until condition;
wait for time expression;
wait on signal;
wait;
```

The Xilinx Foundation Express has implemented only the first form of the `wait` statement. The syntax is as follows:

```
wait until signal = value;
wait until signal'event and signal = value;
wait until not signal'stable and signal = value;
```

The condition in the `wait until` statement must be true for the process to resume, for example:

```
wait until CLK='1';
wait until CLK='0';
wait until CLK'event and CLK='1';
wait until not CLK'stable and CLK='1';
```

For the first example, the process will wait until a positive-going clock edge occurs, while for the second example, the process will wait until a negative-going clock edge arrives. The last two examples are equivalent to the first (positive-edge or 0-1 transitions). The hardware implementation for these three statements will be identical. It should be noted that a process that contains a `wait` statement cannot have a sensitivity list. If a process uses one or more `wait` statements, the Foundation Express synthesizer will use sequential logic. The results of the computations are stored in flip-flops.

The `null` statement states that no action will occur. The syntax is as follows:

```
null;
```

It can be useful in a `case` statement where all choices must be covered, even if some of them can be ignored. As an example, consider a control signal `cntl` in the range 0 to 31. When the value of `cntl` is 3 or 15, the signals A and B will be `xor`ed; otherwise, nothing will occur [12–14].

```
library IEEE;
use IEEE.STD_LOGIC_1164.ALL;
entity ex_wait is
port (cntl: in integer range 0 to 31:=0;
a, b: in std_logic_vector(7 downto 0):="00000000";
z: out std_logic_vector(7 downto 0):="00000000");
end ex_wait;
architecture arch_wait of ex_wait is
begin
p_wait: process (cntl)
begin
z <=a;
case cntl is
when 3 | 15 =>
z <= a xor b;
```

```
when others =>
null;
end case;
end process p_wait;
end arch_wait;
```

DESIGN EXAMPLE 2.12

Design and develop a sequential digital system using the algorithmic design flow as given below:

```
  i. Initial state is s0, input is x and output is y,
 ii. s0 to s1 if x=1 with y=1; otherwise, state is s0 with y=0,
iii. s1 to s2 if x=0 with y=1; otherwise, state is s1 with y=0,
 iv. s2 to s3 if x=1 with y=0; otherwise, state is s2 with y=1 and
  v. s3 to s0 if x=0 with y=0; otherwise, state is s3 with y=1,
```

Solution

```
library ieee;
use ieee.std_logic_1164.all;
entity algor_des is
port (clk,x: in std_logic:='0';
z: out std_logic:='0');
end algor_des;
architecture exa of algor_des is
type state_type is (s0,s1,s2,s3);
signal st: state_type:=s0;
begin
p0: process (clk)
begin
if clk'event and clk = '1' then
case st is
when s0 =>
if x='0' then st<=s0;
else st<=s1;
end if;
when s1 =>
if x='1' then st<=s1;
else st<=s2;
end if;
when s2 =>
if x='0' then st<=s2;
else st<=s3;
end if;
when s3 =>
if x='1' then st<=s3;
else st<=s0;
end if;
when others => null;
end case;
end if;
end process;
z <='1' when (st=s0 and x='1') else
'1' when (st=s1 and x='0') else
'1' when (st=s2 and x='0') else
'1' when (st=s3 and x='1') else
'0';
end exa;
```

DESIGN EXAMPLE 2.13

Develop and implement a decoder using sequential design flow.

Solution

```
library ieee;
use ieee.std_logic_1164.all;
use ieee.std_logic_arith.all;
use ieee.std_logic_unsigned.all;
entity sequential_decoder is
port(a,b,enable:in std_logic:='0';
i:out std_logic_vector(3 downto 0):="0000");
end sequential_decoder;
architecture behavioral of sequential_decoder is
signal abar,bbar:std_logic:='0';
begin
process(a,b,enable)
begin
abar<=not a;
bbar<=not b;
if (enable='1')then
i(0)<=(abar and bbar);
i(1)<=(abar and b);
i(2)<=(a and bbar);
i(3)<=(a and b);
else
i<="1111";
end if;
end process;
end behavioral;
```

DESIGN EXAMPLE 2.14

Design and develop a digital system to realize a single normalizing digital circuit.

Solution

```
library ieee;
use ieee.std_logic_1164.all;
use ieee.std_logic_arith.all;
use ieee.std_logic_unsigned.all;
entity divider is
port(dividend_in:in std_logic_vector(7 downto 0);
divisor:in std_logic_vector(3 downto 0);
st,clk:in std_logic;
quotient:out std_logic_vector(3 downto 0);
remainder:out std_logic_vector(3 downto 0);
overflow:out std_logic);
end divider;
architecture behavioral of divider is
signal state,nextstate:integer range 0 to 5;
signal clk_sig,c,load,su,sh:std_logic;
signal subout:std_logic_vector(4 downto 0);
signal dividend:std_logic_vector(8 downto 0);
begin
subout<=dividend(8 downto 4)-('0'& divisor);
c<=not subout(4);
remainder<=dividend(7 downto 4);
quotient<=dividend(3 downto 0);
------------------------------------------
p1:process(clk)
variable cnt:integer;
```

```
begin
if rising_edge(clk) then
if(cnt=20000)then
clk_sig<= not(clk_sig);
cnt:=0;
else
cnt:= cnt + 1;
end if;
end if;
end process;
-------------------------------------------
state_graph:process(state,st,c)
begin
load<='0';overflow<='0';sh<='0';su<='0';
case state is
when 0=>
if(st='1')then load<='1';nextstate<=1;
else nextstate<=0;
end if;
when 1=>
if(c='1')then overflow<='1';nextstate<=0;
else sh<='1';nextstate<=2;
end if;
when 2 |3| 4=>
if(c='1')then su<='1';nextstate<=state;
else sh<='1';nextstate<=state+1;
end if;
when 5=>
if(c='1')then su<='1';end if;
nextstate<=0;
end case;
end process state_graph;
update:process(clk_sig)
begin
if clk_sig'event and clk_sig='1' then --rising edge of clk
state<=nextstate;
if load='1' then dividend<='0'& dividend_in;end if;
if load='1' then dividend(8 downto 4)<=subout;dividend(0)<='1';end if;
if sh='1' then dividend<=dividend(7 downto 0)&'0';end if;
end if;
end process update;
end behavioral;
```

2.7 Flip-Flops and Their Conversions

In many applications, FFs are used for storing data sequences, delaying data flow, controlling the output latch-out, designing clock dividers, and setting/resetting the circuit and so on. There are four FFs, namely the (i) set-reset (SR) FF, (ii) Jack-Kilby (JK) FF, (iii) delay (D) FF, and (iv) toggle (T) FF. We discuss the characteristic functions and state change control of all the FFs, FF conversions and their applications in this section.

2.7.1 Flip-Flops

A flip-flop is a bistable multivibrator. The circuit can be made to change the state by signals applied to one or two control inputs, and it will have two outputs. It is a sequential logic

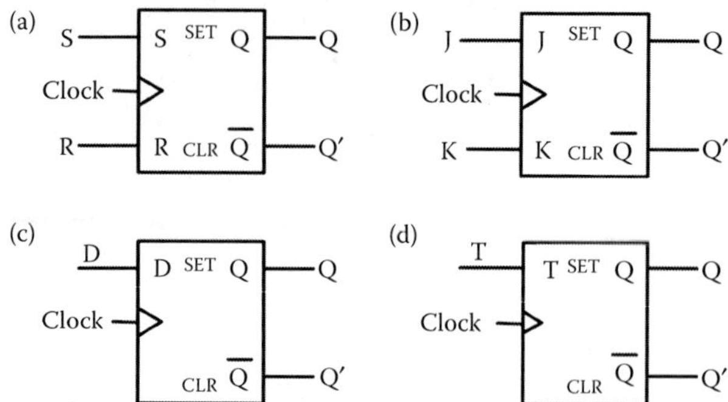

FIGURE 2.1
Symbolic representation of FFs: SR-FF (a), JK-FF (b), D-FF (c), and T-FF (d).

device and storage element. A latch is level sensitive, whereas a flip-flop is edge sensitive. In digital electronics, the flip-flop is a special type of edge-gated circuit. A symbolic representation of all the FFs is shown in Figure 2.1. The internal design of all the FFs with the basic gates is out of the scope of this book since it can be found in many eBooks or on websites. Our main aim in this part is understanding the characteristic function of all the FFs and deriving the state change input control. We discuss these one by one below.

SR-flip-flop: This is similar to an SR latch. Besides the clock input, the SR-FF has two inputs, labeled set (S) and reset (R). If the set input is at logic 1 when the clock is triggered, the output (Q) is logic 1. If the reset input is at logic 1 when the clock is triggered, the output (Q) is logic 0. Note that in an SR-FF, the set and reset inputs should not both be at logic 1 when the clock is triggered. This is considered an invalid input condition, and the resulting output is not predictable if this condition occurs as given in the left part of Table 2.4. Therefore, the next state (Q^+) will be decided by the FF inputs S and R and the present state (Q).

Upon solving the characteristic table of SR-FF using the K-map [17,18], the characteristic equation for the Q^+ can be obtained as

$$Q^+ = S + R'Q \tag{2.1}$$

TABLE 2.4
Characteristic and State Change Control of SR-FF

Characteristic Function					State Change Control I/P		
Set (S)	Reset (R)	Present State (Q)	Next State (Q^+)	Remark	State Change (Q to Q^+)	S	R
0	0	0	0	No Change	0 to 0	0	X
0	0	1	1		0 to 1	1	0
0	1	0	0	Reset	1 to 0	0	1
0	1	1	0		1 to 1	X	0
1	0	0	1	Set			
1	0	1	1				
1	1	0	X	Invalid			
1	1	1	X				

Where Q^+ is the next state, S and R are the input of the SR-FF and Q is the present state. Now, we will discuss deriving the control inputs for the required state changes, which is most important for the FF conversion we discuss in a subsequent section. Let us assume that the present state of the SR-FF is 0; that is, $Q = 0$, and we wish to maintain the same state for next clock pulse as well. This requirement is given as "0 to 0" in the right part of Table 2.4. Now, we need to identify all the inputs to SR-FF to maintain the state "0 to 0"; that is, the present state is 0 and the next state is also 0. We need to refer the characteristic table, that is, the left part of Table 2.4, of this FF to identify those inputs. The input `S` = `R` = `0` and `S` = `0`; `R` = `1` maintains the next state (`Q+`) as 0 when the present state (`Q`) is 0. This means that whenever `S` = `0`, `R` = `X` and `Q` = `0`, then the next state will be 0; therefore, the control inputs for the state change "0 to 0" are `S` = `0` and `R` = `X`. Now, we assume another state change as "0 to 1". Refer to the characteristic table, which shows that whenever `S` = `1`, `R` = `0` and `Q` = `0`, the next state will be 1. Therefore, the control inputs for the state change "0 to 1" are `S` = `1` and `R` = `0`. In the same way, we can obtain the control inputs for all other state changes such that "1 to 0" is `S` = `0`; `R` = `1` and "1 to 1" is `S` = `X`; `R` = `0`, which are given in the left part of Table 2.4. The same logic and analysis can be obtained for all the FFs as well.

JK flip-flop: The JK-FF has two inputs, labeled J and K. The J input corresponds to the set input in an SR-FF, and the K input corresponds to the reset input. The difference between a JK-FF and an SR-FF is that, in the JK-FF, both inputs can be logic 1, which is prohibited in SR-FF. When both the J and K inputs are logic 1, the output (`Q`) is toggled, which means that the output alternates between logic 1 and logic 0. For example, if the output (`Q`) is logic 1 when the clock is triggered and J and K are both logic 1, the output (`Q`) is set to logic 0. If the clock is triggered again while J and K both remain at logic 1, the output (`Q`) is set to logic 1 again and so forth, with the output alternating from logic 1 to logic 0 at every clock pulse.

The characteristic table of JK-FF is given in Table 2.5. Upon solving the characteristic table of JK-FF using the K-map, the characteristic equation for `Q+` can be obtained as

$$Q^+ = JQ' + K'Q \tag{2.2}$$

where Q^+ is the next state, J and K are the input of the JK-FF, and Q is the present state. As discussed in the previous section, the state change control inputs can be obtained as given in the right part of Table 2.5.

TABLE 2.5

Characteristic and State Change Control of JK-FF

Characteristic Function					**State Change Control I/P**		
Jack (J)	**Kilby (K)**	**Present State (Q)**	**Next State (Q^+)**	**Remark**	**State Change (Q to Q^+)**	**S**	**R**
0	0	0	0	No Change	0 to 0	0	X
0	0	1	1		0 to 1	1	X
0	1	0	0	Reset	1 to 0	X	1
0	1	1	0		1 to 1	X	0
1	0	0	1	Set			
1	0	1	1				
1	1	0	1	Complement			
1	1	1	0				

TABLE 2.6

Characteristic and State Change Control of D-FF

Characteristic Function				State Change Control I/P	
Data In (D)	**Present State (Q)**	**Next State (Q^+)**	**Remark**	**State Change (Q to Q^+)**	**D**
0	0	0	Reset	0 to 0	0
0	1	0		0 to 1	1
1	0	1	Set	1 to 0	0
1	1	1		1 to 1	1

D-flip-flop: This FF has just one input in addition to the clock input. This input is called the data input (D). When the clock is triggered, the output (Q) is matched to the data input. Thus, if the data input is logic 1, the output (Q) is logic 1, and if the data input is logic 0, the output is logic 0. This type of FF is mostly used as the delay element. Cascade configuration of this FF will yield a delay for more than one clock pulse.

The characteristic table for this FF is given in the left part of Table 2.6. Upon solving the characteristic table of D-FF using the K-map, the characteristic equation for Q^+ can be obtained as

$$Q^+ = D \tag{2.3}$$

where Q^+ is the next state and D is the input of the D-FF. The state change control input for this FF is given in the right part of Table 2.6.

T flip-flop: This is simply a JK-FF whose output alternates between logic 1 and logic 0 with each clock pulse. Toggles are widely used in logic circuits because they can be combined to form counting circuits that count the number of clock pulses received. This toggling happens whenever the value of T is logic 1. If the value of T is 0, then there is no change in the present and next state values, that is, the "No Change" condition occurs.

The characteristic table for this FF is given in the left part of Table 2.7. Upon solving the characteristic table of T-FF using the K-map [17,18], the characteristic equation for Q^+ can be obtained as

$$Q^+ = TQ^+ + T'Q \tag{2.4}$$

TABLE 2.7

Characteristic and State Change Control of T-FF

Characteristic Function				State Change Control I/P	
Data In (T)	**Present State (Q)**	**Next State (Q^+)**	**Remark**	**State Change (Q to Q^+)**	**T**
0	0	0	No Change	0 to 0	0
0	1	1		0 to 1	1
1	0	1	Complement	1 to 0	1
1	1	0		1 to 1	0

where Q^+ is the next state, T is the input of the T-FF, and Q is the present state. The state change control input for this FF is given in the right part of Table 2.7.

DESIGN EXAMPLE 2.15

Design and develop a digital system to realize all the FFs.

Solution

```
library ieee;
use ieee. std_logic_1164.all;
use ieee. std_logic_arith.all;
use ieee. std_logic_unsigned.all;
entity ffs is
port(s,r,d,j,k,t,clock: in std_logic:='0';
q_sr, q_sr_bar,q_d,q_jk, q_jk_bar,q_t: out std_logic:='0');
end ffs;
architecture behavioral of ffs is
signal tmp: std_logic:='0';
begin
-------sr-ff------
process(clock)
variable tmp: std_logic:='0';
begin
if(clock='1' and clock'event) then
if(s='0' and r='0')then
tmp:=tmp;
elsif(s='1' and r='1')then
tmp:='Z';
elsif(s='0' and r='1')then
tmp:='0';
else
tmp:='1';
end if;
end if;
q_sr<= tmp;
q_sr_bar <= not tmp;
end process;
--------d-ff------
process(clock)
begin
if(clock='1' and clock'event) then
q_d <= d;
end if;
end process;
------jk-ff-----
process(clock)
variable tmp: std_logic:='0';
begin
if(clock='1' and clock'event) then
if(j='0' and k='0')then
tmp:=tmp;
elsif(j='1' and k='1')then
tmp:= not tmp;
elsif(j='0' and k='1')then
tmp:='0';
else
tmp:='1';
end if;
end if;
```

```
q_jk<=tmp;
q_jk_bar<=not tmp;
end process;
---------t-ff--------
process (clock)
begin
if clock'event and clock='1' then
if t='0' then
tmp <= tmp;
elsif t='1' then
tmp <= not (tmp);
end if;
end if;
end process;
q_t <= tmp;
end behavioral;
```

2.7.2 Flip-Flop Conversions

A flip-flop is an electronics circuit that consists of two stable states which are used to store data. FFs are the basic building blocks of a digital electronic system which are used in various applications like communications, computers, control, signal/image processing, and so on. Conversion of one type of FF to another is required for many of these applications and can be done by using a combinational logic circuit. For example, if we need a JK-FF, but we have only an SR-FF, then we need to convert the SR-FF into a JK-FF, as shown in Figure 2.2. The inputs of the JK-FF are given to the combinational circuit, which alters these inputs to another suitable form and applies the state change control inputs to the SR-FF to get the response of the JK-FF. The combinational circuit also gets input from the present state. Thus, the output of the combinational circuit is the function of JK-FF inputs and present state values [17,18].

We need to note one important point at this time: the combination of a combinational circuit and one FF will yield another FF; for example, combinational circuit + SR-FF = JK-FF, combinational circuit + SR-FF = D-FF and combinational circuit + JK-FF = T-FF. Therefore, any FF can be obtained using any other FF; in other words, any FF can be converted to any other FF. We discuss some of the FF conversions below.

SR-FF to JK-FF: This conversion is also called JK-FF from/using SR-FF. In JK-FF, J, and K are given as the external inputs. Both inputs, S and R, of the SR-FF are the outputs of the combinational circuit. The output of the combinational is most important and is actually the unknown that we need to derive. This means that the user will feed JK-FF input into the circuit expecting JK-FF output, but inside the circuit, the SR-FF is actually present; thus,

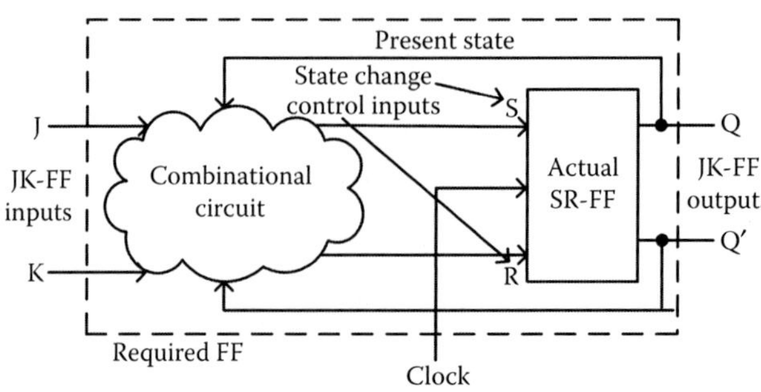

FIGURE 2.2
General design of SR-FF to JK-FF conversion.

TABLE 2.8
SR-FF to JK-FF Conversion Table

JK-FF Input and Output Variables				Remarks	State Change Control Variables	
Jack (J)	Kilby (K)	Present state (Q)	Next state (Q^+)	(Q to Q^+)	Set (S)	Reset (R)
0	0	0	0	0 to 0	0	X
0	0	1	1	1 to 1	X	0
0	1	0	0	0 to 0	0	X
0	1	1	0	1 to 0	0	1
1	0	0	1	0 to 1	1	0
1	0	1	1	1 to 1	X	0
1	1	0	1	0 to 1	1	0
1	1	1	0	1 to 0	0	1

the combinational circuit has to convert the JK-FF input to another form which has to be applied to the actual SR-FF to force it to give the JK-FF output. Now, the question is: What is that combination circuit? That is our next discussion.

First, we take the characteristic table of JK-FF as given in the left side of Table 2.8, where the inputs J and K, the present state Q, and the next state Q^+ values of the JK-FF are given. Here, we need to find what should be the input to S and R to get Q^+ from Q. For example, "$Q = 0$ to $Q^+ = 0$" can be obtained when $S = 0$ and $R = X$, as discussed in the previous section; similarly, state change control variables S and R will be X and 0, respectively, for "1 to 1" and 0 and 1, respectively, for "1 to 0" and so on. In this way, state change control variables can be obtained for all possible inputs and present state values as given in the right part of Table 2.8. This table now has to be resolved using the K-map for S and R, as shown in Figure 2.3. Upon solving the J, K, and Q for S and R, the logical expression for the combinational circuit is

$$S = JQ'$$

$$R = KQ \qquad (2.5)$$

The K-map simplification and logical circuit for the conversion of SR-FF to JK-FF is shown in Figure 2.3, where the JK-FF inputs go to the combinational circuit, whose output goes to the actual SR-FF, and the final output is JK-FF.

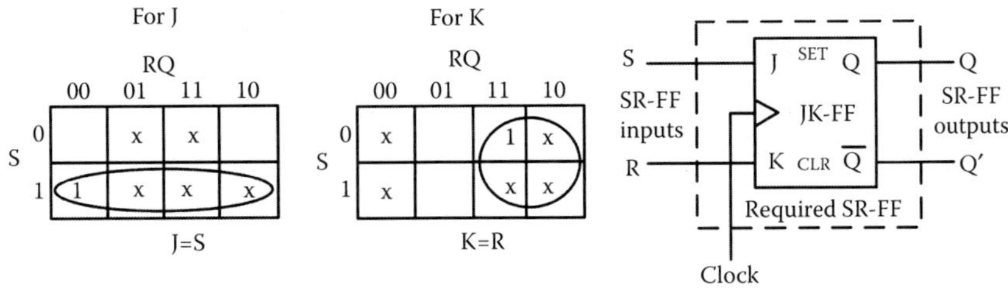

FIGURE 2.3
K-map simplification and circuit diagram for SR-FF to JK-FF conversion.

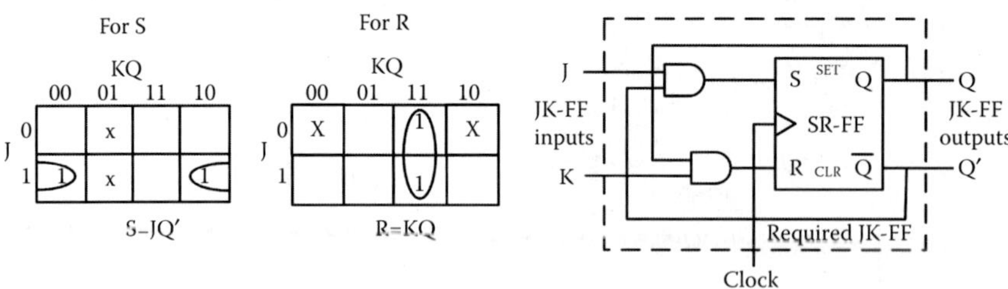

FIGURE 2.4
K-map simplification and circuit diagram for JK-FF to SR-FF conversion.

As shown in Figure 2.3, the combinational circuit consists of two AND gates which convert the actual JK-FF inputs into another convenient form that is suitable to SR-FF to produce the output of JK-FF. In the same way, any FF can be obtained using any other FF. We discuss some FF conversions in the subsequent section.

JK-FF to SR-FF: The conversion of JK-FF to SR-FF is the opposite of SR-FF to JK-FF. Here, S and R will be the external inputs to the combinational circuit, whose output will be the input to the JK-FF as shown in Figure 2.4. Therefore, as we discussed above, the J and K values have to be acquired in terms of S, R and Q. The conversion table for JK-FF to SR-FF is given in Table 2.9.

Upon solving Table 2.9, the logical expressions for J and K can be obtained as given in Equation 2.6:

$$J = S$$

$$K = R \tag{2.6}$$

Thus, there is not any combinational circuit.

SR-FF to D-FF: As shown in Figure 2.5, the actual inputs of the FF are S and R, and the external input is D (Table 2.10). The derivation of state change control variables S and R can be obtained in terms of D and the present state Q using the K-map as given in Figure 2.5.

TABLE 2.9
JK-FF to SR-FF Conversion Table

SR-FF Input and Output Variables				Remarks	State Change Control Variables	
Set (S)	Reset (R)	Present state (Q)	Next state (Q^+)	(Q to Q^+)	Jack (J)	Kilby (K)
0	0	0	0	0 to 0	0	X
0	0	1	1	1 to 1	X	0
0	1	0	0	0 to 0	0	X
0	1	1	0	1 to 0	X	1
1	0	0	1	0 to 1	1	X
1	0	1	1	1 to 1	X	0
1	1	0	X	X	X	X
1	1	1	X	X	X	X

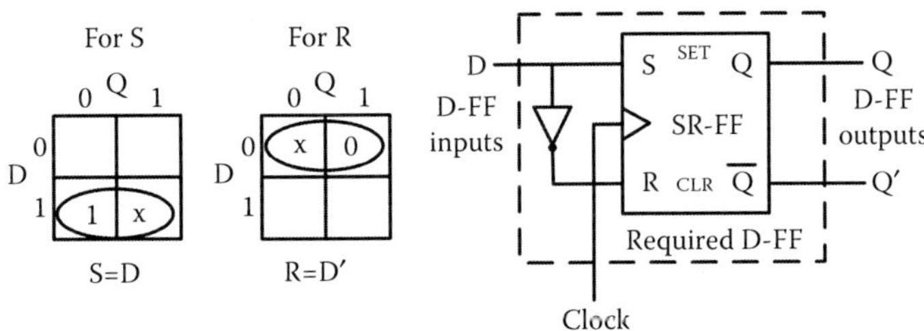

FIGURE 2.5
K-map simplification and circuit diagram for SR-FF to D-FF conversion.

TABLE 2.10
Conversion of SR-FF to D-FF

D-FF Input and Output Variables			Remarks	State Change Control Variables	
D	**Q**	**Q^+**	**(Q to Q^+)**	**Set (S)**	**Reset (R)**
0	0	0	0 to 0	0	X
0	1	0	1 to 0	0	1
1	0	1	0 to 1	1	0
1	1	1	1 to 1	X	0

The values of the characteristic function and state change control variables are given in Table 2.5.

The logical expressions are

$$S = D$$

$$R = D' \tag{2.7}$$

The circuit diagram for this conversion is shown in Figure 2.5; as in that figure, the combinational circuit consists of only one `NOT` gate.

D-FF to SR-FF: In this type of FF conversion, D is the actual the input of the FF and S and R are the external inputs. There are eight possible combinations from the external inputs S, R, and Q. Since the input S = R = 1 is unacceptable, the values of D, Q and Q^+ are taken as "X". The logic diagram of D-FF to SR-FF is shown in Figure 2.6.

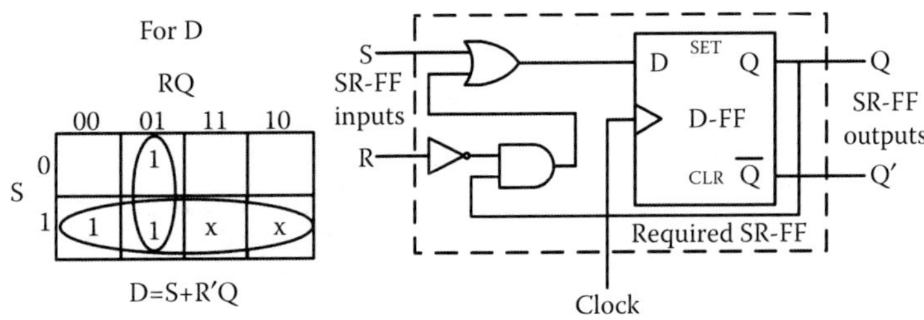

FIGURE 2.6
K-map simplification and circuit diagram for D-FF to SR-FF conversion.

TABLE 2.11

Conversion of D-FF to SR-FF

SR-FF Input and Output Variables				Remarks	State Change Control Variable
Set (S)	Reset (R)	Present State (Q)	Next State (Q^+)	(Q to Q^+)	D
0	0	0	0	0 to 0	0
0	0	1	1	1 to 1	1
0	1	0	0	0 to 0	0
0	1	1	0	1 to 0	0
1	0	0	1	0 to 1	1
1	0	1	1	1 to 1	1
1	1	0	X	X	X
1	1	1	X	X	X

The values of the characteristic function and state change control variables are given in Table 2.11.

Upon solving Table 2.11 using the K-map for the state change variable D, the following logical expressions can be obtained

$$D = S + R'Q \tag{2.8}$$

Thus, the combinational circuit consists of one `or` gate, one `and` gate and one `not` gate.

JK-FF to T-FF: In this type of FF conversion, J, K are the actual the inputs of the FF and T is considered the external input. Four combinations are created by T, Q for finding the expression for J and K, as shown in Figure 2.7.

The characteristic and state change variable controls are given in Table 2.12. Upon solving this table using the K-map, the logic expression can be obtained as

$$J = T$$

$$K = T \tag{2.9}$$

Thus, there is no combinational circuit.

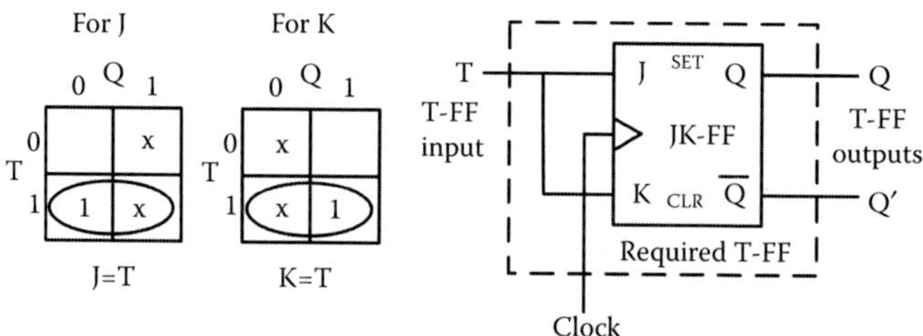

FIGURE 2.7
K-map simplification and circuit diagram for JK-FF to T-FF conversion.

TABLE 2.12

Conversion of JK-FF to T-FF

T-FF Input and Output Variables			Remarks	State Change Control Variables	
T	**Q**	**Q^+**	**(Q to Q^+)**	**Jack (J)**	**Kilby (K)**
0	0	0	0 to 0	0	X
0	1	1	1 to 1	X	0
1	0	1	0 to 1	1	X
1	1	0	1 to 0	X	1

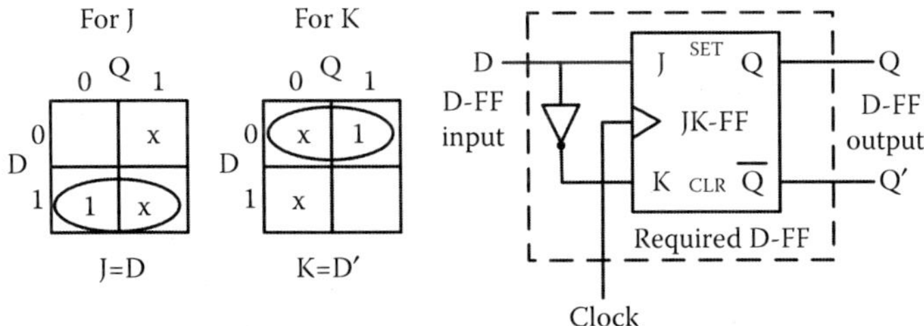

FIGURE 2.8
K-map simplification and circuit diagram for JK-FF to D-FF conversion.

DESIGN EXAMPLE 2.16

Design and develop a digital circuit to convert JK-FF into D-FF and vice versa.

Solution

As given in the design example, we need to convert (i) JK-FF to D-FF and (ii) D-FF to JK-FF. We will design one by one as explained below.

JK-FF to D-FF: In this type of FF conversion, J and K are the actual inputs of the FF, where D is the external input, as shown in Figure 2.8. The four combinations of the FF will be done in terms of D and Q, and the logical expression for J and K can be expressed.

The characteristic and state change variable controls are given in Table 2.13.

TABLE 2.13

Conversion of JK-FF to D-FF

D-FF Input and Output Variables			Remarks	State Change Control Variables	
D	**Q**	**Q^+**	**(Q to Q^+)**	**J**	**K**
0	0	0	0 to 0	0	X
0	1	0	1 to 0	X	1
1	0	1	0 to 1	1	X
1	1	1	1 to 1	X	0

Upon solving this table using the K-map, the logic expression can be obtained as

$$J = D$$
$$K = D' \tag{2.10}$$

D-FF to JK-FF: In this type of FF conversion, J and K are the external inputs of the FF and D is the actual input. The eight combinations one can make using J, K, and Q are shown in Figure 2.9.

The characteristic and state change variable controls are given in Table 2.14. Upon solving this table using the K-map, the logic expression can be obtained as

$$D = K'Q + JQ' \tag{2.11}$$

Thus, the combinational circuit consists of one NOT gate, two AND gates and one OR gate.

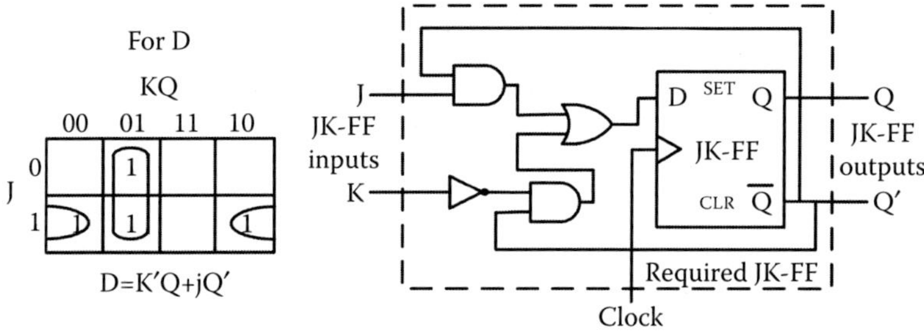

FIGURE 2.9
K-map simplification and circuit diagram for D-FF to JK-FF conversion.

TABLE 2.14
Conversion of D-FF to JK-FF

JK-FF Input and Output Variables				**Remarks**	**State Change Control Variable**
Jack (J)	**Kilby (K)**	**Present State (Q)**	**Next State (Q^+)**	**(Q to Q^+)**	**D**
0	0	0	0	0 to 0	0
0	0	1	1	1 to 1	1
0	1	0	0	0 to 0	0
0	1	1	0	1 to 0	0
1	0	0	1	0 to 1	1
1	0	1	1	1 to 1	1
1	1	0	1	0 to 1	1
1	1	1	0	1 to 0	0

2.8 Data Shift Registers

The registers which will shift bits to left are called shift-left registers. An FF can store a single bit of binary data, that is, 1 or 0, and when we need to store multiple bits of data, we need multiple FFs. As a single FF is used for one bit of storage, n FFs are connected in an order to store n bits of data. In digital electronics, a register, which is constructed using FFs, is a device which is used to store the information. For example, if a computer stores 16-bit data, then it needs a set of 16 FFs. The input and output of the register may be in serial or parallel based on the requirements. The stored information can be transferred within the registers just by shifting; hence, they are called shift registers. A shift register is a sequential circuit which stores the data and shifts it toward the output on every clock cycle. Basically, the shift registers are of four types, namely, (i) serial in serial out (SISO) shift register, (ii) serial in parallel out (SIPO) shift register, (iii) parallel in serial out (PISO) shift register, and (iv) parallel in parallel out (PIPO) shift register.

Serial in Serial out Shift Register: The input to this register is given in serial fashion, that is, one bit after another through a single data line, and the output is collected serially. The data can be shifted only in the left or right direction; hence, it is called the SISO shift register. As the data are input from the left bit by bit, the shift register shifts the data bits to right. A 4-bit SISO shift register consisting of four FFs is shown in Figure 2.10a, where, if we pass the data 1101 to the data input, the same sequence will start coming out after four clock pulses. As the clock signal is connected to all four FFs, the serial data in and

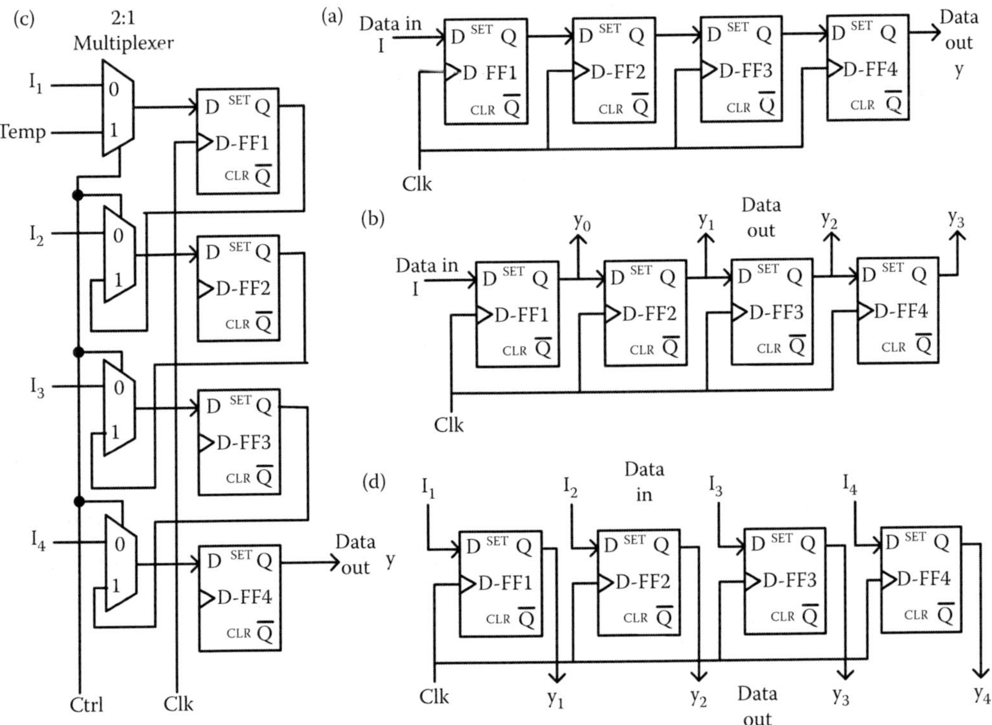

FIGURE 2.10
Design of a SISO shift register (a), SIPO shift register (b), PISO shift register (c), and PIPO shift register (d).

out are connected to the leftmost or rightmost FF, respectively. The output of the first FF is connected to the input of the next FF and so on. As we see in Figure 2.10a, the shifting takes place from left to right whenever the clock signal is applied and the serial data are given. Only one bit will be available at output at a time in the order of the input data. The use of the SISO shift register is to act as temporary data storage device and delay element.

Serial in Parallel out Shift Register: The input to this register is given in serial and the output is collected in parallel. The serial data are connected to the FF at either end depending on the configuration of the shift left registers. All the flip-flops are connected with a common clock. The output of the first FF is connected to the input of the next FF as well as to the output and so on.

Unlike the SISO shift registers, the output is collected at each FF in the SIPO shift register configuration, as shown in Figure 2.10b. The outputs are denoted by y_0, y_1, y_2, and y_3 are the outputs of the first, second, third, and fourth FFs, respectively. The main application of a SIPO shift register is to convert serial data into parallel data. Hence, they are used in communication lines where demultiplexing of a data line into several parallel lines is required.

Parallel in Serial out Shift Register: The input to this register is given in parallel; that is, data are given separately to each FF and the output is collected in serial at the output of the last FF. The clock input is directly connected to all the FFs, but the input data are connected individually to each FF through a digital multiplexer at input of every FF, as shown in Figure 2.10c, where I_1, I_2, I_3, and I_4 are the individual parallel inputs to the shift register, while `y` is the output of the last FF; therefore, the register output is collected in serial. Thus, this type of FF configuration is called a PIPO shift register. The output of the previous FF and the parallel data input are connected to the input of the multiplexer, and its output is connected to the next respective FF. The multiplexer in the PISO shift register multiplexes the actual parallel input (I_1 through I_4) or the output of the previous FF, as shown in Figure 2.10c. The logical expression of the multiplexer in this configuration for the first FF is $D_{in1} = (ctrl)'I_1 + (temp)(ctrl)$. As in this expression, the input to the first D-FF will be either `I1` or `temp` based on the value of control (`ctrl`) input. If the `cnt` is 0, then all the FFs read the inputs in parallel, and, if 1, the FFs shift out the inputted values one by one in serial with outputting the most significant bit (MSB) first. During the shifting process, the first FF will be loaded with the temporary (`temp`) data. Therefore, the PISO shift register converts parallel data to serial data. Hence, they are used in communication lines where a number of data lines are multiplexed into a single serial data line [12,13,19].

Parallel in Parallel out Shift Register: In this shift register, the inputs are given in parallel and the outputs are also collected in parallel. Data are given as the input separately for each FF and, in the same way, output is also collected individually from each FF. Figure 2.10d shows a four-stage PIPO shift register. I_1, I_2, I_3, and I_4 and y_1, y_2, y_3, and y_4 are the inputs and outputs of the PIPO shift registers, respectively. There are no interconnections between any of the four FFs. A PIPO shift register is used as a temporary storage device and also as a delay element.

The entire set of shift registers mentioned above are unidirectional shift registers; that is, they shift the data only to the right or left. To achieve both directional and SIPO, SISO, PIPO, and PISO shifts, the bidirectional and universal shift registers are used respectively. These two shift registers are more commonly used in many applications in digital electronics. We discuss their operations one by one below.

Bidirectional Shift Register: If we shift a binary number to the left by one position, this operation is equivalent to multiplying the original number by 2. Similarly, if we shift

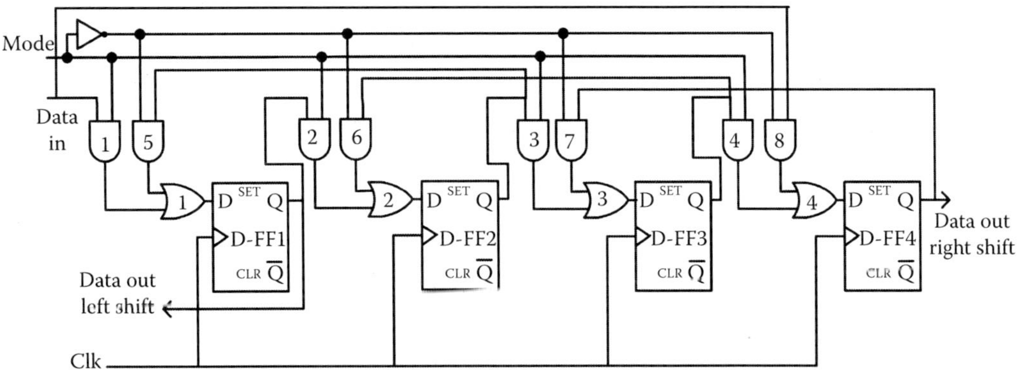

FIGURE 2.11
Design of a bidirectional shift register.

a binary number to the right by one position, this operation is equivalent to dividing the original number by 2. Hence, in order to perform these mathematical operations, we need a shift register that can shift the bits in either direction. This can be achieved by bidirectional shift register (BSR). The BSR can be defined as a register in which the data can be shifted either in the left or right direction based on the status of the mode input. If the mode input is 1, then the data will be shifted right; otherwise, the data will be shifted left. The circuit of a bidirectional shift register using D-FF is shown in Figure 2.11.

The serial input data are connected at two ends of the circuit, that is, to `and` gates 1 and 8. Based on the status of mode input, only one `and` gate is in the active state. When the mode input is "1", then the serial data path is $\texttt{and}_1 \rightarrow \texttt{or}_1 \rightarrow \texttt{FF}_1 \rightarrow Q_1 \rightarrow \texttt{and}_2 \rightarrow \texttt{or}_2 \rightarrow \texttt{FF}_2 \rightarrow Q_2 \rightarrow \texttt{and}_3 \rightarrow \texttt{or}_3 \rightarrow \texttt{FF}_3 \rightarrow Q_3 \rightarrow \texttt{and}_4 \rightarrow \texttt{or}_4 \rightarrow \texttt{FF}_4 \rightarrow Q_4$. When the mode input is low "0", then the serial data path is $\texttt{and}_8 \rightarrow \texttt{or}_4 \rightarrow \texttt{FF}_4 \rightarrow Q_4 \rightarrow \texttt{and}_7 \rightarrow \texttt{or}_3 \rightarrow \texttt{FF}_3 \rightarrow Q_3 \rightarrow \texttt{and}_6 \rightarrow \texttt{or}_2 \rightarrow \texttt{FF}_2 \rightarrow Q_2 \rightarrow \texttt{and}_5 \rightarrow \texttt{or}_1 \rightarrow \texttt{FF}_1 \rightarrow Q_1$. In this way, any length of data can be shifted either in the left or right directions.

Universal Shift Register: The universal shift register (USR) can be defined as a register that can be used to shift the data in both directions (left, right), and it can load parallel data as well. This register can perform three different operations, namely parallel loading, shifting left and shifting right; that means the universal shift register can store and transmit data in parallel. Simply, the universal shift register will load the data either in serial/parallel and will produce output as required, that is, either in serial/parallel. Thus, the USR can be used for left shift, right shift, serial to serial, serial to parallel, parallel to serial, and parallel to parallel operations. The mode input is directly connected to the multiplexer input, and the inverted mode input is connected to the inputs of upper-stage FFs. The inputs $\texttt{I}_1$, $\texttt{I}_2$, $\texttt{I}_3$, and $\texttt{I}_4$ are connected in parallel and the outputs $\texttt{y}_1$, $\texttt{y}_2$, $\texttt{y}_3$, and $\texttt{y}_4$ are collected in parallel. A serial input pin feeds the data into the register for both left shift and right shift operations. The logic diagram of a 4-bit USR is shown in Figure 2.12.

The mode input is connected to "1" to load the data in parallel, and for serial shifting, the mode input is connected to "0". When the mode pin is connected to "0", the USR will operate/behave as a BSR. To shift the data to the right, the input is connected to the `and` gate1 of the first FF through the serial input pin. To shift the data to left, the input is connected to `and` gate8 of the last FF through the input D. When the control selection input

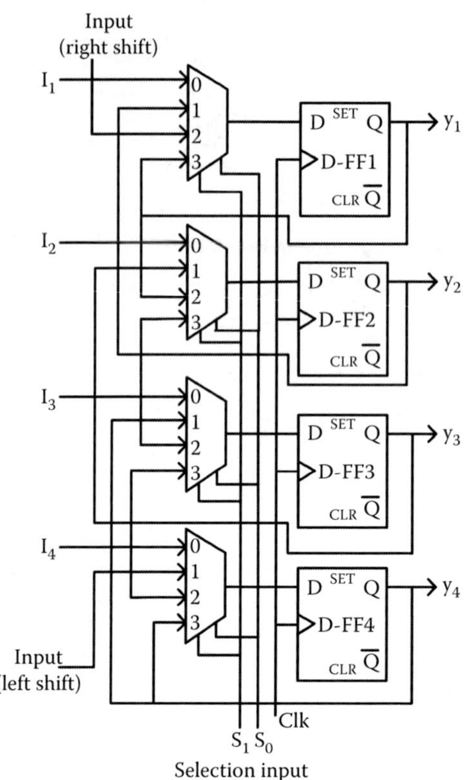

FIGURE 2.12
Circuit diagram of a 4-bit bidirectional universal shift register.

$S_0 = S_1 =$ `'0'`, then the USR will be in a locked state, which means no operation will be performed. When $S_0 = 1$ and $S_1 = 0$, then it will shift the data to the right; that is, the shift right operation is performed. When $S_0 = 0$ and $S_1 = 1$, then it will shift the data to the left; that is, the shift left operation is performed. When $S_0 = S_1 =$ `'1'`, then the register results in a parallel data loading operation. All these selections are performed through a 4:1 multiplexer.

DESIGN EXAMPLE 2.17

Design and develop a digital system to perform the serial to parallel data conversion operation.

Solution

```
library ieee;
use ieee.std_logic_1164.all;
use ieee.std_logic_arith.all;
use ieee.std_logic_unsigned.all;
entity serial_parallel_converter is
generic(n:integer:=8);
port(clk,reset,enable:in std_logic:='0';
data_in:in std_logic:='0';
data_out:out std_logic_vector(n-1 downto 0):="00000000");
end serial_parallel_converter;
```

```
architecture behavioral of serial_parallel_converter is
signal acc:std_logic_vector(n-1 downto 0):=(others=>'0');
type state_type is (rst_state,s0,s1,assign_state);
signal state:state_type:=rst_state;
signal done:std_logic:='0';
signal count:integer:=0;
begin
process(clk,reset)
begin
if(reset='1')then
acc<=(others=>'0');
count<=0;
done<='0';
state<=rst_state;
elsif(clk='1' and clk'event) then
case state is
when rst_state=>
if(enable='1')then
acc<=(others=>'0');
done<='0';
count<=0;
state<=s0;
else
state<=rst_state;
end if;
when s0=>
if(count=n)then
count<=0;
state<=s1;
else
acc<=data_in & acc(n-1 downto 1);
count<=count+1;
state<=s0;
end if;
when s1=>
state<=rst_state;
done<='1';
when others=>null;
end case;
end if;
end process;
data_out<=acc;
end behavioral;
```

DESIGN EXAMPLE 2.18

Design and develop a digital system to perform the parallel to serial data conversion operation.

Solution

```
library ieee;
use ieee.std_logic_1164.all;
use ieee.std_logic_arith.all;
use ieee.std_logic_unsigned.all;
entity parrall_serial_converter is
port(clk,reset,enable:in std_logic;
data_in:in std_logic_vector(7 downto 0);
data_out:out std_logic);
end parrall_serial_converter;
architecture behavioral of parrall_serial_converter is
```

```
signal acc:std_logic_vector(7 downto 0):=(others=>'0');
type state_type is (rst_state,s0,s1,assign_state);
signal state:state_type:=rst_state;
signal done:std_logic:='0';
signal count:integer:=0;
begin
process(clk,reset)
begin
if(reset='1')then
acc<=(others=>'0');
count<=0;
done<='0';
state<=rst_state;
elsif(clk='1' and clk'event)then
case state is
when rst_state=>
if(enable='1')then
acc<=data_in;
done<='0';
count<=0;
state<=s0;
else
state<=rst_state;
end if;
when s0=>
if(count=8)then
count<=0;
state<=s1;
else
data_out<=acc(0);
acc<=('0' & acc(7 downto 1));
count<=count+1;
state<=s0;
end if;
when s1=>
done<='1';
state<=rst_state;
when others=>null;
end case;
end if;
end process;
end behavioral;
```

DESIGN EXAMPLE 2.19

Design and develop a digital system to perform the right to left shift operation on the input data.

Solution

```
library ieee;
use ieee.std_logic_1164.all;
use ieee.std_logic_arith.all;
use ieee.std_logic_unsigned.all;
entity right_left_shift is
generic(n:integer:=8);
port(clk,reset,enable:in std_logic;
data_in:in std_logic_vector(n-1 downto 0);
data_out:out std_logic_vector(n-1 downto 0));
end right_left_shift;
architecture behavioral of right_left_shift is
```

```
signal temp,acc:std_logic_vector(n-1 downto 0):=(others=>'0');
type state_type is (rst_state,s0,s1,shift_state);
signal state:state_type:=rst_state;
signal done:std_logic:='0';
signal count:integer:=8;
begin
process(clk,reset)
begin
if(reset='1')then
temp<=(others=>'0');
acc<=(others=>'0');
count<=8;
done<='0';
state<=rst_state;
elsif(clk='1' and clk'event)then
case state is
when rst_state=>
if(enable='1')then
acc<=(others=>'0');
done<='0';
count<=8;
temp<=data_in;
state<=s0;
else
state<=rst_state;
end if;
when s0=>
if(count=0)then
count<=8;
state<=s1;
else
acc<=(acc(n-2 downto 0) & '0');
count<=count-1;
state<=shift_state;
end if;
when shift_state=>
acc(0)<=temp(count);
state<=s0;
when s1=>
done<='1';
state<=rst_state;
when others=>null;
end case;
end if;
end process;
data_out<=acc;
end behavioral;
```

DESIGN EXAMPLE 2.20

Design and develop a digital system to perform the left to right shift operation on the input data.

Solution

```
library ieee;
use ieee.std_logic_1164.all;
use ieee.std_logic_arith.all;
use ieee.std_logic_unsigned.all;
entity shift_register is
generic(n:integer:=8);
```

```
port(clk,reset,enable:in std_logic;
data_in:in std_logic_vector(n-1 downto 0);
data_out:out std_logic_vector(n-1 downto 0));
end shift_register;
architecture behavioral of shift_register is
signal temp,acc:std_logic_vector(n-1 downto 0):=(others=>'0');
type state_type is (rst_state,s0,s1,shift_state);
signal state:state_type:=rst_state;
signal done:std_logic:='0';
signal count:integer:=0;
begin
process(clk,reset)
begin
if(reset='1')then
temp<=(others=>'0');
acc<=(others=>'0');
count<=0;
done<='0';
state<=rst_state;
elsif(clk='1' and clk'event)then
case state is
when rst_state=>
if(enable='1')then
acc<=(others=>'0');
done<='0';
count<=0;
temp<=data_in;
state<=s0;
else
state<=rst_state;
end if;
when s0=>
if(count=n)then
count<=0;
state<=s1;
else
acc<='0' & acc(n-1 downto 1);
state<=shift_state;
end if;
when shift_state=>
acc(n-1)<=temp(count);
count<=count+1;
state<=s0;
when s1=>
done<='1';
state<=rst_state;
when others=>null;
end case;
end if;
end process;
data_out<=acc;
end behavioral;
```

2.9 Multifrequency Generator

While designing a system with FPGA, there are many situations where we want clock signals with small frequencies, that is, high time periods. For example, liquid crystal display

(LCD)/graphical LCD (GLCD) interfacing, DC motor driving, seven-segment display, stepper motor control, robotic control, serial communication, relay control, and so on. But in most FPGA boards, the frequencies of the onboard crystal oscillators are in the range of tens of MHz [13,19,20]. One solution to the above problem is taking the high-frequency clock available onboard and converting it to a lower-frequency clock. This is called frequency down conversion. In this section, we discuss multiple low-frequency generation, that is, a clock divider circuit to generate multiple down-sampled clock pulses from the onboard crystal clock frequency. The trade-off of the multifrequency generator circuit depends on the maximum operating frequency, power consumption, amount of resources needed, and flexibility. Depending on the specific application, the frequency divider will be designed and used. Typically, in a phase-locked loop (PLL), the output of the voltage controlled oscillator (VCO) is divided by the frequency divider to reduce the frequency. For signals with duty cycles other than 50%, an additional `divideby-2` can be used to generate the 50% duty cycle [19,20]. The basic and very simple clock divider can be designed using the D-FF, as shown in Figure 2.13a, which provides a very easy method of dividing an incoming pulse train by a factor of two. The dividing-by-two logical circuit requires only one D-FF. Simply applying the input pulse train into the clock input of a D-FF, connecting the Q′ output to the D_{in} and taking the output (`clk_out`) from the Q gives a clock divider by a factor of 2, as shown in Figure 2.13a and b. The circuit operates in a simple way; the incoming pulse train acts as a clock to the D-FF, and the data D_{in} are connected to the Q′; thus, the output is Q of the D-FF.

Take a situation when the Q is at logic 1, which means that the Q′ output will be at logic 0. These data, that is, D_{in}, are clocked to the output Q on the next positive rising edge of the incoming pulse train, that is, clock input. At this point, the output changes from logic 1 to logic 0. At the next positive edge of the clock pulse, the data on the Q′ output are again clocked to the previous logical value. As it is now at logic 1 (opposite the Q), this will be transferred to the output, and the output again changes its state. It can be seen that the output of the circuit changes state only at the positive-rising edges of the incoming pulse clock stream, as shown in Figure 2.13b. Each positive edge occurs once every cycle, but as the output of the D-FF requires two changes to complete one cycle, the output of the D-FF changes at half the rate of the incoming pulse train. In other words, the output is divided by two.

Two D-FFs can be configured as a master-slave to design a clock divider circuit, as shown in Figure 2.13a. The first D-FF is commonly called the master and the second one is

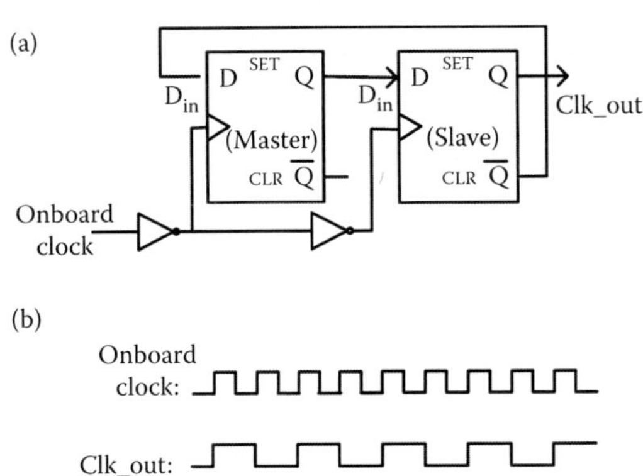

FIGURE 2.13
Clock divider circuit with D-FF (a) and clock divider output (b).

normally referred to as the slave. The inverted output (Q′) of the slave FF is fed back to the input (D_{in}) of master FF. The onboard clock is used to drive both D-FFs with opposite logic. The reason for the inverter is to operate the FFs at both edges; that is, when the onboard clock goes to logic 0, the master FF is driven, and when the onboard clock goes to logic 1, the slave FF is driven. Either master or slave is activated in each clock cycle, but not both D-FFs at the same time because of the inverter between them. Thus, the master and slave complete one full cycle once in every two onboard clocks. Therefore, the output `clk_sig` toggles for every cycle of the onboard clock. This event repeats for every two onboard input clock cycle. Thus, the output frequency is half of the input frequency, as shown in Figure 2.13b.

In these types of designs, we require multiple stages of D-FFs to generate the different rates of down converted frequencies, for example, dividing by 4, 100, 1000, 25,000, and 50,000 requires enormous D-FFs, which increases the complication in the design. To avoid such complications, a simple design called a counter-based clock divider is preferred. This design consists of three blocks, namely a binary/integer counter, comparator, and inverter as shown in Figure 2.14. The onboard clock pulse is directly connected to the counter; hence, it increments its value at that speed/rate/frequency.

This counter is designed to reset itself in accordance with a reset control signal that comes from the comparator. At all instants, that is, either rising or falling edge, the comparator compares the current value of the counter with the scale factor. Once both are equal, it sends a reset signal to the counter to reset it and sends a complement signal to the inverter, which toggles its output (`clk_out`). In accordance with the complement signal, the inverter toggles the `clk_out`; for example, at an instant, the `clk_out` is x, and when the comparator gives a pulse to the inverter, then the logical transitions of x for every comparator pulse are x→x′→(x″=x)→x′→(x″=x) and so on. In this way, `clk_out` produces a clock divided signal. The scale factor is determined as

$$\text{Scale factor} = (\text{frequency of onboard clk}/2\text{x frequency of clk\_out}) \quad (2.12)$$

For example, if we want to convert a 4-MHz (4,000,000-Hz) signal into a 4-Hz signal, as per Equation 2.12, scale factor = (4,000,000/2 × 4) = (4,000,000/8) = 500,000. This means that at the moment the counter value reaches 500,000, the comparator sends a reset signal and complement sign to the counter and inverter, respectively. Hence, the counter resets its value to 0 and the inverter toggles its output. These operations happen once in every 500,000 onboard clock pulses, that is, eight times within a second; thus, the frequency of `clk_out` is 4 Hz. In this way, just by assigning the appropriate scale factors, we can derive any frequency from the onboard clock rate. Thus, the frequency divider is a simple component whose objective is to reduce the input frequency.

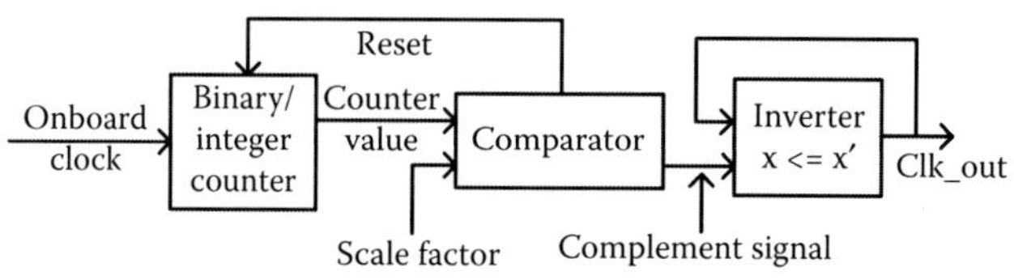

FIGURE 2.14
Counter-based clock divider circuit.

DESIGN EXAMPLE 2.21

Design a digital circuit to generate multiple clock dividers to generate seven different frequencies and display them using the LEDs. Assume that the onboard clock crystal frequency is 4,000,000 Hz, start the frequency generation from 1 Hz and produce further low frequencies.

Solution

As given in the design example, the onboard frequency is 4 MHz and the starting low frequency is 1 Hz. Therefore, the scale factor is (4 MHz/2 × 1) = 2 MHz = 2,000,000. We generate 1 Hz using this scale factor and derive other low frequencies by using 1 Hz.

```
library ieee;
use ieee.std_logic_1164.all;
use ieee.std_logic_arith.all;
use ieee.std_logic_unsigned.all;
entity clkd is
port (clk,reset : in std_logic:='0';
dout,d0out,d1out,d2out,d3out,d4out,d5out: out std_logic:='0');
end clkd;
architecture behavioral of clkd is
signal clk_sig,clk0,clk1,clk2,clk3,clk4,clk5:std_logic:='0';
signal d0cnt,d1cnt,d2cnt,d3cnt,d4cnt,d5cnt:std_logic_vector(3 downto
0):="0000";
begin
-- clock divider for 1hz generater
p1:process(clk,reset)
variable cnt:integer:=0;
begin
if(reset='0') then
clk_sig<='0';
cnt:=0;
elsif rising_edge(clk) then
if(cnt=2000000)then
clk_sig<= not(clk_sig);
cnt:=0;
else
cnt:= cnt + 1;
end if;
end if;
end process;
dout<= clk_sig;
--zeroth display time
p2:process(clk_sig)
begin
if rising_edge(clk_sig) then
if(d0cnt="1001") then
clk0<=not(clk0);
d0cnt<="0000";
else
d0cnt<=d0cnt + 1;
end if;
end if;
end process;
d0out<=clk0;
-- first display time
p3:process(clk0)
begin
if rising_edge(clk0) then
if(d1cnt="1001") then
```

```
clk1<=not(clk1);
d1cnt<="0000";
else
d1cnt<=d1cnt + 1;
end if;
end if;
end process;
d1out<= clk1;
-- second display time
p4:process(clk1)
begin
if rising_edge(clk1) then
if(d2cnt="1001") then
clk2<=not(clk2);
d2cnt<="0000";
else
d2cnt<=d2cnt + 1;
end if;
end if;
end process;
d2out<=clk2;
-- thired display time
p5:process(clk2)
begin
if rising_edge(clk2) then
if(d3cnt="1001") then
clk3<=not(clk3);
d3cnt<="0000";
else
d3cnt<=d3cnt + 1;
end if;
end if;
end process;
d3out<=clk3;
-- fourth display time
p6:process(clk3)
begin
if rising_edge(clk3) then
if(d4cnt="1001") then
clk4<=not(clk4);
d4cnt<="0000";
else
d4cnt<=d4cnt + 1;
end if;
end if;
end process;
d4out<=clk4;
-- fifth display time
p7:process(clk4)
begin
if rising_edge(clk4) then
if(d5cnt="1001") then
clk5<=not(clk5);
d5cnt<="0000";
else
d5cnt<=d5cnt + 1;
end if;
end if;
end process;
d5out<= clk5;
end behavioral;
```

LABORATORY EXERCISES

1. Design and develop a digital system to realize a binary to excess-3 code converter.
2. Design and develop a suitable digital circuit to implement all the data types' converters with multidimensional array declarations.
3. Design and develop a digital system to realize the VHDL special operators `resize` and `casting`.
4. Design and develop a digital system to realize the following:
   a. Three switches, start, SW0, and SW1, and eight LEDs are connected to the FPGA.
   b. The LEDs have to light up if the SW1 gets three 1 s after getting two 0 s at SW0.
   c. The LEDs have to light up if the SW1 gets two 1 s after getting five 0 s at SW0. Repeat this operation as long as the start input is at logic high.
5. Design and sketch a circuit to realize all possible conversion of one FF to other FFs.
6. Design and develop digital circuits to implement conversion of one FF to other FFs.
7. Design and implement a digital system to realize data (i) shifting, (ii) rotation, concatenation, swapping, serial to parallel conversion, and parallel to serial conversion with some logical condition.
8. Design and develop a digital circuit to realize a bidirectional shift register.
9. Design and develop a digital circuit to realize a universal shift register.
10. Design and implement a circuit on the breadboard to increase the frequency of an input pulse using a PLL.

3

Simple System Design Techniques

3.1 Introduction

In the previous chapter, we discussed many VHDL instructions, functions, input/output declarations, data type conversions, arithmetic/logical operators, and special operation functions. In this chapter, we see some design examples for simple system design using all those instructions. As we progress through other chapters, we will see many examples of interfacing and complicated system designs. This chapter begins with a simple design of half-adder and full-adder implementations. Logical expressions, function tables, and implementation of a full adder as a function of two half adders are also discussed. Half subtractor and full subtractor design and implementation using data flow, structural, and behavioral models are discussed, along with Boolean expressions, function tables, and design examples. The differences and advantages of a full adder/subtractor over the half adder/subtractor are highlighted.

The significance of design of a signed magnitude comparator is listed, and the designs of a single-bit magnitude comparator and 4-bit magnitude comparator are explicitly explained. Design examples using different modeling techniques are also explained. Different configurations of seven-segment displays, interfacing techniques, and advantages of driver ICs are detailed, along with interfacing circuit and design examples. The design of various counters (synchronous and asynchronous), their working principles and their limitations are detailed, with suitable design examples. A real-time digital clock design approach, its digital architecture, and the function of each and every submodule are described, with 12-hr and 24-hr digital clock design examples. The significance of pulse width modulation (PWM), generation of PWM signals and techniques of varying the duty cycle and applications are detailed. Methodologies of controlling power electronic switches by PWM signal are also explained, with different design examples. Digital system design techniques using packages, libraries, functions, and procedures are explained, with standard templates and design examples. The design examples are designed to introduce the concepts of subprograms (functions and procedures) and packages in VHDL.

3.2 Half and Full Adder

A common and very useful combinational logic circuit that can be constructed using a few basic logic gates for adding two or more binary numbers together is the known as a binary adder. A basic binary adder circuit can be made using `xor`, `and`, and `or` gates as

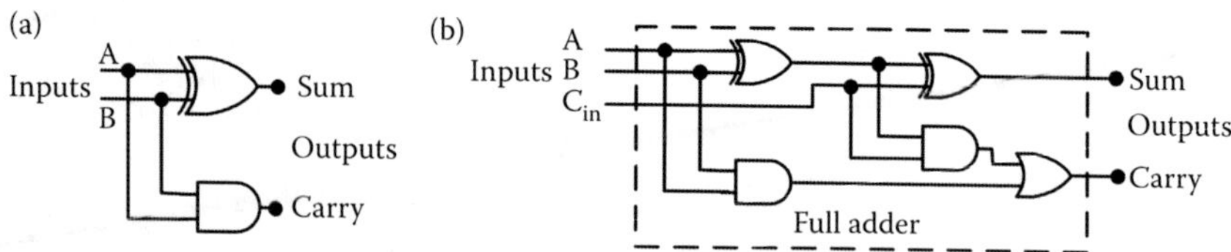

FIGURE 3.1
Combinational circuit of half (a) and full (b) adder.

shown in Figure 3.1. The addition of two or more digits produces an output called the sum and carry bit according to the rules of binary addition. One of the main uses of the binary adder is arithmetic operations and counting circuits. Consider the simple addition of two decimal numbers: if augend A is $(234)_{10}$ and addend B is $(567)_{10}$, then the sum is $(801)_{10}$. As we know, we add both numbers (A and B) together column-wise starting from the right-hand side, and each digit has a weighted value depending upon its column position. When each column is added, a carry is generated if the result is greater than or equal to $(10)_{10}$, which is then added to the result of the addition of the next column on the left side and so on. The adding of binary numbers is exactly the same idea as that for adding decimal numbers, but a carry is generated only when the result in any column is greater than or equal to $(2)_{10}$ since this is the base value of the binary number. These basic rules of decimal addition can also be applicable for binary addition. However, in binary addition, there are only two digits, with the largest value being 1. Hence, the carry is generated whenever two or more 1s are added together, and the carry bit is passed to the subsequent addition in the next column on the left side. Single-bit addition is performed as $(0 + 0 = 0)$, $(0 + 1 = 1)$, $(1 + 0 = 1)$ and $(1 + 1 = 10)$; the 10 is not a $(10)_{10}$, rather it is $(2)_{10}$ as a binary number. Thus, whenever two single binary bits are added together, the addition of $(0 + 0)$ results in a sum and carry of 0, $(0 + 1$ and $1 + 0)$ results in a sum of 1 and carry of 0 and $(1 + 1)$ results in a sum of 0 and carry of 1 [21–24].

A combination of the `xor` gate with the `and` gate results in a simple digital binary adder circuit known commonly as a half adder. A half adder is a logical circuit that performs an addition operation on two binary digits. The half adder produces sum and carry values which are both binary digits. One-bit addition can be performed by a half adder, which is shown in Figure 3.1a, where the sum and carry of the binary addition resemble `xor` and `and` gates. If we label two binary inputs `A` and `B`, then the resulting truth table for the sum and carry is as given in Table 3.1. We can see from that truth table that an `xor` gate produces an output of 1 only when either input is at logic 1, and an `and` gate produces logic 1 when both inputs are 1.

From the truth table of the half adder, we can see that the sum and carry-out (C_{out}) outputs are given by the Boolean expression using the K-map as

$$\begin{aligned} \text{Sum} &= \text{A xor B} = \text{A} \oplus \text{B} \\ C_{out} &= \text{A and B} = \text{A.B} \end{aligned} \tag{3.1}$$

One major disadvantage of the half adder circuit, when used as a binary adder for multiple data bits, that is, more than one bit, is that there is no provision for a carry-in (C_{in}) from the previous addition. For example, suppose we want to add together two 8-bits, that is, a byte of data. Any resulting carry bit would need to be able to "ripple" or move across the bit patterns starting from the least significant bit (LSB). The most complicated operation in the half adder is just "1 + 1", but, as the half adder has no carry input, the resultant

TABLE 3.1
Function Table of Half and Full Adders

Half Adder				Full Adder				
Inputs		Outputs		Inputs			Outputs	
A	B	Carry	Sum	C_{in}	A	B	C_{out}	Sum
0	0	0	0	0	0	0	0	0
0	1	0	1	0	0	1	0	1
1	0	0	1	0	1	0	0	1
1	1	1	0	0	1	1	1	0
				1	0	0	0	1
				1	0	1	1	0
				1	1	0	1	0
				1	1	1	1	1

added value would be incorrect. In order to perform the addition of two numbers, that is, more than two bits, we require one additional bit, called the carry bit (C_{in}). This operation can be done by a full adder, as shown in Figure 3.1b, whose function table is also given in Table 3.1. The main difference between the full adder and the previous half adder is that a full adder has three inputs: the same two single-bit data inputs A and B, as before, plus an additional C_{in} input to receive the carry from a previous stage. Therefore, the full adder is a logical circuit that performs an addition operation on three binary inputs and generates a carry out to the next addition column. Thus, C_{in} is a possible carry from a less significant digit, while a carry out (C_{out}) represents a carry to a more significant digit. In many ways, the full adder can be thought of as two half adders connected together, as shown in Figure 3.2a, with the first half adder passing its carry to the second half adder [21–24].

The truth table for the full adder includes an additional column to take C_{in} into account along with the other inputs A and B. It produces sum and C_{out} as given in Table 3.1 and shown in Figure 3.2a. The Boolean expression for a full adder can be obtained using the K-map as

$$\begin{aligned} \text{Sum} &= ((\text{A xor B}) \text{ xor } C_{in}) = ((A \oplus B) \oplus C_{in}) \\ C_{out} &= (\text{A and B}) \text{ or } (C_{in} \text{ and } (\text{A xor B}) = A.B + Cin(A \oplus B) \end{aligned} \tag{3.2}$$

We have seen above that single 1-bit binary adders can be constructed from basic logic gates. But if we want to add two n-bit numbers together, then one 1-bit half adder and n – 1 one-bit full adders need to be connected/cascaded together to produce the sum and final carry; this configuration is known as a ripple carry adder (RCA). It is called an RCA

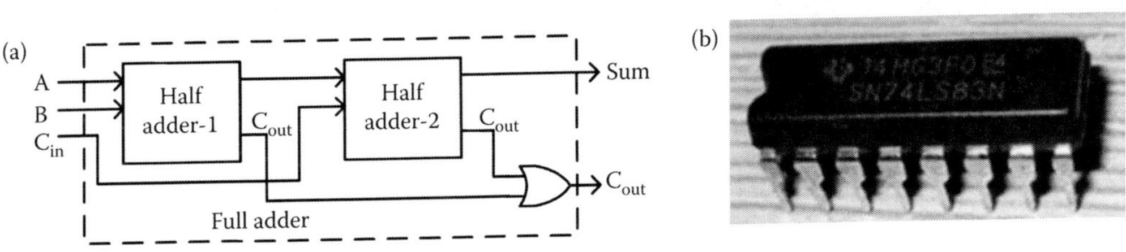

FIGURE 3.2
Construction of a full adder using two half adder circuits (a) and picture of a CLA adder IC74LS83 (b).

because the carry signals produce a ripple effect through the binary adder from right to left, that is, LSB to MSB. For example, suppose we want to add two 4-bit numbers together. The two outputs of the first half adder will provide the first place digit sum (S_0) of the addition plus a carry-out bit that acts as the carry-in digit of the next full adder. The second binary adder in the chain also produces a summed output (S_1) plus another carry-out bit, and we can keep adding more full adders to the combination to add larger numbers, linking the carry bit output from the first full binary adder to the next full adder and so forth, up to the fourth bit.

One main disadvantage of cascading 1-bit binary adders together to add large binary numbers is that if inputs A and B change, the sum at the output will not be valid until any carry input has rippled through every full adder in the chain, because the MSB of the sum has to wait for any changes from the carry input of the LSB. Consequently, there will be a finite delay before the output of the adder responds to any change in its inputs, resulting in an accumulated delay.

When the size of the bits being added is not too large, for example, 4 or 8 bits, or the summing speed of the adder is not important, this delay may neglected. However, when the size of the bits is larger, for example, 32 or 64 bits used in multibit adders, summation is required at a very high clock speed, and this delay may become prohibitively large with the addition processes not being completed correctly within one clock cycle.

This unwanted delay time is called a propagation delay. Also, another problem called overflow occurs when an n-bit adder adds two parallel numbers together whose sum is greater than or equal to 2n. One solution is to generate the carry input signals directly from the A and B inputs rather than using the ripple arrangement; this is another type of binary adder circuit called a carry look-ahead adder (CLAA), where the speed of the parallel adder can be greatly improved using CLA logic. The advantage of carry look-ahead adders is that the length of time a carry look-ahead adder needs in order to produce the correct sum is independent of the number of data bits used in the operation, unlike the cycle time a parallel ripple adder needs to complete the sum, which is a function of the total number of bits in the addend. A 4-bit full adder circuit with CLA features is available in standard IC packages in the form of the TTL 4-bit binary adder 74LS83, as shown in Figure 3.2b, or the 74LS283 and CMOS 4008, which can add together two 4-bit binary numbers and generate a sum and a carry output [21,25].

We have seen above some details about simple binary adders that can add two binary numbers together, producing a sum and carry out. More details and design approaches for half and full adders with various techniques are discussed in Chapter 4.

DESIGN EXAMPLE 3.1

Design and develop a digital system to realize a half and full adder using a data flow model.

Solution

```
library ieee;
use ieee.std_logic_1164.all;
entity half_adder is
port(a,b,cin: in bit:='0';
s_ha,c_ha,s_fa,c_fa: out bit:='0');
end half_adder;
architecture half_adder of half_adder is
begin
s_ha<=(a xor b);
```

```
c_ha<=(a and b);
s_fa<=((a xor b) xor cin);
c_fa<=((a and b) or (b and cin) or (cin and a));
end half_adder;
```

DESIGN EXAMPLE 3.2

Design and implement a full adder using a behavioral model.

Solution

```
library ieee;
use ieee.std_logic_1164.all;
entity full_adder is
port(a, b, cin: in  bit;
sum, cout: out bit);
end full_adder;
architecture rtl of full_adder is
begin
process(a, b, cin)
variable indata: std_logic_vector(2 downto 0);
begin
indata := to_stdulogic(cin) & to_stdulogic(a) & to_stdulogic(b);
case indata is
when "000" =>  cout <= '0'; sum <= '0';
when "001" =>  cout <= '0'; sum <= '1';
when "010" =>  cout <= '0'; sum <= '1';
when "011" =>  cout <= '1'; sum <= '0';
when "100" =>  cout <= '0'; sum <= '1';
when "101" =>  cout <= '1'; sum <= '0';
when "110" =>  cout <= '1'; sum <= '0';
when "111" =>  cout <= '1'; sum <= '1';
when others => cout <= '0'; sum <= '0';
end case;
end process;
end rtl;
```

DESIGN EXAMPLE 3.3

Design and implement a full adder as a function of two half adders using a structural model.

Solution

Component

```
library ieee;
use ieee.std_logic_1164.all;
entity half_adder is
port(a, b: in bit;
sum, carry: out bit);
end half_adder;
architecture dataflow of half_adder is
begin
sum   <= a xor b;
carry <= a and b;
end dataflow;
```

Main Code

```
library ieee;
use ieee.std_logic_1164.all;
```

```
entity full_adder is
port(a, b, cin: in  bit;
sum, cout: out bit);
end full_adder;
architecture struct of full_adder is
component half_adder
port(a, b: in bit;
sum, carry: out bit);
end component;
signal u0_carry, u0_sum, u1_carry: bit;
begin
u0: half_adder port map (a => a, b => b, sum => u0_sum, carry =>
u0_carry);
u1: half_adder port map (a => u0_sum, b => cin, sum => sum, carry =>
u1_carry);
cout <= u0_carry or u1_carry;
end struct;
```

3.3 Half and Full Subtractor

Subtraction is a mathematical operation in which one number is deducted from another to obtain the equivalent quantity. The number from which the other number is to be deducted is called the minuend, and the number subtracted from the minuend is called the subtrahend. In all the operations, each subtrahend bit is deducted from the minuend bit. Similarly to binary addition, a single-bit binary subtraction also has four possible operations, such as (0–1 = 0), (1–0 = 1), (1–1 = 0), and (0–1 = 10–1 = 11). This means that, like decimal subtraction, we can simply subtract a smaller number from a bigger number without borrowing anything as the first three subtractions, whereas in the last case, that is, (0–1), subtraction of "1" from "0" is not possible; hence, we need to get a borrow. Then the "0" becomes "10" ("10"–"1" = 1); thus, the difference is "1" and the borrow is also "1". Similarly to adder circuits, subtraction circuits are also classified as half and full subtractors. A half subtractor is a combinational circuit that does the subtraction of two bits of binary data. It has two input variables and two output variables that correspond to difference and borrow bits. Binary subtraction is performed by the `xor` gate and a `not` with an `and` gate, as shown in Figure 3.3a, which produces the difference and borrow, respectively. Thus, a half subtractor is designed by an `xor` gate including an `and` gate with input (`A`) complemented before being fed to the `and` gate [21–25].

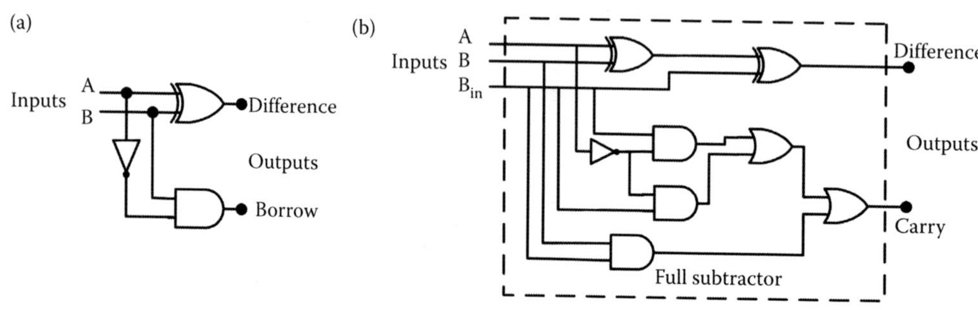

FIGURE 3.3
Combinational circuit of a half (a) and full (b) subtractor.

TABLE 3.2

Function Table of Half and Full Subtractors

Half Subtractor				Full Subtractor				
Inputs		Outputs		Inputs			Outputs	
A	B	Borrow	Difference	A	B	B_{in}	B_{out}	Difference
0	0	0	0	0	0	0	0	0
0	1	1	1	0	0	1	1	1
1	0	0	1	0	1	0	1	1
1	1	0	0	0	1	1	1	0
				1	0	0	0	1
				1	0	1	0	0
				1	1	0	0	0
				1	1	1	1	1

By comparing the adder and subtractor circuits or truth tables, one can observe that the output D in the full subtractor is exactly the same as the output S of the full adder. The only difference is that input variable A is complemented in the full subtractor. Therefore, it is possible to convert the full adder circuit into a full subtractor by simply complementing the input A before it is applied to the gates to produce the final borrow bit output B_{out}. A half subtractor is a logical circuit that performs a subtraction operation on two binary digits. The half subtractor produces a sum and a borrow bit for the next stage. From the truth table given in Table 3.2 of the half subtractor, we can see that the difference (D) output is the result of the xor gate, and the borrow out (B_{out}) is the result of the not-and gate combination. Then, the Boolean expression for a half subtractor can be obtained using the K-map as

$$\begin{aligned} \text{Difference} &= \text{A xor B} = \text{A} \oplus \text{B} \\ \text{Borrow} - \text{out} &= \text{A}' \text{ and B} = \text{A}'\text{B} \end{aligned} \tag{3.3}$$

In the case of multidigit subtraction, subtraction between the two digits must be performed along with a borrow of the previous digit subtraction; hence, a subtractor needs to have three inputs. Therefore, a half subtractor has limited application and strictly is not used in practice. A combinational logic circuit that performs a subtraction between two binary bits by considering the borrow of the lower significant stage is called a full subtractor. In this subtraction, the process is performed by taking into consideration whether a 1 has already been borrowed by the previous adjacent lower minuend bit. It has three input terminals in which two terminals correspond to the two bits to be subtracted (minuend A and subtrahend B) and a borrow bit (B_{in}) corresponds to the borrow operation. There are two outputs; one corresponds to the difference (D) output and the other to the borrow output (B_{out}), as given in Table 3.2 and shown in Figure 3.3b.

Thus, the combinational circuit of a full subtractor performs the operation of subtraction on three binary bits as A-B-B_{in}, producing outputs for the difference (D) and borrow (B_{out}). The truth table for the full subtractor has eight different input combinations, as there are three input variables, the data bits and the B_{in} input. Then, the Boolean expression for a full subtractor can be obtained using the K-map as

$$\text{Difference} = (\text{A}'\text{B}'\text{B}_{in}) + (\text{A}'\text{B}\text{B}'_{in}) + (\text{A}\text{B}'\text{B}'_{in}) + (\text{A}\text{B}\text{B}_{in})$$

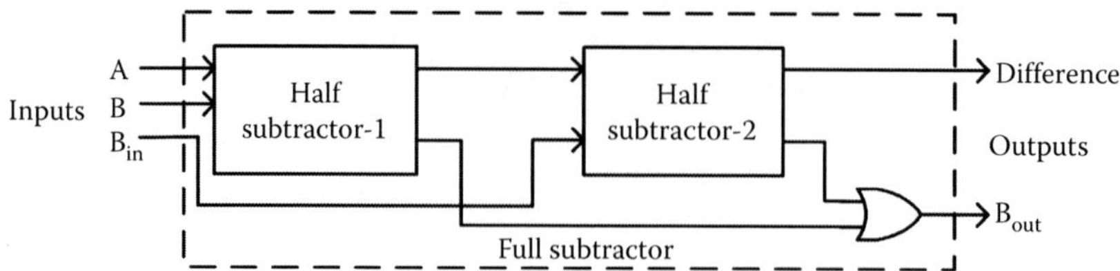

FIGURE 3.4
Construction of a full subtractor using two half subtractor circuits.

This can be simplified as

$$\text{Difference} = (\text{A xor B}) \text{ xor } B_{in} = ((A \oplus B) \oplus B_{in}), \text{ then}$$

$$B_{out} = (A'B'B_{in}) + (A'BB_{in}') + (A'BB_{in}) + (ABB_{in})$$

This can be simplified as

$$B_{out} = (\text{A and B}) \text{ or } (\text{A xor B})B_{in} = AB + (A \oplus B)B_{in} \quad (3.4)$$

As given in Equation 3.4, a full subtractor can be implemented with two half subtractors and an or gate, as shown in Figure 3.4. Just like the binary adder circuit, the full subtractor can also be thought of as two half subtractors connected together, with the first half subtractor passing its borrow to the second half subtractor as shown in Figure 3.4.

Like the binary adder, we can also have n 1-bit full binary subtractors connected/cascaded together to subtract two parallel n-bit numbers from each other [21,25]. We have seen above that simple binary subtractors can subtract two binary numbers, producing a difference and borrow out. More details and design approaches for half and full subtractors with various techniques are discussed in Chapter 4.

DESIGN EXAMPLE 3.4

Design and implement a half and full adder circuit using a data flow model.

Solution

```
library ieee;
use ieee.std_logic_1164.all;
entity subtrs is
port(a,b,Bin: in bit:='0';
diff_hs,Bout_hs,diff_fs,bout_fs: out bit:='0');
end subtrs;
architecture subtractor_dfm of subtrs is
begin
diff_hs<=(a xor b);
Bout_hs<=((not a) and b);
diff_fs<=((a xor b) or (b xor Bin) or (a xor Bin));
Bout_fs<=(((not a) and (b or Bin)) or (b and Bin));
end subtractor_dfm;
```

3.4 Signed Magnitude Comparator

The logic behavior of a binary magnitude comparator is similar to the adder and subtractor, but not exactly; that is, it does not return a sum, carry, difference, or borrow. Instead,

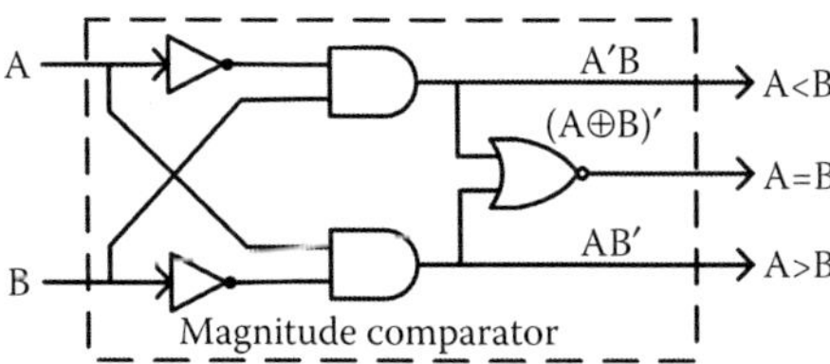

FIGURE 3.5
Construction of a single-bit magnitude comparator.

it indicates that the given number A is larger than, smaller than or equal to another number B. The magnitude comparator is a common and very useful combinational circuit, especially for data and signal processing applications. The comparator circuit is made up of the standard `and`, `nor`, and `not` gates that compare the digital input (data) present at their input terminals and produce an output depending upon the values of those inputs. The digital comparator accomplishes this using several logic gates that operate on the principles of Boolean algebra. There are two main types of digital comparators available, namely (i) the identity comparator, which compares if the inputs A and B are equal ($A = B$) or not ($A \neq B$), and (ii) the magnitude comparator, which has three outputs: equality ($A = B$), greater than ($A > B$) and less than ($A < B$). The purpose of a digital comparator is to compare a set of variables or unknown numbers, for example, A (A1, A2, A3, ... An), against that of a constant or unknown value, such as B (B1, B2, B3, ... Bn), to produce an output condition or flag depending upon the result of the comparison. For example, a magnitude comparator of two 1-bit (A and B) inputs would produce the three outputs, as shown in Figure 3.5, when compared to each other [21,25–27].

The circuit shown in Figure 3.5 is useful if we want to compare two variables (A and B) of 1 bit each and produce an output of whether $A = B$, $A > B$ or $A < B$. The operation of a 1-bit magnitude comparator is given in Table 3.3.

We may notice from Figure 3.5 and Table 3.3 that the magnitude comparator just compares the inputs and takes a decision based on the comparison, as either A is compared with B or B is compared with A, to indicate either $A = B$, $A > B/B > A$ or $A < B/B < A$. The output condition for $A = B$ resembles that of a commonly available logic gate, the `xor-nor`, on each of the n bits, giving $output = (A \oplus B)'$. The magnitude comparators actually use `xor` gates within their design for comparing their respective pairs of bits. When we compare two binary or binary coded decimal (BCD) values or variables against each other, we are comparing the magnitude of these values, logic 0 against logic 1, which is where the term *magnitude comparator* comes from. As well as comparing individual bits, we can

TABLE 3.3
Truth Table of a 1-Bit Magnitude Comparator

Magnitude Comparator				
Inputs		Outputs		
A	B	A > B	A < B	A = B
0	0	0	0	1
0	1	0	1	0
1	0	1	0	0
1	1	0	0	1

design larger bit comparators by cascading together n stages of single-bit magnitude comparators and produce an n-bit comparator just as we did for the n-bit adders/subtractors in the previous sections. Multibit comparators can be constructed to compare whole binary or BCD words to produce an output if one word is larger, equal to or less than the other. A two 4-bit magnitude comparator circuit design is shown in Figure 3.6.

A portion of the function table for a 4-bit magnitude comparator is given in Table 3.4. Now, we will discuss how to analyze this 4-bit magnitude comparator and get its logical expression. Assume two inputs of 4-bit length as `A[3:0]`, that is, `A(3)`, `A(2)`, `A(1)` and `A(0)`, and `B[3:0]`, that is, `B(3)`, `B(2)`, `B(1)` and `B(0)`. When comparing the large binary or BCD numbers for equality, we can just use the `xnor` gate, that is, a combination of `xor` and `not` gates, as xor + not. Here, the `xor` gate will give logic 0 when the inputs `A` and `B` are equal, and an inverter, that is, a `not` gate, complements it. Hence, the `xnor` gate output is logic 1 if both inputs are equal. For `A` > `B`, to save time, the comparator has to start comparing the highest-order bit, that is, the MSB bit, first as `(A(3) and not B(3))`. Here, if `A(3)` is logic 1 and `B(3)` is logic 0, then the result of this expression is logic 1; thus, `A` > `B`. If equality exists at `A(3)` and `B(3)`, then we need to compare the next lowest bit as `[(A(3) xnor B(3)) and (A(2) and not B(2))]`. Here, the first part is for the equality of `A(3)` and `B(3)`, and the second part is for `A(2)` > `B(2)`; thus, the result is `A` > `B`. If the equality still exists, then we have to continue until it reaches the lowest-order bit, i.e., the LSB. In the same way, the logical expression for `A` < `B` can be obtained. The Boolean expressions for `A` = `B`, `A` > `B` and `A` < `B` are:

$$\text{Eq} = (\text{A xnor B})$$

$$\begin{aligned}\text{Ag} = [&(\text{A(3) and not(B(3))}) + ((\text{A(3) xnor B(3)}) \text{ and } (\text{A(2) and not(B(2)))} + ((\text{A(3) xnor B(3)})\\ &\text{and } (\text{A(2) xnor B(2)}) \text{ and } (\text{A(1) and not(B(1)))} + ((\text{A(3) xnor B(3)}) \text{ and } (\text{A(2) xnor B(2)})\\ &\text{and } (\text{A(1) xnor B(1)}) \text{ and } (\text{A(0) and not(B(0)))}]\end{aligned}$$

$$\begin{aligned}\text{Al} = [&(\text{B(3) and not(A(3))}) + ((\text{B(3) xnor A(3)}) \text{ and } (\text{B(2) and not(A(2)))} + ((\text{B(3) xnor A(3)})\\ &\text{and } (\text{B(2) xnor A(2)}) \text{ and } (\text{B(1) and not(A(1)))} + ((\text{B(3) xnor A(3)}) \text{ and } (\text{B(2) xnor A(2)})\\ &\text{and } (\text{B(1) xnor A(1)}) \text{ and } (\text{B(0) and not(A(0)))}]\end{aligned} \tag{3.5}$$

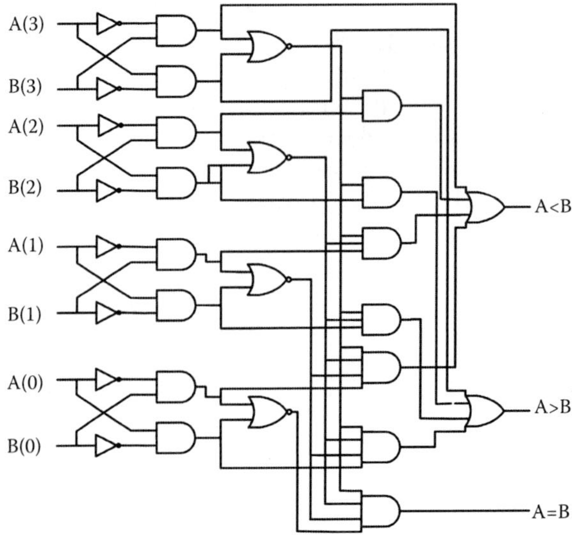

FIGURE 3.6
Circuit for a 4-bit magnitude comparator.

TABLE 3.4

Truth Table of Two 4-Bit Magnitude Comparators

Magnitude Comparator				
Inputs		Outputs		
A[3:0]	B[3:0]	A > B	A = B	A < B
0000	0000	0	1	0
0000	0001	0	0	1
0000	0010	0	0	1
.	.	.	.	.
.	.	.	.	.
1111	0000	1	0	0
1111	0001	1	0	0
.	.	.	.	.
.	.	.	.	.
1111	1111	0	1	0

The logical expressions given in Equation 3.5 can be altered for a magnitude comparator of any number of bits.

Some of the commercially available digital single-package ICs, such as the TTL74LS85 or CMOS4063, can be used as 4-bit magnitude comparators. These ICs have additional input terminals that allow the individual comparators to be cascaded together to compare words larger than 4 bits; hence, magnitude comparators of n bits can be designed. These cascading designs can be used to compare 8, 16, 32, or even more bits. Digital comparators are used widely in analog to digital converters (ADCs) and ALUs to perform a variety of arithmetic and signal-processing operations.

DESIGN EXAMPLE 3.5

Design and implement a single-bit magnitude comparator using the behavioral model.

Solution

```
library ieee;
use ieee.std_logic_1164.all;
entity compare is
port(   num1 : in std_logic_vector(3 downto 0);
num2 : in std_logic_vector(3 downto 0);
less : out std_logic;
equal : out std_logic;
greater : out std_logic);
end compare;
architecture behavioral of compare is
begin
process(num1,num2)
begin
if (num1 > num2 ) then
less <= '0';
equal <= '0';
greater <= '1';
elsif (num1 < num2) then
less <= '1';
```

```
equal <= '0';
greater <= '0';
else
less <= '0';
equal <= '1';
greater <= '0';
end if;
end process;
end behavioral;
```

DESIGN EXAMPLE 3.6

Design and implement a 4-bit magnitude comparator using the data flow model.

Solution

```
library ieee;
use ieee.std_logic_1164.all;
entity mag_comp_4b is
port ( a, b: in std_logic_vector(3 downto 0);
ag, al, eq : out std_logic);
end mag_comp_4b;
architecture beh of mag_comp_4b is
signal s : std_logic_vector(3 downto 0);
begin
s(0)<= a(0) xnor b(0);
s(1)<= a(1) xnor b(1);
s(2)<= a(2) xnor b(2);
s(3)<= a(3) xnor b(3);
eq<=s(3) and s(2) and s(1) and s(0);--equal
ag<=(a(3) and (not b(3))) or (s(3) and a(2) and (not b(2)))or (s(3) and
s(2) and a(1)and (not b(1))) or (s(3) and s(2) and s(1) and a(0) and
(not b(0)));--a greater
al<=(b(3) and (not a(3))) or (s(3) and b(2) and (not a(2)))or (s(3) and
s(2) and b(1)and (not a(1))) or (s(3) and s(2) and s(1) and b(0) and
(not a(0))); -- a lesser
end beh;
```

DESIGN EXAMPLE 3.7

Design and implement a 4-bit magnitude comparator using the structural model.

Solution

Component 1

```
library ieee;
use ieee.std_logic_1164.all;
entity notgate is
port( inport  : in std_logic;
outport : out std_logic);
end notgate;
architecture func of notgate is
begin
outport <= not inport;
end func;
```

Component 2

```
library ieee;
use ieee.std_logic_1164.all;
entity full_adder is
```

```
port( x, y, cin   : in std_logic;
sum, cout   : out std_logic);
end full_adder;
architecture func of full_adder is
begin
sum <= (x xor y) xor cin;
cout <= (x and (y or cin)) or (cin and y);
end func;
```

Component 3

```
library ieee;
use ieee.std_logic_1164.all;
entity orgate is
port( a, b : in std_logic;
f : out std_logic);
end orgate;
architecture func of orgate is
begin
f <= a or b;
end func;
```

Component 4

```
library ieee;
use ieee.std_logic_1164.all;
entity xorgate is
port( a, b : in std_logic;
f : out std_logic);
end xorgate;
architecture func of xorgate is
begin
f <= a xor b;
end func;
```

Component 5

```
library ieee;
use ieee.std_logic_1164.all;
entity norgate is
port( a, b : in std_logic;
f : out std_logic);
end norgate;
architecture func of norgate is
begin
f <= a nor b;
end func;
```

Main Code

```
library ieee;
use ieee.std_logic_1164.all;
entity comparator is
port( x3, x2, x1, x0 : in std_logic;
y3, y2, y1, y0 : in std_logic;
equ, grt, lss : out std_logic);
end comparator;
architecture struct of comparator is
component notgate is
port( inport : in std_logic;
outport : out std_logic);
end component;
component full_adder is
```

```
port( x, y, cin : in std_logic;
sum, cout : out std_logic);
end component;
component orgate is
port( a, b : in std_logic;
f : out std_logic);
end component;
component xorgate is
port( a, b : in std_logic;
f : out std_logic);
end component;
component norgate is
port( a, b : in std_logic;
f : out std_logic);
end component;
--interconnecting wires
signal inv0, inv1, inv2, inv3: std_logic;
signal c1, c2, c3, c4: std_logic;
signal s1, s2, s3, s0: std_logic;
signal ortop, orbot  : std_logic;
signal xor1, xor2    : std_logic;
signal nor1          : std_logic;
signal const : std_logic := '1';
begin
g_i0: notgate port map(y0, inv0);
g_i1: notgate port map(y1, inv1);
g_i2: notgate port map(y2, inv2);
g_i3: notgate port map(y3, inv3);
fa0: full_adder port map(x0, inv0, const, s0, c1); -- s0
fa1: full_adder port map(x1, inv1, c1,  s1, c2);   -- s1
fa2: full_adder port map(x2, inv2, c2,  s2, c3);   -- s2
fa3: full_adder port map(x3, inv3, c3,  s3, c4);   -- s3
g_o1: orgate port map(s0, s1, ortop);
g_o2: orgate port map(s2, s3, orbot);
g_n: norgate port map(ortop, orbot, nor1);
g_x1: xorgate port map(c3, c4, xor1);
g_x2: xorgate port map(s3, xor1, xor2);
---------final outputs
equ <= nor1;
g_grt: norgate port map(nor1, xor2, grt);
lss <= xor2;
end struct;
```

3.5 Seven-Segment Display Interfacing

A seven-segment display is an output display device that provides a way to display information in the form of numbers and text. LED-based seven-segment displays are very popular amongst electronics hobbyists, as they are easy to use and understand. The most commonly used displays, along with the control modules, are LCD, GLCD, light-emitting diodes (LEDs), seven-segment displays and so on. The seven-segment display is the most common display device used in many gadgets and electronic appliances such as digital meters, digital clocks, microwave ovens, electric stoves, and so on. These displays consist of seven segments of LEDs assembled into a structure like the numeral 8, as shown in Figure 3.7a. Actually, seven-segment displays contain about eight segments

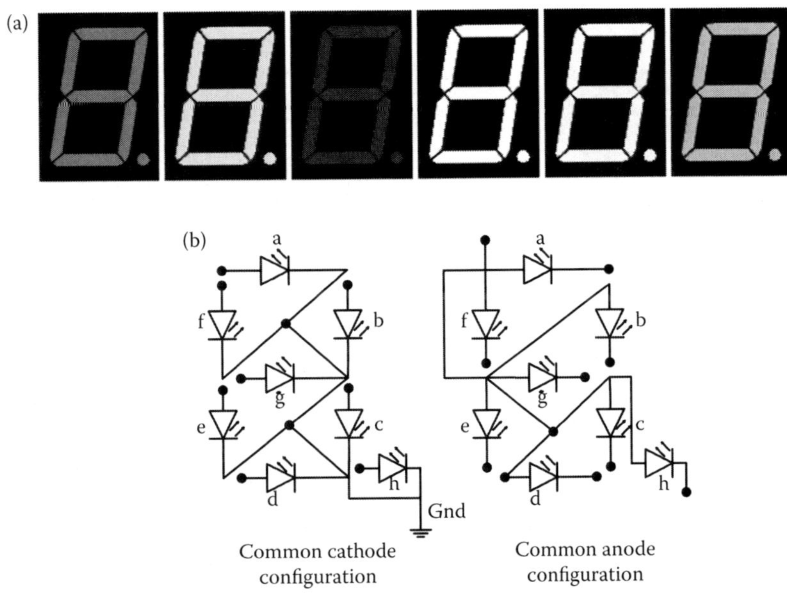

FIGURE 3.7
Seven-segment display array (a) and different configuration of display modules (b).

wherein an extra eighth segment is used to display the dot. This segment is useful while displaying noninteger numbers. Seven segments are indicated as A through g, and the eighth segment is indicated as h, as shown in Figure 3.7b. A seven-segment display is generally available in a 10-pin package, in that eight pins relate to the eight LEDs, and the remaining pins at the middle are internally shorted. These segments come in two outlines, the common cathode and common anode, as shown in Figure 3.7b. In the common cathode configuration, the negative terminals are connected together to the DC ground. When the corresponding pin is given high, then that particular LED glows. In a common anode arrangement, the common pin is given logic high and the pins of the LED are given low to display a number [28].

When power is given to all the segments, then the number 8 will be displayed. If you disconnect the power for segment g, that will result in the number 0. The circuit of the seven-segment display is designed in such a way that the voltage at different pins can be applied at the same time. In the same way, we can form combinations to display numerals from 0 to 9. Both types of display consist of 10 pins. Numeric seven-segment displays can also display some of the hex characters, such as A, B, C, D, E, and F; however, seven-segment displays are commonly used for displaying numeric values only. These types of displays still have a real purpose due to their high illumination, which is more suitable to dark areas like railway stations. Therefore, common anode seven-segment displays are very popular, as many logic circuits can sink more current than they can source. These displays are not a direct replacement in a circuit for a common anode display, as it is the same as connecting the LEDs in reverse; hence, light emission will not take place. Depending upon the decimal number displayed, the particular set of LEDs is forward biased in this configuration. For instance, to display the number 0, we need to light up the remaining segments corresponding to a, b, c, d, e, and f. Then, the digits from 0 through 9 can be displayed using a seven-segment display. There are different types of

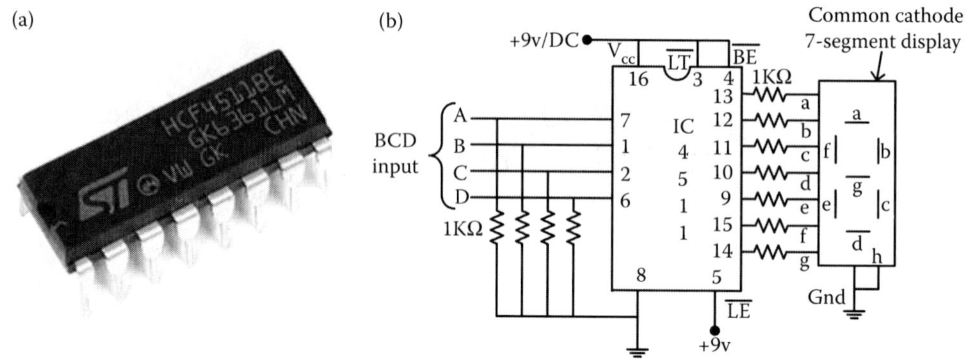

FIGURE 3.8
Picture of a seven-segment display decoder IC (a) and a decoder circuit configuration (b).

controlling techniques that are implemented by interfacing the seven-segment display with the external control device.

In most practical applications, seven-segment displays are driven by a suitable decoder/driver IC such as the CMOS4511 or TTL7447 from a 4-bit BCD input. In the simple interfacing circuit of a common cathode seven-segment display, the anode terminals are connected directly to a BCD to seven-segment driver IC451, shown in Figure 3.8a, and the cathode terminals are connected to ground, as shown in Figure 3.8b. The current from each output passes through a 1-KΩ resistor that limits the current to a safe amount. The binary input to the IC4511 is via the 4-bit inputs that come from the FPGA. Therefore, we can control the LED display using just four inputs instead of eight. Most digital equipment uses seven-segment displays for converting digital signals into a form that can be displayed and understood by the viewer. This information is often numerical data in the form of numbers, characters, and symbols. Thus, common anode or common cathode seven-segment displays produce the required number by illuminating the individual segments in various combinations. Seven-segment displays are commonly used in timers, clock radios, digital clocks, calculators, and wristwatches. These devices can also be found in speedometers, motor-vehicle odometers, radio frequency indicators, and practically any other display that makes use of alphanumeric characters alone. The use of seven-segment displays is very limited due to advancement in display technologies [21,28]. Now, the dot matrix display is mostly used in the place of seven-segment displays, yet seven-segment displays are still a good starting point to read about display technologies.

DESIGN EXAMPLE 3.8

Design and develop a digital system to display 0 through 9 on a seven-segment display.

Solution

```
library ieee;
use ieee.std_logic_1164.all;
use ieee.std_logic_arith.all;
use ieee.std_logic_unsigned.all;
entity clkd is
port ( clk,reset : in  std_logic;
seout:out std_logic_vector(8 downto 0):="000000000";
clkout : inout  std_logic);
end clkd;
architecture behavioral of clkd is
```

```
signal clk_sig:std_logic;
signal discnt:std_logic_vector(3 downto 0):="0000";
begin
--clock divider section
p1:process(clk,reset)
variable cnt:integer;
begin
if(reset='0') then
clk_sig<='0';
cnt:=0;
elsif rising_edge(clk) then
if(cnt=4000000)then
clk_sig<= not(clk_sig);
cnt:=0;
else
cnt:= cnt + 1;
end if;
end if;
end process;
clkout<= clk_sig;
--7 segment display counter section;
p2:process(clkout)
begin
if rising_edge(clkout) then
if(discnt="1001") then
discnt<="0000";
else
discnt<=discnt+1;
end if;
end if;
end process;
--7 segment display section
p3:process(discnt)
begin
case discnt is
-- cshgfedcba 0-on 1- off
when "0000" =>seout<="011000000";
when "0001" =>seout<="011111001";
when "0010" =>seout<="010100100";
when "0011" =>seout<="010110000";
when "0100" =>seout<="010011001";
when "0101" =>seout<="010010010";
when "0110" =>seout<="010000010";
when "0111" =>seout<="011111000";
when "1000" =>seout<="010000000";
when "1001" =>seout<="010011000";
when others=> seout<="xxxxxxxxx";
end case;
end process;
end behavioral;
```

3.6 Counter Design and Interfacing

In the previous chapter, we saw a multifrequency clock generator where the decimal counter was used to reduce the clock rate, that is, frequency, to the required low level from the onboard high frequency. In this section, we discuss the synchronous binary counter.

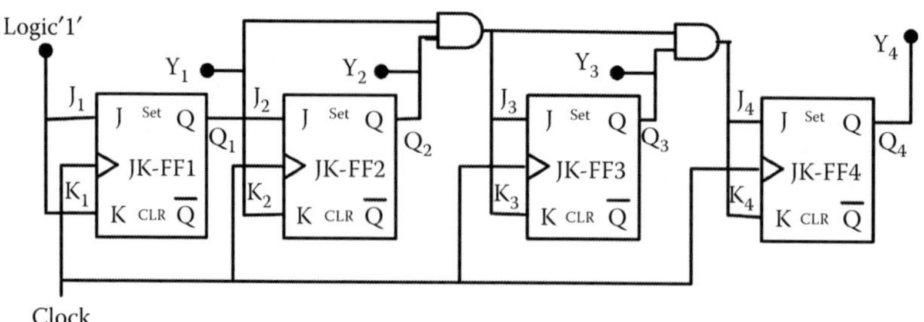

FIGURE 3.9
Binary 4-bit synchronous up counter.

In the synchronous binary counter, the external clock signal is connected to the clock input of all the FFs of the counter so that all the FFs are clocked together simultaneously, that is, in parallel. In other words, changes in the output of the FFs occur synchronously with either the rising or falling edge of the clock signal. The result of this synchronization is that all the individual output bit changes occur at exactly the same time in response to the common clock signal with no ripple effect and therefore with equal propagation delay. A simple synchronous binary up counter constructed using four JK-FFs is shown in Figure 3.9. It can be seen from this figure that the external clock pulses, that is, the pulses to be counted, are fed directly to all the JK-FFs. The inputs of the first FF, that is, J_1 and K_1, are tied together and connected to logic 1; hence, it is now in toggling mode so as to toggle for every clock pulse. Then, the synchronous counter follows a predetermined sequence of states in response to the common clock signal, advancing one state for each pulse. The inputs J_2 and K_2 are connected directly to the output Q_1. The inputs J_3, K_3, J_4 and K_4 are driven from separate and gates which are also supplied with signals from the input and output of the previous stage, as shown in Figure 3.9.

These additional and gates generate the required logic for the JK inputs of the next stage. If we enable each JK-FF to toggle based on all preceding FF outputs (Q), we can obtain the same counting sequence without the ripple effect, since each FF in this circuit will be clocked at exactly the same time. Then, as there is no inherent propagation delay in synchronous counters because all the counter stages are triggered in parallel at the same time, the maximum operating frequency of this type of frequency counter is much higher than that for a similar asynchronous counter circuit [21,22,27,29]. The counting for the clock inputs is "0000", "0001", "0010", "0011", "0100", "0101", "0110", "0111", "1000" … "1111"; then it rolls back to "0000"; because this is a 4-bit synchronous counter that counts sequentially on every clock pulse resulting from the output from $(0)_{10}$, that is, "0000" through $(15)_{10}$, that is, "1111". The circuit shown in Figure 3.9 can easily be altered as a 4-bit synchronous down counter just by connecting the and gates to the $\overline{Q}$ output of the FFs. As the synchronous counters are formed by connecting the FFs together, any number of FFs can be connected/cascaded together to form a divide-by-n binary counter, modulo counter, decade counter, BCD counter, and so on.

We can quite easily rearrange the additional and gates in the above counter circuit to produce other count numbers, such as a mod-12 counter which counts 12 states from "0000" to "1011", that is, $(0)_{10}$ through $(11)_{10}$, and then repeats, making them suitable for clocks and so on. Generally, synchronous counters count on the rising edge, which is the low-to-high transition of the clock signal, and asynchronous ripple counters count on the falling edge, which is the high-to-low transition of the clock signal. It may seem unusual that ripple counters use the falling edge of the clock cycle to change state, but this makes it easier to link counters together because the MSB of one counter can drive the clock input of the next. This works because the next bit must change state when the previous bit changes from high to low, the point at which a carry must occur to the next bit. Synchronous counters usually have a carry-out and carry-in pin for linking counters together without introducing any propagation delays.

DESIGN EXAMPLE 3.9

Design and implement a six-digit parallel synchronous counter. Show the result of the counters on six different seven-segment displays.

Solution

```
library ieee;
use ieee.std_logic_1164.all;
use ieee.std_logic_arith.all;
use ieee.std_logic_unsigned.all;
entity clkd is
port (clk,reset : in  std_logic;
discnt:out std_logic_vector(5 downto 0):="000000";
seout:out std_logic_vector(7 downto 0):="00000000";
dout,d0out,d1out,d2out,d3out,d4out,d5out: out std_logic);
end clkd;
architecture behavioral of clkd is
signal clk_sig,clk_sig1,clk0,clk1,clk2,clk3,clk4,clk5:std_logic:='0';
signal seout0,d0cnt,d1cnt,d2cnt,d3cnt,d4cnt,d5cnt:std_logic_vector
(3 downto 0):="0000";
--------seven-segment display ------------------------
signal discnt0:std_logic_vector(2 downto 0):="000";
begin
-- clock divider for 1hz generater
p1:process(clk,reset)
variable cnt:integer;
begin
if(reset='0') then
clk_sig<='0';
cnt:=0;
elsif rising_edge(clk) then
if(cnt=2000000)then
clk_sig<= not(clk_sig);
cnt:=0;
else
cnt:= cnt + 1;
end if;
end if;
end process;
dout<= clk_sig;
--zeroth display time
p2:process(clk_sig)
begin
if rising_edge(clk_sig) then
if(d0cnt="1001") then
clk0<=not(clk0);
d0cnt<="0000";
else
d0cnt<=d0cnt + 1;
end if;
end if;
```

```
end process;
d0out<=clk0;
-- first display time
p3:process(clk0)
begin
if rising_edge(clk0) then
if(d1cnt="1001") then
clk1<=not(clk1);
d1cnt<="0000";
else
d1cnt<=d1cnt + 1;
end if;
end if;
end process;
d1out<= clk1;
-- second display time
p4:process(clk1)
begin
if rising_edge(clk1) then
if(d2cnt="1001") then
clk2<=not(clk2);
d2cnt<="0000";
else
d2cnt<=d2cnt + 1;
end if;
end if;
end process;
d2out<=clk2;
-- thired display time
p5:process(clk2)
begin
if rising_edge(clk2) then
if(d3cnt="1001") then
clk3<=not(clk3);
d3cnt<="0000";
else
d3cnt<=d3cnt + 1;
end if;
end if;
end process;
d3out<=clk3;
-- fourth display time
p6:process(clk3)
begin
if rising_edge(clk3) then
if(d4cnt="1001") then
clk4<=not(clk4);
d4cnt<="0000";
else
d4cnt<=d4cnt + 1;
end if;
end if;
end process;
d4out<=clk4;
-- fifth display time
p7:process(clk4)
begin
if rising_edge(clk4) then
if(d5cnt="1001") then
clk5<=not(clk5);
d5cnt<="0000";
```

```
else
d5cnt<=d5cnt + 1;
end if;
end if;
end process;
d5out<= clk5;
-----------------------------seven-segment display section-----------------
p12:process(clk)
variable cnt1:integer;
begin
if rising_edge(clk) then
if(cnt1=1000)then
clk_sig1<= not(clk_sig1);
cnt1:=0;
else
cnt1:= cnt1 + 1;
end if;
end if;
end process;
p8:process(clk_sig1)
begin
if rising_edge(clk_sig1) then
if(discnt0="101") then
discnt0<="000";
else
discnt0<=discnt0+1;
end if;
end if;
end process;
--display control  selector
p9:process(discnt0)
begin
case discnt0 is
when "000" =>discnt<="011111";seout0<=d0cnt;
when "001" =>discnt<="101111";seout0<=d1cnt;
when "010" =>discnt<="110111";seout0<=d2cnt;
when "011" =>discnt<="111011";seout0<=d3cnt;
when "100" =>discnt<="111101";seout0<=d4cnt;
when "101" =>discnt<="111110";seout0<=d5cnt;
when others=>discnt<="xxxxxx";
end case;
end process;
-- 7 segment value
p11:process(seout0)
begin
case seout0 is
-- hgfedcba 0-on 1- off
when "0000" =>seout<="11000000";
when "0001" =>seout<="11111001";
when "0010" =>seout<="10100100";
when "0011" =>seout<="10110000";
when "0100" =>seout<="10011001";
when "0101" =>seout<="10010010";
when "0110" =>seout<="10000010";
when "0111" =>seout<="11111000";
when "1000" =>seout<="10000000";
when "1001" =>seout<="10011000";
when others=> seout<="xxxxxxxx";
end case;
end process;
end behavioral;
```

DESIGN EXAMPLE 3.10

Design and implement a synchronous counter to count from $(0)_{10}$ through $(999999)_{10}$. The frequency of the counting speed has to be 1 Hz and show the counter values on six different seven-segment displays.

Solution

```
library ieee;
use ieee.std_logic_1164.all;
use ieee.std_logic_arith.all;
use ieee.std_logic_unsigned.all;
entity clkd is
port (clk,reset : in  std_logic;
discnt:out std_logic_vector(5 downto 0):="000000";
seout:out std_logic_vector(7 downto 0):="00000000";
dout,d0out,d1out,d2out,d3out,d4out,d5out: out std_logic);
end clkd;
architecture behavioral of clkd is
signal clk_sig,clk_sig1:std_logic:='0';
signal clk0,clk1,clk2,clk3,clk4,clk5:std_logic:='1';
signal seout0,d0cnt,d1cnt,d2cnt,d3cnt,d4cnt,d5cnt:std_logic_vector
(3 downto 0):="0000";
--------seven-segment display ------------------------
signal discnt0:std_logic_vector(2 downto 0):="000";
begin
-- clock divider for 1hz generater
p1:process(clk,reset)
variable cnt:integer;
begin
if(reset='0') then
clk_sig<='0';
cnt:=0;
elsif rising_edge(clk) then
if(cnt=2000000)then
clk_sig<= not(clk_sig);
cnt:=0;
else
cnt:= cnt + 1;
end if;
end if;
end process;
dout<= clk_sig;
--zeroth display time
p2:process(clk_sig)
begin
if rising_edge(clk_sig) then
if (d0cnt="0101")then
clk0<=not(clk0);
end if;
if(d0cnt="1001") then
clk0<=not(clk0);
d0cnt<="0000";
else
d0cnt<=d0cnt + 1;
end if;
end if;
end process;
d0out<=clk0;
-- first display time
p3:process(clk0)
```

```
begin
if rising_edge(clk0) then
if (d1cnt="0101")then
clk1<=not(clk1);
end if;
if(d1cnt="1001") then
clk1<=not(clk1);
d1cnt<="0000";
else
d1cnt<=d1cnt + 1;
end if;
end if;
end process;
d1out<= clk1;
-- second display time
p4:process(clk1)
begin
if rising_edge(clk1) then
if (d2cnt="0101")then
clk2<=not(clk2);
end if;
if(d2cnt="1001") then
clk2<=not(clk2);
d2cnt<="0000";
else
d2cnt<=d2cnt + 1;
end if;
end if;
end process;
d2out<=clk2;
-- thired display time
p5:process(clk2)
begin
if rising_edge(clk2) then
if (d3cnt="0101")then
clk3<=not(clk3);
end if;
if(d3cnt="1001") then
clk3<=not(clk3);
d3cnt<="0000";
else
d3cnt<=d3cnt + 1;
end if;
end if;
end process;
d3out<=clk3;
-- fourth display time
p6:process(clk3)
begin
if rising_edge(clk3) then
if (d4cnt="0101")then
clk4<=not(clk4);
end if;
if(d4cnt="1001") then
clk4<=not(clk4);
d4cnt<="0000";
else
d4cnt<=d4cnt + 1;
end if;
end if;
end process;
```

```
d4out<=clk4;
-- fifth display time
p7:process(clk4)
begin
if rising_edge(clk4) then
if (d5cnt="0101")then
clk5<=not(clk5);
end if;
if(d5cnt="1001") then
clk5<=not(clk5);
d5cnt<="0000";
else
d5cnt<=d5cnt + 1;
end if;
end if;
end process;
d5out<= clk5;
------------------------------seven-segmentdisplaysection-----------------
p12:process(clk)
variable cnt1:integer;
begin
if rising_edge(clk) then
if(cnt1=1000)then
clk_sig1<= not(clk_sig1);
cnt1:=0;
else
cnt1:= cnt1 + 1;
end if;
end if;
end process;
p8:process(clk_sig1)
begin
if rising_edge(clk_sig1) then
if(discnt0="101") then
discnt0<="000";
else
discnt0<=discnt0+1;
end if;
end if;
end process;
--display control  selector
p9:process(discnt0)
begin
case discnt0 is
when "000" =>discnt<="011111";seout0<=d0cnt;
when "001" =>discnt<="101111";seout0<=d1cnt;
when "010" =>discnt<="110111";seout0<=d2cnt;
when "011" =>discnt<="111011";seout0<=d3cnt;
when "100" =>discnt<="111101";seout0<=d4cnt;
when "101" =>discnt<="111110";seout0<=d5cnt;
when others=>discnt<="xxxxxx";
end case;
end process;
-- 7 segment value
p11:process(seout0)
begin
case seout0 is
--                          hgfedcba       0-on  1- off
when "0000" =>seout<="11000000";
when "0001" =>seout<="11111001";
when "0010" =>seout<="10100100";
```

```
when "0011" =>seout<="10110000";
when "0100" =>seout<="10011001";
when "0101" =>seout<="10010010";
when "0110" =>seout<="10000010";
when "0111" =>seout<="11111000";
when "1000" =>seout<="10000000";
when "1001" =>seout<="10011000";
when others=> seout<="xxxxxxxx";
end case;
end process;
end behavioral;
```

3.7 Digital Clock Design and Interfacing

There are four main blocks in the design of a digital clock, namely (i) pulse per second (PPS) generator, (ii) timing counter, (iii) display manager, and (iv) seven-segment display interfacing, as shown in Figure 3.10. Now, we will discuss the design and functions of these blocks one by one. The PPS generator block consists of a DCM, 1-Hz generator, and timing control circuit. The DCM converts the onboard high frequency to low frequency so as to slowly drive the 1-Hz generator module. As discussed in the previous chapter, the DCM is designed with a decimal counter, equality comparator, and complementing circuit. In this design, the onboard frequency is 4 MHz; thus, the DCM converts it to 100 Hz, which drives the 1-Hz generator to again lower the frequency to 1 Hz. This type of cascaded frequency reduction circuit does not require large bit counters or large bit equality comparators. The PPS output of the 1-Hz counter is applied to the timing control circuit which controls all the functions of timing counters [30–33].

The timing counter block consists of second, minute, and hour counters. The second counter has two digits, one modulo 10 digit, and the other modulo 6 digit. That means the first digit rolls back to 0 once it reaches 10 by incrementing the second digit by a value of 1. Then, once the second digit reaches a value of 6, it rolls back to 0 by incrementing the

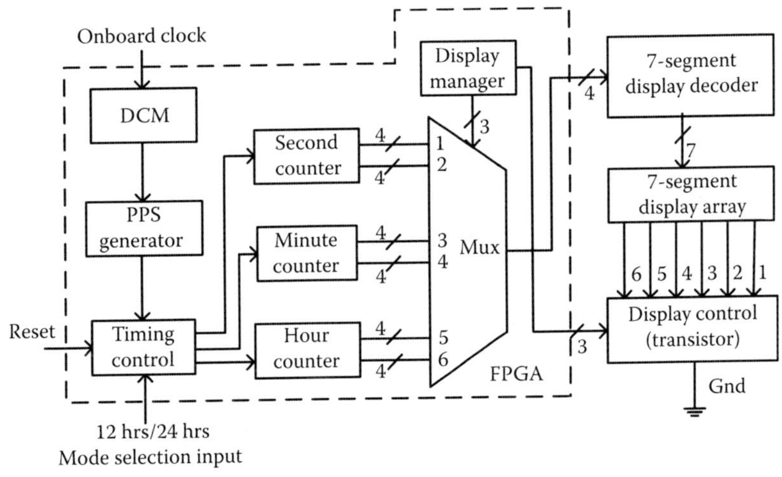

FIGURE 3.10
Schematic design of digital real-time clock.

minute counter. In the same way, the minute counter is also designed with two digits; the first counts from 0 through 9 and rolls back to 0 once it reaches 10 by incrementing the second digit. The second digit rolls back to 0 once it reaches 6, like the second counter. Thus, in general, the second and minute counters' count values vary from 00 to 59 for every cycle of rolling back. Based on the value of the minute counter, a trigger pulse is applied to the hour counter, which also has two digits that vary from 00 through 12 or 00 through 24 based on the 12/24 mode selection input. The timing control unit can reset the values of all the counters in accordance with the reset input. Time management of all these operations is performed by the timing control unit. The counter values are sent to the data multiplexer as 4 bits for each digit, that is, sec_0`[3:0]`, sec_1`[3:0]`, min_0`[3:0]`, min_1`[3:0]`, hr_0`[3:0]` and hr_1`[3:0]`, in parallel so as to route any one of those values to the external seven-segment display.

The display manager block is operated at a relatively higher frequency since it has to update the display values on six different seven-segment displays within a second. The display manager selects the channels of the multiplexer in order as `sec0`, `sec1`, `min0`, `min1`, `hr0`, and `hr1` by sending the values `1`, `2`, `3`, `4`, `5`, and `6` to the selection lines so as to send the appropriate counter value to the respective seven-segment display. The display manager also performs another important task: it manages the display control circuit, which enables only one segment to be on at a given instant of time. Since these two operations are performed by the display manager, it is perfectly possible to synchronize multiplexer channel selection and enable a particular seven-segment display to display a value corresponding to the selected digit.

The seven-segment display interfacing block is an external circuit, not a part of digital circuit built inside the FPGA. This module consists of a seven-segment display decoder, display array and display control circuit. The display decoder converts the BCD input `[7:0]` into the form `[6:0]`, which is required to display the BCD number on a seven-segment display module. In this design, the display array is constructed with six seven-segment displays, and two modules are used for each digit (hr, min, sec) displays. The display control circuit is designed with six transistors (each display module has one transistor), and their base terminals are controlled by the display manager. A particular display module can be switched just by appropriately passing logic 1 to the respective transistor. The base input logic 1 drives the transistor on; thus, the collector terminal of the transistor gets shorted with the emitter terminal, that is, ground. Therefore, the particular seven-segment display gets powered with proper grounding; thus, it would show the number/character on the enabled segment. Control synchronization between the display data loading and controlling the display control circuit are taken care of by the display manager. All the above operations have to be performed with effectively managing the time and logical operations. In this way, any digital real-time clock can be designed [30,33].

DESIGN EXAMPLE 3.11

Design and implement a digital system in FPGA to realize a digital clock. Display the hours, minutes, and seconds on six seven-segment displays (two for each).

Solution

```
library ieee;
use ieee.std_logic_1164.all;
use ieee.std_logic_arith.all;
use ieee.std_logic_unsigned.all;
entity clkd is
port (clk,reset : in  std_logic;
```

```
discnt:out std logic_vector(5 downto 0):="000000";
seout:out std_logic_vector(7 downto 0):="00000000";
dout,d0out,d1out,d2out,d3out,d4out: out std_logic);
end clkd;
architecture behavioral of clkd is
signal clk_sig,clk_sig1:std_logic:='0';
signal clk0,clk1,clk2,clk3,clk4:std_logic:='1';
signal seout0,d0cnt,d1cnt,d2cnt,d3cnt,d4cnt,d5cnt:std_logic_vector
(3 downto 0):="0000";
--------seven-segment display -----------------------
signal discnt0:std_logic_vector(2 downto 0):="000";
begin
-- clock divider for 1hz generater
p1:process(clk,reset)
variable cnt:integer;
begin
if(reset='0') then
clk_sig<='0';
cnt:=0;
elsif rising_edge(clk) then
if(cnt=2000000)then
clk_sig<= not(clk_sig);
cnt:=0;
else
cnt:= cnt + 1;
end if;
end if;
end process;
dout<= clk_sig;
----------------------------sec calculation--------------------
--zeroth display time
p2:process(clk_sig)
begin
if rising_edge(clk_sig) then
if (d0cnt="0101")then
clk0<=not(clk0);
end if;
if(d0cnt="1001") then
clk0<=not(clk0);
d0cnt<="0000";
else
d0cnt<=d0cnt + 1;
end if;
end if;
end process;
d0out<=clk0;
-- first display time
p3:process(clk0)
begin
if rising_edge(clk0) then
if (d1cnt="0100")then
clk1<=not(clk1);
end if;
if(d1cnt="0101") then
clk1<=not(clk1);
d1cnt<="0000";
else
d1cnt<=d1cnt + 1;
end if;
end if;
end process;
```

```
d1out<= clk1;
--------------------------------min calculation-------------------
-- second display time
p4:process(clk1)
begin
if rising_edge(clk1) then
if (d2cnt="0101")then
clk2<=not(clk2);
end if;
if(d2cnt="1001") then
clk2<=not(clk2);
d2cnt<="0000";
else
d2cnt<=d2cnt + 1;
end if;
end if;
end process;
d2out<=clk2;
-- thired display time
p5:process(clk2)
begin
if rising_edge(clk2) then
if (d3cnt="0100")then
clk3<=not(clk3);
end if;
if(d3cnt="0101") then
clk3<=not(clk3);
d3cnt<="0000";
else
d3cnt<=d3cnt + 1;
end if;
end if;
end process;
d3out<=clk3;
-----------------------------hr calculation-----------------------
-- fourth display time
p6:process(clk3)
begin
if rising_edge(clk3) then
if (d5cnt="0000") then
if (d4cnt="0101")then
clk4<=not(clk4);
end if;
if(d4cnt="1001") then
clk4<=not(clk4);
d4cnt<="0000";
else
d4cnt<=d4cnt + 1;
end if;
else
if (d4cnt="0001")then
clk4<=not(clk4);
end if;
if(d4cnt="0010") then
clk4<=not(clk4);
d4cnt<="0001";
else
d4cnt<=d4cnt + 1;
end if;
end if;
end if;
```

```
end process;
d4out<=clk4;
-- fifth display time
p7:process(clk4)
begin
if rising_edge(clk4) then
if(d5cnt="0001") then
d5cnt<="0000";
else
d5cnt<=d5cnt + 1;
end if;
end if;
end process;
-----------------------------seven-segmentdisplaysection-----------------
p12:process(clk)
variable cnt1:integer;
begin
if rising_edge(clk) then
if(cnt1=1000)then
clk_sig1<= not(clk_sig1);
cnt1:=0;
else
cnt1:= cnt1 + 1;
end if;
end if;
end process;
p8:process(clk_sig1)
begin
if rising_edge(clk_sig1) then
if(discnt0="101") then
discnt0<="000";
else
discnt0<=discnt0+1;
end if;
end if;
end process;
--display control  selector
p9:process(discnt0)
begin
case discnt0 is
when "000" =>discnt<="011111";seout0<=d0cnt;
when "001" =>discnt<="101111";seout0<=d1cnt;
when "010" =>discnt<="110111";seout0<=d2cnt;
when "011" =>discnt<="111011";seout0<=d3cnt;
when "100" =>discnt<="111101";seout0<=d4cnt;
when "101" =>discnt<="111110";seout0<=d5cnt;
when others=>discnt<="xxxxxx";
end case;
end process;
-- 7 segment value
p11:process(seout0)
begin
case seout0 is
-- hgfedcba 0-on 1- off
when "0000" =>seout<="11000000";
when "0001" =>seout<="11111001";
when "0010" =>seout<="10100100";
when "0011" =>seout<="10110000";
when "0100" =>seout<="10011001";
when "0101" =>seout<="10010010";
when "0110" =>seout<="10000010";
```

```
when "0111" =>seout<="11111000";
when "1000" =>seout<="10000000";
when "1001" =>seout<="10011000";
when others=> seout<="xxxxxxxx";
end case;
end process;
end behavioral;
```

DESIGN EXAMPLE 3.12

Design and implement a digital system in FPGA to realize a digital clock. Display the hours, minutes, and seconds on six seven-segment displays (two for each). Indicate AM/PM using the decimal dot.

Solution

```
library ieee;
use ieee.std_logic_1164.all;
use ieee.std_logic_arith.all;
use ieee.std_logic_unsigned.all;
entity clkd is
port (clk,reset : in  std_logic;
discnt:out std_logic_vector(5 downto 0):="000000";
seout:out std_logic_vector(7 downto 0):="00000000";
dout,d0out,d1out,d2out,d3out,d4out: out std_logic);
end clkd;
architecture behavioral of clkd is
signal clk_sig,clk_sig1:std_logic:='0';
signal clk5:std_logic:='0';
signal clk0,clk1,clk2,clk3,clk4:std_logic:='1';
signal seout0,d0cnt,d1cnt,d2cnt,d3cnt,d4cnt,d5cnt:std_logic_vector
(3 downto 0):="0000";
--------seven-segment display ------------------------
signal discnt0:std_logic_vector(2 downto 0):="000";
begin
-- clock divider for 1hz generater
p1:process(clk,reset)
variable cnt:integer;
begin
if(reset='0') then
clk_sig<='0';
cnt:=0;
elsif rising_edge(clk) then
if(cnt=2000000)then
clk_sig<= not(clk_sig);
cnt:=0;
else
cnt:= cnt + 1;
end if;
end if;
end process;
dout<= clk_sig;
-----------------------------sec calculation--------------------
--zeroth display time
p2:process(clk_sig)
begin
if rising_edge(clk_sig) then
if (d0cnt="0101")then
clk0<=not(clk0);
end if;
if(d0cnt="1001") then
```

```
clk0<=not(clk0);
d0cnt<="0000";
else
d0cnt<=d0cnt + 1;
end if;
end if;
end process;
d0out<=clk0;
-- first display time
p3:process(clk0)
begin
if rising_edge(clk0) then
if (d1cnt="0100")then
clk1<=not(clk1);
end if;
if(d1cnt="0101") then
clk1<=not(clk1);
d1cnt<="0000";
else
d1cnt<=d1cnt + 1;
end if;
end if;
end process;
d1out<= clk1;
--------------------------------min calculation------------------
-- second display time
p4:process(clk1)
begin
if rising_edge(clk1) then
if (d2cnt="0101")then
clk2<=not(clk2);
end if;
if(d2cnt="1001") then
clk2<=not(clk2);
d2cnt<="0000";
else
d2cnt<=d2cnt + 1;
end if;
end if;
end process;
d2out<=clk2;
-- thired display time
p5:process(clk2)
begin
if rising_edge(clk2) then
if (d3cnt="0100")then
clk3<=not(clk3);
end if;
if(d3cnt="0101") then
clk3<=not(clk3);
d3cnt<="0000";
else
d3cnt<=d3cnt + 1;
end if;
end if;
end process;
d3out<=clk3;
-----------------------------hr calculation----------------------
-- fourth display time
p6:process(clk3)
begin
```

```
if rising_edge(clk3) then
if (d4cnt ="0001" and d5cnt="0001") then
clk5<= not(clk5);
end if;
if (d5cnt="0000") then
if (d4cnt="0101")then
clk4<=not(clk4);
end if;
if(d4cnt="1001") then
clk4<=not(clk4);
d4cnt<="0000";
else
d4cnt<=d4cnt + 1;
end if;
else
if (d4cnt="0001")then
clk4<=not(clk4);
end if;
if(d4cnt="0010") then
clk4<=not(clk4);
d4cnt<="0001";
else
d4cnt<=d4cnt + 1;
end if;
end if;
end if;
end process;
d4out<=clk4;
-- fifth display time
p7:process(clk4)
begin
if rising_edge(clk4) then
if(d5cnt="0001") then
d5cnt<="0000";
else
d5cnt<=d5cnt + 1;
end if;
end if;
end process;
------------------------------Seven-segment display section-----------------
p12:process(clk)
variable cnt1:integer;
begin
if rising_edge(clk) then
if(cnt1=1000)then
clk_sig1<= not(clk_sig1);
cnt1:=0;
else
cnt1:= cnt1 + 1;
end if;
end if;
end process;
p8:process(clk_sig1)
begin
if rising_edge(clk_sig1) then
if(discnt0="101") then
discnt0<="000";
else
discnt0<=discnt0+1;
end if;
end if;
```

```
end process;
--display control  selector
p9:process(discnt0)
begin
case discnt0 is
when "000" =>discnt<="011111";seout0<=d0cnt;
when "001" =>discnt<="101111";seout0<=d1cnt;
when "010" =>discnt<="110111";seout0<=d2cnt;
when "011" =>discnt<="111011";seout0<=d3cnt;
when "100" =>discnt<="111101";seout0<=d4cnt;
when "101" =>discnt<="111110";seout0<=d5cnt;
when others=>discnt<="xxxxxx";
end case;
end process;
-- 7 segment value
p11:process(seout0)
begin
if (clk5='1' and (discnt0="000" or discnt0="010" or discnt0="100")) then
case seout0 is
-- hgfedcba 0-on 1- off
when "0000" =>seout<="01000000";
when "0001" =>seout<="01111001";
when "0010" =>seout<="00100100";
when "0011" =>seout<="00110000";
when "0100" =>seout<="00011001";
when "0101" =>seout<="00010010";
when "0110" =>seout<="00000010";
when "0111" =>seout<="01111000";
when "1000" =>seout<="00000000";
when "1001" =>seout<="00011000";
when others=> seout<="xxxxxxxx";
end case;
else
case seout0 is
-- hgfedcba 0-on 1- off
when "0000" =>seout<="11000000";
when "0001" =>seout<="11111001";
when "0010" =>seout<="10100100";
when "0011" =>seout<="10110000";
when "0100" =>seout<="10011001";
when "0101" =>seout<="10010010";
when "0110" =>seout<="10000010";
when "0111" =>seout<="11111000";
when "1000" =>seout<="10000000";
when "1001" =>seout<="10011000";
when others=> seout<="xxxxxxxx";
end case;
end if;
end process;
end behavioral;
```

3.8 Pulse Width Modulation Signal Generation

Modern FPGAs provide very good hardware design flexibility. This section presents details of the generation of PWM signals and techniques that are needed to vary the duty cycle. The requirements of PWM signals are found in a large number of applications as a voltage control signal. It is mainly used for controlling the output voltage via power electronic switches in most high-voltage applications. PWM control-based inverters are one of the power converters which particularly use the PWM concept for their control operation. Recently, PWM inverters have gained great popularity in industrial applications because of their linearity and superior performance. The PWM has a fixed frequency and a variable average voltage via varying the duty cycle. The voltage level of the PWM signal normally falls between 2 and 5 V. A PWM signal can be generated using few counters. The advantage of the counter-based method is that it can be used to generate a high-frequency variable duty cycle PWM signal. The center of each pulse occurs at the PWM frequency, and the pulse width varies around the center according to the duty cycle of interest. If set to multiple phases, the component generates one PWM signal for each phase with even spacing among them. For example, when set to three phases, it generates three PWM outputs 120° out of phase with respect to one another. Figure 3.11 shows a typical example of the PWM signals of 50%, 10%, and 90% duty cycles. The onboard clock is divided by the suitable decimal value to get the required PWM frequency. Counters define the PWM period for each phase. There is one counter for each PWM phase, with their values offset by the phase. Each counter gets incremented based on the onboard clock and is cleared once it reaches the maximum value. The duty cycle determines the points during the period when the PWM signal's rising and falling edges occur. Once the counter reaches each of these positions, the PWM signal is toggled appropriately [34].

In general, there are two types of PWM techniques, namely analog and digital. The disadvantages of the analog technique are that it is easily affected by noise and changes with respect to voltage and temperature. The advantages of the digital technique are that it is good for designing variable PWM signals, very flexible and less sensitive to environmental noise. The construction of a PWM signal using the digital technique is easy to implement in a very fast manner. The reprogramming capability of the FPGA makes it suitable to develop any design using FPGA. In many applications, the PWM signal is not constant and the main parameter of it is the duty cycle. The frequency range of the PWM signal can be calculated as `frang = (fh-fl)`, where, `fh` is the possible operating high frequency and `fl` is the possible operating low frequency [34]. Then, the resolution of the operating

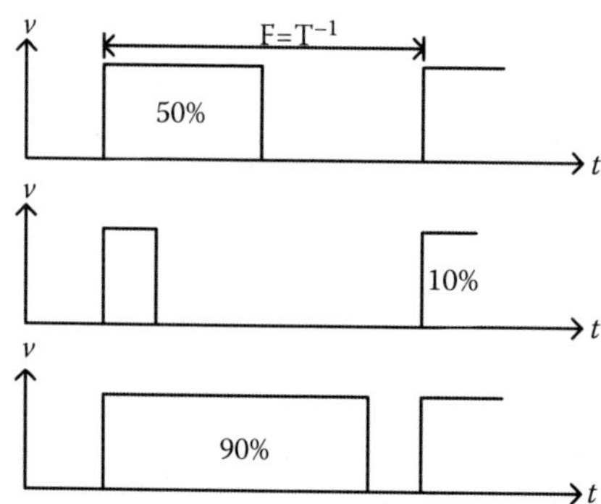

FIGURE 3.11
Waveform of an in-phase PWM signal.

frequency (`freso`) is calculated by (`frang`/resolution count). We now need to find the counter value for the frequency divider to generate the PWM signal at this calculated PWM frequency. In this way, any PWM signal can be generated for any control application.

DESIGN EXAMPLE 3.13

Design and develop a digital system to generate a PWM signal using data-flow modeling.

Solution

```
library ieee;
use ieee.std_logic_1164.all;
use ieee.std_logic_unsigned.all;
entity pwm is
generic(sys_clk : integer := 50_000_000;
pwm_freq : integer := 100_000;
bits_resolution : integer := 8;
phases : integer := 1);
port(clk, reset_n,ena: in  std_logic;
duty : in  std_logic_vector(bits_resolution-1 downto 0);
pwm_out, pwm_n_out : out std_logic_vector(phases-1 downto 0));
end pwm;
architecture logic of pwm is
constant period  : integer := sys_clk/pwm_freq;
type counters is array (0 to phases-1) of integer range 0 to period - 1;
signal count     : counters := (others => 0);
signal half_duty : integer range 0 to period/2 := 0;
begin
process(clk, reset_n)
begin
if(reset_n = '0') then
count <= (others => 0);
pwm_out <= (others => '0');
pwm_n_out <= (others => '0');
elsif(clk'event and clk = '1') then
if(ena = '1') then
half_duty <= conv_integer(duty)*period/(2**bits_resolution)/2;
end if;
for i in 0 to phases-1 loop
if(count(0) = period - 1 - i*period/phases) then
count(i) <= 0;
else
count(i) <= count(i) + 1;
end if;
end loop;
for i in 0 to phases-1 loop
if(count(i) = half_duty) then
pwm_out(i) <= '0';
pwm_n_out(i) <= '1';
elsif(count(i) = period - half_duty) then
pwm_out(i) <= '1';
pwm_n_out(i) <= '0';
end if;
end loop;
end if;
end process;
end logic;
```

DESIGN EXAMPLE 3.14

Design and develop a digital system to generate a PWM signal using structural modeling.

Solution

Component 1

```
library ieee;
use ieee.std_logic_1164.all;
entity clk64khz is
port (clk    : in  std_logic;
reset   : in  std_logic;
clk_out: out std_logic);
end clk64khz;
architecture behavioral of clk64khz is
signal temporal: std_logic;
signal counter : integer range 0 to 780 := 0;
begin
freq_divider: process (reset, clk) begin
if (reset = '1') then
temporal <= '0';
counter  <= 0;
elsif rising_edge(clk) then
if (counter = 780) then
temporal <= not(temporal);
counter  <= 0;
else
counter <= counter + 1;
end if;
end if;
end process;
clk_out <= temporal;
end behavioral;
```

Component 2

```
library ieee;
use ieee.std_logic_1164.all;
use ieee.numeric_std.all;
entity servo_pwm is
port (clk   : in  std_logic;
reset : in  std_logic;
pos   : in  std_logic_vector(6 downto 0);
servo : out std_logic);
end servo_pwm;
architecture behavioral of servo_pwm is
signal cnt : unsigned(10 downto 0);
signal pwmi: unsigned(7 downto 0);
begin
pwmi <= unsigned('0' & pos) + 32;
counter: process (reset, clk) begin
if (reset = '1') then
cnt <= (others => '0');
elsif rising_edge(clk) then
if (cnt = 1279) then
cnt <= (others => '0');
else
cnt <= cnt + 1;
end if;
end if;
```

```
end process;
servo <= '1' when (cnt < pwmi) else '0';
end behavioral;
```

Main Code

```
library ieee;
use ieee.std_logic_1164.all;
entity servo_pwm_clk64khz is
port(clk   : in  std_logic;
reset: in  std_logic;
pos  : in  std_logic_vector(6 downto 0);
servo: out std_logic);
end servo_pwm_clk64khz;
architecture behavioral of servo_pwm_clk64khz is
component clk64khz
port( clk     : in  std_logic;
reset   : in  std_logic;
clk_out: out std_logic);
end component;
component servo_pwm
port ( clk    : in  std_logic;
reset : in  std_logic;
pos   : in  std_logic_vector(6 downto 0);
servo : out std_logic );
end component;
signal clk_out : std_logic := '0';
begin
clk64khz_map: clk64khz port map(clk, reset, clk_out);
servo_pwm_map: servo_pwm port map(clk_out, reset, pos, servo );
end behavioral;
```

3.9 Special System Design Techniques

In this section, we discuss digital system design using some of the special types of functions, namely packages, libraries, functions, and procedures. Design examples are also given to understand the advantages of all these special functions.

3.9.1 Packages and Libraries

Packages and libraries provide a convenient way of referencing frequently used functions and components. Packages are the only language mechanism to share objects among different design units. Usually, they are designed to provide standard solutions for specific designs, for example, data types, subprograms, and type conversion functions. A package consists of a package declaration and an optional package body [12–14]. The package declaration contains a set of declarations which may be shared by several design units, for example, types, signals, components, functions, and procedure declarations. The body package usually contains the functions and procedure bodies. The general syntax rule for a package declaration is

```
package identifier is {package declarations}
begin
{sequential_statement} end [package] [identifier];
```

A package is analyzed separately and placed in the working library by the analyzer. Each package declaration includes function and/or procedure declarations that must have a corresponding package body. The syntax rule for a package body declaration is

```
package body identifier is {package body declarations}
end [package body] [identifier];
```

The package can be created like a new project from the new source wizard. The file name and location have to be given before beginning the package design. Upon creating a package, all the templates will be opened and available for editing. We need to use the necessary template, and other unused templates may be deleted.

DESIGN EXAMPLE 3.15

Design and develop a digital system to realize an 8-bit adder using the package and library.

Solution

Package Code

```
library ieee;
use ieee.std_logic_1164.all;
package my_package is
function add4_func(a,b:std_logic_vector(3 downto 0);
carry : std_logic) return std_logic_vector;
procedure add4_proc
(a, b: in std_logic_vector(3 downto 0);
carry: in std_logic;
signal sum: out std_logic_vector(3 downto 0);
signal cout: out std_logic);
end package my_package;
package body my_package is
function add4_func(a, b: std_logic_vector(3 downto 0);
carry: std_logic) return std_logic_vector is
variable cout: std_logic; variable cin: std_logic;
variable sum: std_logic_vector(4 downto 0);
begin
cin := carry; sum := "00000";
loop1: for i in 0 to 3 loop
cout := (a(i) and b(i)) or (a(i) and cin) or (b(i) and cin); sum(i) :=
a(i) xor b(i) xor cin;
cin := cout;
end loop loop1;
sum(4) := cout; return sum;
end add4_func;
procedure add4_proc
(a, b: in std_logic_vector(3 downto 0);
carry: in std_logic;
signal sum: out std_logic_vector(3 downto 0);
signal cout: out std_logic) is
variable c: std_logic;
begin
c := carry;
for i in 0 to 3 loop
sum(i) <= a(i) xor b(i) xor c;
c := (a(i) and b(i)) or (a(i) and c) or (b(i) and c);
end loop;
cout <= c;
end add4_proc;
end package body my_package;
```

Main Code

```
library my_library;
use my_library.my_package.all;
library ieee;
use ieee.std_logic_1164.all;
package my_package is
function add4_func(a, b : std_logic_vector(3 downto 0);
carry : std_logic) return std_logic_vector;
procedure add4_proc
(a, b : in std_logic_vector(3 downto 0);
carry: in std_logic;
signal sum: out std_logic_vector(3 downto 0);
signal cout: out std_logic);
end package my_package;
package body my_package is
function add4_func(a, b : std_logic_vector(3 downto 0); carry: std_
logic) return std_logic_vector is
variable cout: std_logic; variable cin: std_logic;
variable sum: std_logic_vector(4 downto 0);
begin
cin := carry; sum := "00000";
loop1: for i in 0 to 3 loop
cout := (a(i) and b(i)) or (a(i) and cin) or (b(i) and cin);
sum(i) := a(i) xor b(i) xor cin;
cin := cout;
end loop loop1;
sum(4) := cout;
return sum;
end add4_func;
procedure add4_proc
(a, b : in std_logic_vector(3 downto 0); carry: in std_logic;
signal sum: out std_logic_vector(3 downto 0); signal cout: out std_
logic) is
variable c: std_logic;
begin
c := carry;
for i in 0 to 3 loop
sum(i) <= a(i) xor b(i) xor c;
c := (a(i) and b(i)) or (a(i) and c) or (b(i) and c);
end loop;
cout <= c;
end add4_proc;
end package body my_package;
```

DESIGN EXAMPLE 3.16

Design and develop a digital system using the package for displaying the integers 0 through 9 on a seven-segment display.

Solution

Package Code

```
library IEEE;
use IEEE.STD_LOGIC_1164.all;
package mypackage  is
--procedure <procedure_name>(<type_declaration> <constant_name>: in
<type_declaration>);
procedure ssg_decode(signal hexcode:in std_logic_vector (3 downto 0);
signal ssg_out:out std_logic_vector (7 downto 0));
end mypackage;
```

```
package body mypackage is
-- Procedure Example
--procedure <procedure_name> (<type_declaration> <constant_name> : in
<type_declaration>) is
procedure ssg_decode(signal hexcode:in std_logic_vector (3 downto 0);
signal ssg_out:out std_logic_vector (7 downto 0)) is
variable temp :std_logic_vector (7 downto 0);
begin
case hexcode is
when "0000" => temp(6 downto 0) :="1000000";
when "0001" => temp(6 downto 0) :="1111001";
when "0010" => temp(6 downto 0) :="0100100";
when "0011" => temp(6 downto 0) :="0110000";
when "0100" => temp(6 downto 0) :="0011001";
when "0101" => temp(6 downto 0) :="0010010";
when "0110" => temp(6 downto 0) :="0000010";
when "0111" => temp(6 downto 0) :="1111000";
when "1000" => temp(6 downto 0) :="0000000";
when "1001" => temp(6 downto 0) :="0011000";
when others =>  temp(6 downto 0) :="0000000";
end case;
ssg_out<= not temp;
ssg_out(7)<='1';
end ssg_decode;
end mypackage;
```

Main Code

```
library IEEE;
use IEEE.STD_LOGIC_1164.ALL;
use IEEE.STD_LOGIC_ARITH.ALL;
use IEEE.STD_LOGIC_UNSIGNED.ALL;
use work.mypackage.all;
entity ssegprocedure is
Port ( mclk,sw0 : in  STD_LOGIC:='0';
anodes : out  STD_LOGIC_vector(3 downto 0):="0000";
sseg : out  STD_LOGIC_vector(7 downto 0):="00000000");
end ssegprocedure;
architecture Behavioral of ssegprocedure is
signal clkdiv: std_logic_vector(2 downto 0):="000";--(24 downto
0):="00000000  00000000000000000";
signal cntr: std_logic_vector(3 downto 0):="0000";
signal cclk: std_logic:='0';
begin
ssg_decode(hexcode=>cntr,ssg_out=>sseg);
p0:process(sw0)
begin
if sw0='1' then
anodes<="0000";
else
anodes<="1111";
end if;
end process;
p1:process(mclk)
begin
if mclk='1' and mclk'event then
clkdiv<=clkdiv + 1;
end if;
end process;
cclk<=clkdiv(2); --24
```

```
p2:process(cclk)
begin
if cclk='1' and cclk'event then
if cntr="1001" then
cntr<="0000";
else
cntr<=cntr+1;
end if;
end if;
end process;
end Behavioral;
```

3.9.2 Functions and Procedures

A function executes a sequential algorithm and returns a single value to the calling program. We can think of a function as a generalization of expressions. The syntax rule for a function declaration is

```
[pure | impure] function identifier [(parameter_interface_list)] return
type_mark is
{subprogram declarations}
begin
{sequential statements} end [function] [identifier];
```

By default (i.e., if no keyword is given), functions are declared as pure. A pure function does not have access to a shared variable, because shared variables are declared in the declarative part of the architecture and pure functions do not have access to objects outside of their scope. Only parameters of mode `in` are allowed in function calls, and they are treated as `constant` by default. Functions may be used wherever an expression is necessary within a VHDL statement. Subprograms themselves, however, are executed sequentially like processes. Similar to a process, it is also possible to declare local variables. These variables are initialized with each function call with the leftmost element of the type declaration (`Boolean: false, bit: '0'`). The leftmost value of integers is guaranteed to be at least $-2^{31} - 1$, i.e., zeros must be initialized to 0 at the beginning of the function body. It's recommended to initialize all variables in order to enhance the clarity of the code. A general structure of the function declaration is:

```
function rising_edge(signal clock: std_logic)
return Boolean is
--
-- declarative region: declare variables local to the function
--
begin--
-- body
--
return (value);
end function rising_edge;
```

For example, the following VHDL code describes a simple function that adds two 4-bit vectors and a carry in and returns a 5-bit sum.

```
function add4_func(a, b : std_logic_vector(3 downto 0); carry: std_logic)
return std_logic_vector is
```

```
variable cout : std_logic; variable cin : std_logic;
variable sum : std_logic_vector(4 downto 0); begin
cin := carry; sum := "00000";
loop1 : for i in 0 to 3 loop
cout := (a(i) and b(i)) or (a(i) and cin) or (b(i) and cin); sum(i) :=
a(i) xor b(i) xor cin;
cin := cout; end loop loop1;
sum(4) := cout; return sum; end add4_func;
```

VHDL procedures, in contrast to functions, are used like any other statement in VHDL. Consequently, they do not have a return value, although the keyword `return` may be used to indicate the termination of the subprogram. Depending on their position within the VHDL code, either in an architecture or process, the procedure as a whole is executed concurrently or sequentially, respectively. Procedures facilitate decomposition of VHDL code into modules [12–14]. They can return any number of values using output parameters. The default mode of a parameter is `in`; the keyword `out` or `inout` is necessary to declare output signals/variables. The syntax rule for a procedure declaration is

```
procedure identifier [(parameter_interface_list)] is {subprogram
declarations}
begin
{sequential statements} end [procedure] [identifier];
```

DESIGN EXAMPLE 3.17

Design and develop a digital system to realize an 8-bit binary parallel adder using the VHDL function.

Solution

```
library ieee;
use ieee.std_logic_1164.all;
use ieee.std_logic_arith.all;
use ieee.std_logic_unsigned.all;
use IEEE.numeric_bit.all;
entity func02 is
port(data: in std_logic_vector(7 downto 0):="00000000";
parity: out std_logic:='0';
compl: out std_logic_vector(7 downto 0):="00000000");
end entity func02;
architecture beh of func02 is
function parity_func(signal data: std_logic_vector)
return std_logic is
variable parity: std_logic:='0';
variable compl: std_logic_vector(7 downto 0);
begin
for i in data'range loop -- -for i 0 to 7 loop
parity := parity xor data(i);
end loop;
return (parity);
end function parity_func;
function parity_func1(signal data: std_logic_vector)
return std_logic_vector is
variable compl: std_logic_vector(7 downto 0);
begin
compl:= not data;
return (compl);
```

```
end function parity_func1;
begin
p0:process(data) is
begin
parity <= parity_func(data);
compl<=parity_func1(data);
end process;
end architecture beh;
```

DESIGN EXAMPLE 3.18

Design and develop a digital system to realize a binary parallel adder using the VHDL function.

Solution

```
library ieee;
use ieee.std_logic_1164.all;
use ieee.std_logic_arith.all;
use ieee.std_logic_unsigned.all;
use IEEE.numeric_bit.all;
entity dff is
port(a1,b1: in std_logic_vector (3 downto 0):="0000";
car: in std_logic:='0';
sum: out std_logic_vector(4 downto 0):="00000");
end entity dff;
architecture beh of dff is
function add4_func(a, b : std_logic_vector(3 downto 0);
carry: std_logic)
return std_logic_vector is
variable cout : std_logic;
variable cin : std_logic;
variable sum : std_logic_vector(4 downto 0);
begin
cin := carry;
sum := "00000";
loop1 : for i in 0 to 3 loop
cout := (a(i) and b(i)) or (a(i) and cin) or (b(i) and cin);
sum(i) := a(i) xor b(i) xor cin;
cin := cout;
end loop loop1;
sum(4) := cout;
return sum;
end add4_func;
begin
p0: process (a1,b1) is
begin
sum<=add4_func(a1,b1,car);
end process;
end architecture beh;
```

DESIGN EXAMPLE 3.19

Design and give the procedure parts necessary to realize a 4-bit adder.

Solution

```
procedure add4_proc
(a, b : in std_logic_vector(3 downto 0); carry : in std_logic;
signal sum : out std_logic_vector(3 downto 0); signal cout : out std_
logic) is
variable c : std_logic; begin
```

```
c := carry;
for i in 0 to 3 loop
sum(i) <= a(i) xor b(i) xor c;
c := (a(i) and b(i)) or (a(i) and c) or (b(i) and c); end loop;
cout <= c;
end add4_proc;
```

LABORATORY EXERCISES

1. Design and develop a digital system to realize a 16 × 2 multiplexer and 4 × 32 demultiplexer using the VHDL code in the data flow model.
2. Design and implement a full subtractor using the behavioral and structural models.
3. Design and develop a digital system to realize the simple, priority, octal to binary, and decimal to BCD encoders and their corresponding decoders using the VHDL.
4. Design and develop a digital system to realize a 16-bit signed magnitude comparator using the data flow model.
5. Design and develop a digital system to realize a modulo N counter using generic commands.
6. Design and develop a digital system to realize Johnson and up/down counters. Show the results on a seven-segment display.
7. Design and develop a digital system to realize an object counter. An IR Tx and Rx are aligned in line of sight (LoS). Assume that the digital system receives logic 0 when the object crosses the IR link. The digital system has to increment the object counter in accordance with this input pulse. Illustrate the counter value on the four seven-segment display array.
8. Assume that 8-bit samples are arriving in the digital system built inside the FPGA. Realize a simple delay line canceller (DLC) as `y[n] = x[n] - x[n - 1]`.
9. Design and develop a scoreboard display with score value feed options (inputs) and seven-segment displays.
10. Design and develop a digital system to realize a digital stopwatch.

4

Arithmetic and Logical Programming

4.1 Introduction and Preview

In almost all digital signal-processing techniques, many arithmetic and logical computations are involved, especially in the fields of signal, speech, and image processing. Thus, discussing the algorithms involved in performing several arithmetic and logical operations and designing digital systems to implement them becomes significant, which is dealt with in this chapter. Even though there are several advantages in designing hardware systems using the FPGA, there are also some limitations, even today, in building digital systems of a higher degree of arithmetic and logical computations. This chapter begins with the design and implementation of serial, parallel, and pipelined adders. A complete algorithm requirement for designing and implementing a classical subtractor, 2's complement subtractor, and signed/unsigned subtractors are discussed. The classical multiplication algorithm and FAM algorithm are explained, with necessary numerical examples and VHDL designs. As we know, the design of binary dividers is important as well as complicated, especially from a digital point of view; its classical design approach and a popular technique of left shift and the 2's complement addition algorithm are described, with many numerical examples of different binary lengths and their digital designs. Due to the development of the CORDIC algorithm and high packing density of the gates inside an FPGA, it became possible to design and implement trigonometric computations directly in an FPGA just through the shift and add methods.

The algorithm involved in CORDIC computation and its digital design approaches are described, with several numerical examples. VHDL code design and implementation in the FPGA are also given. The MAC unit is an essential unit, especially in digital filter designs. The MAC unit consists of a multiplier of any kind, as noted above, and an accumulation register, which are clearly explained, with design examples, in this chapter. Another very popular unit, not only in the VLSI area but also in MSI, is the ALU. The design and application of the ALU is detailed, along with the functionality and implementation of a popular ALU chip, IC74181, in the FPGA. The design of ROM and logical function programming/implementation using the ROM unit are detailed. The operational difference between PAL and PLA and logical function implementation using them are also discussed. Finally, the design and implementation of RAM are discussed, along with the design of a binary cell, read/write operations, enable control, and address decoding logic. Design examples are given for all the arithmetic and logical algorithms discussed in this chapter.

4.2 Arithmetic Operations: Adders and Subtractors

In the modern scientific world, many algorithms are emerging for a wide spectrum of applications. These algorithms require almost all the arithmetic and logical operations to complete the computations. The modern version of FPGAs contains provisions to implement almost all the arithmetic and logical operations either directly or through some algorithms. In this section, we discuss some direct and standard algorithms for realizing arithmetic and logical operations. The arithmetic operators are given in Table 4.1. The limitations of all the operators are also given under the heading remarks.

We realize all the above arithmetic operators with design examples below.

DESIGN EXAMPLE 4.1

Develop a digital system to realize all the direct arithmetic operators. Assume that the system gets inputs A and B in 8-bit from the switches. Clearly give all the logic to deal with the input as the signal, integer, and variable in the design. Use the last bit as a sign bit in the subtraction to indicate the result with its sign.

Solution

```
library IEEE;
use IEEE.std_logic_1164.all;
use IEEE.numeric_std.all;
use IEEE.std_logic_unsigned.all;
use IEEE.std_logic_arith.all;
entity arith_ope is
port(A,B: in std_logic_vector(7 downto 0):="00000000";
O1,O2: out std_logic_vector(8 downto 0):="000000000";
O3: out std_logic_vector(15 downto 0):="0000000000000000";
O4,O5,O6,O7:out std_logic_vector(7 downto 0):="00000000");
end arith_ope;
architecture sss of arith_ope is
signal temp1,temp2,temp3,temp4: integer:=0;
begin
p0:process(A,B)
begin
O1<=(('0'& A) + ('0' & B));
if (A>B) then
O2<=('0' & (A-B));
else
O2<=('1' & (not(A+ (not(B) +1)) +1));
end if;
O3<=(A*B);
temp1<=(conv_integer(A)/2);
O4<=conv_std_logic_vector(temp1,8);
temp2<=(9 mod 2);
O5<=conv_std_logic_vector((9 mod 2),8);
temp3<=(11 rem 6);
O6<=conv_std_logic_vector(temp3,8);
temp4<=(3 ** 3);
O7<=conv_std_logic_vector(temp4,8);
end process;
end sss;
```

Upon simulating the above code, we will get all the results. Some of the outputs will appear with one unit time delay due to the pipelined stage introduced in the above code.

TABLE 4.1
Arithmetic Operators

Operator	Description	Remarks
+	Addition	numeric + numeric = numeric unary + numeric = numeric
–	Subtraction	numeric – numeric = numeric unary – numeric = numeric
*	Multiplication	numeric * numeric = numeric
/	Division	numeric/numeric = numeric The denominator should be 2^n
mod	Modulo	integer mod integer = integer
rem	Remainder	integer rem integer = integer
abs	Absolute	numeric abs numeric = numeric
**	Exponentiation	numeric ** integer = numeric

We discuss pipelining and data type conversions subsequently. As we in see in the simulation, the above code does not work for fractional numbers. Chapter 10 deals with computations for different forms of fractional numbers.

4.2.1 Serial Adder

A serial adder is a digital circuit that can add any two arbitrarily large numbers of bits using a full adder circuit. If the computation speed is not of great importance, a cost-effective option is to use a serial adder. The serial adder adds a set of bits that consists of two inputs and one carry at a time in one cycle. Just as humans do, the serial adder operates on one set of bits/digits at a time. For example, when we add two 4-digit radix-10 numbers as 7852 + 1974, we typically start by adding 2 + 4 = 6 (no carry), then 5 + 7 = 12, place the 2 and carry the 1, then 8 + 9 + 1 = 18, place the 8 and carry the 1 and so on. Similarly, for two 4-bit radix-2 numbers as "1011" + "0110", the serial adder starts by adding 1 + 0 = 1, no carry, then 1 + 1 = 10, place the 0 and carry the 1, then 1 + 1 + 0 = 10, place the 0, carry the 1 and so on. Generally, both a human and a serial adder follow the same sequential method to add the digits/bits. A schematic diagram of a serial adder is shown in Figure 4.1, which is designed by using one full adder and one delay element, that is, a D-FF. The full adder adds three inputs (x_i, y_i, and c_i) and produces an output sum (`S`) and carry (`c`). The `c` will be stored within a D-FF and will come out after one clock for the next subsequent addition, as shown in Figure 4.1. Whenever a clock is applied to the register `X`, `Y`, and delay element, the values of the respective `X`, `Y`, and D-FF, that is, x_i, y_i, and c_i, will enter into the full adder, which generates `S` and `c`. This operation has to be continued until all the bits are added. Assume that there are two parallel in serial out shift registers (PISO-SRs) `X` and `Y` of size `[n:0]`. The contents of `X` and `Y` can be given by $(x_n, x_{n-1}, \ldots x_0)$ and $(y_n, y_{n-1}, \ldots y_0)$. Once the data loading into these registers are over, then serial shifting and serial addition will be performed in synchronization with the clock. The serial adder adds $x_0 + y_0 + 0$ in the first (i.e., LSB) addition, then the sum result is S_0 and the carry bit is stored in the delay element; second, it adds $x_1 + y_1 +$ `carry` $= S_1$; third $x_2 + y_2 +$ `carry` $= S_2$ and so on, where `S` represents the elements of the sum `S` $= [S_n\ S_{n-1}, \ldots S_0]$ and the final carry sits with the sum as its MSB. Hence, the actual sum after adding all the bits is `S` $=$ `[carry` $S_n\ S_{n-1}, \ldots S_0]$. In all the stages, the carry is not one of the inputs like `x` and `y`; instead, it is an input

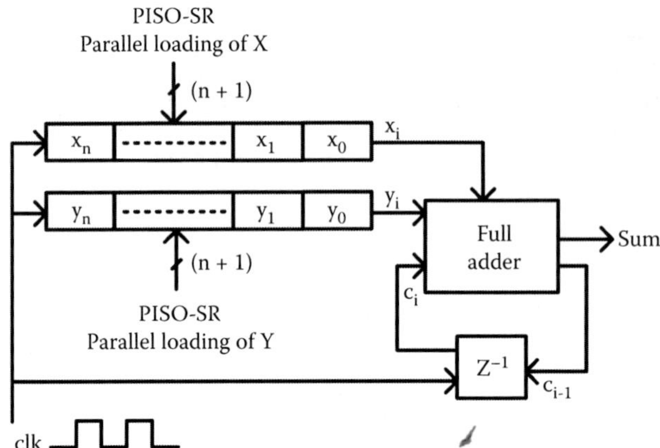

FIGURE 4.1
Schematic diagram of a serial adder.

generated from the previous addition. Further, the value of `carry` does not depend on the current inputs; for example, while adding x_3 and y_3, the carry is of addition x_2 and y_2 and the carry of $x_3 + y_3$ will be added with the addition of x_4 and y_4.

The speed of addition depends on the frequency of the clock input. The full adder can be implemented using the following logical expressions.

$$\begin{aligned} \text{Sum (S)} &= (\text{x xor y xor c}) \\ \text{carry (c)} &= (\text{xc} + \text{yc} + \text{xy}) \end{aligned} \tag{4.1}$$

If the carry is not considered, then this design is said to be a half adder. The carry variable can be either "1" or "0", and a full adder's behavior when the carry is "0" and "1" is not the same; that is, the output of a serial adder will not be the same in both cases when a new input reaches it. For example, at the beginning, if the inputs are "10", the sum is "1" and the carry is "0"; the next input is "11", the sum is "0" and the carry is "1"; and if the third input is again "10", the sum is "0" and the carry is "1", which is dissimilar from the above first case. Therefore, the carry in the serial adder is a decision function and the design has to be operated in two different states: one for `carry` = 0 and another for

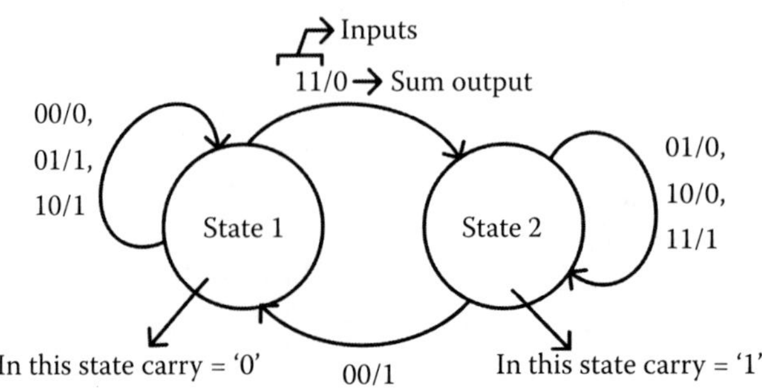

FIGURE 4.2
State transition diagram of a serial adder.

TABLE 4.2

Carry Function Table of a Full Adder

x	y	c	c_{out}	Remarks
0	0	0	0	If x=y then c_{out}=x
0	0	1	0	
0	1	0	0	If x≠y then c_{out}=c
0	1	1	1	
1	0	0	0	
1	0	1	1	
1	1	0	1	If x=y then c_{out}=x
1	1	1	1	

carry = 1, as shown in Figure 4.2. The process in state-1 corresponds to carry = '0', and state-2 is for carry = '1'. Now, if the inputs, that is, x and y, are "00", then the sum is "0", the carry is "0" and the process will be in the same state. If the input is "10", then the sum is "1", the carry is "0" and it will be in the same state; or, if the input is "01", then the sum will be "1", the carry will be "0" and it will be in the same state; or, if the input is "11", the sum will be "0" and it cannot be in the same state because here the carry is "0"; hence, the carry value is now actually "1"; therefore, we need to move to another state named state-2, as shown in Figure 4.2. After reaching state-2, we need to do the same process again and should come back to state-1 whenever we get carry "0". All these transitions are shown in Figure 4.1b.

This state diagram can be carried further to represent it in the state table or excitation table, obtain the logical expressions for any FFs, and design a sequential circuit. This is left to the reader in the laboratory exercises at the end of this chapter. Design of a serial adder using the VHDL code for the FPGA implementation is given below. A full adder can also be implemented using its function table directly using the look up table (LUT). Consider the function table of a full adder as given in Table 4.2. In that table, the method of generating the carry without passing the variables into the logical expression of the carry is also given.

Instead of designing a digital system to generate the carry as per Equation 4.1, Table 4.2 can be used just by a LUT. The throughput rate can be increased and the latency can be decreased using the LUT-based design approach [35]. For example, c_{out} is c whenever both inputs are unequal, and c_{out} is x when both inputs are equal. This type of implementation can be done simply with a multiplexer.

DESIGN EXAMPLE 4.2

Develop a digital system to serially add 4-bit binary numbers. Display the sum and final carry using the LEDs. Initiate the loading and shifting operations using an external input.

Solution

```
library ieee;
use ieee.std_logic_1164.all;
entity seri_adder is
port(start : in std_logic:='0';
m_clk : in std_logic:='0';
a_out : out std_logic_vector(3 downto 0):="0000";
fin_carry: out std_logic:='0';
```

```
x,y: in std_logic_vector(3 downto 0):="0101");
end seri_adder;
architecture ser_adder of seri_adder is
signal shift,load,clk : std_logic:='0';
signal Cin, Cout : std_logic:='0';
signal sum_in : std_logic:='0';
signal state, next_state : integer range -1 to 4:=-1;
signal a,b: std_logic_vector(3 downto 0):="0000";
begin
p0: process(m_clk)
variable cnt: integer:=1;
begin
if rising_edge(m_clk) then
if (cnt=2) then
clk<=not clk;
cnt:=1;
else
cnt:=cnt + 1;
end if;
end if;
end process;
sum_in <= a(0) xor b(0) xor Cin;
Cout <= (Cin and a(0))or(Cin and b(0))or(a(0) and b(0));
a_out <= a;
P1: process(start,state)
begin
case state is
when -1=> fin_carry<='0';
if start='1' then
load<='1';
next_state<=0;
else
load<='0';
next_state<=-1;
end if;
when 0 => load<='0';
shift <= '1';
next_state <= 1;
when 1 => shift <= '1';
next_state <= 2;
when 2 => shift <= '1';
next_state <= 3;
when 3 => shift <= '1';
next_state<=4;
when 4 => fin_carry<=Cout;
if start ='1' then
next_state<=4;
shift <= '0';
else
next_state <= -1;
shift <= '0';
end if;
end case;
end process;
p2:process(clk)
begin
if rising_edge(clk) then
state <= next_state;
if load='1' then
a<=x; b<=y;
else
```

```
if shift = '1' then
a <= sum_in & a(3 downto 1);
b <= b(0) & b(3 downto 1);
Cin <= Cout;
end if;
end if;
end if;
end process;
end ser_adder;
```

4.2.2 Parallel and Pipelined Adders

In the preceding section, we discussed how two binary bits can be added with a carry using a serial adder. We have also seen that a full adder, that is, a 1-bit binary adder, can be constructed from the basic logic gates. However, when we want to add two n-bit numbers together, then n 1-bit full adders need to be cascaded together to produce an n-bit parallel adder. A parallel adder is simply n 1-bit full adders cascaded together, with each full adder representing a single weighted column in a long binary addition, as shown in Figure 4.3. It is called a ripple carry adder because the carry signals produce a ripple effect through the binary adder from right to left, that is, LSB to MSB. For example, suppose we want to add together two 4-bit numbers; the two outputs of the first full adder are the first-place digit in the sum (S_0) of the addition and a carry-out bit that acts as the carry-in to the next binary adder. The second binary adder in the chain also produces a summed output (S_1) and another carry-out bit, and we can keep adding more full adders to the combination to add larger numbers, linking the carry bit output from the first full adder to the next full adder and so forth.

In practical situations, it is required to add two data that contain more than one bit. Two binary numbers of n bits can be added by means of a parallel adder circuit. For example, two 4-bit binary numbers x3x2x1x0 and y3y2y1y0 are to be added with an initial carry input `c0`. This can be done by cascading four full adder circuits. The least significant bits `x0`, `y0` and `c0` are added to the produced sum output `S0` and carry output `c1`, as shown in Figure 4.3. Carry output `c1` is then added to the next significant bits `x1` and `y1`, producing sum output `S1` and carry output `c2`. `c3` is then added to `x3` and `y3` and so on. Thus, it finally produces the 4-bit sum output `S3S2S1S0` and final carry output `C4`, which is actually the MSB of the sum. For example, the results of the addition "1101" + "0111" are "10100". Such four-bit binary adder MSI single-package ICs are commercially available.

One main disadvantage of cascading 1-bit binary adders together to add large binary numbers is that if inputs `A` and `B` change, the sum at the output will not be valid until any

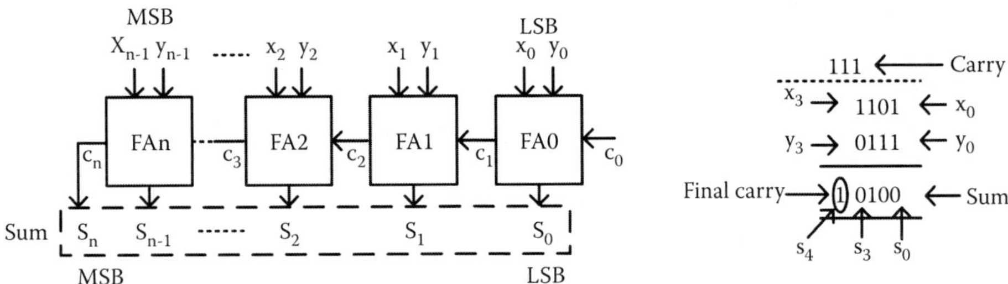

FIGURE 4.3
General n-bit parallel adder with example addition.

carry input has rippled through every full adder in the chain because the MSB of the sum has to wait for any changes from the carry input of the LSB. Consequently, there will be a finite delay before the output of the adder responds to any change in its inputs, resulting in an accumulated delay. Further, the delay associated with the traveling of the carry bit is called carry propagation delay and is found to worsen with an increase in the length of the binary numbers required to be added. For example, if each full adder is considered to have a delay of 10 ns, then the total delay required to produce the output of a 4-bit parallel adder would be $4 \times 10 = 40$ ns. When the size of the bits being added is not too large, for example 4, 8, 16, and 32 bits, the summing speed of the adder may be negligible, but for larger cases, it provides a considerable delay. However, when the size of the bits is larger, for example 64, 128, and 256 bits and so on, summation requires a very high clock speed, and this delay may become prohibitively large, with the addition processes not being completed correctly within one clock cycle [25,36–39].

To reduce the propagation delay, another type of binary adder circuit called a carry look-ahead binary adder can be used. In this adder, the speed of the parallel addition is significantly improved using carry look-ahead logic. The advantage of carry look-ahead adders is that the length of time a carry look-ahead adder needs in order to produce the correct `SUM` is independent of the number of data bits used in the operation, unlike the cycle time a parallel ripple adder needs to complete the `SUM`, which is a function of the total number of bits in the addend [25,39]. From the discussion presented, we can say that in the case of an n-bit parallel adder, each adder has to wait for the carry term to be generated from its preceding adder in order to finish its task of adding. This can be visualized as if the carry term propagates along the chain in the fashion of a ripple.

DESIGN EXAMPLE 4.3

Develop a digital system to add 31 bits of two data "a" and "b" in parallel using a `for` loop. Display the results using the LEDs.

Solution

```
library IEEE;
use IEEE.STD_LOGIC_1164.ALL;
use IEEE.STD_LOGIC_ARITH.ALL;
use IEEE.STD_LOGIC_UNSIGNED.ALL;
entity para_add is
Port (a,b : in std_logic_vector(31 downto 0):="0000000000000000000000000
0000000";
s : out std_logic_vector(31 downto 0):="00000000000000000000000000000000";
c : out std_logic:='0';
cin : in std_logic:='0');
end para_add;
architecture sss of para_add is
begin
process(a,b,cin)
variable u:std_logic:='0';
begin
u:=cin;
for i in 0 to 31 loop
s(i)<=a(i) xor b(i) xor u;
u:=(a(i) and b(i))or(b(i) and u) or(u and a(i));
end loop;
c<=u;
end process;
end sss;
```

DESIGN EXAMPLE 4.4

Develop a digital system to add 8 bits of two data "a" and "b" in parallel using the structural model. Display the results using the LEDs.

Solution

As given in this design example, a structural model has to be followed. Therefore, a full adder has to be designed in the component part, which has to be accessed by the main structure to pass the variables and read the values. The component and main code parts are given below.

Component Code

```
library ieee;
use ieee.std_logic_1164.all;
entity fa is
port(a,b,cin:in std_logic;sum,cout:out std_logic);
end fa;
architecture behv of fa is
begin
sum<=a xor b xor cin;
cout<=(a and b) or (b and cin) or (a and cin);
end behv;
```

Main Code

```
library ieee;
use ieee.std_logic_1164.all;
entity para_Add is
port(a,b:in std_logic_vector(7 downto 0):="00000000";
s:out std_logic_vector(7 downto 0):="00000000";
c:out std_logic:='0');
end para_Add;
architecture sss of para_Add is
signal temp:std_logic:='0';
signal car:std_logic_vector(6 downto 0);
begin
fa1:entity work.fa port map(a(0),b(0),temp,s(0),car(0));
fa2:entity work.fa port map(a(1),b(1),car(0),s(1),car(1));
fa3:entity work.fa port map(a(2),b(2),car(1),s(2),car(2));
fa4:entity work.fa port map(a(3),b(3),car(2),s(3),car(3));
fa5:entity work.fa port map(a(4),b(4),car(3),s(4),car(4));
fa6:entity work.fa port map(a(5),b(5),car(4),s(5),car(5));
fa7:entity work.fa port map(a(6),b(6),car(5),s(6),car(6));
fa8:entity work.fa port map(a(7),b(7),car(6),s(7),c);
end sss;
```

Another important design is the pipelined parallel adder design. Before proceeding into the actual design of the pipelined parallel adder, we should understand the significance of introducing pipelining in digital system designs. Pipelining is a technique that implements a form of parallelism called instruction-level parallelism within a single processor. It therefore allows faster throughput, that is, the number of instructions that can be executed in a unit of time than would otherwise be possible at a given clock rate [25,39]. Pipelining is an implementation technique where multiple instructions are overlapped in execution. The computer pipeline, that is, the entire process, is divided into multiple stages. Each stage completes a part of an instruction in parallel. Pipelining is one way of improving the overall processing performance of a processor. This architectural approach allows the simultaneous execution of several instructions. Pipelining is transparent to the

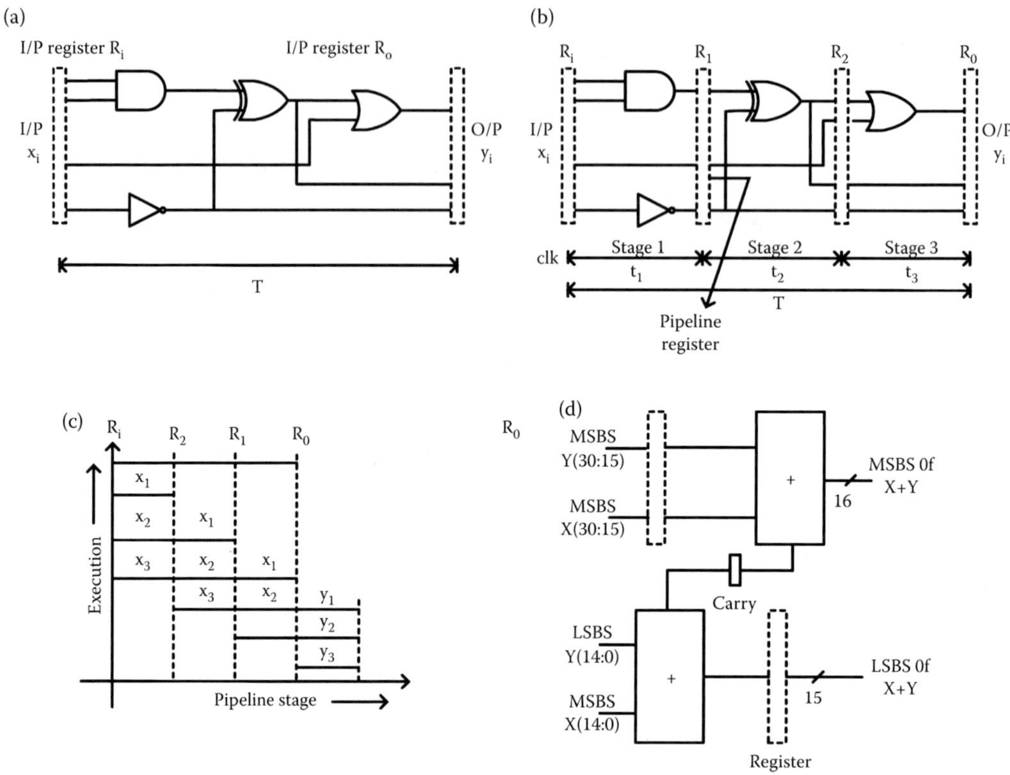

FIGURE 4.4
Digital circuit: standard (a) and pipelined (b), input processing in pipelined digital circuit (c), and schematic diagram of a 31-bit pipelined parallel adder (d).

programmer; it exploits parallelism at the instruction-level by overlapping the execution process of instructions. It is analogous to an assembly line where workers perform a specific task and pass the partially completed product to the next worker. The pipeline design technique decomposes a sequential process into several subprocesses, called stages or segments. A stage performs a particular function and produces an intermediate result. It consists of an input latch, also called a register or buffer, followed by a processing circuit, which can be a combinational or sequential circuit. The processing circuit of a given stage is connected to the input latch of the next stage, as shown in Figure 4.4. A clock signal is connected to each input latch. At each clock pulse, every stage transfers its intermediate result to the input latch of the next stage. In this way, the final result is produced after the input data have passed through the entire pipeline, completing one stage per clock pulse. The period of the clock pulse should be large enough to provide sufficient time for a signal to traverse through the slowest stage, which is called the bottleneck or latency, that is, the stage needing the longest amount of time to complete. In addition, there should be enough time for a latch to store its input signals.

Now, we proceed to design a pipelined digital circuit. Consider Figure 4.4a, which is a simple digital circuit with one input register (`Ri`) and one output register (`Ro`). Once the input is applied, then the circuit should wait for some time to get the output. The applied input will go through different paths in the circuit according to the logic, as shown in Figure 4.4a, and finally, the correct output will be available at the output register. The time delay depends on the logic of the circuit. Typically, it is equal to the processing plus the propagation delay. For example, as in Figure 4.4a, the third output will be available quickly because it is just the output of a not gate. At this instant of time, another two outputs will not be available because the logical computation is not yet over. Therefore, the available output is incorrect; hence, we need to wait for some time to latch out the final output and apply the new set of input. The maximum time delay, that is, latency, say, `T` sec, is decided by the longest logical path. This operation clearly means that the circuit can accept new input once every `T` sec.

Now, we convert this standard circuit into a pipelined circuit just by introducing two registers at the intermediate-level of the circuit, as shown in Figure 4.4b. Two additional registers, R_1 and R_2, are introduced and connected to the clock input as well. Therefore, now, the waiting time is reduced to t_1, t_2, and t_3 sec. It is possible to say now that `T` sec is divided by 3; however, all the delays may not be the same, since it depends on the amount of logic involved in each stage. In general, the registers will be introduced in the longer circuit/logic such as to equally divide the latency [25,37–39]. Assume that $t_1 = t_2 = t_3$. Now, the circuit has no need to wait for `T` sec; instead, it has to wait only for t_1 sec, so latency is decreased from `T` to t_1 and throughput is hence increased. Further, now the circuit can work for multiple inputs, which means: suppose input x_1 is applied into the circuit. After t_1 sec, the output of stage 1 for x_1 will be available at R_1. Now, we can apply the input x_2; hence, stage 1 will work for x_2 while stage 2 works for x_1. In the same way, stage 1, stage 2 and stage 3 will work for x_3, x_2 and x_1, respectively, at one instant in time. The execution of three inputs through this pipelined digital circuit is shown in Figure 4.4c. We understand how latency reduction happens by introducing the pipeline registers with a simple calculation. Let us assume $T = 3$ sec; thus, to process x_1, x_2 and x_3, we require 9 sec. After introducing the pipeline registers, the delay is just 1 sec; now, as shown in Figure 4.4c, when $t = 1$ sec: x_1; when $t = 2$ sec: x_2, x_1; when $t = 3$ sec: x_3, x_2, x_1; when $t = 4$ sec: -, x_3, x_2 and y_1; when $t = 5$ sec: -, -, x_3 and y_2 and when $t = 6$ sec: -,-,- and y_3; hence, in total, we require only 6 sec. Just by introducing two stages of pipeline registers in the circuit, we reduced the latency by 33.33%. A top-level schematic of a 31-bit parallel pipelined adder is shown in Figure 4.4d. The digital implementation of this design is given in the design example below. The introduction of the register actually breaks the larger computation into small computations; hence, obviously, one process becomes the dependent of its previous process. The data flow from one stage to another is controlled by the pipeline register and clock pulse.

DESIGN EXAMPLE 4.5

Develop a digital system to realize a 31-bit pipelined parallel adder. Split the parallel adder into two stages and display the result using the LEDs.

Solution

```
library ieee;
use ieee.std_logic_1164.all;
use ieee.std_logic_arith.all;
use ieee.std_logic_unsigned.all;
entity add1p is
generic (width : integer := 31;
width1 : integer := 15;
width2 : integer := 16);
port (x,y : in std_logic_vector(width-1 downto 0):="11010101010101010101
01010101111";
sum : out std_logic_vector(width downto 0);
```

```
lsbs_carry : out std_logic;
clk : in std_logic:='0');
end add1p;
architecture fpga of add1p is
signal l1, l2, s1 : std_logic_vector(width1-1 downto 0);
signal r1 : std_logic_vector(width1 downto 0);
signal r2 : std_logic_vector(width2 downto 0);
signal l3, l4: std_logic_vector(width2-1 downto 0);
signal s2: std_logic_vector(width2 downto 0);
begin
p0: process (clk,x,y)
begin
if rising_edge(clk) then
l1 <= x(width1-1 downto 0);
l2 <= y(width1-1 downto 0);
l3 <= x(width-1 downto width1);
l4 <= y(width-1 downto width1);
r1 <= ('0' & l1) + ('0' & l2);
r2 <= ('0' & l3) + ('0' & l4);
s1 <= r1(width1-1 downto 0);
s2 <= r1(width1) + r2;
end if;
end process;
lsbs_carry <= r1(width1);
sum <= s2 & s1;
end fpga;
```

4.2.3 Subtractors

The binary subtractor is another type of combinational arithmetic circuit, and it is the opposite of the binary adder. As the name implies, a binary subtractor subtracts two binary numbers like minuend (X) – subtrahend (Y) to find the difference and borrow between these two numbers. Unlike the binary adder, which produces a sum and carry, the binary subtractor produces difference and borrow bits. Then, obviously, the operation of subtraction is the opposite of addition. The truth table of a half subtractor is given in Table 4.3, where, 0 – 0 means no difference and no need for any borrow; 0 – 1 means a borrow is required, and after getting the borrow, the 0 becomes "10"; hence, "10" − 1 = 1, which means the difference is 1. Thus, in this subtraction, the difference as well as the borrow is 1, and in 1 – 1, the difference is 0 and there is no need of any borrow; thus, it is 0, as given in Table 4.3. If the borrow bit is ignored, then the difference of the binary subtraction resembles the modulo 2 adder; that is, the difference can also be generated using a modulo 2 adder, like the sum in a half adder. From Table 4.3, we can draw the K-map and get logical expressions for the difference and borrow as given in Equation 4.2. These logical expressions can be implemented with basic logic gates, as shown in Figure 4.5a. Just by combining the modulo 2 adder with NOT and AND gates, the borrow can be generated.

$$\begin{aligned} \text{Difference} &= (x \oplus y) \\ \text{Borrow} &= (x'y) \end{aligned} \tag{4.2}$$

The above half subtractor can be altered to design a full subtractor to also take the borrow bit into account. The truth table of a full subtractor is given in Table 4.3, where x and y are the inputs and B_{in} is the borrow input. Thus, the subtraction in the full subtractor is x-y-B_{in}; therefore, 0 – 0 – 0; no difference and no need of any borrow, 0 – 0 – 1; as said above, now the difference is 1 and the borrow is also 1, 0 – 1 – 0; the difference is 1 and the borrow

TABLE 4.3

Truth Table of Half and Full Subtractor

Half Subtractor				Full Subtractor				
x	y	D	B	x	y	B_{in}	D	B
0	0	0	0	0	0	0	0	0
0	1	1	1	0	0	1	1	1
1	0	1	0	0	1	0	1	1
1	1	0	0	0	1	1	0	1
				1	0	0	1	0
				1	0	1	0	0
				1	1	0	0	0
				1	1	1	1	1

is 1, 0 – 1 – 1; now, the borrow is required to subtract 0 – 1; then it becomes "10" – 1, whose difference is 1 and borrow is 1; then this difference is –1, i.e., B_{in}, then 1 – 0; the difference is 1; therefore, the difference is 1 and the borrow is 1. In the same way, the subtraction will be continued as given in Table 4.3. The logical expressions for a full adder can also be obtained just by applying the K-map on the truth table given in Table 4.3. The logical expressions for a full adder are given in Equation 4.3. Just like the binary adder circuit, the full subtractor can also be thought of as two half subtractors connected together with the first half subtractor passing its difference to the second half subtractor, as shown in Figure 4.5b. A full subtractor can also be designed using general logical expressions, as shown in Figure 4.5c.

The Boolean expression for the difference and borrow of a full subtractor is given by

$$\text{Difference} = x'y'B_{in} + x'yB'_{in} + xy'B_{in}' + xyB_{in}$$
$$\text{Barrow} = x'B_{in} + x'y + yB_{in}$$

These logical expressions can also be simplified using Demorgan's theorem [38–40]

$$\begin{aligned}\text{Difference} &= ((x \oplus y) \oplus B_{in})\\ \text{Borrow} &= (x'y + (x \oplus y)'B_{in})\end{aligned} \quad (4.3)$$

This expression gives the design of a full subtractor using two half subtractors.

Thus, the combinational circuit of a full subtractor performs the operation of subtraction on three binary bits, producing outputs for the difference and borrow. Like a serial adder, a serial subtractor can also be designed, which is given to the readers as a laboratory exercise.

Here, we discuss the parallel binary adder. A parallel binary subtractor can be implemented by cascading several full adders. Unlike full adders, full subtractors cannot be used due to borrowing issues [38–40]. Suppose we wish to design a 4-bit parallel binary subtractor; then, the inputs can be given by $x_3x_2x_1x_0$ and $y_3y_2y_1y_0$, which will give a 4-bit difference output with a borrow output. We need to follow a 2's complement technique to design a parallel subtractor. The technique is simple, as the minuend and 2's complement of the subtrahend have to be added together with some proceeding result manipulations. The circuit for a 4-bit parallel subtractor is shown in Figure 4.6. As shown in that figure, the n-bit parallel subtractor design

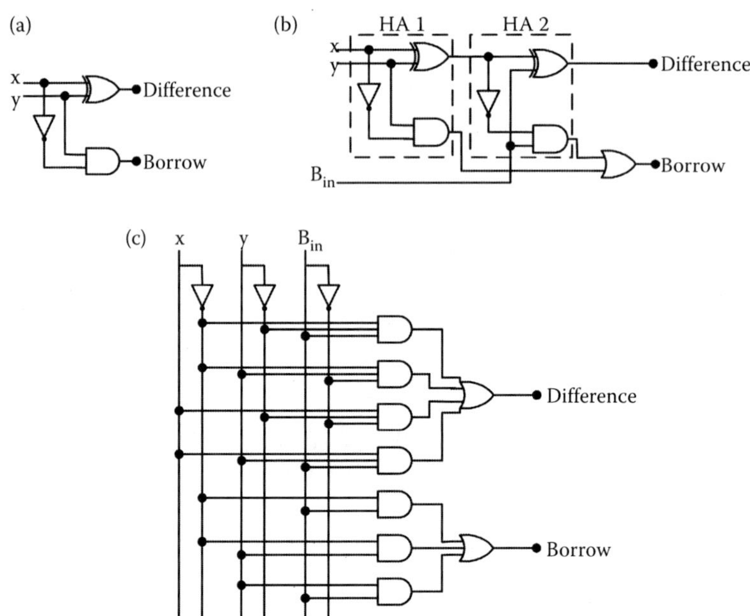

FIGURE 4.5
Half subtractor (a), full subtractor using two half subtractors (b), and a typical full subtractor (c).

requires n-bit parallel adders. Using n adders and n `NOT` gates, the process of n-bit parallel subtraction can be dealt with as the n-bit parallel adder because we use the two's complement technique, that is, 1's complement on all the bits of the subtrahend and setting the LSB carry input to logic 1. For example, consider subtraction as `X - Y`. Here, the `X` is directly applied to the adders, as shown in Figure 4.6. Instead of passing the `Y` directly, we send it through the `NOT` gate; hence, the 1's complement of `Y` is performed. The carry of the first full adder is set to logic 1; hence, the 2's complement of `Y` is also performed. Now, the parallel adder adds `X` and `Y` as `X + (Y' + 1)`, which means the 2's complement of `Y` is added to `X`.

We can understand this computation with one simple example. Consider `X` = $(13)_{10}$ and `Y` = $(7)_{10}$. Then, the first step is the 1's complement of `Y`: $(7)_{10}$ = "0111" will be "1000", then add 1, which gives "1001". This is the 2's complement of `Y`, which is then added to X as

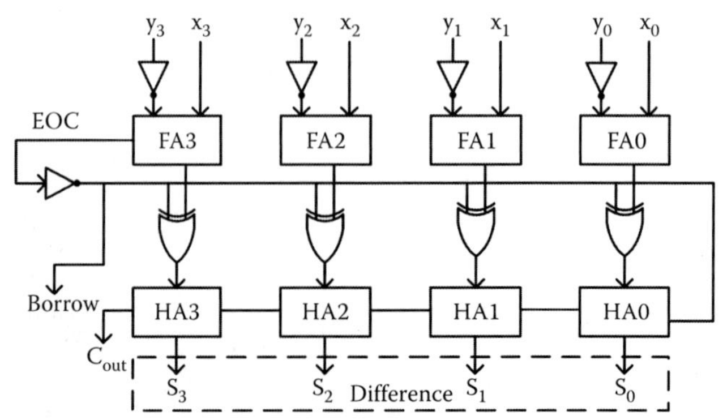

FIGURE 4.6
Design of a 4-bit parallel subtractor.

"1101" + "1001" = "10110". Now the sum (actually the difference) is "0110", which is equal to $(6)_{10,}$ the correct answer. The additional carry bit is "1", which is called the end of carry (EoC) and can be omitted. The important point to be noted here is that if X $\geq$ Y, then the result will be positive and correct and the EoC will be "1"; otherwise, the result will be incorrect and the EoC will be "0". Therefore, whenever the EoC is "0", that is, X < Y, we need to take the 2's complement on the sum to get the correct result, as shown in Figure 4.6; hence, the sign bit will be negative. For example, let us assume X = $(8)_{10}$ and Y = $(14)_{10}$. Then, the 2's complement of Y is "0010". Add to the X, then "1000" + "0010" = "01010". Here, the EoC is "0" and the result is "1010", which is incorrect; hence, once again, take the 2's complement only on the sum, that is, "1010". Then the result is "0110", that is, $(-6)_{10}$. As shown in Figure 4.6, if EoC = '1', then the result of the first parallel adder will be passed to the output as it is, and if EoC = '0', then the modulo 2 adder array provides the 1's complement and the half adder array provides the 2's complement as well as parallel addition and finally gives the correct result. Further, the sum outputs of each and every adder actually correspond to the difference bits (the expected result), while the final carry will be nothing but the resultant borrow [39,40].

Now, we discuss a design which functions as a parallel adder as well as a parallel subtractor. The operations of both addition and subtraction can be performed by one common binary adder. Such a binary circuit can be designed by adding the modulo 2 adders with every full adder circuit, as shown in Figure 4.7. The mode selection input (M) is connected with the carry input of the first full adder as well as one input of all the modulo 2 adders. When M = 1, the circuit is a subtractor, and when M = 0, the circuit becomes an adder. One input of the modulo 2 adder is connected to Y; therefore, when M = 0, the modulo 2 adder's output is Y; otherwise, it is Y′. Then full adders add either Y (when M = 0) or the complement of Y, i.e., Y′ (when M = 1), with X and the LSB carry; which is either 1 (when M = 1) or 0 (when M = 0). When M = 1, the modulo 2 adder produces the 1's complement of Y and, because of adding 1 at LSB carry, the 2's complement of Y is performed. Therefore, the parallel adder does the subtraction operation.

We can see the examples to understand the working principle of the parallel adder/subtractor shown in Figure 4.7. Let us assume two examples: X = 10, Y = 4 and X = 9, Y = 14. The data computation, that is, addition/subtraction flow through the circuit shown in Figure 4.7 is given in Table 4.4.

This is the way this design works for n bits of inputs X and Y. Thus, a combinational circuit can function as either a parallel adder or subtractor by selecting the appropriate mode of operation.

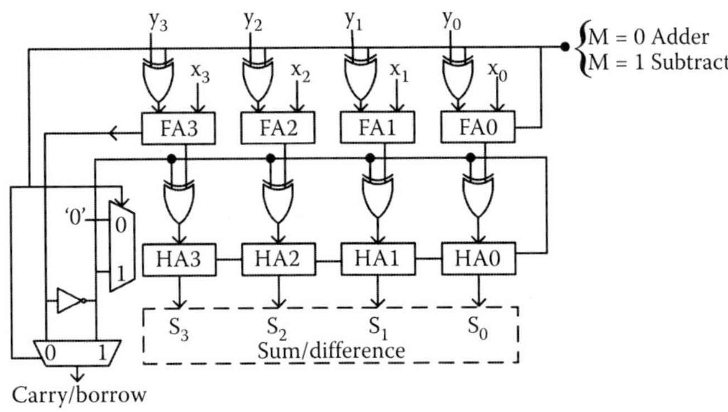

FIGURE 4.7
Design of a 4-bit parallel adder/subtractor.

TABLE 4.4

Function of a Parallel Adder/Subtractor Circuit

X	Y	When M = 0	When M = 1	Remarks
$(10)_{10}$ = "1010"	$(4)_{10}$ = "0100"	–	–	Binary representation of X and Y
–	–	Y = "0100"	Y′ = "1011"	Input to the full adder array
–	–	LSB carry = 0	LSB carry = 1	LSB carry is equal to M
–	–	X + Y + 0 = "1010" + "0100" = "1110" = $(14)_{10}$	X + (Y′ + 1) = "1010" + ("1011" + 1) = "1010" + "1100" = "10110"; EoC=1 and the result is $(6)_{10}$	Addition or subtraction based on the value of M. EoC = 1, no need of 2's complement.
$(9)_{10}$ = "1001"	$(14)_{10}$ = "1110"	–	–	Binary representation of X and Y
–	–	Y = "1110"	Y′ = "0001"	Input to the full adder array
–	–	LSB carry = 0	LSB carry = 1	LSB carry is equal to M
–	–	X + Y + 0 = "1001" + "1110" = "10111" = $(23)_{10}$	X + (Y′ + 1) = "1001" + ("0001" + 1) = "1001" + "0010" = "01011"; EoC = 0 and result is "1011"	Addition or subtraction based on the value of M. EoC = 0, need 2's complement.
–	–	–	Result = "0100" + 1 = "0101"=$(-5)_{10}$	Final result.

DESIGN EXAMPLE 4.6

Design a digital system to realize the logical function of half subtractor and full subtractor. Display the results in parallel, that is, the difference and borrow of a half and full subtractor separately.

Solution

```
library IEEE;
use IEEE.std_logic_1164.all;
use IEEE.numeric_std.all;
use IEEE.std_logic_unsigned.all;
use IEEE.std_logic_arith.all;
entity full_subtractor is
port(a,b,Bin : in STD_LOGIC:='0';
diff_hal,borrow_hal,diff_full, borrow_full: out STD_LOGIC:='0');
end full_subtractor;
architecture full_subtractor_arc of full_subtractor is
begin
diff_hal<=(a xor b);
borrow_hal<=((not a)and b);
diff_full <= (a xor b xor Bin);
borrow_full <= ((not a) and Bin) or ((not a) and b) or (Bin and b);
end full_subtractor_arc;
```

DESIGN EXAMPLE 4.7

Design a digital system to realize a 4-bit parallel subtractor using the 2's complement technique. Display the final result, difference, and borrow with sign bit using the LEDs.

Solution

```
library ieee;
use ieee.std_logic_1164.all;
entity fullsub is
port(x,y:in std_logic_vector(3 downto 0);
s:out std_logic_vector(3 downto 0);
m:in std_logic;
bout:out std_logic);
end entity;
architecture mark0 of fullsub is

signal c,c1:std_logic_vector(3 downto 0); -- intermediate carry
signal y1:std_logic_vector(3 downto 0); -- complimented number
signal cn:std_logic;
signal d:std_logic_vector(3 downto 0); -- intermediate subtraction
signal d1:std_logic_vector(3 downto 0); -- sencondary answer
begin
-- 1st stage --
y1 <= not y;
d(0) <= (x(0) xor y1(0)) xor '1';
c(0) <= x(0) or y1(0);
d(1) <= (x(1) xor y1(1)) xor c(0);
c(1) <= (x(1) or y1(1)) or (x(1) or c(0)) or (c(0) or y1(1));
d(2) <= (x(2) xor y1(2)) xor c(1);
c(2) <= (x(2) or y1(2)) or (x(2) or c(1)) or (c(1) or y1(2));
d(3) <= (x(3) xor y1(3)) xor c(3);
c(3) <= (x(3) or y1(3)) or (x(3) or c(2)) or (c(2) or y1(3));
----------------- 2nd stage --
cn <= not c(3);
d1 <= (cn&cn&cn&cn) xor d;
----------------- 3rd stage --
s(0) <= cn xor d1(0);
c1(0) <= cn and d1(0);
s(1) <= c1(0) xor d1(1);
c1(1) <= c1(0) and d1(1);
s(2) <= c1(1) xor d1(2);
c1(2) <= c1(1) and d1(2);
s(3) <= c1(2) xor d1(3);
c1(3) <= c1(2) and d1(3);
bout <= cn;
end architecture;
```

DESIGN EXAMPLE 4.8

Design a digital system to realize a 4-bit parallel adder/subtractor in a circuit. Use the 2's complement technique for realizing the subtractor. Select the operating mode of the circuit, that is, either adder or subtractor, by the user mode selection input. Display the final result, sum and carry, or difference and borrow with a sign bit using the LEDs.

Solution

```
library ieee;
use ieee.std_logic_1164.all;
entity fullsub is
port(x,y:in std_logic_vector(3 downto 0);
```

```
s:out std_logic_vector(3 downto 0);
m:in std_logic;
bout:out std_logic);
end entity;
architecture mark1 of fullsub is
signal c,c1:std_logic_vector(3 downto 0); -- intermediate carry
signal y1:std_logic_vector(3 downto 0); -- complimented number
signal cn,cd:std_logic;
signal d:std_logic_vector(3 downto 0); -- intermediate subtraction
signal d1:std_logic_vector(3 downto 0); -- secondary answer
begin
-- 1st stage --
y1 <= (m&m&m&m) xor y;
d(0) <= (x(0) xor y1(0)) xor '1';
c(0) <= x(0) or y1(0);
d(1) <= (x(1) xor y1(1)) xor c(0);
c(1) <= (x(1) or y1(1)) or (x(1) or c(0)) or (c(0) or y1(1));
d(2) <= (x(2) xor y1(2)) xor c(1);
c(2) <= (x(2) or y1(2)) or (x(2) or c(1)) or (c(1) or y1(2));
d(3) <= (x(3) xor y1(3)) xor c(3);
c(3) <= (x(3) or y1(3)) or (x(3) or c(2)) or (c(2) or y1(3));
---------------- 2nd stage --
cd <= not c(3);
cn <= cd when (m='1') else '0';
d1 <= (cn&cn&cn&cn) xor d;
----------------- 3rd stage --
s(0) <= cn xor d1(0);
c1(0) <= cn and d1(0);
s(1) <= c1(0) xor d1(1);
c1(1) <= c1(0) and d1(1);
s(2) <= c1(1) xor d1(2);
c1(2) <= c1(1) and d1(2);
s(3) <= c1(2) xor d1(3);
c1(3) <= c1(2) and d1(3);
bout <= cd when(m='0') else cn;
end architecture;
```

4.3 Arithmetic Operations: Multipliers

A binary multiplier is a combinational logic circuit used in digital systems to perform the multiplication of two binary signed or unsigned numbers. These are most commonly used in various applications, especially in the field of DSP, to perform signal/image processing algorithms and computations. Commercial applications like computers, mobiles, high-speed calculators, general-purpose processors, and controllers require binary multipliers. Compared with addition and subtraction, multiplication is a complex process [25,39]. In the multiplication process, the number to be multiplied by the other number is called the multiplicand (`Md`) and the number multiplied is called the multiplier (`Mr`). Similar to the multiplication of decimal numbers, binary multiplication follows the same process for producing a product result of the two binary numbers. Binary multiplication is much easier, as it contains only 0s and 1s. The multiplication of two binary numbers can be performed by using various methods, and we discuss many algorithmic methods in later chapters. Here, we concentrate only on two basic methods, namely the partial product method and the add and shift method. Let us start with the unsigned partial product binary multiplication

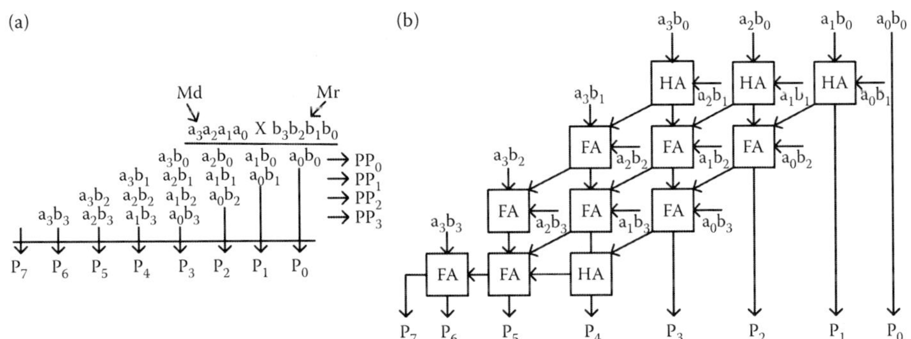

FIGURE 4.8
A 4-bit unsigned binary multiplication (a) and its implementation using half and full adders (b).

process: consider two 4-bit binary numbers given by "$a_3a_2a_1a_0$" and "$b_3b_2b_1b_0$". The partial product multiplication is shown in Figure 4.8a; there, bitwise multiplication is performed.

As shown in Figure 4.8a, partial products (PPs), that is, PP_0 through PP_3, are generated for each bit in the multiplier (`Mr`). Then, all these partial products are added to produce the final product value, that is, P_0 through P_7. In partial product multiplication, when the multiplier bit is zero, the partial product is zero, and when the multiplier bit is 1, the resulting partial product is the multiplicand. Similar to decimal numbers, each successive partial product is shifted one position left relative to the preceding partial product before summing all partial products. The combinational circuit implemented to perform this method of multiplication is called an array multiplier or combinational multiplier, which is shown in Figure 4.8a and b. The first partial product "a_0b_0", that is, the LSB of the multiplication result (P_0) can be obtained by an `AND` gate. Since the second partial product is shifted to the left position, the first partial second term "a_1b_0"and second partial first term "a_0b_1" are added by a half adder (HA) and produce the sum output, which is P_1, along with the carry out. This carry out is passed to the next full adder (FA) as an input, as shown in Figure 4.8b. Likewise, the addition operation, with the appropriate inputs and carry, is performed, and finally, it produces the multiplication result, that is, P_0 through P_7, of two binary numbers by using the simple HA and FA circuit configurations.

Another method to multiply two binary numbers is the add and shift method. This method follows the manual multiplication approach and can be implemented by using an add and shift control logic, a few registers (`Mr`, `Md`, `Acc`, and `C`) and one n-bit parallel adder, as shown in Figure 4.9, where the n-bit multiplier (`Mr`) is loaded into the `Mr` register, the 4 bit multiplicand (`Md`) is loaded into the `Md` register and the registers `C` and `Acc` are initially cleared to zero. It actually stores the sum result upon applying the clock pulse, keeping the final carry in the `C` register, as shown in Figure 4.9. The multiplication process starts with checking whether the LSB of `Mr` is 0 or 1. If `Mr[0]` is 1, then, the value of the multiplicand (`Md`) is added with the value of the `Acc` register, then the combined `C` and `Acc` values are shifted to the right by one bit. If the bit `Mr[1]` is 0, then the combined `C` and `Acc` registers are shifted to the right by one bit without performing any addition. This process is repeated n times for n-bit numbers. This method of binary multiplication is also called a parallel multiplier. The control commands for loading the `Md` and `Mr` values, clearing the `C` and `Acc` registers, controlling the adding and shifting operations or only shifting are given by the `add` and `shift` control logic to the processor that executes the operations.

Now, we will see an example to understand multiplication using the add and shift method. Consider Table 4.5, which gives all the multiplication steps of the add and shift

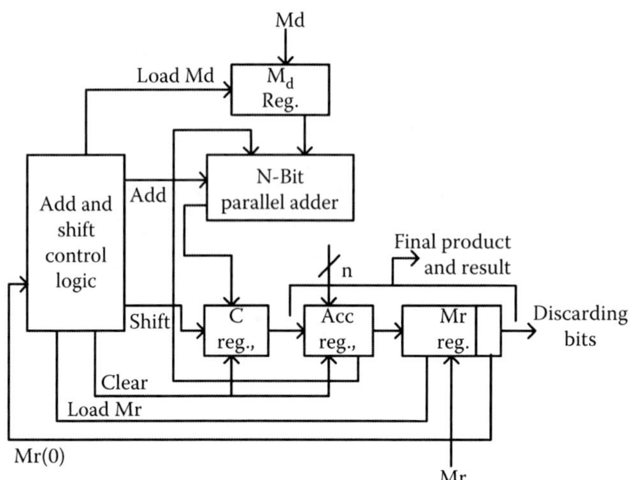

FIGURE 4.9
Design of an add and shift method-based multiplier.

method for the Md and Mr of $(13)_{10}$ = "1101" and $(11)_{10}$ = "1011", respectively, which are loaded into the Md and Mr registers, respectively. Initially, the C and Acc registers, which will actually store the sum and carry results of the addition, are cleared. Clearing the C and Acc registers and loading the values of Md and Mr into the design are taken care of by the control logic block, as shown in Figure 4.9. Now, Mr [0] is 1; hence, the value of Md has to be added to the value of Acc. Then, the addition result will be "0000" + "1101" = "1101", as given in Table 4.5, and it is stored in the Acc register. The values of the C, Acc and Mr registers are serially shifted to right by one bit; thus, the new values after this shifting are "0", "0110" and "1101", respectively, and the last 1 is discarded.

After shifting, now Mr [0] is 1; hence, the above step is repeated. Then, "0110" + "1101" = "1 0011", as shown in Table 4.5; then shift right. After shifting, the values will be "0", "1001"

TABLE 4.5
Add and Shift-Based Multiplication: Md = $(13)_{10}$ and Mr = $(11)_{10}$

Clock (t)	C	Acc	Mr	Discarding Values	Remarks
t_0	0	0000	1011	–	Initialization
t_1	0	**1101**	1011	–	Mr [0] = 1; hence add Md to Acc
	0	1101			
t_2	0	0110	1101	1	Shift right
t_3	0	**1101**	1101	1	Now Mr [0] = 1; hence add Md to Acc
	1	0011			
t_4	0	1001	1110	11	Shift right
t_5	0	0100	1111	011	Now Mr [0] = 0; hence, shift right
t_6	0	**1101**	1111	011	Now Mr [0] = 1; hence add Md to Acc
	1	0001			
t_7	0	1000	1111	1011	Shift right
Completed, the result is "10001111" = $(143)_{10}$					

and "1110". Now, `Mr[0]` is 0; hence, addition is not required and only the right shift is performed, as given in Table 4.5. Then, the results will be "0", "0100" and "1111". This process has to be repeated four times to perform 4-bit multiplication. The results of all the steps are given in Table 4.5. The final multiplication result will be available in the `Acc` and `Mr` registers as "10001111" = $(143)_{10}$. A 4 × 4 unsigned binary multiplier takes two 4-bit inputs and produces an output of 8 bits. Similarly, an 8 × 8 multiplier accepts two 8-bit inputs and generates an output of 16 bits. Thus, n-bit by n-bit multiplication requires a 2n-bit register to store the product result bits. Many advanced and algorithmic signed and unsigned multipliers are discussed in the following chapters.

DESIGN EXAMPLE 4.9

Design and develop an 8-bit add and shift method-based unsigned binary multiplier. Display the results using the LEDs.

Solution

```
library IEEE;
use IEEE.STD_LOGIC_1164.ALL;
use IEEE.STD_LOGIC_ARITH.ALL;
use IEEE.STD_LOGIC_UNSIGNED.ALL;
entity mult8x8 is
Port (clk,st : in STD_LOGIC:='0';
Mplier, Mcand:in STD_LOGIC_VECTOR(7 downto 0):="00000000";
Done:out std_logic:='0';
product: out STD_LOGIC_VECTOR (15 downto 0):="0000000000000000");
end mult8x8;
architecture Behavioral of mult8x8 is
signal state, nextstate:integer range 0 to 3;
signal count:std_logic_vector(2 downto 0):= "000"; -- 3bit counter
signal A:std_logic_vector(8 downto 0):="000000000";
signal B:std_logic_vector(7 downto 0):="00000000";
alias M:std_logic is B(0);
signal addout:std_logic_vector(8 downto 0):="000000000";
signal k,load, Ad, Sh,clk_sig:std_logic:='0';
begin
product <=A(7 downto 0)& B;
addout<='0' & A(7 downto 0)  + Mcand;
k<= '1' when count = 7 else '0';
p1:process (St,State,K,M)
begin
Load <='0';
sh<='0';
Ad<='0';
done<='0';
case state is
when 0=>
if St='1' then
Load <='1';
NextState<=1;
else
Nextstate<=0;
end if;
when 1=>
if M='1' then
Ad <='1';
nextstate<=2;
else
if k='0' then
```

```
Sh<='1'; nextstate <=1;
else
Sh <='1';
Nextstate <=3;
end if;
end if;
when 2=>
if k='0' then
Sh <='1';
nextstate<=1;
else
Sh <='1';
Nextstate <=3;
end if;
when 3=>
Done <='1';
Nextstate <=0;
end case;
end process;
p2:process(clk)
variable cnt1:integer:=1;
begin
if rising_edge(clk) then
if(cnt1=2)then
clk_sig<= not(clk_sig);
cnt1:=1;
else
cnt1:= cnt1 + 1;
end if;
end if;
end process;
p3:process (clk_sig)
begin
if rising_edge(clk_sig) then
if load ='1' then
A<= "000000000";
count<="000";
B<= Mplier;
end if;
if Ad='1' then
A<= addout;
end if;
if sh='1' then
A<='0' & A(8 downto 1);
B<=A(0)&B(7 downto 1);
count <= count + 1;
end if;
state<=Nextstate;
end if;
end process;
end Behavioral;
```

DESIGN EXAMPLE 4.10

Design and develop a digital system to multiply 8-bit binary numbers. Display the results in a 16 × 2 LCD.

Solution

```
library IEEE;
use IEEE.STD_LOGIC_1164.ALL;
use IEEE.STD_LOGIC_ARITH.ALL;
```

```
use IEEE.STD_LOGIC_UNSIGNED.ALL;
entity testc is
Port (clk: in STD_LOGIC:='0';
Mplier, Mcand:in STD_LOGIC_VECTOR(7 downto 0):="00000000";
lcd_cont: out STD_LOGIC_VECTOR (2 downto 0):="000";
lcd_data_out:out std_logic_vector(7 downto 0):="00000000");
end testc;
architecture Behavioral of testc is
--Multiplier
signal state, nextstate:integer range 0 to 3:=0;
signal count:std_logic_vector(2 downto 0):= "000"; -- 3bit counter
signal A:std_logic_vector(8 downto 0):="000000000";
signal B:std_logic_vector(7 downto 0):="00000000";
alias M:std_logic is B(0);
signal addout:std_logic_vector(8 downto 0):="000000000";
signal k,load, Ad, Sh:std_logic:='0';
signal Done:std_logic:='0';
signal st:std_logic:='1';
signal product: STD_LOGIC_VECTOR (15 downto 0):="0000000000000000";
--binary to bcd
signal pst,nst: integer range 0 to 66:=0;
signal count_shift: std_logic_vector(3 downto 0):="0000";
signal lcd_data1,lcd_data2,lcd_data3,lcd_data4,lcd_data5: std_logic_
vector(7 downto 0):="00000000";
signal bcd:std_logic_vector(19 downto 0):="00000000000000000000";
signal b1:STD_LOGIC_vector(35 downto 0):="000000000000000000000000000000
000000";
signal mr_lcd,md_lcd,pr_lcd,lcd_data_in:std_logic_vector(15 downto
0):="0000000000000000";
signal de_mux: std_logic_vector(1 downto 0):="00";
--lcd
signal clk_sig: std_logic:='0';
begin
mr_lcd<="00000000" & Mplier;
md_lcd<="00000000" & Mcand;
product <=A(7 downto 0)& B;
addout<='0' & A(7 downto 0) + Mcand;
k<= '1'
when
count = 7
else '0';
p0:process (St,State,K,M)
begin
Load <='0';
sh<='0';
Ad<='0';
done<='0';
case state is
when 0=>
if St='1' then
Load <='1';
NextState<=1;
else
Nextstate<=0;
end if;
when 1=>
if M='1' then
Ad <='1';
nextstate<=2;
else
if k='0' then
```

```
Sh<='1';
nextstate <=1;
else
Sh <='1';
Nextstate <=3;
end if;
end if;
when 2=>
if k='0' then
Sh <='1';
nextstate<=1;
else
Sh <='1';
Nextstate <=3;
end if;
when 3=>
Done <='1';
Nextstate <=0;
end case;
end process;
p1:process (clk)
begin
if rising_edge(clk) then
if load ='1' then
A<= "000000000";
count<="000";
B<= Mplier;
end if;
if Ad='1' then
A<= addout;
end if;
if sh='1' then
A<='0' & A(8 downto 1);
B<=A(0)&B(7 downto 1);
count <= count + 1;
end if;
state<=Nextstate;
end if;
end process;
p2: process(done)
begin
if rising_edge(done) then
pr_lcd <= product;
end if;
end process;
p3:process(de_mux)
begin
case de_mux is
when "00"=> lcd_data_in<=mr_lcd;
when "01"=> lcd_data_in<=md_lcd;
when "10"=> lcd_data_in<=pr_lcd;
when others => null;
end case;
end process;
p4:process(clk)
variable cnt1:integer:=0;
begin
if rising_edge(clk) then
if(cnt1=200)then
clk_sig<= not(clk_sig);
```

```
cnt1:=0;
else
cnt1:= cnt1 + 1;
end if;
end if;
end process;
p5:process(clk_sig)
begin
if rising_edge(clk_sig) then
pst<=nst;
end if;
end process;
p6:process(pst)
begin
case pst is
when 0=> lcd_data_out<="00111000";lcd_cont<="100"; nst<=pst+1;--38
when 1|3|5|7|9|28|49|59=> lcd_cont<="000"; nst<=pst+1;
when 2=> lcd_data_out<="00001111";lcd_cont<="100"; nst<=pst+1;--0F
when 4=> lcd_data_out<="00000001";lcd_cont<="100"; nst<=pst+1;--01
when 6=> lcd_data_out<="00000110";lcd_cont<="100"; nst<=pst+1;--06
when 8=> lcd_data_out<="10000010";lcd_cont<="100"; nst<=pst+1;--80
when 10 => bcd<="00000000000000000000";nst<=pst+1;
when 11 => b1<= bcd & lcd_data_in;nst<=pst+1;
when 22 => b1<=b1(34 downto 0) & b1(35);nst<=pst+1;
-- first digit
when 13=> if (b1(19 downto 16)>"0100") then
b1(19 downto 16)<=b1(19 downto 16) + "0011";
end if;nst<=pst+1;
-- second digit
when 14=> if (b1(23 downto 20)>"0100") then
b1(23 downto 20)<=b1(23 downto 20) + "0011";
end if;nst<=pst+1;
-- third digit
when 15=> if (b1(27 downto 24)>"0100") then
b1(27 downto 24)<=b1(27 downto 24) + "0011";
end if;nst<=pst+1;
-- fourth digit
when 16=> if (b1(31 downto 28)>"0100") then
b1(31 downto 28)<=b1(31 downto 28) + "0011";
end if;nst<=pst+1;
-- fifth digit
when 17=> if (b1(35 downto 32)>"0100") then
b1(35 downto 32)<=b1(35 downto 32) + "0011";
end if;nst<=pst+1;
when 18=> count_shift<=count_shift +"0001";nst<=pst+1;
when 19=> if (count_shift="1111") then
count_shift<="0000";nst<=pst+1;
else nst<=22;
end if;
when 20=> b1<=b1(34 downto 0) & b1(35);nst<=pst+1;
when 21=> lcd_data5<="0011" & b1(19 downto 16);nst<=pst+1;
when 22=> lcd_data4<="0011" & b1(23 downto 20);nst<=pst+1;
when 23=> lcd_data3<="0011" & b1(27 downto 24);nst<=pst+1;
when 24=> lcd_data2<="0011" & b1(31 downto 28);nst<=pst+1;
when 25=> lcd_data1<="0011" & b1(35 downto 32);nst<=pst+1;
--Mr
when 26=> if (de_mux="00") then nst<=pst+1;
else nst<=47;
end if;
when 27=> lcd_data_out<="10000100";lcd_cont<="100"; nst<=pst+1;--84
```

```
when 29=> lcd_data_out<="01010100";lcd_cont<="110"; nst<=pst + 1;--M
when 30|32|34|37|39|41|43|45|51|53|55|61|63|65=>
lcd_cont<="010";nst<=pst + 1;
when 31=> lcd_data_out<="01010100";lcd_cont<="110"; nst<=pst + 1;--r
when 33=> lcd_data_out<="01010100";lcd_cont<="110"; nst<=pst + 1;--=
--de_mux increment
when 35=> if (de_mux="10") then
de_mux<="00";nst<=pst + 1;
else
de_mux<=de_mux + "01";nst<=pst + 1;
end if;
--lcd outut
when 36=> lcd_data_out<=lcd_data5;lcd_cont<="110";nst<=pst + 1;
when 38=> lcd_data_out<=lcd_data4;lcd_cont<="110";nst<=pst + 1;
when 40=> lcd_data_out<=lcd_data3;lcd_cont<="110";nst<=pst + 1;
when 42=> lcd_data_out<=lcd_data2;lcd_cont<="110";nst<=pst + 1;
when 44=> lcd_data_out<=lcd_data1;lcd_cont<="110";nst<=pst + 1;
when 46=> nst<=10;
--Md
when 47=> if (de_mux="01") then nst<=pst + 1;
else nst<=57;
end if;
when 48=> lcd_data_out<="10000100";lcd_cont<="100"; nst<=pst + 1;--8A
when 50=> lcd_data_out<="01010100";lcd_cont<="110"; nst<=pst + 1;--M
when 52=> lcd_data_out<="01010100";lcd_cont<="110"; nst<=pst + 1;--d
when 54=> lcd_data_out<="01010100";lcd_cont<="110"; nst<=pst + 1;--=
when 56=> nst<=35;
--Pr
when 57=> if (de_mux="10") then nst<=pst + 1;
else nst<=10;
end if;
when 58=> lcd_data_out<="10000100";lcd_cont<="100"; nst<=pst + 1;--C4
when 60=> lcd_data_out<="01010100";lcd_cont<="110"; nst<=pst + 1;--P
when 62=> lcd_data_out<="01010100";lcd_cont<="110"; nst<=pst + 1;--r
when 64=> lcd_data_out<="01010100";lcd_cont<="110"; nst<=pst + 1;--=
when 66=> nst<=35;
end case;
end process;
end Behavioral;
```

We will discuss another type of multiplier, the FAM. As the name implies, there is an array to multiply the numbers, and it is really a fast multiplication because of using the LUT (Table 4.6). To understand this concept, we see some mathematical simplifications as follows. The product (P) of two 8-bit binary numbers A and X is given by:

$$P = A \times X$$

$$P = \sum_{k=0}^{N-1} a_k 2^k \times X$$

When $N = 7$

$$P = \sum_{k=0}^{7} a_k 2^k \times = (a_0 2^0 + a_1 2^1 + a_2 2^2 + \cdots + a_7 2^7) \times X$$

TABLE 4.6

Look-Up Table of an 8-Bit Fast Array Multiplier

LUT Values for the Position a_0, a_1, a_2, a_3, a_4, a_5, a_6 and a_7							
X	2X	4X	8X	16X	32X	64X	128X
0	0	0	0	0	0	0	0
1	2	4	8	16	32	64	128
2	4	8	16	32	64	128	256
3	6	12	24	48	96	192	384
4	8	16	32	64	128	256	512
5	10	20	40	80	160	320	640
6	12	24	48	96	192	384	768
7	14	28	56	112	224	448	896
8	16	32	64	128	256	512	1024
9	18	36	72	144	288	576	1152
10	20	40	80	160	320	640	1280
11	22	44	88	176	352	704	1408
12	24	48	96	192	384	768	1536
13	26	52	104	208	416	832	1664
14	28	56	112	224	448	896	1792
.	.	.	.	.	.	.	.
.	.	.	.	.	.	.	.
247	494	988	1976	3952	7904	15808	31616
248	496	992	1984	3968	7936	15872	31744
249	498	996	1992	3984	7968	15936	31872
250	500	1000	2000	4000	8000	16000	32000
251	502	1004	2008	4016	8032	16064	32128
252	504	1008	2016	4032	8064	16128	32256
253	506	1012	2024	4048	8096	16192	32384
254	508	1016	2032	4064	8128	16256	32512
255	510	1020	2040	4080	8160	16320	32640

$$P = (a_0 1 + a_1 2 + a_2 4 + a_3 8 + a_4 16 + a_5 32 + \cdots + a_7 128) \times X$$

$$P = a_0 X + a_1 2X + a_2 4X a_3 8X + a_4 16X + \cdots + a_7 128X \tag{4.4}$$

According to Equation 4.4, the multiplication of two binary bits can be done just by direct addition of X, 2X, 4X, 8X, 16X, 32X, 64X and 128X based on the values of a_0, a_1, a_2, a_3, a_4, a_5, a_6, and a_7. The digital architecture for this fast array multiplier is shown in Figure 4.10. Per Equation 4.4, a LUT can be prepared with the terms X, 2X, 4X, … 128X. For example, suppose, we take $X = 4$. Then, $2X = 8$, $4X = 16$, $8X = 32$, $16X = 64$, $32X = 128$, $64X = 256$, and $128X = 512$. Thus, the LUT will have [4 8 16 32 64 128 256 512] for the X value of 4. In the same way, we can prepare the LUT for any value of X from 1 to 255 since we are interested in designing an 8-bit multiplier. One example demonstrates this concept.

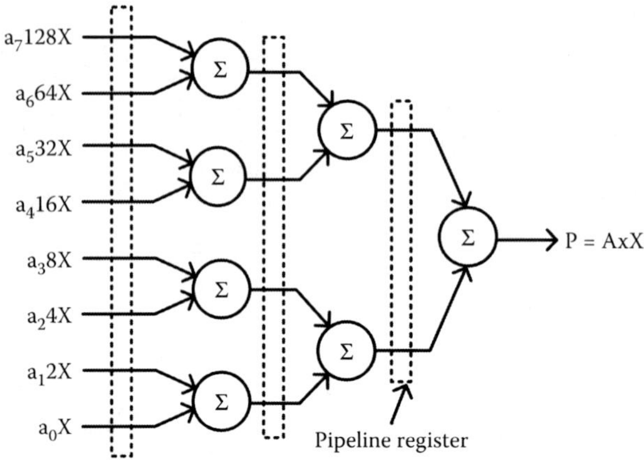

FIGURE 4.10
Fast array multiplier.

Assume that we need to multiply $(149)_{10}$ by $(4)_{10}$, that is, in binary "10010101" × "00000100"; the value of the multiplier has to be compared in the LUT, and we take the corresponding values. For 4, it will be [4 8 16 32 64 128 256 512]. Now, the value of the multiplicand gives $a_0 = 1$, $a_1 = 0$, $a_2 = 1$, $a_3 = 0$, $a_4 = 1$, $a_5 = 0$, $a_6 = 0$, and $a_7 = 1$. According to Equation 4.4, we need to compare these two results and add the valid values to get the product result. That means $(4 + 16 + 64 + 512) = 596$. In this way, multiplication of any length of bits can be done with just a LUT at a faster rate. The only issue is designing a suitable LUT, but since we have many resources in modern FPGAs, this is not a major issue. As shown in Figure 4.10, pipeline registers can be suitably incorporated in the design to increase the computation speed still further [25,39,41,42].

DESIGN EXAMPLE 4.11

Write MATLAB® code to realize the computation of a fast array multiplier for 8-bit multiplication. Prepare a LUT and store the results in an Excel file to read the values whenever needed.

Solution

```
clc;
clear all;
x=0:1:255;
for i=1:1:256,
temp1(i)=x(i);
temp2(i)=2*x(i);
temp4(i)=4*x(i);
temp8(i)=8*x(i);
temp16(i)=16*x(i);
temp32(i)=32*x(i);
temp64(i)=64*x(i);
temp128(i)=128*x(i);
end
lut=[temp1' temp2' temp4' temp8' temp16' temp32' temp64' temp128'];
xlswrite('C:\Users\DIAT\Documents\MATLAB\T_F_book.xls',lut,5,'E3');
X1=[1 1 0 1 0 0 1 0];
X=bi2de(X1);
y1=[1 1 0 1 1 0 1 1];
Y=bi2de(y1);
```

```
mul_val=X + 1;
mul=(y1(1)*lut(mul_val,1) +  y1(2)*lut(mul_val,2) +....
y1(3)*lut(mul_val,3) +  y1(4)*lut(mul_val,4) +.....
y1(5)*lut(mul_val,5) +  y1(6)*lut(mul_val,6) +....
y1(7)*lut(mul_val,7) +  y1(8)*lut(mul_val,8));
Fina_produ=[X Y mul]
```

This code will give the LUT values and multiply two binary numbers using FAM. The digital design and implementation of a fast array multiplier is left to the readers as a laboratory exercise.

The multiplier result for A = 75 and X = 219 is as follows:

```
Fina_produ = 75 219 16425.
```

4.4 Arithmetic Operations: Dividers

In this section, we discuss signed and unsigned binary dividers. There are many algorithms for binary division; a few of them are discussed in this section, with necessary design examples. Before proceeding to division algorithms, we should understand some terminology related to them. Suppose we wish to divide x by y, that is, (x/y), where x and y are the binary inputs, called the dividend and divisor, respectively, for example, (10/3). Here, 10 is the dividend and 3 is the divisor. In the actual division of (10/3), the quotient is 3 and the remainder is 1. As this is a decimal number division, binary division can also be carried out. The binary division of (135/13), that is, ("10000111"/"1101") is given in Table 4.7. The binary divisor is used to divide the binary dividend like the classical decimal divider. In the case where the intermediate remainder is smaller, we need to add 0 to the quotient and subtract with that intermediate value.

TABLE 4.7

Classical Binary Division for 135/13

Divisor	Dividend	Quotient	Remarks
1101	10000111	–	Initialization
	1101	**1**	One time
	011	–	Subtract
	0111	–	Take 1 from dividend
	0000	**0**	Smaller value, hence, subtract "0000"
	1111	–	Take another 1
	1101	**1**	One time
	010	–	Subtract
	0101	–	Take 1 from dividend
	0000	**0**	Smaller value, hence, subtract "0000"
	0101	–	**Complete**
The quotient is "1010" = 10 and the remainder is "101" = 5			

In this method, the division is done by just shift and subtract operations, and it has to be continued until getting the final remainder, that is, until assigning all the bits of the dividend for the shift and subtract operations. Another type of dividing algorithm is the parallel binary divider, for which the basic principles are left shift and add or subtract operations. We will discuss the division algorithm step by step below with examples and see the digital architecture of this algorithm subsequently. Let assume that we divide x by y, that is, (x/y), where x is the dividend and y is the divisor. We will denote the dividend and divisor by the classical variables Q and M, respectively, that is, Q = x and M = y. For example, assume (x/y) = (24/4); here, x = $(24)_{10}$ = "11000" = Q and y = $(4)_{10}$ = "100" = M. The value of `count` is the width of Q, `count` = 5 in this example.

1. Let us assume n_1 and n_2 are the width of Q and M, respectively. Now, we need to design a temporary register, called an accumulator and denoted as A, with 0 s according to the following conditions: (i) the length of A has to be $(n_1 + 1)$ if $n_1 \neq n_2$ and (ii) the length of A has to be $(n_1 + 2)$ if $n_1 = n_2$. In this example, $n_1 = 5$ and $n_2 = 3$; therefore, the register A is "000000".
2. Now, we need to formulate a table with the values of A and Q as given in Table 4.8a.

TABLE 4.8a

Initialization of Division Algorithm

Operation	A Register	Q Register	Count
Initialization	000000	11000	5

3. Left shift the A and Q registers by 1 bit. Then, Q[0] becomes empty; thus, the registers A and Q become as given in the third row of Table 4.8b.

TABLE 4.8b

Division Algorithm

Operation	A Register	Q Register	Count
Initialization	000000	11000	5
Left Shift	000001	1000-	

4. Now, we need to make the width of M equal to the width of the A register; hence, in this example, M = "000100". In this algorithm, we need to subtract M using the 2's complement whenever needed; hence, we require the 2's complement of M because, as we know, the subtraction of M can be performed just by adding the 2's complement of M, which is actually –M. In this example, –M = "111011" + 1 = "111100".
5. Based on the value of the MSB of the A register, we need to decide whether we should add M to the A register or subtract M from it. As we have seen already in the preceding sections, subtraction is performed in the 2's complement by adding –M to the A register. Addition is to be done if the MSB of the A register is "1"; otherwise, we use subtraction. In this example, the MSB of the A register is "0"; hence, we need to subtract. That is, we perform the addition of the A register to –M,

as `A + (−M)` = ("000001" + "111100") = "111101", and this result has to be entered in the table appropriately. Then, the above table becomes

TABLE 4.8c

Division Algorithm

Operation	`A` Register	`Q` Register	Count
Initialization	000000	11000	5
Left Shift	000001	1000-	
`A + (−M)`	111101	1000-	

6. Based on the value of the MSB of the `A` register, we need to determine the value for `Q[0]` as if the MSB is "0", then `Q[0]` = '1', and otherwise '0'. In this example, the MSB of the `A` register is "1"; hence, `Q[0]` = '0'. Then, the above table becomes

TABLE 4.8d

Division Algorithm

Operation	`A` Register	`Q` Register	Count
Initialization	000000	11000	5
Left Shift	000001	1000-	
`A + (−M)`	111101	1000-	
`Q[0]`	111101	10000	

7. The above steps (6) and (7) have to be repeated until the count value is 0, as given in the table below.

TABLE 4.8e

Division Algorithm

Operation	`A` Register	`Q` Register	Count
Initialization	000000	11000	5
Left Shift	000001	1000-	
`A + (−M)`	111101	10000	
Left Shift	111011	0000-	4
`A + M`	111111	0000-	
`Q[0]`	111111	00000	
Left Shift	111110	0000-	3
`A + M`	000010	0000-	
`Q[0]`	000010	00001	
Left Shift	000100	0001-	2
`A + (−M)`	000000	0001-	
`Q[0]`	000000	00011	
Left Shift	000000	0011-	1
A + (−M)	111100	0011-	
Q[0]	111100	00110	
Division is Completed			0

8. Now, the quotient value is `Q` = "00110" = $(6)_{10}$. If the MSB of the `A` register is "1", then we need to add `M` to the `A` register to get the correct remainder value; otherwise, the remainder = the `A` register. In this example, the MSB of the `A` register is "1"; hence, add the `A` register to `M` as ("111100" + "000100") = "000000" = $(0)_{10}$. The actual result of (24/4) is `Q` = 6 and `Rem` = 0.

Thus, the division algorithm is completed. In this way, division can be performed for any binary data. Signed numbers can also be divided in the same way by adjusting the sign bit at the final stage. Now, we discuss the digital architecture of this algorithm, which is shown in Figure 4.11. The inputs, the dividend and divisor, are fed into the Q-register and data formatter, respectively. The data formatter takes care of manipulating the value of the divisor, as discussed above, with respect to the length and value of the `A` register, and generates `M` and `-M`, which will be stored in the `M` and `-M` registers, respectively. The 2:1 multiplexer passes either `M` or `-M` to the `add/sub` (+/−) module in accordance with the controller commend. The controller decides either `M` or `-M` will be passed based on the value of the MSB of the `A` register. The left shift at the `A` and `Q` registers is performed by the command `left_shift` from the controller. The controller writes (`Wr`) the current value of the `A` register to the +/− module to perform either (`A` + `M`) or (`A` + `(-M)`) and then reads (`Rd`) and stores the result in the `A` register. Thus, the (+/−) modules get input from the `A` register and multiplexer to perform their operation, then send the result to the `A` register. The controller repeats all these operations until the counter value reaches 0 [39,43]. Finally, the values of the quotient and remainder can be collected from the `Q` and `A` registers. We also see some design numerical examples and VHDL code design subsequently.

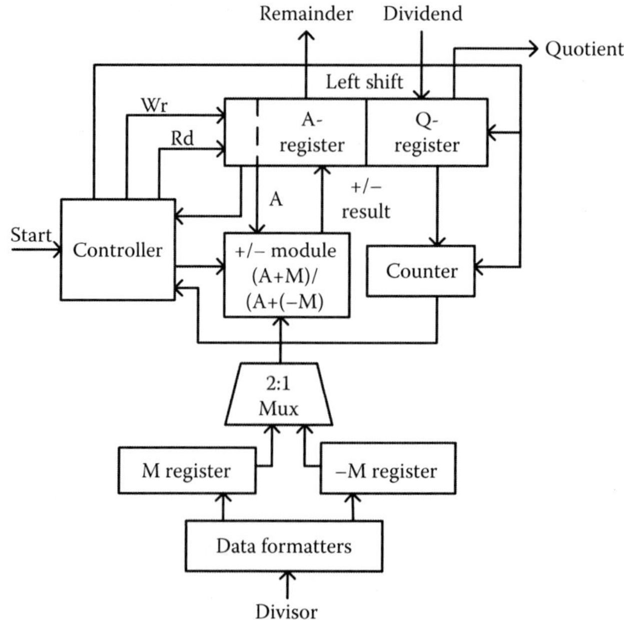

FIGURE 4.11
Digital architecture for division operation.

DESIGN EXAMPLE 4.12

Explain an 8-bit by 4-bit binary divider algorithm of the shift and subtract method with one example.

Solution

We take one 8-bit × 4-bit division example as (234/15), that is, ("11101010"/"1111"); then, `Q` = "11101010", `M` = "1111", `count` = 8, $n_1 = 8$ and $n_2 = 4$. Since, $n_1 \neq n_2$, the length of the `A` register is (n_1 + 1). Thus, the `A` register is "000000000". According to this `A` register, we need to alter the width of `M`, as now, `M` = "000001111"; then, −`M` = "111110001". With these values, we can start iterative computations as given in Table 4.9.

Here, the MSB of the `A` register is "0"; therefore, the quotient and remainder values can be directly taken as `Q` = "00001111" = $(15)_{10}$ and `Rem` = "000001001" = $(9)_{10}$.

TABLE 4.9

Division Algorithm for 8-Bit/4-Bit Data

Operation	`A` Register	`Q` Register	Count
Initialization	000000000	11101010	8
Left Shift	000000001	1101010-	8
`A + (−M)`	111110010	1101010-	
`Q[0]`	111110010	11010100	
Left Shift	111100101	1010100-	7
`A + M`	111110100	1010100-	
`Q[0]`	111110100	10101000	
Left Shift	111101001	0101000-	6
`A + M`	111111000	0101000-	
`Q[0]`	111111000	01010000	
Left Shift	111110000	1010000-	5
`A + M`	111111111	1010000-	
`Q[0]`	111111111	10100000	
Left Shift	111111111	0100000-	4
`A + M`	000001110	0100000-	
`Q[0]`	000001110	01000001	
Left Shift	000011100	1000001-	3
`A + (−M)`	000001101	1000001-	
`Q[0]`	000001101	10000011	
Left Shift	000011011	0000011-	2
`A + (−M)`	000001100	0000011-	
`Q[0]`	000001100	00000111	
Left Shift	000011000	0000111-	1
`A + (−M)`	000001001	0000111-	
`Q[0]`	000001001	00001111	
Division is Completed			0

DESIGN EXAMPLE 4.13

Explain an 8-bit by 8-bit binary divider algorithm of the shift and subtract method with one example.

Solution

We take one 8-bit division example as (253/172), i.e., ("11111101"/"10101100"), then, Q = "11111101", M = "10101100", count = 8, $n_1 = 8$ and $n_2 = 8$. Since $n_1 = n_2$, the length of the A register is $(n_1 + 2)$. Thus, the A register is "0000000000". According to this A register, we need to alter the width of M because now, M = "0010101100"; then, −M = "1101010100". With these values, we can start our iterative computations as given in Table 4.10.

Here, the MSB of the A register is "0"; therefore, the quotient and remainder values can be directly taken as Q = "00000001" $= (1)_{10}$ and Rem = "0001010001" $= (81)_{10}$.

TABLE 4.10

Division Algorithm for 8-Bit/4-Bit Data

Operation	A Register	Q Register	Count
Initialization	0000000000	11111101	8
Left Shift	0000000001	1111101-	8
A + (−M)	1101010101	1111101-	
Q[0]	1101010101	11111010	
Left Shift	1010101011	1111010-	7
A + M	1101010111	1111010-	
Q[0]	1101010111	11110100	
Left Shift	1010101111	1110100-	6
A + M	1101011011	1110100-	
Q[0]	1101011011	11101000	
Left Shift	1010110111	1101000-	5
A + M	1101100011	1101000-	
Q[0]	1101100011	11010000	
Left Shift	1011000111	1010000-	4
A + M	1101110011	1010000-	
Q[0]	1101110011	10100000	
Left Shift	1011100111	0100000-	3
A + M	1110010011	0100000-	
Q[0]	1110010011	01000000	
Left Shift	1100100110	1000000-	2
A + M	1111010010	1000000-	
Q[0]	1111010010	10000000	
Left Shift	1110111101	0000000-	1
A + M	0001010001	0000000-	
Q[0]	0001010001	00000001	
Division is Completed			0

DESIGN EXAMPLE 4.14

Develop a simple digital system to realize a 24-bit divider q = (Xx4095)/8192, where X is the input of width 24 bits.

Solution

```
library ieee;
use ieee.std_logic_1164.all;
use ieee.numeric_std.all;
entity divbyr is
port (clk : in std_logic:='0';
rst : in std_logic:='0';
d : in unsigned (23 downto 0):="000000000000000000000000";
q : out unsigned (23 downto 0):="000000000000000000000000");
end;
architecture rtl of divbyr is
begin
process (rst,clk)
begin
if rst='1' then
q <= (others=>'0');
elsif rising_edge (clk) then
q <= to_unsigned((to_integer(d) * 4095 / 8192),24);
end if;
end process;
end rtl;
```

DESIGN EXAMPLE 4.15

Develop a digital system to build an unsigned binary divider of 8 bits by 4 bits in the FPGA. Use a start input to enable the division process. Display the results using the LEDs.

Solution

```
library ieee;
use ieee.std_logic_1164.all;
use ieee.std_logic_arith.all;
use ieee.std_logic_unsigned.all;
entity divider is
port (dividend_in:in std_logic_vector (7 downto 0):="00000000";
divisor:in std_logic_vector (3 downto 0):="0000";
st, clk: in std_logic:='0';
quotient: out std_logic_vector(3 downto 0):="0000";
remainder: out std_logic_vector(3 downto 0):="0000";
overflow: out std_logic:='0');
end divider;
architecture behavioral of divider is
signal state, nextstate: integer range 0 to 5:=0;
signal c, load, su, sh: std_logic:='0';
signal subout: std_logic_vector(4 downto 0):="00000";
signal dividend: std_logic_vector(8 downto 0):="000000000";
begin
subout <= dividend(8 downto 4)-('0'& divisor);
c <= not subout(4);
remainder<=dividend(7 downto 4);
quotient <= dividend(3 downto 0);
state_graph:process (state,st,c)
begin
load <='0';
overflow <='0';
```

```
sh <='0';
su <='0';
case state is
when 0=>
if (st = '1') then
load <= '1';
nextstate <= 1;
else
nextstate <= 0;
end if;
when 1 =>
if (c ='1') then
overflow <='1';
nextstate <= 0;
else
sh <= '1';
nextstate <= 2;
end if;
when 2 | 3 | 4 =>
if (c ='1') then
su <='1';
nextstate <=state;
else
sh <='1';
nextstate <= state + 1;
end if;
when 5 =>
if (c ='1') then
su <='1';
end if;
nextstate <= 0;
end case;
end process state_graph;
update: process (clk)
begin
if clk'event and clk ='1' then
state <= nextstate;
if load ='1' then
dividend <='0' & dividend_in;
end if;
if su ='1' then
dividend(8 downto 4) <= subout;
dividend(0)<='1';
end if;
if sh = '1' then
dividend <= dividend(7 downto 0) & '0';
end if;
end if;
end process update;
end behavioral;
```

4.5 Trigonometric Computations Using the COordinate Rotation DIgital Computer (CORDIC) Algorithm

Most engineers and scientists aim to implement trigonometric functions such as sine, cosine, tan, cosecant, secant, and cot in FPGA, which they initially thought of accomplishing

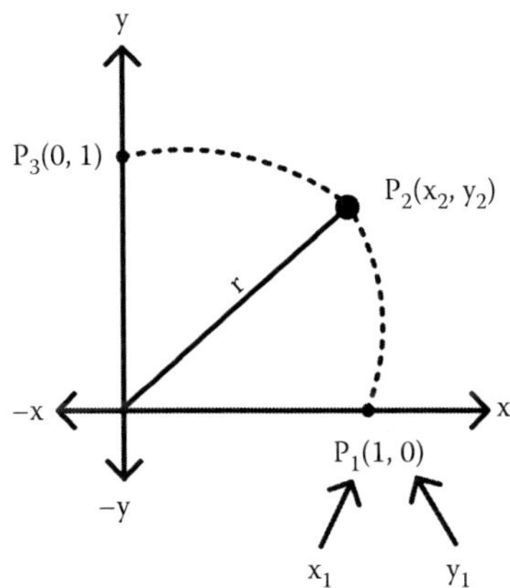

FIGURE 4.12
Illustration of coordinate rotations.

by means of a lookup table, possibly combined with linear interpolation or a power series if multipliers are available. The CORDIC algorithm is one of the most important, and a simple one was designed to calculate trigonometric functions just by rotation and additions. This algorithm was invented by Jack Volder in 1959 while designing a new navigation computer [39,44,45]. The real beauty of this algorithm is that we can implement it with a very small FPGA footprint because we require only a shifter and adder. The CORDIC requires only a small lookup table, along with the logic to perform shift and add operations. Importantly, the algorithm requires no dedicated multipliers or dividers. This algorithm is one of the most important for many modern scientific applications. For example, designers use CORDIC in many controllers to implement mathematical transfer functions, true RMS measurement, signal processing, image processing, and biomedical applications. The CORDIC algorithm can operate in one of three configurations: linear, circular, or hyperbolic [39,45]. Within each of these configurations, the algorithm functions in one of two modes: rotation or vectoring. In rotation mode, the input vector is rotated by a specified angle, whereas in vectoring mode, the algorithm rotates the input vector to the x axis while recording the angle of rotation required.

Before proceeding into discussing the CORDIC algorithm, we need to understand something about the coordinate rotations. Let us assume a 2D plot as shown in Figure 4.12, where x, −x, y, and −y are the axis variables. We see the coordinate rotation in the first quadrant since the same rotation and computation can be applicable in other quadrants, just with the appropriate sign for the respective variables. That means in the second quadrant, the value of x and its associated variables will have a negative sign and y will be positive, and in the fourth quadrant, both will have a negative sign. Assume a point P_1 at (1,0), as shown in Figure 4.12, and this point can be rotated up, that is, in an anti-clockwise direction, to point P_2 at (x_2,y_2). If the rotation is continued further, it is possible to reach point P_3 at (0,1). An important point to be noted here is that the rotation is not random; instead, it is done with a constant radius r from the origin, that is, (0,0). That means all the above points move away from the origin by the radius of r only. This is the reason point shifting is called rotation. Since the radius is constant, the rotation can also be dealt with in terms of theta (θ), which means that P_1 is at 0° (0 rad), P_2 is at 45° (π/4 rad) and P_3 is at 90° (π/2 rad) or any point from 0° through 90° in the first quadrant. This is the reason this is called coordinate rotation.

Another important point is while rotating in an anti-clockwise direction, the value of x decreases towards 0 and the value of y increases toward 1. Suppose we wish to rotate from point P_3 to P_2 in a clockwise direction, that is, the downward direction. The variation of the values of x and y are vice versa; that is, x increases while y decreases. Therefore, in coordinate rotation, we need to know three variables, namely theta (θ), x and y, during all the rotations. To know the relation among these variables, we need to understand the coordinate systems related to the CORDIC algorithm. Consider the following simple MATLAB code:

```
clc;
clear all;
f=figure;
si=0:0.01:2*pi;
subplot(2,2,1);
plot(cos(si),sin(si),'lineWidth',1.5);
set(gca,'fontsize',10);
grid on;
title('(a)');
subplot(2,2,2);
plot(2.*cos(si),4.*sin(si),'lineWidth',1.5);
set(gca,'fontsize',10);
grid on;
x0=2;
y0=2;
xd= x0.*cos(si)-y0.*sin(si);
yd= y0.*cos(si) +x0.*sin(si);
title('(b)');
subplot(2,2,3);
plot(xd,yd,'lineWidth',1.5);
set(gca,'fontsize',10);
grid on;
xlim([-3 3]);
ylim([-3 3]);
x0=1;
y0=2.5;
xd= x0.*cos(si)-y0.*sin(si);
yd= y0.*cos(si) +x0.*sin(si);
title('(c)');
subplot(2,2,4);
plot(xd,yd,'lineWidth',1.5);
set(gca,'fontsize',10);
grid on;
title('(d)');
```

In the above MATLAB code, θ varies from 0° through 360°, that is, 0 rad through 2π rad at the increment resolution of 0.01 rad. In the above MATLAB code, there are four sections, that is, four subplots. The first plots a unit circle as a function of Cos(θ) and Sin(θ) values; that means a unit circle can be generated, as shown in Figure 4.13a, when the Cos(θ) and Sin(θ) are taken as the x- and y-axis variables, respectively. In the second subplot, the same is followed, but with different amplitudes for x and y; that is, $x = 2$ and $y = 4$. The result is shown in Figure 4.13b. The third part is slightly different, because we do not directly take Cos(θ) and Sin(θ) values; instead, we take them in terms of coordinate expressions as $X_d = X_0\text{Cos}(\theta) - Y_0\text{Sin}(\theta)$ and $Y_d = Y_0\text{Cos}(\theta) + X_0\text{Sin}(\theta)$, which will also give the circle, as shown in Figure 4.13c. When X_0 and Y_0 are unequal, the result will not be a circle; instead, it will look like a tilted ellipse, as

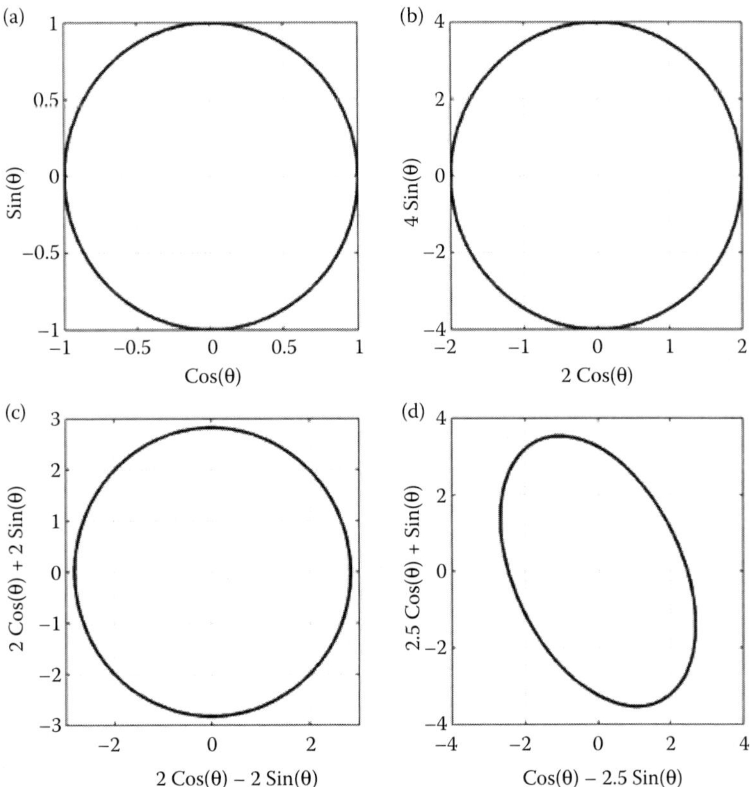

FIGURE 4.13
Results of coordinate rotation with different values of x and y.

shown in Figure 4.13b and d. Our main discussion of the CORDIC algorithm starts from these coordinate expressions. Actually, we have no need to think in terms of 360° because of the reasons above, as well as our interest not being in plotting a circle. Further, our main focus is only on the first quadrant. Therefore, we slightly alter these expressions as:

$$\begin{aligned} X_{i+1} &= X_i \,\mathrm{Cos}(\theta_i) - Y_i \,\mathrm{Sin}(\theta_i) \\ Y_{i+1} &= Y_i \,\mathrm{Cos}(\theta_i) + X_i \,\mathrm{Sin}(\theta_i) \end{aligned} \tag{4.5}$$

where i is the rotation or iteration index, which varies from 0 through N – 1, and we will see what the value of N – 1 is later.

Equation 4.5 can be altered further as

$$\begin{aligned} X_{i+1} &= \mathrm{Cos}(\theta_i)\left(X_i - Y_i \frac{\mathrm{Sin}(\theta_i)}{\mathrm{Cos}(\theta_i)}\right) \\ Y_{i+1} &= \mathrm{Cos}(\theta_i)\left(Y_i + X_i \frac{\mathrm{Sin}(\theta_i)}{\mathrm{Cos}(\theta_i)}\right) \\ X_{i+1} &= \mathrm{Cos}(\theta_i)\left(X_i - Y_i \,\mathrm{Tan}(\theta_i)\right) \\ Y_{i+1} &= \mathrm{Cos}(\theta_i)\left(Y_i + X_i \,\mathrm{Tan}(\theta_i)\right) \end{aligned} \tag{4.6}$$

Now, we need to understand Equation 4.6 more clearly. While rotating a point in the first quadrant, the values of X and Y vary, and they can be estimated from their previous points or positions. That means X_{i+1} and Y_{i+1} can be calculated from X_i and Y_i, respectively. We continue our discussion about X and Y now and discuss theta subsequently; therefore, assume $K_i = \text{Cos}(\theta_i)$. Then, Equation 4.6 becomes:

$$\begin{aligned} X_{i+1} &= K_i(X_i - Y_i \ \text{Tan}(\theta_i)) \\ Y_{i+1} &= K_i(Y_i + X_i \ \text{Tan}(\theta_i)) \end{aligned} \tag{4.7}$$

In the above equation, everything is fine except K_i and $\text{Tan}(\theta_i)$, because other variables are just related to the previous position. To simplify the above equation, we take the $\text{Tan}(\theta_i)$ as $(1/2^i) = 2^{-i}$. As we know that whenever the values of X and Y are changed on a 2D plane, that is, a point is shifted from one point to another, the theta also will be varied; that is why the index i is included. Further, $\text{Tan}(\theta_i) = 2^{-i}$; hence, $\theta = \text{Tan}^{-1}(2^{-i})$. This assumption limits the theta variations with some order, which means the theta variation should be on the order of $\text{Tan}^{-1}(2^{-i})$. The advantage of this assumption is (1), (1/2), (1/4), … , (1/8), … $(1/2^i)$ can be performed just by right shifting the binary data. Further, all the values of θ can be stored in a LUT, as given in Table 4.11. As in the table, the order of theta rotation is restricted to 45°, 26.565°, 14.036°, 7.125°, and so on. Every theta is almost half the value of the previous theta.

Now, we will see the theta rotation. Assume a point P_1, as in Figure 4.14, at $\theta = 0°$. Our intention is to shift the P_1 to 60.183°. We are restricted from directly going to 60.183°; instead, we can go only in the order of 45°, 26.565°, 14.036°, 7.125°, and so on. Because of this limitation, we need to rotate the point up and down to reach the expected theta, that is, 60.183°. At the beginning, the theta is 0°; therefore, we need to add, that is, going up, the first theta, that is, 45° to the 0°. Then, the new theta is 45°, as indicated by point P_2 in Figure 4.14. Still, this point is below the theta we wish to reach; hence, add the next theta, that is, 26.565° to 45°, which will be 71.565° and is shown as P_3 in Figure 4.14. Now, the theta is above the

TABLE 4.11

Order of Rotation Angle in the CORDIC Algorithm

i	$\theta_i = \tan^{-1}(2^{-i})$	$\tan(\theta_i)$
0	$\tan^{-1}(1) = 45°$	1
1	$\tan^{-1}(2^{-1}) = 26.565°$	1/2
2	$\tan^{-1}(2^{-2}) = 14.036°$	1/4
3	$\tan^{-1}(2^{-3}) = 7.125°$	1/8
4	$\tan^{-1}(2^{-4}) = 3.576°$	1/16
5	$\tan^{-1}(2^{-5}) = 1.79°$	1/32
6	$\tan^{-1}(2^{-6}) = 0.895°$	1/64
7	$\tan^{-1}(2^{-7}) = 0.448°$	1/128
8	$\tan^{-1}(2^{-8}) = 0.224°$	1/256
9	$\tan^{-1}(2^{-9}) = 0.112°$	1/512
.	.	.
.	.	.

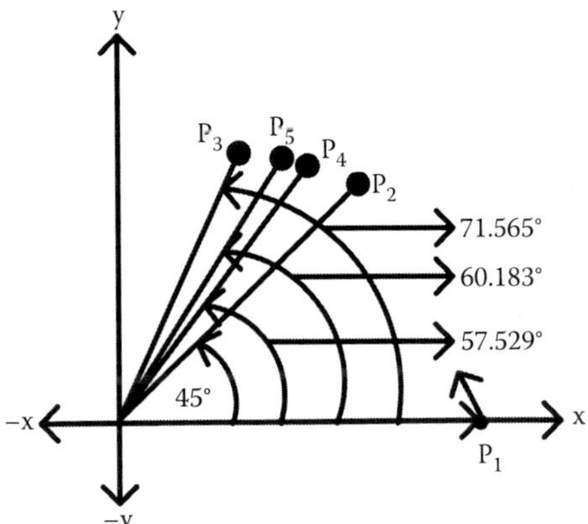

FIGURE 4.14
Illustration of coordinate rotations.

theta we wish to reach; hence, we need to come down; that is, subtract the third theta from this value as 71.565° – 14.036° = 57.529°, which is shown as P_4 in Figure 4.14. This theta is below the actual theta; hence, add the next theta as 57.529° + 7.125° = 64.654°. This is above the actual theta; hence, subtract the next theta as 64.654° – 3.576° = 61.078°.

Again, subtract the next theta as 61.078° – 1.79° = 59.288° and add the next theta as 59.288° + 0.895° = 60.183°, which is shown by point P_5. Now, the rotation of theta is converged to the actual theta, that is, 60.183°. Actually, we have done the coordinate rotations to converge the theta to the decided/actual theta (or point) via the precalculated (or particular specified) order of thetas as given in Table 4.11. We need to decide that either addition or subtraction is required just by comparing the current theta with the actual theta, as discussed above. After seven (N = 7) iterations, that is, i = 0,1,2,3, … 6, the theta converged to 60.183°. Based on this information, we need to formulate a recursive equation to find the theta for N – 1 iterations as

$$\theta_{i+1} = \theta_i - d_i \tan^{-1}(2^{-i}) \tag{4.8}$$

where θ_{i+1} is the future theta, θ_i is the present theta, and d_i is the direction function. For every i, the value of $\tan^{-1}(2^{-i})$ will be taken from the LUT as in Table 4.11. The d_i will be either positive or negative based on the direction we want to rotate. That means if the actual theta is above the present theta, then we need to add the LUT theta to the present theta to go up; therefore, d_i should be negative (–1). If the present theta is above the actual theta, then we need to subtract the LUT theta from the actual theta; hence, the d_i should be positive (+ 1). We will see this logic again in the design examples. Now, we go back to Equation 4.7 and alter it in light of the above discussions as

$$\begin{aligned} X_{i+1} &= X_i - d_i\, Y_i 2^{-i} \\ Y_{i+1} &= Y_i + d_i\, X_i 2^{-i} \end{aligned} \tag{4.9}$$

where X_{i+1} is the future value of x; X_i is the present x value; Y_{i+1} is the future value of y; Y_i is the present value of y; d_i is the decision function; and 2^{-i} is the $\mathrm{Tan}(\theta_i)$, as we assumed. Here, the d_i will have either a positive or negative sign based on the point rotation. For example, if we rotate the point down, the value of x will increase and y will decrease;

TABLE 4.12

Recursive Equations of the CORDIC Algorithm

Coordinate Rotation	Iterative or Recursive Equation	Decision Function (d_i)
Theta(θ)	$\theta_{i+1} = \theta_i - d_i \tan^{-1}(2^{-i})$	$d_i = -1$ if $\theta_{actual} > \theta_i$ $d_i = +1$ if $\theta_{actual} < \theta_i$
X value	$X_{i+1} = X_i - d_i Y_i 2^{-i}$	$d_i = +1$ if $\theta_{actual} > \theta_i$ $d_i = -1$ if $\theta_{actual} < \theta_i$
Y value	$Y_{i+1} = Y_i + d_i X_i 2^{-i}$	

therefore, d_i should be negative (–1). If the point moves up, the value of x decreases and y increases; hence, d_i is positive (+1). The recursive equations for the coordinate rotation of theta, X and Y are summarized in Table 4.12. As in Equation 4.9, because of rotating theta in the order of precalculated values, as given in Table 4.12, $(Yi/2^{-i})$ and $(Xi/2^{-i})$ can be simply performed just by appropriately right shifting the Y_i and X_i, which does not require any multiplication or division.

Note that the value of d_i is different for theta, X and Y. The CORDIC algorithm in coordinate rotation is almost over, except K_i in Equation 4.7, which we purposely left to form Equation 4.8 K_i is equal to Cos(θ). Since the theta varies at every rotation, the value of K_i can also be written as $K_i = \text{Cos}(\theta_i)$. We know that θ is assumed to be $\tan^{-1}(2^{-i})$; hence, $K_i = \cos(\tan^{-1}(2^{-i}))$. This can be simplified further as

$$\cos(\tan^{-1}(2^{-i})) = \left(\frac{1}{\sqrt{1+\tan^2(\tan^{-1}(2^{-i}))}}\right) \tag{4.10}$$

As in Equation 4.10, $\tan^2(\tan^{-1}(2^{-i}) = \tan(\tan^{-1}(2^{-i})) \times \tan(\tan^{-1}(2^{-i})$, the tan and $\tan^{-1}$ will be cancelled, then $\tan^2(\tan^{-1}(2^{-i}) = (2^{-i}) \times (2^{-i}) = (2^{-i})^2 = 2^{-2i}$. Substituting this result in Equation 4.10, it becomes:

$$\cos(\tan^{-1}(2^{-i})) = \left(\frac{1}{\sqrt{1+(2^{-2i})}}\right) \tag{4.11}$$

As in Equation 4.11, the value of K_i is only in terms of i; therefore, K_i becomes an independent factor of the rotation of theta, X as well as Y, because it depends only on the iteration index, that is, i. Further, K_i is not a function of the direction of rotation and is a constant for the rotation in any direction. Therefore, we can do the multiplication of K_i with the left-hand side of Equation 4.8 as in Equation 4.7. That is how Equation 4.8 is formed.

Now, we should know what value we need to multiply at the last and how to find that value. To get these answers, we analyze the variation profile of K_i for different values of the iteration index (i), as shown in Figure 4.15, where the values of K_i are 0.7071 when $i = 0$, 0.8944 when $i = 1$, 0.9701 when $i = 2$, 0.9923 when $i = 3$, 0.9981 when $i = 4$, 0.9995 when $i = 5$, 0.9999 when $i = 6$, 1 when $i = 7$, 1 when $i = 8$, 1 when $i = 9$, and so on, which means that K_i increases for the first seven iterations and then saturates at a value of 1. That means that K_i is unity after seven iterations. Therefore, the multiplication is

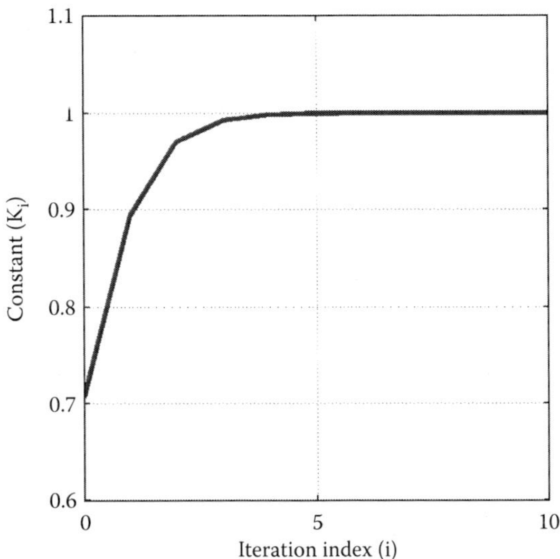

FIGURE 4.15
Profile of the CORDIC scaling factor (K_i).

$$K_i = \prod_{i=0}^{N-1}\left(\frac{1}{\sqrt{1+(2^{-2i})}}\right) \tag{4.12}$$

This value is actually $K_i = (0.7071) \times (0.8944) \times (0.9701) \times (0.9923) \times (0.9981) \times (0.9995) \times (0.9999) \times (1) \times (1) \ldots = 0.6073$. This is the value we need to multiply by the right-hand side values of Equation 4.8 as in Equation 4.7 if the convergence occurs after seven iterations. Otherwise, the multiplication of respective values has to be multiplied at the end. That is all for the CORDIC algorithm based on coordinate rotation.

However, we need to implement the CORDIC algorithm in the FPGA to perform trigonometric operations. To realize it, we need to implement the LUT given in Table 4.1, recursive equations given in Table 4.12 and final X and Y result multiplication by 0.6073. The digital architecture for this implementation is shown in Figure 4.16. The architecture shows the iterative stages from $i = 0$ through $i = N - 1$. The precalculated theta values that are required to rotate the point are indicated as $\alpha_0, \alpha_1, \alpha_2, \alpha_3, \ldots \alpha_{N-1}$, and these values will be taken from a LUT. Initial values of X, Y, and θ are given to the respective adder/subtractor. The sign for the decision function (d_i) will be found by comparing the actual and present theta at every iteration. Then, the appropriate d_i will be passed into the adder/subtractor, based on which it performs addition or subtraction, i.e., rotating up or down. The division of $X_i/2^i$ and $Y_i/2^i$ is performed just by right shifting X_i and Y_i appropriately, which is indicated by $\gg$0 (no right shift because $i = 0$), $\gg$1 (one right shift because $i = 1$), $\gg$2 (two right shift because $i = 2$), ... $\gg N - 1$ ($N - 1$ right shift because $i = N - 1$).

Finally, after converging the theta to the actual theta, we need to multiply the result of X_{N-1} and Y_{N-1} by 0.6073, as shown in Figure 4.16, and these values will be equal to the Cos(θ) and Sin(θ) values, respectively. All the other trigonometric operations like tan(x), cosec (x), cot(x), tan(A + B) and $\tan^2(x)$ can be performed in digital form using the FPGA in terms of sin(x) and cos(x) just by doing some simple mathematical operations at the end. Further, any values in the other quadrants can also be found by doing some easy simplification;

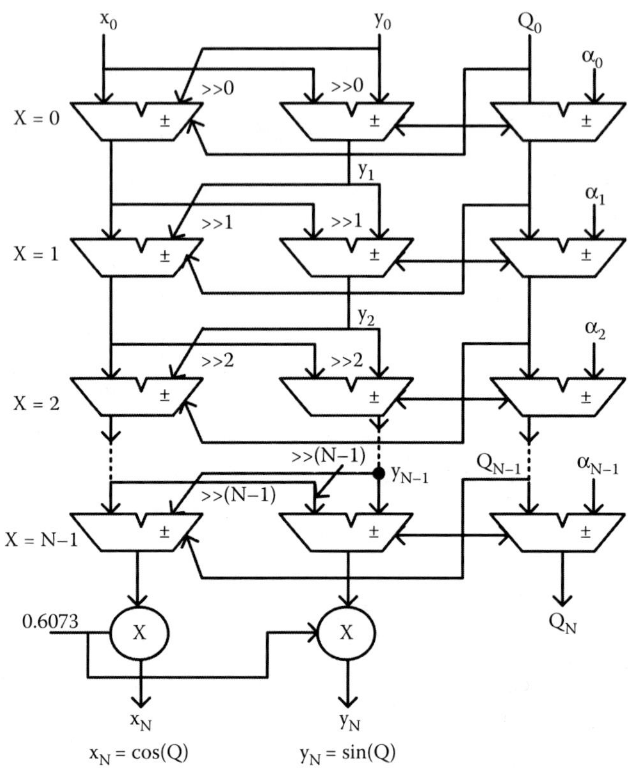

FIGURE 4.16
Digital design for implementation of CORDIC [44,45].

for example, sin (135°) and cos (135°) can be found as sin (45°) and –cos (45°), respectively. In this way, the sin and cos functions can be found for any theta using the CORDIC algorithm. We realize the algorithm with some design examples below.

DESIGN EXAMPLE 4.16

Find the Sine (5.0385°) and Cos (5.0385°) using the CORDIC algorithm. Clearly explain all the steps for theta, X, and Y.

Solution

The value of theta for which we need to find the Sine and Cos is $\theta = 5.0385°$. We start our iteration from (1,0); therefore, the initial theta is 0°; that is, $X_i = 1$, $Y_i = 0$ and $\theta_i = 0°$. We start our iteration with these values as given in Table 4.13 below.

In the above steps, the Sine values for the decision function (d_i) of theta, X_i and Y_i are selected based on the values of the actual theta (θ) and the present theta (θ_i); that is, in which direction we need to rotate the coordinates. For example, θ_i is initially 0°, but we need to reach the actual θ, which is 5.0385°; therefore, we need to add the first value of the LUT, that is, 45°, to θ_i to move up towards 5.0385°. After this iteration, θ_i will be 45°, which is above the actual theta; hence, we need to rotate down. Therefore, we subtract the second value of LUT from 45°. In this way, the iterations are continued. The decision function for X and Y is also chosen based on the direction of rotation. Finally, when θ_i converges to the actual theta with acceptable accuracy, we need to multiply the final values of X and Y by the value of K_i, that is, 0.6073. Then, the result will be the Cos and Sine values of the actual theta. All these steps are clearly given in Table 4.13, with necessary remarks and computations.

TABLE 4.13

CORDIC Iteration for $\theta = 5.0385°$

i	Present Values (i)			Remarks	Iterative Computation	Next Values (i + 1)		
	θ_i	X_i	Y_i			θ_{i+1}	X_{i+1}	Y_{i+1}
0	0°	1	0	Since, $\theta > \theta_i$ d_i for theta is −1 d_i for X and Y is +1	$\theta_{i+1} = \theta_i + 45° = 0° + 45° = 45°$ $X_{i+1} = X_i - (Y_i/1) = 1 - (0/1) = 1$ $Y_{i+1} = Y_i + (X_i/1) = 0 + (1/1) = 1$	45°	1	1
1	45°	1	1	Since, $\theta < \theta_i$ d_i for theta is +1 d_i for X and Y is −1	$\theta_{i+1} = \theta_i - 26.56° = 45° - 26.56° = 18.4349°$ $X_{i+1} = X_i + (Y_i/2) = 1 + (1/2) = 1.5$ $Y_{i+1} = Y_i - (X_i/2) = 1 - (1/2) = 0.5$	18.4349°	1.5	0.5
2	18.4349°	1.5	0.5	Since, $\theta < \theta_i$ d_i for theta is +1 d_i for X and Y is −1	$\theta_{i+1} = \theta_i - 14.0362° = 18.4349° - 14.036° = 4.3987°$ $X_{i+1} = X_i + (Y_i/4) = 1.5 + (0.5/4) = 1.625$ $Y_{i+1} = Y_i - (X_i/4) = 0.5 - (1.5/4) = 0.125$	4.3987°	1.625	0.125
3	4.3987°	1.625	0.125	Since, $\theta > \theta_i$ d_i for theta is −1 d_i for X and Y is +1	$\theta_{i+1} = \theta_i + 7.1250° = 4.3897° + 7.1250° = 11.5237°$ $X_{i+1} = X_i - (Y_i/8) = 1.625 - (0.125/8) = 1.609$ $Y_{i+1} = Y_i + (X_i/8) = 0.125 + (1.625/8) = 0.3281$	11.5237°	1.609	0.3281
4	11.5237°	1.609	0.3281	Since, $\theta < \theta_i$ d_i for theta is +1 d_i for X and Y is −1	$\theta_{i+1} = \theta_i - 3.5763° = 11.5237° - 3.5763° = 7.9474°$ $X_{i+1} = X_i + (Y_i/16) = 1.609 + (0.3281/16) = 1.6295$ $Y_{i+1} = Y_i - (X_i/16) = 0.3281 - (1.609/16) = 0.2275$	7.9474°	1.6295	0.2275
5	7.9474°	1.6295	0.2275	Since, $\theta < \theta_i$ d_i for theta is +1 d_i for X and Y is −1	$\theta_{i+1} = \theta_i - 1.7899° = 7.9474° - 1.7899° = 6.1575°$ $X_{i+1} = X_i + (Y_i/32) = 1.6295 + (0.2275/32) = 1.6366$ $Y_{i+1} = Y_i - (X_i/32) = 0.2275 - (1.6295/32) = 0.1765$	6.1575°	1.6366	0.1765
6	6.1575°	1.6366	0.1765	Since, $\theta < \theta_i$ d_i for theta is +1 d_i for X and Y is −1	$\theta_{i+1} = \theta_i - 0.8952° = 6.1575° - 0.8952° = 5.2623°$ $X_{i+1} = X_i + (Y_i/64) = 1.6366 + (0.1765/64) = 1.6393$ $Y_{i+1} = Y_i - (X_i/64) = 0.1765 - (1.6366/64) = 0.1509$	5.2623°	1.6393	0.1509
7	5.2623°	1.6393	0.1509	Since, $\theta < \theta_i$ d_i for theta is +1 d_i for X and Y is −1	$\theta_{i+1} = \theta_i - 0.4476° = 5.2623° - 0.4476° = 4.8147°$ $X_{i+1} = X_i + (Y_i/128) = 1.6393 + (0.1509/128) = 1.6404$ $Y_{i+1} = Y_i - (X_i/128) = 0.1509 - (1.6393/128) = 0.1380$	4.8147°	1.6404	0.1380
8	4.8147°	1.6404	0.1380	Since, $\theta > \theta_i$ d_i for theta is −1 d_i for X and Y is +1	$\theta_{i+1} = \theta_i + 0.2238° = 4.8147° + 0.2238° = 5.0385°$ $X_{i+1} = X_i - (Y_i/256) = 1.6404 - (0.13809/256) = 1.6398$ $Y_{i+1} = Y_i + (X_i/256) = 0.13809 + (1.6404/256) = 0.1444$	**5.0385°**	**1.6398**	**0.1444**
We need to multiply the constant K_i, that is, 0.6073, with the values of X_i and Y_i.					$Cos(5.0385°) = 1.6398 \times 0.6073 = 0.996$ $Sin(5.0385°) = 0.1444 \times 0.6073 = 0.087$	–		

TABLE 4.14

CORDIC Iteration for $\theta = 28.027°$

	Present Values (i)					Next Values (i + 1)		
i	θ_i	X_i	Y_i	Remarks	Iterative Computation	θ_{i+1}	X_{i+1}	Y_{i+1}
0	0°	1	0	Since, $\theta > \theta_i$ d_i for theta is −1 d_i for X and Y is +1	$\theta_{i+1} = \theta_i + 45° = 0° + 45° = 45°$ $X_{i+1} = X_i - (Y_i/1) = 1 - (0/1) = 1$ $Y_{i+1} = Y_i + (X_i/1) = 0 + (1/1) = 1$	45°	1	1
1	45°	1	1	Since, $\theta < \theta_i d_i$ for theta is +1 d_i for X and Y is −1	$\theta_{i+1} = \theta_i - 26.565° = 45° - 26.565° = 18.435°$ $X_{i+1} = X_i + (Y_i/2) = 1 + (1/2) = 1.5$ $Y_{i+1} = Y_i - (X_i/2) = 1 - (1/2) = 0.5$	18.435°	1.5	0.5
2	18.4349°	1.5	0.5	Since, $\theta > \theta_i$ d_i for theta is −1 d_i for X and Y is +1	$\theta_{i+1} = \theta_i + 14.0362° = 18.435° + 14.036° = 32.471°$ $X_{i+1} = X_i - (Y_i/4) = 1.5 - (0.5/4) = 1.375$ $Y_{i+1} = Y_i + (X_i/4) = 0.5 + (1.5/4) = 0.875$	32.471°	1.375	0.875
3	32.471°	1.375	0.875	Since, $\theta < \theta_i$ d_i for theta is +1 d_i for X and Y is −1	$\theta_{i+1} = \theta_i - 7.125° = 32.471° - 7.125° = 25.346°$ $X_{i+1} = X_i + (Y_i/8) = 1.375 + (0.875/8) = 1.4843$ $Y_{i+1} = Y_i - (X_i/8) = 0.875 - (1.375/8) = 0.7031$	25.346°	1.4843	0.7031
4	25.346°	1.4843	0.7031	Since, $\theta > \theta_i$ d_i for theta is −1 d_i for X and Y is +1	$\theta_{i+1} = \theta_i + 3.576° = 25.346° + 3.576° = 28.922°$ $X_{i+1} = X_i - (Y_i/16) = 1.4843 - (0.7031/16) = 1.4403$ $Y_{i+1} = Y_i + (X_i/16) = 0.7031 + (1.4843/16) = 0.7958$	28.922°	1.4403	0.7958
5	28.922°	1.4403	0.7958	Since, $\theta < \theta_i$ d_i for theta is +1 d_i for X and Y is −1	$\theta_{i+1} = \theta_i - 1.79° = 28.922° - 1.79° = 27.132°$ $X_{i+1} = X_i + (Y_i/32) = 1.4403 + (0.7958/32) = 1.4651$ $Y_{i+1} = Y_i - (X_i/32) = 0.7958 - (1.4403/32) = 0.75079$	27.132°	1.4651	0.75079
6	27.132°	1.4651	0.75079	Since, $\theta > \theta_i$ d_i for theta is −1 d_i for X and Y is +1	$\theta_{i+1} = \theta_i + 0.895° = 27.132° + 0.895 = 28.027°$ $X_{i+1} = X_i - (Y_i/64) = 1.4651 - (0.75079/64) = 1.4533$ $Y_{i+1} = Y_i + (X_i/64) = 0.75079 + (1.4651/64) = 0.7736$	**28.027°**	**1.4533**	**0.7736**
We need to multiply the constant K_i, that is, 0.6073, with the value of X_i and Y_i.					$\cos(28.027°) = 1.4533 \times 0.6073 = 0.8826$ $\sin(28.027°) = 0.7736 \times 0.6073 = 0.4698$	–		

DESIGN EXAMPLE 4.17

Find the Sine (28.027°) and Cos (28.027°) using the CORDIC algorithm. Clearly explain all the steps for theta, `X`, and `Y`.

Solution

The value of theta for which we need to find the Sine and Cos is $\theta = 28.027°$. We start our iteration from (1,0); therefore, the initial theta is 0°; that is, $X_i = 1$, $Y_i = 0$ and $\theta_i = 0°$. We start our iteration with these values, as given in Table 4.14.

DESIGN EXAMPLE 4.18

Design MATLAB code for the CORDIC algorithm to realize Sine and Cos functions. Plot the following:

i. LUT, that is, $\tan^{-1}(2^{-i})$, values
ii. Theta convergence
iii. `X` and `Y` variation profile
iv. Error values

Solution

As given in this design example, we need to write MATLAB code to realize the CORDIC algorithm. Finally, we will plot all the responses. The error will be computed as Error = (CORDIC result – MATLAB function result). At the end, we print `X`, `Y`, `K`, Sine(), and Cos() values for the given theta.

```
clc;
clear all;
clear screen;
format short;
K=1;
Q=input ('Enter the theta: ');
N=input('Enter the number of iteration: ');
i=0:1:N-1;
for a=0:1:length(i)-1,
k(a+1) =(1./sqrt(1 + (1./2^(2.*a))));
K=K*k(a+1);
end
theta=atand(2.^-i);
figure(1);
subplot(2,2,1);
plot(theta,'o-','lineWidth',1.5);
set(gca,'fontsize',10);
xlabel('No of iterations (i)');
ylabel('LUT: tan^{-1}(2^{-i})');
text(2,40,{'Each angle is roughly';'half the previous one'});
grid on;
z0=0;
z00=0;
x0=1;
x00=x0;
y0=0;
y00=y0;
d=1;
d0=d;
for i=0:length(theta)-1,
if z0<Q,
D(i+1)=-1;
elseif z0>=Q,
```

```
D(i + 1)=1;
end
d=D(i + 1);
zn(i + 1)= z0-(d*theta(i + 1));
z0=zn(i + 1);
d=-1*d;
xn(i + 1)= x0-(d* (y0*(1/2^i)));
yn(i + 1)= y0 + (d* (x0*(1/2^i)));
x0=xn(i + 1);
y0=yn(i + 1);
end
subplot(2,2,2);
zn=[z00 zn];
plot(zn,'*-','lineWidth',1.5);
set(gca,'fontsize',10);
xlabel('No of iterations (i)');
ylabel('\theta_{i + 1}');
grid on;
xlim([0 N]);
subplot(2,2,3);
xn=[x00 xn];
plot(xn,'*-','lineWidth',1.5);
set(gca,'fontsize',10);
hold on;
yn=[y00 yn];
plot(yn,'o-','lineWidth',1.5);
set(gca,'fontsize',10);
grid on;
xlabel('No of iterations (i)');
ylabel('X,Y Variation Profile');
legend('Y','X');
xlim([0 N]);
D=[d0 D];
ans=[ (0:1:N-1)' k' theta' D(1:length(D)-1)' zn(1:length(zn)-1)'
xn(1:length(xn)-1)' yn(1:length(yn)-1)']K;
sin_Q=K*yn(length(yn)-1);
cos_Q=K*xn(length(xn)-1);
fin_ans=[xn(length(xn)-1) yn(length(yn)-1) K sin_Q cos_Q]
sin_err=(sind(Q)-sin_Q);
cos_err=(cosd(Q)-cos_Q);
err=[sin_err cos_err];
subplot(2,2,4);
stem(err,'lineWidth',1.5);
set(gca,'fontsize',10);
grid on;
ylabel('Error');
text(1.3,0,{'A - sin(\theta)';'B - cos(\theta)'});
vertical_line_gap=1;
vertical_lines=0:vertical_line_gap:2*vertical_line_gap;
vertical_line_label= [' ';'A';'B';' '];
set(gca,'Xlim',[1,2],'XTick',vertical_lines,'XTicklabel',
vertical_line_label);
xlim([0.5 2.5])
axes('Position',[0 0 1 1],'Visible','off');
text(0.3,0.97,'CORDIC iterations for SIN and COS computations');
fprintf('Y is %.4f\n',yn(length(yn)-1));
fprintf('X is %.4f\n',xn(length(xn)-1));
fprintf('K is %.4f\n',K);
fprintf('Actual theta is %.4f\n',Q);
fprintf('sin value is %.4f\n',sin_Q);
fprintf('cos value is %.4f\n',cos_Q);
```

All the simulation results are shown in Figure 4.17. The first plot is the values of LUT, the second is theta convergence, the third is the variation profile of X and Y, and the last is the error. The last plot clearly shows the error values for the Sine and Cos functions, which are 1.3×10^{-9} and -0.9×10^{-9}, respectively. This result is evident from the computation accuracy of the CORDIC algorithm.

Final results for the given angle 35.56° are:

```
Y is 0.9577
X is 1.3397
K is 0.6073
Actual theta is 35.5600
Sine value is 0.5816
Cos value is 0.8135
```

DESIGN EXAMPLE 4.19

Develop a digital system to implement the CORDIC algorithm in the FPGA. Use the standard add and shift methods to implement the CORDIC algorithm [25,44,45].

Solution

```
library ieee;
use ieee.std_logic_1164.all;
use ieee.numeric_std.all;
entity synth_cordic is port(
clk : in std_logic;
resetn : in std_logic;
z_ip : in std_logic_vector(16 downto 0); --1,16
x_ip : in std_logic_vector(16 downto 0);--1,16
y_ip : in std_logic_vector(16 downto 0);--1,16
cos_op : out std_logic_vector(16 downto 0); --1,16
sin_op : out std_logic_vector(16 downto 0)); --1,16
end entity synth_cordic;
architecture rtl of synth_cordic is
type signed_array is array (natural range<>) of signed(17 downto 0);
--arctan array format 1,16 in radians
constant tan_array : signed_array(0 to 16):=
(to_signed(51471,18),to_signed(30385,18),to_signed(16054,18),to_
signed(8149,18),
to_signed(4090,18), to_signed(2047,18),to_signed(1023,18),
to_signed(511,18),
to_signed(255,18), to_signed(127,18),to_signed(63,18), to_signed(31,18),
to_signed(15,18),to_signed(7,18),to_signed(3,18),to_signed(1,18), to_
signed(0, 18));
signal x_array : signed_array(0 to 14) :=(others => (others =>'0'));
signal y_array : signed_array(0 to 14) :=(others => (others =>'0'));
signal z_array : signed_array(0 to 14) :=(others => (others =>'0'));
begin
rocess(resetn, clk)
begin
if resetn = '0' then
x_array <= (others => (others =>'0'));
z_array <= (others => (others =>'0'));
y_array <= (others => (others =>'0'));
elsif rising_edge(clk) then
if signed(z_ip)< to_signed(0,18)
then
x_array(x_array'low) <=signed(x_ip) + signed('0' & y_ip);
```

```
y_array(y_array'low) <=signed(y_ip) - signed('0' & x_ip);
z_array(z_array'low) <=signed(z_ip) + tan_array(0);
else
x_array(x_array'low) <=signed(x_ip) - signed('0' & y_ip);
y_array(y_array'low) <=signed(y_ip) + signed('0' & x_ip);
z_array(z_array'low) <=signed(z_ip) - tan_array(0);
end if;
for i in 1 to 14 loop
if z_array(i-1) < to_signed(0,17)
then
x_array(i) <= x_array(i-1) +(y_array(i-1)/2**i);
y_array(i) <= y_array(i-1) - (x_array(i-1)/2**i);
z_array(i) <= z_array(i-1) +tan_array(i);
else
x_array(i) <= x_array(i-1) - (y_array(i-1)/2**i);
y_array(i) <= y_array(i-1) +(x_array(i-1)/2**i);
z_array(i) <= z_array(i-1) -tan_array(i);
end if;
end loop;
end if;
end process;
cos_op <=std_logic_vector(x_array(x_array'high)(16 downto 0));
sin_op <=std_logic_vector(y_array(y_array'high)(16 downto 0));
end architecture rtl;
```

DESIGN EXAMPLE 4.20

Develop a digital system to implement coordinate conversion and the CORDIC algorithm in the FPGA [25,44,45].

Solution

```
package eight_bit_int is -- user-defined types
subtype byte is integer range -128 to 127;
type array_byte is array (0 to 3) of byte;
end eight_bit_int;
library work;
use work.eight_bit_int.all;
library ieee;
use ieee.std_logic_1164.all;
use ieee.std_logic_arith.all;
entity cordic is port (clk : in std_logic:='0';
x_in, y_in : in byte;
r, phi, eps : out byte);
end cordic;
architecture fpga of cordic is
signal x, y, z : array_byte:= (0,0,0,0);
begin
process
begin
wait until clk = '1'; r <= x(3);
phi <= z(3);
eps <= y(3);
if y(2) >= 0 then
x(3) <= x(2) + y(2) /4;
y(3) <= y(2) - x(2) /4;
z(3) <= z(2) + 14;
else
x(3) <= x(2) - y(2) /4;
y(3) <= y(2) + x(2) /4;
z(3) <= z(2) - 14;
```

```
end if;
if y(1) >= 0 then
x(2) <= x(1) + y(1) /2;
y(2) <= y(1) - x(1) /2;
z(2) <= z(1) + 26;
else
x(2) <= x(1) - y(1) /2;
y(2) <= y(1) + x(1) /2;
z(2) <= z(1) - 26;
end if;
if y(0) >= 0 then
x(1) <= x(0) + y(0);
y(1) <= y(0) - x(0);
z(1) <= z(0) + 45;
else
x(1) <= x(0) - y(0);
y(1) <= y(0) + x(0);
z(1) <= z(0) - 45;
end if;
if x_in >= 0 then
x(0) <= x_in;
y(0) <= y_in;
z(0) <= 0;
elsif y_in >= 0 then
x(0) <= y_in;
y(0) <= - x_in;
z(0) <= 90;
else
x(0) <= - y_in;
y(0) <= x_in;
z(0) <= -90;
end if;
end process;
end fpga;
```

4.6 Multiply-Accumulation Circuit

As integrated circuit technology has significantly improved to allow more and more components on a single chip, digital systems have continued to grow for the design of complex systems. Many digital signal processors have been designed to perform various mathematical operations with digitally represented signals, that is, 1 s and 0 s. Almost all DSP applications require critical operations, including many multiplications and accumulations. Therefore, high-throughput MAC units have become significant as a key element to achieve the intended real-time DSP applications at the desired speed. An architecture of a MAC unit is shown in Figure 4.18. In the last few years, the main concentration in MAC unit design was on improving its computation speed and reducing the area, because the throughput rate and size of the device are the main concerns for any DSP system. Most DSP applications require computation as the SOP of a series of successive multiplications and additions. In order to implement such functions, MAC units are preferred [25,39,46–48]. A MAC unit consists of a multiplier, adder/subtractor, accumulation register, and output latch. MAC units are used to implement functions of the type $y = \sum_{i=0}^{N-1} x_i y_i$, where y is the final result and x_i and y_i are the inputs. Although addition and multiplication are two different operations, they can be performed in parallel because the MAC is

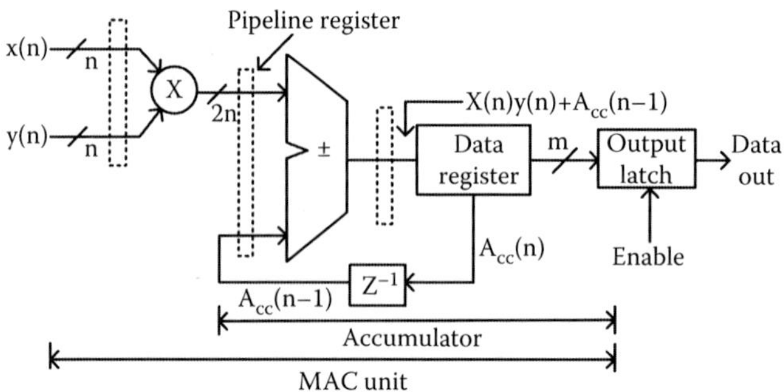

FIGURE 4.17
A typical MAC unit with pipelined register.

designed with pipelined registers. By the time the multiplier has computed the nth product, the adder/subtractor adds the (n−1)th values and the accumulator stores the (n − 2)nd value. Thus, if N products need to be done, N − 1 multiplications can overlap with N − 1 additions. During the very first multiplication, the accumulator will be idle, and during the last accumulation, the multiplier will be idle. Thus, N + 1 clock pulses are required

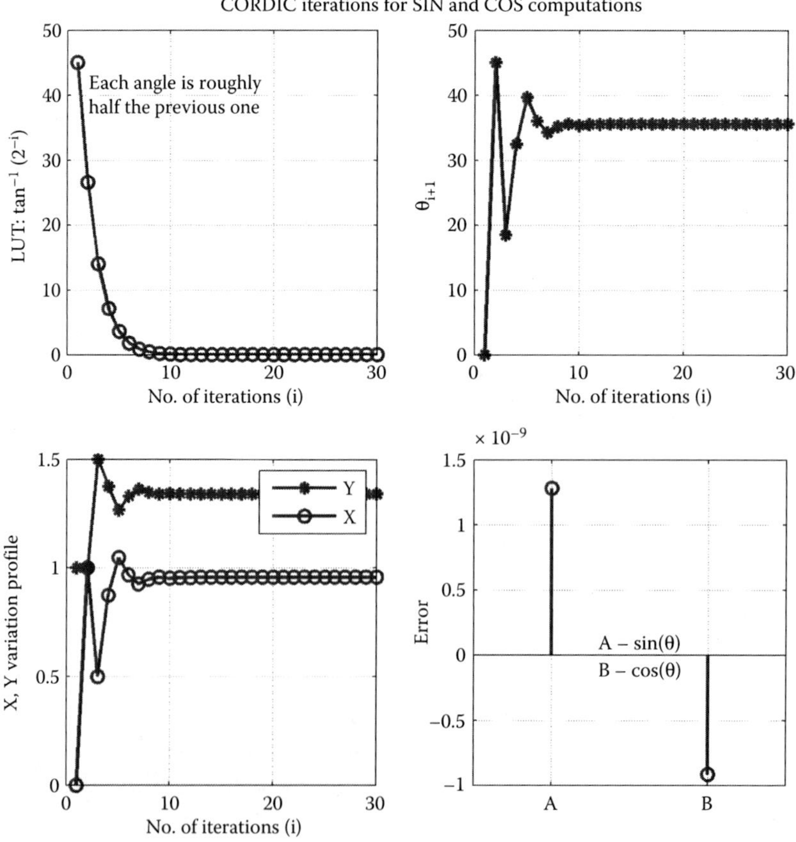

FIGURE 4.18
Simulation result of the CORDIC algorithm.

to compute the sum of N products. The processing time depends on the amount of hardware involved in the MAC unit design. For example, suppose we wish to add the result of 256 multiplications using a MAC unit. If the MAC execution time for one full cycle is 100 nsec, then the total time required to complete the operation is $(N + 1) \times 100E{-}9 = (257 \times 100E{-}9) = 25.7\ \mu sec$.

Another important aspect of designing a MAC unit is the word sizes that we encounter at the input of the multiplier and the sizes of the add/subtract unit and the accumulator, as there is a possibility of overflow and underflow. Over/underflow can be avoided by using any of the following methods:

1. Using level shifters at the input/output of the MAC unit: shifters can be provided at the input of the MAC unit to normalize the data and at the output to denormalize the same.
2. Providing guard bits in the accumulator: as the normalization process does not yield accurate results, it is not desirable for some applications. In such cases, we have another alternative: providing additional bits called guard bits in the accumulator so that there will not be any overflow error. Here, the add/subtract unit also has to be modified appropriately to manage the additional bits of the accumulator. Consider a MAC unit whose inputs are 16-bit numbers. If 256 products are to be summed up in the MAC unit, the number of guard bits necessary for the accumulator to prevent the overflow condition for the sum of 256 multiplications of 16 bits is $(16 + \log_2 256) = 24$ bits. Hence, the accumulator should be capable of handling these 24 bits. Thus, the guard bits required will be $24 - 16 = 8$ bits.
3. Using saturation logic: over/underflow will occur if the result goes beyond the most positive number or below the least negative number that the accumulator can handle. Thus, the over/underflow error can be resolved just by loading the accumulator with the most positive number it can handle at the time of overflow and the least negative number it can handle at the time of underflow [46–48]. This method is called saturation logic.

The result of the adder/subtractor will sit in the data register and come out to the output latch as well as the delay element. The output latch holds the data until getting the enable control signal. Once the multiplication, addition/subtraction and accumulation are completed for the given samples, the Enable goes logic 1, which drives the output latch to bring out the accumulated values, as shown in Figure 4.18. Thus, considering all these aspects, a suitable MAC unit can be designed for the intended DSP applications.

DESIGN EXAMPLE 4.21

Design and develop a MAC unit to multiply and accumulate N samples. Assume $N = 10$ and reset the accumulator once the final value is latched out. Display the final result using the LEDs.

Solution

```
library ieee;
use ieee.std_logic_1164.all;
use ieee.numeric_std.all;
```

```
use ieee.std_logic_unsigned.all;
entity muladd is
generic(N:integer:=10);
port(in1,in2:in std_logic_vector(7 downto 0);
clk,nrst:in std_logic;
macout:out std_logic_vector(N+16 downto 0));
end entity;
architecture mark0 of muladd is
signal mulreg:integer range 0 to 65537;
signal addreg,val:std_logic_vector(N+16 downto 0);
signal temp:std_logic_vector(N-1 downto 0);
signal accm:std_logic_vector(16 downto 0);
signal cnt:integer range 0 to N;
begin
mulreg <= to_integer(unsigned(in1)) * to_integer(unsigned(in2));
accm <= std_logic_vector(to_unsigned(mulreg,17));
addreg <= val + (temp&accm);
process(clk,nrst)
begin
if(nrst='0') then
val <= (others=>'0');
temp <= (others=>'0');
cnt <= 0;
else
if(falling_edge(clk)) then
val <= addreg;
if(cnt=N) then
macout <= addreg;
cnt <= 0;
val <= (others=>'0');
else
cnt <= cnt + 1;
end if;
end if;
end if;
end process;
end architecture;
```

4.7 Arithmetic and Logical Unit

An ALU is a combinational circuit that performs either logic or arithmetic operations with its inputs based on the status of its selection inputs. A simplified representation of an ALU is shown in Figure 4.19a. In this section, we learn how to design an ALU with its necessary digital subsystems. In some digital controllers and processors, the ALU is divided into two units, namely an arithmetic unit (AU) and a logic unit (LU). Some processors contain more than one AU, for example, fixed-/floating-point processors. Typically, the ALU has direct input/output access options in many processors and controllers, like memory and input/output devices. The input consists of an instruction word that contains an operation/format code. The operation code tells the ALU what operation it needs to perform; for example, two operands may be added or compared logically. The format code is combined with the operation code and indicates whether the process is in a fixed-point or floating-point resolution. The output consists of a final result that is stored in a register with a provision to latch it out whenever the entire computation, for example,

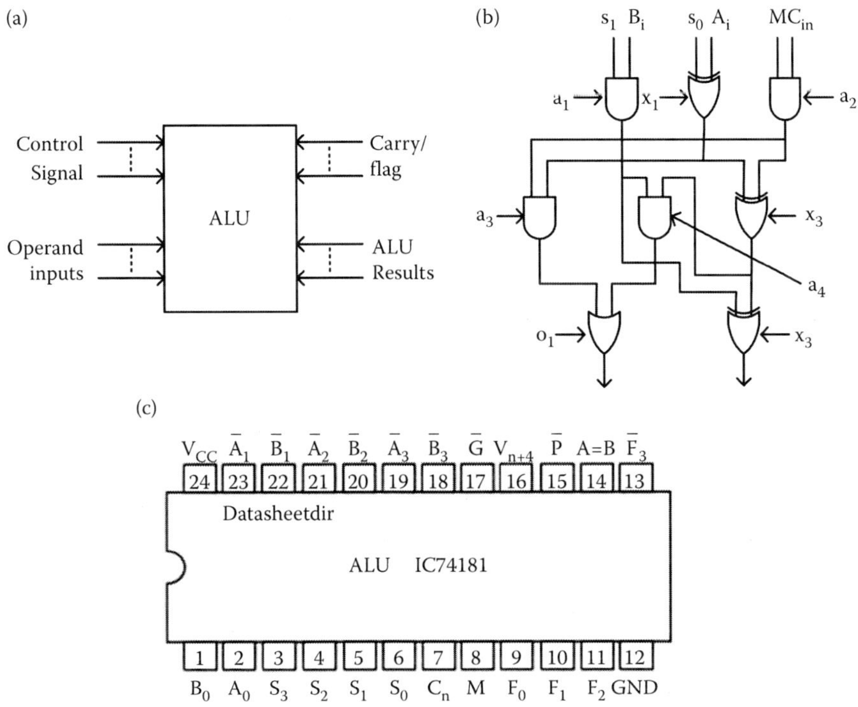

FIGURE 4.19
Entity of a basic ALU (a), single-bit ALU circuit module (b), and pin details of a standard ALU: IC74181 (c).

multiplication and accumulation, is over. In some ALUs, there is a provision to check the current status of the device; that is, whether the computation is still on or completed, which is called the machine status bit. In many processors/controllers, the ALU has storage elements for storing the input operands that are being added and accumulated or shifted and so on. The flow of input bits and the operations being performed with them in the subunits of the ALU are controlled by pipelined registers or gated circuits [39,47–49], as shown in Figure 4.19b. The gates in these circuits are controlled by a sequence logic unit that uses a particular algorithm or sequence for each operational code. In the AU, multiplication and division are performed by a series of additions and subtractions, respectively, with shifting operations.

Now, we start our discussion of the design of an ALU. An n-bit ALU typically has two input words, denoted by $A_i = (A_{n-1}, \ldots, A_0)$ and $B_i = (B_{n-1}, \ldots, B_0)$, where 0 and (n − 1) represent the LSB and MSB of the inputs `A` and `B`. The value of `i` varies from 0 through (n − 1) as $i = 0, 1, 2, \ldots, n - 1$, and it is called the sample index. The output word is denoted by $F_i = F_n, F_{n-1}, \ldots, F_0$, where the high-order output bit `Fn` is actually the final carry. Besides the data inputs and outputs, an ALU must have selection inputs to choose the operations to be performed. One selection input is mode selection (`M`). When `M` = 0, the operation is a logic function, and when `M` = 1, it is the arithmetic operation. In addition to mode selection, there are some other inputs called operation selection inputs (`Si`) which determine the specific operations, either in logical or arithmetic functions with respect to the value of `M`. Consider Table 4.15, which gives some specific operations of a simple ALU for `M` = 0 and `M` = 1 with the selective inputs "00" through "11". We will analyze the arithmetic and

TABLE 4.15
Design of a Single-Bit ALU Module

		Function Selection Control			Example – O/Ps (I/Ps: $A_i = 1$ and $B_i = 0$)	
Mode (M)	**Carry in (C_{in})**	S_0	S_1	**ALU Function**	C_{i+1}	F_i
0	X	0	0	F_i<=A_i	0	1
		0	1	F_i<=not A_i	0	0
		1	0	F_i<=A_i xor B_i	0	1
		1	1	F_i<=A_i xnor B_i	0	0
1	0	0	0	F_i<=A_i	0	1
		0	1	F_i<=not A_i	0	0
		1	0	F_i<=$A_i + B_i$	0	1
		1	1	F_i<=not (A_i) + B_i	0	0
1	1	0	0	F_i<=$A_i + C_{in}$	1	0
		0	1	F_i<=not (A_i) + C_{in}	0	1
		1	0	F_i<=$A_i + B_i + C_{in}$	1	0
		1	1	F_i<=not (A_i) + $B_i + C_{in}$	0	1

logical operations of the ALU shown in Figure 4.19b in accordance with different values of M. The list of operations that can be performed using the ALU circuit shown in Figure 4.19b shows the logical and arithmetic operations with carry-in (C_{in}) 0 and 1. ALUs are relatively simple to implement by designing a 1-bit ALU module and cascading as many as we need to build a multiple-bit structure. Of course, in a multiple-bit structure, the performance is limited because of the propagation of carries among the ALU stages. A single-bit ALU module, shown in Figure 4.19b, has six inputs: A_i, B_i, C_i, M, S_1, and S_0, and two outputs: F_i and C_{i+1}. The logical and arithmetic operations of this single-bit ALU module for different values of M, S_1, and S_0 are given in Table 4.15, where the outputs C_{i+1} and F_i are given against all the logical and arithmetic functions for the specified inputs $A_i = 1$ and $B_i = 0$ [49].

We will discuss and analyze data flow in a single-bit ALU module for its logical and arithmetic operations. The different logical and arithmetic functions for the possible values of M, C_{in}, S_1, and S_0 are given in Table 4.15. According to the design, the logical expressions for the output and C_{i+1} can be deduced from the gates x_3 and o_1 as

$$F_i = S_iB_i \oplus ((S_0 \oplus A_i) \oplus MC_{in}) \text{ and}$$
$$C_{i+1} = MC_{in}(S_0 \oplus A_i) + S_1B_i(S_0 \oplus A_i \oplus MC_{in}) \quad (4.13)$$

which can be implemented by the following VHDL code.

```
library IEEE;
use IEEE.STD_LOGIC_1164.ALL;
use IEEE.STD_LOGIC_ARITH.ALL;
```

```
use IEEE.STD_LOGIC_UNSIGNED.ALL;
entity clkd is
Port (Ai,Bi,Cin,M,S0,S1:in STD_LOGIC:='0';
Fi,Cip1: out std_logic:='0');
end clkd;
architecture Behavioral of clkd is
begin
p1:process(Ai,Bi,Cin,M,S0,S1)
begin
Fi<=((S1 and Bi)xor ((S0 xor Ai) xor (M and Cin)));
Cip1<=(((M and Cin) and (S0 xor Ai)) or ((S1 and Bi) and ((S0 xor Ai)
xor (M and Cin))));
end process;
end Behavioral;
```

This implementation shows that when S_1 is 0, B_i is blocked from affecting the outputs by the and gate a_1; hence, the operation depends on only A_i, as given in Equation 4.13. The same is true for C_{in} as when M is 0 at and gate a_2. Addition is indicated whenever M = 1. In this case, B_i (assuming $S_1 = 1$) and C_i are passed to the inputs of the xor gates x_3 and x_2, respectively. When S0 is 0, the topmost xor gate x_1 simply passes A_i. When S_0 is 1, it passes the complement of A_i, that is, A_i'. Thus, the three cascaded xor gates form a circuit to produce the final result based on the values of the control selection and actual inputs. The first product term is formed from gate A_3, the second from A_4 and finally an or gate produces the carry C_{i+1}. If $S_1 = 0$, we simply replace B_i with 0 in Equation 4.13 to obtain the correct carry function. When the ALU is in logic mode, M = 0, we can concentrate on the cascaded xor gates. When S_0 and S_1 are both 0, A_i is passed through to the output. If S_0 is 1 while S_1 is 0, the complement of A_i is passed. When S0 is 0 while S_1 is 1, the inputs to X_3 are A_i and B_i, and their xor is computed. In the last case, $S_0 = 1$, $S_1 = 1$, the inputs to x_3 are (A_i' xor B_i') when M = 0. This function is equivalent to the XOR function.

The most widely used ALU in TTL designs is IC74181, whose schematic (pin details) is shown in Figure 4.19c. This ALU IC has two groups of 4-bit data inputs (A_0, A_1, A_2, A_3 and B_0, B_1, B_2, B_3), a carry-in (C_n), a single 4-bit data output (F_0, F_1, F_2, F_3) and a carry-out (C_{n+} 4). Many of the input and output logics in this ALU are active low, as shown in Figure 4.19c. This ALU also has four selection inputs (S_0, S_1, S_2, S_3) to choose the appropriate function we wish to perform and a mode bit M to select either logical or arithmetic function. This allows the ALU to compute 32 ($2^4 + 2^1 = 32$) different logical and arithmetic functions. The IC74181 can also be used as a comparator, and it has an open-collector A = B output for easy cascading across multiple stages.

The various logical and arithmetic operations of this ALU are given in Table 4.16, where the inputs and outputs are assumed as the negative logic data. The ALU implements the logic function `nand`, `nor`, `and`, `or`, `xor`, and `xnor` gates, as well as arithmetic plus and minus operations. The ALU chip contains internal logic that implements a 4-bit carry look-ahead adder. Its outputs can be propagated to cascade ALU stages and carry look-ahead units to construct arithmetic units of larger bit widths. Very often, few of the functions available in the IC74181 are used in practice. In other words, to select the first column of arithmetic functions in positive logic, the carry in (C_{in}) must be set to 1. More information about the working principles and interfacing of this ALU IC can be found in References 39 and 49]. We discuss the design of the ALU inside the FPGA with some design examples in the subsequent part.

TABLE 4.16

Function Table of ALU: IC74181

Selection Inputs				Logical Function	Arithmetic Function	
					M = 0	
S_3	S_2	S_1	S_0	M = 1	C_{in}=0	C_{in} = 1
0	0	0	0	F_i<= not Ai	F_i<=Ai-1	F_i<=Ai
0	0	0	1	F_i<= Ai nand Bi	F_i<=AiBi-1	F_i<=AiBi
0	0	1	0	F_i<= (not Ai) + Bi	F_i<=Ai(not Bi)-1	F_i<=Ai(not Bi)
0	0	1	1	F_i<=1	F_i<=-1	F_i<=0
0	1	0	0	F_i<=Ai nor Bi	F_i<=Ai+(Ai+not Bi)	F_i<=Ai + (Ai + not Bi) + 1
0	1	0	1	F_i<=not Bi	F_i<=AiBi+(Ai+not Bi)	F_i<=AiBi + (Ai + not Bi) + 1
0	1	1	0	F_i<= Ai xnor Bi	F_i<=Ai-Bi-1	F_i<=(Ai + not Bi) + 1
0	1	1	1	F_i<=Ai + not Bi	F_i<=Ai+not Bi	F_i<=Ai-Bi
1	0	0	0	F_i<=(not Ai)Bi	F_i<=Ai+(Ai + Bi)	F_i<=Ai + (Ai + Bi) + 1
1	0	0	1	F_i<=Ai xor Bi	F_i<=Ai+Bi	F_i<=Ai + Bi + 1
1	0	1	0	F_i<=Bi	F_i<=Ai(not Bi)+(Ai+Bi)	F_i<=Ai(not Bi) + (Ai + Bi) + 1
1	0	1	1	F_i<=Ai+Bi	F_i<=Ai+Bi	F_i<=(Ai + Bi) + 1
1	1	0	0	F_i<=0	F_i<=Ai	F_i<=Ai + 1
1	1	0	1	F_i<=Ai (not Bi)	F_i<=AiBi+Ai	F_i<=AiBi + Ai + 1
1	1	1	0	F_i<=AiBi	F_i<=Ai(not Bi)+Ai	F_i<=Ai(not Bi) + Ai + 1
1	1	1	1	F_i<=Ai	F_i<=Ai	F_i<=Ai + 1

DESIGN EXAMPLE 4.22

Design a digital circuit to implement an ALU in FPGA. The design has to be in the structural model, and the width of the two inputs is 8 bits.

Solution

As given in the design objective, we need to follow the structural model design; hence, we build the components as an arithmetic unit, logical unit and multiplexer unit. Finally, we link all these components in the main code as given below.

Component – Arithmetic Unit

```
entity arith_unit is
port (a, b : in std_logic_vector(7 downto 0);
sel: in std_logic_vector(2 downto 0);
cin: in std_logic_vector;
x: out std_logic_vector (7 downto 0));
end arith_unit;
architecture arith_unit of arith_unit is
signal arith, logic : std_logic_vector (7 downto 0);
begin
with sel select
x<= a when "000"
```

```
a + 1 when "001",
a-1 when "010",
b when "011",
b + 1 when "100",
b-1 when "101",
a + b when "110",
a + b + cin when others;
end arith_unit;
```

Component – Logic Unit

```
entity logic_unit is
port (a, b : in std_logic_vector (7 downto 0);
sel : in std_logic_vector (2 downto 0);
x : out std_logic_vector (7 downto 0));
end logic unit;
architecture logic_unit of logic_unit is
begin
with sel select
x <=      not a when "000",
not b when "001",
a and b when "010",
a or b when "010",
a nand b when "100",
a nor b when "101",
a xor b when "110",
not (a xor b) when others;
end logic_unit;
```

Component – Multiplexer

```
entity mux is
port (a, b: in std_logic _vector (7 downto o);
sel: in std_logic;
x: out std_logic_vector (7 downto 0));
end mux;
architecture mux of mux is
begin
with sel select
x <= a when '0',
b when others;
end mux;
```

Main Code

```
entity alu is
port (a, b: in std_logic_vector (7 downto 0);
cin: in std_logic;
sel: in std_logic_vector(3 downto 0);
y: out std_logic_vector(7 downto 0));
end alu;
architecture alu of alu is
component arith_unit is
port (a, b: in std_logic_vector (7 downto 0);
cin: in std_logic;
sel: in std_logic_vector(2 downto 0);
x: out std_logic_vector(7 downto 0));
end component;
component logic_unit is
port (a, b: in std_logic_vector (7 downto 0);
sel: in std_logic_vector(2 downto 0);
x: out std_logic_vector(7 downto 0));
end component;
```

```
component mux is
port (a, b: in std_logic_vector (7 downto 0);
sel: in std_logic;
x: out std_logic_vector(7 downto 0));
end component;
signal x1,x2 : std_logic_vector(7 downto 0);
begin
u1 : arith_unit port map (a, b, cin, sel (2 downto 0), x1);
u2 : logic_unit port map (a, b, sel(2 downto 0), x2);
u3 : mux port map (x1, x2, sel(3), y);
end alu;
```

4.8 Read-Only Memory Design and Logic Implementations

ROM is essentially a memory device in which permanent binary information can be stored. The binary information must be specified by the designer and then implemented in the ROM to form the required interconnection pattern. Once the pattern is created, it stays within the unit even when power is turned off and on again [50]. Roughly similar to ROM, the other two combinational devices are PLA and PAL. ROM, PLA, and PAL are storage components. It may be noticed that there is a contradiction because we learned that combinational devices do not have memory, for which we find the answer below. A block diagram of a simple ROM consisting of k inputs and n outputs is shown in Figure 4.20a. The inputs provide the address for memory, that is, the decoder, and the outputs give the data bits of the stored word that is selected by the given address. As in the figure, 2^k possibilities of address can be given to the ROM decoder. The ROM decoder enables the appropriate decoded line among 2^k possible lines in accordance with the decoder address. The selected, that is, decoded, line produces the output based on the logic embedded with it at the programmable output logic section. The number of outputs depends on the number of logical functions we wish to embed. The ROM unit does not have any data inputs other than enable input, because it does not have a write operation. Now, we will discuss how to embed a function within a ROM unit. Consider Figure 4.20b, which corresponds to the internal logic construction of a ROM unit of (k − 1) address

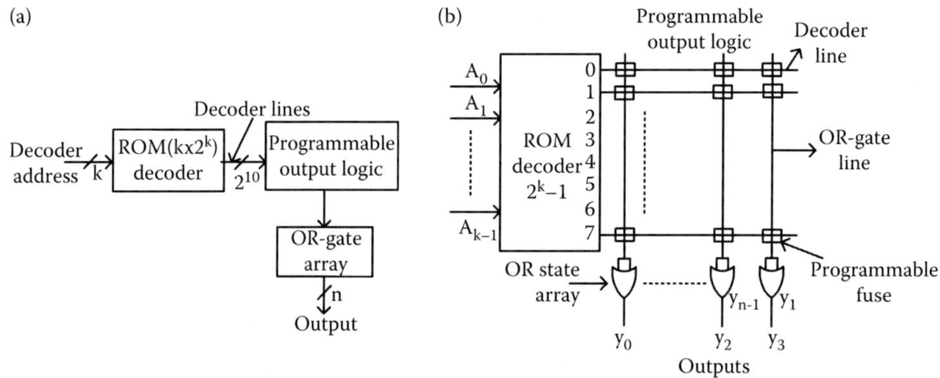

FIGURE 4.20
Entity of a ROM unit (a) and internal logic of (k−1) x (n−1) ROM unit (b).

lines, 2^k decoded lines and (n – 1) output lines. The (k – 1) inputs are decoded into 2^k distinct decoder lines by the decoder. Each line of the decoder represents a memory address in the output logic. Finally, the output of the output logic is connected to the or gate as per the logical functions in which we are interested. Each or gate can have maximum of 2^k connection/words/inputs. Thus, inside the output logic unit, a maximum of ($2^k \times n$) internal connections exists.

All these interconnections, shown by a square box in Figure 4.20b, are programmable. A programmable connection between two lines, that is, a decoder line and or gate line, is logically equivalent to a switch that can be altered to be either closed (two lines are connected) or open (two lines are opened). The internal binary storage at the output logic, that is, creating a connection between the decoder and or gate lines, of a ROM is specified by a truth table that shows the word content of each address. Now, we will demonstrate this logic implementation with one example. For example, consider a ROM with k = 3, that is, 0 through 2, and n = 4, that is, 0 through 3. Because k = 4, the decoder lines are $2^k = 8$, that is, 0 through 7. The logic we wish to implement in this ROM is given in Table 4.17. The truth table gives the three inputs under which are listed all seven addresses. Each address stores a word of 4 bits, which is listed in the output columns. The hardware procedure that programs the ROM blows fuse links in accordance with a given truth table. Programming the ROM according to the truth table given in Table 4.17 results in the output logic configuration shown in Figure 7.21. Every 0 listed in the truth table specifies the absence of a connection and every 1 listed specifies a path that is obtained by a connection. For example, the table specifies the 4-bit word "0110" for permanent storage at address 7. The two 0s in the word are programmed by blowing the fuse links between output 7 of the decoder, and the first and fourth inputs of the `or` gates associated with outputs y_3 and y_0. The two 1s in the word are marked with an "x" to denote a temporary connection between the seventh output of the decoder and the second and third inputs of the or gate corresponding to the outputs y_1 and y_2. When the address input of the ROM is "111", all the outputs of the decoder are 0 except for output 7, which is at logic 1. The signal equivalent to logic 1 at decoder output 7 propagates through the connections to the or gate outputs of y_1 and y_2, producing logic 1.

The other two outputs remain at logic 0. The result is that the stored word "0110" is applied to the four data outputs. This is the way this combinational circuit acts as a

TABLE 4.17

ROM Truth Table

Address Inputs (k = 3)			Outputs (n = 4)			
A_2	A_1	A_0	Y_3	Y_2	Y_1	Y_0
0	0	0	1	1	1	1
0	0	1	0	1	1	0
0	1	0	1	0	0	1
0	1	1	1	1	0	0
1	0	0	1	0	0	0
1	0	1	1	1	1	0
1	1	0	0	0	1	1
1	1	1	0	1	1	0

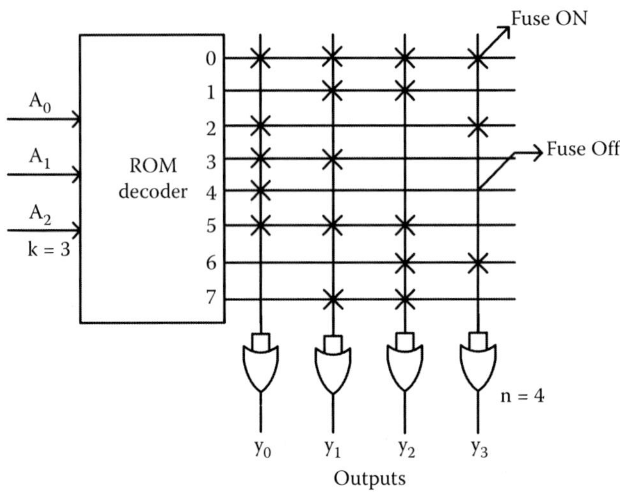

FIGURE 4.21
Programming the ROM according to Table 4.17.

memory device. The `or` gates are inserted to sum the minterms of the Boolean functions we implemented at the output logic, through which we are able to generate any desired combinational logic. In general, the ROM is a device that includes both the decoder and the `or` gates within a single device to form a minterm generator. By choosing connections for those minterms which are included in the function, the ROM outputs can be programmed to represent the Boolean functions of the output variables in a combinational circuit [50,51]. Each output terminal, that is, y_0 through y_3, is considered separately as the output of a Boolean function expressed as an SOP minterm. For example, the ROM shown in Figure 4.21 may be considered a combinational circuit with four outputs, each a function of the three input variables. Output y_2 can be expressed in a SOP form as y_2`(A2, A1, A0)` $=\sum$`(0, 1, 3, 5, 7)`. The connection marked with "x" in the figure produces a minterm for the sum. All other crosspoints are not connected and are not included in the sum, i.e., for y_2. The `AND` array is the portion of the ROM device that decodes the inputs. The `AND` array determines the minterms decoded by the device. A ROM decodes all possible minterms. The `OR` array is the portion of the device that combines the minterms for the definition of a function.

The advantages of implementing the logical functions, i.e., storing the data, in ROM is that there is a need of only one IC to purchase and fewer connections, unlike implementing logical functions using SSI devices, where we need several basic gates. A ROM can be programmed at the factory or in the field. ROMs programmed at the factory are called mask ROMs because during fabrication, the circuit patterns are determined by a mask. There are several different types of field-programmable ROMs, namely, programmable read-only memory (PROM), erasable programmable read-only memory (EPROM), electrically erasable programmable read-only memory (EEPROM), and flash memory. There are another two types of popular programmable logic devices, namely PLAs and PALs. A PLA has a programmable `AND` array and a programmable `OR` array. A PLA with n inputs has fewer than 2^n `and` gates, unlike the ROM decoder, which is the advantage of PLA over a ROM implementation of the same function. A PLA only needs to have enough `AND` gates to

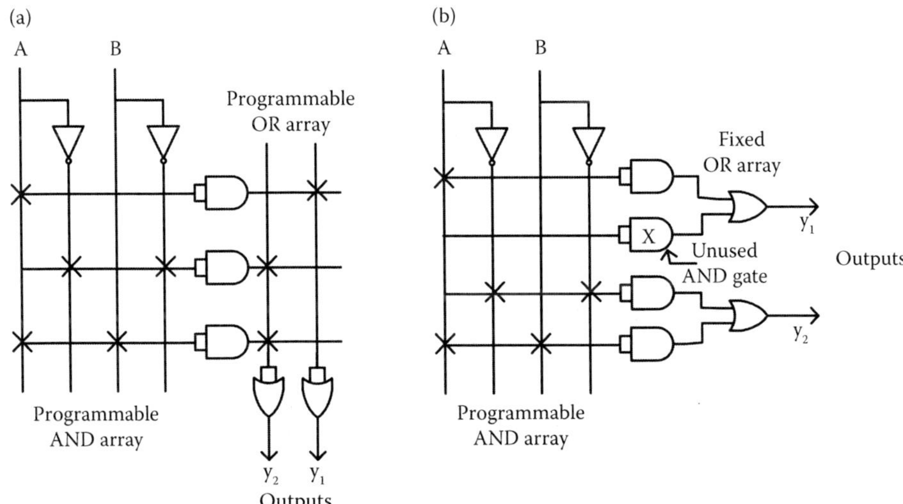

FIGURE 4.22
Function implementation using PLA (a) and using PAL (b).

decode as many unique terms as there are in the functions we wish to implement. Because we can control the and array and there is a limit to the number of minterm terms that can be specified in the and array. Now, we implement two functions, $y_1 = A$ and $y_2 = A'B' + AB$, using the PLA as shown in Figure 4.22a. Notice that we need only three and gates because there are only three unique minterms in the functions y_0 and y_1. Also, notice that since we need to program the OR array, we can share the minterms AB and A′B′ as per the definitions of both functions, as shown in Figure 4.22a. Next, we see the implementation of the same function using the PAL.

The PAL has a programmable AND array and a fixed OR array. Here, the designer cannot program the or array. A fixed OR array makes the device less expensive to manufacture. On the other hand, because of the fixed OR array, we can't share the minterm terms between functions. For example, implementing the above functions y_1 and y_2 using a PAL is shown in Figure 4.22b. However, we need to confirm sufficient or gates are available in the fixed or array to implement the function we wish to embed.

DESIGN EXAMPLE 4.23

Design and implement a circuit that accepts a three-bit number and outputs a binary number equal to the square of the input number.

Solution

As given in the design example, there are 3 bits at the input; hence, its maximum value is 7; thus, the maximum square value is 49, so we need 6 bits of output. That means that here, $k = 3$, $2^k = 8$, and $n = 6$. Now, we prepare the truth table, which is given in Table 4.18 below.

The interesting observations from Table 4.18 are (i) the output y_0 is equal to A_0 and (ii) y_1 is always zero; thus, there is no need to program the or array for two outputs: the y_0 can be directly taken from the input A_0, and y_1 can permanently be assigned to logic 0. Then, the ROM design becomes as shown in Figure 4.23.

TABLE 4.18

ROM Truth Table for Design Example 4.23

Address (k = 3)			Outputs (n = 6)					
A_2	A_1	A_0	y_5	y_4	y_3	y_2	y_1	y_0
0	0	0	0	0	0	0	0	0
0	0	1	0	0	0	0	0	1
0	1	0	0	0	0	1	0	0
0	1	1	0	0	1	0	0	1
1	0	0	0	1	0	0	0	0
1	0	1	0	1	1	0	0	1
1	1	0	1	0	0	1	0	0
1	1	1	1	1	0	0	0	1

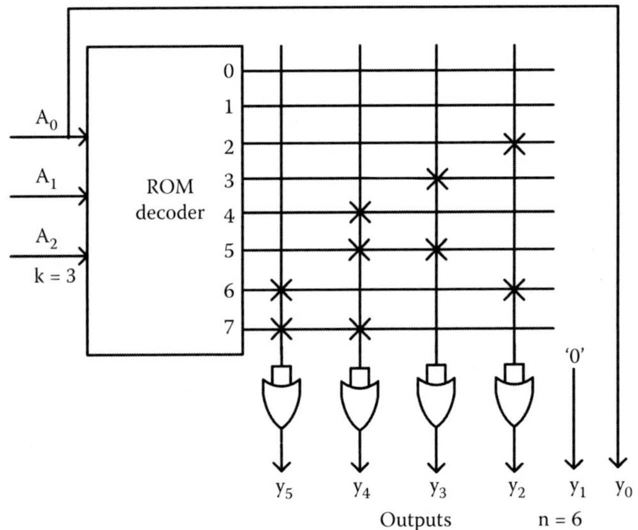

FIGURE 4.23
ROM implementation of Design Example 4.23.

DESIGN EXAMPLE 4.24

Design and develop a digital system to realize a simple ROM unit in the FPGA.

Solution

```
library ieee;
use ieee.std_logic_1164.all;
use ieee.std_logic_arith.all;
use ieee.std_logic_unsigned.all;
entity ROM_Des is
port(clock : in std_logic;
read : in std_logic;
address : in std_logic_vector(1 downto 0);
data_out: out std_logic_vector(7 downto 0));
end ROM_Des;
```

```
architecture beha of ROM_Des is
type rom_array is array (0 to 3)of std_logic_vector(7 downto 0);
constant content: rom_array := (
0 => "10000001",
1 => "10000010",
2 => "10000011",
3 => "10000100",
others => "11111111");
begin
process(clock)
begin
if(clock'event and clock = '0') then
if(read = '1') then
data_out <= content(conv_integer(address));
else
data_out <= "ZZZZZZZZ";
end if;
end if;
end process;
end beha;
```

DESIGN EXAMPLE 4.25

Design and develop a digital system to realize a simple ROM unit in the FPGA using the case statement with enable input.

Solution

```
library ieee;
use ieee.std_logic_1164.all;
use ieee.std_logic_unsigned.all;
use ieee.std_logic_arith.all;
entity ROM_Desi is
port (ce:in std_logic;
read_en :in std_logic;
address :in std_logic_vector (3 downto 0);
data :out std_logic_vector (7 downto 0) );
end entity;
architecture behavior of ROM_Desi is
begin
process (read_en, address) begin
if (read_en = '1') then
case (address) is
when x"0" => data <= conv_std_logic_vector(10,8);
when x"1" => data <= conv_std_logic_vector(55,8);
when x"2" => data <= conv_std_logic_vector(244,8);
when x"3" => data <= (others=>'0');
when x"4" => data <= conv_std_logic_vector(1,8);
when x"5" => data <= x"ff";
when x"6" => data <= x"11";
when x"7" => data <= x"01";
when x"8" => data <= x"10";
when x"9" => data <= x"00";
when x"a" => data <= x"10";
when x"b" => data <= x"15";
when x"c" => data <= x"60";
when x"d" => data <= x"90";
when x"e" => data <= x"70";
when x"f" => data <= x"90";
when others => data <= x"00";
end case;
```

```
end if;
end process;
end architecture;
```

DESIGN EXAMPLE 4.26

Design and develop a digital system to realize a simple ROM unit in the FPGA using a constant array.

Solution

```
library ieee;
use ieee.std_logic_1164.all;
use ieee.std_logic_unsigned.all;
use ieee.std_logic_arith.all;
entity ROM_Desi is
port (read_en :in std_logic;
address :in std_logic_vector (3 downto 0);
data :out std_logic_vector (7 downto 0) );
end entity;
architecture behavior of ROM_Desi is
subtype rom_word is std_logic_vector (7 downto 0);
subtype rom_addr is integer range 0 to 15;
type rom is array (rom_addr) of rom_word;
constant rom_table :rom := (
conv_std_logic_vector(10,8),
conv_std_logic_vector(55,8),
conv_std_logic_vector(244,8),
"00000000",
conv_std_logic_vector(1,8),
x"ff",
x"11",
x"01",
x"10",
x"00",
x"10",
x"15",
x"60",
x"70",
x"90",
x"00");
begin
process (read_en, address) begin
if (read_en = '1') then
data <= rom_table(conv_integer(address));
end if;
end process;
end architecture;
```

4.9 Random Access Memory Design

RAM is a type of digital data storage device which is mainly used in computers as main memory. RAM allows access of data in any order, that is, random; thus, any piece of data can be returned at any time regardless of its physical location and previous operation. It is possible to access any memory cell directly if we know the address of the row and column that intersect at that cell. Most RAM chips are volatile types of memory, where the information is lost after the power is switched off. Two popular RAM types are static random access memory (SRAM) and dynamic random access memory (DRAM). SRAM uses bistable latching circuitry to store each bit, and it is volatile memory. Each bit in SRAM is stored on four transistors that form two cross-coupled inverters. This storage cell has two stable states which are used to denote 0 and 1. Two additional access transistors serve to control the access to a storage cell during read and write operations [39,50,51]. As SRAM does not need to be refreshed, it is faster than other types, but each cell uses at least six transistors, which is very expensive. Thus, SRAM is used for faster access to memory units in the CPU. In DRAM, a transistor and a capacitor are paired to create a memory cell which represents a single bit of data, where the capacitor holds the bit of information. The transistor acts as a switch that lets the control circuitry on the memory chip read the capacitor or change its state. As capacitors leak charge, the information eventually fades unless the capacitor charge is refreshed periodically. Because of this refresh process, it is dynamic memory [50]. The advantage of DRAM is its structural simplicity. As it requires only one transistor and one capacitor per one bit, high density can be achieved. Hence, DRAM is cheaper and slower when compared to SRAM. Some of the other types of memories are fast page mode (FPM) DRAM, extended data out (EDO) DRAM, synchronous dynamic (SD) RAM, double data rate (DDR) SDRAM, and double data rate two (DDR2) SDRAM. Discussion and design of these memory types are out of the scope of this book. Therefore, we now concentrate only on RAM design.

An entity of a RAM unit is shown in Figure 4.24a, where the unit has k bits of address lines, m bits of data lines, clock input, read (Rd) and write (Wr) inputs and m bits of data output lines. The first process involves reading data from the input lines and storing them in the RAM unit. Then, the data are read out again from the written location. The internal construction of a RAM is shown in Figure 4.24b, and it consists of the address decoder, storage cell (also called binary cell [BC]), and `or` array. The BC is the basic building block of memory units and stores 1 bit of information within it. The storage part of the cell can be modeled by an SR latch with associated gates to form a D latch [39,50,51]. Every BC is designed with four inputs: data bit, read/write bit, address bit, and one output: data bit, as shown in Figure 4.24b. A binary storage cell must be very small in order to pack as many cells as possible in the small area available in the integrated circuit chip. The memory

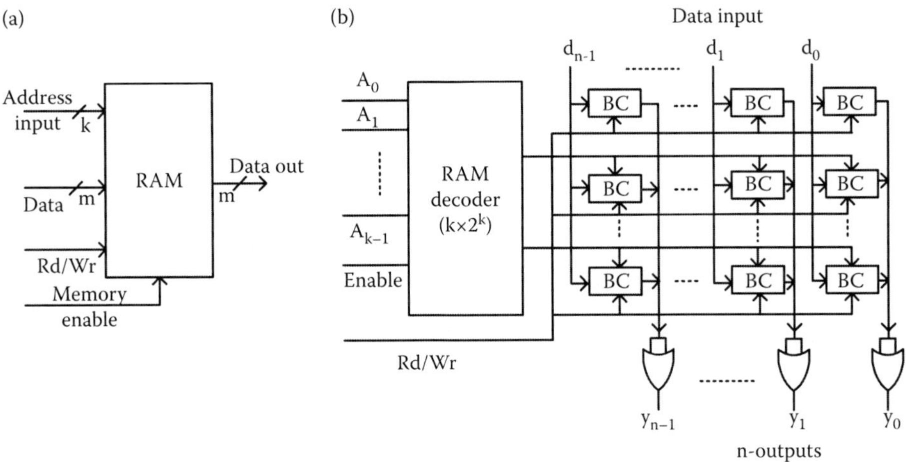

FIGURE 4.24
Entity of a RAM module (a) and internal logic of a RAM cell (b).

enable input of the BC enables the BC for reading or writing, and the read/write input determines the operation of the BC when it is selected. Logic 1 in the read/write input provides the read operation by forming a path from the BC to the output terminal. Logic 0 in the read/write input provides the write operation by forming a path from the input terminal to the BC, as shown in Figure 4.24b. If the RAM consists of four words of four bits each, it will have a total of 16 binary cells.

A memory with four words needs two address lines. The two address inputs go through a 2^2 decoder to select one of the four words. The decoder is enabled with the memory-enable input. When the memory enable is 0, all outputs of the decoder are 0 and none of the memory words are selected. With the memory select at logic 1, one of the four words is selected, dictated by the value in the two address lines. Once a word has been selected, the read/write input determines the operation. During the read operation, the four bits of the selected word go through or gates to the output terminals. During the write operation, the data available in the input lines are transferred into the four BCs of the selected word. The binary cells that are not selected are disabled and their previous binary values remain unchanged. Commercial RAMs may have a capacity of thousands of words, and each word may range from 1 to 64 bits [39,50,51].

DESIGN EXAMPLE 4.27

Design and develop a digital system to realize a RAM unit in the FPGA.

Solution

```
library ieee;
use ieee.std_logic_1164.all;
use ieee.std_logic_arith.all;
use ieee.std_logic_unsigned.all;
entity RAM_Desi is
generic(width: integer:=8;
depth: integer:=4;
addr: integer:=2);
port( clock,write,read :in std_logic;
addrs : in std_logic_vector(addr-1 downto 0);
data_out: out std_logic_vector(width-1 downto 0);
data_in: in std_logic_vector(width-1 downto 0));
end RAM_Desi;
architecture beha of RAM_Desi is
type ram_type is array (0 to depth-1) of std_logic_vector(width-1 downto
0);
signal tmp_ram: ram_type;
begin
process(clock)
begin
if (clock'event and clock='1') then
if write='1' then
tmp_ram(conv_integer(addrs)) <= data_in;
data_out <= "ZZZZZZZZ";
elsif read = '1' then
data_out <= tmp_ram(conv_integer(addrs));
else
data_out <= "ZZZZZZZZ";
end if;
end if;
end process;
end beha;
```

LABORATORY EXERCISES

1. Develop a digital system to implement a serial adder to add 10 bits sequentially and give the final sum and carry values. The design should follow the logic shown in Figure 4.1.
2. Design and develop a digital system to realize a 10-bit carry look-ahead adder.
3. Design and develop a digital system to implement a serial subtractor for n bits of subtraction.
4. Design and develop a digital system to realize an 8-bit fast array multiplier. The logic explained in Design Example 4.11 needs to be followed while designing the multiplier.
5. Design a digital system to realize the logical and arithmetic functions of an arithmetic logic unit chip IC 74181 in the FPGA.
6. Design and develop a digital system to realize a simple calculator with all the arithmetic and trigonometric functions.
7. Design and develop a digital system to realize a digital calendar. This calendar should display the day on an LCD corresponding to valid date, month and year inputs. If the input data are wrong, a message "wrong input" should be displayed.
8. Design and develop a digital system to realize the function SQRT $(a^2 + b^2)$ using the CORDIC algorithm.
9. Design and develop a digital system to realize the logarithmic and exponential operations.
10. Simplify and build the following functions: (i) $y_0 = AB' + AC + A'BC'$ and $y_1 = (AC + BC)'$ and (ii) $y_0 = \sum(2, 12, 13)$, $y_1 = \sum(7, 8, 9, 10, 11, 12, 13, 14, 15)$, $y_2 = \sum(0, 2, 3, 4, 5, 6, 7, 8, 10, 11, 15)$, and $y_3 = \sum(1, 2, 8, 12, 13)$ using the PLA and PAL, respectively, and implement them in the FPGA.

Part II

Custom Input/Output Peripheral Interfacing

5

Input/Output Bank Programming and Interfacing

5.1 Introduction and Preview

Any real-world applications can be accomplished through interfacing many external peripheral devices as needed for the intended requirements. There are many input and output devices through which real-time input signals can be passed into the FPGA whose outputs can be used for control and display applications through some other suitable device. This chapter discusses the working principles and interfacing techniques of many popular input-output devices. Following the discussion, VHDL code design examples are also given to help readers easily understand the concepts. All the design examples given in this chapter are tested in the development board, which facilitates direct implementation, and, at the end, readers will be able to develop VHDL code for any relevant new applications. This chapter is organized as follows: first of all, optical display interfacing techniques are discussed. The working principles of LEDs and multisegment (scrolling text) displays are discussed. The design circuits and assembly of a dot-matrix LED display are also detailed, followed by VHDL code. The working principles of a piezo buzzer and different methods of interfacing it with the FPGA are explained. A design example to generate different tones using a piezo buzzer is also given. Interfacing protocols of LCD/GLCD are detailed, along with their driver circuits. All the commands/addresses needed for the preparation of LCDs/GLCDs to display characters as we wish are also listed. The functionalities of many general-purpose input switches are discussed. The construction of DIPs and reed switches is explained, with their equivalent circuits. Applications of these switches are described, with design examples. The significance of making a general purpose input output (GPIO) bit/port as a bidirectional switch is explained, along with an application. The different logics behind controlling forward and reverse data transmission in a bidirectional switch are detailed. Applications of tristate switches are also described, along with design examples. The scanning principles of a matrix keypad and its interfacing advantages are explained. Algorithms behind the identification of key presses, their reorganization and key number assignment are detailed. A design example of interfacing a 4 × 4 matrix keypad and displaying the pressed key in an eight-LED array is given. Telephone number dialling and getting connectivity are based on DTMF signaling. That means the sum of two different frequencies will be sent whenever a key is pressed. Hence, decoding digital data (4 bits) at the receiver (the called phone) for a key pressed at the transmitter (the calling phone) is possible using a DTMF decoder. This decoded data can be used to do any operation at the receiver, keeping control at the transmitter. An example design for controlling electrical appliances from a remote telephone is explained. Types of optical sensors and their interfacing techniques are explained. Different configurations of IR Tx/Rx and their basic design circuits are also discussed. An example of measuring the speed of a rotating disc using IR-Tx/Rx is given. Applications of the IR-Tx/Rx proximity arrangement and methods of using it for obstacle finding/counting, line-following robot design, height/depth finding, and so on are also discussed, with a design example. The working principles of pyrometric sensors and IR profile changes due to a person's thermal influence are described, along with a person-counting design example. Various applications of the metal detector and its working principles are explained, with an application. Characteristic responses and applications of LDRs are discussed. As a design example, an automatic solar panel steering system is given. Standard wind-speed measurement techniques are listed, through which the advantages of using the cup anemometer are highlighted. The design of a cup anemometer and direction-finding vane are explained, along with a real-time wind-speed/direction measurement design example. Finally, laboratory practice exercises are listed.

5.2 Optical Display Interfacing

In this section, the operating principles of two types of optical displays, (i) light emitting diodes and (ii) multisegment LEDs, and their FPGA interfacing techniques are discussed, along with interfacing circuit and VHDL code design examples.

5.2.1 Light-Emitting Diode Displays

LEDs are known to be the best optoelectronic device for many real-world applications. An LED is capable of emitting a fairly narrow bandwidth of visible or invisible (infrared) light when its PN junction attains a forward electric potential, the wind direction is measured, forward current or voltage. This phenomenon is generally called electroluminance, which can be defined as the emission of light from a semiconductor under the influence of an electric field. Due to the supply voltage (V_{cc}), some part of the energy must be dissipated at the LED junction in order to recombine the electrons and the holes. This recombining energy is emitted in the form of heat and light. Once the LED is configured with forward bias, the current will increase at a greater rate in accordance with a small increase in supply voltage. A series current limiting resistor "R" is connected, as shown in Figure 5.1a, to the LED to prevent it from damage due to current overflow. Forcing a logic 1, typically +5 V, into the

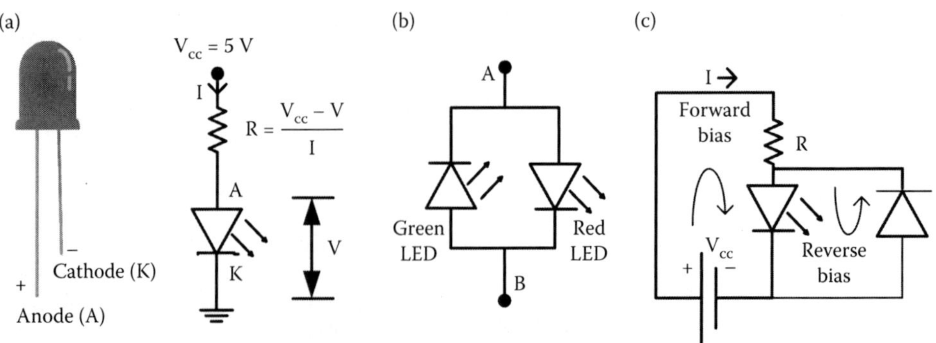

FIGURE 5.1
Appearance and typical wiring of LED (a), color LEDs switching circuit (b), and LED wiring with reverse bias protection diode (c).

anodes of the LED would make it glow if its cathode is already grounded. This type of wiring is called common cathode configuration. Alternatively, forcing a logic 0 would make the LED glow in the case where it is already wired in common anode configuration [52,53].

Commercially used LEDs have a typical voltage drop between 1.5 and 2.5 V or current between 10 and 50 mA. The exact voltage drop depends on the LED current, color, and tolerance. Two different-colored LEDs (e.g., green and red) can be wired to indicate two different results, as shown in Figure 5.1b, where the LEDs are "off" when the control pin "AB" is "00"/"11", only the green LED is "on" (forward bias) when "AB" is "01", and only the red LED is "on" (forward bias) when "AB" is "10". In another method, when the PN junction is reverse biased, no light will be produced by the LED, and the device may also be damaged in the case where a slightly high voltage is applied. The circuit shown in Figure 5.1c is designed by wiring an LED with an inverse parallel PN junction diode, typically 1N4007, to prevent the LED from reverse-biased damage. A single-color LED usually emits orange, red, yellow, or green, whereas a multicolored LED emits light at different colors over time based on the state of the internal switching circuit. The response time of the LED is known to be very fast, on the order of 0.1 μsec, when compared with a tungsten lamp, which is on the order of 100 msec. The lifespan of an LED supersedes the short life of an incandescent bulb by thousands of hours. Tiny LEDs are replacing picture tubes to make dramatically thinner LED/HD televisions.

DESIGN EXAMPLE 5.1

Sixteen LEDs are connected to the FPGA in common cathode configuration. Develop VHDL code to make the first eight LEDS glow in a downward direction while the other eight glow in an upward direction. This has to happen five times; then all the LEDs have to blink two times.

Solution

```
library IEEE;
use IEEE.STD_LOGIC_1164.ALL;
use IEEE.STD_LOGIC_ARITH.ALL;
use IEEE.STD_LOGIC_UNSIGNED.ALL;
entity LED is
Port (m_clk : in STD_LOGIC:='0';
led_out1,led_out2:inout std_logic_vector(7 downto 0):="00000000");
end LED;
architecture Behavioral of LED is
signal clk_sig: std_logic:='0';
begin
p0: process(m_clk)
variable cnt: integer:=0;
begin
if rising_edge(m_clk) then
if (cnt=2) then
clk_sig<= not clk_sig;
cnt:=0;
else
cnt:=cnt+1;
end if;
end if;
end process;
p1:process(clk_sig)
variable cnt1,cnt2:integer:=0;
begin
if rising_edge(clk_sig) then
```

```
if (cnt1>7)then
if (cnt2<4) then
cnt1:=0;
cnt2:=cnt2+1;
else
if (cnt2=4) then
led_out1<="11111111"; led_out2<="11111111";
cnt2:=cnt2+1;
elsif (cnt2=5) then
led_out1<="00000000"; led_out2<="00000000";
cnt2:=cnt2+1;
elsif (cnt2=6) then
led_out1<="11111111"; led_out2<="11111111";
cnt2:=cnt2+1;
elsif (cnt2=7) then
led_out1<="00000000"; led_out2<="00000000";
cnt2:=0;
cnt1:=0;
end if;
end if;
else
if (cnt1=0)then
led_out1<="10000000"; led_out2<="00000001";
cnt1:=cnt1+1;
else
led_out1<=(led_out1(0) & led_out1(7 downto 1));
led_out2<=(led_out2(6 downto 0) & led_out2(7));
cnt1:=cnt1+1;
end if;
end if;
end if;
end process;
end Behavioral;
```

5.2.2 Multisegment Display

The multisegment display, also called the dot-matrix display, is currently used for displaying scrolling text, numerical digits, symbols, messages, digital signatures, tables, plots, advertising signs, school signs, restaurant signs, custom street signs, business signs, church signs, animated signs, banners, safety signs, political signs, hand paintings, and so on, unlike the well-known seven-segment display. Character scrolling and displaying images using multisegment displays are found in almost all kinds of public display devices. However, multisegment displays can be used for various display applications where resolution is not a big concern. One method of driving a multisegment display is organizing LEDs in rows (R) and columns (C) and then selecting individual rows and columns by the intersection of the row and column lines. A commercially available multisegment display chip and a configuration of LEDs for 7×5 grid display is shown in Figure 5.2. A logic 1 (high state) is required to activate a row, and a logic 0 (low state) is required to activate a column, because the LED anodes are connected to row lines and the cathodes are connected to column lines.

In general, the construction of a dot-matrix display consists of a matrix of LEDs arranged in a rectangular configuration. Any individual LED or group of LEDs in the matrix can be activated by switching the required number of rows and columns. The character or graph can be displayed in a dot-matrix display just by switching particular LEDs on/off with the proper time delay. Commercially available dot-matrix displays are 7×5, 8×8, 7×15 and

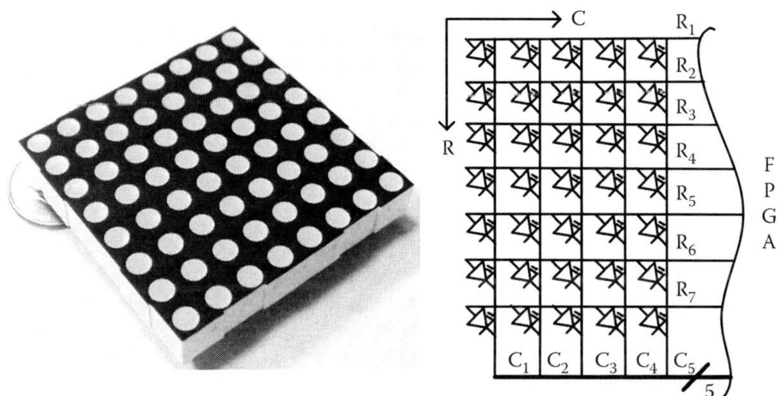

FIGURE 5.2
A multisegment display chip and arrangement of LEDs for a 7 × 5 dot-matrix display.

so on. For example, in Figure 5.2, if R_1 is made high and C_1 is made low, only the top-left LED (addressed R_1C_1) will glow. As a demonstration, let us see how we can display a character "A" using this dot-matrix display. The state diagram required to send the necessary row and column data (logic levels) to display this character is given in Figure 5.3.

As in the above state diagram, only one LED in a row will be "on" at a time, but any number of LEDs can be "on" in a column. This means that two processing modules have to be built in the FPGA to directly drive the row and the column lines. The schematic diagram for interfacing more than one dot-matrix display (M_0, M_1...M_{N-1}) is shown in Figure 5.4. A high-current Darlington transistor array (HCDTA) is used to drive the column lines of the display. Typically, ULN2803 is used due to its high voltage—(50 V) and current—(500 mA per channel) handling capability. Each IC has eight channels with individual output clamp diodes. ULN2803 is an active high device, which means a logic high must be applied to the input to make the corresponding output high. Scrolling messages in the dot-matrix display can be shown using SIPO shift registers [52,53].

All dot-matrix displays are driven by display data sets via a PNP transistor, typically SK100, array. Each display is connected to separate SIPO shift register via a separate

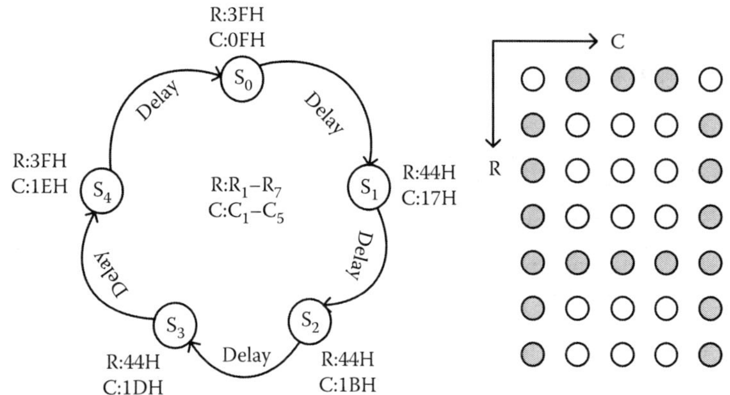

FIGURE 5.3
State diagram to display "A" and illustration of character "A" in a 7 × 5 dot-matrix display.

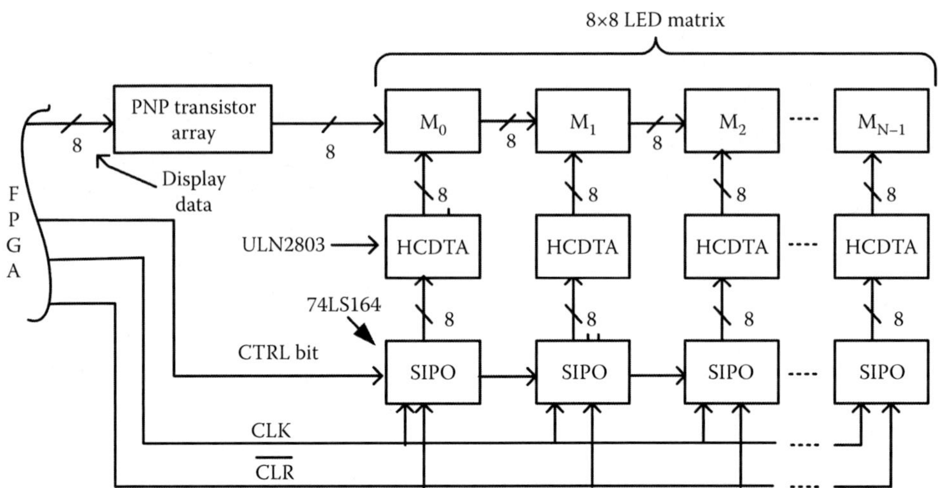

FIGURE 5.4
Schematic diagram of a scrolling text display system.

current sinker, for example, ULN2803, to allow larger current to be sunk. Separate modules are built in the FPGA to provide the display data to the transistor array and control bits and clock/clear signals to the SIPO shift register, as shown in Figure 5.4. If a SIPO shift register provides the data sequence "10000000", that is, 80H, to an HCDTA, the first column in the corresponding dot-matrix display will be connected to ground via ULN2803, and the remaining other columns will be left unconnected. Therefore, LEDs in that column will glow according to the display data. In the same way, the next columns will be driven one by one with the corresponding display data set and control sequence. By building the suitable architecture in the FPGA, we can make the text display scroll. As per the persistence of vision phenomenon, which scientifically determined that a rate of less than 16 frames per second causes the mind to see a flashing character/images, the refresh rate should be equal to at least 24 frames per second. One frame takes 1/24 second to scan all the columns in the matrix. The character scrolling speed can be varied using a potentiometer (POT) and A/D convertor. Details about A/D interfacing and more of its applications can be found in Chapter 7.

DESIGN EXAMPLE 5.2

Develop a digital system to display "DIAT" on a multisegment LED display panel. The scrolling of the word has to be in the right-to-left direction. Assume each segment of the LED display panel has eight rows and eight columns.

Solution

The general circuit diagram needed for any scrolling display is shown in Figure 5.5. Since there are eight rows and eight columns, the display data required to drive the rows are 44H, 81H, FFH, 81H, 81H, 81H, 42H, 3CH, 81H, 81H, 81H, FFH, 81H, 81H, 81H, 81H, 3FH, 44H, 84H, 84H, 84H, 84H, 44H, 3FH, 10H, 10H, 10H, FFH, 10H, 10H, 10H, and 10H. The SK100 transistor is used to drive the rows; hence, these display data have to be inverted before pumping them one by one to the segment. The `CTRL` bit has to be shifted from left to right to scroll the text from right to left. The SIPO shift register has to be cleared before writing any display data to the segment.

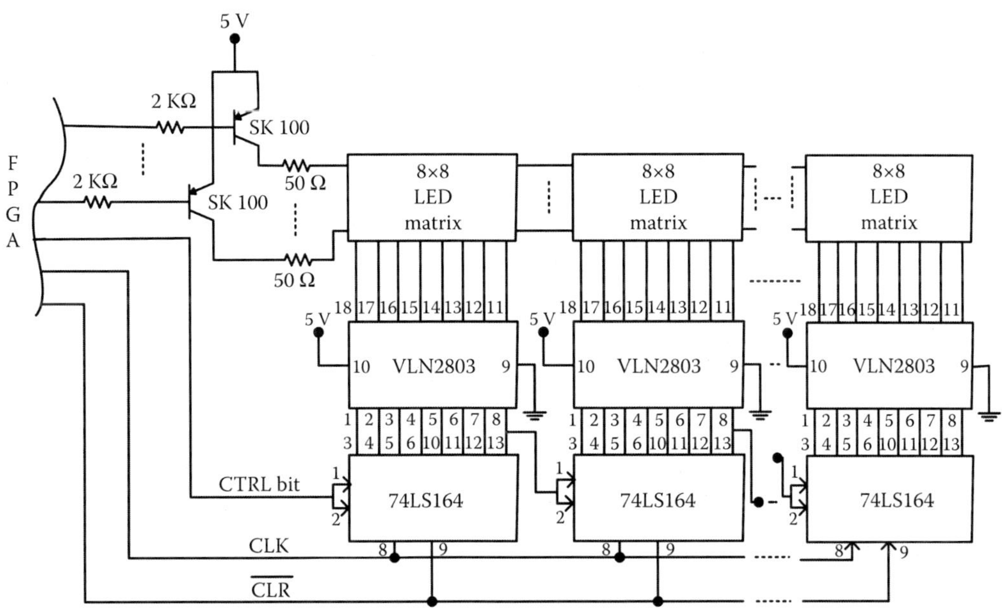

FIGURE 5.5
General circuit diagram for a scrolling display system.

The VHDL code is given below.

```
LIBRARY ieee;
USE ieee.std_logic_1164.all;
USE ieee.numeric_std.all;
use ieee.std_logic_unsigned.all;
use IEEE.STD_LOGIC_ARITH.ALL;
entity scro_disp is
Port (m_clk: in  STD_LOGIC:='0';
ctrl_bit: out std_logic:='1';
clk: out std_logic:='0';
clr: out std_logic:='0';
disp_data_out: out std_logic_vector(7 downto 0):="00000000");
end scro_disp;
architecture Behavioral of scro_disp is
signal clk_sig: std_logic:='0';
signal sta,nsta:integer range 0 to 36:=0;
type pix_data is array (0 to 31) of std_logic_vector(7 downto 0);
constant data_d : pix_data
:=(x"44",x"81",x"FF",x"81",x"81",x"81",x"42",x"36",x"81",x"81",x"81",x"FF",
x"81",x"81",x"81",x"81",x"3F",x"44",x"84",x"84",x"84",x"84",
x"44",x"3F",x"10",x"10",x"10",x"FF",x"10",x"10",x"10",x"10");
signal disp_mux,shft_cnt:std_logic_vector(4 downto 0):="00000";
signal disp_data: std_logic_vector(7 downto 0):="00000000";
begin
disp_data_out<= not (disp_data);
p0:process(m_clk)
variable cnt:integer:=1;
begin
if rising_edge(m_clk) then
if(cnt=2)then
clk_sig<= not(clk_sig);
cnt:=1;
```

```
else
cnt:= cnt + 1;
end if;
end if;
end process;
p1: process(disp_mux)
begin
case disp_mux is
when "00000"=> disp_data<=data_d(0);
when "00001"=> disp_data<=data_d(1);
when "00010"=> disp_data<=data_d(2);
when "00011"=> disp_data<=data_d(3);
when "00100"=> disp_data<=data_d(4);
when "00101"=> disp_data<=data_d(5);
when "00110"=> disp_data<=data_d(6);
when "00111"=> disp_data<=data_d(7);
when "01000"=> disp_data<=data_d(8);
when "01001"=> disp_data<=data_d(9);
when "01010"=> disp_data<=data_d(10);
when "01011"=> disp_data<=data_d(11);
when "01100"=> disp_data<=data_d(12);
when "01101"=> disp_data<=data_d(13);
when "01110"=> disp_data<=data_d(14);
when "01111"=> disp_data<=data_d(15);
when "10000"=> disp_data<=data_d(16);
when "10001"=> disp_data<=data_d(17);
when "10010"=> disp_data<=data_d(18);
when "10011"=> disp_data<=data_d(19);
when "10100"=> disp_data<=data_d(20);
when "10101"=> disp_data<=data_d(21);
when "10110"=> disp_data<=data_d(22);
when "10111"=> disp_data<=data_d(23);
when "11000"=> disp_data<=data_d(24);
when "11001"=> disp_data<=data_d(25);
when "11010"=> disp_data<=data_d(26);
when "11011"=> disp_data<=data_d(27);
when "11100"=> disp_data<=data_d(28);
when "11101"=> disp_data<=data_d(29);
when "11110"=> disp_data<=data_d(30);
when "11111"=> disp_data<=data_d(31);
when others=> null;
end case;
end process;
p2:process(clk_sig)
begin
if rising_edge(clk_sig) then
sta<=nsta;
end if;
end process;
p5:process(sta)
variable i,j:integer:=0;
begin
case sta is
when 0=>   clr<='0';
nsta<=sta+1;
when 1=>   clr<='1';
nsta<=sta+1;
when 2=>   disp_mux<=shft_cnt;
nsta<=sta+1;
when 3=>  if (i=32) then
i:=0;
```

```
j:=0;
nsta<=7;
else
if(j=0) then
ctrl_bit<='1';
nsta<=sta+1;
else
ctrl_bit<='0';
nsta<=sta+1;
end if;
end if;
when 4=>  clk<='1';
nsta<=sta+1;
when 5=>  clk<='0';
nsta<=sta+1;
when 6=>  j:=1;
i:=i+1;
shft_cnt<=shft_cnt+"00001"; nsta<=sta+1;
nsta<=2;
when 7=> shft_cnt<=shft_cnt+"00001";
nsta<=2;
when others => nsta<=0;
end case;
end process;
end Behavioral;
```

5.3 Buzzer Control

The piezo buzzer is an electronic device that is commonly used to produce sound. Lightweight, simple construction and low price make it usable in various real-world applications such as car/truck reverse indicators, computers, over voltage/current indicators, doorbells, and so on. Piezoceramic is a class of composite material that poses the piezo electric effect, which is used in the piezo buzzer. When the buzzer is subjected to an alternating electric field, stretching, or compression happens in accordance with the frequency of the signal, which results in a sound. The buzzer can also be driven using a DC source or digital pulses. Typical application circuits to make use of a buzzer are shown in Figure 5.6.

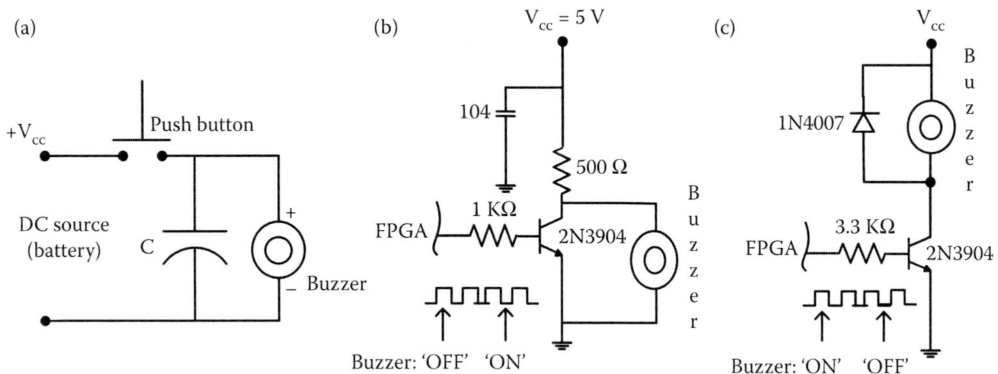

FIGURE 5.6
Buzzer driver circuits using DC source and digital pulses.

When the push button is pushed and held down, the buzzer will emit a steady tone. This is because of a DC source (battery) that delivers steady supply voltage to the buzzer as well as charging the capacitor.

If the push button is released, then the capacitor starts supplying power to the buzzer and discharges, which results in rapidly rising sound that then decreases. If the push button is pressed/released in rapid succession, the buzzer will not make a steady tone because it is not receiving constant voltage from the battery, which results in a wavering sound. Transistor-based wiring is required to automatically control the buzzer by any electronic system, as shown in Figure 5.6. As in the middle figure, whenever a logic 0 is applied from the FPGA, the transistor goes "off" (open circuit), then V_{cc} is grounded via the buzzer; hence, it produces sound [54]. Alternatively, if logic 1 is applied, the transistor goes "on", then V_{cc} is grounded through the transistor; hence, the buzzer does not produce a sound. The reverse operation happens in the rightmost circuit in Figure 5.6 since the buzzer is connected in the collector of the transistor. A freewheeling diode (1N4148) is used to make the buzzer immediately respond to logic changes at the base of the transistor.

DESIGN EXAMPLE 5.3

Develop a digital system in FPGA to drive a buzzer at different duty cycles 0%, 40%, 80%, and 100% whenever the input (`data _ in`) is (`0` ≤ `data _ in` ≤ `10`), (`10` < `data _ in` ≤ `100`), (`100` < `data _ in` ≤ `200`) and (`200`< `data _ in`), respectively. Assume that the inputs are given via a port of eight bits.

Solution

```
library IEEE;
use IEEE.STD_LOGIC_1164.ALL;
USE ieee.numeric_std.ALL;
entity buz_cnt is
port(m_clk:in std_logic:='0';
data_in:in std_logic_vector(7 downto 0):="00000000";
buz_ctrl:out std_logic:='0');
end buz_cnt;
architecture Behavioral of buz_cnt is
signal clk_sig: std_logic:='0';
signal cnt40, cnt80: std_logic:='1';
signal data_int:integer:=0;
begin
data_int <= to_integer(unsigned(data_in));
po:process(m_clk)
variable cnt:integer:=0;
begin
if rising_edge(m_clk) then
if (cnt=2) then --For 1 Hz, this value has to be decided based on
onboard clock
clk_sig<=not clk_sig;
cnt:=1;
else
cnt:=cnt+1;
end if;
end if;
end process;
p1: process(clk_sig) --Duty Cycle section
variable cnt1:integer:=1;
begin
if rising_edge(clk_sig) then
if (cnt1=4) then
```

```
cnt40<='0';
cnt1:=cnt1+1;
elsif (cnt1=8) then
cnt80<='0';
cnt1:=cnt1+1;
elsif (cnt1=10) then
cnt40<='1';
cnt80<='1';
cnt1:=1;
else
cnt1:=cnt1+1;
end if;
end if;
end process;
p2:process(data_int,cnt40,cnt80) --Duty cycle selection
begin
if(data_int >= 0 and data_int<=10) then
buz_ctrl<='0';
elsif(data_int >10 and data_int<=100) then
buz_ctrl<= cnt40;
elsif(data_int >100 and data_int<=200) then
buz_ctrl<= cnt80;
elsif(data_int >200) then
buz_ctrl<='1';
end if;
end process;
end Behavioral;
```

5.4 Liquid Crystal Display Interfacing and Programming

In this section, the interfacing protocols of LCDs and GLCDs are discussed, along with programming techniques for displaying ASCII characters and real-time graphs.

5.4.1 Liquid Crystal Display

The LCD is a very common electronic display module, which has a wide range of applications, such as digital voltmeters/ammeters, digital clocks, home automation displays, status indicator displays, digital code locks, digital speedometers/odometers, displays for music players, calculators, laptops, mobile phones, and so on. A 16 × 2 LCD can display two lines of 16 characters. Each character is typically displayed using 5 × 7 pixel resolution. LCDs can be interfaced with the FPGA in 4-bit or 8-bit mode, which differ in how the data is sent to the LCD. In 8-bit mode, to write a character, 8 bits of ASCII data have to be sent through the data lines D_0–D_7, and a data strobe is given through enable (E) of the LCD. Other LCD commands, such as register select (RS) and read/write (R/W), are also 8 bit and have to be written to the LCD in a similar way.

Four-bit mode uses only four data lines, D_4–D_7, whereas 8-bit ASCII character or command data are divided into two parts and sent sequentially through the first four data lines. The idea of 4-bit communication is used to save the pins. Four-bit communication is a bit slower than 8-bit communication, but this speed difference can be neglected since LCDs are slow-speed devices. Thus, 4-bit-mode data transfer is preferred whenever eight pins are not available for LCD interfacing. In order to understand the interfacing protocol, first we need to understand the 16 × 2 LCD module, which is available in a 16-pin package, as shown

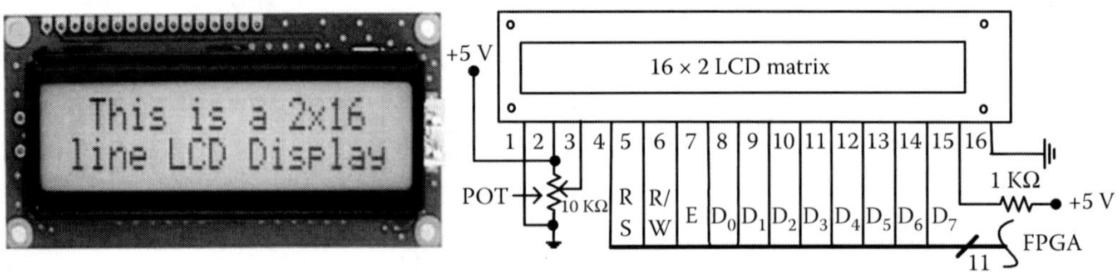

FIGURE 5.7
Picture and basic wiring of 16 × 2 LCD.

in Figure 5.7, with backlight and contrast adjustment provisions. The pin numbers, their names and corresponding functions are also given in Figure 5.7 and detailed in Table 5.1.

The V_{EE} pin is meant for adjusting the contrast of the LCD display, which can be achieved by varying the voltage at that pin using a POT, as shown in Figure 5.7. The LCD has two built-in registers, namely the command register and data register. The command register is to place the commands to the LCD to prepare it for displaying the character as the designer likes, and the data register is for placing the ASCII data that need to display on the LCD. The 16 × 2 LCD module has a set of commands, as given in Table 5.2, each meant for doing a particular job to prepare the LCD. A low logic level at the RS pin will select the command register, and high logic selects the data register. Making the RS pin low and then passing data to the 8-bit data line (D_0–D_7) activates the LCD module to recognize the data as a command to prepare it. Otherwise, if the data are passed when the RS is high, the data will be taken as an ASCII character. The R/W pin is meant for selecting either a read or write operation. A high level at the R/W pin enables read mode, and a low level enables write mode. A high-to-low transition at the E pin will enable the command/ASCII data in the LCD module. LED+ is the anode of the background LED light, and this pin must be connected to V_{cc} through a suitable series current-limiting resistor. LED− is the cathode of the background LED light, and this pin must be connected to ground [55].

The steps that must be done for initializing the LCD display are given below. These steps are common for almost all LCDs: (i) send 38H to the 8-bit data line for initialization; (ii) send 0FH to turn the LCD "on", cursor "on" and cursor blinking "on"; (iii) send 01H for clearing the display and return the cursor; (iv) send 06H for incrementing the cursor

TABLE 5.1
Pin Details of a 16 × 2 LCD

Pin No.	Name	Functions
1	V_{SS}	Gnd
2	V_{CC}	Positive supply voltage pin (5 V)
3	V_{EE}	Contrast adjustment
4	RS	Register selection: command (RS = 0) or ASCII (RS = 1)
5	R/W	Read (R/W = 1) or write (R/W = 0)
6	E	Enable (falling edge)
7–14	D_0–D_7	Command or ASCII data
15	LED+	Backlight LED+
16	LED−	Backlight LED− (gnd)

TABLE 5.2

16 × 2 LCD Module Command Data and the Corresponding Functions

Command	Functions
01	Clear display screen
02	Return to home
04	Decrement cursor: shift cursor right to left
06	Increment cursor: shift cursor left to right
05	Shift display right
07	Shift display left
08	Display "off", cursor "off"
0A	Display "off", cursor "on"
OC	Display "on", cursor "off"
0E	Display "on", and cursor "on", blinking "off"
0F	LCD "on", cursor "on", and cursor blinking "on"
10	Shift cursor position to left
14	Shift cursor position to right
18	Shift entire display to the left
1C	Shift entire display to the right
80–8F	Starting and ending address for first line
82	Jump to position 3 in first line
C0–CF	Starting and ending address for second line
C5	Jump to position 6 in second line
38	Use 8-bit, two lines and 5 × 7 resolution matrix
30	Use 8-bit, one line and 5 × 7 resolution matrix
20	Use 4-bit, one line and 5 × 7 resolution matrix
28	Use 4-bit, two line and 5 × 7 resolution matrix
00	Row 1 address in 8-bit, four lines and 5 × 8 resolution matrix
40	Row 2 address in 8-bit, four lines and 5 × 8 resolution matrix
10	Row 3 address in 8-bit, four lines and 5 × 8 resolution matrix
50	Row 4 address in 8-bit, four lines and 5 × 8 resolution matrix

position; and (v) send 80H to begin the display at row1 and column1. During this preparation time, the values of RS and R/W have to be "0" and the enable signal has to be applied appropriately (high-to-low transition) for every passing of command data. The designers can chose any command data based on the way they want to display the character in the LCD. Once passing of the command data is over, then make RS = '1' and pass the data to be displayed. Therefore, LCD control pins can be designed as a 3-bit output vector as [2:0], that is, [E,RS,R/W], and the data: command/display can be designed as an 8-bit output vector as [7:0], that is, [D_7,D_6...D_0]. Almost all ASCII characters, for example, 20H for a blank space and 30H for printing 0, can be displayed in the LCD.

DESIGN EXAMPLE 5.4

Interface a 16 × 2 LCD with the FPGA and develop VHDL code to print "The counter" in the first row and "value is:" in the second row. Also, construct a six-digit counter to count the values from "000000" to "999999" and print the values of the counter every second in the second line followed by the text. The counter counting speed has to be 1 Hz. Whenever the reset is "0", the counter values have to set to "000000"; otherwise, the counter has to continue counting.

Solution

```
library IEEE;
use IEEE.STD_LOGIC_1164.ALL;
use IEEE.STD_LOGIC_ARITH.ALL;
use IEEE.STD_LOGIC_UNSIGNED.ALL;
entity LCD is
Port (clk,reset : in STD_LOGIC:='0';
c:out std_logic_vector(2 downto 0):="000";
d:out std_logic_vector(7 downto 0):="00000000");
end LCD;
architecture Behavioral of LCD is
type state is(s0,s1,s2,s3,s4,s5,s6,s7,s8,s9,s10,s11,s12,s13,s14,s15,s16,
s17,s18,s19,s20,s21,s22,s23,
s24,s25,s26,s27,s28,s29,s30,s31,s32,s33,s34,s35,s36,s37,s38,s39,s40,s41,
s42,s43,s44,s45,
s46,s47,s48,s49,s50,s51,s52,s53,s54,s55,s56,s57,s58,s59,s60,s61,s62,s63,
s64,s65);
signal ps,ns:state:=s0;
signal d0cnt,d1cnt,d2cnt,d3cnt,d4cnt,d5cnt:std_logic_vector(7 downto
0):="00110000";
signal clk_sig,
clk_sig0,clk_sig1,clk0,clk1,clk2,clk3,clk4:std_logic:='1';
begin
-------Low Frequency clock generator for LCD control--------
p1:process(clk,reset)
variable cnt:integer;
begin
if rising_edge(clk) then
if(cnt=2000)then          ----This value of the counter has to
be chosen based on
clk_sig0<= not(clk_sig0); ----the master clock frequency.
cnt:=0;
else
cnt:= cnt + 1;
end if;
end if;
end process;
------------State updating----------------
p2:process(clk_sig0)
begin
if rising_edge(clk_sig0) then
ps<=ns;
end if;
end process;
------------------LCD Command and Data ---------------------
p3:process(ps,d0cnt,d1cnt,d2cnt,d3cnt,d4cnt,d5cnt)
begin
case ps is
when s0=> d<="00111000";c<="100"; ns<=s1;--38
when s1=> c<="000";ns<=s2;
when s2=> d<="00001111";c<="100"; ns<=s3;--0F
when s3=> c<="000";ns<=s4;
when s4=> d<="00000001";c<="100"; ns<=s5;--01
when s5=> c<="000";ns<=s6;
when s6=> d<="00000110";c<="100"; ns<=s7;--06
```

```
when s7=> c<="000";ns<=s8;
when s8=> d<="10000000";c<="100"; ns<=s9;--80
when s9=> c<="000";ns<=s10;
when s10=> d<="01010100";c<="110"; ns<=s11;--T
when s11=> c<="010";ns<=s12;
when s12=> d<="01101000";c<="110"; ns<=s13;--h
when s13=> c<="010";ns<=s14;
when s14=> d<="01100101";c<="110"; ns<=s15;--e
when s15=> c<="010";ns<=s16;
when s16=> d<="00100000";c<="110"; ns<=s17;-- 20H for a space
when s17=> c<="010";ns<=s18;
when s18=> d<="01100011";c<="110"; ns<=s19;--c
when s19=> c<="010";ns<=s20;
when s20=> d<="01101111";c<="110"; ns<=s21;--o
when s21=> c<="010";ns<=s22;
when s22=> d<="01110101";c<="110"; ns<=s23;--u
when s23=> c<="010";ns<=s24;
when s24=> d<="01101110";c<="110"; ns<=s25;--n
when s25=> c<="010";ns<=s26;
when s26=> d<="01110100";c<="110"; ns<=s27;--t
when s27=> c<="010";ns<=s28;
when s28=> d<="01100101";c<="110"; ns<=s29;--e
when s29=> c<="010";ns<=s30;
when s30=> d<="01110010";c<="110"; ns<=s31;--r
when s31=> c<="010";ns<=s32;
when s32=> d<="11000000";c<="100"; ns<=s33;--C0
when s33=> c<="000";ns<=s34;
when s34=> d<="01110110";c<="110"; ns<=s35;--v
when s35=> c<="010";ns<=s36;
when s36=> d<="01100001";c<="110"; ns<=s37;--a
when s37=> c<="010";ns<=s38;
when s38=> d<="01101100";c<="110"; ns<=s39;--l
when s39=> c<="010";ns<=s40;
when s40=> d<="01110101";c<="110"; ns<=s41;--u
when s41=> c<="010";ns<=s42;
when s42=> d<="01100101";c<="110"; ns<=s43;--e
when s43=> c<="010";ns<=s44;
when s44=> d<="00100000";c<="110"; ns<=s45;-- 20H for a space
when s45=> c<="010";ns<=s46;
when s46=> d<="01101001";c<="110"; ns<=s47;--i
when s47=> c<="010";ns<=s48;
when s48=> d<="01110011";c<="110"; ns<=s49;--s
when s49=> c<="010";ns<=s50;
when s50=> d<="00111010";c<="110"; ns<=s51;--:
when s51=> c<="010";ns<=s52;
when s52=> d<=d5cnt;c<="110"; ns<=s53;
when s53=> c<="010";ns<=s54;
when s54=> d<=d4cnt;c<="110"; ns<=s55;
when s55=> c<="010";ns<=s56;
when s56=> d<=d3cnt;c<="110"; ns<=s57;
when s57=> c<="010";ns<=s58;
when s58=> d<=d2cnt;c<="110"; ns<=s59;
when s59=> c<="010";ns<=s60;
when s60=> d<=d1cnt;c<="110"; ns<=s61;
when s61=> c<="010";ns<=s62;
when s62=> d<=d0cnt;c<="110"; ns<=s63;
when s63=> c<="010";ns<=s64;
when s64=> d<="11001001";c<="100"; ns<=s65;--C9
when s65=> c<="001";ns<=s52;
when others=> null;
end case;
```

```
end process;
--------------1 Hz clock generator for the counters----------
p4:process(clk,reset)
variable cnt:integer;
begin
if(reset='0') then
clk_sig<='0';
cnt:=0;
d0cnt<="00110000";d1cnt<="00110000";d2cnt<="00110000";
d3cnt<="00110000";
d4cnt<="00110000";d5cnt<="00110000";
elsif rising_edge(clk) then
if(cnt=2000000)then ---This value of the counter has to be chosen based
on the master
clk_sig<= not(clk_sig);----clock frequency.
cnt:=0;
else
cnt:= cnt + 1;
end if;
end if;
end process;
-----------------------0th display counter
p5:process(clk_sig)
begin
if rising_edge(clk_sig) then
if (d0cnt="00110101")then
clk0<=not(clk0);
end if;
if(d0cnt="00111001") then
clk0<=not(clk0);
d0cnt<="00110000";
else
d0cnt<=d0cnt + 1;
end if;
end if;
end process;
-- --------------------1st display counter
p6:process(clk0)
begin
if rising_edge(clk0) then
if (d1cnt="00110101")then
clk1<=not(clk1);
end if;
if(d1cnt="00111001") then
clk1<=not(clk1);
d1cnt<="00110000";
else
d1cnt<=d1cnt + 1;
end if;
end if;
end process;
-------------------------2nd display counter
p7:process(clk1)
begin
if rising_edge(clk1) then
if (d2cnt="00110101")then
clk2<=not(clk2);
end if;
if(d2cnt="00111001") then
clk2<=not(clk2);
d2cnt<="00110000";
```

```
else
d2cnt<=d2cnt + 1;
end if;
end if;
end process;
-- -------------------------3rd display counter
p8:process(clk2)
begin
if rising_edge(clk2) then
if (d3cnt="00110101")then
clk3<=not(clk3);
end if;
if(d3cnt="00111001") then
clk3<=not(clk3);
d3cnt<="00110000";
else
d3cnt<=d3cnt + 1;
end if;
end if;
end process;
-- --------------------4th display counter
p9:process(clk3)
begin
if rising_edge(clk3) then
if (d4cnt="00110101")then
clk4<=not(clk4);
end if;
if(d4cnt="00111001") then
clk4<=not(clk4);
d4cnt<="00110000";
else
d4cnt<=d4cnt + 1;
end if;
end if;
end process;
-- ----------------------5th display counter
p10:process(clk4)
begin
if rising_edge(clk4) then
if(d5cnt="00111001") then
d5cnt<="00110000";
else
d5cnt<=d5cnt + 1;
end if;
end if;
end process;
end Behavioral;
```

5.4.2 Graphical Liquid Crystal Display

The GLCD is a type of LCD which is used to display characters and graphs at relatively higher resolutions. A thin profile, low power consumption, easy interfacing, reasonable resolution and more pixels are the unique features of the GLCD. The graphical representation of any data presents a better understanding than just characters and/or numbers. A GLCD of size 128 × 64 and its pin details are shown in Figure 5.8. The illumination contrast of the GLCD can be adjusted through a POT. GLCD interfacing requires 8 data bits (D_0–D_7) and six control lines as register select (RS), read/write (R/W),

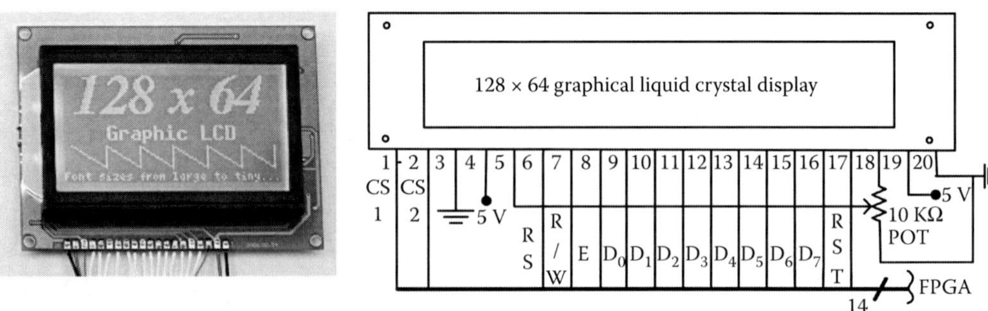

FIGURE 5.8
Picture and pin wiring of a GLCD.

enable (E), reset (RST) and segment select 1 and 2 (CS1 and CS2), whose function details are given in Table 5.3.

We need to send 8-bit data to turn the GLCD pixels, consisting of 8 bits/dots vertically, "on" or "off". A byte of data like FFH, that is, "11111111", turns all dots of the column "on", and F0H, that is, "11110000", turns the first four dots [d_0–d_3] "off" and the last four dots [d_4–d_7] "on". The designers can make text and/or images by turning these pixels "on" or "off". The combination of all this logic makes a picture. The GLCD has a graphic RAM where each bit in RAM corresponds to one pixel on screen. This means writing to the graphic RAM and modifying its content will accordingly result in changes on the screen. There are various types of GLCD modules available in the market depending on pixel size, such as 120 × 64, 128 × 128, 240 × 64, 128 × 64, and 240 × 128. The two halves of the display (left and right segments) can be individually accessed through the chip select pins (CS1 and CS2). Each half consists of eight horizontal pages (p_0–p_7), which are 8 bits (1 byte), as shown in Figure 5.9. The pixel data should be sent through the bus by choosing the data mode [55,56].

TABLE 5.3
GLCD Pin Details and Their Functions

Pin No.	Name	Functions
1	CS1	Chip select 1: left segment (CS1 = 0 and CS2 = 1)
2	CS2	Chip select 2: Right segment (CS1 = 1 and CS2 = 0)
3	V_{SS}	Gnd
4	V_{DD}	+5 V
5	CST	Contrast
6	RS	Register select: command (RS = 0) and data (RS = 1)
7	R/W	Read/write: read (R/W = 1) and write (R/W = 0)
8	E	Enable
9–16	D_0-D_7	GLCD – command/data lines
17	RST	Reset
18	V_{EE}	Negative voltage
19	LED+	LED anode
20	LED–	LED cathode

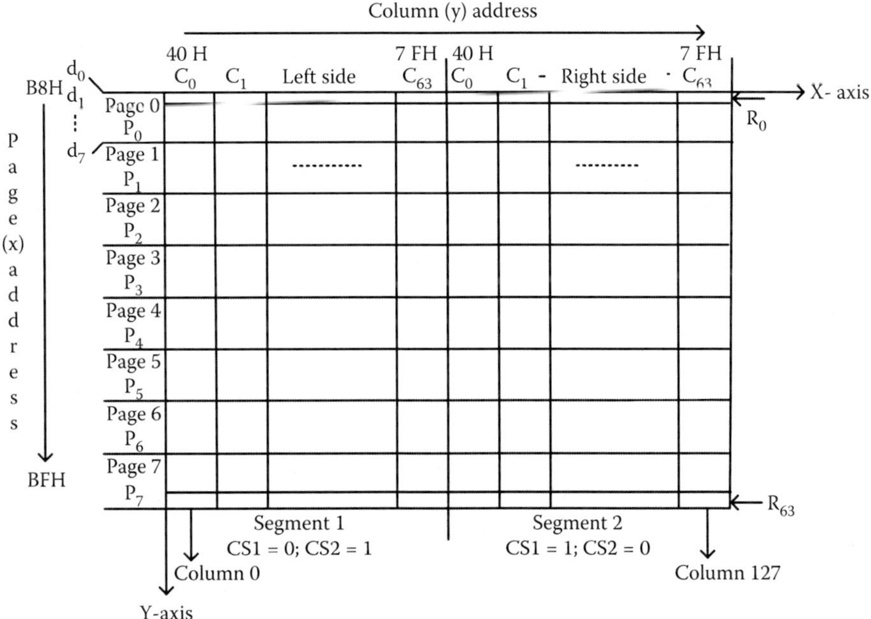

FIGURE 5.9
Page and byte mapping of a 128 × 4 GLCD.

Before sending display data, we need to select the exact page and the column from where we wish to begin the display. Eight rows (R_0–R_7) are combined as a page, thus forming eight pages supporting 64 rows, as indicated in Figure 5.9. Eight-bit pixel data sent to the GLCD module would be displayed in a single vertical column as one pixel in each row of the selected page. The display control instructions (protocol) of a GLCD are given in Table 5.4. Starting from page 0, whose address is B8H, on the left half (CS1 = 0 and CS2 = 1), if we pass 1 data byte, it will appear on the first column of page 0. If you repeat this 64 times, then switch to the second half and repeat until the last (128th) position is reached, the first eight rows (R_0–R_7) will now be plotted. The next eight rows (R_8–R_{15}) can be plotted similarly by switching to page 1, whose address is B9H. The total number of bytes

TABLE 5.4
Display Control Instructions

Command	Functions
3E	Display "off"
3F	Display "on"
40–7F	Columns (y) address for both segments
B8–BF	Pages (x) address for both segments
C0–FF	Columns (y) address for both segments to display the start line at the top of the screen
D_7	Ready (0) and busy (1)
D_5	Display off (1) and display on (0)
D_4	Reset (1) and normal (0)

needed for a complete display frame is $128 \times 8 = 1024$ pixels. Communication to the GLCD module from the FPGA is classified into two transaction modes: (i) control mode: sending commands to initialize/configure the GLCD and (ii) data mode: sending the pixel information that needs to be displayed on the GLCD. The GLCD module has to be configured/ initialized with a set of commands by the FPGA before beginning any writing operations.

The steps to communicate with the GLCD module are described below: (i) select the segment, (ii) enable the display "on", (iii) send the column address, (iv) send the page address, (v) send the column address to display at the top of the screen (optional), and (vi) send the data sequences. The appropriate logic levels/transitions have to be applied to the all control pins such as RS, R/W and E during the above initialization and data display operations. The display data RAM of the GLCD automatically increment the column (y) address once a byte of data is printed on the screen.

Please note that the pin details and required logic levels at CS1 and CS2 to select the segments differ in some GLCDs. Designers have to check them before wiring and programming. However, regardless of the GLCD models, an output of 6 bits as [5:0] has to be alerted for control pins, that is, [CS1, CS2, RS, R/W, E, RST] and another output of 8 bits as [7:0] for the command/data lines [55,56].

DESIGN EXAMPLE 5.5

Develop a digital system to display digital pulses on pages 1 and 6 of a GLCD. Use both segments and display two cycles in each page. Use the onboard DIP switch for resetting the GLCD.

Solution

```
LIBRARY ieee;
USE ieee.std_logic_1164.all;
USE ieee.numeric_std.all;
use ieee.std_logic_unsigned.all;
use IEEE.STD_LOGIC_ARITH.ALL;
entity GLCD is
Port (m_clk: in STD_LOGIC:='0';
GLCD_cnt: out std_logic_vector(2 downto 0):="100"; --E,RS,RW
GLCD_data: out std_logic_vector(7 downto 0):="00000000"; --d7,d6....d0
CS:out std_logic_vector(1 downto 0):="00"); --CS1,CS2
end GLCD;
architecture Behavioral of GLCD is
signal clk_sig: std_logic:='0';
signal sta,nsta:integer range 0 to 36:=0;
type com_data is array (1 to 4) of std_logic_vector(7 downto 0);
constant data_c : com_data :=(x"3F",x"40", "10111001",x"BE");
type disp_data is array (1 to 2, 1 to 3) of std_logic_vector(7 downto
0); --(1 to 6) is row;
signal data_d: disp_data:=((x"FF", x"01", x"80"), ---(1 to 3) is column
(x"FF", x"80", x"01"));
begin
p0:process(m_clk)
variable cnt:integer:=1;
begin
if rising_edge(m_clk) then
if(cnt=2)then
clk_sig<= not(clk_sig);
cnt:=1;
else
cnt:= cnt + 1;
end if;
```

```
end if;
end process;
p4:process(clk_sig)
begin
ıt rising_edge(clk_sig) then
sta<=nsta;
end if;
end process;
p5:process(sta)
variable cnt,cnt1:integer:=1;
begin
case sta is
when 0=> CS<="01"; nsta<=sta+1; --CS1, CS2
when 1|20=> GLCD_data<=data_c(1); GLCD_cnt<="100"; nsta<=sta+1;
--3FH
when 2|4|6|21|23|25=> GLCD_cnt<="000";nsta<=sta+1;
when 3|22=> GLCD_data<=data_c(2);GLCD_cnt<="100"; nsta<=sta+1;--40H
when 5|24=> if (cnt1=1) then
GLCD_data<=data_c(3);GLCD_cnt<="100"; nsta<=sta+1;--B9H
cnt1:=cnt1+1;
elsif (cnt1=2) then
GLCD_data<=data_c(4);GLCD_cnt<="100"; nsta<=sta+1;--BEH
cnt1:=1;
end if;
when 7=> GLCD_data<=data_d(1,1); GLCD_cnt<="110"; nsta<=sta+1; --FFH
(Column 0)
when 8|10|13|15|18|27|30|32|35=> GLCD_cnt<="010";nsta<=sta+1;
when 9=> GLCD_data<=data_d(1,2); GLCD_cnt<="110"; nsta<=sta+1; --01H
(Column 1 to 31)
when 11=> if (cnt=31) then
nsta<=sta+1;
cnt:=1;
else
cnt:=cnt+1;
nsta<=9;
end if;
when 12=> GLCD_data<=data_d(1,1); GLCD_cnt<="110"; nsta<=sta+1; --FFH
(Column 32)
when 14=> GLCD_data<=data_d(1,3); GLCD_cnt<="110"; nsta<=sta+1; --00H
(Column 33 to 62)
when 16=> if (cnt=30) then
nsta<=sta+1;
cnt:=0;
else
cnt:=cnt+1;
nsta<=14;
end if;
when 17=> GLCD_data<=data_d(1,1); GLCD_cnt<="110"; nsta<=sta+1; --00H
(Column 63)
when 19=> CS<="10"; nsta<=sta+1; --CS1, CS2
when 26=> GLCD_data<=data_d(1,2); GLCD_cnt<="110"; nsta<=sta+1; --00H
(Column 0 to 31)
when 28=> if (cnt=31) then
nsta<=sta+1;
cnt:=1;
else
cnt:=cnt+1;
nsta<=26;
end if;
when 29=> GLCD_data<=data_d(1,1); GLCD_cnt<="110"; nsta<=sta+1; --00H
(Column 32)
```

```
when 31=> GLCD_data<=data_d(1,3); GLCD_cnt<="110"; nsta<=sta+1; --00H
(Column 33 to 62)
when 33=> if (cnt=30) then
nsta<=sta+1;
cnt:=1;
else
cnt:=cnt+1;
nsta<=31;
end if;
when 34=> GLCD_data<=data_d(1,1); GLCD_cnt<="110"; nsta<=sta+1; --00H
(Column 63)
when 36=> nsta<=0;
when others => nsta<=0;
end case;
end process;
end Behavioral;
```

5.5 General-Purpose Switch Interfacing

In this section, we discuss various input switches and their different possibilities of interfacing with the FPGA. The necessary circuit and design examples are also given to provide the reader with a clear idea of the usages of manual and automatic input switches.

5.5.1 Dual Inline Package Switch

The simplest and most common type of input switch is the DIP switch, which is available in different forms as a mechanical on/off toggle switch, push-button switch, rocker switch, key switch, reed switch, and so on. All these DIP switches are popular because of their low cost and easy interfacing to any circuit. The operator can change the state of an input simply by operating a switch knob or moving a magnet over a reed switch. More than one DIP switches together form a keypad, as shown in Figure 5.10a, which can be interfaced with the FPGA. This allows DIP switches to be inserted into standard IC sockets or printed circuit boards. Each switch in a keypad normally indicates one of two conditions, on or off status, and a four-switch DIP package will have four outputs and indicate 16 different statuses. DIP switches or push buttons are mechanical devices that have two or more sets of electrical contacts. When the switch is open, the contacts are open circuited, and when closed, these contacts are shorted together. The most common way of interfacing a DIP switch or push button to an electronic circuit is via a pull-up resistor with the supply voltage, as shown in Figure 5.10b. When the two left switches are open, 5 V, that is, a logic 1, is given as the output signal. When those switches are closed, the output is grounded and 0 V, that is, a logic 0, is given as the output. The reverse operation happens for the two right-side switches. Therefore, depending upon the position of the switch, a "high" or "low" output is produced. A pull-up resistor is necessary to hold the output voltage level at the required value when the switch is open and also to prevent the switch from shorting out the supply when closed. Therefore, digital switches are entirely different from analog (electrical) switches [57].

In addition to manually operated switches, there is one called the reed switch that can be operated by an electronic (or) mechanical system, which means human intervention is not needed to operate it. The reed switch comes in two varieties, called normally open

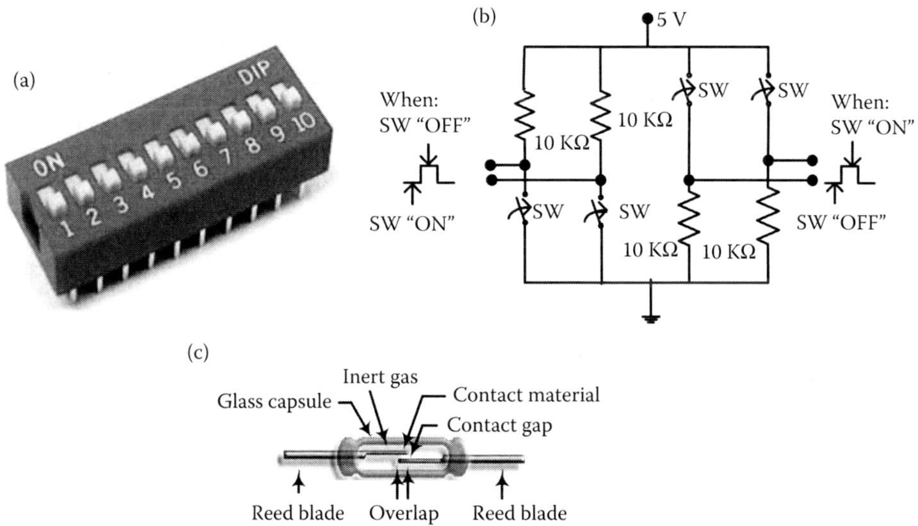

FIGURE 5.10
Picture of a DIP switch (a), application circuit (b), and reed switch (c).

and normally closed, with two leads (which look like metal reeds) made from conduction materials housed inside a thin glass envelope, as shown in Figure 5.10c. In a normally opened switch, as we bring a magnet closer to the switch on top of it, the contact metals are attracted; hence, each of them makes contact. Therefore, the reed blades are connected to each other, the switch acts as a short circuit and current flows through them. Further, as we take the magnet away, the contact metals go back to their original positions and the switch becomes an open circuit. In a normally closed switch, one of the contacts is a magnetic north pole, while the other is a south pole. As we bring a magnet up to the switch, it affects the contacts in opposite ways, that is, attracting one and repelling the other; hence, they become an open circuit. When the magnet is taken away, the contacts go back to their original positions. This switch was first invented by Bell Laboratory, and its length today is just a few millimetres. Reed switches are used in numerous applications, including coffee makers, rain gauges, cup anemometers, wind direction finders and satellite TV positioners.

DESIGN EXAMPLE 5.6

Develop a digital system to do different operations based on the magnitude of a set of 4-bit DIP switches. A reed switch is also an input to the system. Increment (0 through F) the seven-segment display and force high to only the green LED when the magnitude at the first four DIP switches is greater than that of the second four DIP switches. Decrement the display and force high to only the red LED if the second switch inputs are less than the first four inputs. In the case where both inputs are equal, maintain the counter as it is, look at the reed switch and complement the status of the LEDs if the reed switch is at "1"; otherwise, turn both LEDs off.

Solution

```
LIBRARY ieee;
USE ieee.std_logic_1164.all;
USE ieee.numeric_std.all;
use ieee.std_logic_unsigned.all;
```

```
use IEEE.STD_LOGIC_ARITH.ALL;
entity dip_swi is
Port (m_clk,reed_sw: in STD_LOGIC:='0';
dip_sw: in std_logic_vector(7 downto 0):="00000000";
led_r,led_g: inout std_logic:='0';
sev_Seg_data: out std_logic_vector(7 downto 0):="00000000");
end dip_swi;
architecture Behavioral of dip_swi is
signal clk_sig: std_logic:='0';
signal count: std_logic_vector(3 downto 0):="0000";
begin
p0:process(m_clk)
variable cnt:integer:=1;
begin
if rising_edge(m_clk) then
if(cnt=2)then
clk_sig<= not(clk_sig);
cnt:=1;
else
cnt:= cnt + 1;
end if;
end if;
end process;
p1:process(clk_sig,dip_sw,reed_sw)
begin
if rising_edge(clk_sig) then
if (dip_sw(7 downto 4)> dip_sw(3 downto 0)) then
count<=count+"0001";
led_g<='1';
led_r<='0';
elsif (dip_sw(7 downto 4)< dip_sw(3 downto 0)) then
count<=count-"0001";
led_g<='0';
led_r<='1';
elsif (dip_sw(7 downto 4)= dip_sw(3 downto 0)) then
count<=dip_sw(7 downto 4);
if (reed_sw='1') then
led_g<=not led_g;
led_r<=not led_r;
else
led_g<='0';
led_r<='0';
end if;
end if;
end if;
end process;
p2: process(count)
begin
case count is
when "0000"=> sev_Seg_data<="11000000";
when "0001"=> sev_Seg_data<="11111001";
when "0010"=> sev_Seg_data<="10100100";
when "0011"=> sev_Seg_data<="10110000";
when "0100"=> sev_Seg_data<="10011001";
when "0101"=> sev_Seg_data<="10010010";
when "0110"=> sev_Seg_data<="10000010";
when "0111"=> sev_Seg_data<="11111000";
when "1000"=> sev_Seg_data<="10000000";
when "1001"=> sev_Seg_data<="10011000";
when "1010"=> sev_Seg_data<="10001000";
```

```
when "1011"=> sev_Seg_data<="10000000";
when "1100"=> sev_Seg_data<="11000110";
when "1101"=> sev_Seg_data<="11000000";
when "1110"=> sev_Seg_data<="10000110";
when "1111"=> sev_Seg_data<="10001110";
when others=> sev_Seg_data<="11000000";
end case;
end process;
end Behavioral;
```

5.5.2 Bidirectional Port/Switch Design

Bidirectional switches have to be designed inside the FPGA whenever we need to make use of a port as a bidirectional (input-output) terminal. This type of switch is mainly used whenever we need to pass data to a port from the FPGA as well as receive data to the FPGA through the same port. Therefore, it is called a bidirectional switch. A simple design of a bidirectional switch is shown in Figure 5.11a. A bidirectional switch has two inputs, input (x) and enable (`E`); one output (`r`) and one inout (`y`). The `x` is routed to `y` when `E` is equal to "0"; otherwise, `y` will be set at high impedance `Z`. That means the input (x) is connected to the input terminal of inout whenever `E` is "0". The input terminal of the inout (`y`) is totally detached from getting input from the input (x) and is kept at state `Z` whenever `E` is "1". However, the output terminal of inout (`y`) is always connected to the output (`r`). Hence, the external input of the inout (`y`) can always be collected at `r`. Therefore, `y` acts as an output terminal when `E` is set at logic 0 and vice versa. This operation is illustrated in Figure 5.11 by the dotted lines. More than one switch can be designed in parallel inside the FPGA to design a port of bidirectional switches.

There is another switch, called a tristate switch, whose output (`y`) can also be electrically disconnected from its input (x) when required. The enable signal (`E`) can be either at logic 0 or logic 1, which results in connecting the output (`y`) either with input (x) or high impedance (`Z`). Making a short circuit between the input (x) and output (`y`) would be attained either by forcing a logic 1 (active high) or a logic 0 (active low) at the enable (`E`). However, in both cases, the output will be at high impedance for one logic level of the enable. For example, in Figure 5.11b, the input (x) is routed to output (`y`) when the enable is at logic 0; otherwise, the high impedance will be forced to the output, which means the switch becomes an open circuit; hence, the output (`y`) does not get anything from the input (x) [57,58].

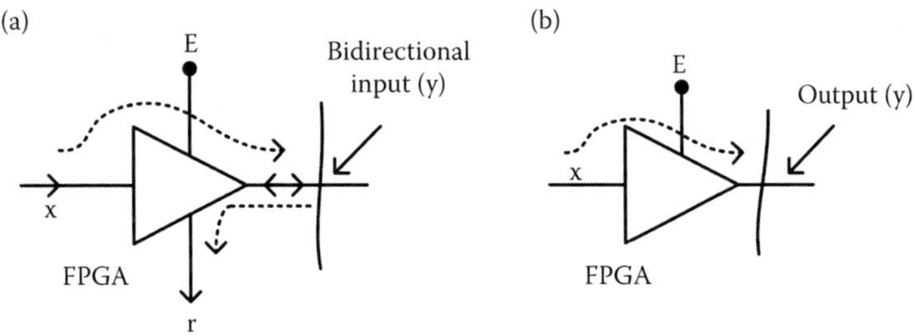

FIGURE 5.11
Bidirectional (a) and tristate switches (b).

DESIGN EXAMPLE 5.7

Develop a bidirectional switch to make use of a pin as output as well as input based on an enable control.

Solution

```
library ieee;
use ieee.std_logic_1164.all;
entity clkdcd1 is
port (Data_w,clk,reset : in std_logic:='0';
Data_r,clk_sig_out : out std_logic;
Data_rw : inout std_logic);
end clkdcd1;
architecture RTL of clkdcd1 is
signal data_rw_cnt :std_logic:='0';
signal clk_sig0: std_logic:='0';
begin
p1:process(clk,reset)
variable cnt:integer:=0;
begin
if(reset='0') then
clk_sig0<='0';
cnt:=0;
elsif rising_edge(clk) then
if(cnt=2)then
clk_sig0<= not(clk_sig0);
data_rw_cnt<=not data_rw_cnt;
cnt:=0;
else
cnt:= cnt + 1;
end if;
end if;
clk_sig_out<=clk_sig0;
end process;
Data_rw <= Data_w when (data_rw_cnt = '1') else
'Z' when (data_rw_cnt='0');
Data_r <= Data_rw;
end RTL;
```

DESIGN EXAMPLE 5.8

Develop two tristate switches to route one input to an output while keeping another output at high impedance. Swap this operation with enable control.

Solution

```
LIBRARY ieee;
USE ieee.std_logic_1164.ALL;
ENTITY trist IS
PORT(
w,r,en: in STD_LOGIC:='0';
w1,r1 : out STD_LOGIC:='0');
END trist;
ARCHITECTURE switch OF trist IS
begin
p1:process(en,r,w)
begin
if (en='0') then
w1<=w;
r1<='Z';
elsif (en='1') then
```

```
w1<='Z';
r1<=r;
end if;
end process;
END switch;
```

DESIGN EXAMPLE 5.9

Develop a tristate switch to route an input to two outputs while the enable is set at logic 1; otherwise, keep the outputs at `Z`. The input and outputs have to be of N width.

Solution

```
library ieee;
use ieee.std_logic_1164.all;
entity Test_trisw is
generic (width : natural := 2);
port (Data_In : in std_logic_vector (width downto 0);
Data_Out : out std_logic_vector (width downto 0);
Data : inout std_logic_vector (width downto 0);
E : in std_ulogic);
end Test_trisw;
architecture RTL of Test_trisw is
begin
Data <= Data_in when (E = '1') else
(others => 'Z');
Data_out <= Data;
end RTL;
```

5.5.3 Matrix Keypad Interfacing

Matrix keypads are commonly used in calculators, telephones, weighing scales, vending machines, and so on, where a number of input switches are required. The matrix keypad is made by arranging push-button switches in rows and columns, as shown in Figure 5.12. To connect 4 × 4 push buttons to the FPGA in a straightforward way, we need 16 input pins. By connecting the switches in a matrix, we need only eight pins [39,57]. We can read the status of each switch using a port of eight pins in the FPGA.

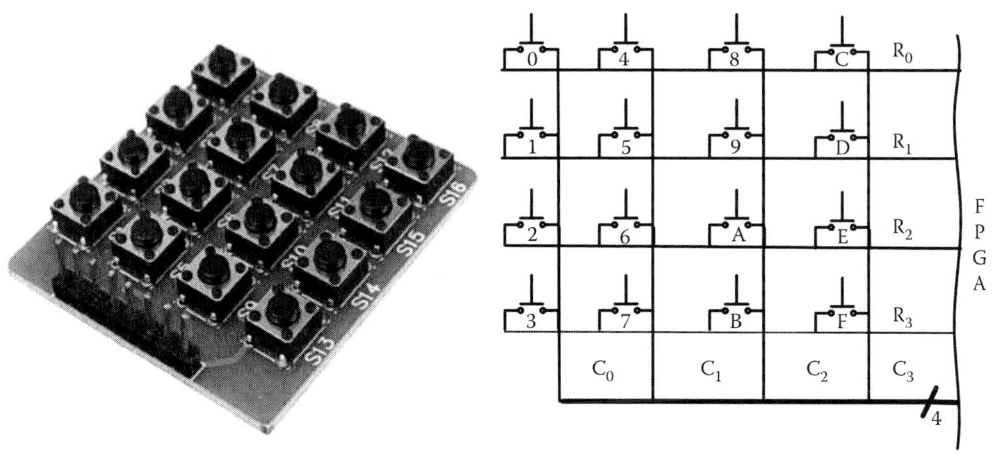

FIGURE 5.12
Construction of a matrix keypad.

Now, we see how it is practically possible. The status of each key can be determined by a process called scanning. For the sake of understanding, let us assume that all the column pins (C_0–C_3) and all the row pins (R_0–R_3) are connected to the FPGA as the output and input, respectively. To detect a pressed key in the first column (C_0), the FPGA sends "1110" to the columns of the keypad and then immediately reads the values of rows. If the data read from the rows are "1111", that is, no bit is "0", it means that no key has been pressed. If the read data are "0111", then the pressed key is "R_0C_0"; if "1011", the key is "R_1C_0"; if "1101", the key is "R_2C_0" and if "1110", the key is "R_3C_0". After a key press is detected, the FPGA will go through the process of assigning the key number. Suppose the address of a identified key is "R_3C_0". Then the key number is "3", as shown in Figure 5.12. Since all the scanning, key identification and key number assignment happen at a high frequency, missing any key press is avoided. This process continues until all the columns have been scanned, then subsequently scans the rows. Therefore, a 4-bit input and another 4-bit output are required to interface matrix-keypad switches with FPGA. All processes like scanning, detection of pressed keys, and assigning switch numbers can be synchronized with an independent universal counter.

DESIGN EXAMPLE 5.10

Develop a digital system to interface a 4 × 4 matrix keypad to the FPGA. Display the pressed key using sufficient LEDs.

Solution

```
library IEEE;
use IEEE.STD_LOGIC_1164.ALL;
use IEEE.STD_LOGIC_ARITH.ALL;
use IEEE.STD_LOGIC_UNSIGNED.ALL;
entity exc is
Port (clk : in STD_LOGIC;
c:out std_logic_vector(3 downto 0);
r:in std_logic_Vector(3 downto 0);
d: out std_logic_vector(3 downto 0));
end exc;
architecture Behavioral of exc is
signal clkcont:std_logic_vector(29 downto 0);
begin
p1:process(clk)is
begin
if rising_edge (clk) then
clkcont<=clkcont+1;
case clkcont(10 downto 8)is
when "000" => c<="1110";
when "001" =>
if(r="1110") then
d<="0000";--0
elsif (r="1101") then
d<="0001";--1
elsif (r="1011") then
d<="0010";--2
elsif (r="0111") then
d<="0011";--3
end if;
when "010" => c<="1101";
when "011" =>
if(r="1110") then
d<="0100";--4
```

```
elsif (r="1101") then
d<="0101";--5
elsif (r="1011") then
d<="0110";--6
elsif (r="0111") then
d<="0111";--7
end if;
when "100" => c<="1011";
when "101" =>
if(r="1110") then
d<="1000";--8
elsif (r="1101") then
d<="1001";--9
elsif (r="1011") then
d<="1010";--A
elsif (r="0111") then
d<="1011";--B
end if;
when "110" => c<="0111";
when "111" =>
if(r="1110") then
d<="1100";--C
elsif (r="1101") then
d<="1101";--D
elsif (r="1011") then
d<="1110";--E
elsif (r="0111") then
d<="1111";--F
end if;
when others=>
end case;
end if;
end process;
end Behavioral;
```

5.6 Dual-Tone Multifrequency Decoder

This section explains using the DTMF decoder to do some operations based on a telephone signal. Some of the more common applications of the DTMF are interactive voice response (IVR), digital telephony, caller ID display, device status monitoring and control, and so on. Tones can be sent only over an active/connected voice calls. DTMF provides two selected output frequencies (high and low band) for a duration of 100 ms whenever a number is pressed.

The CM8870 is a full DTMF decoder that has both a frequency band split filter and tone decoder in a single 18-pin package, as shown in Figure 5.13. In the filter section, a switched capacitor technology is used for high/low band filtering as well as for dial tone rejection. Delayed steering output (StD) is used to indicate the valid frequencies during the guard time. The FPGA will start reading the decoded data once the StD goes high. The decoder uses digital counting techniques to detect and decode all 16 DTMF tones as a 4-bit code. When the DTMF input is given at pin 2, the equivalent binary data will be available at the output in 4 bits: Q_0–Q_3. The key number pressed by the caller can be identified at the receiver, through which many operations would be performed at the receiver by the caller. Therefore, the input of five bits as [$\mathtt{StD}Q_3Q_2Q_1Q_0$] is the inputs to the FPGA. The number of bits for the output has to be decided based on the subsequent operations.

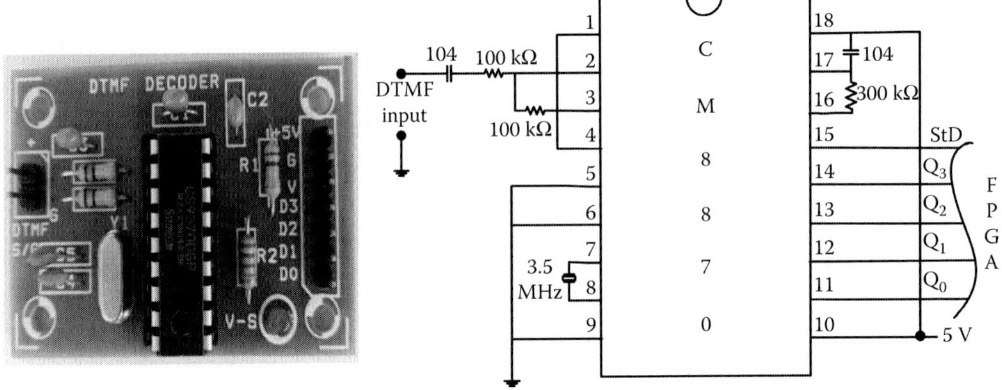

FIGURE 5.13
Appearance and application circuit of a DTMF decoder.

DESIGN EXAMPLE 5.11

Develop a digital system in the FPGA to control appliances from a remote station using DTMF tones as per the following requirements.

i. There are nine electrical appliances that have to be controlled.
ii. DTMF tone "0" has to be used to switch off all the appliances.
iii. DTMF tone data (1–9) have to be used to select a particular appliance.
iv. "*" and "#", that is, A ("1010") and C ("1100"), have to be used to switch the appliances "on" and "off", respectively.

Solution

```
library IEEE;
use IEEE.STD_LOGIC_1164.ALL;
entity dtmf is
Port (StD,m_clk : in STD_LOGIC:='0';
dtmf_data : in STD_LOGIC_vector(3 downto 0):="0000";
ctrl_out:out std_logic_vector(9 downto 1):="000000000");
end dtmf;
architecture Behavioral of dtmf is
signal clk_sig: std_logic:='0';
type state is(s0,s1,s2,s3,s4,s5,s6,s7,s8,s9);
signal nst:state:=s0;
begin
p0:process(m_clk)
variable cnt: integer:=0;
begin
if rising_edge(m_clk) then
if (cnt=2) then
clk_sig<= not clk_sig;
cnt:=0;
else
cnt:=cnt+1;
end if;
end if;
end process;
p1:process(clk_sig,dtmf_data,StD)
begin
if (StD='1') then
if rising_edge(clk_sig) then
```

```
case dtmf_data is
when "0000"=> nst<=s0;
when "0001"=> nst<=s1;
when "0010"=> nst<=s2;
when "0011"=> nst<=s3;
when "0100"=> nst<=s4;
when "0101"=> nst<=s5;
when "0110"=> nst<=s6;
when "0111"=> nst<=s7;
when "1000"=> nst<=s8;
when "1001"=> nst<=s9;
when others=> null;
end case;
end if;
if falling_edge(clk_sig) then
case nst is
when s0=> ctrl_out<="000000000";
when s1=> if (dtmf_data="1010") then
ctrl_out(1)<='1';
elsif (dtmf_data="1100") then
ctrl_out(1)<='0';
end if;
when s2=> if (dtmf_data="1010") then
ctrl_out(2)<='1';
elsif (dtmf_data="1100") then
ctrl_out(2)<='0';
end if;
when s3=> if (dtmf_data="1010") then
ctrl_out(3)<='1';
elsif (dtmf_data="1100") then
ctrl_out(3)<='0';
end if;
when s4=> if (dtmf_data="1010") then
ctrl_out(4)<='1';
elsif (dtmf_data="1100") then
ctrl_out(4)<='0';
end if;
when s5=> if (dtmf_data="1010") then
ctrl_out(5)<='1';
elsif (dtmf_data="1100") then
ctrl_out(5)<='0';
end if;
when s6=> if (dtmf_data="1010") then
ctrl_out(6)<='1';
elsif (dtmf_data="1100") then
ctrl_out(6)<='0';
end if;
when s7=> if (dtmf_data="1010") then
ctrl_out(7)<='1';
elsif (dtmf_data="1100") then
ctrl_out(7)<='0';
end if;
when s8=> if (dtmf_data="1010") then
ctrl_out(8)<='1';
elsif (dtmf_data="1100") then
ctrl_out(8)<='0';
end if;
when s9=> if (dtmf_data="1010") then
ctrl_out(9)<='1';
elsif (dtmf_data="1100") then
ctrl_out(9)<='0';
```

```
end if;
when others=> null;
end case;
end if;
end if;
end process;
end Behavioral;
```

5.7 Optical Sensor Interfacing

In this section, the operating principles of optical sensors are discussed. Some of the most important interfacing application circuits are explained. Different possibilities of wiring to photodiode (PD) and phototransistor (PT) are also discussed, along with a rotating disc speed measurement example.

5.7.1 Infrared Sensors

An IR LED is a special-purpose LED that transmits (Tx) infrared rays (not visible to the naked eye) in the range of the 760 nm wavelength. An infrared LED operates like a normal LED, but uses different materials to produce infrared light. Such LEDs are usually made of gallium arsenide or aluminium gallium arsenide. They are commonly used as sensors along with IR receivers (Rx), that is, photodiodes or phototransistors. Since the human eye cannot see infrared radiation, it is not possible for a person to identify whether the IR LED is working, unlike a common LED. To overcome this problem, the camera on a cell phone can be used in IR mode. The camera can show us the IR rays being emanated from the IR LED [39,52].

Depending upon the type and amount of semiconductor doping, some photodiodes respond to visible light, and some only to IR light. When there is no incident light, the reverse/dark current flows in the circuit, which is negligible. An increase in the amount of light intensity increases the reverse current. Then, we can see that a photodiode allows reverse current to flow in one direction only, which is the opposite of a standard rectifying diode, for example, 1N4007. The picture and application circuits of IR Tx/Rx are shown in Figure 5.14a and b, respectively. As shown in the figure, the circuit can be used to measure the amount of light falling on its junction. The output of a photodiode can be given to a comparator from which an output of either logic 0 or 1 can be obtained with respect to the reference voltage divider circuit (VDC) input, as in Figure 5.14c. The appearance and application circuit of a phototransistor are also shown in Figure 5.14d and e, respectively. Optical switching, also called opto-switching, is another application of IR Tx/Rx which can be used for finding objects and counting the speed of a rotating disc.

A DC voltage is used to drive the LED, which converts the input signal into infrared light. This light travels in the isolation gap and falls on the phototransistor, which converts it back to an electrical output signal. For understanding, assume that a slotted disc is made to rotate within the isolation gap, as shown in Figure 5.15. During one revolution of the disc, the infrared light from the LED strikes the phototransistor through the slot and is then blocked as the disk rotates. This results in turning the transistor "on" and "off" in each pass of the slot. The collector of the phototransistor is connected to the

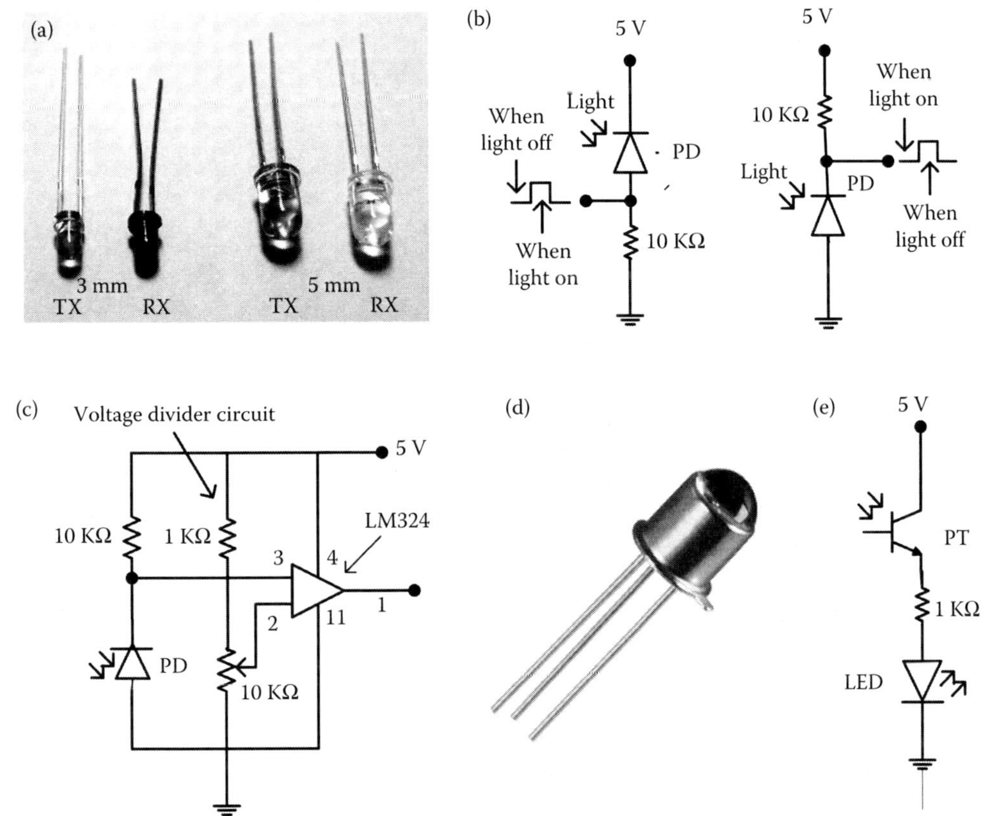

FIGURE 5.14
Appearance of IR Tx/Rx (a), IR Rx application circuits (b), photodiode with a comparator (c), appearance of a phototransistor (d), and application circuit of a phototransistor (e).

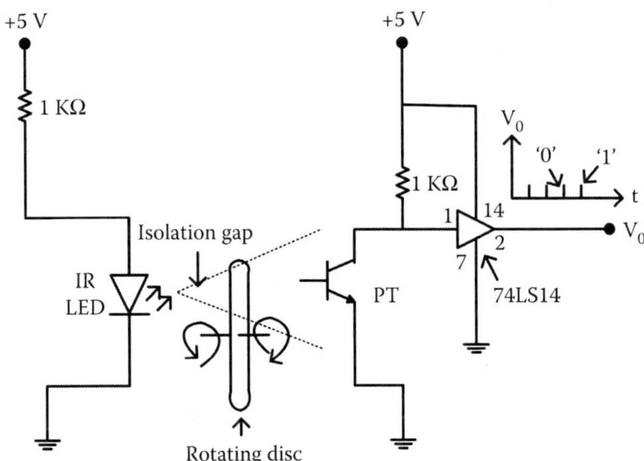

FIGURE 5.15
Circuit for rotating disc speed measurement.

input of a Schmitt inverter (IC74LS14), which produces a logic 0 whenever the transistor is "off" and a logic 1 whenever the transistor is "on". If the inverter output is given to a digital counter, then it would be possible to determine the rotating speed of the disc per minute [52,53].

DESIGN EXAMPLE 5.12

Develop a digital system in FPGA to meet the following requirements. A parcel conveyer belt system consists of a scanner, three IR sensors and four solenoid plungers. Every pair of IR-Tx and Rx is placed opposite at different heights such as 3, 10, and 18 cm beside the belt so as to have the IR link blocked by parcels of different heights. For example, if the height of the parcel falls below 10 cm, the IR-Rx output will be "110", "100" if the height is below 18 cm, and "000" if the height is above 18 cm. The scanner default value is "10", and it gives "00" if the parcel has defective material and "11" for good material. To sort out the parcels according to height, the respective solenoid plunger has to be activated. If defective material is identified, a buzzer has to be enabled for some time before activating the respective solenoid plunger. Separate counters are required to count the number of good material (parcels) passing the IR-Tx/Rx and display the counter values using LEDs. Use maximum of eight LEDs per counter. Assume that the speed of the conveyer belt is constant at a predecided value; hence, all these operations would be done according to the time needed for each and every action. There should be a provision for the operator to reset the system.

Solution

```
library IEEE;
use IEEE.STD_LOGIC_1164.ALL;
use IEEE.STD_LOGIC_ARITH.ALL;
use IEEE.STD_LOGIC_UNSIGNED.ALL;
entity ir_tx_rx is
Port (m_clk,reset : in STD_LOGIC:='0';
ir_rx: in std_logic_vector(2 downto 0):="000";
scan : in STD_LOGIC_vector(1 downto 0):="10";
buzz_defe: out std_logic:='0';
sole_plun: out std_logic_vector(3 downto 0):="0000";
count3,count10,count18: inout std_logic_vector(7 downto 0):="00000000");
end ir_tx_rx;
architecture Behavioral of ir_tx_rx is
signal clk_sig: std_logic:='0';
signal ir_rx_data: std_logic_vector(2 downto 0):="000";
begin
p0:process (m_clk)
variable cnt:integer:=1;
begin
if rising_edge(m_clk) then
if (cnt=2) then
clk_sig<=not clk_sig;
cnt:=1;
else
cnt:= cnt+1;
end if;
end if;
end process;
p1:process(clk_sig,ir_rx,scan,reset)
variable cnt:integer:=1;
begin
if (reset='1') then
count3<="00000000";
```

```
count10<="00000000";
count18<="00000000";
buzz_defe<='0';
sole_plun<="0000";
else
if rising_edge(clk_sig) then
if (cnt=1) then
if (scan="10") then
buzz_defe<='0';
sole_plun<="0000";
cnt:=1;
elsif (scan="00") then
buzz_defe<='1';
cnt:=2;
elsif (scan="11") then
buzz_defe<='0';
cnt:=3;
end if;
elsif (cnt=2) then
buzz_defe<='0';
sole_plun<="0001";---plunger for defected parcel
cnt:=8;
elsif (cnt=3) then
ir_rx_data<=not (ir_rx);
cnt:=cnt+1;
elsif (cnt=4) then
if (ir_rx_data="001") then
count3<=count3+"00000001"; ---3cm counter
cnt:=cnt+1;
elsif (ir_rx_data="011") then
count10<=count10+"00000001"; --10cm counter
cnt:=cnt+1;
elsif (ir_rx_data="111") then
count18<=count18+"00000001";--18 cm counter
cnt:=cnt+1;
end if;
elsif (cnt=5) then
if (ir_rx_data="001") then
sole_plun<="0010";
cnt:=8;
else
cnt:=cnt+1;
end if;
elsif (cnt=6) then
if (ir_rx_data="011") then
sole_plun<="0100";
cnt:=8;
else
cnt:=cnt+1;
end if;
elsif (cnt=7) then
if (ir_rx_data="111") then
sole_plun<="1000";
cnt:=8;
end if;
elsif(cnt=8) then
sole_plun<="0000";
if (scan/="10") then
cnt:=8;
else
cnt:=1;
```

```
end if;
end if;
end if;
end if;
end process;
end Behavioral;
```

5.7.2 Proximity Sensor

Another type of optical switching device is a proximity sensor or reflective opt-switching sensor, which uses an IR LED and photodetector (PD or PT) to detect an object, marked path/line, height of an object and so on. The reflective opto-switch can detect the presence of the object by receiving reflected infrared light from the surface of the object being sensed. The basic arrangement of a reflective opto-sensor is shown in Figure 5.16.

The phototransistor has a very high "off" resistance and a low "on" resistance, which are controlled by the amount of light striking its base from the IR LED, that is, actually from the surface of the object [39,52–54]. If there is no object in front of the sensor, then the infrared light will shine forward as a single beam. When there is an object in close proximity to the sensor, the infrared light is reflected back, falls on the IR Rx and hence is detected by the phototransistor. The amount of reflected light sensed by the phototransistor and the degree of transistor saturation will give information about how close the reflective object is.

DESIGN EXAMPLE 5.13

Develop a digital system to continuously monitor and automatically fill the chemical fluid in a tank. A proximity sensor is placed inside the tank so as to get IR radiation reflected back to the phototransistor from the chemical fluid surface. The output of the phototransistor is given to an 8-bit A/D convertor. Hence, the fluid level is measured by A/D and given to the FPGA via 8-bit data. The system has to indicate the fluid level in the first row of an LCD and switch on the fluid inlet motor if the level goes below "00001111".

Solution

```
library IEEE;
use IEEE.STD_LOGIC_1164.ALL;
use IEEE.STD_LOGIC_ARITH.ALL;
use IEEE.STD_LOGIC_UNSIGNED.ALL;
entity prox_Sen is
```

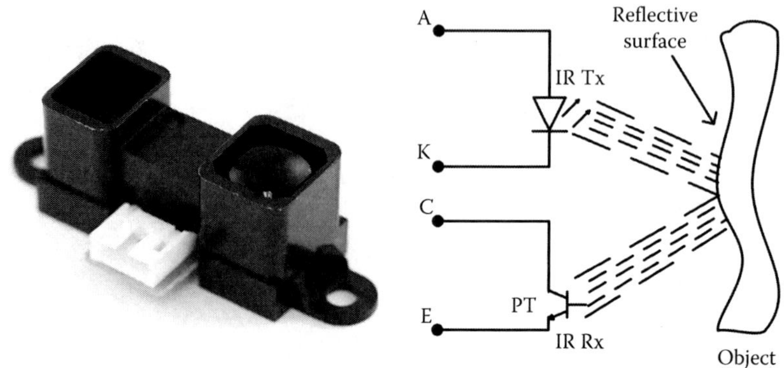

FIGURE 5.16
IR proximity or obstruction sensor.

```
Port (m_clk : in STD_LOGIC:='0';
wat_lev: in std_logic_vector(7 downto 0):="00000000";
lcd_cnt: out std_logic_vector(2 downto 0):="000";
lcd_data: out std_logic_vector(7 downto 0):="00000000";
wat_moto: out std_logic:='0');
end prox_Sen;
architecture Behavioral of prox_Sen is
signal clk_sig,clk_sig1: std_logic:='0';
signal wat_lev_int: integer:=0;
signal lcd_lev,data_temp: std_logic_vector(7 downto 0):="10000000";
type state is(s0,s1,s2,s3,s4,s5,s6,s7,s8,s9,s10,s11,s12,s13);
signal ps,ns:state:=s0;
begin
p0:process (m_clk)
variable cnt:integer:=1;
begin
if rising_edge(m_clk) then --Proximity sensor clock
if (cnt=2) then
clk_sig<=not clk_sig;
cnt:=1;
else
cnt:= cnt+1;
end if;
end if;
end process;
wat_lev_int<=conv_integer(unsigned(wat_lev));
p1:process(clk_sig,wat_lev)
begin
if rising_edge(clk_sig) then
if (wat_lev<=15) then
wat_moto<='1';
lcd_lev<="10000000";
elsif (wat_lev>=16 and wat_lev<=31) then
lcd_lev<="10000001";
elsif (wat_lev>=32 and wat_lev<=47) then
lcd_lev<="10000010";
elsif (wat_lev>=48 and wat_lev<=63) then
lcd_lev<="10000011";
elsif (wat_lev>=64 and wat_lev<=79) then
lcd_lev<="10000100";
elsif (wat_lev>=80 and wat_lev<=95) then
lcd_lev<="10000101";
elsif (wat_lev>=96 and wat_lev<=111) then
lcd_lev<="10000110";
elsif (wat_lev>=112 and wat_lev<=127) then
lcd_lev<="10000111";
elsif (wat_lev>=128 and wat_lev<=143) then
lcd_lev<="10001000";
elsif (wat_lev>=144 and wat_lev<=159) then
lcd_lev<="10001001";
elsif (wat_lev>=160 and wat_lev<=175) then
lcd_lev<="10001010";
elsif (wat_lev>=176 and wat_lev<=191) then
lcd_lev<="10001011";
elsif (wat_lev>=192 and wat_lev<=207) then
lcd_lev<="10001100";
elsif (wat_lev>=208 and wat_lev<=223) then
lcd_lev<="10001101";
elsif (wat_lev>=224 and wat_lev<=239) then
lcd_lev<="10001110";
elsif (wat_lev>=240) then
```

```
wat_moto<='0';
lcd_lev<="10001111";
end if;
end if;
end process;
p2:process (m_clk) --LCD low frequwency....
variable cnt:integer:=1;
begin
if rising_edge(m_clk) then
if (cnt=4) then
clk_sig1<=not clk_sig1;
cnt:=1;
else
cnt:= cnt+1;
end if;
end if;
end process;
p3:process(clk_sig1)
begin
if rising_edge(clk_sig1) then
ps<=ns;
end if;
end process;
-------------LCD Command and Data ---------------
p4:process(ps,lcd_lev)
begin
case ps is
when s0=> lcd_data<="00111000";lcd_cnt<="100"; ns<=s1;--38
when s1=> lcd_cnt<="000";ns<=s2;
when s2=> lcd_data<="00001111";lcd_cnt<="100"; ns<=s3;--0F
when s3=> lcd_cnt<="000";ns<=s4;
when s4=> lcd_data<="00000001";lcd_cnt<="100"; ns<=s5;--01
when s5=> lcd_cnt<="000";ns<=s6;
when s6=> lcd_data<="00000110";lcd_cnt<="100"; ns<=s7;--06
when s7=> lcd_cnt<="000";ns<=s8;
when s8=> lcd_data<="10000000";lcd_cnt<="100"; ns<=s9;--80
when s9=> lcd_cnt<="000";ns<=s10;
when s10=>data_temp<="10000000"; ns<=s11;
when s11=> lcd_data<="11011011";lcd_cnt<="110"; ns<=s12;--#
when s12=> lcd_cnt<="010";ns<=s13;
when s13=> if (data_temp=lcd_lev) then
ns<=s0;
else
data_temp<=data_temp+"00000001";
ns<=s11;
end if;
when others=> ns<=s0;
end case;
end process;
end Behavioral;
```

5.8 Special Sensor Interfacing

In this section, we study some of the special sensors, namely the PIR sensor, metal detector and LDR. Working principles of all these sensors, standard application circuits, pictures and VHDL code design examples are also discussed.

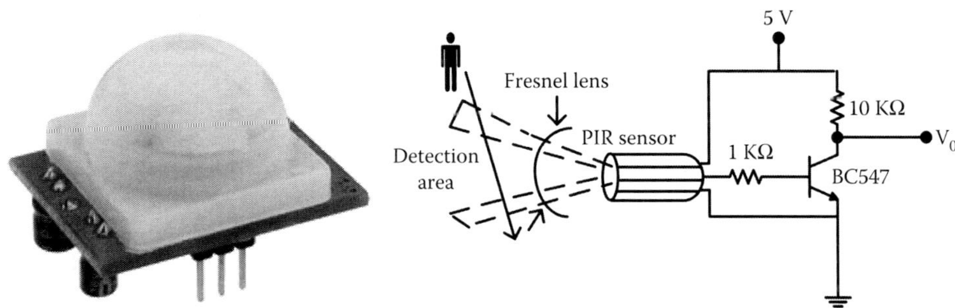

FIGURE 5.17
Appearance and working principles of a PIR sensor.

5.8.1 Passive Infrared Sensor

The PIR sensor, also called a human intruder detector or pyrometric sensor, can be used to detect the presence of a person. Certain crystalline materials have the property to generate a surface electric charge when exposed to thermal infrared radiation. This phenomenon is known as pyroelectricity. The PIR sensor module works on the same principle. The front part of the sensor module has a Fresnel lens to focus the infrared light to the sensor element, as shown in Figure 5.17. The human body radiates heat in the form of infrared radiation, which is about 9.4 um. The presence of the human body creates a sudden change in the IR profile of the surroundings that is sensed by the sensor. The PIR sensor module has an instrumentation circuit on board that amplifies this signal to the appropriate voltage level to indicate detection of a person.

A PIR sensor is more complicated than many other sensors. The PIR sensor itself has two slots; each slot is made of a special material that is sensitive to IR, as shown in Figure 5.17. When the sensor is idle, both slots detect the same amount of IR radiation, that is, the ambient radiation within the room. When a warm body, like a human, passes by, it first intercepts half of the PIR sensor, which causes a positive differential change between the two halves. When the warm body leaves the sensing area, the reverse happens, whereby the sensor generates a negative differential change. These change pulses are detected to identify the person's presence. The sensor output is connected to a NPN transistor; hence, the whole unit has a single output that goes low when motion is detected. A module in the FPGA continuously monitors the output of this unit and activates the concerned operation once a person is detected [52–54]. The PIR sensor has a typical range of 20 feet and is designed to adjust to slowly changing conditions such as a gradual change in the thermal profile of the surroundings as the day passes. A continuous motion will give repeated high/low pulses.

DESIGN EXAMPLE 5.14

Develop a digital system in the FPGA to display the total number of persons present in an auditorium using two PIR sensors. The system design has to be as follows.

i. The PIR sensor module is installed at entry and exit.
ii. Person entry as well as exit have to be counted independently.
iii. The total number of persons present in the room has to be displayed on a two-digit seven-segment display. Therefore, the display may vary from 00 to 99.

Solution

```
library IEEE;
use IEEE.STD_LOGIC_1164.ALL;
use IEEE.STD_LOGIC_ARITH.ALL;
use IEEE.STD_LOGIC_UNSIGNED.ALL;
entity pir is
Port (m_clk,pir_ent,pir_ext : in STD_LOGIC:='0';
sev_seg_out:out std_logic_vector(7 downto 0):="00000000";
tr_ctrl: out std_logic_vector(1 downto 0):="00");
end pir;
architecture Behavioral of pir is
signal clk_sig,clk_sig1: std_logic:='0';
signal count_ent,count_ext,count_seg:std_logic_vector(6 downto
0):="0000000";
type state is(s0,s1,s2,s3,s4,s5,s6,s7,s8,s9,s10,s11,s12,s13);
signal pst,nst:state:=s0;
signal bcd: std_logic_vector (14 downto 0):="000000000000000";
signal seg_mux:std_logic_vector(3 downto 0):="0000";
begin
p0:process(m_clk)
variable cnt: integer:=0;
begin
if rising_edge(m_clk) then
if (cnt=6) then ---Clock for Entry and Exit detection
clk_sig<= not clk_sig;
cnt:=0;
else
cnt:=cnt+1;
end if;
end if;
end process;
p1:process(clk_sig, pir_ent,pir_ext)
variable cnt1,cnt2:integer:=1;
begin
if rising_edge(clk_sig) then
if (cnt1=1) then
if (pir_ent='1') then
if (count_ent="1100011") then
count_ent<="0000000";
cnt1:=1;
else
count_ent<=count_ent+ "0000001"; --Entry Counter
cnt1:=2;
end if;
else
cnt1:=1;
end if;
elsif (cnt1=2) then
if (pir_ent='1') then
cnt1:=2;
else
cnt1:=1;
end if;
end if;
if (cnt2=1) then
if (pir_ext='1') then
if (count_ext="1100011") then
count_ext<="0000000";
cnt2:=1;
else
```

```
count_ext<=count_ext+ "0000001"; --Exit Counter
cnt2:=2;
end if;
else
cnt2:=1;
end if;
elsif (cnt2=2) then
if (pir_ext='1') then
cnt2:=2;
else
cnt2:=1;
end if;
end if;
end if;
if falling_edge(clk_sig) then
if (count_ent > count_ext) then
count_seg <=(count_ent-count_ext); --Difference b/w counters
elsif (count_ent < count_ext) then
count_seg <=(count_ext - count_ent);
elsif (count_ent = count_ext) then
count_seg<="0000000";
end if;
end if;
pst<=nst; --Updating of State
end process;
p2: process(pst,count_seg)---Binary to BCD conversion
begin
case pst is
when s0=> bcd<=("00000000" & count_seg); --loading
nst<=s1;
when s1=> bcd<=(bcd(13 downto 0) & bcd(14));--LR1
nst<=s2;
when s2=> bcd<=(bcd(13 downto 0) & bcd(14));--LR2
nst<=s3;
when s3=> bcd<=(bcd(13 downto 0) & bcd(14));--LR3
nst<=s4;
when s4=> if bcd(10 downto 7) > "0100" then
bcd(10 downto 7)<= bcd(10 downto 7) + "0011";--Add 3 if need
end if;
nst<=s5;
when s5=> bcd<=(bcd(13 downto 0) & bcd(14)); --LR4
nst<=s6;
when s6=> if bcd(10 downto 7) > "0100" then
bcd(10 downto 7)<= bcd(10 downto 7) + "0011";--Add 3 if need
end if;
nst<=s7;
when s7=> bcd<=(bcd(13 downto 0) & bcd(14)); --LR5
nst<=s8;
when s8=> if bcd(10 downto 7) > "0100" then
bcd(10 downto 7)<= bcd(10 downto 7) + "0011"; --Add 3 if need
end if;
nst<=s9;
when s9=> bcd<=(bcd(13 downto 0) & bcd(14)); --LR6
nst<=s10;
when s10=>if bcd(10 downto 7) > "0100" then
bcd(10 downto 7)<= bcd(10 downto 7) + "0011"; --Add 3 if need
end if;
nst<=s11;
when s11=>bcd<=(bcd(13 downto 0) & bcd(14)); --LR7
nst<=s12;
```

```
when s12=>seg_mux<=bcd(14 downto 11);
tr_ctrl<="10";
nst<=s13;
when s13=>seg_mux<=bcd(10 downto 7);
tr_ctrl<="01";
nst<=s0;
when others=> nst<=s0;
end case;
end process;
p3:process(m_clk)
variable cnt: integer:=0;
begin
if rising_edge(m_clk) then
if (cnt=2) then ---Clock for 7-segment display
clk_sig1<= not clk_sig1;
cnt:=0;
else
cnt:=cnt+1;
end if;
end if;
end process;
p4: process(clk_sig1,seg_mux)
begin
if rising_edge(clk_sig1) then --7 Segment display
case seg_mux is
when "0000" => sev_seg_out<="11000000";
when "0001" => sev_seg_out<="11111001";
when "0010" => sev_seg_out<="10100100";
when "0011" => sev_seg_out<="10110000";
when "0100" => sev_seg_out<="10011001";
when "0101" => sev_seg_out<="10010010";
when "0110" => sev_seg_out<="10000010";
when "0111" => sev_seg_out<="11111000";
when "1000" => sev_seg_out<="10000000";
when "1001" => sev_seg_out<="10011000";
when others=> sev_seg_out<="11000000";
end case;
end if;
end process;
end Behavioral;
```

5.8.2 Metal Detector

A security system can be developed by using a metal detector. The metal detector is used to identify/avoid any illegal or unauthorized entry of metallic objects such as bombs, knives and guns, especially in public places like theaters, shopping malls, parks, airports, hotels, military/defense areas, and railway stations. Metal detectors are also used to sense weapons, identify steel reinforcement in concrete structures and detect the condition of pipes/wires buried in walls/floors. A metal detector is an electronic device that comprises an oscillator, which generates an AC current that passes via a coil generating an alternating magnetic field, as shown in Figure 5.18. When a part of the metal is near the coil, an eddy current will be induced in the metal object, and it generates a magnetic field of its own. This magnetic field can be sensed and measured using the coil to detect the metal object [59].

When the output pin is high, the base resistor will offer a positive voltage to the transistor, which turns "on" the LED; hence, the output (V_0) goes low. The logic level change at the collector of the transistor is continuously monitored by a module built

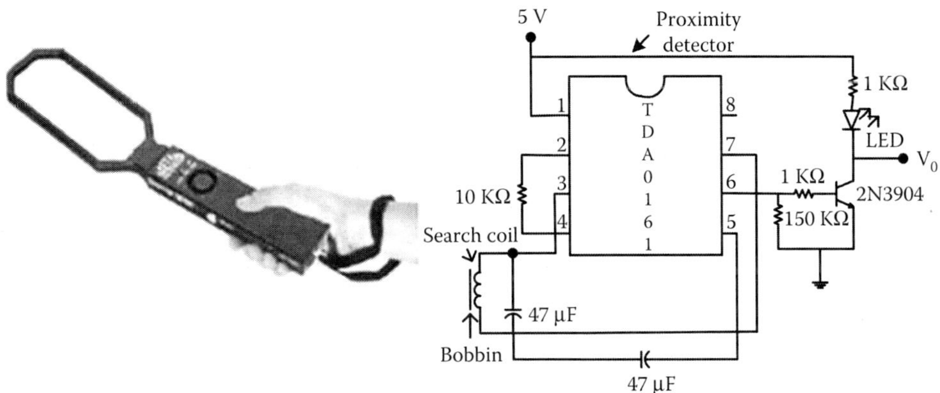

FIGURE 5.18
Appearance and application circuit of a metal detector.

inside the FPGA. Once a logic 0 is detected, the module immediately initiates the necessary actions.

DESIGN EXAMPLE 5.15

Develop a digital system for a metal detector stick with the following features.

i. The metal detector circuit has to give logic 1 whenever metal is detected.
ii. A provision has to exist to enter the location code (0 through 7) before starting the scanning process.
iii. There must be a provision to choose the stick for detection (logic 1) or display (logic 0) mode. That means the system has to increment the appropriate counter whenever metal is detected under detection mode. The system has to display the counter value in eight LEDs, corresponding to any location chosen by the operator.
iv. The detection also has to be alarmed via a buzzer.

Solution

```
library IEEE;
use IEEE.STD_LOGIC_1164.ALL;
use IEEE.STD_LOGIC_ARITH.ALL;
use IEEE.STD_LOGIC_UNSIGNED.ALL;
entity met_det is
Port (m_clk, met_det,op_mode : in STD_LOGIC:='0';
loc_id: in std_logic_vector(2 downto 0):="000";
met_cnt_valu: out std_logic_vector(7 downto 0):="00000000";
buz_out: out std_logic:='0');
end met_det;
architecture Behavioral of met_det is
signal clk_sig: std_logic:='0';
signal
met_cnt_temp0,met_cnt_temp1,met_cnt_temp2,met_cnt_temp3,met_cnt_temp4,
met_cnt_temp5,met_cnt_temp6,met_cnt_temp7: std_logic_vector(7 downto
0):="00000000";
begin
p0:process (m_clk)
variable cnt:integer:=1;
begin
if rising_edge(m_clk) then
```

```
if (cnt=2) then
clk_sig<=not clk_sig;
cnt:=1;
else
cnt:= cnt+1;
end if;
end if;
end process;
p1:process(clk_sig,met_det,op_mode,loc_id)
variable cnt:integer:=1;
begin
if rising_edge(clk_sig) then
if (op_mode='1') then --detection mode
buz_out<=met_det;
met_cnt_valu<="00000000";
if (cnt=1) then
if (met_det='0') then
cnt:=1;
else
if (loc_id="000") then
met_cnt_temp0<=met_cnt_temp0+"00000001";
cnt:=cnt+1;
elsif (loc_id="001") then
met_cnt_temp1<=met_cnt_temp1+"00000001";
cnt:=cnt+1;
elsif (loc_id="010") then
met_cnt_temp2<=met_cnt_temp2+"00000001";
cnt:=cnt+1;
elsif (loc_id="011") then
met_cnt_temp3<=met_cnt_temp3+"00000001";
cnt:=cnt+1;
elsif (loc_id="100") then
met_cnt_temp4<=met_cnt_temp4+"00000001";
cnt:=cnt+1;
elsif (loc_id="101") then
met_cnt_temp5<=met_cnt_temp5+"00000001";
cnt:=cnt+1;
elsif (loc_id="110") then
met_cnt_temp6<=met_cnt_temp6+"00000001";
cnt:=cnt+1;
elsif (loc_id="111") then
met_cnt_temp7<=met_cnt_temp7+"00000001";
cnt:=cnt+1;
end if;
end if;
elsif (cnt=2) then
if (met_det='1') then
cnt:=2;
else
cnt:=1;
end if;
end if;
elsif (op_mode='0') then --display mode
case loc_id is
when "000"=>met_cnt_valu<=met_cnt_temp0;
when "001"=>met_cnt_valu<=met_cnt_temp1;
when "010"=>met_cnt_valu<=met_cnt_temp2;
when "011"=>met_cnt_valu<=met_cnt_temp3;
when "100"=>met_cnt_valu<=met_cnt_temp4;
when "101"=>met_cnt_valu<=met_cnt_temp5;
when "110"=>met_cnt_valu<=met_cnt_temp6;
```

```
when "111"=>met_cnt_valu<=met_cnt_temp7;
when others=> null;
end case;
end if;
end if;
end process;
end Behavioral;
```

5.8.3 Light-Dependent Resistor

An LDR or photo-resistor is a device whose resistivity is a function of the incident light. They are also called photo conductors, photo-conductive cells or simply photocells. They are made up of semiconductor materials having high resistance. The appearance and characteristic response of an LDR are shown in Figure 5.19. Photo conductivity is an optical phenomenon in which the material's conductivity increases when light falls on it. The photons in the incident light should have energy greater than the band gap of the semiconductor material to make the electrons jump from the valence band to the conduction band. Hence, when light having enough energy strikes the device, more and more electrons are excited to the conduction band, which results in a large number of charge carriers. The result of this process makes more and more current flow through the device when the circuit is closed; hence, it is said that the resistance of the device has been decreased. If a constant voltage is applied to it and the intensity of light is increased, the current flow starts increasing. However, the sensitivity varies as a function of the wavelength of light incident on the device. Some photocells are not at all responsive to a certain range of wavelengths [60].

The intensity of light falling on the LDR can be estimated using the amount of resistance across the LDR, that is, (R_{LDR}). The model equation for this estimation is lux $= 1.25 \times 10^{7}$ $R_{LDR}^{-1.4059}$. The nonlinear characteristic response of an LDR obtained using this equation is shown in Figure 5.19b, and the corresponding MATLAB® code is as below.

```
clc;
clear all;
r=0:1:100000;
lux=1.25E7 * r.^-1.4059;
```

(a)

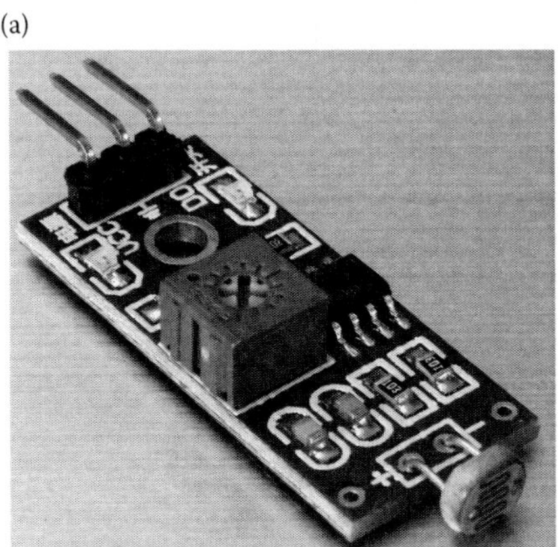

(b)

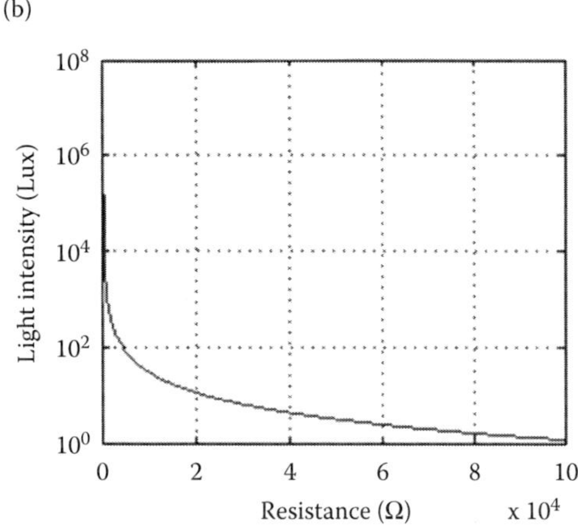

FIGURE 5.19
Appearance (a) and typical nonlinear characteristic response of an LDR (b).

```
subplot(2,2,1)
semilogy(r,lux);
xlabel('Resistance (\Omega)');
ylabel ('Light Intensity (Lux)');
grid on;
```

Therefore, LDRs can be used with a series resistor (R) to form a VDC across the supply, as shown in Figure 5.20a. In the dark, the resistance of the LDR is much greater than that of the resistor (R) which connects the LDR from the supply or to ground. This circuit can be used as a light or dark detector as well. When the light level is low, the resistance of the LDR is high. This prevents current flow to the base of the transistors. Consequently, the LED does not light. When more light shines on the LDR, its resistance decreases and then current flows into the base of the first transistor and then the second transistor, and now the LED glows, i.e., the output (V_0) produces a logic 0. The R can be turned up/down to increase/decrease resistance through which the voltage level across R would be adjusted more/less [60].

LDRs can also be connected to an op-amp voltage comparator circuit to produce data for interfacing to digital circuits, as shown in Figure 5.20b. Each comparator has inputs from VDC (reference voltage) and LDR (sensor voltage). Whenever the sensor voltage exceeds the reference voltage, the outputs (y_0–y_3) of the op-amp comparator go low. For example, the reference voltage of the second comparator is 2 V; as long as another input is less than 2 V, the output is at logic 1. If another input is equal to or greater than 2 V, the output produces logic 0. Therefore, as the light intensity increases, the resistance of the LDR decreases; hence, the voltage across R_1 increases, which results in producing digital data at the outputs as "1111", "1110", "1100", "1000", and "0000" over time.

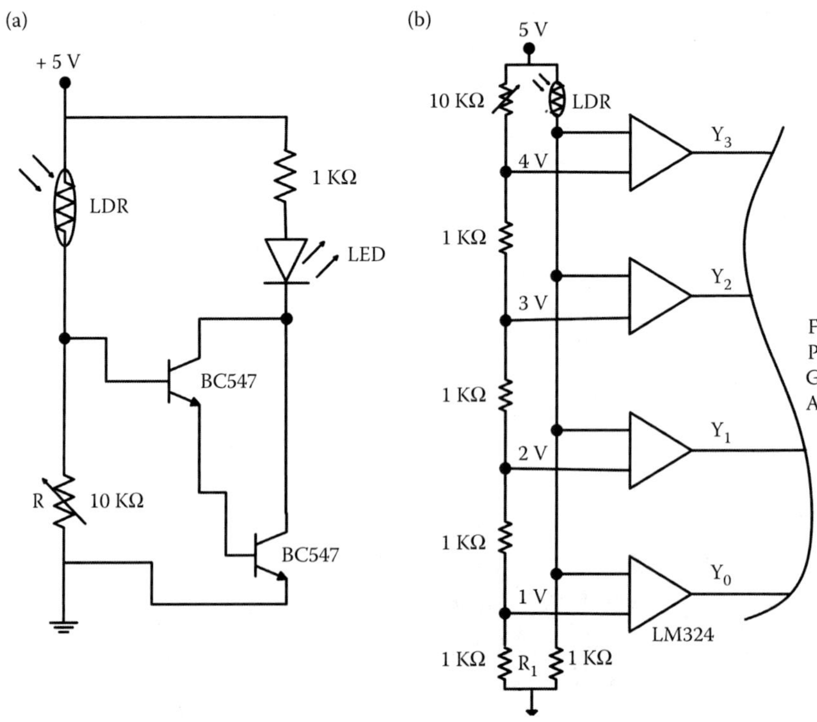

FIGURE 5.20
LDR-based light/dark detector circuit (a) and LDR with an op-amp comparator (b).

DESIGN EXAMPLE 5.16

Develop a digital system to automatically steer a solar panel according to the movement of the sun. The movement of the sun is measured using eight LDRs. A wood patch of size 30 cm × 30 cm × 1 cm is vertically positioned on a wooden base of size 40 cm × 40 cm × 1 cm on which four LDRs are placed, vertical to the patch, one by one on both sides. This whole setup is deployed in the East–West direction, so that the voltage values of LDRs can be used to measure the movement of the sun. Output of eight (four east side and another four west) LDRs are connected to the FPGA through LM324. The digital system built inside FPGA has to steer a stepper motor so as to tilt the solar panel toward the sun. The output of the solar panel is given to a comparator, based on whose output the delivery of power from a battery is enabled whenever the comparator value is less than "1111". Once a full steering is completed, bring the panel back to its initial position.

Solution

```
library IEEE;
use IEEE.STD_LOGIC_1164.ALL;
use IEEE.STD_LOGIC_ARITH.ALL;
use IEEE.STD_LOGIC_UNSIGNED.ALL;
entity ldr_sol_track is
Port (    m_clk : in STD_LOGIC:='0';
ldr_E,ldr_W: in std_logic_vector(3 downto 0):="0000";
sol_vol: in std_logic_vector(3 downto 0):="0000";
step_mot: out std_logic_vector(3 downto 0):="0000";
sol_bt_ctrl: out STD_LOGIC:='0');
end ldr_sol_track;
architecture Behavioral of ldr_sol_track is
signal clk_sig:std_logic:='0';
signal tra_W: std_logic_vector(3 downto 0):="1000";
signal tra_E: std_logic_vector(3 downto 0):="0001";
signal step_mux:std_logic_vector(1 downto 0):="00";
begin
p0:process (m_clk,sol_vol)
variable cnt:integer:=1;
begin
if rising_edge(m_clk) then
if (cnt=2) then ----Stepper motor
clk_sig<=not clk_sig;
cnt:=1;
else
cnt:= cnt+1;
end if;
if (sol_vol < "1111") then
sol_bt_ctrl<='1';
else
sol_bt_ctrl<='0';
end if;
end if;
end process;
p1:process(clk_sig,ldr_W,ldr_E)
variable cnt:integer:=0;
begin
if rising_edge(clk_sig) then
tra_W<=(tra_W(0) & tra_W(3 downto 1));
tra_E<=(tra_E(2 downto 0) & tra_E(3));
end if;
if falling_edge(clk_sig) then
if (cnt=1) then
if (ldr_E="1111" and ldr_W="0000") then
```

```
cnt:=1;
else
cnt:=cnt+1;
end if;
elsif (cnt=2) then
step_mux<="01";
cnt:=cnt+1;
elsif (cnt=6) then
step_mux<="00";
if (ldr_E="1111" and ldr_W="0001") then
cnt:=6;
else
cnt:=cnt+1;
end if;
elsif (cnt=7) then
step_mux<="01";
cnt:=cnt+1;
elsif (cnt=11) then
step_mux<="00";
if (ldr_E="1111" and ldr_W="0011") then
cnt:=11;
else
cnt:=cnt+1;
end if;
elsif (cnt=12) then
step_mux<="01";
cnt:=cnt+1;
elsif (cnt=16) then
step_mux<="00";
if (ldr_E="1111" and ldr_W="0111") then
cnt:=16;
else
cnt:=cnt+1;
end if;
elsif (cnt=17) then
step_mux<="01";
cnt:=cnt+1;
elsif (cnt=21) then
step_mux<="00";
if (ldr_E="1111" and ldr_W="1111") then
cnt:=21;
else
cnt:=cnt+1;
end if;
elsif (cnt=22) then
step_mux<="01";
cnt:=cnt+1;
elsif (cnt=26) then
step_mux<="00";
if (ldr_E="1110" and ldr_W="1111") then
cnt:=26;
else
cnt:=cnt+1;
end if;
elsif (cnt=27) then
step_mux<="01";
cnt:=cnt+1;
elsif (cnt=31) then
step_mux<="00";
if (ldr_E="1100" and ldr_W="1111") then
cnt:=31;
```

```
else
cnt:=cnt+1;
end if;
elsif (cnt=32) then
step_mux<="01";
cnt:=cnt+1;
elsif (cnt=36) then
step_mux<="00";
if (ldr_E="1000" and ldr_W="1111") then
cnt:=36;
else
cnt:=cnt+1;
end if;
elsif (cnt=37) then
step_mux<="01";
cnt:=cnt+1;
elsif (cnt=41) then
step_mux<="00";
if (ldr_E="0000" and ldr_W="1111") then
cnt:=41;
else
cnt:=cnt+1;
end if;
elsif (cnt=42) then
if (ldr_E="0000" and ldr_W="0000") then
step_mux<="10";
cnt:=cnt+1;
else
cnt:=41;
end if;
elsif (cnt=62) then
step_mux<="00";
cnt:=1;
else
cnt:=cnt+1;
end if;
end if;
end process;
p2:process(step_mux, clk_sig)
begin
if rising_edge(clk_sig) then
case step_mux is
when "01" => step_mot<=tra_W;
when "10"=> step_mot<=tra_E;
when others=> null;
end case;
end if;
end process;
end Behavioral;
```

5.9 Wind-Speed Sensor Interfacing

A cup anemometer is a common wind sensor that can be used to measure wind speed. The cup anemometer assembly consists of hemispherical cups that rotate according to the wind speed, a drive shaft, magnetic bars, and a single-pole (normally opened) reed switch, as shown in Figure 5.21. The cup anemometer has the reed switch and magnets at the

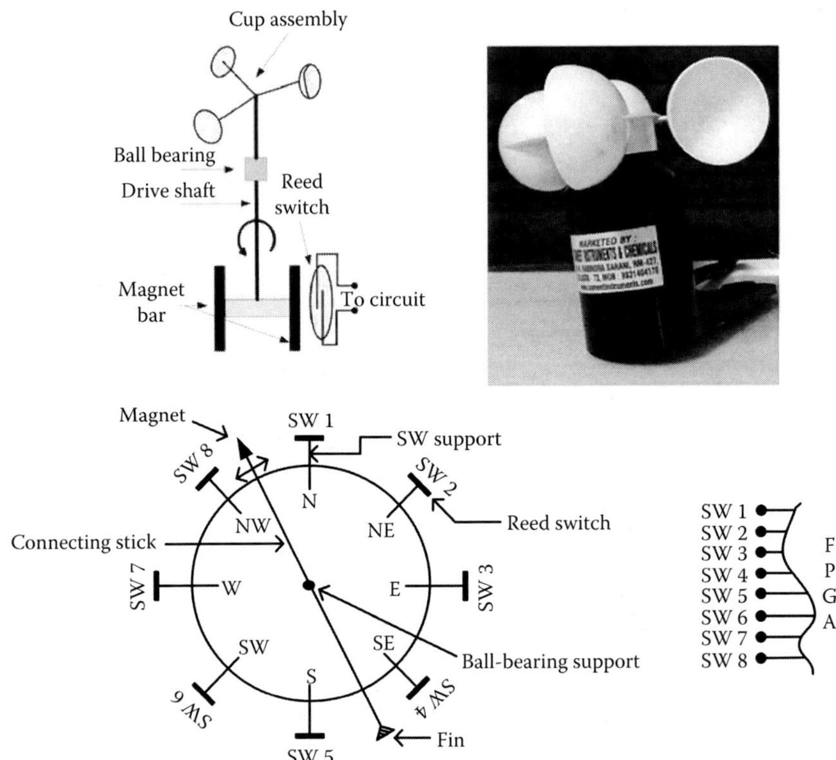

FIGURE 5.21
Cup anemometer assembly and wind-direction-finding vane.

bottom of the rotating shaft. This produces a switching output signal with frequency proportional to the wind speed. This means the angular movement of the magnet is directly proportional to the rotational speed of the cups, that is, wind speed. The reed switch is closed when any one of the magnetic poles comes closer to the centre part of it and opens when the pole moves away. This kind of anemometer effectively makes a series of electric pulses at a rate proportional to the wind speed. By counting how often the pulses come in, we can figure out the wind speed. The number of contacts (pulses) per second is the measurement of wind speed per second. There is a minimum wind speed that will set the cup in motion which depends on the shaft bearing friction and the design of the instrument. In a steady wind, the cup performs well from almost 0.27–60 ms^{-1}. The wind speed $\mu_a(t)$ is related to the angular velocity of the anemometer by $Cs(t)+U_0$, where C is a calibration constant (0.6201 m), $s(t)$ is the angular velocity of the device in Hz and U_0 is the offset speed (0.27 ms^{-1}). An architecture to measure the wind speed using the cup-anemometer is shown in Figure 5.22.

The wind direction is measured with a simple circuit, as shown in Figure 5.21. A stick arrow is placed at the center, on top of a circular disc with a small ball bearing on which the stick can freely roll. One end of the arrow is a fin to turn the entire stick along the direction of the wind. A magnet piece is placed at the other end of the stick arrow. Eight reed switches are distributed at equal radii from the center of the disc. Arrow turning now depends the wind direction; hence, the magnet moves toward any one of the reed switches or between any two switches. Therefore, either one or two reed switches will be "on" based

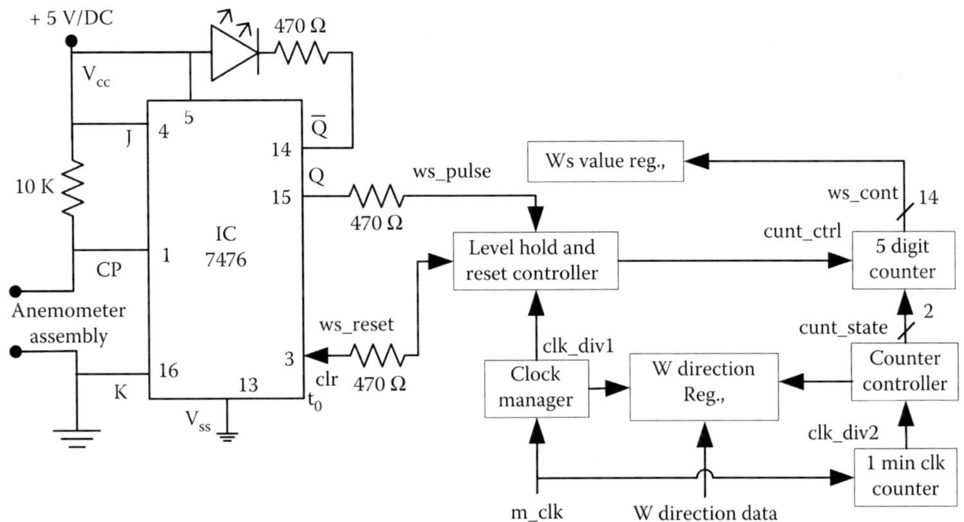

FIGURE 5.22
Circuit and architecture for cup anemometer interfacing.

on the position of the stick, that is, the direction of the wind. The circuit has eight outputs, corresponding to wind direction, that are connected to the FPGA from which the wind direction is measured [61–64]. More reed switches and a smaller magnet size (magnetic flux) increase the accuracy of wind-direction measurement. These outputs can also be connected to a data logging system in real time to perform on/offline data analysis. Therefore, we need to use one input of a bit for wind speed input (Pulse), one output of a bit for FF reset and another input of an 8-bit vector [7:0] for the wind direction data.

DESIGN EXAMPLE 5.17

Develop a digital system in FPGA to measure the wind speed and its direction using a cup anemometer and a direction-finding vane, respectively. Assume that the vane consists of eight reed switches. Measure and log those data once every second in the registers.

Solution

A JK-FF is used to hold the pulse coming from the anemometer and to reset it after reading the pulse. Every polling of the anemometer is actually a clock pulse to the JK-FF, and its output (`Q`) goes high for all the rising edges of the clock pulse. The level hold and reset controller continuously monitors the voltage changes at `Q` and increments the value of the five-digit counter at the rising edge of `Q`. After one millisecond, it resets the JK-FF by sending a low signal to the `clr` pin.

The LED glows every time the anemometer sends a signal to the FPGA. A 1-second clock counter is designed to trigger the counter controller unit once every second, which keeps the counter in counting, shifting or reset mode. The clock manager generates the control and trigger signal at a 5-kHz rate in order to synchronize the measurement. The counter value $(000000010000)_2$ for a given second is equal to 16 in radix 10. The reed switch of the anemometer polls four times per cycle; hence, the angular velocity `s(t)` of the anemometer is given by counter value/4. The VHDL code is as below.

```
library IEEE;
use IEEE.STD_LOGIC_1164.ALL;
```

```
use IEEE.STD_LOGIC_ARITH.ALL;
use IEEE.STD_LOGIC_UNSIGNED.ALL;
entity ws_se is
Port (ws_data_in,clk,ws_reg_reset : in STD_LOGIC:='0';
w_dir: in std_logic_vector(7 downto 0):="00000000";
ws_ff_reset: out std_logic:='0';
ws_out : out STD_LOGIC_vector(13 downto 0):="00000000000000");
end ws_se;
architecture Behavioral of ws_se is
signal ws_reg:std_logic_vector(13 downto 0):="00000000000000";
signal w_dir_reg:std_logic_vector(7 downto 0):="00000000";
signal ws_clk_sig:std_logic:='0';
begin
p0:process(clk)
variable ws_cnt1:integer:=0;
begin
if rising_edge(clk) then
if(ws_cnt1=400)then
ws_clk_sig<= not(ws_clk_sig);
ws_cnt1:=1;
else
ws_cnt1:= ws_cnt1 + 1;
end if;
end if;
end process;
p1:process(ws_clk_sig,ws_data_in,ws_reg_reset,w_dir)
variable ws_cnt2: integer:=0;
begin
if (ws_reg_reset='1') then
ws_reg<="00000000000000";
ws_cnt2:=0;
ws_ff_reset<='1';
elsif (rising_edge(ws_clk_sig))then
if(ws_cnt2=0) then
ws_ff_reset<='1';
ws_cnt2:=ws_cnt2+1;
elsif(ws_cnt2=1) then
if (ws_data_in ='0') then
ws_ff_reset<='1';
ws_cnt2:=ws_cnt2;
else ws_cnt2:=ws_cnt2+1;
end if;
elsif(ws_cnt2=3) then
ws_reg<=ws_reg+ 1;
ws_ff_reset<='1';
ws_cnt2:=ws_cnt2+1;
elsif(ws_cnt2=20) then
ws_ff_reset<='0';
ws_cnt2:=ws_cnt2+1;
elsif(ws_cnt2=50) then
ws_ff_reset<='1';
ws_cnt2:=0;
else
ws_cnt2:=ws_cnt2+1;
w_dir_reg<=w_dir;
end if;
end if;
end process;
ws_out<=ws_reg;
end Behavioral;
```

LABORATORY EXERCISES

1. Design a simple pendulum and develop a digital system to measure the total number of oscillations of it using an IR sensor. Display the result in a 4 × 16 LCD.
2. Develop a digital system to produce different waveforms like sine, cosine, ramp and square wave forms with the x–y axis in a GLCD. Use DIP switches to select the wave form to be shown on a GLCD. Use 2D array techniques for storing and accessing the display data.
3. Design a simple line-following robot using IR Tx/Rx and other robot accessories. Develop a digital system to run the robot properly on the drawn line based on the IR sensor data.
4. Design a circuit and develop a digital system to make a telephone caller ID. The system should have the provision to see the history of accepted and missed calls.
5. Develop a digital system to act as a voltmeter using a few magnitude comparators. The measured value has to be displayed in a GLCD. There should be a provision for the user to select either analog (using a needle) or digital display (using a seven-segment display).
6. Develop a digital system to act as a smart water irrigation system. This system should have ground-level sensors, decision making, and delivery controls.
7. Develop a digital system for an escalator controller application using reed switches. Also, design an automatic car parking system for a business complex.
8. Design a simple circuit and develop a digital system to interface a TV remote, sensor, and display. Show the pressed numbers in dot-matrix displays.
9. A mechanical assembly throws a small ball in a straight line. Develop a digital system to measure the speed of that ball and show the result in an LCD.
10. Develop a water solenoid valve control system using a proximity sensor.

6

System Design with Finite and Algorithmic State Machine Approaches

6.1 Introduction and Preview

FSM- and ASM-based designs are widely used for sequential control logic, which is the core of many digital systems. State machines are required in a variety of applications, including peripheral interfacing, bus arbitration and timing generation, data encryption and decryption, control signal generation, transmission protocols, and so on. In this chapter, we will examine the significance and functionality of FSMs and ASMs and their implementations in the FPGA for different applications. This chapter begins with Moore and Mealy model-based FSM design as well as conversion of Moore to Mealy machines and vice versa. Design examples, data flow model-based and behavioral model-based implementations of these FSM models are also given. Based on the FSM design approaches, code classification and binary-to-binary-coded-decimal (BCD) conversion are developed and explained. A binary data sequence detector and sequence recognizer are designed with different state machine models (Moore and Mealy), and the design methodologies are detailed with numerical and VHDL code design examples. Some popular applications of FSMs are vending machine controllers and traffic light controllers, which are discussed in this chapter, with state diagrams and several design examples. The working principles and interfacing techniques of an escalator, dice game, and model electronic train controller are also described, with several design examples. The significance of an ASM chart, the difference between an FSM diagram and ASM chart, ASM block sets, ASM chart design flow and conversion of FSM to ASM are explained, followed by several ASM-based real-world design examples. ASM-based digital system design is introduced, with several design approaches and VHDL code design examples. Design methodologies, state diagrams, and VHDL codes for all the above topics are given and explained clearly so as to help the reader get complete information about them to apply to any other suitable applications.

6.2 Finite State Machine Design: Moore and Mealy Models

Up to now, most of the circuits presented have used combinatorial designs. That means their outputs are dependent only on their current inputs. Previous inputs for that type of circuit have no effect on the output. However, there are many applications where there is a need for the circuits to have memory (a storage element) to remember previous inputs and calculate their outputs according to them. A circuit whose output depends not only on the present input but also on the history of the input is called a sequential circuit [39,65]. In this section, we will learn how to design and build such sequential circuits. In order to understand how this procedure works, we will study the design of an FSM. For this purpose, assume that we need to design a digital system that reads an input from an external button/switch. Once the button is pressed, the system has to transmit a logic high pulse that is, logic 1, with a duration of only one cycle of 1 sec and will not transmit another pulse until the button is released and then pressed. Now, we need to convert this requirement into a state diagram. The states will actually be implemented by memory or storage elements, that is, FFs. The state diagram describes the operation of interest in a sequence; hence, it is called a sequential circuit. We require some finite number of states in order to get this work done; hence, it is also called an FSM. As shown in Figure 6.1, a circle represents a state with information about the: (i) state ID, (ii) output at that state, and (iii) next state for all the possible input values. Therefore, each state should have information for what it is intended to do from that state.

We start drawing the state diagram. At the first state, that is, circle S_0, the circuit waits for the input (x), that is, for the button to be pressed. If the input is logic 0; that is, the button is not pressed, the same state has to be retained (S_0), and the output should be logic 0. When the button is pressed, that is, the input is logic 1, the present state has to move to another state (S_1), where the output will be logic 1. Therefore, the second circle is the condition where the button has just been pressed and our circuit needs to transmit a high pulse. After producing a logic high output for a duration of 1 sec, the output has to be logic 0; hence, we should not be in the same state (S_1). If the button is released, we can go back to state S_0; instead, if it is still pressed, we can't move to S_0 because it will come back to S_1 immediately, which is prohibited in this case. Hence, we need to move to a new state (S_2), where the output will be logic 0 if the input is logic 1, and stay there as long as the same input is maintained, as shown in Figure 6.1. Therefore, the third circle is the condition where our circuit waits for the button to be released before it returns to the initial state (S_0). Within the circle, in the lower part, the output of our circuit is given as if we want our circuit to transmit a logic high at a specific state, and we put a 1 on that state; otherwise, we put a 0. Every arrow represents a transition from one state to another. A transition happens once every clock cycle. Depending on the current input, we may go to a different state each time.

Now, we need to convert this state diagram into a state or transition table. For this purpose, we need to replace the IDs of the states with unique numbers. Normally, we start the enumeration from $(0)_{10}$, which has to be assigned for the initial state (S_0). We then continue the enumeration with any state we like until all states have been given their numbers. In this example, the state numbers are: S_0 is represented by $(0)_{10}$, S_1 by $(1)_{10}$ and S_2 by $(2)_{10}$,

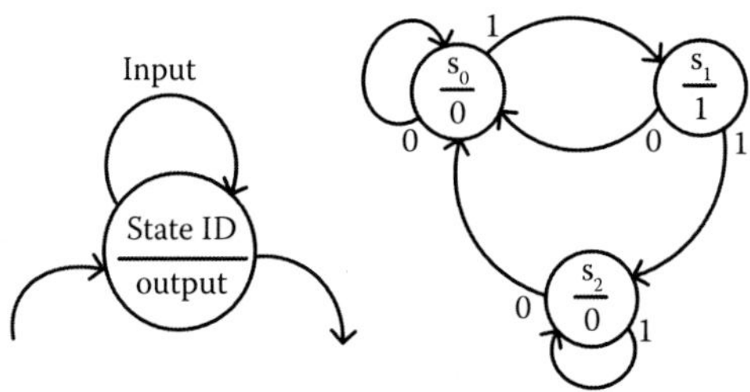

FIGURE 6.1
State diagram for sensing a button and producing a logic high pulse for only one cycle.

TABLE 6.1

State or Transition Table

Present State (Q)	Next State (Q^+)		
	Input x = 0	Input x = 1	
Q_BQ_A	$Q_B^+Q_A^+$	$Q_B^+Q_A^+$	Output (y)
0 0	0 0	0 1	0
0 1	0 0	1 0	1
1 0	0 0	1 0	0
1 1	x x	x x	x

which shows we need two binary bits to represent these states in binary form, that is, S_0 by $(00)_2$, S_1 by $(01)_2$, and S_2 by $(10)_2$. The $(11)_2$ is not required in this work; hence, it has to be treated as a "don't care" term. We need to allot n-bits to represent $(2^{n-1} + 1)$ through 2^n states. If we had five states, we would have to use the binary numbers $(000)_2$ through $(100)_2$. We will prepare a state or transition table with all these entries (Table 6.1). Since we allot one memory or storage element, that is, one FF for 1 bit, in the state representation, here we need two FFs for the 2 binary bits we have in this example. This table has a very specific form.

The first column in Table 6.1 describes the current state [Q(t) or simply Q] of our circuit. To the right of the current state column, we have the next state [Q(t + 1) or simply Q^+] column for all the possibilities of the input. Finally, we have the output column. The output column is filled by the output of the corresponding current state in the state diagram. All the entries in the state table, given in Table 6.1, are made completely based on the state diagram; hence, it describes the behavior of our circuit as fully as the state diagram does. In this example, the output is dependent on only the current input states, that is, the not the input (x). This is the reason the output column has two 1s. We take this point as it is now and discuss it in more depth in subsequent sections.

Now, we need to proceed further for implementing the state table using digital circuits, that is, FSM. As we know, we need two FFs to implement this design. In this section, we discuss the implementation of our design using both D-FFs as well as JK-FFs. The selection of the FF to use is arbitrary and is usually determined by cost factors. The best choice is to both perform analysis and decide which type of FF results in a minimum number of logic gates and lower cost [39,65,66]. First, we discuss how to implement our FSM using D-FFs. To proceed further, we need to include two columns, one for each FF, in Table 6.1 to get the excitation table as given in Table 6.2. The new column that corresponds to each FF describes what input we must give the FF in order to get the expected, that is, next, state (Q^+) from the present state (Q). As discussed in Chapter 2, for the D-FF, this is easy, because the input will be the next state. That means, as in Table 6.2, the expected state values, that is, the next state values, are equal to the FF inputs $D_A = Q_A^+$ and $D_B = Q_B^+$.

Now, we need to determine the Boolean functions to implement in the digital machine to produce the inputs for the FFs and output based on the input and present state values. We need to extract one Boolean function for each FF input we have. This can be done with a Karnaugh map (K-map). The input variables of this map are the variables used in Table 6.2. The preparation and simplification of a K-map are shown in Figure 6.2. In this example, the first two columns, that is, Q and x, are the input to the Boolean function, while the third and fifth columns, that is, FF inputs and y, are the output of it.

TABLE 6.2

Excitation Table with D-FFs

Present State (Q)		FF Inputs		Next State (Q^+)	
Q_BQ_A	Input (x)	D_B	D_A	$Q_B^+Q_A^+$	Output (y)
0 0	0	0	0	0 0	0
	1	0	1	0 1	
0 1	0	0	0	0 0	1
	1	1	0	1 0	
1 0	0	0	0	0 0	0
	1	1	0	1 0	
1 1	0	x	x	x	x
	1	x	x	x	

Upon simplifying the K-map as shown in Figure 6.2, the Boolean functions for first the FF input, second FF input and output are

$$\begin{aligned} D_A &= Q_B'Q_A'x \\ D_B &= (Q_A + Q_B)x \text{ and} \\ Y &= Q_A \end{aligned} \tag{6.1}$$

We implement the FSM as per Equation 6.1 using the D-FFs and logic gates to form the Boolean functions that we calculated, which will look like Figure 6.3. The combinational circuits at the left of the FFs take input from the present state as well as the input of the circuit (x). This combinational circuit actually generates the necessary inputs into the FFs to force it to the next expected state. In the same way, a combinational circuit at the right of the FFs generates the output (y) as the function of only the present state.

We will follow the same approach using JK-FFs, which will cause some differences in the third column of Table 6.2, as given in Table 6.3. We need to use the function table of JK-FF, as discussed in Chapter 2, to find the suitable inputs we must apply to the JK-FF to force it to the expected next state. Based on the present state and expected next state, the values for the variables (J_A, K_A and J_B, K_B) in the third column of Table 6.3 are obtained using the function table of JK-FF. The JK-FF has two inputs; therefore, we need to derive four equations, two for each FF.

Table 6.3 says that if we want to go from state "10" to state "00", we need to use the specific input values for the FFs, that is, $J_A = 0$, $K_A = x$ and $J_B = x$, $K_B = 1$. Then, we need to

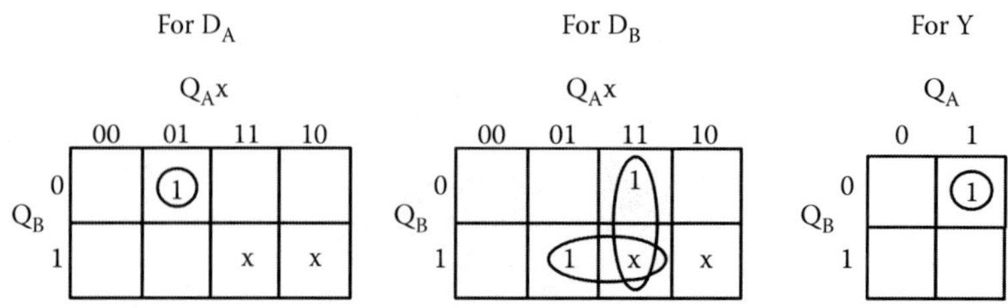

FIGURE 6.2
K-map and logical simplification with D-FF.

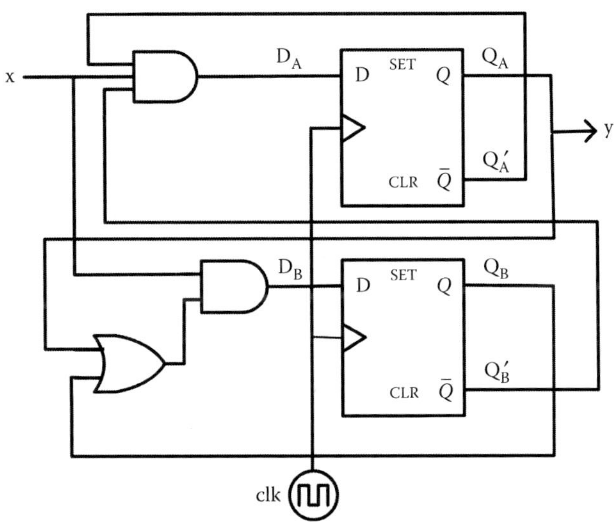

FIGURE 6.3
FSM realization of Equation 6.1.

find the Boolean function to generate these inputs. Therefore, we prepare the K-map and simplify, as shown in Figure 6.4. The same output function can be used.

Upon simplifying Figure 6.4, the Boolean functions can be obtained as

$$\begin{aligned} J_A &= Q'_B x \\ K_A &= 1 \\ J_B &= Q_A x \text{ and} \\ K_B &= x' \end{aligned} \tag{6.2}$$

This is the function equation of FSM we wish to implement. We will implement it using JK-FFs and gates, which will look like Figure 6.5.

Make a note that these two designs are a Moore FSM. Its output is a function of only its current state, not its input. That is in contrast to the Mealy FSM, where the input affects

TABLE 6.3

Excitation Table with JK-FFs

Present State (Q)		FF Input				Next State (Q^+)	Output
Q_BQ_A	Input (x)	J_B	K_B	J_A	K_A	$Q_B^+Q_A^+$	(y)
0 0	0	0	x	0	x	0 0	0
	1	0	x	1	x	0 1	
0 1	0	0	x	x	1	0 0	1
	1	1	x	x	1	1 0	
1 0	0	x	1	0	x	0 0	0
	1	x	0	0	x	1 0	
1 1	0	x	x	x	x	x	x
	1	x	x	x	x	x	

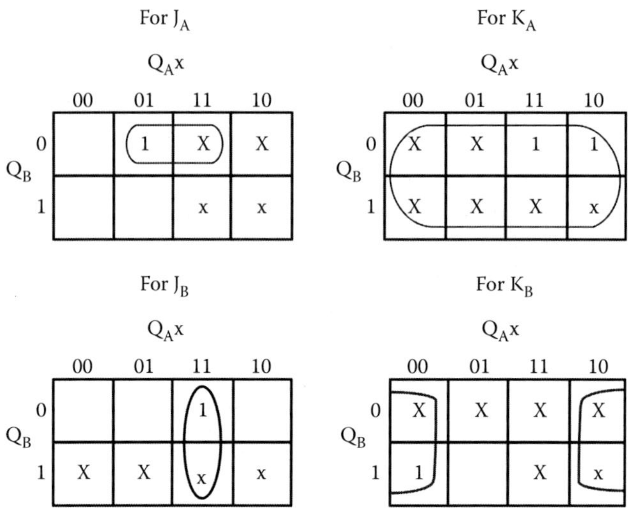

FIGURE 6.4
K-map and logical simplification with JK FF.

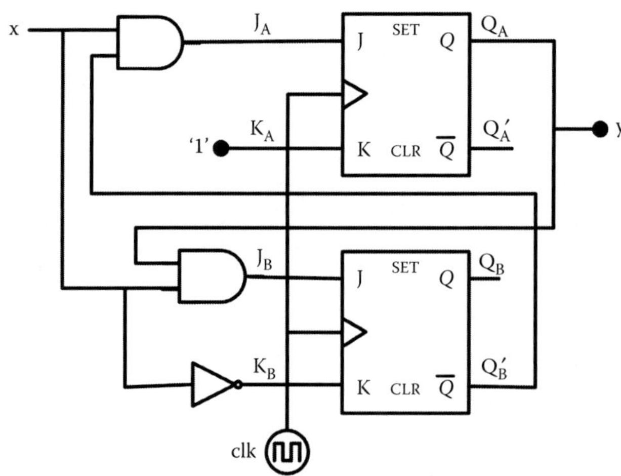

FIGURE 6.5
FSM realization of Equation 6.2.

the output. In later sections, we discuss Mealy FSMs as well as their conversions, that is, Moore to Mealy and vice versa.

DESIGN EXAMPLE 6.1

Design a sequential circuit using SR FFs to convert non return to zero (NRZ) binary input (x) to Manchester code output (y).

Solution

In the NRZ form, logic 1 and logic 0 are represented by positive and zero voltage, respectively. This is the data pattern of the input in this design example. In Manchester code, logic 1 and logic 0 are represented by high-to-low and low-to-high transitions, respectively. Therefore, the frequency of the sequential circuit must be twice that of the

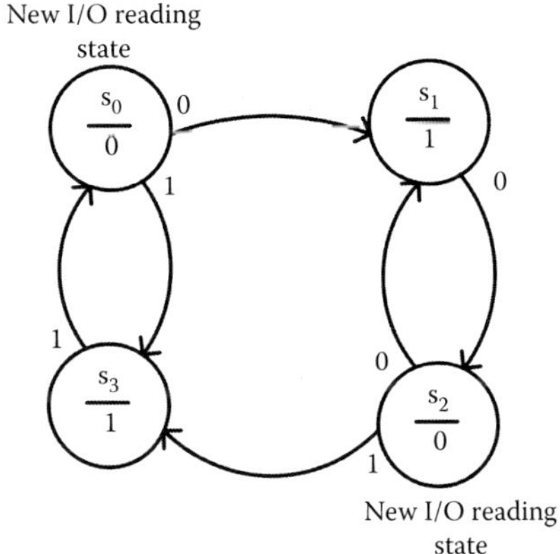

FIGURE 6.6
State diagram for NRZ to Manchester code conversion.

input sequence frequency. The state diagram and excitation table with SR-FFs for this design example are given below.

As mentioned in Figure 6.6, the new input will be read only at the states s_0 and s_2; hence, there is no need to consider the transition from other states, that is, s_1 and s_3, for the inputs $x = 1$ and $x = 0$, respectively. Therefore, they are the intermediate states, that is, not the state from where we need to read new input and take the decision to move; hence they are taken as "don't care" states for that input, as marked in Table 6.4. As we discussed above, the excitation table (Table 6.4) is prepared for implementing the given task using SR-FFs.

Upon solving Table 6.4 using the K-map, we will get the following Boolean functions for the input variables of FFs, that is, S_A, R_A, S_B, R_B, and output (y).

$$\begin{aligned} S_A &= Q'_A \\ R_A &= Q_A \\ S_B &= (Q_A \oplus x) \\ R_B &= (Q_A \oplus x)' \text{ and} \\ y &= Q_B \end{aligned} \tag{6.3}$$

TABLE 6.4

Excitation Table with SR-FFs

Present State (Q)		FFs Input				Next State (Q^+)	
Q_BQ_A	Input (x)	S_B	R_B	S_A	R_A	$Q_B{}^+Q_A{}^+$	Output (y)
0 0	0	0	x	1	0	0 1	0
	1	1	0	1	0	1 1	
0 1	0	1	0	0	1	1 0	1
	1	x	x	x	x	x	
1 0	0	0	1	1	0	0 1	1
	1	x	0	1	0	1 1	
1 1	0	x	x	x	x	x	1
	1	0	1	0	1	0 0	

VHDL code developed as per these logical equations (Equation 6.3) to get Manchester code from NRZ input is given below.

```
library ieee;
use ieee.std_logic_1164.all;
use ieee.std_logic_arith.all;
use ieee.std_logic_unsigned.all;
entity FSM is
port ( clk,x : in std_logic:='0';
y : out std_logic:='0');
end entity;
architecture beh1 of FSM is
signal SA,RA,SB,RB: std_logic:='0';
signal QA,QAB,QB,QBB: std_logic:='0';
begin
SA<=QAB;
RA<=QA;
SB<=(QA xor x);
RB<=not (QA xor x);
p0:process (clk,SA,RA,SB,RB)
variable tmpA,tmpB: std_logic:='0';
begin
if rising_edge(clk) then
if(SA='0' and RA='0')then
tmpA:=tmpA;
elsif(SA='1' and RA='1')then
tmpA:='Z';
elsif(SA='0' and RA='1')then
tmpA:='0';
else
tmpA:='1';
end if;
if(SB='0' and RB='0')then
tmpB:=tmpB;
elsif(SB='1' and RB='1')then
tmpB:='Z';
elsif(SB='0' and RB='1')then
tmpB:='0';
else
tmpB:='1';
end if;
QA<=tmpA;
QAB <= not tmpA;
QB<=tmpB;
QBB <= not tmpB;
end if;
end process;
y<=QB;
end beh1;
```

The FSM can be implemented for any application in two different techniques or a combination of them. The main difference between them is how they generate the output. The two types are (i) Moore FSM and (ii) Mealy FSM. The following points highlight the differences between a Moore machine and Mealy machine [39,65,67].

Moore FSM:

i. Output depends upon only the present state.
ii. It has more states than the Mealy machine in general.
iii. The output changes only at the clock edges.
iv. More logic is needed to decode the outputs since it has more circuit delays.

Mealy FSM:

i. The output depends upon both the present state and present input.
ii. It has fewer states than the Moore machine in general.
iii. Input change can cause a change in output as soon as the logic is done.
iv. Mealy machines react faster to inputs.

In the subsequent section, we discuss these models one by one in detail with some design examples.

6.2.1 Moore Finite State Machine Design

As we stated already, the Moore machine is an FSM whose output depends on only the present state. The input changes at any instant in time do not make any impact on the output that exists. The output of this FSM is the logical relation of only the present state. A standard block diagram of a Moore FSM model is shown in Figure 6.7. There are three important sections: two combinational circuits and one sequential circuit. The input is applied to the first combinational circuit, which also gets input from the output of the state memory, that is, the present state. Therefore, the first combinational circuit generates output based on input changes as well as present state changes. If the input is changed at an instant in time, the output of the first combinational circuit also gets changed. The first combinational circuit is designed so as to generate suitable input, based on the input and present state values, to the state memory to force it to the expected next state. Therefore, this combinational circuit is also called a next state-generating circuit. The number of inputs to the next state-generating circuit depends on the number of bits in the input and present state; however, the number of outputs depends the number of FFs and their type. For example, three D-FFs are used in the state memory circuit, so the number of outputs is three, that is, there are three logical equations to generate three outputs as the function of the input and present state. If three JK-FFs are used, six outputs, that is, six logical equations, are required to generate the outputs.

The state memory consists of registers, that is, FFs. The input of the FFs decides the next state after applying the clock pulse. Even though the inputs of the state memory are sensitive to input all the time, there is no impact on the present state until the clock input is applied. Once the clock is applied, then state transition occurs and the state memory locks with the new state values. In this model, the third combinational circuit generates the

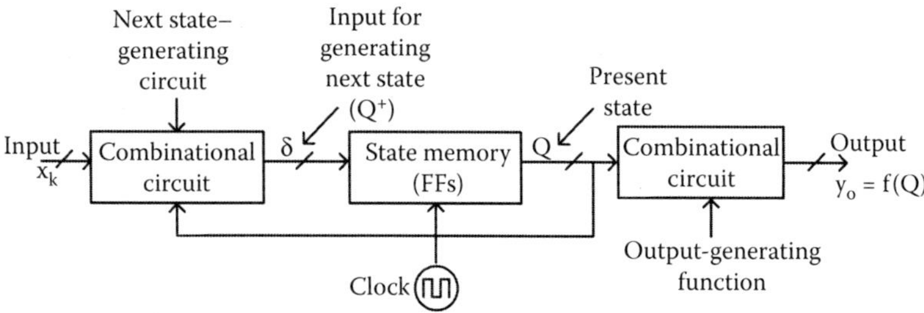

FIGURE 6.7
Standard Moore FSM model.

TABLE 6.5
General State Transition Table—Moore Model

Present State (Q)	Next State (Q^+)				Output (y)
	Input x = 00	Input x = 01	Input x = 10	Input x = 11	
a	b	c	b	d	01
b	b	d	c	c	11
c	c	d	c	a	10
d	d	d	a	b	00

output only as a function of the present state values. The processing speed of this model depends on the frequency of the clock pulse. A Moore machine can be described by six variables: input (x_k), state memory input (δ), next state (Q), output (y_o); where "k" and "o" are elements of vectors of total input and output variables, respectively. A simple state transition table for the Moore FSM model is given in Table 6.5. We can deduce the width of all the model parameters for Table 6.5 as $x_k = 2$ bits, $\delta = 2$ bits if D-FFs or T- FFs are preferred; otherwise, $\delta = 4$ bits, $Q = 2$ bits, and $y = 2$ bits [65,67].

The state diagram of the above Moore machine given in Table 6.5 is shown in Figure 6.8. As we discussed above, the outputs of this FSM are marked inside the state circle because $y_o = f(Q)$. The state transitions for different possible inputs are marked in the state diagram.

Therefore, all the designs we discussed in the previous section use the Moore FSM model. We need to properly generate the state transition table and state diagram for the given task before beginning the implementation. Once the state diagram is ready, we can proceed our design step to the excitation table and consider the type of FFs we wish to use. Based on the excitation table, we would derive the logical functions for both combinational circuits. Then, the design can be implemented either with digital components on PCB or in the FPGA using HDL codes. We discuss some design examples below.

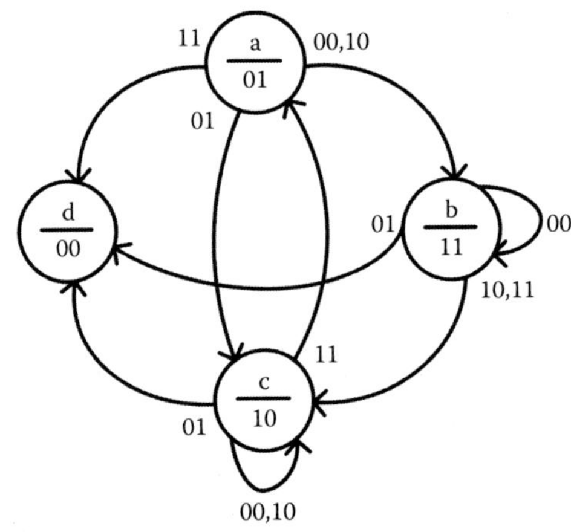

FIGURE 6.8
State diagram for a simple Moore FSM model.

DESIGN EXAMPLE 6.2

Develop a Moore FSM digital system to implement the following state transitions in the FPGA using VHDL code. Assume that there are four states, S_0 through S_3, and one input (x). The outputs (y) at the states S_0, S_1, S_2, and S_3 are "1", "1", "0", and "0", respectively. The state transition sequences are

i. S_0 to S_1 if x=1; otherwise to S_2.
ii. S_1 to S_3.
iii. S_2 to S_3.
iv. S_3 to S_0.

Solution

The given state transitions show that the transition from S_0 is a conditional transition and all the other transitions are unconditional transitions. That means we do not need to check the status of the input (x). The VHDL code for the given task is as follows.

```
library ieee;
use ieee.std_logic_1164.all;
use ieee.std_logic_arith.all;
use ieee.std_logic_unsigned.all;
entity FSM is
port ( clk, reset, x : in std_logic:='0';
y : out std_logic:='0');
end entity;
architecture beh1 of FSM is
type state_type is (s0,s1,s2,s3);
signal state: state_type:=s0;
begin
p0:process (clk,reset)
begin
if (reset ='1') then
state <=s0;
y<='1';
elsif (clk='1' and clk'event) then
case state is
when s0 => if x='1' then state <= s1;
else     state <= s2;
end if;
y<= '1';
when s1 => state <= s3; y<= '1';
when s2 => state <= s3; y<= '0';
when s3 => state <= s0; y<= '0';
end case;
end if;
end process;
end beh1;
```

6.2.2 Mealy Finite State Machine Design

A Mealy model machine is an FSM whose output depends on the present state as well as the present input. Other than this, there is no difference between the general structure of the Moore and Mealy FSM models, as shown in Figure 6.9. Based on the application, we need to decide whether Moore or Mealy or both FSM models are required. A Mealy machine can also be described by four variables: input (x_k), state memory input (δ), next state (Q), output (y_o); where the "k" and "o" are elements of the vectors of the total input and output variables, respectively. Since there is a difference in the output-generating function

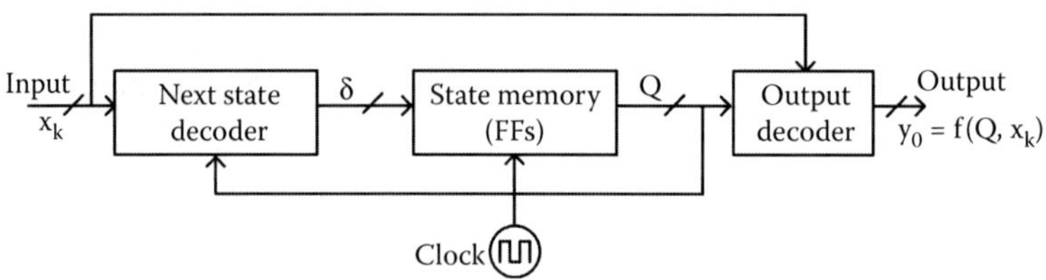

FIGURE 6.9
Structure of a Mealy FSM model.

in this model, $y_0 = f(x_k, Q)$, there is a difference in the state transition table as well as the state diagram. A standard state transition table is given in Table 6.6. There are five main columns, one for the present state, two for the next state, and the last two for output.

We need to clearly understand the state table of the Mealy FSM model and how it differs from the state transition table of the Moore FSM model. The first three columns in Table 6.6 are the same as the columns in the Moore FSM machine. The fourth and fifth columns are different from the Moore FSM model. In the Moore FSM model, the output depends upon only the present state; hence, one column is sufficient to indicate the output, whereas in the Mealy FSM model, the output depends upon the present input as well as present state; hence, more than one column is required based on the possible input values. For example, in Table 6.6, when the present state is S_2, if $x = 0$, the next state and output will be S_0 and "1", respectively; otherwise, the next state and output will be S_3 and "0", respectively. This is the method of preparing a state transition table for the Mealy FSM model [65,67]. Now, we proceed to the state diagram for the Mealy model.

The state diagram corresponding to Table 6.6 is shown in Figure 6.10. The methods of representing the state ID, input and output are different from the Moore model. Inside the circle, only the state ID will be marked, and the input and output will be marked over the transition line (arrow line). The input and output are represented as the numerator and denominator factors, as indicated in Figure 6.10. For example, from state S_2, if $x = 0$, then the next state and output will be S_0 and "1", respectively, which is given by "0/1"; otherwise, if $x = 1$, the next state and output will be S_3 and "0", respectively, which is given by "1/0".

Once the state diagram generation is done, we need to prepare the excitation table and logical equations for the next state-generating function and output function, as we discussed in the preceding sections. Now, we see an example to understand the design flow for the Mealy FSM model. We take a BCD-to-excess-3 code converter, where the BCD numbers 0 through 9 are represented by 3 through 12, respectively. Assume that the BCD numbers arrive at the code converter bit by bit, keeping the LSB first. That means the BCD number 5, that is, "0101", arrives the code converter as "1", "0", "1", and "0". The corresponding output

TABLE 6.6

State Transition Table for Mealy Model

Present State (Q)	Next State (Q^+)		Output (y)	
	$x = 0$	$x = 1$	$x = 0$	$x = 1$
S_0	S_1	S_0	0	0
S_1	S_3	S_4	1	1
S_2	S_0	S_3	1	0
S_3	S_4	S_0	1	0
S_4	S_0	S_1	0	1

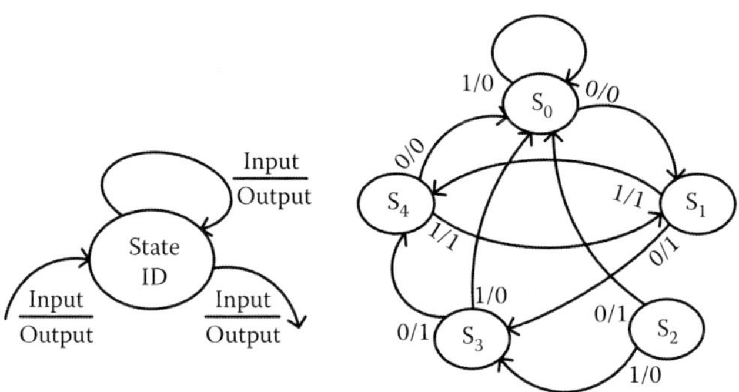

FIGURE 6.10
Mealy model state diagram for Table 6.6.

is 8, that is, "1000", which will be "0", "0", "0" and "1". Based on this information, we start drawing the state diagram, which will look like Figure 6.11. The decision for the first bit, that is, the LSB of the BCD, is taken at S_0, as if the LSB of the BCD is "0"; then the output is "1". Otherwise, if the LSB of the BCD is "1", the output is "0". In the same way, we continue the analysis for three more bits of BCD inputs.

Now, we directly proceed to the excitation table for this state diagram. We assume that we wish to use D-FFs. There are six different states in all; hence, we need a minimum of three D-FFs. When using D-FFs, we require three logical equations for generating the next state-generating inputs and one equation for generating the output. Then, the excitation table for this design will be as given in Table 6.7.

The expected next state is equal to the input of the D-FFs. Therefore, they are copied under the FFs' input as it is in Table 6.7. Upon solving Table 6.7 using the K-map for D_A, D_B, D_C, and output (y), the following logical expressions can be obtained.

$$
\begin{aligned}
D_A &= Q'_B \\
D_B &= Q_A \\
D_C &= Q_A Q_B Q_C + x' Q_A Q'_C + x Q'_A Q'_B \\
y &= x' Q'_C + x Q_C
\end{aligned}
\tag{6.4}
$$

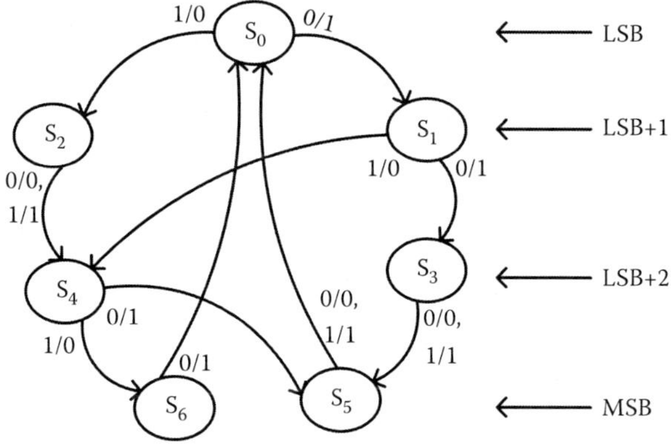

FIGURE 6.11
Mealy model state diagram for BCD to excess-3 converter.

TABLE 6.7

Excitation Table for BCD to Excess-3 Converter

Present State (Q)		FF Inputs			Next State (Q^+)	
$Q_CQ_BQ_A$	Input (x)	D_C	D_B	D_A	$Q_C^+Q_B^+Q_A^+$	Output (y)
0 0 0	0	0	0	1	0 0 1	1
	1	0	1	0	0 1 0	0
0 0 1	0	0	1	1	0 1 1	1
	1	1	0	0	1 0 0	0
0 1 0	0	1	0	0	1 0 0	0
	1	1	0	0	1 0 0	1
0 1 1	0	1	0	1	1 0 1	0
	1	1	0	1	1 0 1	1
1 0 0	0	1	0	1	1 0 1	1
	1	1	1	0	1 1 0	0
1 0 1	0	0	0	0	0 0 0	0
	1	0	0	0	0 0 0	1
1 1 0	0	0	0	0	0 0 0	1
	1	x	x	x	x	x
x	x	x	x	x	x	x
	x	x	x	x	x	x

These logical equations can be implemented using either the digital components or in the FPGA using HDL code to get a BCD-to-excess-3 code converter. If the same design is done with JK-FFs, we need seven logical equations, as given in Equation 6.5.

$$\begin{aligned} J_A &= Q_B' \\ K_A &= Q_B \\ J_B &= Q_A \\ K_B &= Q_A' \\ J_C &= (x \oplus Q_A) \\ K_C &= Q_A' + Q_B' \\ Y &= x'Q_C' + xQ_C \end{aligned} \tag{6.5}$$

In this way, any Mealy FSM model can be designed and implemented. The above logical expressions can be implemented using HDL code, as explained in Design Example 6.1. We will discuss a design example to understand how to implement a Mealy FSM model with state machines using HDL code in the FPGA below.

DESIGN EXAMPLE 6.3

Design a Mealy FSM model using the state machine approach for the following state transitions and outputs. Assume that there are four states: S_0, S_1, S_2 and S_3. The output and state transition details are:

i. S0 to S1: 0/0 and S0 to S2: 1/1
ii. S1 to S3: 0/0 and S1 to S1: 1/0

iii. S2 to S2: 0/1 and S2 to S3: 1/0
iv. S3 to S3: 0/1 and S3 to S0: 1/1

Solution

```
library ieee;
use ieee.std_logic_1164.all;
use ieee.std_logic_arith.all;
use ieee.std_logic_unsigned.all;
entity mealy_FSM is
port (clk : in std_logic;
reset : in std_logic;
input : in std_logic;
output : out std_logic);
end mealy_FSM;
architecture behavioral of mealy_FSM is
type state_type is (s0,s1,s2,s3);
signal current_s,next_s: state_type:=s0;
begin
p0:process (clk,reset)
begin
if (reset='1') then
current_s <= s0;
elsif rising_edge(clk) then
current_s <= next_s;
end if;
end process;
p1:process (current_s,input)
begin
case current_s is
when s0 =>
if(input ='0') then
output <= '0'; next_s <= s1;
else
output <= '1'; next_s <= s2;
end if;
when s1 =>
if(input ='0') then
output <= '0'; next_s <= s3;
else
output <= '0'; next_s <= s1;
end if;
when s2 =>
if(input ='0') then
output <= '1'; next_s <= s2;
else
output <= '0'; next_s <= s3;
end if;
when s3 =>
if(input ='0') then
output <= '1'; next_s <= s3;
else
output <= '1'; next_s <= s0;
end if;
end case;
end process;
end behavioral;
```

6.2.3 Finite State Machine Model Conversion

In the previous section, we discussed the Moore and Mealy FSM models separately. In this section, we discuss the conversion from one to the other, that is, Mealy FSM to Moore FSM and vice versa. First, we see how to get the Mealy FSM from the Moore FSM. We need to follow the following procedure to complete this conversion.

i. Take a state transition table (shaded potion of Table 6.8) corresponding to a Moore FSM mode from which we wish to get the Mealy FSM model.
ii. Insert output columns for all possibilities for the input.
iii. Check the present states and their corresponding outputs in the Moore FSM model and copy the output values into the output columns of the Mealy FSM model, where that state will appear as the next state.
iv. Ignore the actual output column that is in the state transition table of the Moore FSM model.
v. Now, the table will look like Table 6.8. The shaded column is the actual output of the Moore FSM model.

We will examine Table 6.8 now. Before alteration, the state transition table for the Moore FSM model consists of the present state, the next state for $x = 0$ and $x = 1$ and output (y) (shaded potion) only. Two columns (for $x = 0$ and $x = 1$) are introduced as the first step. Consider a present state S_2 whose output is 0; a next state for $x = 0$ is S_2 and the state for $x = 1$ is S_3. Copy the actual output of S_2, that is, 0, and keep it as the output for the next state S_2 under $x = 0$. Copy the actual output of S_3 and keep is as the output of the next state S_3 under $x = 1$ as given in Table 6.8. This step is followed for all columns. Now we have a perfect state transition table of a Mealy FSM model obtained from a Moore FSM model. We can proceed further to devise the excitation table based on the FFs we wish to use for implementing this Mealy FSM model [65,67].

We continue our discussion to understand the conversion of a Mealy FSM model to a Moore FSM model. The steps that need to be followed for this conversion are as given below:

i. Take a state transition table, as in Table 6.9, corresponding to a Mealy FSM model from which we wish to get the Moore FSM model.
ii. Introduce a column at the rightmost side for the output (y) of the Moore FSM model.
iii. Calculate the number of different outputs for each state that are in the state table of the Mealy FSM model under $x = 0$ and $x = 1$. If all the outputs of the

TABLE 6.8

State Transition Table for Moore FSM Conversion

	x = 0		x = 1		
Present State	**Next State**	**Output (y)**	**Next State**	**Output (y)**	**Output (y)**
S_0	S_3	1	S_1	0	1
S_1	S_0	1	S_2	0	0
S_2	S_2	0	S_3	1	0
S_3	S_1	0	S_0	1	1

TABLE 6.9

State Transition Table for Mealy FSM Conversion

Present State	x = 0		x = 1		
	Next State	Output (y)	Next State	Output (y)	Output (y)
S_0	S_3	0	S_1/S_1*	1	1
S_1	S_0	1	S_2	0	0
S_1*	S_0	–	S_2	–	1
S_2	S_2/S_2*	1	S_3	0	0
S_2*	S_2*	–	S_3	–	1
S_3	S_1	0	S_0	1	0

state are the same, copy and keep it under the output (y) of the Moore FSM model. If it has different outputs, break that state into the substates and keep the output (y) of the Moore FSM model accordingly.

iv. Ignore the actual output column that is in the state transition table of the Mealy FSM model.

v. Now, the table will look like Table 6.9. The last column is the actual output of the Mealy FSM model [65,67].

We will examine Table 6.9 first. Here, the outputs for the next states S_0 and S_3 are "1" and "0", respectively, regardless of the value of x. Hence, we can keep these output values as they are under the output (y) of the Moore model (last column) against the respective present state. That is why the outputs in the first and last rows are "1" and "0", respectively, for S_0 and S_3 in Table 6.9. Next, the next states S_1 and S_2 get different values with respect to the value of x; for example, look at the shaded next states in Table 6.9: the output of the next state S_1 when $x = 0$ is "0", while for $x = 1$, it is "1". In the same way, the output value of the next state S_2 when $x = 0$ is "1" and when $x = 1$ is "0". Hence, we need to split these two states into four different states to generate all the required outputs. That means states S_1 and S_2 have to be split into S_1, S_1* and S_2, S_2*, respectively, as given in the first column of Table 6.9. Now, we need to fill in the appropriate next state and output for the newly introduced states and alter the state transitions accordingly. For example, the output "0" is allotted for the present state S_1, and the output "1" is allotted for the present state S_1*. To accomplish this state transition, we need to move the present state S_0 to S_1* when $x = 1$, as given in Table 6.9. Therefore, at S_1, the output is "0", and for S_1*, the output is "1". In the same way, one can understand the S_2 and S_2* state transition as well as the output. Finally, now we have six states: S_0, S_1, S_2, S_3, S_4 and S_5. We can proceed further to devise the excitation table based on the FFs we wish to use for implementing this Moore FSM model.

6.3 Code Classifier and Binary to Binary-Coded Decimal Converters

In this section, we discuss serial and parallel data reading and subsequently classifying received codes under various categories. For example, computing the even parity, odd parity, number of ones, number of zeros and frequency of samples in received code is one of the popular examples for the code classifier. At the end of this section, we discuss the conversion of binary code into equivalent BCD code, which is most important for many real-world applications like sensor data reading and displaying on LCDs.

6.3.1 Input Code Classifier

The main purpose of this section is helping readers understand the various operations of the input sequence/binary-code classifier. The design in the sequence classification circuit receives a sequence of binary codes and performs some of the specified operations with the number of ones and zeros as even parity, odd parity, gray code, gold code, and so on. Receiving the data sequence and performing the operation process stops when a specific sequence of input numbers is encountered. The system has to maintain enough counters to perform the various operations needed in the code classifier. For example, in a simple application where we wish to compute how many ones and zeros are received over a specified duration, the code classifier circuit has to maintain three counters: one for storing the number of codes received, another for storing the number of ones received, and one for storing the number of zeros received. Another method of stopping the operation of the code classifier, instead of stopping after receiving a specified number of bits, is using a password or termination sequence. In this case, the receiving, counting and classification of codes in the input sequence continues until a specific five-digit sequence (password) is received [22,39,68,69]. Then, the process of receiving, counting and classifying stops. The code classifier also classifies the received sequence as binary code, BCD, gray code, gold code, and so on.

We will discuss one example to understand this concept more clearly. Assume a binary sequence "11001001001110100001110010111001100100110011001110101010", which serially arrives at the code classifier built inside the FPGA. Now, we need to perform the following operations to find (i) total number of bits; (ii) number of zeros, number of ones; (iii) whether the sequence is binary, BCD or gray code; and (iv) whether it has even or odd parity. The answers are: (i) total number of bits: 56; (ii) number of zeros: 27, so the number of ones: $56 - 27 = 29$; (iii) since every 4 bits consist of randomly varying ones as well as more than "1001", the received sequence is binary, not gray code or BCD; and (iv) the number of ones is 29; therefore, the sequence is a odd parity.

DESIGN EXAMPLE 6.4

Develop a digital system using VHDL code to perform the following operations:

i. Receive 8 bits of binary code in parallel from the external switches.
ii. Count how many ones are present in the received binary sequence.
iii. Indicate either that the number of ones is greater than the number of zeros or vice versa using two LEDs.
iv. If both are equal in number, display logic 1 at both LEDs.

Solution

```
library ieee;
use ieee.std_logic_1164.all;
use ieee.std_logic_arith.all;
use ieee.std_logic_unsigned.all;
entity code_clas is
port (data_in : in std_logic_vector(7 downto 0):="00000000";
out0s, out1s: out std_logic:='0');
end code_clas;
```

```
architecture behavioral of code_clas is
signal cou1s: integer range 0 to 8:=0;
begin
p0:process (data_in)
variable n: integer range 0 to 8:=0;
begin
n:=0;
for i in 0 to 7 loop
if data_in(i)='1' then
n:=n+1;
end if;
end loop;
cou1s<=n;
end process;
p1: process(cou1s)
begin
if cou1s>4 then
out0s<='0';
out1s<='1';
elsif cou1s<4 then
out0s<='1';
out1s<='0';
elsif cou1s=4 then
out0s<='1';
out1s<='1';
end if;
end process;
end behavioral;
```

DESIGN EXAMPLE 6.5

Develop a digital system to find and display the detection rate of a given 4-bit sequence, number of ones and zeros present in the received sequence, and parity of the received sequence. Assume that the system has four inputs, namely reset, clock, serial data input, and reference pattern [3:0], and the output width for the number of ones and zeros is 6 bits, logic 1, for indicating odd parity, logic 0 for indicating even parity, and 4 bits for indicating the count rate. All these results have to be found once in every 52 bits of data input. The data input has to be serially taken into the system for every rising of the clock.

Solution

```
library ieee;
use ieee.std_logic_1164.all;
use ieee.std_logic_unsigned.all;
entity led is
port (    d : in  std_logic;
clk,rst : in  std_logic;
ones,zeros : out  std_logic_vector(5 downto 0);
parity:out std_logic;
x : in  std_logic_vector (3 downto 0);
cnt_dec : out std_logic_vector(3 downto 0));
end entity;
architecture behavioral of led is
signal c0,c1,c2,c3,c4,c5,c6,c7,c8,c9,c10,c11,c12,c13,c14,c15: std_logic_
vector(3 downto 0);
signal cntr:std_logic_vector(1 downto 0);
signal cnt0,cnt1:std_logic_vector(5 downto 0);
signal cntreg :integer range 0 to 52:=0;
signal reg : std_logic_vector(3 downto 0);
```

```
signal stop : std_logic;
begin
stop <= '0' when (cntreg <= 51) else '1';
p0: process(clk,rst)
begin
if(rst='0') then
cntreg <= 0;
cnt0 <= "000000";
cnt1 <= "000000";
else
if rising_edge(clk) then
if(cntreg <= 51) then
cntreg <= cntreg + 1;
cnt0 <= cnt0 + (not d);
cnt1 <= cnt1 + d;
else
cntreg <= 0;
cnt0 <= "000000";
cnt1 <= "000000";
end if;
end if;
end if;
end process;
zeros <= cnt0;
ones <= cnt1;
parity<= cnt1(0);

p1: process(clk,rst)
begin
if(rst='0') then
cntr <= "00";
else
if rising_edge(clk) then
if (stop='0') then
reg <= d & reg(3 downto 1);
cntr <= cntr + 1;
end if;
end if;
end if;
end process;
p2:process(rst,cntr(1),stop)
begin
if(rst='0') then
c0 <= "0000";
c1 <= "0000";
c2 <= "0000";
c3 <= "0000";
c4 <= "0000";
c5 <= "0000";
c6 <= "0000";
c7 <= "0000";
c8 <= "0000";
c9 <= "0000";
c10 <= "0000";
c11 <= "0000";
c12 <= "0000";
c13 <= "0000";
c14 <= "0000";
c15 <= "0000";
else
if(stop='1') then
```

```
c0 <= "0000";
c1 <= "0000";
c2 <= "0000";
c3 <= "0000";
c4 <= "0000";
c5 <= "0000";
c6 <= "0000";
c7 <= "0000";
c8 <= "0000";
c9 <= "0000";
c10 <= "0000";
c11 <= "0000";
c12 <= "0000";
c13 <= "0000";
c14 <= "0000";
c15 <= "0000";
else
if(rising_edge(cntr(1))) then
case(reg) is
when "0000"=>
c0 <= c0 + '1';
when "0001"=>
c1 <= c1 + '1';
when "0010"=>
c2 <= c2 + '1';
when "0011"=>
c3 <= c3 + '1';
when "0100"=>
c4 <= c4 + '1';
when "0101"=>
c5 <= c5 + '1';
when "0110"=>
c6 <= c6 + '1';
when "0111"=>
c7 <= c7 + '1';
when "1000"=>
c8 <= c8 + '1';
when "1001"=>
c9 <= c9 + '1';
when "1010"=>
c10 <= c10 + '1';
when "1011"=>
c11 <= c11 + '1';
when "1100"=>
c12 <= c12 + '1';
when "1101"=>
c13 <= c13 + '1';
when "1110"=>
c14 <= c14 + '1';
when "1111"=>
c15 <= c15 + '1';
when others=>
null;
end case;
end if;
end if;
end if;
end process;
p_seq_cnt:process(stop,rst)
begin
if(rst='0') then
```

```
cnt_dec <= "0000";
else
if(rising_edge(stop)) then
case(x) is
when ("0000")=> cnt_dec<=c0;
when ("0001")=> cnt_dec<=c1;
when ("0010")=> cnt_dec<=c2;
when ("0011")=> cnt_dec<=c3;
when ("0100")=> cnt_dec<=c4;
when ("0101")=> cnt_dec<=c5;
when ("0110")=> cnt_dec<=c6;
when ("0111")=> cnt_dec<=c7;
when ("1000")=> cnt_dec<=c8;
when ("1001")=> cnt_dec<=c9;
when ("1010")=> cnt_dec<=c10;
when ("1011")=> cnt_dec<=c11;
when ("1100")=> cnt_dec<=c12;
when ("1101")=> cnt_dec<=c13;
when ("1110")=> cnt_dec<=c14;
when ("1111")=> cnt_dec<=c15;
when others => null;
end case;
end if;
end if;
end process;
end behavioral;
```

6.3.2 Binary-to-Binary-Coded-Decimal Converter and Its Arithmetic

The decimal number (radix-10) system is fine for calculations done by humans, but it is not the easiest system for a computer to use. A digital computer contains elements that can be in either of two states: "on" or "off", that is, in other words, "magnetized" or "not magnetized". For such devices, calculations are most conveniently done using the binary (radix-2) number system. Binary numbers are more useful in a digital computer, where each binary digit (bit) can be represented by one state of a binary switch that is either "on" or "off". However, binary numbers are hard to read, partly because of their great length. Hexadecimal, octal, and BCD number systems allow us to express binary numbers more compactly, and they make the transfer of data between computers and people much easier. The BCD is one of the most important ways to represent binary numbers. The advantage of the BCD number system is that each decimal digit is represented by a group of four binary digits. Therefore, a 4-bit group represents each displayed decimal digit from 0000 for a zero to 1001 for a nine. Although 16 numbers (2^4) can be represented using four binary digits in the BCD numbering system, the six binary codes "1010", that is, $(10)_{10}$, through "1111", that is, $(15)_{10}$, are classed as forbidden numbers and cannot be used. The main advantage of the BCD number system is that it allows easy conversion between decimal and binary form. However, the disadvantage is that BCD code is wasteful, as the states between "1010" and "1111" are not used. Nevertheless, BCD has many important applications, especially for digital displays. For example, $(357)_{10}$ in BCD is "0011 0101 0111" and $(8579)_{10}$ is "1000 0101 0111 1001".

The conversion of decimal-to-BCD is a straightforward task, but we need to remember that BCD numbers are decimal numbers and not binary numbers, even though they are represented using bits. However, while BCD is easy to code and decode, it is not an efficient way to store numbers. For example, in binary, a three-digit decimal number from $(0)_{10}$ through $(999)_{10}$ requires only 10 bits, "0000000000" through "1111100111", whereas in BCD, the same number requires a minimum of 12 bits, "0000 0000 0000" through "0011 1110

0111", for the same representation. Nevertheless, the use of a BCD coding system in both microelectronic and computer systems is particularly useful in situations where the BCD is intended to be displayed on one or more seven-segment LED or LCD displays, and there are many popular integrated circuits available that are configured to give BCD outputs.

Now, we discuss the conversion of an 8-bit binary number into the equivalent 12-bit BCD number. The procedures for this conversion are as follows:

i. Load the 8-bit binary number into the binary register. Assume the width of the binary number is "n".
ii. Reset the BCD register.
iii. Concatenate both the registers.
iv. Perform the left shift by 1 position.
v. If the value of any column (every 4 bits) in the BCD register exceeds or is equal to 5, then add 3 to that column. Otherwise, continue the left shift.
vi. If "n" left shifts have been performed, evaluate each column of the BCD, which is equal to the BCD value.

The same procedure can be continued for any value of a binary number to convert it into the equivalent BCD number. A conversion example based on the above procedure is given in Table 6.10. If the width of the binary number is "n", then the left shift of the entire register has to be performed n times with the addition of 3 as required.

TABLE 6.10

Binary to BCD Conversion

BCD			Binary [MSB...LSB]	Operation Happened
0000	0000	0000	11111110	Load binary value and initialize BCD register
0000	0000	0001	1111110	Left shift 1
0000	0000	0011	111110	Left shift 2
0000	0000	0111	11110	Left shift 3
0000	0000	0011	11110	Add 3
0000	0001	1010		
		0101	1110	Left shift 4
0000	0001	0011	1110	Add 3
0000	0011	1000	110	Left shift 5
		0001		
0000	0110	0011	10	Left shift 6
0000	0011	0011	10	Add 3
	1001			
0001	0010	0111	0	Left shift 7
0001	0010	0011	0	Add 3
		1010		
0010	0101	0100		Left shift 8
2	**5**	**4**		**BCD value**

Now, we will discuss some of the arithmetic related to the BCD number system. In BCD addition, if the result is greater than 9, the result is not a valid BCD number. We have to add 6, that is, "0110", to the result of addition. Then, what we get will be a valid BCD number. For example, we add two valid BCD numbers "1001" and "0101", and the result is "1110", which is an invalid BCD number. Then, add "0110" with the result as "1110" + "0110" = "0001 0100", which is a valid BCD number. If the result itself were valid, then we would not need to add "0110" to it, for example, "0011" + "0100" = "0111", which is a valid BCD number. Now, a question may arise as to why "0110" is being added to the result instead of any other number. It is done to skip the six invalid BCD numbers, that is, 10 to 15. Like addition, BCD subtraction is also important. There are several methods of BCD subtraction, namely the 1's compliment method, 9's compliment method and 10's compliment method. Among all these methods, the 9's compliment method and 10's compliment method are the easiest ones. We will discuss the first two methods of BCD subtraction with examples below. The third method is left to the readers as a laboratory exercise.

To understand the 1's complement method, consider a BCD subtraction as 541 – 216 = 325, that is, in BCD, "0101 0100 0001" – "0010 0001 0110" = "0011 0010 0101". We discuss this subtraction step by step now: (i) take the 1's complement for the subtrahend. The 1's complement of "0010 0001 0110" is "1101 1110 1001"; (ii) add this value to "0101 0100 0001" as "0101 0100 0001" + "1101 1110 1001" = "1 0011 0010 1010". Here, the EoC is "1"; hence, the result is a positive number and this EoC is added to the LSB of the result as "0011 0010 1010" + "0000 0000 0001" = "0011 0010 1011"; (iii) in the above additions, the second and third groups (second and third 4 bits) generated auxiliary carries, while the first group does not; hence, we have to take the result of the second and third group as it is, that is, adding "0000 0000" while adding "1010" to the first group. Then, the result is "0011 0010 1011" + "0000 0000 1010" = "0011 0010 1 0101", and we have to ignore the auxiliary carry that occurred in this addition, that is, "1". Then, the result will be "0011 0010 0101", which is equal to 325 in BCD form. As another example, 62 – 97 = –35. According to the subtraction procedure, this can be written as "0110 0010" – "1001 0111". With the 1's complement, it becomes "0110 0010" + "0110 1000". The sum is "1100 1010", and here, the EoC is not generated; hence, the result is a negative number and we need to once again take the 1's complement to get the correct result, which is "0011 0101" = 35 in BCD form. We take one more example: 69 – 97 = –28. This can be written in BCD form as "0110 1001" – "1001 0111", with a 1's complement "0110 1001" + "0110 1000" = "1101 0001". Here, there is no EoC; hence, the result is a negative number and we take the 1's complement once again. Then, the result will be "0010 1110". The first group, i.e., "1110", is an invalid BCD number, so add "0110" to it, and the result will be "0010 1110" + "0000 1010" = "0010 1 1000". Ignore the auxiliary carry, and the result is "0010 1000", which is 28 in BCD form.

The second method is the 9's complement method. We will examine this method through some examples. Assume an example 87 – 39 = 48. We will write this in BCD form as "1000 0111" – "0011 1001". With the 9's complement (99 – 39 = 60 = "0110 0000"), this can be written as "1000 0111" + "0110 0000" = "1110 0111". The second group is not a BCD number, so add "0110" to it. Then, "1110 0111" + "0110 000" = "1 0100 0111", and an EoC is generated. Add the EoC to the LSB of the result as "0100 0111" + "1" = "0100 1000", which is 48 in BCD form. Consider another example, 18 – 72 = –54. We write it in BCD form as "0001 1000" – "0111 0010" and in 9's complement form (99 – 72 = 27) as "0001 1000" + "0010 0111" = "0011 1111". Here, the first group is not a valid BCD number, so add "0110" to it as "0011 1111" + "0000 0110" = "0100 0101". Here, no EoC is generated, so the result is a negative number, which is –45 in BCD form.

In the same way, 10's complement–based BCD subtraction can be done, which is left to the readers as a laboratory exercise.

DESIGN EXAMPLE 6.6

Develop a digital system to convert 8-bit binary data into equivalent BCD data using the `for-loop` statement.

Solution

```
library ieee;
use ieee.std_logic_1164.all;
use ieee.std_logic_arith.all;
use ieee.std_logic_unsigned.all;
entity bin2bcd is
port(bin:in std_logic_vector(7 downto 0);
bcd:out std_logic_vector(11 downto 0));
end entity;
architecture mark0 of bin2bcd is
begin
p0:process(bin)
variable obcd:std_logic_vector(19 downto 0);
begin
obcd := "000000000000"& bin;
for i in 0 to 7 loop
if(obcd(11 downto 8)>4) then
obcd(11 downto 8) := obcd(11 downto 8) + 3;
end if;
if(obcd(15 downto 12)>4) then
obcd(15 downto 12) := obcd(15 downto 12) + 3;
end if;
if(obcd(19 downto 16)>4) then
obcd(19 downto 16) := obcd(19 downto 16) + 3;
end if;
obcd := obcd(18 downto 0)&'0';
end loop;
bcd <= obcd(19 downto 8);
end process;
end architecture;
```

DESIGN EXAMPLE 6.7

Develop a digital system to convert 16-bit binary data into equivalent BCD data. The BCD data has to be in a form that is convenient for LCD display.

Solution

As given in this design example, there are 16 bits to represent the binary input, so we need five digits (4-bit groups of 20 bits) to represent the BCD form. The final BCD result should have "0011" as the MSB word with each digit to have data convenient for LCD display. The VHDL code for this conversion is given below:

```
library IEEE;
use IEEE.STD_LOGIC_1164.ALL;
use IEEE.STD_LOGIC_ARITH.ALL;
use IEEE.STD_LOGIC_UNSIGNED.ALL;
entity testc is
Port ( clk:in std_logic:='0';
data_in : in  STD_LOGIC_vector(15 downto 0):="0000000000000000";
b : out  STD_LOGIC_vector(19 downto 0):="00000000000000000000");
end testc;
architecture Behavioral of testc is
signal pst,nst: integer range 0 to 15:=0;
signal count: std_logic_vector(3 downto 0):="0000";
```

```
signal lcd_data1,lcd_data2,lcd_data3,lcd_data4,lcd_data5: std_logic_
vector(7 downto 0):="00000000";
signal bcd:std_logic_vector(19 downto 0):="00000000000000000000";
signal b1:STD_LOGIC_vector(35 downto 0):="000000000000000000000000000000
000000";
begin
p0:process(clk)
begin
if rising_edge(clk)then
pst<=nst;
end if;
end process;
p1:process(pst)
begin
case pst is
when 0 => bcd<="00000000000000000000";nst<=pst+1;
when 1 => b1<= bcd & data_in; nst<=pst+1;
when 2 => b1<=b1(34 downto 0) & b1(35);nst<=pst+1;
-- first digit
when 3=> if (b1(19 downto 16)>"0100") then
b1(19 downto 16)<=b1(19 downto 16)+"0011";
end if;nst<=pst+1;
-- second digit
when 4=> if (b1(23 downto 20)>"0100") then
b1(23 downto 20)<=b1(23 downto 20)+"0011";
end if;nst<=pst+1;
-- third digit
when 5=> if (b1(27 downto 24)>"0100") then
b1(27 downto 24)<=b1(27 downto 24)+"0011";
end if;nst<=pst+1;
-- fourth digit
when 6=> if (b1(31 downto 28)>"0100") then
b1(31 downto 28)<=b1(31 downto 28)+"0011";
end if;nst<=pst+1;
-- fifth digit
when 7=> if (b1(35 downto 32)>"0100") then
b1(35 downto 32)<=b1(35 downto 32)+"0011";
end if;
nst<=pst+1;
when 8=> count<=count +"0001";nst<=pst+1;
when 9=> if (count="1111") then count<="0000";nst<=pst+1;
else nst<=2;
end if;
when 10=> b1<=b1(34 downto 0) & b1(35);nst<=pst+1;
when 11=> lcd_data5<="0011" & b1(19 downto 16);nst<=pst+1;
when 12=> lcd_data4<="0011" & b1(23 downto 20);nst<=pst+1;
when 13=> lcd_data3<="0011" & b1(27 downto 24);nst<=pst+1;
when 14=> lcd_data2<="0011" & b1(31 downto 28);nst<=pst+1;
when 15=> lcd_data1<="0011" & b1(35 downto 32);nst<=0;
end case;
end process;
end Behavioral;
```

6.4 Binary Sequence Recognizer

There are lot of applications of and requirements for the sequence recognizer/detector. This is a most important circuit for various applications like digital locks, bit error rate (BER) tester design, wireless sensor networks, multichannel data logging systems, digital signature verification, network security, and so on [22,39,48,68,69]. This section explains the implementation of a sequence detector in FPGA using two Mealy and Moore FSM models. We also discuss the sequence recognizer with and without overlapping of the binary input sequence. We assume that data will enter the system in serial, one bit at a time, and the reference pattern for detecting the sequence is "1010". Assume that the LSB, that is, logic 0, is arriving in the system first. If the pattern is recognized, then the system has to issue an output signal "found", that is, logic 1; otherwise, it gives logic "0", as shown in Figure 6.12a. If the sequence detector's output signal will be a function of both the input and the current state, then our design will be implemented as a Mealy machine. Otherwise, if the system's output depends upon only the present state, then it will be implemented as a Moore machine.

We will start the design of the sequence detector with the state diagram. The reference sequence is "1010" and the FSM is the Mealy model. In this reference pattern and sequence detector, there is only a single input, so there can be only two possible transitions from each state. In the idle state, that is, the first state S_0, the input data can be either logic 0 or logic 1. If the input bit is logic 0, the state diagram will reflect that fact with a transition to a new state, state S_1, as shown in Figure 6.12b, because logic 0 is a valid input as per the reference frame. Instead of logic 0, if logic 1 arrives, the process remains in S_0 as long as it receives logic 1. Again, the input can be either logic 0 or logic 1. If the input data bit is logic 0, the system remains in state S_1, which implies that a valid sequence may begin from this logic 0. If the input data bit is logic 1, then the system enters a new state, state S_2. When the process is at S_2, the system receives a valid sequence of "10"; however, the output is logic 0, since the value for the output can be decided upon receiving the fourth bit only. In state S_2, the input can be either logic 0 or logic 1. If the input data bit is logic 0, then the system enters a new state, state S_3, because the sequence "010" is a valid pattern. If logic 1 is received in this state (S_2), the pattern we received before now is "01" and it becomes invalid; hence, we need to begin the receiving process again; therefore, we need to move back to S_0.

In state S_3, once again, the input can be either logic 0 or logic 1. Since we are designing the sequence detector as a Mealy FSM model, we need to produce the appropriate output upon receiving the input bit. If the input data bit is logic 0, the system must return to state S_1. Then, the pattern we received before now, "010", becomes invalid. If logic 1 is received, then the system returns to state S_0 and the output is a logic 1, as shown in Figure 6.12b, to start the complete detection process again for a subsequent new sequence. The

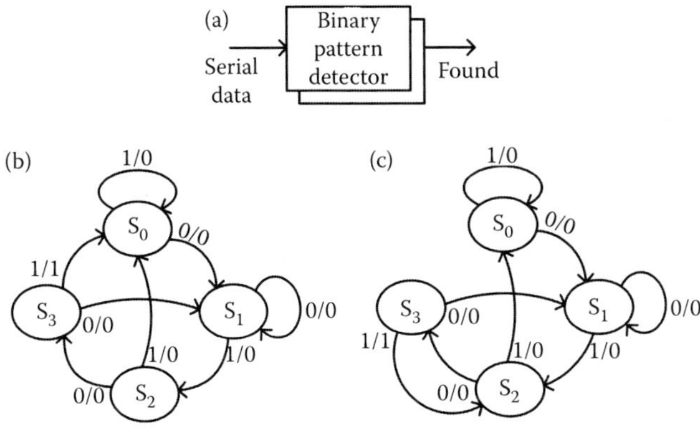

FIGURE 6.12
Sequence detector-high level block diagram (a), state diagram without overlapping (b), and state diagram with overlapping (c).

pattern we received while coming back to state S_0 is "1010", which is a valid sequence; hence, the output also goes high to indicate the detection of a correct (valid) sequence. An example of a long input and output sequence as per the state diagram shown in Figure 6.12b is given below:

Input sequence: "1100100**1010**1101**1010**11001**10101010**001001"
Output sequence: "0000000000**1**00000000**1**000000000**1**000**1**000000"

In the above sequence, the valid inputs and corresponding outputs are bolded. As given in the above sequences, input bit overlapping is not allowed. That means that in the input sequence, two valid patterns of "10101010" are subsequently received by the sequence detector. The corresponding output is "00010001". The input sequence may also be taken, when overlapping is allowed, as "1010", "1010" and "1010", with the corresponding output of "00010101". Here, the first 4 bits are used to form a valid pattern, the last 2 bits in the first valid pattern and a new 2 bits are used to form the another valid pattern and, finally, the last s bits of the second group and a new 2 bits are used to form a third group. The state diagram has to be slightly altered, as shown in Figure 6.12c, to design the sequence detector for overlapping the input sequence. In the state diagram, shown in Figure 6.12c, state transition occurs to S_2 instead of S_0 from state S_3 when a valid bit is received. The input and output as per Figure 6.12c are given below:

Input sequence: "1100100**1010**1100**1010**11000**10101010**001001"
Output sequence: "0000000000**1**00000000**1**000000000**1**0**1**0**1**000000"

This state diagram can be used to devise the respective excitation table from which this sequence detector would be implemented using digital combinational and sequential circuits.

DESIGN EXAMPLE 6.8

Develop a sequence detector to indicate the reception of a valid binary sequence. Assume that the reference pattern has to be given to the circuit in parallel, while the input has to be given serially. Write general VHDL code to give the value for the desired length of the sequence.

Solution

```
library IEEE;
use IEEE.STD_LOGIC_1164.ALL;
entity gen_sequence2 is
generic(N: integer:= 4);
Port ( x,clk,rst : in  STD_LOGIC:='0';
sq : in  STD_LOGIC_VECTOR (N-1 downto 0);
y : out  STD_LOGIC:='0');
end gen_sequence2;
architecture Behavioral of gen_sequence2 is
signal x_reg: std_logic_vector (N-1 downto 0);
begin
p0:process (clk)
begin
if (rst/='1') then
if rising_edge(clk) then
x_reg<=x_reg(N-2 downto 0) & x;
end if;
```

```
end if;
end process;
p1:process(rst,x_reg)
begin
if ((x_reg=sq) and (rst='0')) then
  y<='1';
else
  y<='0';
end if;
end process;
end Behavioral;
```

DESIGN EXAMPLE 6.9

Develop a digital system to accomplish a sequence-recognizing task as detailed below:

i. The width of the input sequence is 2 bits, that is, x[1:0].
ii. If the input at state S_0 is either "01" or "11", the process has to move to another state (S_1) and display 1 in a seven-segment display. Assume that the display is a common anode seven-segment display.
iii. At state S_1, if the same input status, i.e., "01" or "11", is retained, then the process has to be in the same state, that is, S_1.
iv. The process has to come back to S_0 once either "00" or "10" is received.
v. Display valid sequence detection, that is, every state transition, on the seven-segment display.
vi. The maximum detection limit is FH, then reset the display.

Solution

The VHDL design for this design example is given below:

```
library ieee;
use ieee.std_logic_1164.all;
use ieee.std_logic_arith.all;
use ieee.std_logic_unsigned.all;
entity code_recog is
port (clock: in std_logic:='0';
x: in std_logic_vector(1 downto 0):="00";
sev_seg: out std_logic_vector(6 downto 0):="0000000");
end code_recog;
architecture fsm of code_recog is
type state_type is (s0,s1);
signal state : state_type:=s0;
signal cnt_sig: std_logic_vector(3 downto 0):="0000";
begin
process (clock,x,state)
begin
if rising_edge(clock) then
case state is
when s0 =>
if (x="00" or x="10") then
state <= s0; cnt_sig<=cnt_sig;
elsif (x="01" or x="11") then
state <= s1; cnt_sig<=cnt_sig+1;
end if;
when s1 =>
if (x="01" or x="11") then
state <= s1;cnt_sig<=cnt_sig;
elsif (x="00" or x="10") then
state <= s0; cnt_sig<=cnt_sig+1;
```

```
end if;
when others => state <= state;
end case;
end if;
end process;
process(cnt_sig)
begin
case cnt_sig is
when "0000"=> sev_seg<="0000001";
when "0001"=> sev_seg<="1001111";
when "0010"=> sev_seg<="0010010";
when "0011"=> sev_seg<="0000110";
when "0100"=> sev_seg<="1001100";
when "0101"=> sev_seg<="0100100";
when "0110"=> sev_seg<="0100000";
when "0111"=> sev_seg<="0001111";
when "1000"=> sev_seg<="0000000";
when "1001"=> sev_seg<="0000110";
when "1010"=> sev_seg<="0001001";
when "1011"=> sev_seg<="0000000";
when "1100"=> sev_seg<="0110001";
when "1101"=> sev_seg<="0000001";
when "1110"=> sev_seg<="0110000";
when "1111"=> sev_seg<="0111000";
when others=> null;
end case;
end process;
end fsm;
```

DESIGN EXAMPLE 6.10

Design a Mealy FSM model to detect a sequence "1011" with overlapping. Assume that the LSB arrives the detector circuit first.

Solution

The given reference pattern is "1011", and overlapping is allowed in this design; hence, overlapping would be done only with the MSB, i.e., "1", in this design. The Mealy FSM model state diagram for this design is shown in Figure 6.13.

The VHDL code for implementing this sequence detector is given below:

```
library IEEE;
use IEEE.std_logic_1164.all;
```

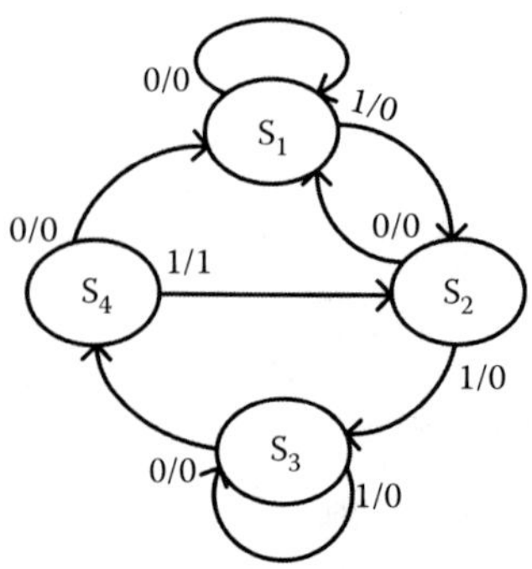

FIGURE 6.13
Mealy FSM model for recognizing a pattern "1011" with overlapping.

```
use IEEE.std_logic_arith.all;
use IEEE.std_logic_unsigned.all;
entity sequence_detector is
port (clk,rst,x: in STD_LOGIC:='0';
z: out STD_LOGIC:='0');
end sequence_detector;
architecture sequence_detector_arch of sequence_detector is
type seq_detect_type is (S1, S2, S3, S4);
signal seq_detect: seq_detect_type:=S1;
begin
p0: process (clk)
begin
if clk'event and clk = '1' then
if rst='1' then
seq_detect <= S1;
else
case seq_detect is
when S1 =>
if x = '1' then
seq_detect <= S2;
elsif x = '0' then
seq_detect <= S1;
end if;
when S2 =>
if x = '1' then
seq_detect <= S3;
elsif x = '0' then
seq_detect <= S1;
end if;
when S3 =>
if x = '1' then
seq_detect <= S3;
elsif x = '0' then
seq_detect <= S4;
end if;
when S4 =>
if x = '1' then
seq_detect <= S2;
elsif x = '0' then
seq_detect <= S1;
end if;
when others =>  null;
end case;
end if;
end if;
end process;
z <= '0' when (seq_detect = S1 and x = '1') else
'0' when (seq_detect = S1 and (x = '0' and not (x = '1'))) else
'0' when (seq_detect = S2 and x = '1') else
'0' when (seq_detect = S2 and (x = '0' and not (x = '1'))) else
'0' when (seq_detect = S3 and x = '1') else
'0' when (seq_detect = S3 and (x = '0' and not (x = '1'))) else
'1' when (seq_detect = S4 and x = '1') else
'0' when (seq_detect = S4 and (x = '0' and not (x = '1'))) else   '0';
end sequence_detector_arch;
```

DESIGN EXAMPLE 6.11

Develop a digital system using the Moore FSM model to detect a sequence "10010" without overlapping.

Solution

The given sequence is "10010", the required FSM is the Moore model and overlapping is not permitted. Based on this information, we can begin generating the state diagram, which will be as shown in Figure 6.14.

The VHDL code for the implementation of this Moore FSM model is given below:

```
library ieee;
use ieee.std_logic_1164.all;
use ieee.std_logic_arith.all;
use ieee.std_logic_unsigned.all;
entity moore is
port (clock,x,reset: in std_logic:='0';
y : out std_logic:='0');
end moore;
architecture moore_fsm of moore is
type state_type is (s0,s1,s2,s3,s4,s5);
signal state : state_type:=s0;
begin
process (clock, reset)
begin
if (reset = '1') then
state <= s0;
elsif rising_edge(clock) then
case state is
when s0 =>
if x='1' then
state <= s1;
else
state <= s0;
end if;
when s1 =>
if x='0' then
state <= s2;
else
state <= s1;
```

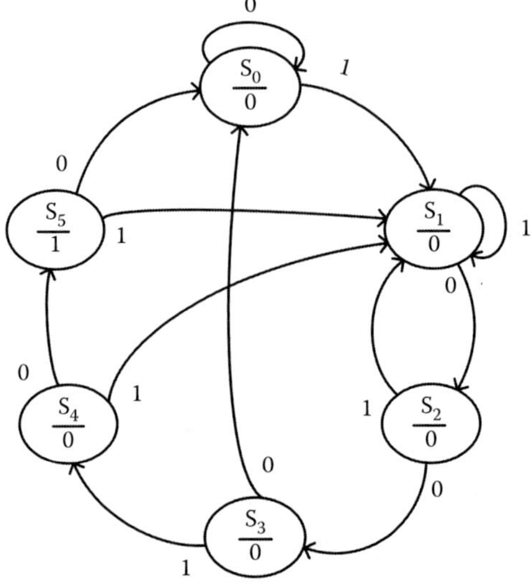

FIGURE 6.14
Moore FSM model for recognizing a pattern "10010" without overlapping.

```
end if;
when s2 =>
if x='0' then
state <= s3;
else
state <= s1;
end if;
when s3=>
if x='1' then
state <= s4;
else
state <= s0;
end if;
when s4=>
if x='1' then
state <= s1;
else
state <= s5;
end if;
when s5=>
if x='1' then
state <= s1;
else
state <= s0;
end if;
when others => state <= state;
end case;
end if;
end process;
y<='1' when state=s5 else '0';
end moore_fsm;
```

6.5 Vending Machine Controller

A vending machine controller is a relatively complex design that will give us a clear idea about various electro-mechanical system interfacing/control from an FSM. The vending machine has a coin-in mechanism that can accept coins like Rs. 2/- (item A), Rs.5/- (item B), Rs.10/- (item C) and currency notes. Assume that a pulse indicating the value of the coin or currency note will be generated when a coin/currency is input. A sufficient number of seven-segment LED displays will display the total amount input and the balance amount to be collected. An LCD or GLCD will be used to display the menu card, that is, price of the items, their availability and the current process. The vending machine may deliver items/products like coffee, tea, milk, cold drinks, and candies. Ordinary DIP switches or a matrix keypad can be used to select the required items. If the input amount exceeds the actual rate of the selected item, the vending machine should provide the product by turning on the corresponding relay/solenoid valve/conveyer and an appropriate LED to indicate that the product is being dispensed and then, after the vending machine dispenses the product, it also releases the balance amount as change. If the user attempts to select an item without having input the correct amount or more for that item, the vending machine simply does nothing until a valid item is selected or the required amount is input. The vending machine can accept coins up to the price of the most expensive item. In other words, if the user inputs an excess amount, the machine has to wait for the user to perform an action, that is, selection of item A, B, C, or D, and then the appropriate balance has to be issued to the user.

All this functionality has to be considered while designing the state diagram. The state diagram mainly consists of five states: (i) start, (ii) money insertion, (iii) item selection, (iv) money sufficiency validation and balance calculation, and (v) item/product delivery. We will briefly see the operations at all the states below. Initially, when the start button is pressed, the vending machine will be ready for the user to input the money. This state is the initial state of the design. After this, the control unit will move to the waiting state, where it will wait for the money to be inserted. Then, the user will input the money as required for the item being purchased. The machine can accept only a few types of coins/notes. Then, the user will select the product to be dispensed. This state may get any one ID value from one of the items A, B, C, D, and so on available at the vending machine. The machine will first check whether the inserted amount is too low, sufficient or too high for the selected item. If too low, the machine will display "the amount is insufficient for the selected items" on the LCD, terminate the entire operation and come back to the initial state. If the amount is sufficient, the machine will activate the dispensing system. If the amount is too high, then the machine will calculate the balance and deliver the item as well as issuing the balance. A `mon _ cnt` signal is used for calculating the total amount of money inserted in the machine. If the money inserted is more than the amount of the product, then the extra change will be returned to the user. There is also an additional feature of withdrawing the request if the user does not want to take the product. When the cancel button is pressed, the money inserted will be returned to the user through the return output. The total amount of a product taken at a time is shown by the money signal. Similarly, the user can select and get other products following the above procedure.

Now, we will discuss the different operations of a vending machine control system at the top level one by one, with appropriate states.

i. State S_0: In this state, the process waits for the start input. As long as the start input is logic 0, the process stays in this state. Once logic 1 is applied to the start input, the process moves to the next state, S_1, opening the coin inlet valve to allow the user to insert the coin/currency.

ii. State S_1: In this state, the process waits for insertion of the money for some time. Once the waiting time has elapsed, the process will move to next state, S_2. A local counter in this state decides the duration of the waiting time. Suppose no coin/currency is deposited until the waiting time elapses. In this case, the process moves back to the initial state, that is, S_0, and terminates the current process.

iii. State S_2: In this state, once the money deposit time is over, the process evaluates the amount deposited by the user and waits for selection of the item. When item selection is done by the user, the process immediately checks the price of the selected item against the amount deposited. If a sufficient amount is deposited, the process moves to next state, S_3. If the amount is too low, the process terminates the operation and moves back to the initial state, S_0. If the amount is too high, the process computes the balance that needs to be issued to the user, and with this information, it moves to the next state, S_3. This operation may also be done in many different ways. Assume that the price of an item is Rs.50/-. If 10/- rupees is inserted, then the process will go to one state and wait until the desired money, that is, Rs. 40/-, is inserted. Suppose Rs. 20/- is inserted. Then, the process will move to a different state and then wait until Rs. 30/- is inserted into the machine. This pattern would continue until the required amount or more is deposited.

iv. State S_3: In this state, the process activates the item delivery assembly and also the balance issuing system, if any.

Apart from all these functions, there are many more operations like maintaining stock history, running a real-time clock, buzzer control, balance-issuing control, and maintaining the necessary temperature/cooling inside the vending machine as required for the stored items.

Now, we discuss the complete design of a vending machine controller. Assume a vending machine that has two provisions for loading coins and selecting the item of interest. It has eight outputs of different widths for indicating the current action of the vending machine, item delivery control and balance-issuing system. The permitted coins, as available in India, are Rs.1/-, Rs.2/-, Rs.5/- and Rs.10/- only. There are five different items, namely (i) Appy, (ii) Mangosip, (iii) Bovonto, (iv) Frooti, and (v) Kalimark, and their approximate prices per cup are Rs.5/-, Rs.10/-, Rs.7/-, Rs.3/-, and Rs.8/-, respectively. An item can be selected using the available switches. The switches are normally in the logic low, that is, "00000", condition and the user has to push up any one of the switches to select the item of interest. In real time, the value of the deposited coin will be given by a circuit as 4-bit data; hence, in this design, we take the value of the deposited coin directly from the four switches. There are six LEDs to indicate the state of various operations, such as the system being ready for service (`ready_led`), deposit of the coin (`coin_in_led`), selection of the item (`item_sel_led`), amount validation (`amo_vali_led`), item delivery (`item_deli_led`), and balance indication (`bala_buzzer`). Finally, we have item delivery control (`item_deliv_ctrl`) of width [4:0] and balance issuing system (`bal_deliv`) of length [2:0]. Other than these inputs and outputs, there are two inputs, one for start and another for the clock, as shown in Figure 6.15a. The `item_deliv_ctrl` output is connected directly to the cool drink flow control driver circuit so as to open the flow valve for a prespecified time duration [39,70].

The state diagram for this design is shown in Figure 6.15b. The state 0 checks the status of the start switch and keeps all the outputs at logic 0. Once the start goes high, the control moves to state 1, giving `ready_led` logic 1. After reaching state 1, logic 1 will be sent to `coin_in_led` to indicate to the user that he or she should insert a coin. This also opens the coin inlet valve through which the user can deposit the coin. In this design, we have to select the switch to input the amount as "0001" for Rs. 1/-, "0010" for Rs. 2/-, "0101" for Rs. 5/-, and "1010" for Rs. 10/-. This binary value goes into the variable `amount`, which we will use subsequently for various decision making and computations. After this operation, the FSM moves to state 2, where we will turn off some LEDs from previous operations and allow the user to select the item. Five switches are available to select the item: "00001" for Appy, "00010" for Mangosip, "00100" for Bovonto, "01000" for Frooti, and "10000" for Kalimark.

The actual rate or amount, as mentioned above, for all these items will be taken from a look-up table. That means once the item is selected, the actual amount of the selected item will immediately be known to the FSM. This operation is indicated by turning on `item_sel_led`, and the selection input is stored in a variable `item_temp`. Then, the FSM moves to state 3.

In state 3, the value of `amount` will be added to a temporary variable `amount_temp`, whose initial value is "0000", and the LEDs of previous operations will be turned off. Then, the FSM moves to state 4. In state 4, the received amount and actual rate of the item selected

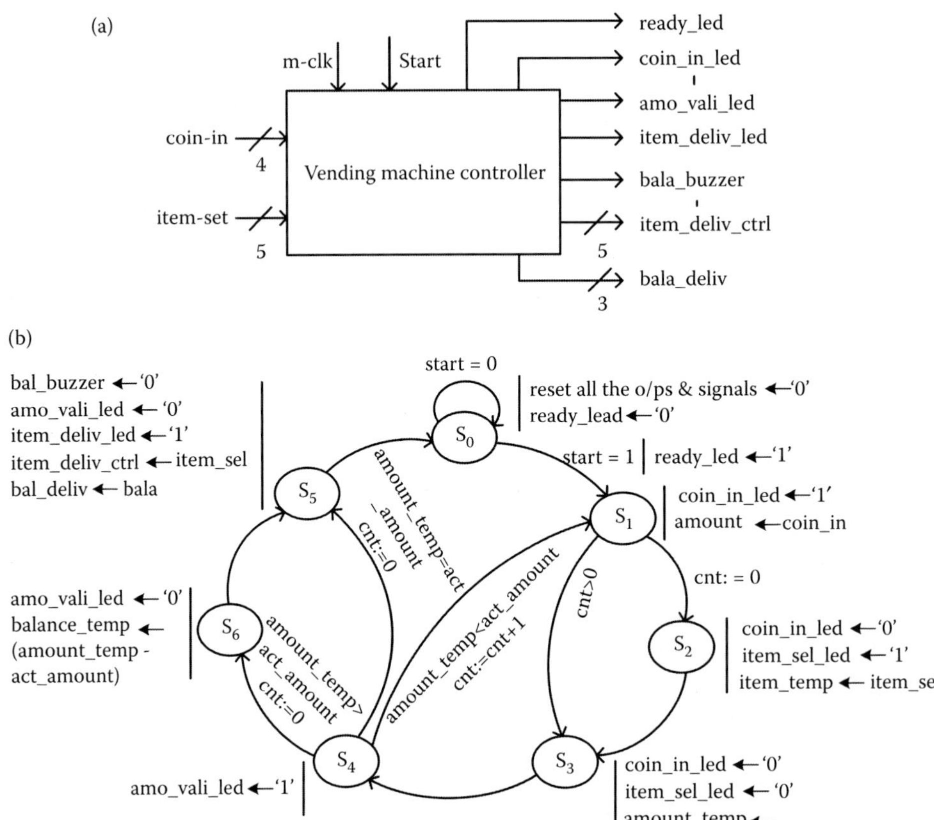

FIGURE 6.15
FSM design for a vending machine control – top-level illustration (a) and control flow state diagram (b).

will be validated to decide that the user has either deposited a sufficient amount or an amount greater or less than the price. If the deposited amount is equal, then the delivery system can be driven with its associated other operations; if greater, the FSM has to calculate the balance to be returned to the user and if lower, the FSM has to allow the user to deposit the required amount. The state transitions are as follows: if sufficient, to state 5; if greater, to state 6 and if lower, to state 1, as shown in Figure 6.15b. That means the FSM does not initiate the delivery process until it gets a sufficient amount or more. The state transition from state 1 either to state 2 or state 3 is decided with the help of a local counter. State 6 calculates the balance and moves to state 5, which drives the delivery of the selected item and the balance-issuing system. After completing a delivery, the FSM moves back to state S_0 and waits for the start input to begin the next delivery. In this way, any vending machine controller can be designed.

DESIGN EXAMPLE 6.12

Develop a digital system to realize simple vending machine controller functions. Display the value of the input amount, the remaining amount that the user has to insert and the balance that needs to be returned to the user on a 16 × 2 LCD. Assume that the inserted amount is given to the vending machine via three DIP switches, the frequency of the onboard clock is 4 MHz and the LCD is connected in 8-bit mode.

Solution

```
library IEEE;
use IEEE.STD_LOGIC_1164.ALL;
use IEEE.STD_LOGIC_ARITH.ALL;
use IEEE.STD_LOGIC_UNSIGNED.ALL;
entity vmcc is
Port ( a : in  STD_LOGIC_vector(2 downto 0):="000";
clk:in std_logic:='0';
c:out std_logic_vector(2 downto 0):="000";
d:out std_logic_vector(7 downto 0):="00000000";
oclk,sdv:out std_logic:='0'  );
end vmcc;
architecture Behavioral of vmcc is
signal st,nst:integer range 0  to 5:=0;
signal depo,bal,f: STD_LOGIC_vector(7 downto 0):="00110000";
signal t:integer range 0 to 3:=0;
signal clk0,clk_sig0,clk_sig1,qin:std_logic:='0';
signal sta,nsta:integer range 0 to 70:=0;
begin
p1:process(clk_sig0)
begin
if rising_edge(clk_sig0) then
sta<=nsta;
end if;
end process;
p2:process(sta,bal,depo)
begin
case sta is
when 0=> d<="00111000";c<="100"; nsta<=1;--38
when 1|3|5|7|9|33|65|69=> c<="000";nsta<=sta+1;
when 2=> d<="00001100";c<="100"; nsta<=3;--0C
when 4=> d<="00000001";c<="100"; nsta<=5;--01
when 6=> d<="00000110";c<="100"; nsta<=7;--06
when 8=> d<="10000000";c<="100"; nsta<=9;--80
when 10=> d<="01000100";c<="110"; nsta<=11;--D
when 11|13|15|17|19|21|23|25|27|29|31|35|37|39|41|43|45|47|
49|51|53|55|57|59|61|63|67=> c<="010";nsta<=sta+1;
when 12|42=> d<="01100101";c<="110"; nsta<=sta+1;--e
when 14=> d<="01110000";c<="110"; nsta<=15;--p
when 16|36=> d<="01101111";c<="110"; nsta<=sta+1;--o
when 18|28|58=> d<="01110011";c<="110"; nsta<=sta+1;--s
when 20=> d<="01101001";c<="110"; nsta<=21;--i
when 22|46=> d<="01110100";c<="110"; nsta<=sta+1;--t
when 24|48|54=> d<="00100000";c<="110"; nsta<=sta+1;--
when 26|56=> d<="01010010";c<="110"; nsta<=sta+1;--R
when 30|60=> d<="00111010";c<="110"; nsta<=sta+1;--:
when 32=> d<="11000000";c<="100"; nsta<=33;--C0
when 34=> d<="01000011";c<="110"; nsta<=35;--C
when 38|40=> d<="01101100";c<="110"; nsta<=sta+1;--l
when 44=> d<="01100011";c<="110"; nsta<=45;--c
when 50=> d<="01000010";c<="110"; nsta<=51;--B
when 52=> d<="01100001";c<="110"; nsta<=53;--a
when 62|70=> d<=bal;c<="110";nsta<=63;
when 64=> d<="10001011";c<="100";nsta<=65;
when 66=> d<=depo;c<="110";nsta<=67;
when 68=> d<="11001110";c<="100";nsta<=69;
end case;
end process;
p3:process(clk)
variable cnt,cnt1,dd: integer:=0;
```

```
begin
if rising_edge(clk) then
if(cnt1=4)then   --40000
clk_sig0<= not(clk_sig0);
cnt1:=0;
else
cnt1:= cnt1 + 1;
end if;
if(cnt=15) then --16000000
clk0<=not clk0;
cnt:=0;
else
cnt:=cnt+1;
end if;
end if;
oclk<=not clk0;
end process;
p4:process(clk0)
begin
if rising_edge(clk0) then
st<= nst;
end if;
end process;
p5:process(clk0)
begin
if rising_edge(clk0) then
if(a="000" or a="011" or a="101" or a="110" or a="111") then
t<=0;
elsif (a="001")then
t<=1;
elsif (a="010")then
t<=2;
elsif (a="100")then
t<=3;
end if;
end if;
end process;
p6:process(t,st)
begin
case t is
when 0=>
case st is
when 0|1|2|3|4|5=>
nst<=st;depo<=(depo or f);bal<=(bal or f);sdv<='0';
end case;
when 1=>
case st is
when 0=>
nst<=1;depo<="00110101";sdv<='0';bal<="00110000";
when 1=>
nst<=2;depo<="00110100";sdv<='0';bal<="00110000";
when 2=>
nst<=4;depo<="00110011";sdv<='0';bal<="00110000";
when 3=>
nst<=0;depo<="00110000";sdv<='1';bal<="00110000";
when 4=>
nst<=5;depo<="00110010";sdv<='0';bal<="00110000";
when 5=>
nst<=3;depo<="00110001";sdv<='0';bal<="00110000";
end case;
when 2 =>
```

```
case st is
when 0=>
nst<=2;depo<="00110100";sdv<='0';bal<="00110000";
when 1=>
nst<=4;depo<="00110011";sdv<='0';bal<="00110000";
when 2=>
nst<=5;depo<="00110010";sdv<='0';bal<="00110000";
when 3=>
nst<=0;depo<="00110000";sdv<='1';bal<="00110001";
when 4=>
nst<=3;depo<="00110001";sdv<='0';bal<="00110000";
when 5=>
nst<=0;depo<="00110000";sdv<='1';bal<="00110000";
end case;
when 3 =>
case st is
when 0=>
nst<=3;depo<="00110001";sdv<='0';bal<="00110000";
when 1=>
nst<=0;depo<="00110000";sdv<='1';bal<="00110000";
when 2=>
nst<=0;depo<="00110000";sdv<='1';bal<="00110001";
when 3=>
nst<=0;depo<="00110000";sdv<='1';bal<="00110100";
when 4=>
nst<=0;depo<="00110000";sdv<='1';bal<="00110010";
when 5=>
nst<=0;depo<="00110000";sdv<='1';bal<="00110011";
end case;
end case;
end process;
end Behavioral;
```

DESIGN EXAMPLE 6.13

Develop a digital system to realize a vending machine controller, as given in Figure 6.15. In this design, the count value has to be chosen based on the onboard clock value to maintain sufficient delay between all the processes at different states.

Solution

```
library IEEE;
use IEEE.STD_LOGIC_1164.ALL;
use IEEE.STD_LOGIC_ARITH.ALL;
use IEEE.STD_LOGIC_UNSIGNED.ALL;
entity VMC_FSM is
port (m_clk, start : in std_logic:='0';
coin_in : in std_logic_vector (3 downto 0):="0000";
item_sel : in std_logic_vector(4 downto 0):="00000";
ready_led,coin_in_led,item_sel_led,amo_vali_led,item_deli_led,bala_
buzzer: out std_logic:='0';
item_deliv_cntl: out std_logic_vector (4 downto 0):="00000";
bal_deliv: out std_logic_vector (2 downto 0):="000");
end VMC_FSM;
architecture behavior of VMC_FSM is
type state_type is (s0,s1,s2,s3,s4,s5,s6);
signal pst,nst: state_type:=s0;
signal amount,act_amount,amount_temp:std_logic_vector(3 downto 0):="0000";
signal item_temp:std_logic_vector(4 downto 0):="00000";
signal balance_temp:std_logic_vector(3 downto 0):="0000";
signal clk_sig:std_logic:='0';
```

```
begin
p0:process(m_clk)
variable cnt:integer:=0;
begin
if rising_edge(m_clk) then
if (cnt=2) then ---as per the on-board clock
clk_sig<= not clk_sig;
cnt:=0;
else
cnt:=cnt+1;
end if;
end if;
end process;
p1:process(clk_sig)
begin
if(clk_sig'event and clk_sig='1') then
pst<=nst;
end if;
end process;
p2:process(start,pst,coin_in,item_Sel)
variable cnt:integer:=0;
begin
case pst is
when s0 =>
coin_in_led<='0';
item_Sel_led<='0';
amo_vali_led<='0';
item_deliv_cntl<="00000";
bal_deliv<="000";
bala_buzzer<='0';
amount_temp<="0000";
item_deli_led<='0';
if(start='0') then
ready_led<='0';
nst<=s0;
else
ready_led<='1';
nst<=s1;
end if;
when s1=>
amo_vali_led<='0';
coin_in_led<='1';
amount<=coin_in;
if cnt>0 then
nst<=s3;
else
nst<=s2;
end if;
when s2=>
coin_in_led<='0';
item_Sel_led<='1';
item_temp<=item_Sel;
nst<=s3;
when s3=>
coin_in_led<='0';
item_Sel_led<='0';
amount_temp<=amount+amount_temp;
nst<=s4;
when s4=>
amo_vali_led<='1';
if (amount_temp=act_amount) then
```

```
cnt:=0;
nst<=s5;
elsif (amount_temp>act_amount) then
cnt:=0;
nst<=s6;
else
nst<=s1;
cnt:=cnt+1;
end if;
when s5=>
amo_vali_led<='0';
item_deli_led<='1';
item_deliv_cntl<=item_sel;
bal_deliv<=balance_temp(2 downto 0);
if (balance_temp/="000") then
bala_buzzer      <='1';
else
bala_buzzer      <='0';
end if;
nst<=s0;
when s6=>
amo_vali_led<='0';
balance_temp<=(amount_temp-act_amount);
nst<=s5;
end case;
end process;
p3:process(m_clk,item_sel)
begin
if rising_edge(m_clk) then
case item_Sel is
when "00001"=> act_amount<="0101"; --Rs.5/-
when "00010"=> act_amount<="1010"; --Rs.10/-
when "00100"=> act_amount<="0111"; --Rs.7/-
when "01000"=> act_amount<="0011"; --Rs.3/-
when "10000"=> act_amount<="1000"; --Rs.8/-
when others=> act_amount<="0000";
end case;
end if;
end process;
end behavior;
```

6.6 Traffic Light Controller

Traffic lights are used to control vehicular traffic. In the modern era, every family has more than one vehicle, resulting in a rapid increment in the usage of number of vehicles. That is why traffic lights are mandatory to avoid traffic jams and accidents. In general, there are three lights in a traffic signal, as shown in Figure 6.16a, each having a different message for the drivers: the red light (upper one) asks the driver to stop at the intersection, the green light (lowest one) permits the driver to drive through the intersection, and the yellow light (middle one) alerts the driver to wait if the next light is red or get ready to go/turn on the engine if the next light is green. The traffic light system has proved to be a remarkable technique to stop vehicular collisions and control traffic jams [39,71]. The fundamental idea of the traffic light system is just controlling the group (more than one) of green, yellow, and red LEDs on a time-sharing basis with a simple electronic circuit, as shown in Figure 6.16b.

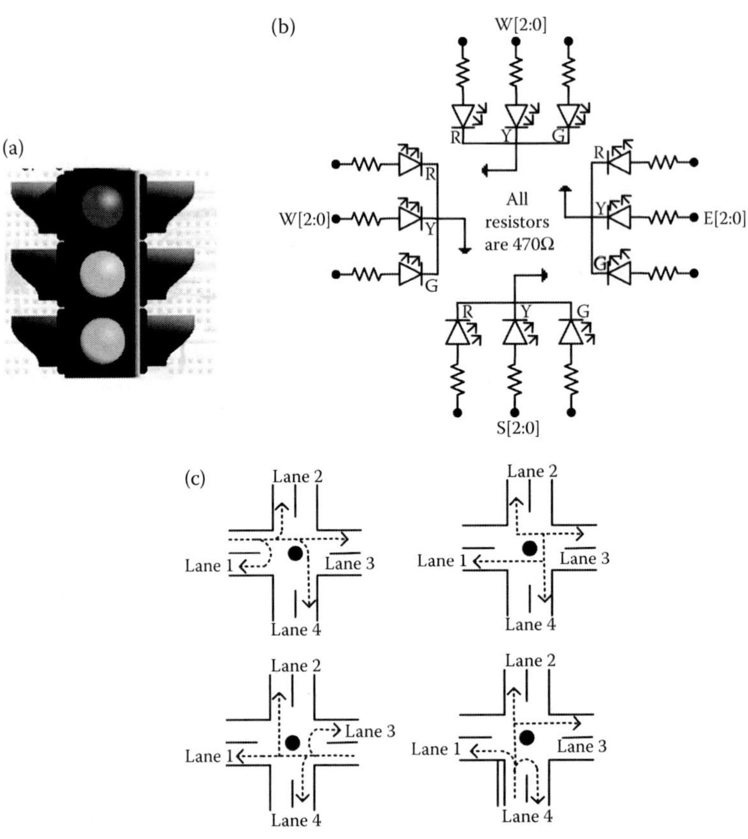

FIGURE 6.16
Photograph of a traffic light (a), circuit for simple traffic light wiring (b), and illustration of traffic control sequences (c).

Four groups of three LEDs (green, yellow and red) are deployed at the four sides of an intersection. All the LEDs in each group are individually wired through separate current-limiting resisters. In practice, an array of LEDs of all the colors is used to increase visibility to longer distances. A common anode or cathode configuration would be used to turn on/off the LEDs and, based on this configuration, control logic 1 or 0 has to be chosen. In Figure 6.16b, the common cathode configuration is used, and the anodes of all the LEDs (W[2:0], N[2:0], E[2:0] and S[2:0]) are connected to the control circuit that drives the LEDs over time. For example, if the control data are "001" for W, "100" for N, "100" for E, and "100" for S, it signals the driver to go from the west to the north, east and south. After some specified amount of time, this status will be "010" for W, "010" for N, "100" for E, and "100" for S to alert the vehicles in the west that the signal is about to change to red and the signal is about to change to green at the north side. Again, after some amount of time, the status will be "100" for W, "001" for N, "100" for E, and "100" for S to signal the drivers to go from the north to the west, south and east. This will cyclically continue for other directions as well. This traffic light control system is also capable of operating in manual mode, with a dedicated sequence of buttons for each of the LEDs.

Consider Figure 6.16c: lane 1 (the west side) has its green light turned on; hence, in all the other lanes, the corresponding red lights are turned on. This display indicates that the vehicle can go from the west to all other directions, as marked by dotted lines in Figure 6.16c. After a time delay of a predefined time, say, 120 seconds, the green light in lane 2 must be turned on and the green light in lane 1 must be turned off. As a warning indicator, the yellow light in lane 1 is turned on, indicating that the red light is about to light up. Similarly, the yellow light in lane 2 is also turned on as an indication that the green light is about to be turned on. The yellow lights in lanes 1 and 3 are turned on for a small duration, say, 5 seconds, after which the red light in lane 1 is turned on and the green light in lane 2 is also turned on. The green light in lane 2 is also turned on for a predefined time, and the process moves forward to lane 3 and finally lane 4. The system then loops back to lane 1, where the process mentioned above will be repeated all over again.

In the modern scientific world, wireless traffic light control systems have become popular. Most wireless traffic light control systems work with RF wireless technology. They are fully automatic with provisions for timing adjustments and reprogrammable options for altering the function sequences. This type of system is more suitable for multiroad junctions and different traffic sequences. Normally, this system consists of a programmable master control unit for passing the commands and multiple slave units for executing the commands at the traffic lights. The slave units have multiple relay output points that are connected to the LEDs. The master control unit can be programmed to control a variety of different light on/off sequences for various directions of traffic. The individual timing parameters can be programmed as per requirements or traffic load conditions and rules.

Now, we will discuss the design of a wireless traffic light control system. Assume that there is one master unit and four slave units. The master unit is designed with an FSM (built inside the FPGA) and RF 433 MHz encoder (HT12E) which transmits an 8-bit address and 4-bit data, as shown in Figure 6.17a, and is deployed at the centre, marked by a black dot in Figure 6.17b, of the intersection of the four cross roads. The slave unit is designed with a decoder (HT12D) and four single-pole single-through (SPST) relays with its driver circuits and array of green, yellow, and red LEDs. The slave units are installed at the appropriate west-north (WN), north-east (NE), east-south (ES), and south-west (SW) points, marked by the circles in Figure 6.17b, with respect to the master unit. More details about the wiring and working principles of HT12E and HT12D can be found in Chapter 5. In our design, the slave unit controls the traffic displays in accordance with the command that comes from the master unit. Each slave unit controls (i) its actual direction display (d_{act}), that is, forward, right, and U-turn; (ii) its free left display (d_l); (iii) its actual pedestrian display (p_{act}); and (iv) its adjacent side pedestrian display (p_{adj}). These variables are marked in Figure 6.17b for clear understanding.

Four data that come from the master unit are used to control all these terminals. For example, if the data are "1010" (note that the MSB denotes the d_{act} and so on, as given in Table 6.11), then $d_{act} = 1$, $p_{act} = 0$, $p_{adj} = 1$ and $d_l = 0$, which displays (i) green for its actual direction that is, a-b-c, a-b-d and a-b-e display; (ii) red for its actual pedestrian; (iii) green for its adjacent pedestrian; and (iv) red on its free left direction, that is, a b-f display. The logic 1 at d_{act} turns on the green display and also enables a down counter (60 through 00) with two seven-segment displays to indicate the timing for the next control. The down counting operation is performed by a simple circuit kept along with every slave unit. Logic 0 at d_l turns on the red light, that is, the "X" symbol, for 20 seconds, that is, during the counting of 60 through 40, and at this duration, logic 1 at the p_{adj} turns on the green light. Once the counter value reaches 40, p_{adj} goes low and d_l goes high. Therefore, the complete operation and switching time of all the displays of the slave units are controlled by the master unit.

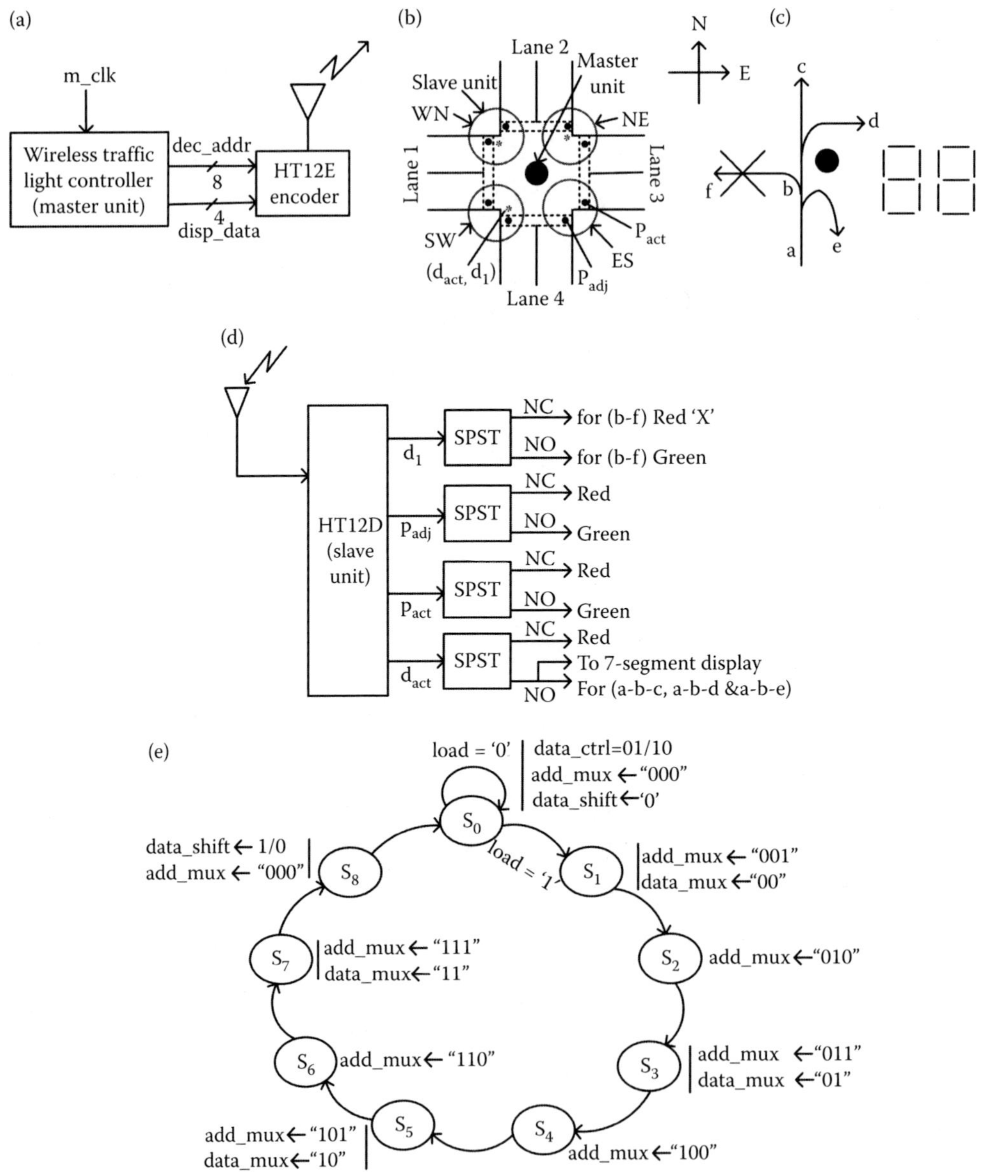

FIGURE 6.17
Master unit in wireless traffic light system (a), geometrical installation of slave units (b), direction and seven-segment (down counter) display at the slave unit (c), configuration of light displays with slave unit (d), and state diagram of control engine (e).

The display pattern for the forward, right, U-turn, "X" sign, and left are shown in Figure 6.17c, and the LEDs and display configurations are shown in Figure 6.17d. Now, we need to frame the data sequences which need to be sent to the slave units for controlling the displays. All the data patterns are given in Table 6.11. The master unit (transmitter) sets the address of the slave unit first, then sets the data so as to communicate only with a particular slave unit. For example, if the master wants to enable the slave deployed at the WN, then it sets address 05H first, then data "1010"; hence, the WN slave receives the data

TABLE 6.11

Master to Slave Control and Display Data Pattern

Slave Unit Position	Slave Unit Address	Control data (d_{act} p_{act} p_{adj} d_l)							
		c_1	c_2	c_3	c_4	c_5	c_6	c_7	c_8
WN	05H	1010	1001	0000	0000	0000	0000	0100	0000
NE	0AH	0100	0000	1010	1001	0000	0000	0000	0000
ES	50H	0000	0000	0100	0000	1010	1001	0000	0000
SW	A0H	0000	0000	0000	0000	0100	0000	1010	1001

and turns on d_{act} and p_{adj}. Immediately, the master sets the address as 00H; hence, no slave will be synchronized. Then, it sets the address to 0AH and data "0100"; hence, p_{act} gets turned on. Since by this time, p_{adj} of the WN slave and p_{act} of the NE slave have turned on, pedestrians are allowed to cross lane 2. The data of the complete sequence of all the cycles (c_1 through c_8) are given in Table 6.11. In each cycle, four 4-bit data are transmitted to control the display operation of all the slave units.

The state diagram of the FSM designed in the FPGA for transmitting the address and control data sequences is shown in Figure 6.17e. The entire operation of the wireless traffic light controller designed with HT12E and HT12D is controlled with eight states. We discuss some design examples below to understand these concepts.

DESIGN EXAMPLE 6.14

Develop a digital system to realize the working principles of a simple traffic light control system. Assume that there are four green, yellow, and red LEDs at each side of the intersection of the four cross roads. The system has to cyclically control the traffic lights as green to yellow to red for all the sides whenever their turn comes. Different timing has to be chosen for turn-on and turn-off of green to yellow and yellow to red LEDs. The system must have a reset option to initialize its operation.

Solution

```
library ieee;
use ieee.std_logic_1164.all;
use ieee.std_logic_unsigned.all;
entity trafc is
port(clk:in std_logic:='0';
rst:in std_logic:='0';
lr,ly,lg:out std_logic_vector(3 downto 0):="0000");
end entity;
architecture mark0 of trafc is
signal cnt:std_logic_vector(4 downto 0):="00000";
signal clock, clks,clks3,x:std_logic:='0';
signal lrs1,lrs2,lys1,lys2,lgs1,lgs2:std_logic_vector(3 downto
0):="0000";
signal cnt30:std_logic_vector(2 downto 0):="000";
begin
p0:process(clk,rst)
begin
if(rst='0') then
cnt <= "00000";
else
if(falling_edge(clk)) then
cnt <= cnt + 1;
```

```
end if;
end if;
end process;
clks <= cnt(4);
p1:process(clks,rst)
begin
if(rst='0') then
cnt30 <= "000";
else
if(falling_edge(clks)) then
cnt30 <= cnt30 + 1;
end if;
end if;
end process;
x <= cnt30(2) and cnt30(1) and cnt30(0);
clks3 <= cnt30(2);
p2:process(rst,clks3)
begin
if(rst='0') then
lrs1 <= "0111";
lrs2 <= "0011";
lys1 <= "0000";
lys2 <= "1100";
lgs1 <= "1000";
lgs2 <= "0000";
else
if(falling_edge(clks3)) then
lrs1 <= lrs1(0) & lrs1(3 downto 1);
lrs2 <= lrs2(0) & lrs2(3 downto 1);
lys1 <= lys1(0) & lys1(3 downto 1);
lys2 <= lys2(0) & lys2(3 downto 1);
lgs1 <= lgs1(0) & lgs1(3 downto 1);
lgs2 <= lgs2(0) & lgs2(3 downto 1);
end if;
end if;
end process;
with x select lr <= lrs1 when '1', lrs2 when '0', "ZZZZ" when others;
with x select ly <= lys1 when '1', lys2 when '0', "ZZZZ" when others;
with x select lg <= lgs1 when '1', lgs2 when '0', "ZZZZ" when others;
end architecture;
```

DESIGN EXAMPLE 6.15

Develop a digital system to realize a wireless traffic light control system, as given in Figure 6.17. In this design, the values of the local counters have to be chosen based on the onboard clock speed (frequency) to maintain sufficient time delay between the control and display operations of the traffic light system.

Solution

```
library ieee;
use ieee.std_logic_1164.all;
use ieee.std_logic_unsigned.all;
entity wireless_tlc is
port(m_clk:in std_logic:='0';
disp_data: out std_logic_vector(3 downto 0):="0000";
dec_addr:out std_logic_vector(7 downto 0):="00000000");
end wireless_tlc;
architecture behavior of wireless_tlc is
signal clk_sig,clk_sig1,data_shift:std_logic:='0';
signal load:std_logic:='1';
```

```
type state_type is (s0,s1,s2,s3,s4,s5,s6,s7,s8);
signal pst,nst: state_type:=s0;
signal disp_data_temp1:std_logic_vector(15 downto 0):=
"1010010000000000";
signal disp_data_temp2:std_logic_vector(15 downto 0):=
"1001000000000000";
signal add_mux: std_logic_vector(2 downto 0):="000";
signal data_mux,data_ctrl: std_logic_vector(1 downto 0):="00";
begin
p0:process(m_clk)
variable cnt:integer:=0;
begin
if(rising_edge(m_clk)) then
if (cnt=2) then
clk_sig<= not clk_sig;
cnt:=0;
else
cnt := cnt + 1;
end if;
end if;
end process;
p1:process(clk_sig)
variable cnt:integer:=0;
begin
if rising_edge(clk_sig) then
if (cnt=8) then
load<='0';
cnt:=cnt+1;
elsif (cnt=20) then
load<='1';
cnt:=cnt+1;
elsif (cnt=28) then
load<='0';
cnt:=cnt+1;
elsif (cnt=50) then
load<='1';
cnt:=1;
else
cnt:=cnt+1;
end if;
end if;
end process;
clk_sig1<=(clk_sig and load);
p2:process(clk_sig1)
begin
if(rising_edge(clk_sig1)) then
pst<=nst;
end if;
end process;
p3:process(pst,load)
begin
case pst is
when s0=> if (load='0') then
add_mux<="000";
data_shift<='0';
nst<=s0;
else
if (data_ctrl="10") then
data_ctrl<="01";
else
data_ctrl<=data_ctrl+1;
```

```
end if;
add_mux<="000";
data_shift<='0';
nst<=s1;
end if;
when s1=>
add_mux<="001";
data_mux<="00";
nst<=s2;
when s2=>
add_mux<="010";
nst<=s3;
when s3=>
add_mux<="011";
data_mux<="01";
nst<=s4;
when s4=>
add_mux<="100";
nst<=s5;
when s5=>
add_mux<="101";
data_mux<="10";
nst<=s6;
when s6=>
add_mux<="110";
nst<=s7;
when s7=>
add_mux<="111";
data_mux<="11";
nst<=s8;
when s8=>
if (data_ctrl="10") then
data_shift<='1';
else
data_shift<='0';
end if;
add_mux<="000";
nst<=s0;
when others=> add_mux<="000";
nst<=s0;
end case;
end process;
p4:process(add_mux)
begin
case add_mux is
when "000"=> dec_addr<=x"00";
when "001"=> dec_addr<=x"05";
when "010"=> dec_addr<=x"00";
when "011"=> dec_addr<=x"0A";
when "100"=> dec_addr<=x"00";
when "101"=> dec_addr<=x"50";
when "110"=> dec_addr<=x"00";
when "111"=> dec_addr<=x"A0";
when others => dec_addr<=x"00";
end case;
end process;
p5:process(data_mux,data_ctrl,disp_data_temp1,disp_data_temp2)
begin
if (data_ctrl="01") then
case data_mux is
when "00"=> disp_data<=disp_data_temp1(15 downto 12);
```

```
when "01"=> disp_data<=disp_data_temp1(11 downto 8);
when "10"=> disp_data<=disp_data_temp1(7 downto 4);
when "11"=> disp_data<=disp_data_temp1(3 downto 0);
when others => disp_data<="0000";
end case;
elsif (data_ctrl="10") then
case data_mux is
when "00"=> disp_data<=disp_data_temp2(15 downto 12);
when "01"=> disp_data<=disp_data_temp2(11 downto 8);
when "10"=> disp_data<=disp_data_temp2(7 downto 4);
when "11"=> disp_data<=disp_data_temp2(3 downto 0);
when others => disp_data<="0000";
end case;
end if;
end process;
p6:process(data_shift)
begin
if rising_edge(data_shift) then
disp_data_temp1<=(disp_data_temp1(3 downto 0) & disp_data_temp1(15
downto 4));
disp_data_temp2<=(disp_data_temp2(3 downto 0) & disp_data_temp2(15
downto 4));
end if;
end process;
end behavior;
```

6.7 Escalator, Dice Game and Model Train Controller Designs

In this section, we discuss some real-world and day-to-day life examples to understand system design concepts more clearly. The working mechanism and interfacing logic of an escalator, a dice game, and model train controllers are discussed below. Some design examples related to the above designs to perform the given specific tasks are also given.

6.7.1 Escalator Controller Design

We will understand the escalator's working principles first before proceeding to designing a model escalator controller. An escalator is a conveyor transport device with a pair of rotating chain loops that pull a series of stairs in a constant cycle to carry people between two floors. An escalator works almost in a way that a conveyor belt does and, in most cases, the moving stairs are actually on a belt that rotates around a set of gears at a certain fixed speed. The gears tend to be large and typically sit just below the steps. They are electrically powered, and as they turn, the steps move [39,72]. In practice, the stairs themselves are just grooved metal that lies flat as it travels down the back side, beneath the floor and back around again. At the top of the machine, housed in the truss, is an electric motor that runs the four primary gears all models have, and these four gears include two drive gears on either side at the top and two return gears on either side at the bottom. Chains loop around the gears and run down each side, and they are connected to each step and help each make their way up or down at a speed that is set by the motor, often through an electronic control panel. The way the steps flatten out at the tops and bottoms has to do with how each step is constructed as a unit [39,72]. A motion sensor might be connected at either end and is used to regulate the speed of the motor powered by electric motors. The maximum angle of inclination of an

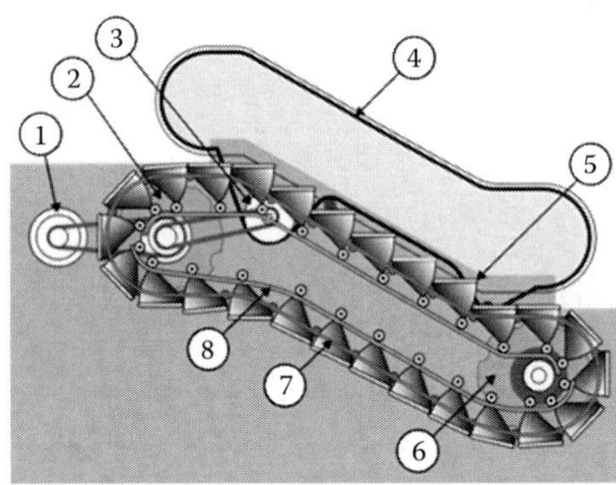

FIGURE 6.18
Photograph and design layout of a single escalator unit: 1. electric motor, 2. drive gear, 3. handrail drive, 4. handrail, 5. step, 6. return wheel, 7. chain guide, and 8. inner rail. (Adapted from http://www.newworldencyclopedia.org/entry/Escalator.)

escalator to the horizontal floor level is 30° and could be typically up to maximum of 60°. The appearance and design layout of an escalator are shown in Figure 6.18. The use and working mechanism of the main parts of the escalator system are briefed below:

i. The top and bottom landing platforms house the curved sections of the tracks, as well as the gears and motors that drive the stairs. The top platform contains the motor assembly and the main drive gear, while the bottom holds the step return idler sprockets. In addition, the platforms contain a floor plate and a comb plate. The floor plate provides a place for the passengers to stand before they step onto the moving stairs. This plate is flush with the finished floor and is either hinged or removable to allow easy access to the machinery below. The comb plate is the piece between the stationary floor plate and the moving step.
ii. The truss is a hollow metal structure that bridges the lower and upper landings. It is composed of two side sections joined together with cross braces across the bottom and just below the top. The ends of the truss are attached to the top and bottom landing platforms via steel or concrete supports. The truss carries all the straight track sections connecting the upper and lower sections [39,72].
iii. The track system is built into the truss to guide the step chain, which continuously pulls the steps from the bottom platform and back to the top in an endless loop. There are actually two tracks: one for the front wheels of the steps and another for the back wheels of the steps. The relative positions of these tracks cause the steps to form a staircase as they move out from under the comb plate. Along the straight section of the truss, the tracks are at their maximum distance apart. This configuration forces the back of one step to be at a 90° angle relative to the step behind it. This right angle bends the steps into a stair shape.
iv. The steps themselves are solid, one-piece, die-cast aluminum. Rubber mats may be affixed to their surfaces to reduce slippage, and yellow demarcation lines may be added to clearly indicate their edges. The leading and trailing edges of each step are cleared with comb-like protrusions that mesh with the comb plates on the top and bottom platforms. The steps are linked by a continuous metal chain so they form a closed loop, with each step able to bend in relation to its neighbors.

v. The railing provides a convenient handhold for passengers while they are riding the escalator. It is constructed of four distinct sections. At the centre of the railing is a slider, also known as a glider ply, which is a layer of a cotton or synthetic textile. The purpose of the slider layer is to allow the railing to move smoothly along its track. The next layer, known as the tension member, consists of either steel cable or flat steel tape. It provides the handrail with the necessary tensile strength and flexibility. Finally, the outer layer, the only part that passengers actually see, is the rubber cover, which is a blend of synthetic polymers and rubber. This cover is designed to resist degradation from environmental conditions, mechanical wear and tear, and human vandalism [72].

Now, we will discuss a simple design of an escalator control system. Assume that all the electro-mechanical assembly needed for an escalator is available to us and we need to develop only the controller to operate it with some specific conditions. Therefore, the electronic control section begins now. We will keep one IR Tx-Rx pair just before the first step of the escalator, that is, at the entrance, to detect the entry of a person. Let us assume that at any instant of time, 20 steps exist in the escalator. Therefore, 20 IR Tx-Rx pairs have been kept at the steps-rail-guide along the direction of the escalator so as to detect the person standing on the step. When one or more people are standing on the step, the respective IR Tx-Rx pair goes logic low. That means one pair or more than one pair would go logic low. Once any IR Tx-Rx kept at the step-rail-guide goes low after getting logic low at the entrance IR Tx-Rx pair, the system has to start driving the motors. The motor driving has to be continued until getting logic high from all the step-rail-guide IR LEDs as well as the entry IR LED.

There is a direction display which is constructed using green LEDs for showing the direction of the escalator; in our design, it will display the upward direction. The LEDs are arranged row-wise; hence, we need a separate 8 bits to control them to illustrate the moving-forward light display. A red LED has to glow while the motors are in the turned-off condition. A beep has to be produced using a buzzer for a few seconds before starting to drive the escalator. All these control and display logics have been considered and a state diagram is designed, as shown in Figure 6.19, where the state transition from S_0 to S_1 occurs whenever the start button is on. At state S_1, the buzzer is enabled and produces a beep as specified for some duration. To give a gap, the next operation is designed from state S_5 onwards. This means that after triggering the buzzer at S_1, we disable it at S_5; therefore, it produces the beep for some notable duration. Then, the subsequent operations

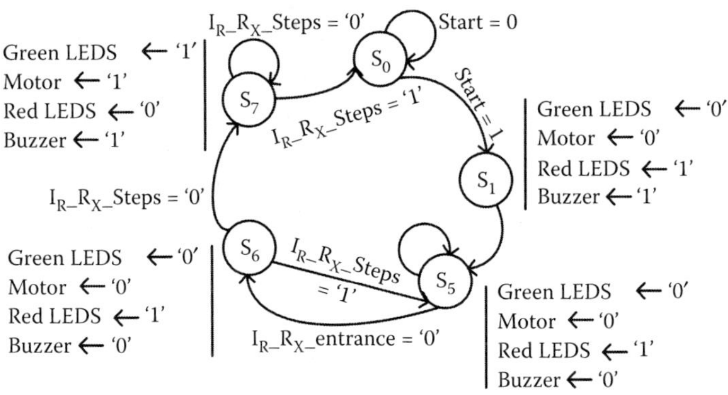

FIGURE 6.19
State diagram of an escalator controller.

are continued as specified at different states with respect to the IR-Rx inputs. In a similar way, any controller can be designed to control the operation of the escalator as required.

DESIGN EXAMPLE 6.16

Develop a digital system to turn on and off the escalator motor based on the number of persons entering and leaving the escalator. Assume that there are two IR sensors: one for entry detection and another for exit detection. A counter has to increment for every detection of person entry and subtract the number of people leaving so as to decide if the escalator motor has to be turned on or off.

Solution

```
library ieee;
use ieee.std_logic_1164.all;
entity escalator_case is
port ( x,clk : in  std_logic;
y: in std_logic;
z : out  std_logic);
end escalator_case;
architecture behavioral of escalator_case is
signal cnt: integer:=0;
signal cnt1, cnt2: integer;
signal s: integer:=0;
begin
process(clk)
begin
if rising_edge(clk) then
case (s) is
when (0)=>
if (x='1' and y='1') then
s<=0;
elsif ( x='0' and y='1') then
s<=1;
elsif( x='0' and y='0' ) then
s<=2;
elsif( x='1' and y='0') then
s<=3;
end if;
when (1) =>
if (x='0' and y='1') then
s<=1;
elsif (x='1' and y='1') then
s<=0;
cnt<= cnt +1;
end if;
when (2) =>
if (x='0' and y='0') then
s<=2;
elsif (x='1' and y='1') then
s<=0;
cnt<= cnt;
end if;
when (3) =>
if (x='1' and y='0') then
s<=2;
elsif (x='1' and y='1') then
s<=0;
cnt<= cnt -1;
end if;
```

```
when others =>
end case;
cnt1<=cnt;
end if;
if (cnt/=0) then
z<='1';
else
z<='0';
end if;
end process;
end behavioral;
```

DESIGN EXAMPLE 6.17

Develop a digital system to control escalator operations and display all the indications, as mentioned in Figure 6.19.

Solution

```
library ieee;
use ieee.std_logic_1164.all;
entity escalator is
port ( ir_rx : in  std_logic;
ir_steps : in  std_logic_vector (3 downto 0);
clk: in  std_logic;
motor1,motor2 : out  std_logic;
led : out  std_logic_vector (3 downto 0));
end escalator;
architecture behavioral of escalator is
signal clk_1hz : std_logic:='0';
signal en,check : std_logic;
signal state: integer range 0 to 1;
begin
check<= ir_steps(3) and ir_steps(2) and ir_steps(1) and ir_steps(0);
p0: process(clk)
begin
if(rising_edge(clk)) then
if(ir_rx='1' and check='1') then
en<='0';
motor1<='0';
motor2<='0';
end if;
if(ir_rx='0' and check='1') then
en<='1';
motor1<='0';
motor2<='0';
end if;
if(ir_rx='1' and check='0') then
en<='1';
motor1<='1';
motor2<='1';
end if;
if(ir_rx='0' and check='0') then
en<='1';
motor1<='1';
motor2<='1';
end if;
end if;
end process;
p1: process(clk)
variable cnt: integer range 0 to 1;
```

```
begin
if(rising_edge(clk)) then
if(cnt=1) then
clk_1hz<= not clk_1hz;
cnt:= 0;
else
cnt:= cnt+1;
end if;
end if;
end process;
p2: process(clk_1hz,en)
variable led1 : std_logic_vector(3 downto 0);
begin
if(en='0') then
led<="0000";
elsif(en='1' and ir_rx='0') then
led<="1111";
elsif(en='1' and ir_rx='1') then
if(rising_edge(clk_1hz)) then
case state is
when 0 =>
led1:="1000";
state<=1;
when 1 =>
led1:=led1(0) & led1(3 downto 1);
state<=1;
end case;
led<=led1;
end if;
end if;
end process;
end behavioral;
```

6.7.2 Dice Game Controller Design

Dice games are some of the most popular games and are simple to play with family members and friends. Dice games are versatile, as they can be played with any number of players, on the floor or table or in the car and with people of any age. Dice games teach numbers and counting to children, which increases their ability to remember and mentally add numbers. Dice games introduce kids to strategic thinking and planning. Dice games teach social skills such as taking turns and winning and losing gracefully. A dice game box is portable, as we can keep it in the glove box, handbag, pocket, or backpack. Dice are inexpensive and readily available in stores. The digital system design to play a dice game is an interesting one, which will display a random number from 1 to 9 on a seven-segment display or more than 9 using multiple seven-segment displays. Digital dice are an alternative electronic gadget that can be used to replace the traditional dice we used to play with, as shown in Figure 6.20. In this section, we discuss the strategy that we need to follow while playing as well as designing the digital dice game gadget. The electronic digital dice game can be designed with the help of seven-segment displays. In our design, we consider only two players; each player gets his or her turn to roll the dice, and the player who reaches the target score, for example, 25, fastest is declared the winner. Both players have one push button which freezes with a number when pressed, showing it on the respective seven-segment display. Each player has to submit the score by pressing the submit-score button. The scores of both the players and the status of who leads up to that turn are displayed on two seven-segment displays [73].

FIGURE 6.20
Photographs of different dice.

We will build the electronic dice in such a way that the pulse generator and the pulse counter will be enabled when we press the player's push button. Since the pulse generator works at the rate of the onboard clock frequency, the pulse counter will start counting very rapidly; as a result, binary bits in the counter will have quick succession. Hence, the corresponding seven-segment display looks as if all seven LEDs are permanently lit up. As soon as the player releases the push button, the pulse counter stops counting, and it is impossible to say in advance at what number it will stop, as it simply changes too fast. The counter stops in a certain combination and we can see the corresponding decimal number on the seven-segment display. This operation is equal/alternative to rolling an actual die. In actual die rolling, the number-printed square will be rotated manually. After leaving manual rotation, the square's rotation is slowed down and then stopped. Now, any one number appears exactly or close to the needle position, which is the score of the player. In digital dice, rolling is performed by the binary counter that randomly rolls a number from 1 to 9 or more based on the number of seven-segment displays we use, and immediately stops when the switch is released, unlike the conventional mechanical rolling system.

With this knowledge, now we start designing a digital dice game controller. A schematic diagram of the dice game controller is shown in Figure 6.21, which consists of several submodules, namely a clock divider, pulse generator, pulse counter, game turn counter, score-computing module, and data routing controller. Other than these internal modules,

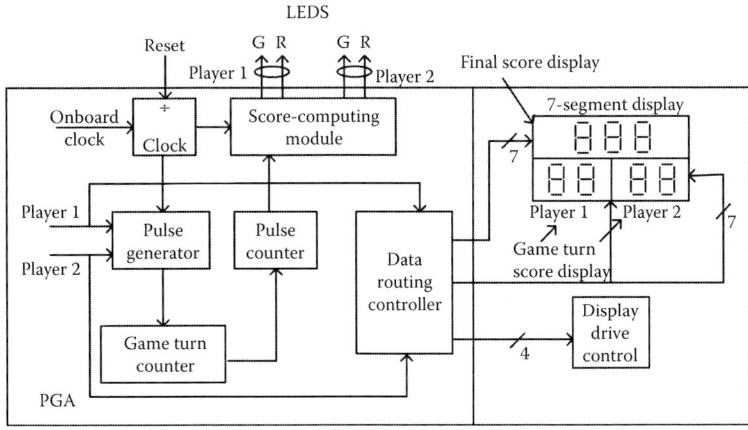

FIGURE 6.21
Top-level schematic diagram of a dice game controller.

a display driver circuit and seven-segment display modules are externally used. The main operation and dice game strategies are as follows:

i. The clock divider divides the onboard clock frequency into the necessary sub-clocks and applies it in various internal modules.
ii. When the player presses the button, the pulse generator is enabled and subsequently gives pulses to the pulse counter, which counts at a high speed.
iii. The data routing controller passes the counter value into the appropriate seven-segment display unit.
iv. Once the player releases the push button, the counter abruptly stops and retains the value, which is score of the player at that turn.
v. The same methodology has to be followed by the other player.
vi. The game turn counter increments its value by "1" for every cycle of game playing.
vii. During all the cycles, the score is stored in the score computing module. This module adds the score differences of subsequent cycles if the value is greater than the previous score; otherwise, it subtracts. For example, if the cycle game scores of player-1 are {9, 13, 6, 2, 23}, the actual score is $(9 + 4 - 7 - 4 + 21 = 23)$. In the same way, the scores will be computed in the score-computing module for all the players with the cycle scores of prefixed numbers of game turns.
viii. Finally, based on the value of the score computing module, the decision will be made as to the winner, runner-up or draw. These decisions will be shown by green (winner), red (runner-up) and both green (draw) LEDs, as shown in Figure 6.21.
ix. The one who gets the higher score is declared the winner.

In this way, a digital dice game controller can be built.

DESIGN EXAMPLE 6.18

Develop a digital system to have an electronics gadget perform as electronic digital dice. Assume that a random number has to be displayed on a seven-segment display. The digital rolling has to happen as long as a switch gets pressed; once it is released, the decimal value of the counter has to be displayed. A reset input will reset the entire circuit to its initial state.

Solution

```
library ieee;
use ieee.std_logic_1164.all;
use ieee.std_logic_unsigned.all;
entity dice is
port(roll:in std_logic;
seg7:out std_logic_vector(6 downto 0);
clk:in std_logic;
nrst:in std_logic);
end entity;
architecture mark0 of dice is
signal rng1,rng2,rng3:std_logic_vector(7 downto 0);
signal x,y,z:std_logic;
signal lcg_sig:std_logic_vector(2 downto 0);
signal cnt:std_logic_vector(1 downto 0);
begin
```

```
x <= (((rng1(0)) xor rng1(2)) xor rng1(3)) xor rng1(4);
y <= (((rng2(1)) xor rng2(3)) xnor rng2(5)) xor rng2(6);
z <= (((rng3(0)) xnor rng3(2)) xor rng3(6)) xnor rng3(7);
p0:process(clk,nrst,roll)
begin
if(nrst='0') then
rng1 <= "00100101";
elsif(roll='1') then
if(falling_edge(clk)) then
rng1 <= x & rng1(7 downto 1);
end if;
end if;
end process;
p1:process(x,nrst,roll)
begin
if(nrst='0') then
rng2 <= "10000101";
else
if(x='1') then
if(roll='1') then
if(falling_edge(clk)) then
rng2 <= y & rng2(7 downto 1);
end if;
end if;
end if;
end if;
end process;
p2:process(y,nrst,roll)
begin
if(nrst='0') then
rng3 <= "10010101";
else
if(y='1') then
if(roll='1') then
if(falling_edge(clk)) then
rng3 <= z & rng3(7 downto 1);
end if;
end if;
end if;
end if;
end process;
p3:process(clk)
begin
if(nrst='0') then
lcg_sig <= "000";
cnt <= "00";
else
if(rising_edge(clk)) then
cnt <= cnt + 1;
case(cnt) is
when "00" =>
lcg_sig <= rng1(7 downto 5);
when "01" =>
lcg_sig <= rng2(3 downto 1);
when "11" =>
lcg_sig <= rng3(2 downto 0);
when others =>
lcg_sig <= (rng1(4 downto 2) xor rng2(4 downto 2)) xor rng3(4 downto
2);
end case;
end if;
```

```
end if;
end process;
seg7 <= "1011000" when(lcg_sig="110") else
"1111001" when(lcg_sig="000") else
"0101100" when(lcg_sig="010") else
"0110000" when(lcg_sig="011") else
"1011000" when(lcg_sig="101") else
"0010010" when(lcg_sig="111") else
"0000010" when(lcg_sig="100") else
"1011000" when(lcg_sig="001") else
"ZZZZZZZ";
end architecture;
```

DESIGN EXAMPLE 6.19

Design and develop a digital system to play a dice game similar to the design shown in Figure 6.21. Display all the results on seven-segment displays and LEDs.

Solution

```
library ieee;
use ieee.std_logic_1164.all;
use ieee.std_logic_unsigned.all;
entity temp1 is
port (    clk,reset:in std_logic;
play,score : in  std_logic;
seg7,score_seg1,score_seg2,score_seg3,score_seg4 : out std_logic_vector
(6 downto 0);
win, draw: out std_logic:='0');
end entity;
architecture behavioral of temp1 is
-- score deciders --
signal score1,score2,score3,score4 : std_logic_vector (3 downto 0);
-- player control --
signal player: std_logic_vector(1 downto 0);
signal ref1,ref2,ref3,ref4,s1,s2,s3,s4 : std_logic_vector(3 downto 0);
signal en:std_logic;
signal round:integer range 0 to 10;
signal in_rst:std_logic;
--------------------
-- dice control --
signal num : std_logic_vector(3 downto 0);
signal rn : std_logic_vector(6 downto 0);
signal x: std_logic;
------------------
begin
-- random number generato (dice) --
x <= (((rn(0)) xor rn(2)) xor rn(3)) xor rn(4);
p0:process(clk,reset,play,in_rst)
begin
if(reset='1' or in_rst='1') then
rn <= "0100101";
num<= "0000";
elsif (reset='0' and play='1') then
if(rising_edge(clk)) then
rn <= x & rn(6 downto 1);
num<= rn(6 downto 3);
end if;
end if;
end process;
----------------------------------
```

```
-- switch between players --
p1:process(play,reset,in_rst)
begin
if(reset='1' or in_rst='1') then
player <= "11";
en <= '0';
round <= 0;
in_rst <= '0';
else
if(rising_edge(play)) then
case player is
when "00" => player <= "01";
when "01" => player <= "10";
when "10" => player <= "11";
when "11" => player <= "00";
en <= '1';
if(round=10) then
in_rst <= '1';
round <= 0;
else
in_rst <= '0';
end if;
round <= round + 1;
when others=> player <= "00";
end case;
end if;
end if;
end process;
-----------------------------
-- decide the scores --
p2:process(player,reset,num,play,in_rst)
begin
if(reset='1' or in_rst='1') then
s1 <= "0000";
s2 <= "0000";
s3 <= "0000";
s4 <= "0000";
ref1 <= "1111";
ref2 <= "1111";
ref3 <= "1111";
ref4 <= "1111";
elsif(en='1') then
case(player) is
when "00"=>
if(play='0') then
if(num>ref1) then
s1<=s1+1;
else
ref1<=num;
end if;
end if;
when "01"=>
if(play='0') then
if(num>ref2) then
s2<=s2+1;
else
ref2<=num;
end if;
end if;
when "10"=>
if(play='0') then
```

```
if(num>ref3) then
s3 <= s3+1;
else
ref3 <= num;
end if;
end if;
when "11"=>
if(play='0') then
if(num>ref4) then
s4 <= s4+1;
else
ref4 <= num;
end if;
end if;
when others=>
s1 <= "0000";
s2 <= "0000";
ref1 <= "1111";
ref2 <= "1111";
ref3 <= "1111";
ref4 <= "1111";
end case;
else
s1 <= "0000";
s2 <= "0000";
ref1 <= "1111";
ref2 <= "1111";
ref3 <= "1111";
ref4 <= "1111";
end if;
end process;
-------------------------
-- score output --
p3:process(score,reset,in_rst)
begin
if(reset='1') then
score1 <= "0000";
score2 <= "0000";
score3 <= "0000";
score4 <= "0000";
else
if(in_rst='1') then
if((s1>s2 and s1>s3 and s1>s4) or (s2>s1 and s2>s3 and s2>s4) or
(s3>s1 and s3>s2 and s3>s4) or (s4>s1 and s4>s2 and s4>s3)) then
win <= '1'; draw <= '0';
else
win <= '0'; draw <= '1';
end if;
else
if(score='1') then
score1<=s1;
score2<=s2;
score3<=s3;
score4<=s4;
end if;
end if;
end if;
end process;
-------------------
-- display unit --
seg7 <= "1111110" when (num="0000") else
```

```
"0110000" when (num="0001")else
"1101101" when (num="0010")else
"1111001" when (num="0011")else
"0110011" when (num="0100")else
"1011011" when (num="0101")else
"1011111" when (num="0110")else
"1110000" when (num="0111")else
"1111111" when (num="1000")else
"1111011" when (num="1001")else
"1110111" when (num="1010")else
"0011111" when (num="1011")else
"1001110" when (num="1100")else
"0111101" when (num="1101")else
"1001111" when (num="1110")else
"1000111" when (num="1111")else
"0000000";
score_seg1 <= "1111110" when (score1="0000") else
"0110000" when (score1="0001")else
"1101101" when (score1="0010")else
"1111001" when (score1="0011")else
"1110111" when (score1="0100")else
"1011011" when (score1="0101")else
"1011111" when (score1="0110")else
"1110000" when (score1="0111")else
"1111111" when (score1="1000")else
"1111011" when (score1="1001")else
"0000000";
score_seg2 <= "1111110" when (score2="0000") else
"0110000" when (score2="0001")else
"1101101" when (score2="0010")else
"1111001" when (score2="0011")else
"0110011" when (score2="0100")else
"1011011" when (score2="0101")else
"1011111" when (score2="0110")else
"1110000" when (score2="0111")else
"1111111" when (score2="1000")else
"1111011" when (score2="1001")else
"0000000";
score_seg3 <= "1111110" when (score3="0000") else
"0110000" when (score3="0001")else
"1101101" when (score3="0010")else
"1111001" when (score3="0011")else
"0110011" when (score3="0100")else
"1011011" when (score3="0101")else
"1011111" when (score3="0110")else
"1110000" when (score3="0111")else
"1111111" when (score3="1000")else
"1111011" when (score3="1001")else
"0000000";
score_seg4 <= "1111110" when (score4="0000") else
"0110000" when (score4="0001")else
"1101101" when (score4="0010")else
"1111001" when (score4="0011")else
"0110011" when (score4="0100")else
"1011011" when (score4="0101")else
"1011111" when (score4="0110")else
"1110000" when (score4="0111")else
"1111111" when (score4="1000")else
"1111011" when (score4="1001")else
"0000000";
end architecture;
```

6.7.3 Electronic Model Train Controller Design

Various algorithms are currently being found to control electronic trains to avoid train accidents, indicate the exact arrival time of a train at a particular station, display the running status of a train, alert passengers to arrival at their station, and so on. All these algorithms are being tested with electronic model trains at the beginning level. A simple electronic model train with routes is shown in Figure 6.22. We will discuss and understand a simple train control system first. Assume a control system has to be designed to avoid railway accidents happening at unattended railway gates. For this design, we need to install two IR Tx-Rx pairs: one pair of transmitter and receiver is at the up side and another pair is at the down side, both at a level above a person with exact line-of-sight alignment. One sensor is called the foreside sensor and the other is the aft-side sensor [74]. Sensor activation time is adjusted by calculating the time taken at a certain speed for at least one compartment of the standard minimum size of the train to cross. Typically, we consider 3 seconds in our design. IR Tx-Rx pairs are deployed at 1 kilometer ahead on both sides of a railway gate. When the foreside sensor gets activated, the gate motor is turned on in one direction and produces a beep alarm sound, turns off the green LEDs and turns on the red LEDs. The gate is closed and stays closed until the train crosses the gate and reaches the aft-side sensor pair. When the aft-side receiver gets activated, the motor turns in the opposite direction, the gate opens and the motor stops. The buzzer sound will stop and the green LEDs will turn back on once the motor starts rotating back to open the gate. The buzzer sound alerts the vehicles which are about to cross the railway track at the gate.

The same logic can also be applied for track change-over switch controls. Consider a situation where a local train and an express train are traveling one by one on a train track. The express train has to be allowed to travel on the same track to go to the front platform, and the local train has to be switched to the other track to go to another platform. To accomplish this task, a pair of IR sensors is placed at a point 1 kilometer away from the track change-over switch. Since we know which train is arriving the railway station, if the sensor placed along that direction gets activated, the control unit immediately checks the ID (local or express) of the approaching train. If it is the local train, then it activates the change-over switch. The switching operation is performed using a high torque stepper motor. Assume that within a certain delay or with some suitable mechanism, the train passing can be confirmed. Once the train passes the switching point, the track change-over will be switched back to its original position to allow the next express train without any interruption. This concept of track switching can be applied on both sides of the railway junction.

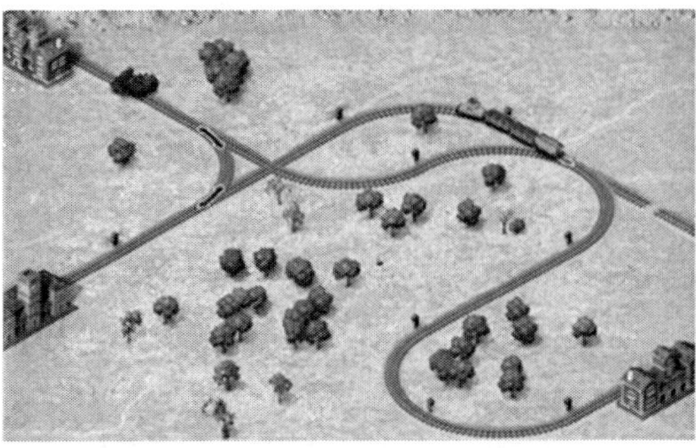

FIGURE 6.22
Illustrations of an electronic model train and its routes.

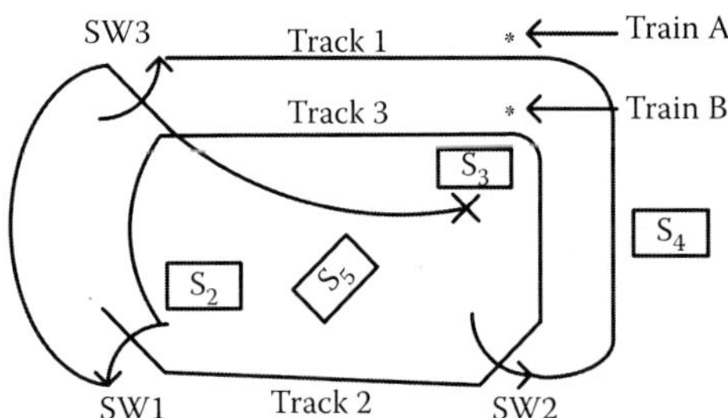

FIGURE 6.23
Track layout for an electronic model training control.

Now, we design a model train controller to run two trains (trains A and B) on the tracks shown in Figure 6.23, where the controller has to run two trains without any collisions. Five sensors (`S1`, `S2`, … , `S5`) are placed close to the track to approximate the current position of the trains. Three switches (`SW1`, `SW2`, `SW3`) are used to change the track so as to route the train to the decided direction/track. The model train is operated on the track using battery-powered DC motors; hence, we need two controls to operate the train in the clockwise or counterclockwise direction on the train track. Therefore, four controls, `DA0` and `DA1` for train A, and `DB0` and `DB1` for train B, respectively, are required. Based on the sensor inputs, that is, the current position of the trains, the controller needs to take the decision to open or close the track change switches. Thus, the inputs are clock, reset, `S1`, `S2`, `S3`, `S4`, and `S5` and the outputs are `SW1`, `SW2`, `SW3`, `DA0`, `DA1`, `DB0`, and `DB1`. We can plan any logic to operate the trains on the track without any collisions. With this information, we can design a controller to perform this operation. More details about this type of design can be found in [74].

DESIGN EXAMPLE 6.20

Develop a digital system to control the electronic model train in the given situation, as discussed in this section above. The input and output details are as given in Figure 6.23 [74].

Solution

```
library ieee;
use ieee.std_logic_1164.all;
use ieee.std_logic_arith.all;
use ieee.std_logic_unsigned.all;
entity tcontrol is
port (reset, clock, sensor1, sensor2,
sensor3, sensor4, sensor5      :in   std_logic;
switch1, switch2, switch3      :out  std_logic;
dira, dirb              :out std_logic_vector(1 downto 0));
end tcontrol;
architecture a of tcontrol is
type state_type is(about, ain, bin, astop, bstop);
signal state: state_type;
signal sensor12, sensor13, sensor24: std_logic_vector(1 downto 0);
begin
sensor12 <= sensor1 & sensor2;
sensor13 <= sensor1 & sensor3;
sensor24 <= sensor2 & sensor4;
switch3 <= '0';
p0:process (reset, clock)
begin
```

```
if reset = '1' then
state <= about;
elsif clock'event and clock = '1' then
case state is
when about =>
case sensor12 is
when "00" => state <= about;
when "01" => state <= bin;
when "10" => state <= ain;
when "11" => state <= ain;
when others => state <=about;
end case;
when ain =>
case sensor24 is
when "00" => state <= ain;
when "01" => state <= about;
when "10" => state <= bstop;
when "11" => state <= about;
when others => state <=about;
end case;
when bin =>
case sensor13 is
when "00" => state <= bin;
when "01" => state <= about;
when "10" => state <= astop;
when "11" => state <= about;
when others => state <=about;
end case;
when astop =>
if sensor3 = '1' then
state <= ain;
else
state <= astop;
end if;
when bstop =>
if sensor4 = '1' then
state <= bin;
else
state <= bstop;
end if;
end case;
end if;
end process;
with state select
switch1 <= '0'   when about,
'0'              when ain,
'1'              when bin,
'1'              when astop,
'0'              when bstop;
with state select
switch2 <= '0'   when about,
'0'              when ain,
'1'              when bin,
'1'              when astop,
'0'              when bstop;
with state select
dira  <= "01"    when about,
"01"             when ain,
"01"             when bin,
"00"             when astop,
"01"             when bstop;
```

```
with state select
dirb <= "01"       when about,
"01"               when ain,
"01"               when bin,
"01"               when astop,
"00"               when bstop;
end a;
```

DESIGN EXAMPLE 6.21

Assume that bidirectional data communication between train and station is established through an optical (laser) beam. As the train leaves a station, the communication will be handed over to the next station. The current station measures the strength and angle of the received communication signal, and once the value approaches the minimum value, it signals the next station to receive the handover of the data link. Develop a digital system and necessary setups to demonstrate this concept. Display the values on an LCD.

Solution

```
library ieee;
use ieee.std_logic_1164.all;
use ieee.std_logic_arith.all;
use ieee.std_logic_unsigned.all;
entity kings_train is
port ( m_clk: in  std_logic:='0';
track_data_in:in std_logic_vector(7 downto 0):="00000001";
train_step_out : inout  std_logic_vector (3 downto 0):="1000";
lcd_control_out:out std_logic_vector(2 downto 0):="000";
handover_laser_control:out std_logic:='0';
stepper_motor_out:out std_logic_vector(3 downto 0):="0000";
lcd_data_out: out std_logic_vector(7 downto 0):="00000000");
end kings_train;
architecture Behavioral of kings_train is
signal
train_step_out_clk_sig,disp_clk_sig,data_read_clk_sig,handover_laser_
control_temp:std_logic:='0';
signal ps,ns:integer range 0 to 79:=0;
signal track_data_temp,voltage_hd,voltage_ld,distance_hd,distance_ld,
angle_hd,angle_ld,junction_ch1,junction_ch2,junction_ch3:std_logic_
vector(7 downto 0):="00000000";
begin
handover_laser_control<=handover_laser_control_temp;
p1:process(data_read_clk_sig)
begin
if rising_edge(data_read_clk_sig) then
track_data_temp<=track_data_in;
case track_data_temp is
when "00000000"=> voltage_hd<=X"30";voltage_ld<=X"30";
distance_hd<=X"30";distance_ld<=X"30";
angle_hd<=X"30";angle_ld<=X"30";
junction_ch1<=X"54";junction_ch2<=X"50";junction_ch3<=X"4C";
stepper_motor_out<="0000";
handover_laser_control_temp<='0';
when "00000001"=> voltage_hd<=X"30";voltage_ld<=X"31";
distance_hd<=X"30";distance_ld<=X"31";
angle_hd<=X"30";angle_ld<=X"31";
junction_ch1<=X"54";junction_ch2<=X"50";junction_ch3<=X"4C";
stepper_motor_out<="1000";
handover_laser_control_temp<='0';
when "00000011"=>  voltage_hd<=X"30";voltage_ld<=X"32";
distance_hd<=X"30";distance_ld<=X"32";
angle_hd<=X"30";angle_ld<=X"32";
```

```
junction_ch1<=X"54";junction_ch2<=X"50";junction_ch3<=X"4C";
stepper_motor_out<="0100";
handover_laser_control_temp<='0';
when "00000111"=> voltage_hd<=X"30";voltage_ld<=X"33";
distance_hd<=X"30";distance_ld<=X"33";
angle_hd<=X"30";angle_ld<=X"33";
junction_ch1<=X"54";junction_ch2<=X"50";junction_ch3<=X"4C";
stepper_motor_out<="0010";
handover_laser_control_temp<='0';
when "00001111"=> voltage_hd<=X"30";voltage_ld<=X"34";
distance_hd<=X"30";distance_ld<=X"34";
angle_hd<=X"30";angle_ld<=X"34";
junction_ch1<=X"54";junction_ch2<=X"50";junction_ch3<=X"4C";
stepper_motor_out<="0001";
handover_laser_control_temp<='0';
when "00011111"=> voltage_hd<=X"30";voltage_ld<=X"35";
distance_hd<=X"30";distance_ld<=X"35";
angle_hd<=X"30";angle_ld<=X"35";
junction_ch1<=X"54";junction_ch2<=X"50";junction_ch3<=X"4C";
stepper_motor_out<="1000";
handover_laser_control_temp<='0';
when "00111111"=> voltage_hd<=X"30";voltage_ld<=X"36";
distance_hd<=X"30";distance_ld<=X"36";
angle_hd<=X"30";angle_ld<=X"36";
junction_ch1<=X"54";junction_ch2<=X"50";junction_ch3<=X"4C";
stepper_motor_out<="0100";
handover_laser_control_temp<='0';
when "01111111"=> voltage_hd<=X"30";voltage_ld<=X"37";
distance_hd<=X"30";distance_ld<=X"37";
angle_hd<=X"30";angle_ld<=X"37";
junction_ch1<=X"54";junction_ch2<=X"50";junction_ch3<=X"4C";
stepper_motor_out<="0010";
handover_laser_control_temp<='0';
when "11111111"=> voltage_hd<=X"30";voltage_ld<=X"38";
distance_hd<=X"30";distance_ld<=X"38";
angle_hd<=X"30";angle_ld<=X"38";
junction_ch1<=X"54";junction_ch2<=X"4E";junction_ch3<=X"4A";
stepper_motor_out<="0001";
handover_laser_control_temp<='1';
when others=> null;
end case;
end if;
end process;
p2:process(m_clk)
variable disp_state_change_clk_cnt:integer:=0;
begin
if rising_edge(m_clk) then
if(disp_state_change_clk_cnt=200)then
disp_clk_sig<= not(disp_clk_sig);
disp_state_change_clk_cnt:=0;
else
disp_state_change_clk_cnt:= disp_state_change_clk_cnt + 1;
end if;
end if;
end process;
p3:process(disp_clk_sig)
begin
if rising_edge(disp_clk_sig) then
ps<=ns;
end if;
end process;
```

```
--------------------------------LCD Command and Data --------------------
p4:process(ps)
begin
case ps is
when 0=> lcd_data_out<=X"38";lcd_control_out<="100"; ns<=ps+1;--38
when 1|3|5|7|9|43=> lcd_control_out<="000";ns<=ps+1;
when 2=> lcd_data_out<=X"0C";lcd_control_out<="100"; ns<=ps+1;--0C
when 4=> lcd_data_out<=X"01";lcd_control_out<="100"; ns<=ps+1;--01
when 6=> lcd_data_out<=X"0E";lcd_control_out<="100"; ns<=ps+1;--0E
when 8=> lcd_data_out<=X"80";lcd_control_out<="100"; ns<=ps+1;--80
when 10=> lcd_data_out<=X"56";lcd_control_out<="110"; ns<=ps+1;--V
when 11|13|15|17|19|21|23|25|27|29|31|33|35|37|39|41|45|47|49|51|53|55|
57|59|61|63|65|67|69|71|73|75=> lcd_control_out<="010";ns<=ps+1;
when 12=> lcd_data_out<=X"6F";lcd_control_out<="110"; ns<=ps+1;--o
when 14=> lcd_data_out<=X"6C";lcd_control_out<="110"; ns<=ps+1;--l
when 16=> lcd_data_out<=X"74";lcd_control_out<="110"; ns<=ps+1;--t
when 18=> lcd_data_out<=X"3A";lcd_control_out<="110"; ns<=ps+1;--:
when 20=> lcd_data_out<=voltage_hd;lcd_control_out<="110";
ns<=ps+1;--d1
when 22=> lcd_data_out<=voltage_ld;lcd_control_out<="110";
ns<=ps+1;--d0
when 24=> lcd_data_out<=X"20";lcd_control_out<="110"; ns<=ps+1;--
when 26=> lcd_data_out<=X"20";lcd_control_out<="110"; ns<=ps+1;--
when 28=> lcd_data_out<=X"44";lcd_control_out<="110"; ns<=ps+1;--D
when 30=> lcd_data_out<=X"69";lcd_control_out<="110"; ns<=ps+1;--i
when 32=> lcd_data_out<=X"73";lcd_control_out<="110"; ns<=ps+1;--s
when 34=> lcd_data_out<=X"74";lcd_control_out<="110"; ns<=ps+1;--t
when 36=> lcd_data_out<=X"3A";lcd_control_out<="110"; ns<=ps+1;--:
when 38=> lcd_data_out<=distance_hd;lcd_control_out<="110";
ns<=ps+1;--d1
when 40=> lcd_data_out<=distance_ld;lcd_control_out<="110";
ns<=ps+1;--d0
when 42=> lcd_data_out<=X"C0";lcd_control_out<="100"; ns<=ps+1;--C0
when 44=> lcd_data_out<=X"41";lcd_control_out<="110"; ns<=ps+1;--A
when 46=> lcd_data_out<=X"6E";lcd_control_out<="110"; ns<=ps+1;--n
when 48=> lcd_data_out<=X"67";lcd_control_out<="110"; ns<=ps+1;--g
when 50=> lcd_data_out<=X"6C";lcd_control_out<="110"; ns<=ps+1;--l
when 52=> lcd_data_out<=X"3A";lcd_control_out<="110"; ns<=ps+1;--:
when 54=> lcd_data_out<=angle_hd;lcd_control_out<="110";
ns<=ps+1;--d1
when 56=> lcd_data_out<=angle_ld;lcd_control_out<="110";
ns<=ps+1;--d0
when 58=> lcd_data_out<=X"20";lcd_control_out<="110"; ns<=ps+1;--
when 60=> lcd_data_out<=X"20";lcd_control_out<="110"; ns<=ps+1;--
when 62=> lcd_data_out<=X"53";lcd_control_out<="110"; ns<=ps+1;--S
when 64=> lcd_data_out<=X"74";lcd_control_out<="110"; ns<=ps+1;--t
when 66=> lcd_data_out<=X"61";lcd_control_out<="110"; ns<=ps+1;--a
when 68=> lcd_data_out<=X"3A";lcd_control_out<="110"; ns<=ps+1;--:
when 70=> lcd_data_out<=junction_ch1;lcd_control_out<="110";
ns<=ps+1;--d3
when 72=> lcd_data_out<=junction_ch2;lcd_control_out<="110";
ns<=ps+1;--d2
when 74=> lcd_data_out<=junction_ch3;lcd_control_out<="110";
ns<=ps+1;--d1
when 76=>data_read_clk_sig<='1';ns<=ps+1;
when 77=>data_read_clk_sig<='1';ns<=ps+1;
when 78=>data_read_clk_sig<='0';ns<=ps+1;
when 79=> ns<=8;
when others=>null;
end case;
end process;
```

```
p5:process(m_clk)
variable train_cnt:integer :=0;
begin
if rising_edge(m_clk) then
if(train_cnt=20000) then
train_step_out_clk_sig<=not train_step_out_clk_sig;
train_cnt:=0;
else
train_cnt:=train_cnt+1;
end if;
end if;
end process;
p6:process(train_step_out_clk_sig)
begin
if rising_edge(train_step_out_clk_sig) then
train_step_out<=(train_step_out(0) & train_step_out(3 downto 1));
end if;
end process;
end Behavioral;
```

6.8 Algorithmic State Machine Charts

Sequential system design using ASM charts is an alternative approach for many applications ranging from simple to highly complicated real-time systems. There are a number of methods to design and realize FSM control units. Simple control units can be designed using the state diagram, also called the state graph, and state/excitation table methods. Complex control units may be designed using algorithmic charts just like flowcharts that are being used in software programs. In this and subsequent sections, we will discuss ASM chart blocks and system design using them. As we have seen in the previous sections, FSM-based digital systems typically consist of control and data path processing modules. The control path is implemented using state machines which can be realized via state graphs. As the control path of the digital system becomes complex, it becomes increasingly difficult to design the control path using the state diagram (graph) technique. The ASM charts technique becomes useful and handy in designing complex and algorithmic circuits. A simple digital system as a combination of the Moore and Mealy FSM models is shown in Figure 6.24a and b. In Figure 6.24a, the digital system is partitioned into two parts as a controller (ASM) to generate the control signals and a controlled architecture or data path processor to process the input and state data. The complete operation of the data path processor is controlled by the ASM controller, which decides the subsequent control operation of the data processor by looking at its status, as specified in the design, for every clock input. In Figure 6.24b, the system produces two outputs; one depends on the input and present state, while the other depends only on the present state; hence, this design has both Moore and Mealy FSM modules [75].

The ASM chart differs from the ordinary flowchart on some specific rules that must be followed for constructing the chart. When those rules are followed, the ASM chart is directly equivalent to the hardware realisation. The ASM chart or block sets are divided into three blocks, namely the (i) state box, (ii) decision box, and (iii) conditional output box,

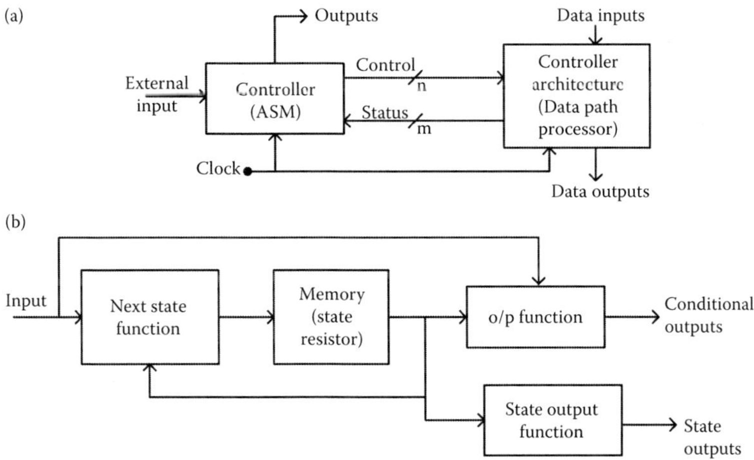

FIGURE 6.24
Top-level partitioning of a digital system (a) and simple complex system, that is, combination of Moore and Mealy models (b).

which are shown in Figure 6.25. We will briefly discuss the function of every ASM block below:

i. The state box represents the state of the system. The state box contains a state name, optional state code and output lists. The optional state code may be placed outside the box at the top.
ii. The decision box is usually a diamond-shaped one and always has true (logic 1) and false (logic 0) branches. The condition placed in the decision box must be a Boolean expression that is evaluated to determine which branch to take to proceed further.
iii. The conditional output box contains a conditional output list. The conditional outputs depend on both the state of the system and the inputs [39,75].

The ASM chart is constructed using SM blocks, which contain exactly one state box together with decision boxes and conditional output boxes associated with that state. An SM block has exactly one entrance path and one or more exit paths. Each SM block describes the machine operation during the time that the machine is in that state. A data path through an SM block from entrance to exit is referred to as a data link path. We will understand the function of an ASM chart before proceeding to actual design of ASM for

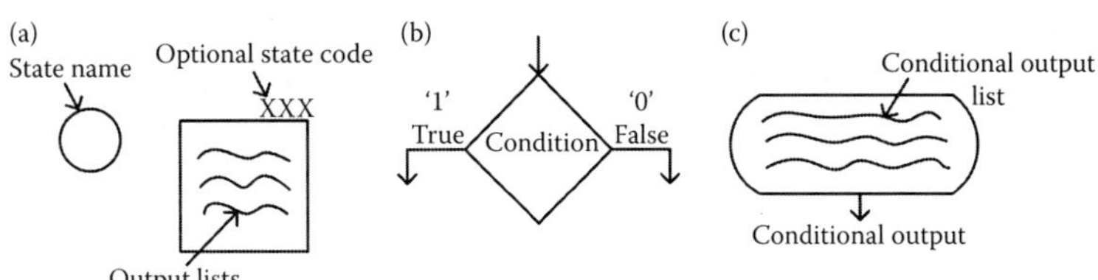

FIGURE 6.25
ASM blocks: state box (a), decision box (b), and conditional output box (c).

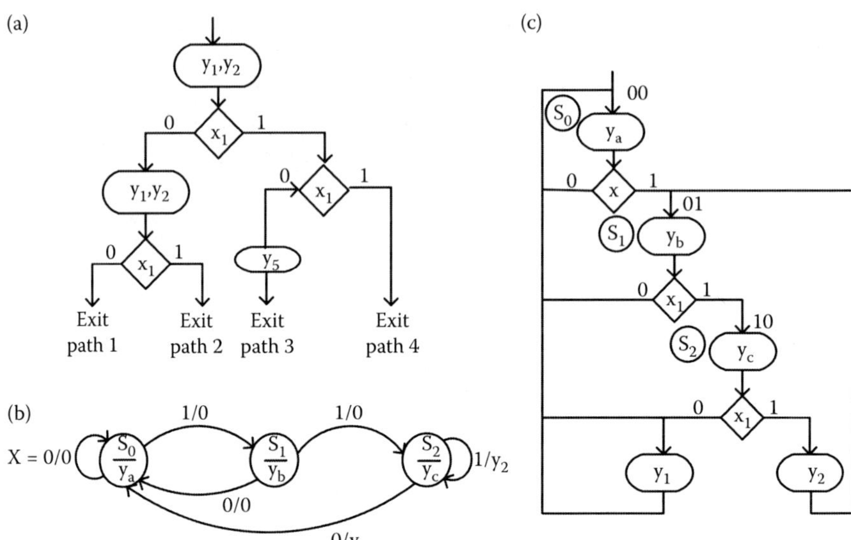

FIGURE 6.26
A simple ASM chart with one entry and four exit paths (a), a complex state diagram (b), and corresponding ASM chart (c).

some real-world applications. Let us consider an ASM chart as shown in Figure 6.26a, through which we will understand the flow of the data sequence. The operation of that ASM chart can be given as

i. When the state S_0 is reached, the outputs y_1 and y_2 become logic 1.
ii. Then, if the inputs x_1 and x_2 are equal to logic 0, the outputs y_3 and y_4 are logic 1 and the machine goes to the next state via exit path 1.
iii. If the inputs x_1 = "0" and x_2 = "1", the outputs y_3 and y_4 are logic 1 and the machine moves to next state via exit path 2.
iv. On the other hand, if x_1 = "1" and x_3="0", the output y_5 is logic 1 and the machine goes to the next state via exit path 3.
v. if x_1 = "1" and x_3 = "1", the machine simply goes to the next state via exit path 4.

Now, we will understand how a state diagram can be converted to an ASM chart. Let us consider a state diagram, as shown in Figure 6.26b, which contains both Mealy and Moore model outputs. The Mealy output depends on the present state as well as input; thus, outputs y_1 and y_2 should be conditional outputs, and they have to reside in the conditional box. Outputs y_a, y_b and y_c are Moore model outputs, and they do not depend on x; thus, they should reside in the state box. The input x can be either "0" or "1"; thus, it should reside in the decision box [39,75].

The ASM chart of the state diagram, that is, Figure 6.26b, is shown in Figure 6.26c. As shown there, after reaching state S_0, the output y_a becomes logic 1, then state S_0 will move to state S_1 only when conditional input x = "1" and output y_b will become logic 1. If x = "0", state S_0 remains unchanged. On the other hand, if x = "0", state S_1 will transition back to state S_0. State S_1 will change to state S_2 only when conditional input x = "1" and output y_c become logic 1. Then, if x = "1" and output y_2 is logic 1, then state S_2 remains unchanged.

If x = "0" and output y_1 is logic 1, the state goes to state S_0. With this knowledge, we proceed to design some digital systems using the ASM chart in the next section.

DESIGN EXAMPLE 6.22

Design and sketch an ASM chart for a digital controller which performs the below-given tasks. Assume that there are two JK-FFs (J_1K_1 and J_2K_2) and a 4-bit binary counter ($d_3d_2d_1d_0$) at the data processor. This processor is controlled by a controller whose operations on various input conditions are as follows:

i. An input "start" begins the operation by clearing the counter and FFs.
ii. At each subsequent clock pulse, the counter is incremented by 1 until the operation stops.
iii. If $d_2 = 0$, then J_1K_1 is set to 0 and the count continues.
iv. If $d_2 = 1$, then J_1K_1 is set to 1, then, if $d_3 = 0$, the count continues, but if $d_3 = 1$, JK_2 is set to 1 on the next clock pulse and the system stops counting.
v. If Start = 0, the system remains in the initial state, but if Start = 1, the operation cycle repeats.

Solution

We will sketch the ASM chart based on the above working conditions, which is shown in Figure 6.27 and consists of three state, three conditional and three output blocks. As long as the start input is 0, the process is at the idle state. Once the start becomes 1, it enters into state 1 by clearing the counter and FF J_2K_2. As long as the third bit of the counter (d_2) is 0, the process clears FF J_1K_1 and stays at state 1; however, the counter is incremented by 1 for every clock pulse. Once d_2 becomes 1, the process clears FF J_1K_1

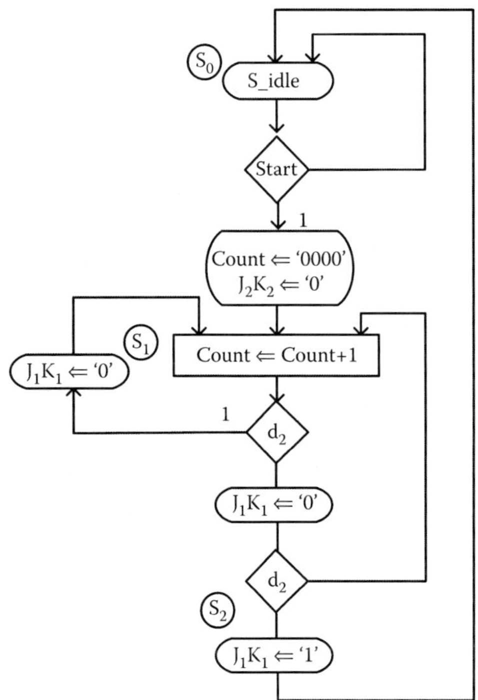

FIGURE 6.27
ASM chart for Design Example 6.22.

and moves into state 2 once the value of d_3 is 1; otherwise, the process moves to state 1 and the counter continues to be incremented.

Since the counter is getting incremented, the process can clear/set FF J_1K_1 as shown in Figure 6.27. When $d_2 = d_3 = 1$, the process reaches state 2, clears FF J_2K_2 and moves to the idle state. Then, this operation continues based on the value of the start input.

6.9 Algorithmic State Machine-Based Digital System Design

In general, in digital system design, we need to deal with two main units, namely the control unit and processor unit. In the processor unit, the data, that is, discrete elements of information, need to be manipulated by performing arithmetic, logic, shift, and other similar data-processing operations. These operations are implemented with digital hardware components such as adders, decoders, multiplexers, counters, and shift registers [75,76]. The control unit provides command signals to the data processor unit to coordinate and execute the various operations in it in predefined order to accomplish the desired data-processing tasks. Thus, any design of a digital system can be divided into two parts, one for performing data-processing operations and the other for control circuits that determine the sequence in which the various data manipulations have to happen. The general relationship between the control unit and the data processor unit in a digital system is shown in Figure 6.28. The control unit decides the next command signal based on the status of the external input and feedback that comes from the processor unit. Thus, the control and processor units are in a closed loop to decide further action by the control unit. The data inputs that need to be processed are directly applied to the processor unit for data manipulation in accordance with the control unit comments.

The control logic that generates the signals for sequencing the operations in the processor unit is an ASM which generates the control commands as functions of the external inputs, feedback status signal and current state of the FSM. In a given state, the outputs of the control unit are the inputs to the processor unit and determine the operations that it will execute. Depending on status conditions and other external inputs, the ASM goes to its next state to initiate other operations. The digital circuits

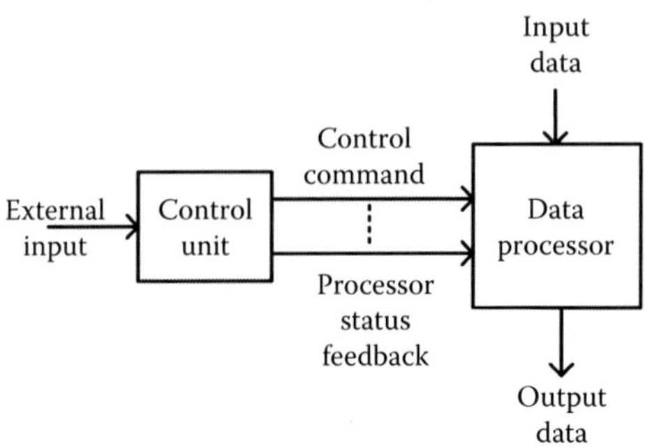

FIGURE 6.28
ASM-based digital system design entity.

that act as the control unit provide a time sequence of signals for initiating the operations in the processor unit and also determine the next state of the control unit. The control sequence and data processor tasks of a digital system are specified by means of a hardware algorithm. A hardware algorithm is a procedure for solving a problem with a given piece of equipment [75,76]. The most challenging and creative part of digital design is the formulation of hardware algorithms for achieving the required application. A flowchart is a convenient way to specify the sequence of procedural steps and decision paths for an algorithm. A flowchart for a hardware algorithm translates the verbal instructions to an information diagram that enumerates the sequence of operations together with the conditions necessary for their execution. We see some design examples with the ASM control unit below.

DESIGN EXAMPLE 6.23

Sketch an ASM chart for a processor that generates the Fibonacci series. As we know, the Fibonacci number is the sum of the two preceding numbers. The simplest is the series 1, 1, 2, 3, 5, 8, etc [76,77].

Solution

The ASM chart required for generating the Fibonacci series number is shown in Figure 6.29. VHDL code corresponding to this ASM chart is given below.

```
library ieee;
use ieee.std_logic_1164.all;
use ieee.numeric_std.all;
entity fibo_desi is
port(clk, reset: in std_logic:='0';
start: in std_logic:='0';
```

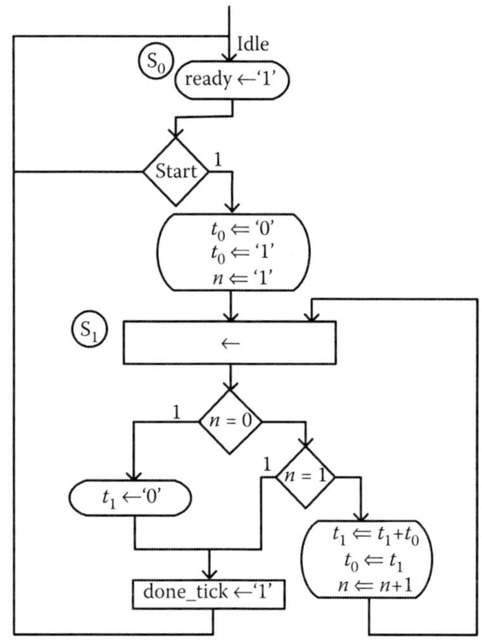

FIGURE 6.29
ASM chart for Design Example 6.23.

```
i: in std_logic_vector(4 downto 0):="00000";
ready, done_tick: out std_logic:='0';
f: out std_logic_vector(19 downto 0):="00000000000000000000");
end fibo_desi;
architecture behav of fibo_desi is
type
state_type is (idle,op,done);
signal state_reg, state_next: state_type;
signal t0_reg, t0_next, t1_reg, t1_next: unsigned(19 downto 0);
signal n_reg, n_next: unsigned(4 downto 0);
begin
p0:process(clk,reset)
begin
if reset='1' then
state_reg
<= idle;
t0_reg <= (others=>'0');
t1_reg <= (others=>'0');
n_reg<= (others=>'0');
elsif
(clk'event and clk='1') then
state_reg<= state_next;
t0_reg <= t0_next;
t1_reg <= t1_next;
n_reg<= n_next;
end if;
end process;
p1:process(state_reg,n_reg,t0_reg,t1_reg,start,i,n_next)
begin
ready <='0';
done_tick <= '0';
state_next<= state_reg;
t0_next <= t0_reg;
t1_next <= t1_reg;
n_next<= n_reg;
case state_reg is
when idle =>ready <= '1';
if start='1' then
t0_next <= (others=>'0');
t1_next <= (0=>'1', others=>'0');
n_next<= unsigned(i);
state_next<= op;
end if;
when op =>
if n_reg=0 then
t1_next <= (others=>'0');
state_next<= done;
elsif n_reg=1 then
state_next<= done;
else
t1_next <= t1_reg + t0_reg;
t0_next <= t1_reg;
n_next<= n_reg-1;
end if;
when done =>done_tick<= '1';
state_next<= idle;
end case;
end process;
f <=std_logic_vector(t1_reg);
end behav;
```

DESIGN EXAMPLE 6.24

Design and sketch an ASM chart of a digital system that functions as an arbiter unit [76,77].

Solution

The ASM chart design corresponding to Design Example 6.24 is shown in Figure 6.30, and its corresponding VHDL code is given below.

```
library ieee;
use ieee.std_logic_1164.all;
entity arbi_desi is
port ( clock, resetn      :in      std_logic:='0' ;
r              : in     std_logic_vector(1 to 3):="000" ;
g              : out    std_logic_vector(1 to 3):="000") ;
end arbi_desi ;
architecture behavior of arbi_desi is
type state_type is (idle, gnt1, gnt2, gnt3) ;
signal y : state_type ;
begin
process ( resetn, clock )
begin
if resetn = '0' then y <= idle ;
elsif (clock'event and clock = '1') then
case y is
when idle =>
if r(1) = '1' then y <= gnt1 ;
elsif r(2) = '1' then y <= gnt2 ;
elsif r(3) = '1' then y <= gnt3 ;
else y <= idle ;
end if ;
when gnt1 =>
if r(1) = '1' then y <= gnt1 ;
else y <= idle ;
end if ;
when gnt2 =>
if r(2) = '1' then y <= gnt2 ;
else y <= idle ;
end if ;
```

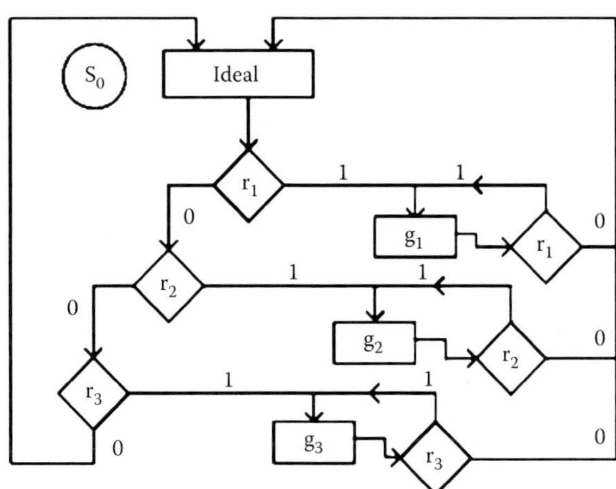

FIGURE 6.30
ASM chart for Design Example 6.24.

```
when gnt3 =>
if r(3) = '1' then y <= gnt3 ;
else y <= idle ;
end if ;
end case ;
end if ;
end process ;
g(1) <= '1' when y = gnt1 else '0' ;
g(2) <= '1' when y = gnt2 else '0' ;
g(3) <= '1' when y = gnt3 else '0' ;
end behavior ;
```

LABORATORY EXERCISES

1. Design a digital system using VHDL to produce a spike pulse whenever an edge is detected at the input using Moore and Mealy FSM models. The width of the input is [1:0].
2. Develop a digital system using VHDL to realize various arithmetic operations like addition, subtraction, multiplication and division using the 9's and 10's complement methods.
3. Develop a vending machine controller with the following interfacing and provisions:
   a. The amount and item selection inputs have to be given by DIP and matrix keypad switches, respectively.
   b. The menu and status display has to be in a GLCD.
   c. Sixteen different items of different rates are available inside the vending machine.
   d. There must be a separate button for displaying the menu with the prices of the items on a GLCD.
   e. Provisions must exist for canceling the request at any point in time.
   f. Provisions must be made for selecting more than one item.
4. Assume that there is a narrow bridge only one vehicle can pass through. There is a wide space at both ends of the bridge, so vehicles can wait in the queue to pass through the bridge. Develop a digital system to automatically control this common-way traffic scheme with proper displays and a signalling system for the drivers.
5. Design an FSM and develop a digital system as a gadget to play the dice game as per the following rules: (i) roll a die; (ii) after the first roll, the sum of scores must be 7 or 11; (iii) once this sum is obtained, the player can begin the game by recording the sum of scores of all rolls; (iv) on any roll, if the sum score is less than the previous score, it is a loss; hence, the present score has to be subtracted from the total score acquired; otherwise, the player wins and adds the scores; and (v) after some turns, the one who has the higher score is declared the winner of the game. Assume there are two needles to indicate the scores on the dice.
6. Design and develop a digital system to have an electronics gadget play a blackjack rummy game. Display all the necessary scores on a GLCD.
7. Develop a digital system to realize the automatic car parking system. Assume that a garage has 10 floors with 50 parking spaces on each floor. The driver will park

the car at the entrance. Then, the digital system has to check the availability of the spaces, give the space details to the driver and operate all the necessary conveyers to properly lift and park the car in the space. When the driver enters the space details, the system has to bring the car back to the entrance.

8. Design an FSM and develop a digital system to find the direction and sequence of five rotating disks. The slotted disks have to be rotated within the covered space; hence, the player(s) cannot see the direction of rotation of their coplayers' disks. The rotating disk is made to block the IR link and allow it through its slots. Therefore, the digital system can identify the direction of rotation of the disk by looking at the input sequence pattern that comes from an individual disk. The more players rotating in any one direction is declared as winning. For example, in an attempt, three players are rotating their disk in clockwise direction while another two are in anti-clockwise. Now the three are winners and the two are runners for that particular attempt, thus, point 1 has to be assigned to winners and 0 to runners by the digital system. In this way, the game has to be continued up to the pre-specified number of attempts. The digital system has to declare the end result of all the five players once all the attempts are completed. The one or more who have obtained maximum score is declared as overall winner(s).
9. Design an ASM chart and digital system to function as an electronic gadget to play the Pallanguli game with four players one by one on a rotation basis. The playing methodology and rules of the Pallanguli game can be found at https://www.youtube.com/watch?v=VDMP-GDjwsA.
10. Design an ASM chart and develop a digital system to function as an electronic gadget to play the Paramapadham game with four players one by one on a rotation basis. The playing methodology and rules of the Paramapadham game can be found at https://www.youtube.com/watch?v=-q4dSDxxZlg.

7

Interfacing Digital Logic to the Real World: Sensors, Analog to Digital, and Digital to Analog

7.1 Introduction and Preview

Sensors, A/D converters and D/A converters are essential electronic devices for many real-world interfaces and applications. Many applications, including signal processing, image processing, industrial control systems, automobiles, and meteorology and security systems, could not be accomplished without real-world sensors and A/D and D/A converters. This chapter explains the sensor interfacing techniques, application circuits, signal processing modules, and data logging systems that are essential for different applications. The basic principles and fundamental concepts associated with A/D and D/A operations are briefed. Working principles and interfacing circuits for optical power, temperature, and strain measurements using different sensors are described, with design examples. The development of a UART in the FPGA, data reading/writing from a computer (MATLAB®) via the RS232 serial port, pin details of the DB9 connector, a voltage-level converter, real-time data logging/plotting, and so on are explained, with many VHDL and MATLAB design examples. Multichannel sensor interfacing, data logging, parallel data reading, and serial data fetching are described, with associated timing diagrams and digital designs.

Estimation of atmospheric turbulence strength (C_n^2) using the PAMELA model [39,61–64,78–80] is detailed, with meteorological parameter measurement techniques along with data writing/reading in/from Excel files. The significance of bidirectional A/D converters is listed, and one important application, OPD interfacing using the AD1674 core, is discussed. Wireless data transmission and reception using a popular encoder and decoder, respectively, are explained, with many control and wireless network applications. Establishing data links using IR and RF transmitters/receivers are also detailed. Applications and generation techniques, serial and parallel, of PRBSs are given, with design examples. The digital architecture required for generating a small-set Kasami sequence is presented. A popular application circuit for analog multiplexing and M-array PAM using it are detailed, with the necessary digital designs. The last section describes D/A interfacing, different waveform generations and low-power high-voltage D/A converter interfacing, followed by laboratory practices.

7.2 Basics of Signal Conditioning for Sensor Interfacing

Sense is an important factor that makes humans different from machines and other animals. To sense, we need sensor(s), which in our case are nose, ears, tongue, eyes, and so on. We can sense happiness, guilt, love, anger, suffocation, wind, cold, hot, taste, smell, and much more. Similarly, in electronics, numerous sensors are available to get the values of many physical parameters like temperature, light, humidity, obstacles, and gas. These sensors play an important role in electronics, since they make the system more intelligent, as it has the ability to make decisions on its own. Every sensor has its own property; for example, an LDR senses light, temperature sensor (LM35) senses temperature, infrared proximity sensor senses obstacles and humidity sensor (SHT11) senses humidity in the air. In most applications, it is necessary to convert an analog signal to accurate digital data and vice versa, as shown in Figure 7.1. For example, in an application where the FPGA is controlling the process, the analog signals coming from the sensors need to be converted into digital data so that they can be bumped into the FPGA. After the process takes place in the FPGA in digital form, the outputs of the FPGA need to be converted back to analog form to interact again with the analog world. Today, in a few sensors, like SHT11, the A/D converter is integrated with sensor chips. However, without using the A/D converter, it is impossible to interface sensors with the FPGA. Therefore, the conversion of A/D as well as D/A becomes significant in many real-world applications.

7.2.1 Analog to Digital Conversion

In this section, we briefly discuss the principles, characteristics, and limitations of the process of A/D conversion. The operation required to convert the voltage generated by the sensor to its equivalent digital data is performed by the A/D converter as a three-stage process: (i) sampling, (ii) quantization, and (iii) encoding. The first stage takes an instantaneous snapshot of the sensor voltage, in accordance with the sampling clock, and freezes it for the duration of the conversion process that is being performed by a sample-and-hold (S/H) circuit, which is located directly at the input of the A/D converter. The S/H briefly opens its aperture window, captures the input voltage on the rising edge of the sampling clock signal, closes its aperture, and holds that voltage level as a newly acquired sample. The sample of the S/H is updated on every rising edge of the sampling clock signal. The second stage assigns a nearest numerical value to the sample present at the output of the S/H circuit, as shown in Figure 7.2. This process, known as quantization, gives the nearest value out of many fixed quantization levels that cover the entire amplitude range.

The third stage is an encoding; that is, once the quantization level (closest discrete value) has been assigned by the quantizer, the encoder converts it to the equivalent binary data in

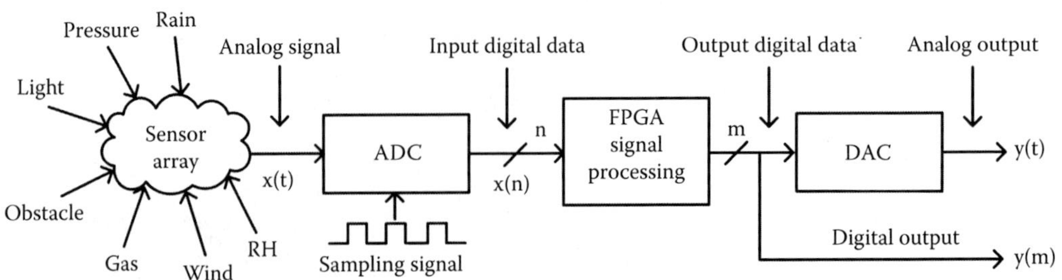

FIGURE 7.1
Basic illustration of sensor interfacing and data processing.

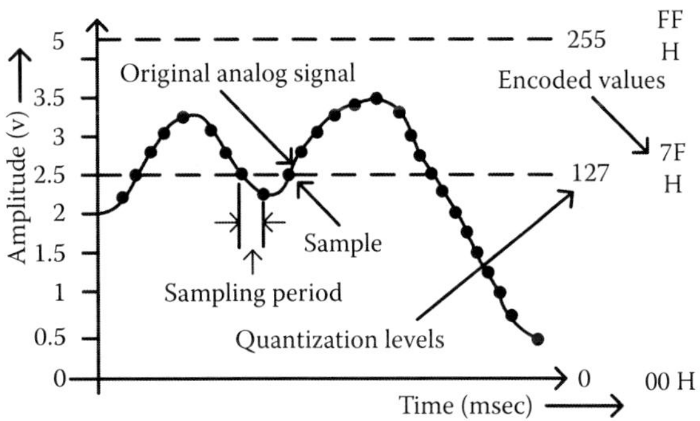

FIGURE 7.2
Illustration of basic process of A/D conversion.

n bits. Therefore, the number of the quantization level is directly proportional to 2^n. There are many advantages to working with digitized signals instead of working in the original analog form, such as (i) a large amount of intrinsic data reduction, (ii) the digital data won't normally be degraded, and (iii) digital data can be stored, transmitted and transformed into other number systems [39,81]. As long as the information remains in a digital form, it can be manipulated and processed using any digital system. This is not possible for analog signals, since it requires dedicated means to process each specific type of signal. The fidelity of processing the analog signal can be improved by converting the analog input to digital form, then processing it by any suitable means in the digital domain.

7.2.2 Digital to Analog Conversion

D/A conversion is a process in which the digital signals are converted into analog signals with a theoretically infinite number of voltage levels. Basically, D/A conversion is the opposite of A/D conversion. In most cases, the D/A converter is placed after an A/D converter; hence, the analog output signal is the processed version of the analog input signal. There are two popular types of D/A conversion, namely (i) binary-weighted resisters and (ii) R/2R ladders. Construction of a D/A converter using binary-weighted resistors is shown in Figure 7.3. The circuit diagram represents a 4-bit (D_0–D_3) D/A converter that yields 16 (2^4) combinations of analog outputs. If only D_0 is connected to logic 1, that is, +5 V, the current that flows through the feedback resistor (R_f) under the assumption that the bias current (I_B) is very low/negligible is (5 V/10 KΩ), that is, 0.5 mA. Thus, the output voltage (V_o) is −(1 kΩ × 0.5 mA), that is, −0.5 V, as shown in Figure 7.3. If D_0 and D_1 are connected to logic 1, then the current flow through R_f is (5 V[(1/10 KΩ) + (1/5 KΩ)]), that is, 1.5 mA. Thus, the output voltage is −(1 KΩ × 1.5 mA), that is, −1.5 V. Therefore, according to the binary weights of the inputs (D_o–D_3), the current will flow through R_f, which will be linearly converted into the output voltage. Thus, the circuit is basically working as a current-to-voltage converter [81].

The output voltage of this digital to analog convertor (DAC) will be at its maximum when all the switches are connected to logic 1, and it can be calculated as

$$V_0 = -5\,R_f\left[\frac{D_0}{10\,K\Omega} + \frac{D_1}{5\,K\Omega} + \frac{D_2}{2.5\,K\Omega} + \frac{D_3}{1.25\,K\Omega}\right] \tag{7.1}$$

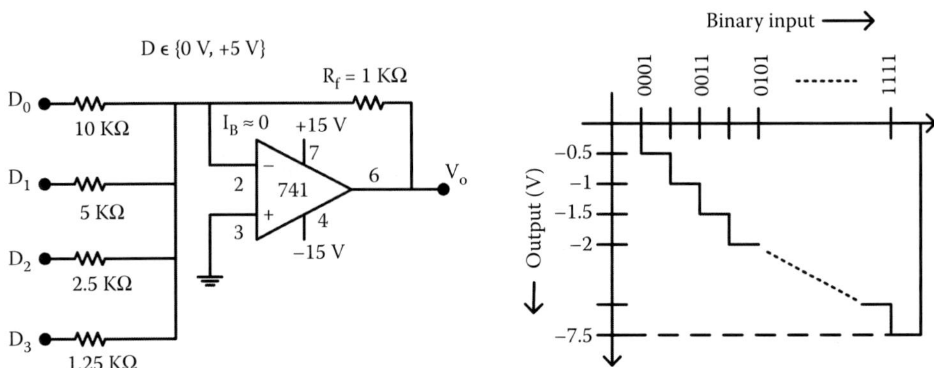

FIGURE 7.3
Working principles of D/A converter using binary weighted resistors.

The output voltages for different possible combinations (weights) of the binary data are shown in Figure 7.3. The output is a negative-going staircase waveform with 15 steps of −0.5 V each. In practice, the variations in the voltage level of logic 1 and the preciseness in the different values of the resistance, as in Equation 7.1, change the step size. Thus, the need for a large range of resistors with high precision is the primary limitation in this type of DACs.

The second type of D/A converter is shown in Figure 7.4a. The R/2R ladder circuit directly converts a parallel digital input (D_0–D_3) into an equivalent analog voltage. Each bit in the digital input adds its own weighted contribution to the analog output. The unique advantages of this type of D/A converter are (i) it is easily scalable to any desired number of bits; (ii) it needs only two values of resistors (R and 2R), which makes the design easy and accurate; (iii) its output impedance is equal to "R" regardless of the number of bits; and (iv) its simplicity for further interfacing and analog signal processing. Thevenin's theorem and the superposition theorem are required here to understand the working principles of the R/2R ladder D/A converter circuit. Thevenin's theorem says that if the circuit contains linear elements like voltage sources, current sources, and resistors, then we can cut the circuit at any point and replace everything on one side of the cut with a voltage source and a single-series resistor. The voltage source is the open-circuit voltage at the cut point, marked by "x" in Figure 7.4a, and the series resistor is the equivalent open-circuit resistance with all voltage sources shorted. The cut points, according to Thevenin's theorem, are shown in Figure 7.4a, through which now we calculate the output impedance of the R/2R ladder circuit, assuming that the digital inputs are connected to logic 0, that is, 0 V. Therefore, the first two 2R resistors on the left side of the first cut point in Figure 7.4a appear in parallel; hence, they can be replaced with an equivalent resistor (R_{eq}) as [81]

$$R_{eq} = \frac{R_1 R_2}{R_1 + R_2} = \frac{2R2R}{2R + 2R} = R \tag{7.2}$$

A resistor R at the right side of the first cut point and R_{eq}, that is, Equation 7.2, now becomes a series; therefore, R+R becomes parallel to the immediate right side resistor, that is, 2R. This process repeats itself each time as we simplify from left to right and, ultimately, the resistance value will be R. Thus, the output impedance of the R/2R resistor circuit is always equal to R, as shown in Figure 7.4b, regardless of the size of the circuit. This simplifies the design of analog filtering, amplification, and signal conditioning circuits that follow this type of D/A converter.

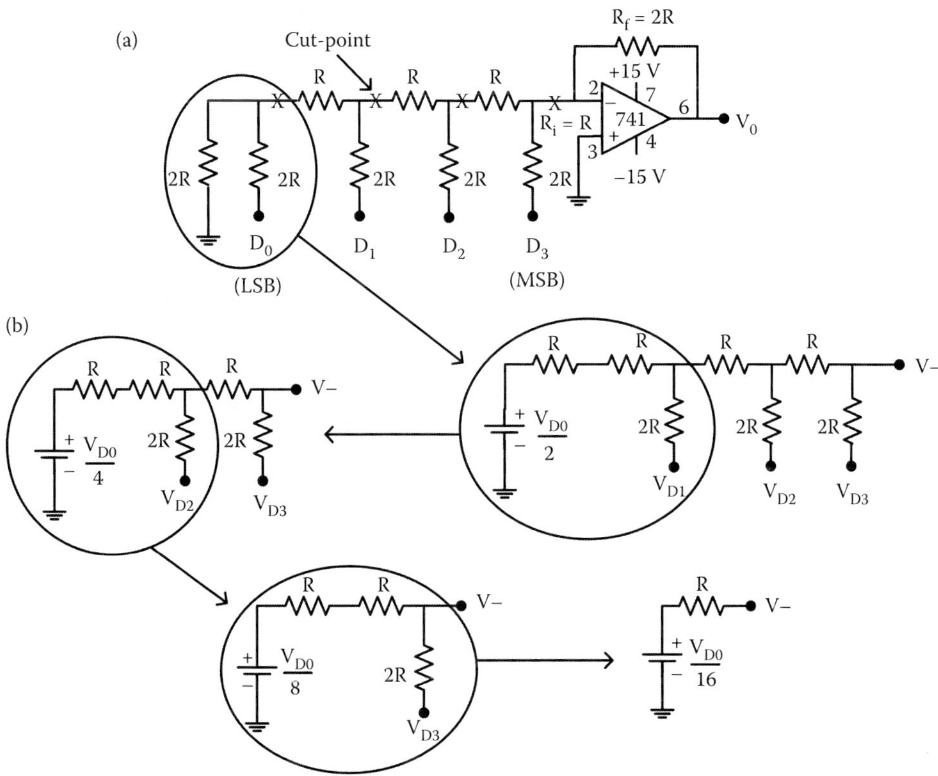

FIGURE 7.4
D/A converter using R/2R ladder (a) and illustration of Thevenin simplification (b).

Now, we calculate the output voltage (V−) of the R/2R ladder circuit for a given digital input. Whenever the digital inputs are connected to logic 1, a voltage ($V_{D0} - V_{D3}$) will typically be applied at the input terminals; hence, current flows through the R/2R ladder circuit. The superposition theorem says that the total current in any part of a linear circuit is equal to the algebraic sum of the currents produced by each source separately. Further, to evaluate the separate currents to be combined, replace all other voltage sources by short circuits and all other current sources by open circuits. We assume that the bit D_0 is logic 1, and D_1, D_2, and D_3 are logic 0; then, we start replacing the circuit by its Thevenin equivalence from the leftmost cut point. Figure 7.4b shows this simplification. Finally, the output voltage of the R/2R ladder circuit is equal to $V_{D0}/16$. That means that the voltage contribution from the bit D_0 is 1/16th of the logic high voltage level (V_{D0}). This analysis can be continued step by step for each possible binary input. In each bit stage, the voltage passing through the cut points contributes by a factor of 2. Therefore, the contribution of each bit to the output of the R/2R ladder circuit is a simple binary weighting function of each bit. Thus, the general form of the equation to calculate the output voltage of R/2R is $[(V_{D0}/16) + (V_{D1}/8) + (V_{D2}/4) + (V_{D3}/2)]$, which is actual input to the operational amplifier (op-amp) circuit. Then, the output voltage (V_o) of the op-amp is calculated by [81,82]

$$V_o = -\frac{R_f}{R} V_{Ref}\left[\frac{D_0}{16} + \frac{D_1}{8} + \frac{D_2}{4} + \frac{D_3}{2}\right] \tag{7.3}$$

where V_{Ref} is the voltage applied for logic 1, that is, $V_{D0} - V_{D3}$. The disadvantage of this type of D/A converter is only a need for more complicated analysis/simplification to understand, especially the high-precision R/2R ladder D/A converter designs.

DESIGN EXAMPLE 7.1

In an R/2R ladder DAC circuit, feedback is designed with 1-KΩ (connected between an inverting terminal and ground) and 7-KΩ (connected between an inverting terminal and output [V_o] of an operational amplifier) resistors. A four-stage R/2R ladder circuit is connected to the noninverting terminal. Find the V_o when the binary input is "1010".

Solution

In general, the input to the noninverting terminal is:

$$V_+ = \frac{1}{2^n}\left[2^{n-1}D_{n-1} + 2^{n-2}D_{n-2} \cdots 2^1D_1 + 2^0D_0\right]$$

For the binary input "1010", the above equation becomes

$$V_+ = \frac{1}{2^4}[2^3 \cdot 1 + 0 + 2^1 \cdot 1 + 0]$$

$$V_+ = \frac{10}{16}\,V$$

Gain of the noninverting amplifier is

$$\frac{V_o}{V_+} = \left(1 + \frac{R_f}{R}\right) = \left(1 + \frac{7\,K}{1\,K}\right)V + V_o = 5\,V$$

7.3 Principles of Sensor Interfacing and Measurement Techniques

In this section, the working principles and interfacing protocols of many sensors, namely optical power sensors, LM35, thermistors, thermocouples, and strain gauges, are explained. Measurement techniques for the physical parameters using those sensors are also explained. The A/D conversion of sensed/measured analog signals using the magnitude comparator and ADC0804 chip is detailed, along with few design examples.

7.3.1 Optical Power Measurement

There are several types of optical measurement sensors (photodetectors), like LDR, PN photodiodes, phototransistors, Avalanche photodiodes, thermopiles, pyroelectric detectors, and PIN photodiodes, which work based on similar basic principles that are briefed below. Optical energy is considered packets of light particles (photons). When a photon of sufficient energy enters the depletion region of a semiconductor diode, it may strike an atom to release an electron from the atomic structure. This creates a free electron and a hole. The electron is negatively charged, while the hole is positively charged. The electron and hole may remain free, or other electrons may combine with the hole to form a complete atom again in the crystal lattice. In this way, a hole-electron pair is generated. Further, it is also possible that the electron and hole may remain free and be pulled away from the depletion region by an external field that causes current flow through the diode, which is known as the photocurrent. The photodiode is operated under reverse bias, which keeps the depletion layer free of any carriers, and normally no current will flow. The hole and electron will then migrate in opposite directions under the action of the electric field across the intrinsic region, and a small current can be seen to flow. Operating diodes under reverse bias increase the sensitivity as they widen the depletion layer where the photo action occurs. In this way, increasing the reverse bias has the effect of increasing the active area of the photodiode. The amount of the current flow is proportional to the amount of light entering the intrinsic region [83].

Although most optical measurement devices display values in dBm or Watts, an optical power meter is only capable of measuring either the current or the voltage generated by a photodetector. For example, when interfacing with a photodiode, the quantity that must be measured is the current, and when interfacing with a thermopile or pyroelectric detector, voltage is the quantity the optical meter must measure. A standard optical power measurement circuit using a photodiode (FDS100) and op-amp (OP07) is shown in Figure 7.5. There are numerous techniques for measuring the current, but the only one which will yield good detectivity, signal-to-noise and accuracy is the semiconductor photodiode. The OP07 has very low input offset voltage (75 μV), which eliminates the need of external nulling. The OP07 also features low input bias current (±4 nA) and high open-loop gain (200 V/mV), which make it particularly useful for high-gain instrumentation applications. In Figure 7.5, one lead of the photodiode is tied to the ground and the other is kept at virtual ground by means of the negative input of the transimpedance amplifier (OP07). The transimpedance amplifier does not bias the photodiode with a voltage as the current starts to flow from the photodiode. The transimpedance circuit not only amplifies, but also converts the current produced by the photodiode to voltage that can be read at V_o. The resultant bias across the photodiode is kept at virtually zero volts, which is a condition that helps minimize dark current and noise and increase linearity and detectivity. Effectively, the transimpedance amplifier causes the photocurrent to flow through the feedback resistor, which creates a voltage of $V_o = iR$ at the output of the OP07. The current produced by the photodiode can be amplified using a transimpedance amplifier. Since the value of the precision feedback resistor (R) is known, the current can be calculated with very good accuracy and transformed into other forms using an appropriate signal-processing circuit to show it in the display device in the required units. The resistance values can be adjusted

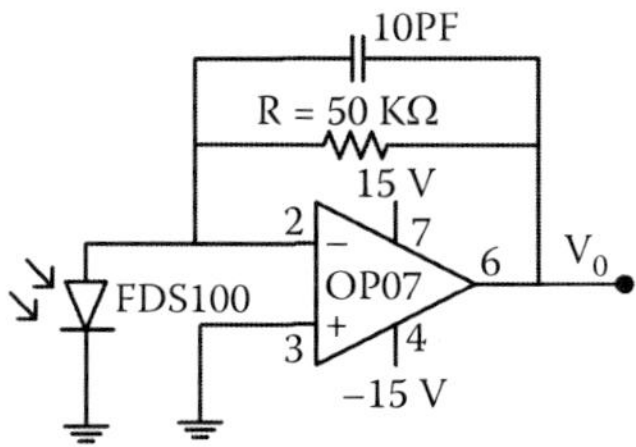

FIGURE 7.5
Optical power measurement-transimpedance amplifier.

to vary the sensitivity of the detector [39,83]. However, this measurement system has to be calibrated against standard master equipment, for example, PD300-IRGP/N7Z02402, to verify the measurement accuracy.

7.3.2 Temperature Measurement

There are many sensors commercially available for measuring temperature. Some of the more commonly used sensors are discussed below. The first is LM35, which has three pins: two for the power supply and one for the analog output, as shown in Figure 7.6. The analog output voltage is linearly proportional to the temperature. The sensing accuracy of LM35 is 10 mV/°C with an accuracy of 0.5°C. This means that at 20°C, the output voltage is 20 × 0.01 = 200 mV, and at 100°C, 100 × 0.01 = 1 V. The LM35 sensor can be powered by DC voltage in the range of 4–30 V. The operating range is −55°C to +150°C. The sensor output voltage can be amplified with sufficient gain, if required, and then fed into an A/D converter. The advantage of using the amplifier is that it considerably reduces circuit-induced electrical noise. The A/D conversion operation can be controlled by an appropriate digital architecture built inside an FPGA. The acquired binary data have to be transformed into a temperature figure that is suitable to show on a display device.

Some of the temperature sensors, like LM335 and LM34, give the temperature in units of °K and °F, respectively. Therefore, temperature conversions from one unit to another are required, which can be done as

$$F = \left(C \times \frac{9}{5}\right) + 32; \quad C = (K - 273.15); \quad F = (K - 273.15)\frac{9}{5} + 32 \tag{7.4}$$

The second temperature sensor is the thermistor, which is an inexpensive, rugged, reliable and simple sensor. The thermistor is preferred for simple temperature measurement applications. They are not used for high temperature measurement. Thermistors are constructed using semiconductor material whose resistivity is especially sensitive to temperature variations. Unlike some other resistive devices, the resistance (R_s) of the thermistor decreases with increasing temperature (T) due to the properties of the semiconductor material the thermistor is made from. Figure 7.7 shows the appearance and characteristic responses of a typical thermistor.

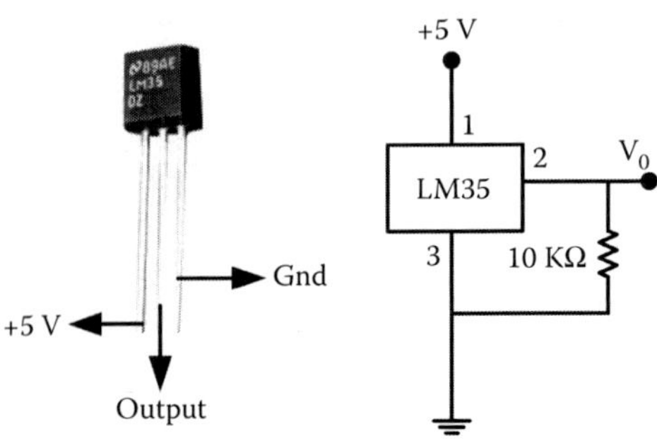

FIGURE 7.6
Appearance and application circuit of an LM35 sensor.

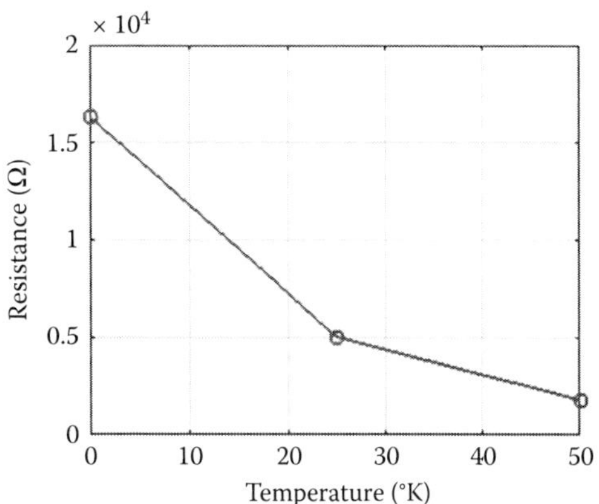

FIGURE 7.7
Appearance of a thermistor and its characteristic response.

The reciprocal of absolute temperature can be calculated as a function of the resistance of a thermistor using the Steinhart-Hart equation as

$$\frac{1}{T} = A + B\ln(R_s) + C\ln(R_s)^3 \tag{7.5}$$

where T is the temperature in °K and R is the sensor resistance in Ω. The constants A, B and C can be determined from the experimental values shown in Figure 7.7 and Equation 7.5. The temperature measured in °C can be converted to °K using Equation 7.4. Substituting the experimental values shown in Figure 7.7 in Equation 7.5 gives three simultaneous linear equations as [84,85]

$$\begin{aligned} C\ln(16,330)^3 + B\ln(16,330) + A &= 0.0036; \quad \text{for } 0°\text{C} \\ C\ln(5000)^3 + B\ln(5000) + A &= 0.0033; \quad \text{for } 25°\text{C} \\ C\ln(1801)^3 + B\ln(1801) + A &= 0.0030; \quad \text{for } 50°\text{C} \end{aligned} \tag{7.6}$$

By solving the above linear equations, the values of the constants A, B, and C are 0.001284, 2.364×10^{-4} and 9.304×10^{-8}, respectively. These values can be substituted in Equation 7.5, and it becomes

$$\frac{1}{T} = 0.001284 + 2.364 \times 10^{-4}\ln(R_s) + 9.304 \times 10^{-8}\ln(R_s)^3 \tag{7.7}$$

Using this equation, we can compute the temperature as a function of the measured resistance. When the thermistor is deployed in an application circuit, for example, a voltage divider or bridge circuit, as shown in Figure 7.8, the measurement electrical quantity is usually voltage, which can be processed further to figure out the measured temperature.

Readers can find the working principles of voltage divider circuits in Chapter 5. In the bridge circuit, there are three resistors, R_2, R_3, and R_4, while the another resistor is a thermistor (R_s). By measuring the output voltage (V_o), the change of resistance can be mapped out and the temperature can be estimated using Equation 7.7. The thermistor can be placed anywhere in the bridge circuit; however, different placements give different polarity to V_o.

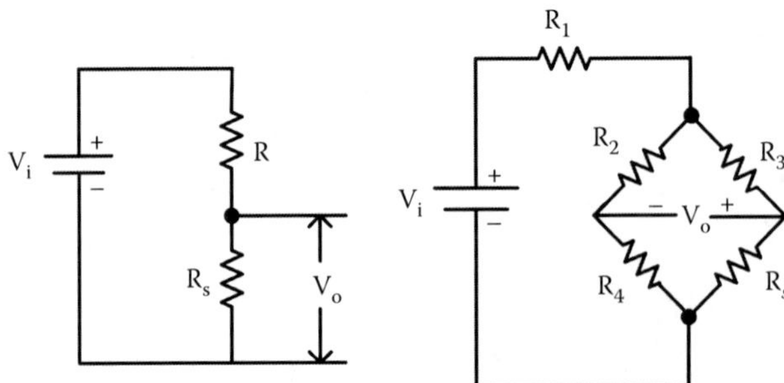

FIGURE 7.8
Application circuits of a thermistor.

The third temperature sensor is the thermocouple, which is a pair of junctions formed by two dissimilar metals. One junction is at a reference temperature, usually 0°C, and the other is at a spot where the temperature has to be measured, as shown in Figure 7.9. Due to the Seebeck effect, the temperature difference between the junctions causes the voltage that depends on the temperature changes. Thermocouples are widely used for measurements over a wide range of temperature variations.

Once we obtain the voltmeter reading, the voltage has to be converted to temperature. The temperature is usually expressed as a polynomial function of the measured voltage or by a decent linear approximation over a limited temperature range. A polynomial equation used to convert thermocouple voltage to temperature over a wide range of temperatures is:

$$T = \sum_{n=0}^{N} a_n v^n \tag{7.8}$$

The coefficients (a_n) in the above equation can be obtained from the manufacturer's data sheet. The values of the coefficient for $0 \leq n \leq 8$ are 0.226584602, 24152.10900, 67233.4248, 2210340.682, −860963914.9, 4.83506×10^{10}, -1.18452×10^{12}, 1.38690×10^{13}, and -6.33708×10^{13}. Therefore, measuring the potential difference and substituting it in Equation 7.8 will give the temperature value. All these measurement systems have to be calibrated against standard master equipment, like testo-810, for validating measurement accuracy [84,85].

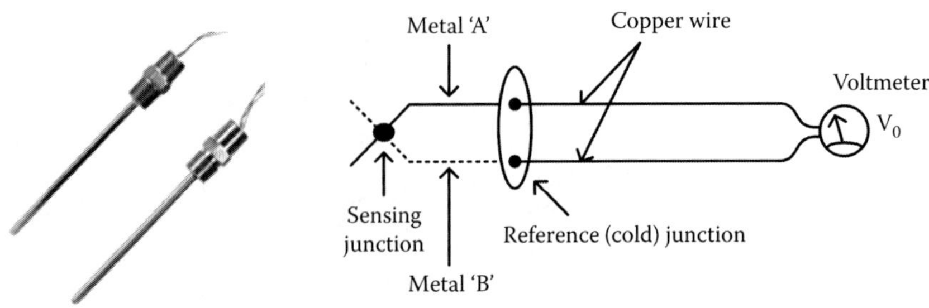

FIGURE 7.9
Appearance and construction of temperature measurement system using a thermocouple.

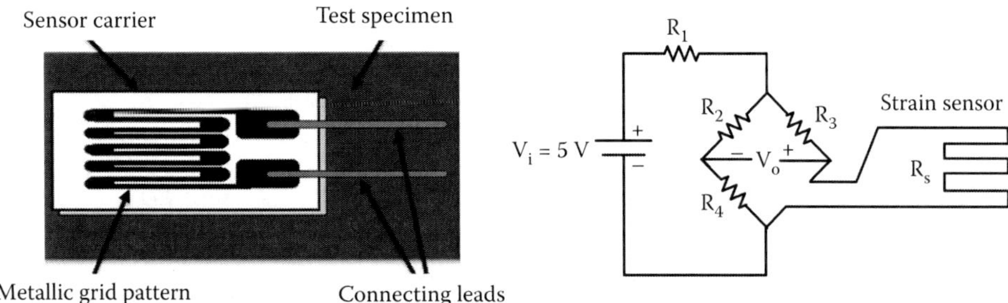

FIGURE 7.10
Appearance and application circuit of a strain gauge.

7.3.3 Strain Measurement

The strain gauge is a thin piece of conducting material whose resistance (R_s) changes with strain and whose shape looks like the drawing shown in Figure 7.10. Assume that a strain gauge is glued on a device, a so called device-under-test (DUT), which is subjected to stress. Now, the strain gauge may elongate or shrink along with the DUT. The strain gauge senses the small geometrical changes. The small geometry change in the strain gauge produces a small change in its resistance (R_s). Even though the resistance change is small, it is what we need to use to get a voltage to figure out the strain applied on the DUT. We need to remember that strain is the fractional change in length in a material when the material is stressed. Normally, on a 100-inch-long piece of material, the strain will not exceed 0.005 inch and elongation will not be more than 0.5 inch. If the strain is 0.005 inch, then the corresponding fractional change in resistance (R_s) will be 1%, which is far larger than we would expect to normally see, since it is an extreme case.

Suppose the strain gauge is connected in a voltage divider circuit; then, the output voltage across the sensor increases as the sensor resistance increases. Assume we have a standard strain gauge sensor of nominal resistance 350 Ω that increases to 350.03 Ω by applying some amount of stress. In this condition, the voltage change is very small and occurs out in the fifth place in a typical display device. We need something that will improve this situation. A bridge circuit, as shown in Figure 7.10, can help in our problem. We will choose R_2 and R_3 to have the same value. That will produce 2.5 V at the middle of the left branch. Since both R_4 and R_s are 350 Ω, the voltage at the midpoint of R_4 and R_s is also 2.5 V. That means $V_o = 0$ V when the strain gauge is unstrained, as in Equation 7.9 [86].

$$V_o = V_i\left[\frac{R_s}{R_s + R_3} - \frac{R_4}{R_4 + R_2}\right] \tag{7.9}$$

There are some implications of this result with the bridge circuit. If the voltage is zero when the gauge is unstrained, the bridge is balanced and the voltage becomes 0.0001071 V when the gauge is strained, then the change is large percentage-wise. That voltage may be small, but we can amplify it using a differential amplifier. Most currently available data acquisition boards have onboard differential amplifiers that will amplify the bridge output voltage. However, this system has to be validated against standard master equipment, like SGT-3F/350-TY41, to conform the measurement accuracy.

7.3.4 Magnitude Comparator

Mostly, as discussed in the previous sections, all sensors' output will be an analog signal; hence, a mechanism to convert it into suitable digital data becomes significant to process the sensor's output signal by any digital devices like the FPGA. One primary and low-cost A/D converter is a magnitude comparator, for example, a LM324. The LM324 series magnitude comparator is a quad operational amplifier with differential inputs (V+ and V−). The quad amplifier can operate at supply voltages of 3–32 V with quiescent currents. The common mode input range includes the negative supply, thereby eliminating the necessity for external biasing components in many applications [87]. The circuits for the two simplest configurations of a voltage comparator and their output response are shown in Figure 7.11a and b. There, the reference voltage (V_{ref}) is fixed at one-half of the supply voltage (V_{cc} = 5 V), while the input voltage (V_{in}) varies from 0 V to the supply voltage.

The output of a magnitude comparator will be logic 0 when the V− is greater than the V+, and the output will be logic 1, when V− < V+. Therefore, when V_{in} increases from 0 V to V_{cc}, as in Figure 7.11a, the output switches to logic 0 at the instant when the $V_{in} = V_{ref}$ and remains there. If we reduce V_{in} back to 0 V, as in Figure 7.11a, then the output switches to logic 1 when again $V_{in} = V_{ref}$ and remains there. A similar, but reverse, operation happens in Figure 7.11b. An application circuit for measuring the temperature level is shown in Figure 7.11c. There, the voltage divider network provides individual reference voltages to all the op-amp comparator circuits. To produce the four reference voltages, we require four resistors. The junction at the bottom produces a reference voltage that is one-fourth the supply voltage ($1/4V_{cc}$) using equal value resistors [39,87]. In the same way, the second pair is $2/4V_{cc}$, the third pair is $3/4V_{cc}$ and so on. Each of these reference voltages is connected directly to the noninverting pin (+) of the operational amplifiers. The output terminal of the temperature sensor (LM35)

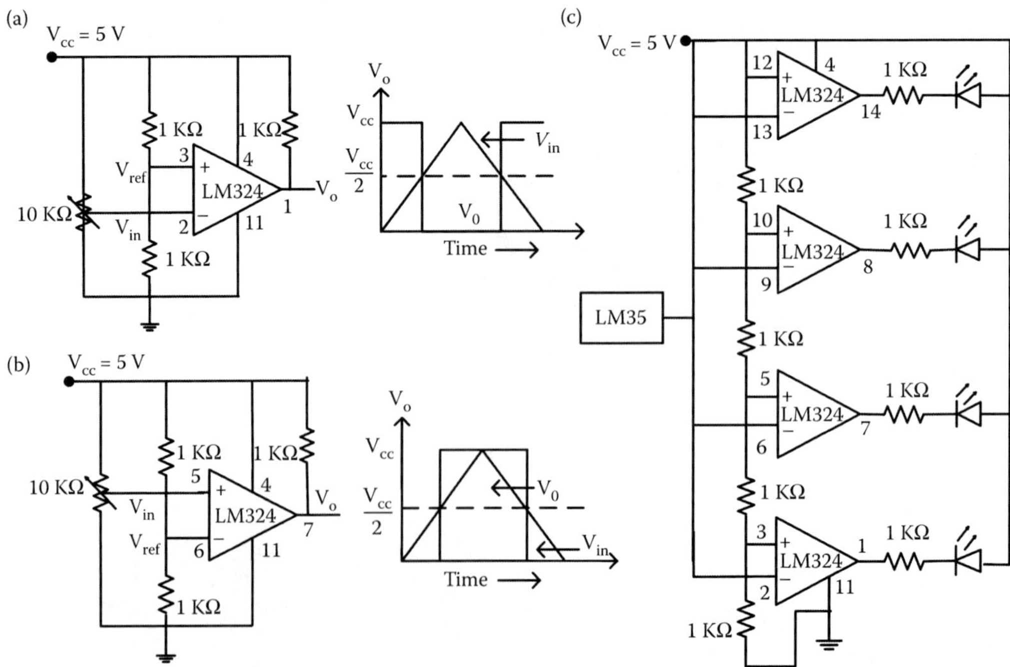

FIGURE 7.11
Basic operation of a voltage comparator: negative input (a), positive input (b), and a 4-bit application circuit (c).

LM324 outputs			Binary outputs	
C	B	A	y	x
0	0	0	0	0
0	0	1	0	1
0	1	1	1	0
1	1	1	1	1

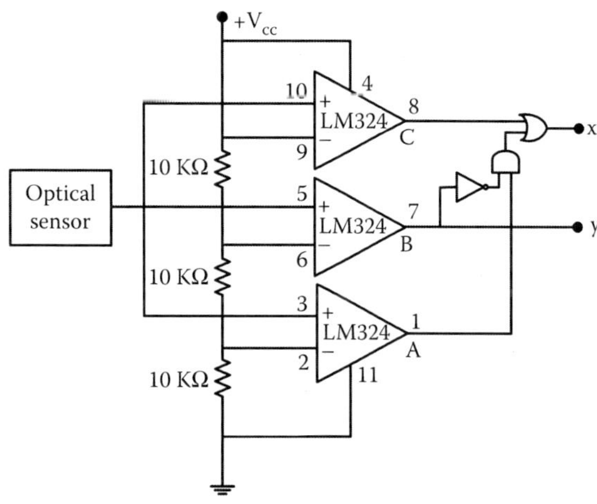

FIGURE 7.12
Two-bit binary A/D converter using LM324.

is directly connected to all inverting terminals (–) of the operational amplifiers. If there is a temperature change, the voltage varies at the output terminal of the LM35 [87]. This voltage is compared to the voltage the comparator has at its noninverting terminal. Therefore, the default value at the outputs of all the comparator is logic 1; hence, the LEDs do not glow. Once the LM35 output crosses the reference voltage of one op-amp, then its output goes logic 0, and the corresponding LED glows. When the temperature is very high, the sensor output voltage will also be very high and all the LEDs will glow.

DESIGN EXAMPLE 7.2

Assume that an optical sensor output voltage is connected to the noninverting terminals of op-amps (LM324). Using this circuit, design a 2-bit A/D converter to get the possible binary outputs for the different values of sensor output voltages.

Solution

The required design is a 2-bit A/D converter, so just three op-amps (magnitude comparator) from an LM324 IC are sufficient. Let us assign A (LSB), B and C (MSB) to the outputs of the first three op-amps, which will be "000", "100", "110", and "111" as the sensor output voltage increases. The corresponding binary outputs x and y (MSB) have to be "00", "01", "10", and "11". Therefore, we need two logical expressions to relate the A, B, and C to x and y, which can be obtained using the K-map. After simplifications, the logical expressions will be x = `AB'` + `C` and y = b.

The truth table and logical circuit diagram for a 2-bit A/D converter using an LM324 are shown in Figure 7.12. This design approach can be applied for any number of A/D converter resolutions; however, this is not preferred for more than 2-bit resolution due to complications in design as well as the requirements of many resistors, op-amps, basic gates, and so on. Dedicated A/D converters are more commonly used in a wide range of applications which are discussed in the following sections.

7.3.5 ADC0804 Interfacing

The A/D converter plays a vital role in many real-world applications, and the following sections cover the working principles of many A/D converters and their interface to the FPGA. The first A/D converter is the ADC0804, which is a single-channel, 8-bit-resolution

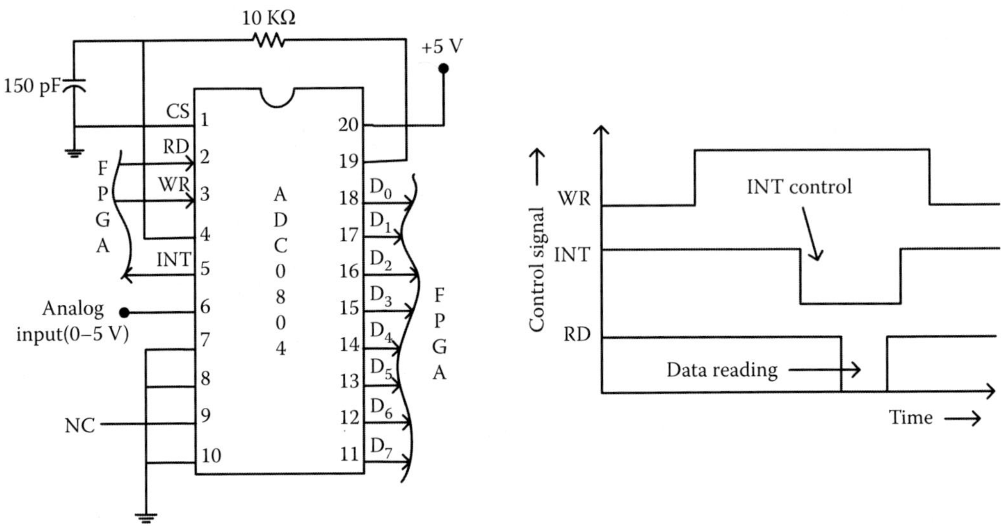

FIGURE 7.13
Application circuit and interfacing timing diagram of ADC0804.

successive approximation A/D converter. The features of the ADC0804 are that it has differential analog voltage inputs, a 0–5 V input voltage range, no need of zero adjustment, and a built-in clock generator; the reference voltage can be externally adjusted to convert smaller analog voltage spans to 8-bit resolution, and so on. The ADC0804 can be interfaced with the FPGA using a minimum of 11 pins: 8 pins for data (D_0–D_7) and 3 pins for control (WR, INT, and RD). A pin chip select (CS) is permanently connected to ground (0 V), as shown in Figure 7.13. The interfacing timing diagram is also shown in Figure 7.13 and shows which signal has to be switched at what time to have proper conversion and data logging. Force the WR pin high and wait for the INT pin to go low, which means the has conversion ended. Once the conversion in the ADC0804 is done, the data will be available in the output latch of the ADC. Force the RD signal low, and read the data from pins through which the ADC0804 is connected to FPGA. Force the RD signal high, and after some time, force the WR signal low. At this instant, one measurement cycle is completed. The possible different reference voltages at pin 9 and the corresponding conversion step size are given in Table 7.1. If the $V_{ref}/2$ is left unconnected, then the input analog voltage span is 0 to 5 V [37,66,88,89].

Since the ADC0804 is a single-channel, 8-bit resolution A/D converter, if the analog input voltage (Vin) is 5 V, then the output bits will be "11111111" in binary, which is the

TABLE 7.1
Reference Input Voltages and Corresponding Conversion Step Size

$V_{ref}/2$ (V)	Input Voltage Span (V)	Step Size (mV)
NC	0–5	19.6
2	0–4	15.69
1.5	0–3	11.76
1.28	0–2.56	10.04
1	0–2	7.84
0.5	0–1	3.92

equivalent of 255 in decimal and FFH in hexadecimal. A resistor (R) and capacitor (C) are associated with the internal clock circuitry of the ADC0804.

The frequency of the internal sampling clock is estimated by f = (1/1.1RC). For example, if R = 10 KΩ and C = 150 pF, then the sampling frequency will be 606 KHz; hence, the sampling time will be 1.65 μsec.

DESIGN EXAMPLE 7.3

The output of a thermocouple (temperature sensor) is connected to an ADC0804 through a signal conditioning circuit. Develop a digital system in the FPGA to control the conversion operation of the ADC0804 and display the acquired data using eight LEDs.

Solution

```
library IEEE;
use IEEE.STD_LOGIC_1164.ALL;
use IEEE.STD_LOGIC_ARITH.ALL;
use IEEE.STD_LOGIC_UNSIGNED.ALL;
entity adc is
Port ( m_clk:in  STD_LOGIC:='0';
int : in  STD_LOGIC:='1';
rd  : out  STD_LOGIC:='1';
wr  :out std_logic:='0';
adc_data_in:in std_logic_vector(7 downto 0):="00000000";
adc_data_out : out  STD_LOGIC_vector(7 downto 0):="00000000");
end adc;
architecture Behavioral of adc is
signal clk_sig1:std_logic:='0';
begin
p0:process(m_clk)
variable cnt1:integer:=1;
begin
if rising_edge(m_clk) then
if(cnt1=2000)then
clk_sig1<= not(clk_sig1);
cnt1:=1;
else
cnt1:= cnt1 + 1;
end if;
end if;
end process;
p1:process(clk_sig1,adc_data_in)
variable cnt:integer:=1;
begin
if rising_edge(clk_sig1) then
if     (cnt=1) then  wr<='1';cnt:=cnt+1;
elsif (cnt=2) then
if int='1' then cnt:=2;
else cnt:=cnt+1;
end if;
elsif (cnt=3) then   rd<='0'; cnt:=cnt+1;
elsif (cnt=4) then   adc_data_out<=adc_data_in; cnt:=cnt+1;
elsif (cnt=5) then rd<='1'; cnt:=cnt+1;
elsif (cnt=6) then wr<='0'; cnt:=cnt+1;
elsif (cnt=7) then cnt:=1;
else  cnt:=1;
end if;
end if;
end process;
end Behavioral;
```

DESIGN EXAMPLE 7.4

The output of a strain gauge is connected to the ADC0804. The sensor output voltage will be 5 V when the applied strain is 0.0051″. Develop a digital system to control the A/D converter and display the measured strain on an LCD in appropriate unit, that is, ″ (inches). The measured strain value has to be displayed after the word "STRAIN:" in the second row of the LCD.

Solution

From the given specifications, it is noted that we need to find the actual displacement in inches (″) from the sensor output voltage. The maximum output voltage, maximum displacement and measurement voltage can be related as

$$\left[\left(\frac{0.0051}{5\,\text{V}}\right)\times\text{Sensor Voltage}\right]$$

which is also given by the output of A/D converter as

$$\left[\left(\frac{0.0051}{255}\right)\times\text{A/D converter value}\right]$$

To simplify the code design process, the above equation is rewritten as

$$\left[\left(\frac{5100\times10^{-6}}{255}\right)\times\text{A/D converter value}\right]=[(20\times\text{A/D converter value})\text{E}-6]$$

Therefore, we can keep the 10^{-6} as it is and just need to multiply the value of A/D converter by 20. Once this value is calculated, then we need to convert that value to BCD to display the measured value on a LCD. Thus, finally, the display in the second row of the LCD will be "STRAIN: ----E-6″". The "-" will be updated by the measured values in real time. The VHDL code for A/D converter interfacing, data manipulation and LCD interfacing is given below:

```
library IEEE;
use IEEE.STD_LOGIC_1164.ALL;
use IEEE.STD_LOGIC_UNSIGNED.ALL;
USE ieee.numeric_std.all;
entity str_adc_lcd is
Port ( m_clk: in STD_LOGIC:='0';
int: in std_logic:='1';
rd  : out  STD_LOGIC:='1';
wr  :out std_logic:='0';
lcd_cnt: out std_logic_vector(2 downto 0):="000";
lcd_data:out std_logic_vector(7 downto 0):="00000000";
adc_data_in: in STD_LOGIC_vector(7 downto 0):="00000000");
end str_adc_lcd;
architecture Behavioral of str_adc_lcd is
signal clk_sig1:STD_LOGIC:='0';
signal adc_data_temp: STD_LOGIC_vector(7 downto 0):="00000000";
signal d,e,f,g,h,i,j: integer:=0;
constant Message0:string(1 to 15):="STRAIN:     E-6";
begin
p0:process(m_clk)
variable cnt1:integer:=1;
begin
if rising_edge(m_clk) then
if(cnt1=1)then --2000
clk_sig1<= not(clk_sig1);
```

```
cnt1:=1;
else
cnt1:= cnt1 + 1;
end if;
end if;
end process;
p1:process(clk_sig1,adc_data_in)
variable cnt:integer:=1;
begin
if rising_edge(clk_sig1) then
if   (cnt=1) then  wr<='1';cnt:=cnt+1;
elsif (cnt=2) then
if int='1' then cnt:=2;
else cnt:=cnt+1;
end if;
elsif (cnt=3) then   rd<='0'; cnt:=cnt+1;
elsif (cnt=4) then   adc_data_temp<=adc_data_in; cnt:=cnt+1;
elsif (cnt=5) then rd<='1'; cnt:=cnt+1;
elsif (cnt=6) then wr<='0'; cnt:=cnt+1;
elsif (cnt=7) then cnt:=cnt+1;
elsif (cnt=8) then lcd_data<="00111000";lcd_cnt<="100";
cnt:=cnt+1;--38
elsif ((cnt=9)or(cnt=11)or(cnt=13)or(cnt=15)or(cnt=17)) then
lcd_cnt<="000"; cnt:=cnt+1;
elsif (cnt=10) then lcd_data<="00001100";lcd_cnt<="100";
cnt:=cnt+1;--0C
elsif (cnt=12) then lcd_data<="00000001";lcd_cnt<="100";
cnt:=cnt+1;--01
elsif (cnt=14) then lcd_data<="00000110";lcd_cnt<="100";
cnt:=cnt+1;--06
elsif (cnt=16) then lcd_data<="11000000";lcd_cnt<="100";
cnt:=cnt+1;--C0
elsif (cnt=18) then
lcd_data<=std_logic_vector(to_unsigned(character'pos(Message0(1)),8));
lcd_cnt<="110";  cnt:=cnt+1;--S
elsif ((cnt=19)or(cnt=21)or(cnt=23)or(cnt=25)or(cnt=27)or
(cnt=29)or(cnt=31)or(cnt=33)or(cnt=35)or(cnt=37)or(cnt=39)or(cnt=41)
or
(cnt=43)or(cnt=45)or(cnt=47)or(cnt=49)) then
lcd_cnt<="010";cnt:=cnt+1;
elsif (cnt=20) then
lcd_data<=std_logic_vector(to_unsigned(character'pos(Message0(2)),8));
lcd_cnt<="110";  cnt:=cnt+1;--T
elsif (cnt=22) then
lcd_data<=std_logic_vector(to_unsigned(character'pos(Message0(3)),8));
lcd_cnt<="110";  cnt:=cnt+1;--R
elsif (cnt=24) then
lcd_data<=std_logic_vector(to_unsigned(character'pos(Message0(4)),8));
lcd_cnt<="110";  cnt:=cnt+1;--A
elsif (cnt=26) then
lcd_data<=std_logic_vector(to_unsigned(character'pos(Message0(5)),8));
lcd_cnt<="110";  cnt:=cnt+1;--I
elsif (cnt=28) then
lcd_data<=std_logic_vector(to_unsigned(character'pos(Message0(6)),8));
lcd_cnt<="110";  cnt:=cnt+1;--N
elsif (cnt=30) then
lcd_data<=std_logic_vector(to_unsigned(character'pos(Message0(7)),8));
lcd_cnt<="110";  cnt:=cnt+1;--:
elsif (cnt=32) then
lcd_data<=std_logic_vector(to_unsigned(character'pos(Message0(8)),8));
lcd_cnt<="110";  cnt:=cnt+1;--
```

```
elsif (cnt=34) then lcd_data<="0011" & std_logic_vector
(to_unsigned(j,4));lcd_cnt<="110"; cnt:=cnt+1;--j
elsif (cnt=36) then lcd_data<="0011" & std_logic_vector
(to_unsigned(i,4));lcd_cnt<="110"; cnt:=cnt+1;--i
elsif (cnt=38) then lcd_data<="0011" & std_logic_vector
(to_unsigned(g,4));lcd_cnt<="110"; cnt:=cnt+1;--g
elsif (cnt=40) then lcd_data<="0011" & std_logic_vector
(to_unsigned(e,4));lcd_cnt<="110"; cnt:=cnt+1;--e
elsif (cnt=42) then
lcd_data<=std_logic_vector(to_unsigned(character'pos(Message0(13)),8));
lcd_cnt<="110";  cnt:=cnt+1;--E
elsif (cnt=44) then
lcd_data<=std_logic_vector(to_unsigned(character'pos(Message0(14)),8));
lcd_cnt<="110";  cnt:=cnt+1;---
elsif (cnt=46) then
lcd_data<=std_logic_vector(to_unsigned(character'pos(Message0(15)),8));
lcd_cnt<="110";  cnt:=cnt+1;--6
elsif (cnt=48) then
lcd_data<="00100010";lcd_cnt<="110";  cnt:=cnt+1;--" (INCH)
elsif (cnt=50) then cnt:=1;
else  cnt:=1;
end if;
end if;
end process;
d<=20*(to_integer(unsigned(adc_data_temp)));
e<=(d mod 10);
f<=(d/10);
g<=(f mod 10);
h<=(f/10);
i<=(h mod 10);
j<=(h /10);
end Behavioral;
```

NB: The above example is only simulatable design and is nonsynthesizable code due to the last few lines: d through j. Altering the above code as a synthesizable design using Design Example 5.14 (p2) is left to the readers as one of the laboratory exercises.

7.4 Universal Asynchronous Receiver-Transmitter Design

The UART is basically a data transmission/reception protocol through which data can be transferred bitwise without any reference clock input. The UART is also called a serial COM port or RS232 standard COM port. While transferring the data, the serial port sends a logical 1 as a negative voltage and a logical 0 as a positive voltage. When no data are being transferred, the serial port transmits logic 1 (i.e., negative voltage) and the port is said to be in a "mark or stop" state. The serial port can also be forced to be in "space or start" state by transferring logic 0 (i.e., positive voltage). The output signal level usually swings between +12 V and −12 V at the output of the COM port, as shown in Figure 7.14a. Logic 0 is defined between +5 V and +12 V, and logic 1 is defined between −5 V and −12 V, with ±3 V for the noise margin. The signal voltage between +3 V and −3 V is also called a "dead area" designed to absorb the line noise. When transmitting a byte, the UART first sends a "start" bit, which is a logic 0 followed by 8 data bits, followed by two "stop" bits, which is a logic 1, as shown in Figure 7.14b. This order of sending the bits has to repeated for each byte that needs to be sent.

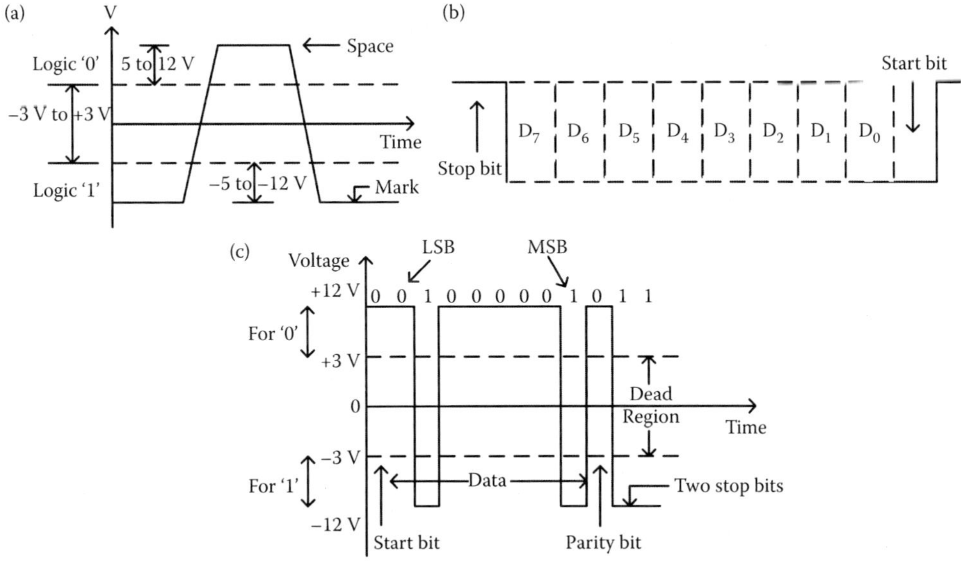

FIGURE 7.14
Data frame format of a serial communication: RS232 voltage level (a), data bit packing for serial transmission (b), and voltage level swing for transferring an ASCII character "A" (c).

At this point, we must know what the duration should be for a bit. In other words, how long does the signal stay in a particular state to send a bit? It is dependent on the baud rate, which decides the number of times that the signal can switch states in one second. For example, if the communication happens at a 9600 baud rate, the line can switch the states 9600 times per second. Serial data communication can be half or full duplex. The data pattern with RS232 voltage levels for sending an ASCII character "A", that is, 41H, is shown in Figure 7.14c. Now, let us see the electrical and mechanical interfacing characteristics of the UART standard. The RS-232 standard 9-pin D-type connector (DB-9) is shown in Figure 7.15a. Its pin details are 1: data carrier detect, 2: receive data, 3: transmit data, 4: data terminal ready, 5: ground, 6: data set ready, 7: request to send, 8: clear to end, and 9: ring indicator [31,89–91].

In almost all digital devices, like the FPGA, the data format is transistor to transistor level (TTL); to interface it to RS232 COM port, an RS-232 line driver between the digital device, and the RS232 port is required. Most devices currently being used for serial data communication require only three wires: Tx (pin 3), Rx (pin 2), and ground (pin 5), as shown in Figure 7.15b. Although many commercial line driver chips are available, the most popular ones are the MAX233 and MAX233A. The only difference between the MAX233 and MAX233A is the data rate; that is, the MAX233 maximum data rate is 115 kbaud, whereas the MAX233A is 230 kbaud. The MAX233 line driver has the capability of simultaneously driving four independent channels, that is, TTL1-RS2321, TTL2-RS2322, RS2321-TTL1, and RS2322-TTL2. The ports/channels corresponding to the independent conversion process are shown/marked in Figure 7.15c as (Tx^1-Rx^1), (Tx^2-Rx^2), (Tx^3-Rx^3), and (Tx^4-Rx^4). Another line driver is the MAX232, which is same as the MAX233 but requires some external capacitors. The RS232-to-universal serial bus (USB) converter cable, as shown in Figure 7.15d, is today commercially available with driver software; hence, a device that has an RS232 COM port can be interfaced to another device that does not have the same port. We discuss serial port data writing/reading from the MATLAB environment, UART (Tx and Rx) design in the FPGA, and real-time data logging/plotting in the following sections.

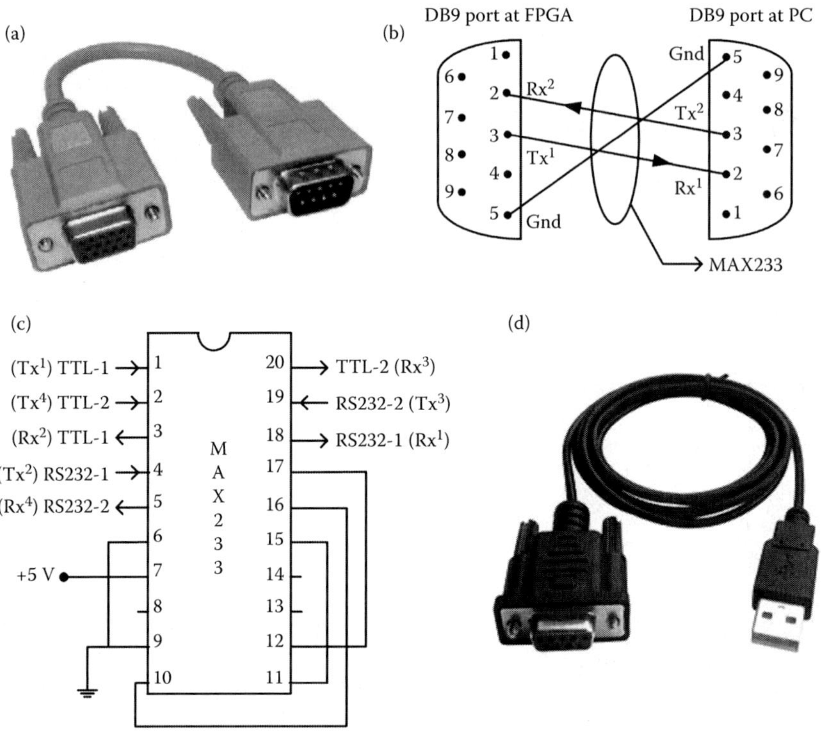

FIGURE 7.15
Electrical and mechanical interfacing of a DB9 connector: RS232 cable (a), three wire connections at the serial port (b), circuit schematic of MAX233 line driver (c), and RS232-to-USB converter cable (d).

7.4.1 Serial Communication: Data Reading/Writing Using MATLAB®

Many external hardware devices have a serial COM port through which the data can be transferred/received to/from the PC either directly or through the RS232-USB converter. We literally send/receive the data over the RS232 cable using two separate pins (wires) as a series of bytes, that is, 8 bits (0–255 or 00H-FFH). Let us assume that data from a sensor are continuously arriving at the serial port of a computer. Code written in MATLAB will read the data and store it in a file specified in the code. On the computer, first the data get stored in a buffer until we read it. New data (the most recent data) that arrive at the serial port get added at the bottom of the buffer. If the code written in MATLAB reads a value from the buffer, it starts reading at the top, that is, the oldest data. Once the code reads a byte of data, it is no longer in the buffer and the data in the second position move up to the top position. The buffer has a finite length, which means there is a limit to how long the buffer can retain the data. Once the buffer is totally full, if new data arrive at the serial port, the oldest data at top of the buffer get discarded forever, and all the previous entries move up to make a room at the bottom for new data. By writing proper code in MATLAB, we can avoid missing data. Data storing/updating in the buffer as data arrive at the serial port is given in Table 7.2. There, (i) create a buffer of length 5 which is initially empty (–); (ii) data "12" enters the buffer; (iii) more data, "6", enters; note that the oldest data are in the first position and new data fill the buffer from the bottom; (iv) we read a byte of data from the buffer; note that once we read a value, it is no longer in the buffer, and we read the top element, which is the oldest data, and the second data jump to the upper position; (v) the buffer is full; and (vi) new

TABLE 7.2

Data Logging in a Serial Port Buffer

Row ID	Action as the data arrives in the buffer (1)	(2)	(3)	(4)	(5)	(6)
1	-	12	12	6	6	16
2	-	-	6	-	16	12
3	-	-	-	-	12	34
4	-	-	-	-	34	129
5	-	-	-	-	129	4

Discarded data

New data

data, "4", arrive in the buffer, and the oldest data in the top entry, that is, "6", are discarded and all the entries shift up by one position to make room for the new entry.

Important attributes related to reading/writing data through the serial communication port from the MATLAB environment are discussed below. The serial communication port details can be read using >> `serialPort = serial('com1')`. This instruction returns some of the default or previously assigned values as given below [90,92].

```
Serial Port Object : Serial-COM1
  Communication Settings
     Port:                 COM1
     BaudRate:             9600
     Terminator:           'LF'
  Communication State
     Status:               closed
     RecordStatus:         off
  Read/Write State
     TransferStatus:       idle
     BytesAvailable:       0
     ValuesReceived:       0
     ValuesSent:           0
```

The above are not the complete set of details about the COM port; however, they are the primary parameters/attributes. Most of the time, we may not need to use the default values. We can view and change any attribute using the instructions >>`get and` >>`set` as

```
>> get(serialPort, 'baudrate')
ans =
        9600
>> set(serialPort, 'BaudRate', 19200)
>> get(serialPort, 'BaudRate')
ans =
       19200
```

A lot of things can be changed while creating a serial object using an instruction. For example, >> `serialPort_new = serial('com1', 'baudrate', 19200,` `'terminator', 'CR')`, which sets the serial port attributes with the values given in the above instruction and returns the following:

```
Serial Port Object : Serial-COM1
  Communication Settings
     Port:                 COM1
     BaudRate:             19200
     Terminator:           'CR'
  Communication State
     Status:               closed
     RecordStatus:         off
  Read/Write State
     TransferStatus:       idle
     BytesAvailable:       0
     ValuesReceived:       0
     ValuesSent:           0
```

In this way, we can edit as many attributes as we want. The names of the attributes have to be in single quotes followed by their values. If the value is in text, then we have to use single quotes; otherwise, we use only numbers. The most common attributes are Terminator, FlowControl, Baudrate, Parity, DataBits, ByteOrder, and so on. If we try to read data from the serial port while there is no data in the buffer, MATLAB will keep trying to read until the timeout, that is, default 10 sec, which can be found by the instruction `>> get(serialPort, 'Timeout')`, which returns

```
ans =
    10
```

Closing, deleting and clearing the serial port from MATLAB is three separate actions, which are performed by

```
>> fclose(serialPort_new)
>> delete(serialPort_new)
>> clear (serialPort_new)
```

Before reading/writing the serial port, we need to open it using `>> fopen(serialPort_new)` in the MATLAB environment. The following instruction can be used to write ASCII data to the serial port.

```
>> fwrite(serialPort_new, [0, 12, 117, 251]);
```

We can also write a mix of text, variables and values as `fprintf(serGPS, 'MOVE %d, PAUSE %d', [moveNum, pauseTime]);`. The serial port buffer data can be quickly cleared by

```
N = serRoomba.BytesAvailable();
while(N~=0)
fread(serRoomba,N);
N = serRoomba.BytesAvailable();
end
```

We can use `>>fread` to read the data (not text). Let us assume the buffer currently has 2 bytes of data in the serial port buffer. Those 2 bytes can be read using

```
a = fread(serialObj, 2);
```

Which will returns the value as

```
a = [1; 8]
```

Using the following instructions, we can read and concatenate 2 bytes, that is, 16 bits, of data, as

```
a = fread(serialObj, 1, 'int16')
ans =
   264
```

which is equivalent to $(0000000100001000)_2$. Note that even though we read only 1 byte (8 bits) of data at a time, one value and two elements were taken out of the buffer since a 16-bit integer is actually composed of two bytes.

The MATLAB code given below in the design examples has been tested in real time. Therefore, the readers can apply it as it is for the applications for which it is designed. Also, the code can be edited according to the reader's requirements.

DESIGN EXAMPLE 7.5

Write MATLAB code to start some operations by the current time (now) and stop them by the specified time. This means the program has to be executed to perform the operation only during the period specified by the current and stop time. Print the results of the operations in the MATLAB command window and a Word file. The operation repeating speed has to be 1 sec.

Solution

```
clc;
clear all;
clear screen;
diary ('e:\Data File\Collected Data.dat'); %% opening a word file….
fprintf ('    ID.No:     Date & Time              SUM          X-DIFF\n');
fprintf ('    ====       ========                 ===          ======\n');
time = now; %%current time
stopTime = 'Oct-15 (Sat) 12:17:00';%% stop time
count=1;
while ~isequal(datestr(now,'mmm-dd (ddd) HH:MM:SS'),stopTime)%% time
comparison
date=datestr(now);
a=255;
b=23;
c=dec2bin(a,8);
d=dec2bin(b,8);
f1=[c d];
sum=bin2dec(f1);
sum_v=sum^2;
xdiff_v=sum_v*0.04;
fprintf('%6d     %s    %0.5E%s    %0.5E%s\n',count,date,sum_v,'V',
xdiff_v,'V');
count = count +1;
pause(1);    %%delay for 1 sec speed.
end
diary off;
```

A portion of the results of this code is shown in Table 7.3.

TABLE 7.3

Screen Snapshot of the Results of Design Example 7.5

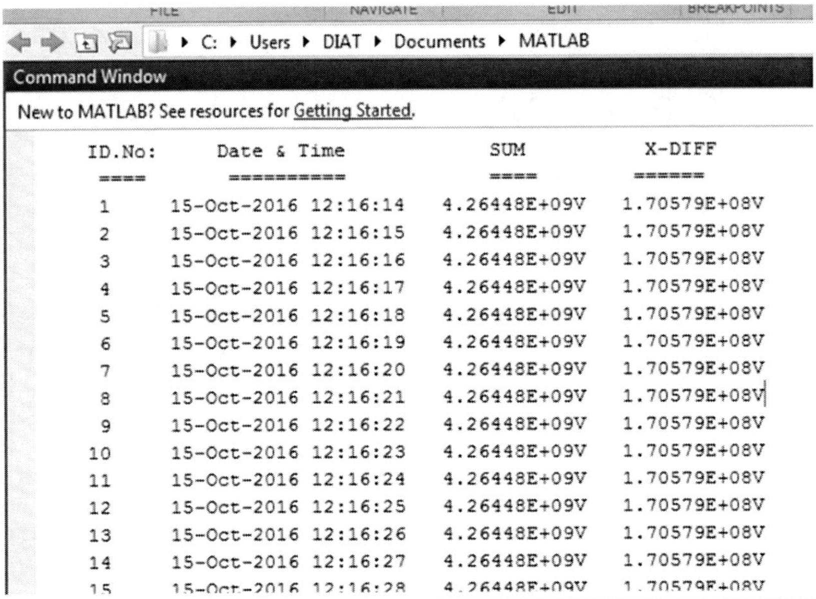

FILE | NAVIGATE | EDIT | BREAKPOINTS

C: ▸ Users ▸ DIAT ▸ Documents ▸ MATLAB

Command Window

New to MATLAB? See resources for Getting Started.

ID.No:	Date & Time	SUM	X-DIFF
====	==========	====	======
1	15-Oct-2016 12:16:14	4.26448E+09V	1.70579E+08V
2	15-Oct-2016 12:16:15	4.26448E+09V	1.70579E+08V
3	15-Oct-2016 12:16:16	4.26448E+09V	1.70579E+08V
4	15-Oct-2016 12:16:17	4.26448E+09V	1.70579E+08V
5	15-Oct-2016 12:16:18	4.26448E+09V	1.70579E+08V
6	15-Oct-2016 12:16:19	4.26448E+09V	1.70579E+08V
7	15-Oct-2016 12:16:20	4.26448E+09V	1.70579E+08V
8	15-Oct-2016 12:16:21	4.26448E+09V	1.70579E+08V
9	15-Oct-2016 12:16:22	4.26448E+09V	1.70579E+08V
10	15-Oct-2016 12:16:23	4.26448E+09V	1.70579E+08V
11	15-Oct-2016 12:16:24	4.26448E+09V	1.70579E+08V
12	15-Oct-2016 12:16:25	4.26448E+09V	1.70579E+08V
13	15-Oct-2016 12:16:26	4.26448E+09V	1.70579E+08V
14	15-Oct-2016 12:16:27	4.26448E+09V	1.70579E+08V
15	15-Oct-2016 12:16:28	4.26448E+09V	1.70579E+08V

DESIGN EXAMPLE 7.6

Develop MATLAB code to read the data (characters) arriving at the serial port and print them in a Word file. The baud rate of the data communication is 9600.

Solution

```
clc;
clear all;
diary ('e:\Data File\Collected Data.dat'); %%opening a word file
s = serial('com3');
s.BaudRate=9600;
fopen(s);
time = now;
stopTime = 'Oct-15 (Sat) 12:37:00';
count = 1;
while ~isequal(datestr(now,'mmm-dd (ddd) HH:MM:SS'),stopTime)
date=datestr(now);
time(count) = datenum(clock);
a=fread(s,1)
fwrite(fid,char(a));
count = count +1;
end
fclose(s);
diary off;
```

DESIGN EXAMPLE 7.7

Write MATLAB code to read the data arriving at the serial port and write them in the command window as well as in a Word file. The data corresponding to different measurements are packed at the transmitter as "Header (FFH), Load-Voltage (LV), Load-Current (LC), Vibration (Vib), and Speed (Spe)".

Solution

The transmitter follows a protocol to send the measurement data to the COM port. Therefore, we need to write receiving and data-logging code according to it. The protocol says that first a header (i.e., FFH) will arrive at the COM port. Since the header itself is "FFH", the measurement data values definitely vary between 00H and FEH. Thus, we need to check whether the received byte is "FFH"; once received, we can read the subsequent four data with proper assignment to the appropriate variables. MATLAB code for this operation is given below.

```
clc;
clear all;
diary ('e:\Data File\Collected Data.dat');  %%opening a file
fprintf('ID.No: Date&Time    LV       LC      Vib      Spe\n');
fprintf ('  === =======    ====     ====    ====     ====\n');
s = serial('com1');
s.BaudRate=9600;
fopen(s);
time = now;
stopTime = 'Oct-15 (Sat) 12::11';
timeInterval = 0.005;
count = 1;
LV=0;
LC=0;
Vib=0;
Spe=0;
while ~isequal(datestr(now,'mmm-dd (ddd) HH:MM:SS'),stopTime)
date=datestr(now);
time(count) = datenum(clock);
a=fread(s,1,'uint8')';
if a ==255 %% header checking
a=fread(s,1,'uint8')';
LV=a;
a=fread(s,1,'uint8')';
LC=a;
a=fread(s,1,'uint8')';
Vib=a;
a=fread(s,1,'uint8')';
Spe=a;
end
fprintf('%6d  %s  %0.3E  %0.3E %0.3E  %0.3E\n',count, date, LV,LC,
Vib,Spe);
count = count +1;
end
fclose(s);
diary off;
```

DESIGN EXAMPLE 7.8

The output voltage of a sensor is connected to a 12-bit parallel out A/D converter whose entire operation is controlled by digital architecture built inside the FPGA. A UART is also designed in the FPGA to send the measured data to the serial port of a computer. Write MATLAB code to read the data and store them in an Excel file as specified below:

Option 1: In sheet 1, the data_ID from the fourth cell in column "A" and the fourth cell in column "B".

Option 2: In sheet 3, the data_ID from the first cell in column "C" and the first cell in column "E".

Option 3: In sheet 2, the data_ID in a row starting from the cell "D2" and data in a row starting from the cell "D4".

Option 4: First 99 data_ID from the cell "B1" to "F99" and first 99 data from the cell "G1" to "K99".

Solution

A 12-bit A/D converter is used in this data-logging system; however, the UART can send only 1 byte at a time. Therefore, we need to receive the first 8 bits and then another 4 bits. While sending the second half of the measurement data (i.e., the last 4 bits), we need to add "0000" (dummy zeros) to form a byte of data. Now, our computer does not know whether the arriving data corresponds to the first or second half of the measurement data, so we need to introduce headers. Here, we require a minimum of two headers, "FFH" and "FFH", to properly synchronize the received data. Once 2 bytes are received appropriately, we need to concatenate them to get the actual value. Finally, we need to store the received data in an Excel sheet as given above. The code for doing this operation is given below:

```
clc;
clear all;
s = serial('Com1');
s.BaudRate=9600;
s.TimeOut=100;
s.Parity='none';
s.InputBufferSize=floor(10000*10);
fopen(s);
time = now;
stopTime = 'Oct-15 (Sat) 18:57:00';
count=1;
temp=0;
data_ID=1;
while ~isequal(datestr(now,'mmm-dd (ddd) HH:MM:SS'),stopTime)
data_ID(count)=count;
date=datestr(now);
time(count) = datenum(clock);
a=fread(s,1,'uint8')';
if a ==255
a=fread(s,1,'uint8')';
if a==255
a=fread(s,2,'uint8')';
b=dec2bin(a(1),8);
c=dec2bin(a(2),8);
f=[c b];
temp(count)=bin2dec(f);
end
end
count = count +1;
end
fclose(s);
clear s;
data_ID_temp=data_ID';
data_temp=temp';
wr_location=input('Ehere do you want to write?: ');
if wr_location==1,
xlswrite('C:\Users\DIAT\Documents\MATLAB\TF_book.xls',data_ID_temp,
1,'A4');
xlswrite('C:\Users\DIAT\Documents\MATLAB\TF_book.xls',data_temp,1,'B4');
elseif wr_location==2,
xlswrite('C:\Users\DIAT\Documents\MATLAB\TF_book.xls',data_ID_temp,
3,'C1');
xlswrite('C:\Users\DIAT\Documents\MATLAB\TF_book.xls',data_temp,3,'E1');
elseif wr_location==3,
```

```
xlswrite('C:\Users\DIAT\Documents\MATLAB\TF_book.xls',data_ID_temp',
2,'D2');
xlswrite('C:\Users\DIAT\Documents\MATLAB\TF_book.xls',data_temp',
2,'D4');
elseif wr_location==4,
xlswrite('C:\Users\DIAT\Documents\MATLAB\TF_book.xls',data_ID_temp,
5,'B1:F99');
xlswrite('C:\Users\DIAT\Documents\MATLAB\TF_book.xls',data_temp,
5,'G1:K99');
end;
```

7.4.2 UART: Transmitter Design

As we discussed in the previous sections, the UART is a kind of serial communication protocol which is mostly used for low-speed and low-cost data exchange between computers and peripherals. Serial communication reduces the distortion of a signal; therefore, it makes data transfer between two systems separated by a great distance possible. The transmitter performs parallel-to-serial conversion on 8-bit data. In order to synchronize the asynchronous serial data transmission and ensure data integrity, the start, parity and stop bits are added to the serial data. The start bit is used to alert the receiver that a byte of data is about to arrive and to force the clock at the receiver to synchronize with the transmission clock. After the start bit, the individual data bits of a byte are sent, with the LSB being sent first. Each bit in the transmission is transmitted for exactly the same amount of time as all of the other bits, and the receiver looks at the wire approximately halfway through the period assigned to each bit to determine whether the bit is a "1" or "0". When the entire data word has been sent, the transmitter has to send a parity bit, if required (not mandatory), which may be used by the receiver to perform simple error checking. Then, at least two stop bits have to be sent by the transmitter to terminate a byte transmission. If another byte is ready for transmission, the start bit for the new byte can be sent as soon as the stop bit for the previous byte has been sent.

The digital architecture of an UART-Tx built in a FPGA is shown in Figure 7.16. There, four sensors are interfaced to the FPGA via a multichannel 12-bit parallel out A/D converter. Thus, we need to properly pack the sensor data with appropriate headers to send them to the

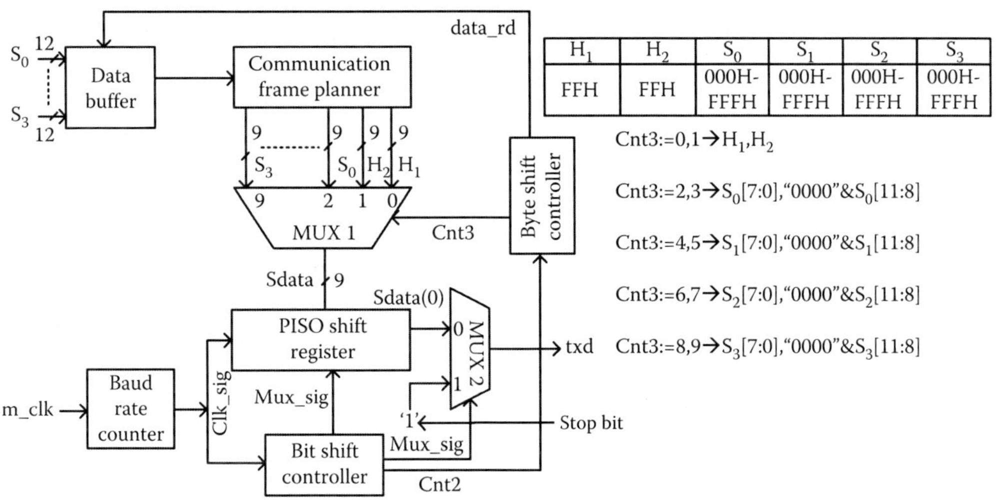

H_1	H_2	S_0	S_1	S_2	S_3
FFH	FFH	000H-FFFH	000H-FFFH	000H-FFFH	000H-FFFH

FIGURE 7.16
Digital architecture of UART-Tx and data transmission frame format.

computer through the serial port. In the architecture shown in Figure 7.16, the output of the MUX1 is forwarded to the parallel-in serial-out shift register (PISOSR), which sends one bit at a time to the MUX2 for the rising edge of the clk_sig signal. The output of the mux2 txd is sdata(0), that is, the LSB of the PISOSR, when the selection line mux_sig is set to "0"; otherwise, it is the stop bit "1". The baud rate counter generates the bit transmission clock, that is, the baud rate clock clk_sig, at the rate of 9600 from the master clock m_clk signal. The transmission of the measurement (M/m) data bit at the PISOSR and the stop bit are controlled by the bit shift controller in accordance with the clk_sig. Further, the bit shift controller conveys the status of the frame transmission to the byte shift controller using cnt2, which switches the selection line of MUX1 cnt3 to choose the next channel (frame). Once all the frames of M/m data corresponding to one sampling cycle are transferred, then the byte shift controller enables the data buffer to read the new set of M/m data by the command data_rd and continues. In order to avoid the occurrence of communication conflict among the frames, two header (H1 and H2) frames are transferred followed by the M/m data frames, with appropriate coordination among the data so that a total of 10 frames needs to be transferred for every data acquisition cycle, as shown in Figure 7.16 [39,42,48,64,93]. A computer is used to log all the M/m data to perform on/offline computations related to the experimental data analysis.

DESIGN EXAMPLE 7.9

Design a UART transmitter in the FPGA to send a word, "Spatan", through the RS232 communication port to the PC. The baud rate has to be 9600. The data frame has to be: 11 stop bits, data bits, and start bit. The onboard frequency is 4 MHz.

Solution

The mentioned baud rate (Br) is 9600, so we need to find the baud rate clock ($\mathtt{Br}_{\mathtt{clk}}$) in accordance with the onboard frequency (f_{clk}). The $\mathtt{Br}_{\mathtt{clk}}$ counter value can be found by

$$Br_{clk} = \left\lfloor \frac{f_{clk}}{9600} \right\rfloor$$

$$Br_{clk}\text{ counter value} = \left(\frac{Br_{clk}}{2} \right)$$

Therefore, in this case, the $\mathtt{Br}_{\mathtt{clk}}$ counter value is 208, which means that the $\mathtt{Br}_{\mathtt{clk}}$ transition has to occur once in every 208 clock pulses of the onboard clock. Further, as mentioned, the data frame has to be: "11111111111 & data & 0". The ASCII values for sending the word "Spatan" are 53H, 70H, 61H, 74H, 61H and 6EH. The VHDL code for doing this operation is given below:

```
library IEEE;
use IEEE.STD_LOGIC_1164.ALL;
use IEEE.STD_LOGIC_ARITH.ALL;
use IEEE.STD_LOGIC_UNSIGNED.ALL;
entity clkd is
Port ( clk : in  STD_LOGIC:='0';
txd : out  STD_LOGIC:='0');
end clkd;
architecture Behavioral of clkd is
constant leda:natural:=19;
signal clk_sig:std_logic:='0';
signal cnt3:std_logic_vector(2 downto 0):="001";
signal sdata:std_logic_vector(leda downto 0):="11111111111010100110";
begin
```

```
p1:process(clk)
variable cnt1:integer:=-2;
begin
if rising_edge(clk) then
if(cnt1=208)then --208 is the baud rate counter value when the on-board
clock is 4MHz.
clk_sig<= not(clk_sig);
cnt1:=1;
else
cnt1:= cnt1 + 1;
end if;
end if;
end process;
p2:process(clk_sig)
variable cnt2:integer:=1;
begin
if rising_edge(clk_sig)then
if (cnt2=leda+1) then
if (cnt3="110") then
cnt3<="000";
else
cnt3<=cnt3+1;
end if;
case cnt3 is
when "000"=>sdata <="11111111111010100110";
when "001"=>sdata <="11111111111011100000";
when "010"=>sdata <="11111111111011000010";
when "011"=>sdata <="11111111111011100100";
when "100"=>sdata <="11111111111011101000";
when "101"=>sdata <="11111111111011000010";
when "110"=>sdata <="11111111111011011100";
when others=> null;
end case;
cnt2:=1;
else
sdata<=('0'& sdata(leda downto 1));
cnt2:=cnt2+1;
end if;
end if;
end process;
txd<=sdata(0);
end Behavioral;
```

DESIGN EXAMPLE 7.10

Design a UART transmitter in the FPGA to send 0 through 255 through the RS232 communication port to the PC. The baud rate has to be 9600.

Solution

```
library IEEE;
use IEEE.STD_LOGIC_1164.ALL;
use IEEE.STD_LOGIC_ARITH.ALL;
use IEEE.STD_LOGIC_UNSIGNED.ALL;
entity ADC is
Port ( clk : in  STD_LOGIC:='0';
txd : out  STD_LOGIC:='0');
end ADC;
architecture Behavioural of ADC is
signal clk_sig:std_logic:='0';
```

```
signal mux_sig:std_logic:='0';
signal txd_temp:std_logic:='0';
signal sdata:std_logic_vector(8 downto 0):="000000000";
begin
p1:process(clk)
variable cnt1:integer:=1;
begin
if rising_edge(clk) then
if(cnt1=208)then  --208 is the baud rate counter value when the on-board
clock is 4MHz.
clk_sig<= not(clk_sig);
cnt1:=1;
else
cnt1:= cnt1 + 1;
end if;
end if;
end process;
p2:process(clk_sig)
variable cnt2:integer:=0;
begin
if rising_edge(clk_sig)then
if (cnt2>7) then
mux_sig<='1';
if(cnt2=8) then
sdata<=(sdata(0)& sdata(8 downto 1));
cnt2:=cnt2+1;
end if;
if (cnt2=14) then -- or (cnt2=11)
mux_sig<='0';
sdata<=sdata+"000000010";
cnt2:=0;
else
cnt2:=cnt2+1;
end if;
else
mux_sig<='0';
sdata<=(sdata(0)& sdata(8 downto 1));
cnt2:=cnt2+1;
end if;
case mux_sig is
when '0'=>txd_temp<=sdata(0);
when '1'=>txd_temp<='1';
when others=> null;
end case;
end if;
end process;
txd<=txd_temp;
end Behavioral;
```

DESIGN EXAMPLE 7.11

Design a UART transmitter in the FPGA to send '00'H through 'FF'H through the RS232 communication port to the PC. The baud rate has to be 9600. The rate of data transfer has to be 1 per second.

Solution

```
library IEEE;
use IEEE.STD_LOGIC_1164.ALL;
use IEEE.STD_LOGIC_ARITH.ALL;
```

```
use IEEE.STD_LOGIC_UNSIGNED.ALL;
entity clkd is
Port ( clk : in  STD_LOGIC:='0';
txd : out  STD_LOGIC:='0');
end clkd;
architecture Behavioral of clkd is
signal clk_sig:std_logic:='0';
signal mux_sig:std_logic:='0';
signal txd_temp:std_logic:='0';
signal sdata:std_logic_vector(10 downto 0):="11000000000";
begin
p1:process(clk)
variable cnt1:integer:=0;
begin
if rising_edge(clk) then
if(cnt1=208)then  --208 is the baud rate counter value when the on-board
clock is 4MHz.
clk_sig<= not(clk_sig);
cnt1:=1;
else
cnt1:= cnt1 + 1;
end if;
end if;
end process;
p2:process(clk_sig)
variable cnt2:integer:=1;
begin
if rising_edge(clk_sig)then
if (cnt2>10) then
mux_sig<='1';
if (cnt2=11) then
sdata<=(sdata(0)& sdata(10 downto 1));
cnt2:=cnt2+1;
end if;
if (cnt2=9690) then
if ( sdata="11111111110") then
sdata<="11000000000";
mux_sig<='0';
cnt2:=1;
else
sdata<=sdata + "00000000010";
cnt2:=1;
mux_sig<='0';
end if;
else
cnt2:=cnt2+1;
end if;
else
mux_sig<='0';
sdata<=(sdata(0)& sdata(10 downto 1));
cnt2:=cnt2+1;
end if;
case mux_sig is
when '0'=>txd_temp<=sdata(0);
when '1'=>txd_temp<='1';
when others=> null;
end case;
end if;
end process;
txd<=txd_temp;
end Behavioral;
```

DESIGN EXAMPLE 7.12

Design a UART the transmitter in the FPGA to read the status of eight DIP switches connected to the FPGA and send them to the PC through the RS232 communication port. The baud rate has to be 9600. The onboard frequency is 1 MHz.

Solution

```
library IEEE;
use IEEE.STD_LOGIC_1164.ALL;
use IEEE.STD_LOGIC_ARITH.ALL;
use IEEE.STD_LOGIC_UNSIGNED.ALL;
entity clkd is
Port ( clk : in  STD_LOGIC:='0';
s_data:in std_logic_vector(7 downto 0):="00000000";
txd:out std_logic:='0');
end clkd;
architecture Behavioral of clkd is
signal clk_sig,txd_temp,mux_sig:std_logic:='0';
signal sdata:std_logic_vector(10 downto 0):="11000000000";
begin
p1:process(clk)
variable cnt1:integer:=1;
begin
if rising_edge(clk) then
if(cnt1=52)then --52 is the baud rate counter value when the on-board
clock is 1MHz.
clk_sig<= not(clk_sig);
cnt1:=1;
else
cnt1:= cnt1 + 1;
end if;
end if;
end process;
p2:process(clk_sig,s_data)
variable cnt2:integer:=1;
begin
if rising_edge(clk_sig)then
if (cnt2>10) then
mux_sig<='1';
if (cnt2=11) then
sdata<=(sdata(0)& sdata(10 downto 1));
cnt2:=cnt2+1;
end if;
if (cnt2=16) then
sdata(8 downto 1)<=s_data;
mux_sig<='0';
cnt2:=1;
else
cnt2:=cnt2+1;
end if;
else
mux_sig<='0';
sdata<=(sdata(0)& sdata(10 downto 1));
cnt2:=cnt2+1;
end if;
case mux_sig is
when '0'=>txd_temp<=sdata(0);
when '1'=>txd_temp<='1';
when others=> null;
end case;
end if;
```

```
end process;
txd<=txd_temp;
end Behavioral;
```

DESIGN EXAMPLE 7.13

A signal conditioning unit sends series of data corresponding to the voltage measurement. The serial communication baud rate is 9600. Write MATLAB code to read the data arriving the serial port and use the received data to immediately update a plot in real time.

Solution

```
clc;
clear all;
s = serial('com1');
s.BaudRate=9600;
fopen(s);
time = now;
voltage = 0;
figureHandle = figure('NumberTitle','off',...
'Name','Voltage Characteristics',...
'Color',[0 0 0],'Visible','off');
axesHandle = axes('Parent',figureHandle,...
'YGrid','on',...
'YColor',[0.9725 0.9725 0.9725],...
'XGrid','on',...
'XColor',[0.9725 0.9725 0.9725],...
'Color',[0 0 0]);
hold on;
plotHandle = plot(axesHandle,time,voltage,'Marker','.','LineWidth',1,
'Color',[0 1 0]);
xlim(axesHandle,[min(time) max(time+0.001)]);
xlabel('Time in Hr:sec','FontWeight','bold','FontSize',14,'Color',[1 1 0]);
ylabel('Amplitude in V','FontWeight','bold','FontSize',14,'Color',[1 1 0]);
title('Arrived Beam Profile','FontSize',15,'Color',[1 1 0]);
stopTime = '10/15 10:00';
timeInterval = 0.005;
count = 1;
while ~isequal(datestr(now,'mm/DD HH:MM'),stopTime)
time(count) = datenum(clock);
a = (fread(s,1));
voltage(count) = a;
set(plotHandle,'YData',voltage,'XData',time);
set(figureHandle,'Visible','on');
datetick('x','mm/DD HH:MM');
pause(timeInterval);
count = count +1;
end
fclose(s);
delete(s);
clear s;
```

7.4.3 Universal Asynchronous Receiver-Transmitter Receiver Design

As we discussed in the previous section, a separate digital architecture for implementing the UART-Rx in the FPGA is required. The UART-Rx literally does the process of serial-to-parallel conversion on the asynchronous data frame received via the serial port. Initially, the data bit din that we receive from the serial port will be assigned to the variable start, and then, based on the status of the start, the process will be either in ready (if start = '1')

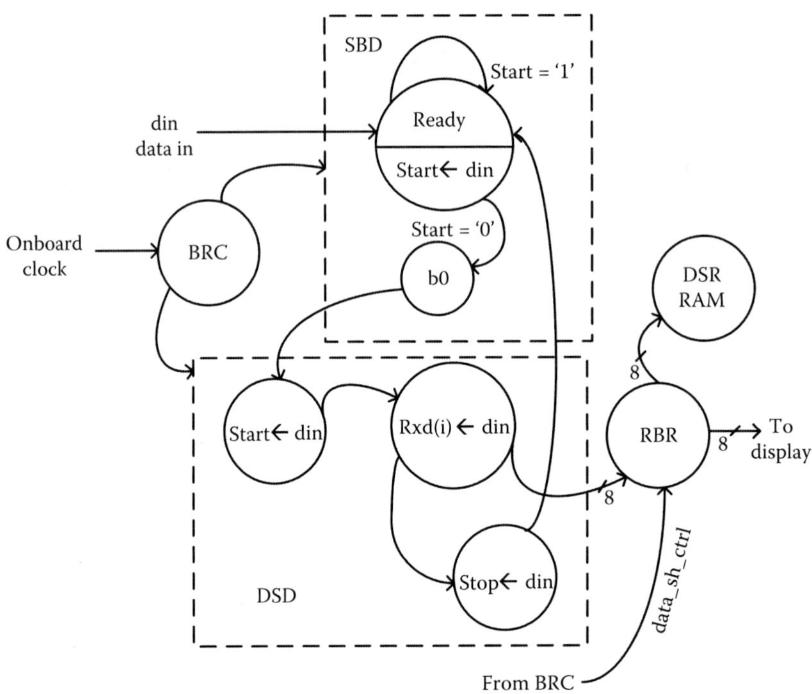

FIGURE 7.17
Data receiving process flow in an UART-Rx.

or bo (if start = '0') state, as shown in Figure 7.17. As long as the start is '1' (stop bit), the state is in ready. Since the serial frame is synchronous only to the receiver clock, a high-to-low transition of din will be treated as the start bit of a frame. All these processes take place in a start bit sequence detector (SBSD) module. The speed of the entire operation of the SBSD module is at the rate of onboard clock. Once a valid start bit is detected, then the state transition occurs from the state ready to state bo in the SBSD, and, after arriving in the state bo, it initiates the data sequence detector (DSD) module to begin the data-receiving process. Three different operations happen in the DSD module in accordance with the values of the baud rate counter (BRC).

First, the DSD module receives the start bit for a duration that is specified by the baud rate of data transmission. Once the duration for the start bit is over, then the DSD starts receiving the data sequence, that is, the subsequent 7 bits. Then, the process will be transferred to another process for receiving the stop bit. Once a valid start bit is received, the data bits will be sampled every clock (based on the receiving baud rate) at the center of the bit duration itself. At the end of the duration of the stop bit, the state will again be forced back to the ready state by the DSD module. Upon completing the receiving process for a byte (8 bits) of data, the DSD module forwards the received data sequence to the receiver buffer register (RBR) module to make the room at the DSD for the subsequent new receptions. The RBR module shifts the data to the data storage register (DSR) RAM as well as to a suitable (as required) data display device in accordance with the synchronized shift control data_sh_ctrl coming from the BRC. Data-storing RAM can be created inside the FPGA, or any external memory chip can be interfaced. The received data are accumulated for further processes, as required, for any applications for which they were sent into the FPGA.

DESIGN EXAMPLE 7.14

Write MATLAB code to write a word, "DIAT", to a device through the RS232 serial port.

SOLUTION

```
s = serial('COM1');
fopen(s)
fprintf(s,'DIAT')
```

DESIGN EXAMPLE 7.15

Write VHDL code to design a UART receiver in the FPGA to receive the data arriving via a serial port from the computer. The baud rate is 9600. Display the received data using LEDs.

Solution

```
library IEEE;
use IEEE.STD_LOGIC_1164.ALL;
use IEEE.STD_LOGIC_ARITH.ALL;
use IEEE.STD_LOGIC_UNSIGNED.ALL;
entity UART_Rx is
port(clk: in std_logic:='0';
din : in std_logic;
rxd : out std_logic_vector(7 downto 0):="00000000");
end UART_Rx;
architecture Behavioral of UART_Rx is
type state is (ready,b0);
signal ps:state:=ready;
signal start,stop:std_logic;
signal trxd: std_logic_vector(7 downto 0):="11111111";
begin
p1:process(clk)
variable i:integer:=0;
begin
if clk'event and clk ='1' then
if ps=ready then
start<= din;
end if;
if start='0' then
ps<= b0;
elsif start<='1' then
ps<=ready;
end if;
if ps=b0 then
i:=i+1;
if i=208 then  ----208 is because of the on-board frequency 4MHz
start<=din;
end if;
if i = 624 then
trxd(0) <=din;
end if;
if i = 1040 then
trxd(1) <=din;
end if;
if i = 1456 then
trxd(2) <=din;
end if;
if i = 1872 then
trxd(3) <=din;
end if;
```

```
if i = 2288 then
trxd(4) <=din;
end if;
if i = 2704 then
trxd(5) <=din;
end if;
if i = 3120 then
trxd(6) <=din;
end if;
if i = 3536 then
trxd(7) <=din;
end if;
if i = 3952 then
stop <=din;
end if;
if i = 4368 then
i:=0;
ps<=ready;
end if;
end if;
end if;
end process;
rxd<=trxd;
end Behavioral;
```

DESIGN EXAMPLE 7.16

Write VHDL code to build a UART receiver in the FPGA to receive the data arriving via the serial port from another device and display them on an LCD. Assume that the device sends four characters at the baud rate of 9600.

Solution

```
library IEEE;
use IEEE.STD_LOGIC_1164.ALL;
use IEEE.STD_LOGIC_ARITH.ALL;
use IEEE.STD_LOGIC_UNSIGNED.ALL;
entity wap is
port(clk: in std_logic:='0';
din : in std_logic;
c:out std_logic_vector(2 downto 0):="000";
d:out std_logic_vector(7 downto 0):="00000000");
end wap;
architecture Behavioral of wap is
type state is (ready,b0);
signal ps:state:=ready;
signal start,stop:std_logic;
signal trxd: std_logic_vector(7 downto 0):="11111111";
signal sta,nsta:integer range 0 to 19:=0;
signal coun:std_logic_vector(12 downto 0):="0000000000000";
begin
p1:process(clk)
variable i:integer:=0;
begin
if clk'event and clk ='1' then
if ps=ready then
start<= din;
end if;
if start='0' then
ps<= b0;
elsif start<='1' then
```

```
ps<=ready;
end if;
if ps=b0 then
i:=i+1;
if i=208 then     ----208 is because of the on-board frequency 4MHz
start<=din;
end if;
if i = 624 then
trxd(0) <=din;
end if;
if i = 1040 then
trxd(1) <=din;
end if;
if i = 1456 then
trxd(2) <=din;
end if;
if i = 1872 then
trxd(3) <=din;
end if;
if i = 2288 then
trxd(4) <=din;
end if;
if i = 2704 then
trxd(5) <=din;
end if;
if i = 3120 then
trxd(6) <=din;
end if;
if i = 3536 then
trxd(7) <=din;
end if;
if i = 3952 then
stop <=din;
end if;
if i = 4368 then
i:=0;
ps<=ready;
end if;
end if;
end if;
end process;
p2:process(clk)
variable cnt: integer:=0;
begin
if rising_edge(clk) then
if(cnt=400)then
sta<=nsta;
cnt:=0;
else
cnt:= cnt + 1;
end if;
end if;
end process;
p3:process(sta,clk)
begin
if clk'event and clk ='1' then
coun<=coun+"000000000001";
case sta is
when 0=> d<="00111000";c<="100"; nsta<=1;--38
when 1|3|5|7|9|18=> c<="000";nsta<=sta+1;
when 2=> d<="00001110";c<="100"; nsta<=3;--0e
```

```
when 4=> d<="00000001";c<="100"; nsta<=5;--01
when 6=> d<="00000110";c<="100"; nsta<=7;--06
when 8|19=> d<="10000000";c<="100"; nsta<=9;--80
when 10|12|14|16=>
if coun="1000100010000" then
d<=trxd;c<="110"; nsta<=sta+1;--Data
coun<="0000000000000";
else
nsta<=sta;
end if;
when 11|13|15|17=> c<="010";nsta<=sta+1;
end case;
end if;
end process;
end Behavioral;
```

DESIGN EXAMPLE 7.17

Write VHDL code to build a UART-Rx in the FPGA to receive the data arriving from the sensor module, which sends the measurement data via serial port. The onboard frequency is 50 MHz and baud rate has to be 9600. Assume four LEDs and one buzzer are connected to the FPGA. Develop digital architecture to perform some operations listed below based on the received character.

1. Make the LEDs glow upward for every reception of a character "R", that is, 52H.
2. Turn the buzzer "on" once the character "Z", that is, 5AH, is received.

Solution

```
library IEEE;
use IEEE.STD_LOGIC_1164.ALL;
use IEEE.STD_LOGIC_ARITH.ALL;
use IEEE.STD_LOGIC_UNSIGNED.ALL;
entity wap is
port(clk: in std_logic;
din : in std_logic;
op1 : out std_logic;
op2 : out std_logic;
op3  : out std_logic;
op4 : out std_logic;
buzzer : out std_logic:='0');
end wap;
architecture Behavioral of wap is
type state1 is (ready1,b01,b02,b03);
signal ps1: state1:=ready1;
type state is (ready,b0);
signal ps: state:=ready;
signal start, stop:std_logic;
signal store: std_logic_vector(7 downto 0):="10101010";
begin
p1:process(clk)
variable i:integer:=0;
begin
if clk'event and clk ='1' then
if ps = ready then
start <= din;
end if;
if start = '0' then
ps<=b0;
elsif start ='1' then
```

```
end if;
if ps =b0 then
i:=i+1;
if i = 2600 then  ----2600 (or  2604 ) is because of the on-board
frequency 50MHz
start <=din;
end if;
if i =7800 then
store(0) <=din;
end if;
if i =13000 then
store(1) <=din;
end if;
if i =18200 then
store(2) <=din;
end if;
if i =23400 then
store(3) <=din;
end if;
if i =28600 then
store(4) <=din;
end if;
if i =33800 then
store(5) <=din;
end if;
if i =39000 then
store(6) <=din;
end if;
if i =44200 then
store(7) <=din;
end if;
if i =49400 then
stop <=din;
end if;
if i =54600 then
i:=0;
ps<=ready;
end if;
end if;
end if;
end process;
p2:process(clk, store)
variable i: integer :=0;
begin
if clk'event and clk='1' then
if store = x"52" then
if ps1=ready1 then
i:= i+1;
if i=800000 then
op1<='0';
op2<='0';
op3<='0';
op4<='0';
buzzer<='0';
ps1<=b01;
i:=0;
end if;
end if;
if ps1=b01 then
i:=i+1;
```

```
if i=800000 then
op1<='0';
op2<='0';
op3<='1';
op4<='0';
buzzer<='0';
ps1<=b02;
i:=0;
end if;
end if;
if ps1=b02 then
i:=i+1;
if i=800000 then
op1<='0';
op2<='1';
op3<='0';
op4<='0';
buzzer<='0';
ps1<=b03;
i:=0;
end if;
end if;
if ps1=b03 then
i:=i+1;
if i=800000 then
op1<='1';
op2<='0';
op3<='0';
op4<='0';
buzzer<='0';
ps1<=ready1;
i:=0;
end if;
end if;
elsif store = x"5A" then
op1<='0';
op2<='0';
op3<='0';
op4<='0';
buzzer<='1';
end if;
end if;
end process;
end Behavioral;
```

7.5 Multichannel Data Logging

In Section 7.3, we dealt with signal digitization and single sensor interfacing techniques. In many real-world applications, interfacing more than one sensor and data acquisition through multiple channels has become significant to understand the behavior of many physical parameters simultaneously. In this section, we discuss multichannel data logging system design using various popular A/D converters. This section also discusses the direct channel addressing, multiplexed channel addressing, and serial data fetching A/D converter interfacing techniques.

7.5.1 ADC0808/ADC0809 Interfacing

Almost all transducers produce an output in the form of voltage, current, charge, capacitance, and resistance. We need to convert these signals to the equivalent voltage in order to send them into the A/D converter. This conversion/modification is commonly called signal conditioning. Therefore, all the sensor outputs, in the case where they are in forms other than voltage or where their voltage level is beyond the input limit of the A/D converter, have to be conditioned before sending them into the multichannel A/D converter. The ADC0808/ADC0809 is a simple 8-channel A/D converter which requires interfacing to a FPGA to control its entire operation to convert an analog signal into digital data. The circuit of the A/D converter configuration around the ADC0808 is shown in Figure 7.18. The ADC0808 is an 8-bit A/D converter that has data output lines D_0–D_7. It operates on the principle of successive approximation–based A/D conversion. It features a total of eight analog input channels, out of which any one can be chosen using the address lines [CBA], as given in Table 7.4.

As given in Table 7.4, the analog input channels CH0 can be chosen by forcing "000" to the address lines [CBA]. Typically, the control signals address latch enable (ALE), EoC, output enable (OE), and start conversion (SC) are interfaced to the FPGA, which controls them over time to perform A/D conversion, data encoding and data fetching. The execution flow of the control operations are: (i) select the analog channel by forcing the appropriate value (3 bits) to the Chan_Sel address lines [CBA]; (ii) force a logic high to the ALE; (iii) force a logic high to the SC; (iv) force a logic low to the ALE as well as the SC simultaneously; (v) once the conversion starts and the EoC signal goes high, wait as long as the EoC is logic high; it goes logic low once the conversion process is over; (vi) wait as long as the EoC is logic low; it goes back to logic high automatically; (vii) force logic high to OE; read the data that correspond to the selected analog channel; and (viii) force logic low to OE. At this instant in time, one measurement cycle for one analog channel is over, as shown in Figure 7.18. The same operation cycle has to be repeated for other channels as well.

Thus, it delivers continuous 8-bit digital output corresponding to the instantaneous value of the analog input signal (V_{in}). If the sensor output exceeds +5 V, it has to be properly scaled down below/equal to the reference level, for example, +5 V. The ADC0808

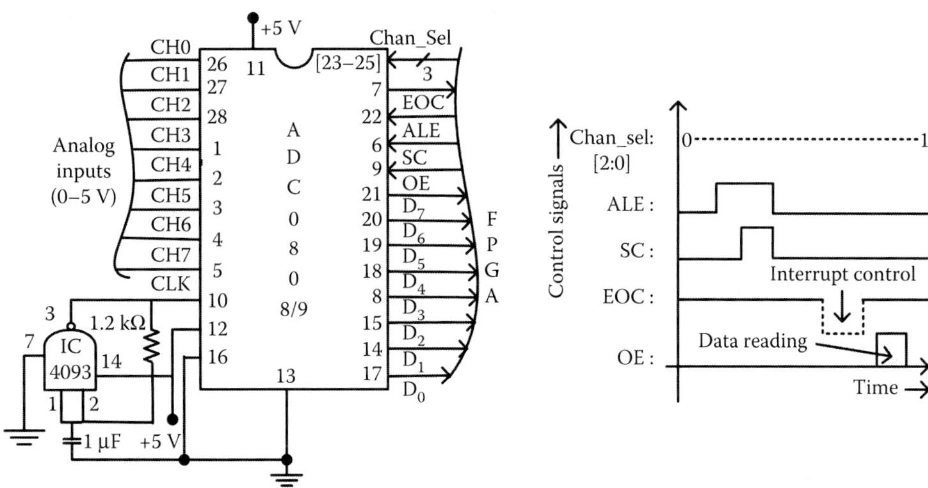

FIGURE 7.18
A/D converter application circuit and timing diagram of control signals.

TABLE 7.4

Analog Channel Selection Control

Analog Channel	Chan_Sel		
	C	B	A
CH0	0	0	0
CH1	0	0	1
CH2	0	1	0
CH3	0	1	1
CH4	1	0	0
CH5	1	0	1
CH6	1	1	0
CH7	1	1	1

normally needs a sampling clock signal of 550 kHz, which can be produced by an astable multivibrator constructed using a hex inverter (7404) or a quad 2-input NAND schematic trigger (4093). The following approximations would help us find many parameters related to A/D conversion operations [94]:

1. Quantization step (Q_{step}) of the A/D converter is ($V_{ref}/2^8$), which is also equal to (($V_{ref+} - V_{ref-})/2^8$) and
2. Digital data (D_2) is the radix-2 of ((V_{in}/Q_{step})−1)

For example, let us assume that the V_{ref+} is 5 V, V_{ref-} is 0 V, A/D converter resolution is 8 bits (parallel out) and V_{in} is 3 V, then

$$Q_{step} = \left(\frac{5-0}{256}\right) = 0.01953 = 19.53\ \text{mV} \quad \text{and}$$

$$D_2 = \left[\text{radix} - 2\left(\frac{3}{0.01953}\right) - 1\right] = \text{radix} - 2(152.6) = \text{radix} - 2(\approx 152) \quad \text{i.e.,}$$

$$D_2 = (10011000)_2 = 98\ \text{H}$$

Thus, the output of the ADC0808 or ADC0809 varies between 00H and FFH in accordance with the instantaneous voltage of the analog signal at the selected channel.

DESIGN EXAMPLE 7.18

Design a digital system in FPGA to interface an 8-bit parallel out A/D converter (ADC0808). Assume that there are three sensors connected at Channel 0, Channel 1 and Channel 3 of the ADC. Display the measurement values using the LEDs.

Solution

```
library IEEE;
use IEEE.STD_LOGIC_1164.ALL;
use IEEE.STD_LOGIC_ARITH.ALL;
use IEEE.STD_LOGIC_UNSIGNED.ALL;
```

```
entity ADC is
Port ( clk : in  STD_LOGIC:='0';
EOC:in std_logic;
ALE,SC,OE:out std_logic:='0';
data_in:in std_logic_vector(7 downto 0):="00000000";
data_out1:out std_logic_vector(7 downto 0):="00000000";
data_out2:out std_logic_vector(7 downto 0):="00000000";
data_out3:out std_logic_vector(7 downto 0):="00000000";
chan_sel:out std_logic_vector(2 downto 0):="000");
end ADC;
architecture Behavioral of ADC is
signal st,nst: integer range 0 to 12:=0;
signal chan_sel_cnt_sig:std_logic:='0';
signal chan_sel_temp:std_logic_vector(2 downto 0):="000";
signal data_in_temp:std_logic_vector(7 downto 0):="00000000";
signal enable:std_logic:='0';
begin
p0:process(clk)
begin
if rising_edge(clk) then
st<=nst;
end if;
end process;
p1:process(st,EOC) is
begin
case st is
when 0=>ALE<='1';nst<=st+1;
when 1=>SC<='1';nst<=st+1;
when 2=>ALE<='0'; SC<='0';nst<=st+1;
when 3=>if EOC='1' then nst<=st;else nst<=st+1;end if;
when 4=>if EOC='0' then nst<=st;else nst<=st+1;end if;
when 5=>OE<='1';nst<=st+1;
when 6=>data_in_temp<=data_in;nst<=st+1;
when 7=>OE<='0';nst<=st+1;
when 8=>enable<='1';nst<=st+1;
when 9=>enable<='0';nst<=st+1;
when 10=>chan_sel_cnt_sig<='1';nst<=st+1;
when 11=>chan_sel_cnt_sig<='0';nst<=st+1;
when 12=>nst<=0;
when others=> nst<=0;
end case;
end process;
p2:process(chan_sel_cnt_sig)
variable cnt1:integer:=0;
begin
if rising_edge(chan_sel_cnt_sig) then
if (cnt1=2) then
cnt1:=0;
else
cnt1:=cnt1+1;
end if;
case cnt1 is
when 0 =>chan_sel_temp<="000";
when 1 =>chan_sel_temp<="001";
when 2 =>chan_sel_temp<="011";
when others=> cnt1:=0;
end case;
end if;
end process;
chan_sel<=chan_sel_temp;
p3:process(enable)
```

```
begin
if rising_edge(enable) then
case chan_sel_temp(1 downto 0) is
when "00" =>data_out1<=data_in_temp;
when "01" =>data_out2<=data_in_temp;
when "11" =>data_out3<=data_in_temp;
when others=>data_out1<="00000000";data_out2<="00000000";
end case;
end if;
end process;
end Behavioral;
```

DESIGN EXAMPLE 7.19

Write VHDL code to build a digital system in the FPGA to interface with an 8-bit A/D converter. Three sensors are connected to channel 0, channel 1 and channel 3 of the ADC0808. Also, build a UART to send the measurement values to the PC through the RS232 serial communication port. Assume that the baud rate is 9600 and the onboard frequency is 4 MHz.

Solution

```
library IEEE;
use IEEE.STD_LOGIC_1164.ALL;
use IEEE.STD_LOGIC_ARITH.ALL;
use IEEE.STD_LOGIC_UNSIGNED.ALL;
entity ADC is
Port ( clk : in  STD_LOGIC:='0';
EOC:in std_logic;
ALE,SC,OE:out std_logic:='0';
txd: out std_logic:='0';
data_in:in std_logic_vector(7 downto 0):="00000000";
chan_sel:out std_logic_vector(2 downto 0):="000");
end ADC;
architecture Behavioral of ADC is
signal st,nst: integer range 0 to 12:=0;
signal chan_sel_cnt_sig:std_logic:='0';
signal chan_sel_temp:std_logic_vector(2 downto 0):="000";
signal data_in_temp:std_logic_vector(7 downto 0):="00000000";
signal enable:std_logic:='0';
signal clk_sig,mux_sig,txd_temp:std_logic:='0';
signal sdata:std_logic_vector(10 downto 0):="11001100000";
signal data_out1,data_out2,data_out3:std_logic_vector(7 downto
0):="00000000";
begin
p0:process(clk)
begin
if rising_edge(clk) then
st<=nst;
end if;
end process;
p1:process(st,EOC) is
begin
case st is
when 0=>ALE<='1';nst<=st+1;
when 1=>SC<='1';nst<=st+1;
when 2=>ALE<='0'; SC<='0';nst<=st+1;
when 3=>if EOC='1' then nst<=st;else nst<=st+1;end if;
when 4=>if EOC='0' then nst<=st;else nst<=st+1;end if;
when 5=>OE<='1';nst<=st+1;
when 6=>data_in_temp<=data_in;nst<=st+1;
```

```
when 7=>OE<='0';nst<=st+1;
when 8=>enable<='1';nst<=st+1;
when 9=>enable<='0';nst<=st+1;
when 10=>chan_sel_cnt_sig<='1';nst<=st+1;
when 11=>chan_sel_cnt_sig<='0';nst<=st+1;
when 12=>nst<=0;
when others=> nst<=0;
end case;
end process;
p2:process(chan_sel_cnt_sig)
variable cnt1:integer:=0;
begin
if rising_edge(chan_sel_cnt_sig) then
if (cnt1=2) then
cnt1:=0;
else
cnt1:=cnt1+1;
end if;
case cnt1 is
when 0 =>chan_sel_temp<="000";
when 1 =>chan_sel_temp<="001";
when 2 =>chan_sel_temp<="011";
when others=> cnt1:=0;
end case;
end if;
end process;
chan_sel<=chan_sel_temp;
p3:process(enable)
begin
if rising_edge(enable) then
case chan_sel_temp(1 downto 0) is
when "00" =>data_out1<=data_in_temp;
when "01" =>data_out2<=data_in_temp;
when "11" =>data_out3<=data_in_temp;
when others=>data_out1<="00000000";data_out2<="00000000";
end case;
end if;
end process;
---------------UART --------------
p4:process(clk)
variable cnt2:integer:=0;
begin
if rising_edge(clk) then
if(cnt2=208)then  ------208 for 9600 baud rate
clk_sig<= not(clk_sig);
cnt2:=1;
else
cnt2:= cnt2 + 1;
end if;
end if;
end process;
p5:process(clk_sig)
variable cnt3,cnt4:integer:=1;
begin
if rising_edge(clk_sig)then
if (cnt3>10) then
mux_sig<='1';
if (cnt3=11) then
sdata<=('0'& sdata(10 downto 1));
cnt3:=cnt3+1;
end if;
```

```
if (cnt3=1610) then
if (cnt4=1)then
sdata<="11000000000";
sdata(8 downto 1)<=data_out1;
cnt3:=1;
mux_sig<='0';
cnt4:=cnt4+1;
elsif (cnt4=2)then
sdata<="11000000000";
sdata(8 downto 1)<=data_out2;
cnt3:=1;
mux_sig<='0';
cnt4:=cnt4+1;
elsif (cnt4=3)then
sdata<="11000000000";
sdata(8 downto 1)<=data_out3;
cnt3:=1;
mux_sig<='0';
cnt4:=1;
end if;
else
cnt3:=cnt3+1;
end if;
else
mux_sig<='0';
sdata<=('0'& sdata(10 downto 1));
cnt3:=cnt3+1;
end if;
case mux_sig is
when '0'=>txd_temp<=sdata(0);
when '1'=>txd_temp<='1';
when others=> null;
end case;
end if;
end process;
txd<=txd_temp;
end Behavioral;
```

7.5.2 ADC0848 Interfacing

The ADC0848 is an eight-channel, 8-bit-resolution A/D converter with a maximum of 256 (2^8) quantization levels. We can scale down the V_{in} of the ADC0848 to produce full-scale output for 2.56 V just by setting the V_{ref} to 2.56 V. The voltage connected to the V_{ref} pin decides the step size of the quantization level; hence, for this A/D converter, the step size is (V_{ref}/256). The circuit configuration around the ADC0848 and the timing diagram of its control signals are shown in Figure 7.19. The control signals of ADC0848 are CS, Write (WR), Interrupt (INTR) and Read (RD). CS and RD are the active low input used to activate the ADC0848 chip. The A/D converter converts the analog input to its equivalent binary data and holds it in an internal register. The RD, also called OE, is used to get the converted data out of the ADC0848. When CS is logic 0, if the RD pin is asserted low, then the 8-bit digital data will be fetched out and available at the D_0–D_7 data pins.

If CS is logic 0 when WR makes a low-to-high transition, the ADC0848 latches the address of the selected channel and starts conversion of the analog input value to 8-bit digital data. The channel address has to be applied when the CS is logic 0 and RD is logic 1; applying a low-to-high transition to WR begins the conversion. The INTR, also called EoC, is an output of ADC0848 and is active low, which is normally logic 1, and when the conversion

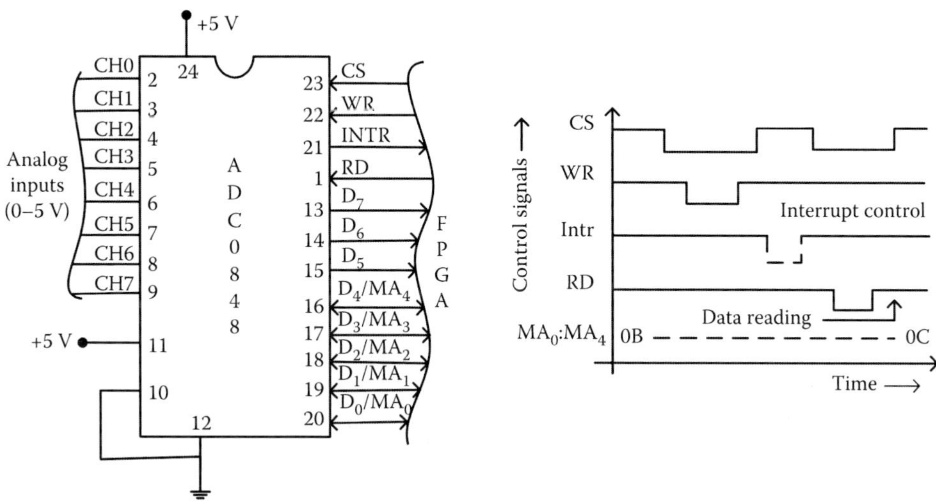

FIGURE 7.19
Application circuit of ADC0848 and timing diagram of control signals.

is finished, it goes logic 0 to signal the FPGA that the converted data is ready to read out. After the INTR goes low, we make CS logic 0 and apply a low pulse to the RD pin to get the binary data out [95].

As shown in Figure 7.19, a portion of the data line (D_0–D_4) pins are also designated as MA_0–MA_4 which are the multiplexed address/data lines that would be used to choose any particular analog channel, as well as to choose the mode of operation of the A/D converter, as given in Table 7.5. CH0–CH7 are the eight channels of analog inputs. D_0–D_4, that is, the MA_0–MA_4 pins, will act as the inputs when we send the channel address. Whenever the converted data is being read, D_0–D_4 are the output pins. This dual operation, that is, use of multiplexed address/data access, saves the interfacing pins. The ADC0848 can be operated in two different modes, namely (i) single-ended mode (mode 1): each of the eight channels (CH0–CH7) can be used for separate analog inputs, and MA_4 and MA_3, for this mode, have to be logic 0 and logic 1, respectively, as in Table 7.5; and (ii) differential mode (mode 2): two channels, such as CH0 and CH1, are paired together for inputting

TABLE 7.5
Multiplexed Channel Selection Controls

Analog Channel	MA4		MA3		MA2	MA1	MA0
	Mod1	**Mod2**	**Mod1**	**Mod2**			
CH0	0	X	1	0	0	0	0
CH1	0	X	1	0	0	0	1
CH2	0	X	1	0	0	1	0
CH3	0	X	1	0	0	1	1
CH4	0	X	1	0	1	0	0
CH5	0	X	1	0	1	0	1
CH6	0	X	1	0	1	1	0
CH7	0	X	1	0	1	1	1

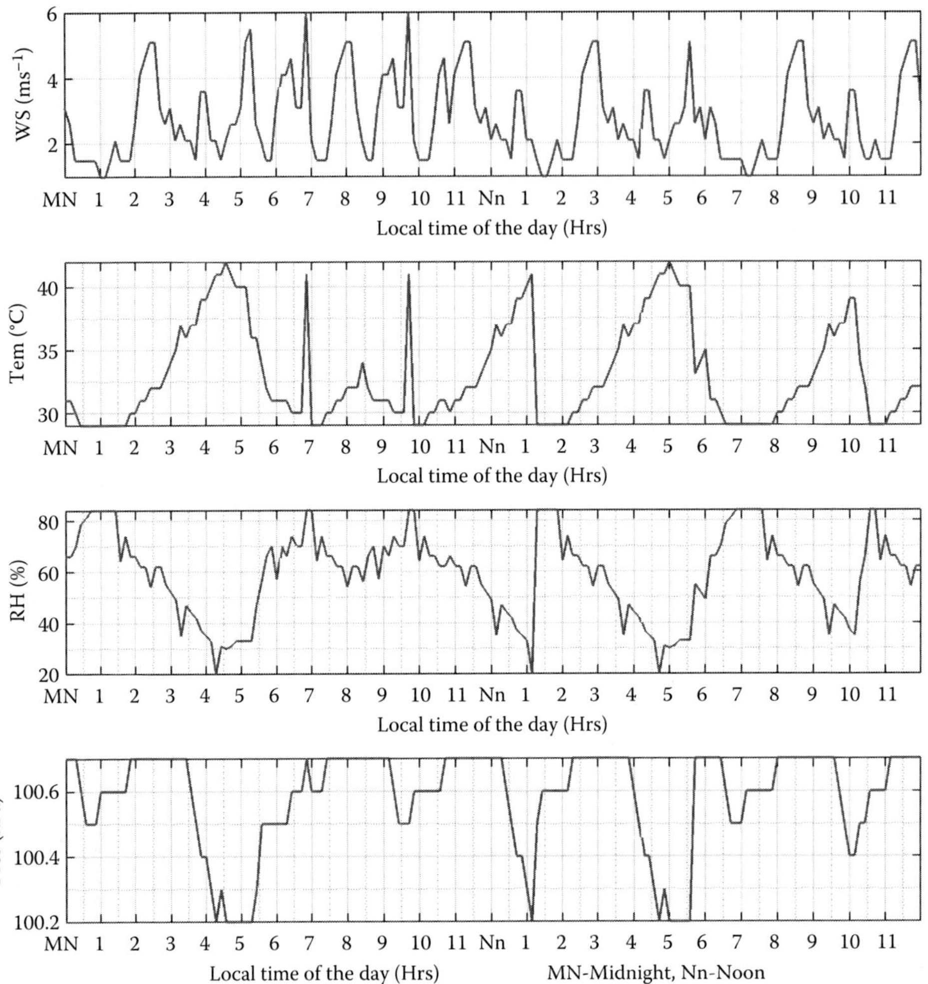

FIGURE 7.20
A diurnal period profile of weather data.

the differential analog input signals as (V_{in+} and V_{in-}), and MA_4 and MA_3, for this mode, are "don't care" (X) and logic 0, as given in Table 7.5. All these processes for channel selection, mode selection, signal conversion, and control signal management have to be repeated for every measurement.

DESIGN EXAMPLE 7.20

Four sensors, namely wind speed (Ws), temperature (T), relative humidity (RH) and pressure (Pr), are connected to the first four channels of an ADC0848. Develop a digital system in the FPGA to control the entire operation of the ADC0848 and send the converted data to a computer through the RS232 serial port. Assume that the sensors value vary from 00H through FAH; hence, the synchronization header may be FFH [39,62,63].

Solution

```
library ieee;
use ieee.std_logic_1164.all;
```

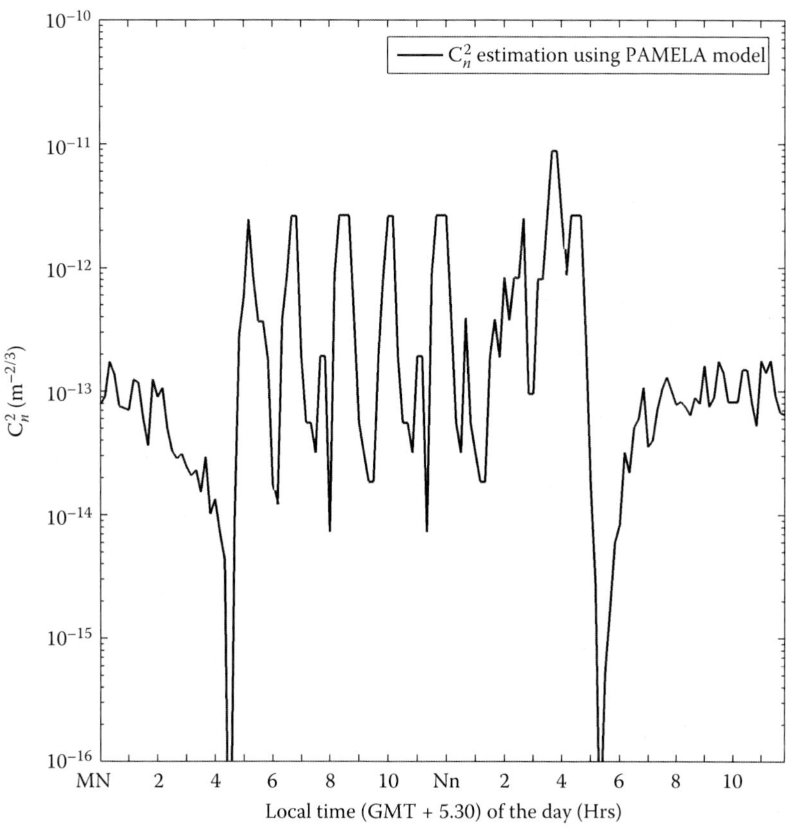

FIGURE 7.21
A diurnal period profile of C_n^2.

```
use ieee.std_logic_arith.all;
use ieee.std_logic_unsigned.all;
entity adc_0848 is
port( CS,WR,RD,txd : out std_logic:='0';
m_clk : in std_logic:='0';
data_d5_d7: in std_logic_vector(2 downto 0):="000";
INTR: in std_logic:='1';
data_add:inout std_logic_vector(4 downto 0):="00000");
end adc_0848;
architecture Pa_ADC of adc_0848 is
signal sw_ctrl, enable,clk_sig,mux_sig,txd_temp : std_logic:='0';
signal add, add_temp,data: std_logic_vector(4 downto 0):="01000";
signal state, next_state : integer range 0 to 19:=0;
signal data_temp,Ws, T, RH, Pr:std_logic_vector(7 downto 0):="00000000";
signal sdata:std_logic_vector(10 downto 0):="11000000000";
begin
data_add <= add when (sw_ctrl='0') else
"ZZZZZ" when (sw_ctrl='1');
data <= data_add;
p0: process(m_clk)
variable cnt: integer:=1;
begin
if rising_edge(m_clk) then
if (cnt=2) then
```

```
state<=next_state;
cnt:=1;
else
cnt:=cnt+1;
end if;
end if;
end process;
P1: process(state,INTR,data,data_d5_d7)
begin
case state is
when 0=> sw_ctrl<='0'; CS<='1'; next_state<=1;
when 1=> RD<='1'; next_state<=2;
when 2=> RD<='1'; next_state<=3;
when 3=> add<=Add_temp; next_state<=4;
when 4=> CS<='0'; next_state<=5;
when 5=> WR<='0'; next_state<=6;
when 6=> CS<='0'; next_state<=7;
when 7=> WR<='1'; next_state<=8;
when 8=> CS<='1'; next_state<=9;
when 9=> sw_ctrl<='1'; next_state<=10;
when 10=> if INTR='1' then next_state<=10;
else next_state<=11;
end if;
when 11=> if INTR='0' then next_state<=11;
else next_state<=12;
end if;
when 12=> CS<='0'; next_state<=13;
when 13=> RD<='0'; next_state<=14;
when 14=> RD<='1'; next_state<=15;
when 15=> data_temp<=(data_d5_d7 & data); next_state<=16;
when 16=> CS<='1'; next_state<=17;
when 17=> enable<='1'; next_state<=18;
when 18=> enable<='0'; next_state<=19;
when 19=> next_state<=0;
when others=> next_state<=0;
end case;
end process;
p2: process(enable)
begin
if rising_edge(enable) then
if (add_temp="01011") then
add_temp<="01000";
else
add_temp<= add_temp + "00001";
end if;
if add_temp(2 downto 0)="000" then
Ws<=data_temp;
elsif add_temp(2 downto 0)="001" then
T<=data_temp;
elsif add_temp(2 downto 0)="010" then
RH<=data_temp;
elsif add_temp(2 downto 0)="011" then
Pr<=data_temp;
end if;
end if;
end process;
---------------UART --------------
p3:process(m_clk)
variable cnt2:integer:=0;
begin
if rising_edge(m_clk) then
```

```
if(cnt2=208)then  ------208 for 9600 baud rate
clk_sig<= not(clk_sig);
cnt2:=1;
else
cnt2:= cnt2 + 1;
end if;
end if;
end process;
p4:process(clk_sig)
variable cnt3,cnt4:integer:=1;
begin
if rising_edge(clk_sig)then
if (cnt3>10) then
mux_sig<='1';
if (cnt3=11) then
sdata<=('0'& sdata(10 downto 1));
cnt3:=cnt3+1;
end if;
if (cnt3=1610) then
if (cnt4=1)then
sdata<="11000000000";
sdata(8 downto 1)<="11111111";
cnt3:=1;
mux_sig<='0';
cnt4:=cnt4+1;
elsif (cnt4=2)then
sdata<="11000000000";
sdata(8 downto 1)<=Ws;
cnt3:=1;
mux_sig<='0';
cnt4:=cnt4+1;
elsif (cnt4=3)then
sdata<="11000000000";
sdata(8 downto 1)<=T;
cnt3:=1;
mux_sig<='0';
cnt4:=cnt4+1;
elsif (cnt4=4)then
sdata<="11000000000";
sdata(8 downto 1)<=RH;
cnt3:=1;
mux_sig<='0';
cnt4:=cnt4+1;
elsif (cnt4=5)then
sdata<="11000000000";
sdata(8 downto 1)<=Pr;
cnt3:=1;
mux_sig<='0';
cnt4:=1;
end if;
else
cnt3:=cnt3+1;
end if;
else
mux_sig<='0';
sdata<=('0'& sdata(10 downto 1));
cnt3:=cnt3+1;
end if;
case mux_sig is
when '0'=>txd_temp<=sdata(0);
when '1'=>txd_temp<='1';
```

```
when others=> null;
end case;
end if;
end process;
txd<=txd_temp;
end Pa_ADC;
```

A portion of measurement of weather data and corresponding turbulence strength (C_n^2) are shown in Figures 7.20 and 7.21, respectively.

DESIGN EXAMPLE 7.21

Assume that the sensors are connected to the first four analog channels of an A/D converter ADC0848. The sensors are Ws, T, RH, and Pr. A data logging system reads the sensors' voltage and sends the digital data corresponding to the instantaneous voltages to a computer through the RS232 serial port. The data logging unit sends the measurement data with a header as "FFHWsTRHPr". Write MATLAB code to read the data and write them in the columns I, J, K, and L of an Excel file (sheet 1) starting from the fourth cell. Once the writing is over, read them and plot all the values in a figure. The length of the data for every measurement cycle is 169.

Solution

```
clc;
clear all;
s = serial('Com1');
s.BaudRate=9600;
s.TimeOut=100;
s.Parity='none';
s.InputBufferSize=floor(10000*10);
fopen(s);
time = now;
stopTime = 'Oct-15 (Sat) 18:57:00';
count=1;
temp=0;
data_ID=1;
Ws_temp=0;
Tinoc_temp=0;
cc_temp=0;
prinkpa_temp=0;
while ~isequal(datestr(now,'mmm-dd (ddd) HH:MM:SS'),stopTime)
data_ID(count)=count;
a=fread(s,1,'uint8')';
if a ==255,
a=fread(s,1,'uint8')';
Ws_temp(count)=a;
a=fread(s,1,'uint8')';
Tinoc_temp(count)=a;
a=fread(s,1,'uint8')';
RH_temp(count)=a;
a=fread(s,1,'uint8')';
prinkpa_temp(count)=a;
count = count +1;
end
end
fclose(s);
clear s;
Ws=Ws_temp';
Tinoc= Tinoc_temp';
RH=RH_temp';
```

```
prinkpa=prinkpa_temp';
%%%%% Writing Part
xlswrite('C:\Users\DIAT\Documents\MATLAB\T_F_book.xls',Ws,1,'I4');
xlswrite('C:\Users\DIAT\Documents\MATLAB\T_F_book.xls',Tinoc,1,'J4');
xlswrite('C:\Users\DIAT\Documents\MATLAB\T_F_book.xls',RH,1,'K4');
xlswrite('C:\Users\DIAT\Documents\MATLAB\T_F_book.xls',prinkpa,1,'L4');
%%%%% Reading and Plotting Part
ws=xlsread('C:\Users\DIAT\Documents\MATLAB\T_F_book.xls','sheet1',
'I4:I172');
Tinoc=xlsread('C:\Users\DIAT\Documents\MATLAB\T_F_book.xls','sheet1',
'J4:J172');
RH=xlsread('C:\Users\DIAT\Documents\MATLAB\T_F_book.xls','sheet1',
'K4:K172');
Prinkpa=xlsread('C:\Users\DIAT\Documents\MATLAB\T_F_book.xls','sheet1',
'L4:L172');
vertical_line_gap=7;
vertical_lines=1:vertical_line_gap:24*vertical_line_gap;
vertical_line_label= ['  MN    ';'   1    ';'   2    ';'   3    ';'   4
';'   5    ';.......
'   6    ';'   7    ';'   8    ';'   9    ';'  10    ';'  11    ';......
'  Nn    ';'   1    ';'   2    ';'   3    ';'   4    ';'   5    ';........
'   6    ';'   7    ';'   8    ';'   9    ';'  10    ';'  11    '];
time=[1:1:169];
figure(1)
subplot(4,1,1)
plot(time,ws,'linewidth',1.2)
grid on;
ylim ([min(ws) max(ws)]);
set(gca,'Xlim',[1,169],'XTick',vertical_lines,'XTicklabel',vertical_
line_label);
xlabel('\fontname{Times New Roman}\fontsize{10} Local time of the day
(Hrs)');
ylabel('\fontname{Times New Roman}\fontsize{10} WS (ms^{-1})');
subplot(4,1,2)
plot(time,Tinoc,'linewidth',1.2);
hold on;
ylim([min(Tinoc) max(Tinoc)]);
set(gca,'Xlim',[1,169],'XTick',vertical_lines,'XTicklabel',vertical_
line_label);
grid on ;
xlabel('\fontname{Times New Roman}\fontsize{10} Local time of the day
(Hrs)');
ylabel('\fontname{Times New Roman}\fontsize{10} Tem (^oC )');
grid minor;
subplot(4,1,3)
plot(time,RH,'linewidth',1.2)
grid on;
ylim ([min(RH) max(RH)]);
set(gca,'Xlim',[1,169],'XTick',vertical_lines,'XTicklabel',vertical_
line_label);
xlabel('\fontname{Times New Roman}\fontsize{10} Local time of the day
(Hrs)');
ylabel('\fontname{Times New Roman}\fontsize{10} RH (%)');
grid minor;
subplot(4,1,4)
plot(time,Prinkpa,'linewidth',1.2)
grid on;
ylim ([min(Prinkpa) max(Prinkpa)]);
set(gca,'Xlim',[1,169],'XTick',vertical_lines,'XTicklabel',vertical_
line_label);
xlabel('\fontname{Times New Roman}\fontsize{10} Local time of the day (Hrs)
```

```
MN-Might Night, Nn- Noon');
ylabel('\fontname{Times New Roman}\fontsize{10} Pres (kPa)');
grid minor;
```

DESIGN EXAMPLE 7.22

Assume that a data-logging system acquired weather data at an uniform interval and stored them in an Excel file. Write MATLAB code to estimate the estimate the atmospheric turbulence strength (C_n^2) using the PAMELA model and also plot the estimated values of C_n^2. Note that all other necessary values are used appropriately in the code itself [39,47,64,78].

Solution

The PAMELA model provides atmospheric turbulence strength (C_n^2) estimation within the surface boundary layer, and it accepts all the parameters of a test field, like geographical location, meteorological values, and optical path, as the inputs. The required geographical inputs are latitude (10°38′46.7334″ and 10°38′52.8468″), longitude (79°3′12.0774″ and 79°2′56.6268″), time of day (diurnal period; GMT+5.30), terrain type (0.03 m-open flat terrain, grass, few isolated obstacles), number of days (as applicable), height above the ground (15.25 m), and meteorological parameters at the desired altitude (15.25 m) of the experiment for estimating C_n^2. The measured meteorological parameters are subsequently applied in the PAMELA model (for which the MATLAB code given below is designed), the turbulence strength is estimated and the necessary plots are updated. A brief tour of the design of PAMELA model is given below, followed by the MATLAB code. Readers can find complete details of the PAMELA model in [89].

The measured solar irradiance R is used to determine the radiation class $c_r = R/300$. For a wind-speed Ws, define the windspeed class $c_w =$ `{0.27 if Ws ≤0.27 else Ws}`, and then the Pasquill stability category P can be determined by

$$P = \frac{-(4 - c_w + c_r)}{2}$$

The length of the surface roughness for open flat terrain, grass and a few isolated obstacles is estimated from tables, $z_r = 0.03$ m, and from this it is possible to calculate the Obukhov buoyancy length scale bl

$$bl = \left[\left(a_1 P + a_2 P^3\right) z_r^{-(a_3 - a_4|P| + a_5 P^2)}\right]^{-1}$$

where $a_1 = 0.004349$, $a_2 = 0.003724$, $a_3 = 0.5034$, $a_4 = 0.231$, and $a_5 = 0.0325$. The mean vertical velocity W and fluctuating part w, horizontal velocity U and fluctuating part u define vertical momentum flux in terms of the eddy viscosity K_m and mean potential temperature Θ and fluctuating part θ, and they also define vertical heat flux in terms of the eddy diffusivity of heat K_h by

$$\overline{uw} = -K_m\left(\frac{\partial U}{\partial Z}\right) \text{ and } \overline{\theta w} = -K_h\left(\frac{\partial \Theta}{\partial z}\right)$$

The mean specific humidity Q and fluctuating part q define the vertical water vapor flux using the eddy diffusivity of water vapor K_w by

$$\overline{qw} = -K_w\left(\frac{\partial Q}{\partial Z}\right)$$

The dimensionless wind shear $\varphi_m(\zeta)$ and the dimensionless potential temperature gradient $\varphi_h(\zeta)$ are expressed as functions of the scaled buoyancy parameter $\zeta = z/L$. The turbulent exchange coefficients for heat K_h and momentum K_m are by

$$K_h = \frac{ku*z}{\varphi_h(\zeta)} \text{ and } K_m = \frac{ku*z}{\varphi_m(\zeta)} \tag{7.10}$$

where $k \cong 0.4$ is von Karman's constant. $K_h = K_m$ as per the optical turbulence model for laser propagation and imaging applications. The friction velocity u* and characteristic temperature T* from the wind speed Ws and the roughness length $z_{r,}$ heat flux H, specific heat c_p and mass density ρ are given by

$$u^* = \frac{kWs}{\ln(z/Z_r)} \text{ and } T^* = \frac{kWs}{c_p\rho u^*} \tag{7.11}$$

The atmospheric refractive index (n) in terms of pressure (Pr) and temperature (T) is

$$n-1 = \frac{77.6\times10^{-6}\,Pr}{T}\left(1+\frac{7.52\times10^{-3}}{\lambda^2}\right) \text{ and } \frac{dn}{dz} = -\frac{77.6\times10^{-6}PrT^*\varphi_h(\varsigma)}{0.4zT^2}$$

The eddy dissipation rate ε and C_n^2 are estimated with the constant $b \approx 2.8$ as

$$\in = \frac{u_*^3(\varphi_m - \varsigma)}{0.4z} \text{ and } C_n^2 = \frac{2.8K_h}{\in^{1/3}}\left(\frac{dn}{dz}\right)^2 \tag{7.12}$$

The direct relationship can be seen by expanding Equations 7.10 to 7.12 as

$$C_n^2 = 5.152\varphi_h\left(\frac{1}{\varphi_m-\zeta}\right)^{0.33}\left(\frac{77.6\times10^{-6}Pr}{T^2}\right)^2 h^{-0.667}\left(\frac{-H}{C_\rho\rho u^*}\right)^2 \tag{7.13}$$

More detailed derivation of the PAMELA model can be found in [39,64]. The MATLAB code for the estimation of C_n^2 in $m^{-2/3}$ using the above models is given below:

```
clc;
clear all;
tstart=18.30;
long=79.049;
lat=10.6472;
hr=0.03;
h=15.25;
Nd=input('Enter the day number in that year: ');
ws_in=xlsread('C:\Users\DIAT\Documents\MATLAB\paper2_plots.xls','sheet1',
'I4:I172');
Temp_in=xlsread('C:\Users\DIAT\Documents\MATLAB\paper2_plots.xls','sheet1',
'J4:J172');
Rh=xlsread('C:\Users\DIAT\Documents\MATLAB\paper2_plots.xls','sheet1',
'K4:K172');
Pr_in=xlsread('C:\Users\DIAT\Documents\MATLAB\paper2_plots.xls','sheet1',
'L4:L172');
time=1:1:length(ws_in);
```

```
gmt1=tstart:0.167:tstart+24;
d=(Nd-1)*(360/365.24);
r=279.93+d;
m=(  12+( 0.12357*sind(d) )-( 0.004289*cosd(d) )+....
( 0.1538*sind(2*d) )+( 0.06078*cosd(2*d) )  );
for i=1:1:144,
Ws=ws_in(i);
if Ws<0.1,
vo=0.1;
else
vo=Ws;
end;
Temp=Temp_in(i);
T=Temp+273.15;
if Rh(i) >=0 && Rh(i)<=10, cc=0;
elseif Rh(i)>=11 && Rh(i)<=20,  cc=1;
elseif Rh(i)>=21 && Rh(i)<=30,  cc=2;
elseif Rh(i)>=31 && Rh(i)<=40,  cc=3;
elseif Rh(i)>=41 && Rh(i)<=50,  cc=1;
elseif Rh(i)>=51 && Rh(i)<=60,  cc=4;
elseif Rh(i)>=61 && Rh(i)<=70,  cc=5;
elseif Rh(i)>=71 && Rh(i)<=80,  cc=6;
elseif Rh(i)>=81 && Rh(i)<=90,  cc=7;
elseif Rh(i)>=91,  cc=8;
end;
Pr=Pr_in(i);
Pa=(Pr*10);
gmt=gmt1(i);
ht=(15*(gmt-m)-long);
beta=( r+ 0.4087*sind(r) +  1.8724*cosd(r) - 0.0182*sind(2*r) .....
- 0.0083*cosd(2*r) ) ;
delta=( asin( 0.39785*sind(beta) ) );
alfa=asin(  (sind(lat)*sin(delta) )+....
cosd(lat)*cos(delta)*cosd(ht)  );
slz=90.57- alfa*57.29;
tc=0.57+0.0045*slz;
itemp=1353*cosd(slz)* tc^(secd(slz));
if itemp< 0,
I=0;
else
I=itemp;
end
B=[1.07 0.89 0.81 0.76 0.72 0.67 0.59 0.45 0.23];
b=B(cc+1);
R=b*I;
if I > 0,
H=0.4*(R-100);
else
H=-40;
end
if H > 0,
crtemp=R/300;
elseif (H <= 0 && cc >= 4);
crtemp=-1;
elseif (H <= 0 && cc < 4);
crtemp=-1 ;
end
if crtemp <= 0
cr=3;
else
cr=crtemp;
```

```
end
if vo<=0.27, %8
cw=0.27;%vo/2;
else % if %vo>8
cw=vo;%4;
end
if cc==8
Pt=0;
else
if H<=0
Pt=0.5*(4-cw+abs(cr));
elseif H>0
Pt=-0.5*(4-cw+abs(cr));
end
end
L=((0.004349*Pt+0.003724*Pt^3)*hr^-(0.5034-0.231*abs(Pt)+
0.0325*Pt^2))^-1;
E=(h/L);
if Pt<=0
pim=(1-16*E)^-0.25;
else
pim=1+5*E;
end
if Pt<=0
y=(1-16*E)^0.25;
sim=log(((1+y^2)/2)*((1+y)/2)^2)-2*atan(y)+pi/2;
else
sim=-5*E;
end;
if Pt<=0
fih=0.74*(1-9*E)^-0.5;
else
fih=0.74+4.7*E;
end;
Ustar=((0.4*vo)/(log(h/hr)-sim));
row=(Pa/(2.87*T));
Tstar=(-H/(1004*row*Ustar));
Kh=(0.4*Ustar*h)/fih;
dndz=-(((77.6*10^-6)*Pa*Tstar*fih)/(0.4*h*T^2));
epsalan=(pim-E)*((Ustar^3)/(0.4*h));
Cn21(i)=( (2.8*Kh*dndz^2)/(epsalan^0.33333) );
end
figure(7);
time=1:length(Cn21);
p1=semilogy(time,Cn21);
set(gca,'fontsize',12);
hold on;
set(p1,'Color','k','LineWidth',1.5);
xlabel({'Local time (GMT+ 5.30) of the day (Hrs)';'( a )'});
ylabel('C_n^{2} (m^{-2/3})');
vertical_line_gap=6;
vertical_lines=1:vertical_line_gap:24*vertical_line_gap;
vertical_line_label= ['  MN    ';'        ';'   2    ';'        ';'   4
';'        ';.......
'   6    ';'        ';'   8    ';'        ';'  10    ';'        ';......
'   Nn   ';'        ';'   2    ';'        ';'   4    ';'        ';.........
'   6    ';'        ';'   8    ';'        ';'  10    ';'        '];
set(gca,'Xlim',[1,144],'XTick',vertical_lines,'XTicklabel',vertical_
line_label);
legend('C_n^2 Estimation using PAMELA model');
ylim([1E-16 1E-10]);
```

7.5.3 Analog to Digital Converter MAX1112 Interfacing and Serial Data Fetching

All the A/D converters we have discussed before are of the parallel type, which means the data (D_0-D_7) are simultaneously fetched by the FPGA from A/D converter. In this way, in a 16-bit parallel A/D converter, we need to reserve 16 pins for the data path and some more pins for the control lines. In many real-time applications, space is a critical issue for using this many pins for data and control lines. For this reason, serial A/D converters are widely used since they reduce the number of pins required for data as well as control. In this section, we discuss the operation principles and interfacing techniques of a serial A/D converter (MAX1112). The MAX1112 is an 8-bit serial A/D converter with eight channels of analog inputs, CH0–CH1. It has a single pin to bring out the digital data after it has been converted [96].

It is compatible with the popular serial peripheral interface (SPI) standard. In single-ended mode, each of the channels can be used for separate analog inputs, whereas in differential mode, we can have a four sets of two-channel differential inputs as CH0 and CH1 together, CH2 and CH3 together, and so on. We select the input channel in single-ended mode, by sending the appropriate control byte via the data input (D_{in}) pin. The structure of the control data pattern, its possible control bits and corresponding operational descriptions are given in Table 7.6. CS is an active low input used to select the MAX1112 chip.

To send the control byte via the D_{in} pin, the CS must be logic 0. Serial clock (SCLK) input is used to bring data out via the D_{out} pin and send in the control byte one bit at a time. The digital data is clocked out on the H-to-L transition (falling edge) of SCLK. Serial data in the control byte is clocked in on the L-to-H transition (rising edge) of SCLK. The serial strobe (SSTRB), which is also called EoC, in internal clock mode, indicates the end of conversion. It goes high when the conversion is complete.

The Shutdown (SHDN) pin is an input and is normally left unconnected or is connected to VDD. If it is at logic 0, the A/D converter will shut down to save power to the A/D converter itself. The D/O bit in the control byte causes shutdown by software, that is, the FPGA. The Reference (R_{efin}) voltage input dictates the step size. The circuit configuration of this serial A/D converter is shown in Figure 7.22. The control bytes has to be send-in, with the MSB going first followed by the other bits, as shown in Figure 7.23. In single-ended mode, with VDD = 5 V, we get 4.096 V for full-scale if the R_{efin} pin is connected to the AGND with

TABLE 7.6

Structure of the Control Byte Register

Start (S) (MSB)	S_2	S_1	S_0	C/B	S/D	P/O	E/I (LSB)
S (MSB)	1—Start bit.						
S_2-S_0	0 through 7 select analog channels.						
C/B	0—2's complement output. 1—Binary output.						
S/D	0—Differential mode of operation. 1—Single-ended mode of operation.						
D/O	0—Power down. 1—Fully operational.						
E/I (LSB)	0—Internal conversion clock. 1—External conversion clock.						

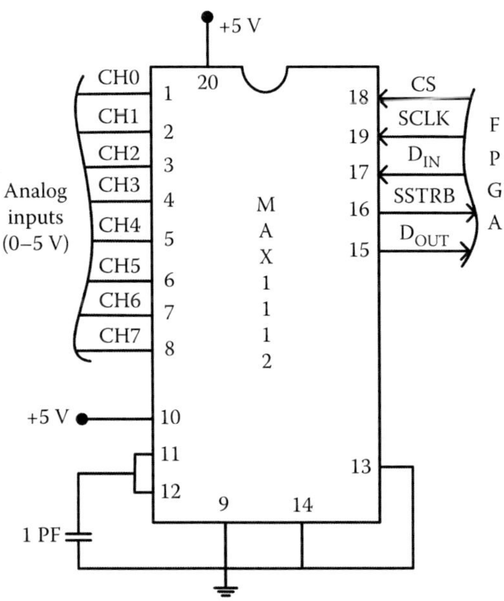

FIGURE 7.22
Application circuit of MAX1112 A/D converter.

a 1-PF capacitor, which gives us a 16-mV step size, as 4.096 V/256 = 16 mV. To get a 10-mV step size, we need to connect the R_{efin} pin to 2.56 V. As shown in Figure 7.23, after the SSTRB pin goes high, the second falling edge of SCLK produces the MSB of converted data, which is available at the D_{out} pin. In other words, we need nine pulses to get data out [96].

DESIGN EXAMPLE 7.23

Develop a digital system in the FPGA to acquire data in single-ended mode from eight sensors connected to the MAX1112 A/D converter. The data of first four channels have to be in 2's complement and the rest in binary form. Interface with the MAX1112 in fully operational mode by applying the conversion clock from the FPGA. Display the data in eight rows of LEDs.

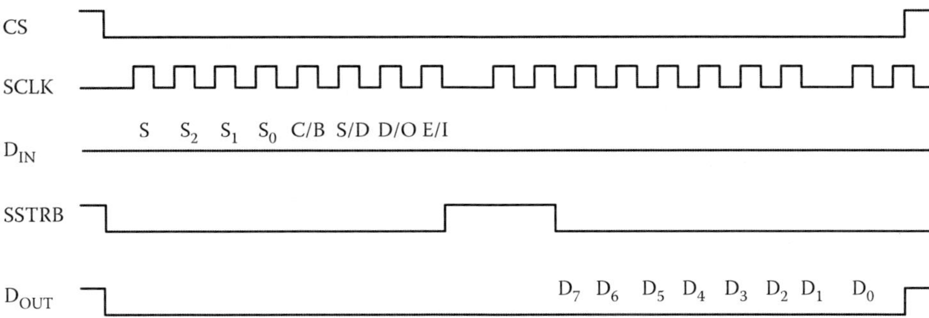

FIGURE 7.23
Timing diagram of single conversion process of MAX1112 A/D converter.

Solution

```
library ieee;
use ieee.std_logic_1164.all;
use ieee.std_logic_arith.all;
use ieee.std_logic_unsigned.all;
use IEEE.numeric_bit.all;
entity adc_max1112 is
port( CS: out std_logic:='1';
sclk,con_bit : out std_logic:='0';
m_clk, data_bit : in std_logic:='0';
sstrb : in std_logic:='1';
d0,d1,d2,d3,d4,d5,d6,d7: out std_logic_vector(7 downto 0):="00000000");
end adc_max1112;
architecture se_ADC of adc_max1112 is
signal sw_ctrl, enable,clk_sig,mux_sig,txd_temp,cb : std_logic:='0';
signal state, next_state : integer range 0 to 19:=0;
signal con_byt,data_temp:std_logic_vector(7 downto 0):="00000000";
signal cha_add:std_logic_vector(2 downto 0):="000";
begin
p0: process(enable)
begin
if rising_edge(enable) then
cha_add<=cha_add + "001";
end if;
if cha_add >= "100" then
cb <='1';
else
cb <='0';
end if;
end process;
p1: process(m_clk)
variable cnt: integer:=1;
begin
if rising_edge(m_clk) then
if (cnt=1) then
state<=next_state;
cnt:=1;
else
cnt:=cnt+1;
end if;
end if;
end process;
P2: process(state,sstrb,data_bit)
variable cnt: integer:=1;
begin
case state is
when 0=> next_state<=state+1;
when 1=> cs<='0'; next_state<=state+1;
when 2=> con_byt <=('1' & cha_Add & cb & "111"); next_state<=state+1;
when 3=> next_state<=state+1;
when 4=> con_bit <= con_byt(7); next_state<=state+1;
when 5=> sclk<='1'; next_state<=state+1;
when 6=> sclk<='0';
con_byt <= con_byt (6 downto 0) & con_byt (7);
if (cnt=8) then
cnt:=1;
next_state<=state+1;
else
cnt:=cnt+1;
next_state<=4;
```

```
end if;
when 7=> next_state<=state+1;
when 8=> next_state<=state+1;
when 9=> if sstrb='0' then next_state<=9;
else next_state<=state+1;
end if;
when 10=> if sstrb='1' then next_state<=10;
else next_state<=state+1;
end if;
when 11=> sclk<='1'; next_state<=state+1;
when 12=> sclk<='0'; next_state<=state+1;
when 13=> sclk<='1'; next_state<=state+1;
when 14=> sclk<='0'; next_state<=state+1;
when 15=> data_temp <=data_temp(6 downto 0) & data_bit;
if cnt=8 then
cnt:=1;
next_state<=state+1;
else
cnt:=cnt+1;
next_state<=13;
end if;
when 16=> case cha_add is
when "000" => d0<=data_temp;
when "001" => d1<=data_temp;
when "010" => d2<=data_temp;
when "011" => d3<=data_temp;
when "100" => d4<=data_temp;
when "101" => d5<=data_temp;
when "110" => d6<=data_temp;
when "111" => d7<=data_temp;
when others=> null;
end case;
next_state<=state+1;
when 17=> enable<='1'; next_state<=state+1;
when 18=> enable<='0'; next_state<=state+1;
when 19=> next_state<=0;
when others=> next_state<=0;
end case;
end process;
end se_ADC;
```

7.6 Bipolar Signal Conditioning and Data Logging

In the previous sections, we discussed many unipolar (single ended) and differential-mode A/D converters of 8-bit resolutions. In this section, we discuss the working principles and interfacing techniques of a bipolar 12-bit-resolution A/D converter. One popular real-time application, optical position detection, which is possible with only a bidirectional A/D converter, is also discussed.

7.6.1 Bidirectional Analog to Digital Converter Interfacing

One popular bidirectional (bipolar) single channel 12-bit resolution A/D converter is the AD1674, which has a built-in sample and hold amplifier. The typical accuracy and resolution requirements for many industrial applications, analog measurements, transducer interfacing, industrial monitoring, and laboratory measurements are mostly met by this type of 12-bit A/D converter. The AD1674 accepts unipolar as well as bipolar analog input signals. A provision to interface eight analog inputs (CH0–CH7) to this A/D converter can be made using an analog multiplexer ADG508A [39].

This analog multiplexer enables us to acquire a signal out of eight different analog sources. The current level that can be dealt with by this A/D converter is 4–20 mA, and by adding a single appropriate resistor in shunt with the inputs, the current signals can also be directly handled as the voltage signals at the inputs of the analog multiplexer. The analog signals can be passed to the multiplexer through a nine-pin terminal strip with a common analog ground. This interface allows us to route any one of the analog channels at a given time to the AD1674 based on the value of the selection lines (chan_sel) of the multiplexer. The industry-standard interface is built around the AD1674 with built-in reference, clock and sample-hold circuit. The external interface module has a 26-pin connector at one edge of the board through which the interface to FPGA can be made with a flat cable connector. External power supplies +12 V and −12 V are connected to the points marked +12 V and −12 V, respectively, on the board. A photograph of this A/D converter board consisting of AD1674, ADG508A, 74LS374, control pins and other input/output ports is shown in Figure 7.24. A full schematic (circuit layout) diagram for eight-channel bipolar analog inputs is shown in Appendix A. The possible different input voltage levels at different operating modes of the AD1674 are 0 to +10 V, 0 to +20 V, −5 to +5 V and −10 to +10 V. The AD1674 can be operated with a +5 V, ±12 V or ±15 V power supply. The voltage levels of the inputs are decided by locating the shorting jumpers JP3 and JP2 appropriately. The JP2 is at "AB" and JP3 is at "BC" for unipolar 0 to +10 V, JP2 is at "AB" and JP3 is at "AB" for bipolar −5 to +5 V and JP2 is at "BC" and JP3 is at "AB" for bipolar −10 to +10 V, as shown in Figure 7.24. On-chip multiple-mode three-state output buffers and its interface logic allow us to have a direct connection to the FPGA.

As shown in Appendix A, a jumper (JP1) decides whether the interface is intended for single-channel or eight-channel operation. When single-channel operation is intended, no multiplexer is needed and JP1 is open, which means the input is directly applied to the A/D converter core. When eight-channel operation is desired, the JP1 is in the closed condition, which means it links the multiplexer output to the AD1674. The analog channel selection at the multiplexer is given in Table 7.7. No channel will be selected once the enable (E) is set at '0'; otherwise, the respective analog input will be routed to the input of the AD1674 based on the values of chan_sel, that is, [C B A]. The interfacing of the A/D converter board to the FPGA is as follows: A bit, SC, is used to commence

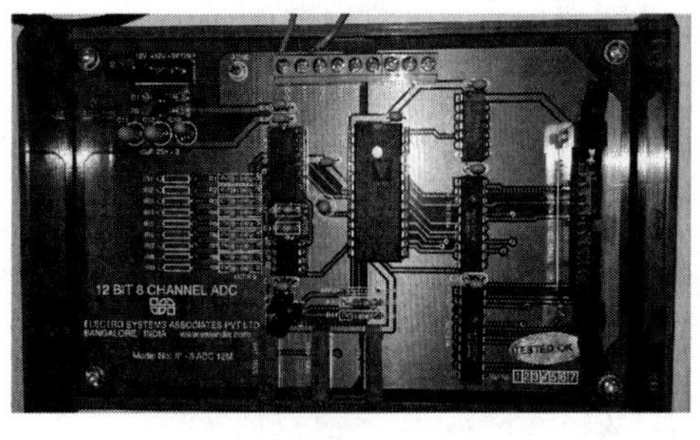

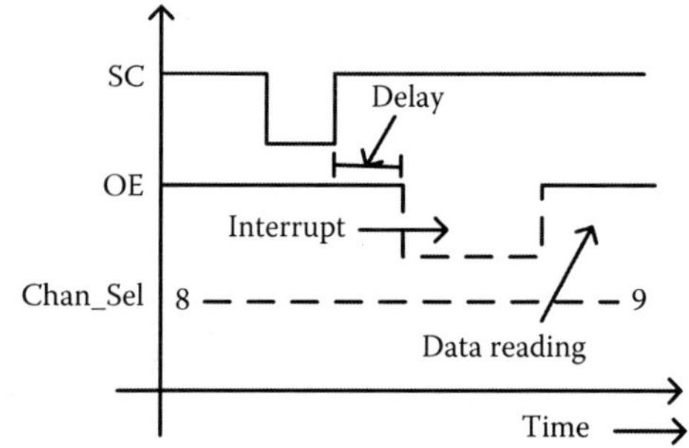

FIGURE 7.24
Photograph of a signal conditioning board using ADC1674 core and its operational timing diagram.

TABLE 7.7
ADC1674 Enable and Channel Selection Details

Enable (E)	`Chan_sel` C	B	A	Selected Channel
0	x	x	x	None
1	0	0	0	0 (8)
1	0	0	1	1 (9)
1	0	1	0	2 (A)
1	0	1	1	3 (B)
1	1	0	0	4 (C)
1	1	0	1	5 (D)
1	1	1	0	6 (E)
1	1	1	1	7 (F)

the conversion. High-to-low transition at SC initiates the conversion, and EoC of AD1674 goes high, indicating that the A/D converter is busy. At the end of the conversion, the EoC line goes low. This transition is used to strobe the converted data into the latches and then to FPGA. The converted data D_0–D_{11} is latched out from a 16 bit D-flip-flop (74LS374). OE controls the operation/condition of this latch. When OE is high, the outputs of the latch are tristated, and when OE is low, output data (D_0-D_{11}) are available at the output ports. We may wait for the status change at the EoC or need to provide some time delay, typically 0.1 μsec, as shown in Figure 7.24, for activating the OE to read out the data from the FPGA.

DESIGN EXAMPLE 7.24

Develop a digital system in the FPGA to acquire data from three sensors that are connected at Ch0, Ch1 and Ch2 using a 12-bit A/D converter (ADC1674). The sensors' output voltage varies from –10 V to +10 V. Assume that three sets of 16 LEDs are already connected to the FPGA onboard.

Solution

```
library IEEE;
use IEEE.STD_LOGIC_1164.ALL;
use IEEE.STD_LOGIC_ARITH.ALL;
use IEEE.STD_LOGIC_UNSIGNED.ALL;
entity ADC is
Port (  clk : in  STD_LOGIC:='0';
chan_sel:out std_logic_vector(3 downto 0):="1000";
data_in:in std_logic_vector(15 downto 0):="0000000000000000";
data_out1,data_out2,data_out3: out std_logic_vector(15 downto
0):="0000000000000000";
sc,oe: out  STD_LOGIC:='1');
end ADC;
architecture Behavioral of ADC is
signal clk_sig0,enable:std_logic:='0';
signal chan_sel_cnt_sig:std_logic:='0';
signal chan_sel_temp:std_logic_vector(3 downto 0):="1000";
signal data_in_temp:std_logic_vector(15 downto 0):="0000000000000000";
```

```
begin
p0:process(clk)
variable cnt1:integer:=0;
begin
if rising_edge(clk) then
if(cnt1=5) then
clk_sig0<= not(clk_sig0);
cnt1:=1;
else
cnt1:= cnt1 + 1;
end if;
end if;
end process;
p1:process(clk_sig0)
variable cnt0:integer:=1;
begin
if rising_edge(clk_sig0) then
if (cnt0=1) then sc<='1';
oe<='1';enable<='0';chan_sel_cnt_sig<='0';cnt0:=cnt0+1;
elsif (cnt0=3) then sc<='0';
oe<='1';enable<='0';chan_sel_cnt_sig<='0';cnt0:=cnt0+1;
elsif (cnt0=5) then sc<='1';
oe<='1';enable<='0';chan_sel_cnt_sig<='0';cnt0:=cnt0+1;
elsif (cnt0=25)then sc<='1';
oe<='0';enable<='0';chan_sel_cnt_sig<='0';cnt0:=cnt0+1;
elsif (cnt0=30)then sc<='1';
oe<='0';enable<='0';chan_sel_cnt_sig<='0';data_in_temp<=data_
in;cnt0:=cnt0+1;
elsif (cnt0=35)then sc<='1';
oe<='1';enable<='1';chan_sel_cnt_sig<='0';cnt0:=cnt0+1;
elsif (cnt0=40)then sc<='1';
oe<='1';enable<='0';chan_sel_cnt_sig<='0';cnt0:=cnt0+1;
elsif (cnt0=45)then sc<='1';
oe<='1';enable<='0';chan_sel_cnt_sig<='1';cnt0:=cnt0+1;
elsif (cnt0=50)then sc<='1';
oe<='1';enable<='0';chan_sel_cnt_sig<='0';cnt0:=cnt0+1;
elsif (cnt0=150)then sc<='1';
oe<='1';enable<='0';chan_sel_cnt_sig<='0';cnt0:=1;
else                                    cnt0:=cnt0+1;
end if;
end if;
end process;
p2:process(enable)
begin
if rising_edge(enable) then
case chan_sel_temp(1 downto 0) is
when "00" =>data_out1<=data_in_temp;
when "01" =>data_out2<=data_in_temp;
when "10" =>data_out3<=data_in_temp;
when others=>data_out1<="0000000000000000";
                        data_out2<="0000000000000000";
                        data_out3<="0000000000000000";
end case;
end if;
end process;
p3:process(chan_sel_cnt_sig)
variable cnt1:integer:=0;
begin
if rising_edge(chan_sel_cnt_sig) then
if (cnt1=2) then
cnt1:=0;
```

```
else
cnt1:=cnt1+1;
end if;
case cnt1 is
when 0 =>chan_sel_temp<="1000";
when 1 =>chan_sel_temp<="1001";
when 2 =>chan_sel_temp<="1010";
when others=> cnt1:=0;
end case;
end if;
end process;
chan_sel<=chan_sel_temp;
end Behavioral;
```

7.6.2 Opto-Electronic Position Detector Interfacing

An OPD is used to measure the centroid position of a circular laser beam. This is mainly used in free space optical (FSO) communication systems, beam steering, adaptive optics, beam focusing, and so on. In all these applications, the reflected or direct beam is made to fall on the OPD. The OPD is constructed with four separate identical silicon photodiodes denoted by A, B, C, and D and arranged in a quadrant geometry, as shown in Figure 7.25. The output voltages of these photodiodes (V_A, V_B, V_C, and V_D) are given to the monopulse arithmetic circuit (MPAC), which is constructed using operational amplifiers for accomplishing the addition and subtraction of V_A, V_B, V_C and V_D to compute the displacement of the arrived beam. The beam centroid displacement errors, azimuth position error (V_{Ex}) and elevation position error (V_{Ey}), along the x and y channels (2D plane) of the OPD are measured as the relative output voltage changes by [39,97,98]

$$\begin{aligned} V_{Ex} &= (V_A + V_C) - (V_B + V_D) \\ V_{Ey} &= (V_A + V_B) - (V_C + V_D) \end{aligned} \tag{7.14}$$

The reference signal (V_{Ref}) is measured by the algebraic sum of all the signals from the quadrants of the OPD as

$$V_{Ref} = (V_A + V_B + V_C + V_D) \tag{7.15}$$

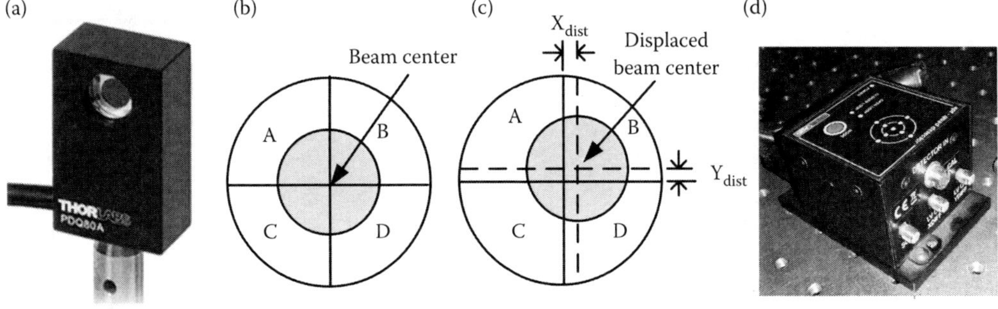

FIGURE 7.25
Appearance of an OPD (a), center beam spot on an OPD surface (b), displaced beam spot on OPD (c), and MPAC (d).

The azimuth and elevation distance are measured by

$$X_{dist} = -2\left(\frac{V_{Ex}}{V_{Ref}}\right)$$
$$Y_{dist} = 2\left(\frac{V_{Ey}}{V_{Ref}}\right) \quad (7.16)$$

where 2 is the maximum radial distance (γ) of the OPD. The minus sign is used due to the formulation of Equation 7.14. The radial distance (γ) from the center of the OPD to the center of the beam can be estimated using Equations 7.14 through 7.16 by

$$\gamma = \sqrt{X_{dist}^2 + Y_{dist}^2} \quad (7.17)$$

The output values of V_{Ex}, V_{Ey} and V_{Ref} vary from −10 to +10 V and from 0 to +10 V, respectively. These signals can be conditioned only using a bidirectional A/D converter and hence applied to the 12-bit parallel data-fetching A/D (AD1674) through an eight-channel analog multiplexer (ADG509A) circuit, as discussed in the previous section. The output of AD1674 is 000H, that is, $(0)_{10}$, for −10 V and FFH, that is, $(4095)_{10}$, for +10 V [37,66,91,97]. The digitized outputs are given to the FPGA via a very-high-density cable interconnect (VHDCI). The information about the beam position displacements on the OPD can be logged in a computer over a longer period to perform on/offline analysis.

DESIGN EXAMPLE 7.25

Develop a digital system in the FPGA to read the wind speed from a cup anemometer and the laser beam centroid displacement error from an OPD. The MPAC of the OPD is connected to the FPGA through a bidirectional A/D converter (AD1674 core). Display the wind speed measurement data on 14 LEDs and send all the measurement (wind speed, x-displacement, y-displacement and reference) values to a computer using a UART at a baud rate of 9600. Assume that the OPD values vary from −10 V to +10 V and the FPGA onboard clock frequency is 4 MHz. Design the UART with an appropriate data synchronisation pilot (header) sequence.

Solution

```
library IEEE;
use IEEE.STD_LOGIC_1164.ALL;
use IEEE.STD_LOGIC_ARITH.ALL;
use IEEE.STD_LOGIC_UNSIGNED.ALL;
entity ADC is
Port (   clk,ws_data_in : in  STD_LOGIC:='0';
chan_sel:out std_logic_vector(3 downto 0):="1000";
data_in:in std_logic_vector(15 downto 0):="0000000000000000";
ws_data_out: out std_logic_vector(13 downto 0):="00000000000000";
txd,sc,oe,ws_ff_reset: out  STD_LOGIC:='1' );
end ADC;
architecture Behavioral of ADC is
signal clk_sig,clk_sig0,tx_cnt:std_logic:='0';
signal mux_sig:std_logic:='0';
signal sdata:std_logic_vector(8 downto 0):="000000000";
signal data_sum,data_sum_temp:std_logic_vector(15 downto
0):="0000000000000000";
```

```
signal data_xdiff,data_xdiff_temp:std_logic_vector(15 downto
0):="0000000000000000";
signal data_ydiff,data_ydiff_temp:std_logic_vector(15 downto
0):="0000000000000000";
signal enable:std_logic:='0';
signal chan_sel_cnt_sig:std_logic:='0';
signal chan_sel_temp:std_logic_vector(3 downto 0):="1000";
signal data_in_temp:std_logic_vector(15 downto 0):="0000000000000000";
signal ws_reg,data_ws_temp:std_logic_vector(13 downto
0):="00000000000000";
signal  ws_clk_sig,ws_reg_reset:std_logic:='0';
begin
p0:process(clk)
variable cnt1:integer:=0;
begin
if rising_edge(clk) then
if(cnt1=5)then
clk_sig0<= not(clk_sig0);
cnt1:=1;
else
cnt1:= cnt1 + 1;
end if;
end if;
end process;
p1:process(clk_sig0)
variable cnt0:integer:=1;
begin
if rising_edge(clk_sig0) then
if (cnt0=1) then
sc<='1';oe<='1';enable<='0';chan_sel_cnt_sig<='0';cnt0:=cnt0+1;
elsif (cnt0=3) then
sc<='0';oe<='1';enable<='0';chan_sel_cnt_sig<='0';cnt0:=cnt0+1;
elsif (cnt0=5) then
sc<='1';oe<='1';enable<='0';chan_sel_cnt_sig<='0';cnt0:=cnt0+1;
elsif (cnt0=25)then
sc<='1';oe<='0';enable<='0';chan_sel_cnt_sig<='0';cnt0:=cnt0+1;
elsif (cnt0=30)then
sc<='1';oe<='0';enable<='0';chan_sel_cnt_sig<='0';data_in_
temp<=data_in;cnt0:=cnt0+1;
elsif (cnt0=35)then
sc<='1';oe<='1';enable<='1';chan_sel_cnt_sig<='0';cnt0:=cnt0+1;
elsif (cnt0=40)then
sc<='1';oe<='1';enable<='0';chan_sel_cnt_sig<='0';cnt0:=cnt0+1;
elsif (cnt0=45)then
sc<='1';oe<='1';enable<='0';chan_sel_cnt_sig<='1';cnt0:=cnt0+1;
elsif (cnt0=50)then
sc<='1';oe<='1';enable<='0';chan_sel_cnt_sig<='0';cnt0:=cnt0+1;
elsif (cnt0=150)then
sc<='1';oe<='1';enable<='0';chan_sel_cnt_sig<='0';cnt0:=1;
else                                          cnt0:=cnt0+1;
end if;
end if;
end process;
p2:process(enable)
begin
if rising_edge(enable) then
case chan_sel_temp(1 downto 0) is
when "00" =>data_sum<=data_in_temp;
when "01" =>data_xdiff<=data_in_temp;
when "10" =>data_ydiff<=data_in_temp;
```

```
when
others=>data_sum<="0000000000000000";data_xdiff<="0000000000000000";
data_ydiff<="0000000000000000";
end case;
end if;
end process;
p3:process(chan_sel_cnt_sig)
variable cnt1:integer:=0;
begin
if rising_edge(chan_sel_cnt_sig) then
if (cnt1=2) then
cnt1:=0;
else
cnt1:=cnt1+1;
end if;
case cnt1 is
when 0 =>chan_sel_temp<="1000";
when 1 =>chan_sel_temp<="1001";
when 2 =>chan_sel_temp<="1010";
when others=> cnt1:=0;
end case;
end if;
end process;
chan_sel<=chan_sel_temp;
p4:process(clk)
variable ws_cnt1:integer:=0;
begin
if rising_edge(clk) then
if(ws_cnt1=400)then
ws_clk_sig<= not(ws_clk_sig);
ws_cnt1:=1;
else
ws_cnt1:= ws_cnt1 + 1;
end if;
end if;
end process;
p5:process(ws_clk_sig,ws_data_in,ws_reg_reset)
variable ws_cnt2: integer:=0;
begin
if (ws_reg_reset='1') then
ws_reg<="00000000000000";
ws_cnt2:=0;
ws_ff_reset<='1';
elsif (rising_edge(ws_clk_sig))then
if(ws_cnt2=0) then  ws_ff_reset<='1';ws_cnt2:=ws_cnt2+1;
elsif(ws_cnt2=1) then if (ws_data_in ='0') then
ws_ff_reset<='1';ws_cnt2:=ws_cnt2; else ws_cnt2:=ws_cnt2+1;end if;
elsif(ws_cnt2=3) then ws_reg<=ws_reg+
1;ws_ff_reset<='1';ws_cnt2:=ws_cnt2+1;
elsif(ws_cnt2=20) then ws_ff_reset<='0';ws_cnt2:=ws_cnt2+1;
elsif(ws_cnt2=50) then ws_ff_reset<='1';ws_cnt2:=0;
else ws_cnt2:=ws_cnt2+1;
end if;
end if;
end process;
ws_data_out<=ws_reg;
p6:process(clk)
variable cnt2:integer:=0;
begin
if rising_edge(clk) then
if(cnt2=208)then
```

```
clk_sig<= not(clk_sig);
cnt2:=1;
else
cnt2:= cnt2 + 1;
end if;
end if;
end process;
p7:process(clk_sig,mux_sig,sdata)
variable cnt3,ws_1mini_cnt:integer:=0;
begin
if rising_edge(clk_sig) then
if    (cnt3=1)then   sdata<=("11111111" &
'0');mux_sig<='1';tx_cnt<='1';cnt3:=cnt3+1;
elsif (cnt3=10)then  mux_sig<='0';tx_cnt<='0';cnt3:=cnt3+1;
elsif (cnt3=860)then  sdata<=("11111111" &
'0');mux_sig<='1';tx_cnt<='1';cnt3:=cnt3+1;
elsif (cnt3=869)then  mux_sig<='0';tx_cnt<='0';cnt3:=cnt3+1;
elsif (cnt3=1719)then  sdata<=("11111111" &
'0');mux_sig<='1';tx_cnt<='1';cnt3:=cnt3+1;
elsif (cnt3=1728)then  mux_sig<='0';tx_cnt<='0';cnt3:=cnt3+1;
elsif (cnt3=2578)then  sdata<=(data_sum_temp(7 downto 0) &
'0');mux_sig<='1';tx_cnt<='1';cnt3:=cnt3+1;
elsif (cnt3=2587)then  mux_sig<='0';tx_cnt<='0';cnt3:=cnt3+1;
elsif (cnt3=3437)then sdata<=(data_sum_temp(15 downto 8) &
'0');mux_sig<='1';tx_cnt<='1';cnt3:=cnt3+1;
elsif (cnt3=3446)then mux_sig<='0';tx_cnt<='0';cnt3:=cnt3+1;
elsif (cnt3=4296)then sdata<=(data_xdiff_temp(7 downto 0) &
'0');mux_sig<='1';tx_cnt<='1';cnt3:=cnt3+1;
elsif (cnt3=4305)then mux_sig<='0';tx_cnt<='0';cnt3:=cnt3+1;
elsif (cnt3=5155)then sdata<=(data_xdiff_temp(15 downto 8) &
'0');mux_sig<='1';tx_cnt<='1';cnt3:=cnt3+1;
elsif (cnt3=5164)then mux_sig<='0';tx_cnt<='0';cnt3:=cnt3+1;
elsif (cnt3=6014)then sdata<=(data_ydiff_temp(7 downto 0) &
'0');mux_sig<='1';tx_cnt<='1';cnt3:=cnt3+1;
elsif (cnt3=6023)then mux_sig<='0';tx_cnt<='0';cnt3:=cnt3+1;
elsif (cnt3=6873)then sdata<=(data_ydiff_temp(15 downto 8) &
'0');mux_sig<='1';tx_cnt<='1';cnt3:=cnt3+1;
elsif (cnt3=6882)then mux_sig<='0';tx_cnt<='0';cnt3:=cnt3+1;
elsif (cnt3=7732)then sdata<=(data_ws_temp(7 downto 0) &
'0');mux_sig<='1';tx_cnt<='1';cnt3:=cnt3+1;
elsif (cnt3=7741)then mux_sig<='0';tx_cnt<='0';cnt3:=cnt3+1;
elsif (cnt3=8591)then sdata<=("00" & data_ws_temp(13 downto 8) &
'0');mux_sig<='1';tx_cnt<='1';cnt3:=cnt3+1;
elsif (cnt3=8600)then mux_sig<='0';tx_cnt<='0';cnt3:=cnt3+1;
mux_sig<='0';tx_cnt<='0';
data_sum_temp<=data_sum;
data_xdiff_temp<=data_xdiff;
data_ydiff_temp<=data_ydiff;
cnt3:=cnt3+1;
elsif (cnt3=9550)then
if (ws_1mini_cnt=59)then
data_ws_temp<=ws_reg;
ws_1mini_cnt:=0;
ws_reg_reset<='0';
cnt3:=9575;
else
ws_1mini_cnt:=ws_1mini_cnt+1;
ws_reg_reset<='0';
cnt3:=9585;
end if;
elsif (cnt3=9575)then
```

```
ws_reg_reset<='1';
cnt3:=cnt3+1;
elsif (cnt3=9585)then
ws_reg_reset<='0';
cnt3:=cnt3+1;
elsif (cnt3=9600)then
ws_reg_reset<='0';
cnt3:=1;
else
cnt3:=cnt3+1;
end if;
if (tx_cnt='1')then
sdata<=('0'& sdata(8 downto 1));
end if;
end if;
case mux_sig is
when '1'=>txd<=sdata(0);
when '0'=>txd<='1';
when others=> null;
end case;
end process;
end Behavioral;
```

DESIGN EXAMPLE 7.26

A data-logging system sends different measurement values to the computer through a serial port. The measurements are OPD reference voltage (12 bits), x-displacement error (12 bits), y-displacement error (12 bits), and wind speed (14 bits). Write MATLAB code to receive the data arriving at the serial port, arrange them back in a correct data frame and convert them to the original measurement values. Write all the measurement values in a Word file under appropriate heads [31,42,48].

Solution

```
clc;
clear all;
diary ('e:\Data File\Collected Data.dat');
fprintf ('ID.No:  Date & Time    SUM     X-DIFF    Y DIFF  Wind Speed\n');
fprintf (  '      ====      =================            =========
===========             ========                ===========\n');
s = serial('Com1');
s.BaudRate=9600;s.DataBits=8;s.Tag = 'FSOC';s.TimeOut=10;
s.InputBufferSize=floor(10000*10);s.OutputBufferSize=floor(10000*10);
s.parity='space';s.Terminator='LF';s.ByteOrder='littleEndian';
fopen(s);
time = now;
stopTime = 'Nov-07 (Sun) 10:05:00';
count = 1;
while ~isequal(datestr(now,'mmm-dd (ddd) HH:MM:SS'),stopTime)
date=datestr(now);
time(count) = datenum(clock);
a=fread(s,1,'uint8')';
if a ==255
a=fread(s,1,'uint8')';
if a==255
a=fread(s,1,'uint8')';
if a==255
a=fread(s,1,'uint8')';
b=dec2bin(a,8);
a=fread(s,1,'uint8')';
```

```
c=dec2bin(a,8);
f1=[c b];
sum=bin2dec(f1);
sum_v=((sum * 0.0048824125)-10);
a=fread(s,1,'uint8')';
b=dec2bin(a,8);
a=fread(s,1,'uint8')';
c=dec2bin(a,8);
f2=[c b];
xdiff=bin2dec(f2);
xdiff_v=((xdiff * 0.0048824125)-10);
a=fread(s,1,'uint8')';
b=dec2bin(a,8);
a=fread(s,1,'uint8')';
c=dec2bin(a,8);
f3=[c b];
ydiff=bin2dec(f3);
ydiff_v=((ydiff * 0.0048824125)-10);
a=fread(s,1,'uint8')';
b=dec2bin(a,8);
a=fread(s,1,'uint8')';
c=dec2bin(a,8);
f4=[c b];
ws=bin2dec(f4);
w_s_f=(ws/60);
w_s=(1/1.45)*w_s_f;
if w_s ~= 0
w_s_actual=2.5+w_s;
else
w_s_actual=w_s;
end
fprintf('%6d     %s     %0.5E%s     %0.5E%s     %0.5E%s       %0.5E%s\n',count,
date,sum_v,'V',xdiff_v,'V',ydiff_v,'V',w_s_actual,'mps');
count = count +1;
end
end
end
end
fclose(s);
diary off;
```

DESIGN EXAMPLE 7.27

Write MATLAB code to read the values of laser beam displacement distance on the x and y axis, calculate the radial distance and plot the corresponding tilted beam of each measurement on a 2D plane.

Solution

```
clc;
clear all;
h_temp=[0 -0.2  0.4  0.5 -0.5 0.8  0.0 -0.6 0.2 1 1.2 -0.6 0.2 -0.6 0.7
-0.8 ];
k_temp=[0  0.4 -0.6 -0.7 -0.2 1.0 -0.3 -0.2 1.0 1 0.2 -0.1 0.5  0.2 0.1
0.2 ];
for k1=1:16,
h=h_temp(k1);
k=k_temp(k1);
r=2;
d=sqrt(h^2 +k^2);
if ( h==0 && k==0 );
```

```
theta2=0;swch=0;
elseif (h>0 && k==0);
theta2=0;swch=1;
elseif (h<0 && k==0);
theta2=3.1415925; swch=2;
elseif (h==0 && k>0);
theta2=1.57079625; swch=3;
elseif (h==0 && k<0);
theta2=4.71238875; swch=4;
elseif (h>0 && k>0);
theta2=atan(k/h);  swch=5;
elseif (h<0 && k>0);
theta2=(-1*atan(k/h))+1.57079625;   swch=6;
elseif (h<0 && k<0);
theta2=(atan(k/h)+ 3.1415925);  swch=7;
elseif (h>0 && k<0);
theta2=(-1*atan(k/h)) + 4.71238875;    swch=8;
end
theta1=(theta2:0.01:((2*pi)+ theta2));
for i1=1:length(theta1)
switch(swch)
case{0},
x = r*cos(theta1);
y = r*sin(theta1);
case{1,2},
x = (r-d)*cos(theta1);
y = r*sin(theta1);
case{3,4},
x = r*cos(theta1);
y = (r-d)*sin(theta1);
case{5,6,7,8},
sweep=theta2+1.57079625;
x = (r-d)*cos(theta1)*cos(sweep) - r*sin(theta1)*sin(sweep);
y = (r-d)*cos(theta1)*sin(sweep) + r*sin(theta1)*cos(sweep);
end
end
figure(1);
subplot(4,4,k1)
plot(x,y)
set(gca, 'FontName','Helvetica','FontSize',8, 'LineWidth', 1.3);
hold on;
fill(x,y,'r');
hold on;
plot(h,k,'rs',....
'MarkerEdgeColor','k',...
'MarkerFaceColor','g',...
'MarkerSize',5);
axis([-4 4 -4 4 ]);
set(gca, 'FontName','Helvetica','FontSize',8, 'LineWidth', 1.3);
axis square;
xlabel('X_{dist}');
ylabel ('Y_{dist}');
plot(-4:1:4,-4:1:4,'--k','linewidth',1.3);
set(gca, 'FontName','Helvetica','FontSize',8, 'LineWidth', 1.3);
hold on;
plot(-4:1:4,4:-1:-4,'--k','linewidth',1.3);
set(gca, 'FontName','Helvetica','FontSize',8, 'LineWidth', 1.3);
hold on;
plot(-4:1:4,[0 0 0 0 0 0 0 0 0],'-k','linewidth',1.3);
set(gca, 'FontName','Helvetica','FontSize',8, 'LineWidth', 1.3);
hold on;
```

```
plot([0 0 0 0 0 0 0 0 0],-4:1:4,'-k','linewidth',1.3);
set(gca, 'FontName','Helvetica','FontSize',8, 'LineWidth', 1.3);
hold off;
end
```

7.7 Encoder/Decoder Interfacing for Remote Control Applications

In this section, we discuss a popular data encoding/decoding system that is essential for various wireless control applications. The working principles and configuration of optical (IR) and RF transmitters (Tx)/receivers (Rx) with encoders/decoders are also discussed.

7.7.1 Optical Tx/Rx Wireless Control

The HT12E and HT12D are two of the more popular encoders and decoders, respectively, being used for different remote control applications. This encoder and decoder are configured with an optical (IR) transmitter and receiver (TSOP1738), shown in Figure 7.26a and b, respectively, for transferring data from HT12E to HT12D. By using this encoder-decoder pair, we can easily transmit 12 bits of parallel data serially. The HT12E simply converts 12-bit parallel data to serial output which can be transmitted through the IR transmitter. The 12-bit parallel data is divided into eight address bits (A_0-A_7) and four data bits (D_0-D_3), as shown in Figure 7.26c. By using these address pins, we can provide an 8-bit security code for each data (D_0-D_3) transmission; hence, multiple receivers can be connected to a single transmitter; that is, 256 (2^8) decoders can be linked to one transmitter. The data can be transferred to all these decoders by selecting their appropriate addresses (synchronization bits) on a time-sharing basis. The HT12E can be operated with a supply voltage of +5 V, and it has a built-in oscillator which requires only one external resistor, typically 750 KΩ. The transmit enable (TE) in HT12E is used for enabling the transmission. This transmission (D_{out}) cycle will continue as long as the TE is enabled. When the TE signal switches to "1", the encoder output completes the current cycle and stops its transmission; that is, it is an active low input. The address pins can be connected to ground or left open if it needs to connect only with one decoder. The data pins are the data inputs through which we would be able to send the data that we wish to the decoder. D_{out} is the serial data output of the encoder HT12E that has to be connected to an optical transmitter through a driver circuit, as shown in Figure 7.26c.

The TSOP1738, as shown in Figure 7.26b and d, is a famous IR remote control receiver. This IR sensor module consists of a PIN diode and a pre-amplifier, which are embedded inside a single package. The output of TSOP is active low, and it gives +5 V in the off state. The output of TSOP1738 goes low when the IR signal of 38 KHz incident on it. The TSOP module has an built-in control circuit for amplifying the coded pulses arriving from the IR transmitter. The received signal is transferred to the decoder HT12D, as shown in Figure 7.26d. The decoder separates the address and data bits from the data frame received through the D_{in} pin and makes the data available at the respective ports once the address is matched. Upon getting a match on the address bits of the encoder and decoder, the LED connected at the valid transmission (V_T) bit of the HT12D goes high to indicate valid reception [99]. The transmitter cannot send data to any receiver if the address bits are different. The decoded data can be used to perform any control operations required within a maximum 100-m range.

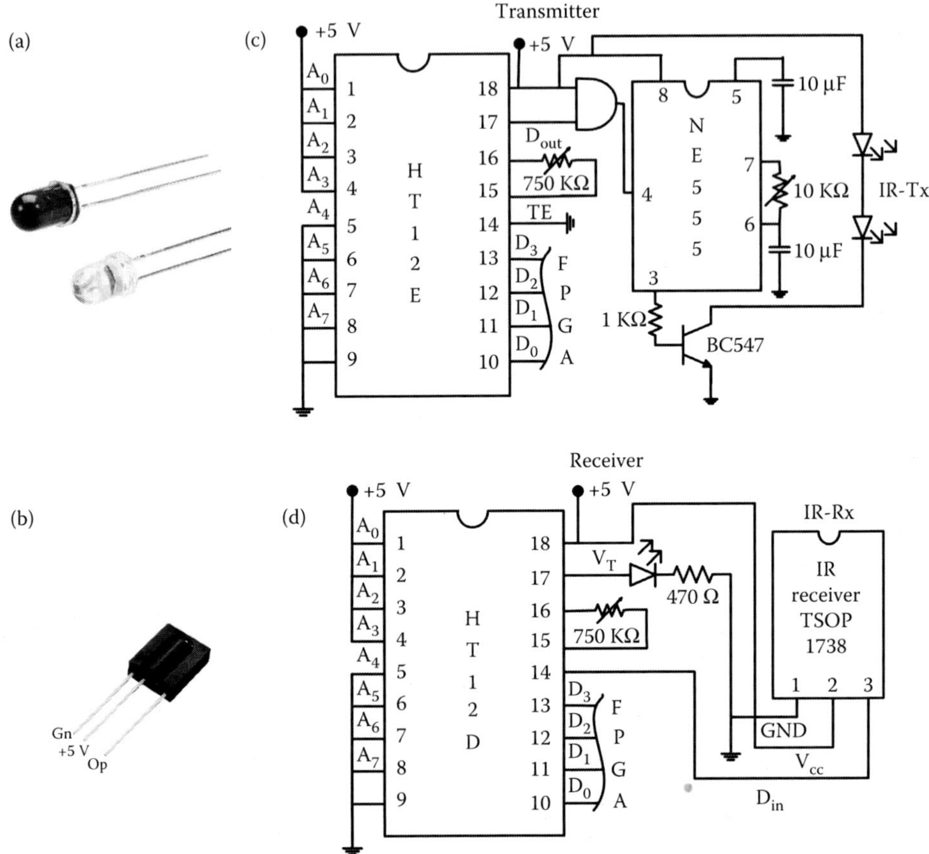

FIGURE 7.26
Optical transmitter (IR transmitter) (a), optical receiver (TSOP1738) (b), HT12E application circuit (c), and HT12D application circuit (d).

7.7.2 Radio Frequency Transmitter/Receiver Wireless Control

The HT12E and HT12D can also be interfaced with radio frequency (RF) transmitter (Tx) and receiver (Rx) modules. RF-Tx/Rx modules work at the radio frequency range of 30 KHz–300 GHz. The RF modules (Tx/Rx) shown in Figure 7.27a and b follow the amplitude shift keying (ASK) modulation technique and work at 433 or 315 MHz. Transmission through RF is better, more reliable and stronger than IR because the RF signal can travel for a longer range as compared to infrared rays. IR mostly supports line-of-sight data transmission, whereas RF signals can travel even if there is an obstructed path. The chosen RF transmitter and receiver pair should have the same frequency. The transmission speed of these modules is 1–10 Kbps [100].

The HT12E encoder will convert the 4-bit parallel data given to pins D_0–D_3 to serial data along with its address and will make them available at D_{out}. This output serial data is given to the ASK RF transmitter module. The external resistor will provide the necessary resistance for the operation of the internal oscillator of the HT12E. The ASK RF receiver receives the data transmitted using the ASK RF transmitter, and the decoder will convert the received serial data to 4-bit parallel data, D_0–D_3. The LED connected in the HT12D circuit glows when valid data transmission occurs from transmitter to receiver. The RF transmitter module has

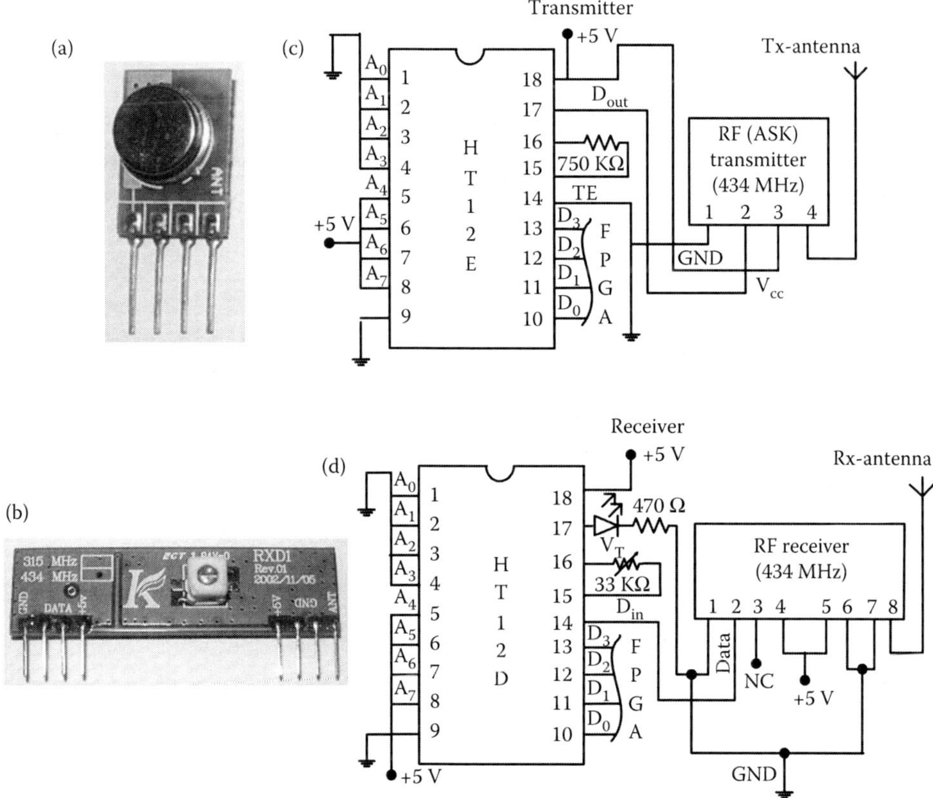

FIGURE 7.27
RF transmitter module (a), RF receiver module (b), encoder application circuit (c), and decoder application circuit (d).

four pins, namely data, V_{cc}, ground (GND), and antenna (RF output). The receiver module has eight pins, namely GND (3 pins), V_{cc} (2 pins), data (2 pins), and antenna (1 pin). This integrated RF receiver module has been tuned to a frequency of 433.92 MHz, exactly the same as for the RF transmitter. The 434 MHz RF receiver module receives an ASK modulation signal and demodulates it to a digital signal for the next decoder stage. A local oscillator and PLL are available in the receiver package. The coded signal transmitted by the transmitter is processed at the receiver side by the decoder HT12D. The HT12D compares the serial input data three times continuously with the local addresses. If no error, that is, matched codes are found, then the input data are decoded and then transferred to the output pins.

DESIGN EXAMPLE 7.28

Develop a simple automatic remote plant control system using RF Tx/Rx modules. A plant consists of 16 RF receiver (HT12D) modules through which the respective process can be controlled by selecting the appropriate address and data (logic "1111" for "ON" and logic "0000" for "OFF") from the control room, that is, Tx station (HT12E). The "ON"/"OFF" order of the process controls are process0 (P0)-P3-P8 (ON)-P15-P2-P7-P9-P1-P8-P14-P4-P5-P6-P10-P11-P8 (OFF)-P12-P13. Once this control chain is over, then reset the plant; while performing the resetting process, indicate the status by forcing a LED "ON". The minimum processing delay is 1 sec. All these operations have to be begun by a start (user input) switch.

Solution

```
library ieee;
use ieee.std_logic_1164.all;
use ieee.std_logic_arith.all;
use ieee.std_logic_unsigned.all;
use IEEE.numeric_bit.all;
entity RF_en is
port( en_add: out std_logic_vector(7 downto 0):="00000000";
en_data: out std_logic_vector(3 downto 0):="1111";
busy_led: out std_logic:='0';
m_clk,start : in std_logic:='0');
end RF_en;
architecture RF of RF_en is
signal clk_sig: std_logic:='0';
begin
p0: process(start, m_clk)
variable cnt:integer:=0;
begin
if (start='0') then
cnt:=0;
else
if rising_edge(m_clk) then
if (cnt=2) then
clk_sig<=not clk_sig;
cnt:=0;
else
cnt:=cnt+1;
end if;
end if;
end if;
end process;
p1:process(clk_sig)
variable con:std_logic_vector(6 downto 0):="0000000";
begin
if rising_edge(clk_sig) then
con:= con + "0000001";
case con is
when "0000000" => en_add<="00000000"; en_data<="1111";
when "0000010" => en_add<="00000011"; en_data<="1111";
when "0000110" => en_add<="00001000"; en_data<="1111";
when "0000111" => en_add<="00001111"; en_data<="1111";
when "0001010" => en_add<="00000010"; en_data<="1111";
when "0010010" => en_add<="00000111"; en_data<="1111";
when "0010100" => en_add<="00001001"; en_data<="1111";
when "0011011" => en_add<="00000001"; en_data<="1111";
when "0011101" => en_add<="00001000"; en_data<="0000";
when "0100110" => en_add<="00001110"; en_data<="1111";
when "0110000" => en_add<="00000100"; en_data<="1111";
when "0111110" => en_add<="00000101"; en_data<="1111";
when "0111111" => en_add<="00000110"; en_data<="1111";
when "1000010" => en_add<="00001010"; en_data<="1111";
when "1000100" => en_add<="00001011"; en_data<="1111";
when "1001000" => en_add<="00001000"; en_data<="1111";
when "1010000" => en_add<="00001100"; en_data<="1111";
when "1010101" => en_add<="00001101"; en_data<="1111";
when "1010110" => en_add<="00000000"; en_data<="0000";
when "1010111" => en_add<="00000001"; en_data<="0000";
when "1011000" => en_add<="00000010"; en_data<="0000";
when "1011001" => en_add<="00000011"; en_data<="0000";
when "1011010" => en_add<="00000100"; en_data<="0000";
```

```
when "1011011" => en_add<="00000101"; en_data<="0000";
when "1011100" => en_add<="00000110"; en_data<="0000";
when "1011101" => en_add<="00000111"; en_data<="0000";
when "1011110" => en_add<="00001000"; en_data<="0000";
when "1011111" => en_add<="00001001"; en_data<="0000";
when "1100000" => en_add<="00001010"; en_data<="0000";
when "1100001" => en_add<="00001011"; en_data<="0000";
when "1100010" => en_add<="00001100"; en_data<="0000";
when "1100011" => en_add<="00001101"; en_data<="0000";
when "1100100" => en_add<="00001110"; en_data<="0000";
when "1100101" => en_add<="00001111"; en_data<="0000";
when others => en_add<="11111111"; en_data<="0000";
end case;
if (con > "1100101") then
busy_led<='1';
else
busy_led<='0';
end if;
end if;
end process;
end RF;
```

7.8 Pseudorandom Binary Sequence Generator and Time-Division Multiple Access

This section deals with many popular techniques of generating random binary sequences. Serial and parallel design of binary random sequence generators are also discussed. Design of a Kasami sequence generator and its implementation in the FPGA are explained.

7.8.1 Serial Pseudorandom Binary Sequence Generator

A PRBS generator is nothing but a random binary sequence generator and is essential for various applications. The PRBS generator is widely used in communications, cryptography, direct sequence spread spectrum (DSSS), image processing, and so on. The word "random" means that the value of an element (FFs) of the sequence is independent of the values of other elements. The word "pseudo" means that the sequence is deterministic and after a length "L", it starts to repeat itself, unlike real random sequences. The implementation of a PRBS generator depends on the wiring of linear feedback shift registers (LFSRs). The PRBS generator produces a predefined sequence of 1s and 0s with the same probability of occurrence. A sequence of consecutive $2^m - 1$ bits, where m is the number of shift registers (FFs) used in the design, comprise one data pattern, and this pattern will repeat itself over time. The maximum number of 1s coming together will be N in the sequence, while the maximum number of zeroes coming together will be (N – 1), and the duration of 1 bit is equal to the period of the clock deriving it. In many applications, we may need a random number generator for generating a sequence of random numbers either in binary or decimal. In MATLAB and other software/languages, there are library functions for this kind of generation. In VHDL, this is achieved by using serially connected shift registers of suitable lengths [22,39,101]. A simple serial PRBS generator circuit is shown in Figure 7.28.

The sequence generated by a PRBS generator is not theoretically random, but for most practical applications, this sequence can be considered random. For example, if $n = 32$,

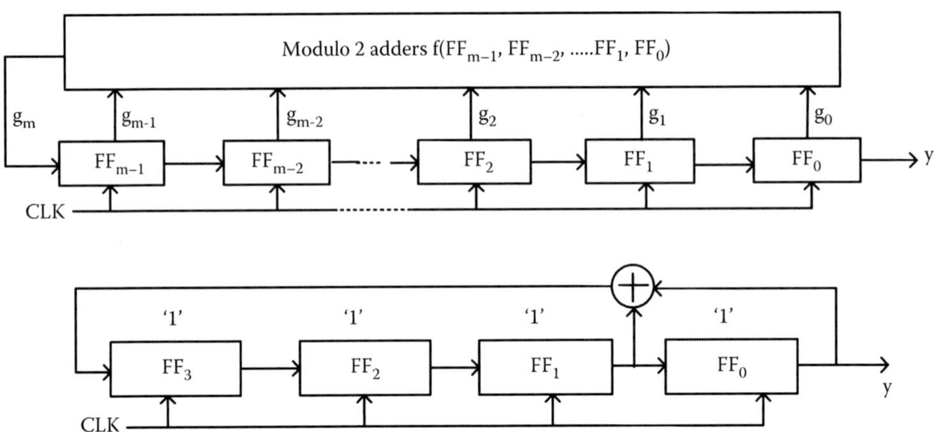

FIGURE 7.28
Simple PRBS generator circuits using serially connected LFSRs.

then the period (L) is 4294967295, which is a long sequence and is large enough for most practical applications. VHDL design modules can be written in a generic way through which the value of n can be specified at the time of synthesizing.

The general polynomial equation for an nth-order PRBS is given by

$$y = g_m x^m + g_{m-1} x^{m-1} + g_{m-2} x^{m-2} + g_{m-3} x^{m-3} + \cdots + g_0 x^0 \quad (7.18)$$

where y represents the PRB sequence, g represents the shift register's output connections to modulo 2 adders and x represents the shift register position. Further, g_n is the MSB and g_0 is the LSM in Equation 7.18. For example, a PRBS generator design is given by y = (110101), which means that the outputs of the second and fourth FFs are not connected to the modulo 2 adders; therefore, modulo 2 addition is not required at those positions. A PRBS generator designed with four shift registers is shown in Figure 7.28, whose polynomial equation is given by

$$y = g_4 x^4 + g_1 x^1 + g_0 x^0 = (10011) \quad (7.19)$$

Assume that all four shift registers are initially filled with a seed pattern of "1111". Then, the subsequent operations of the LFSR for the rising edges of the clock pulses are given in Table 7.8. Before applying the first clock pulse, the output of the modulo 2 adder and LFSR is taken as unknown. This design generate a PRBS of 15 bits as an N-shift register generates $2^N - 1$ bits of PRBS. Right after the final bit pattern, it returns to the top pattern of the bit stream to repeat the same order of sequence. By carefully looking at the bit pattern of the shift registers, we can note that all 4-bit combinations appeared except all 0s. If we feed in the pattern of "0000" at the beginning, the shift register would be stuck and generate only 0s infinitely. A seed pattern must not be all 0s; hence, one of the possible 15 (4-bit) patterns has to be considered as a seed. Interesting characteristics/properties of PRBS can be found in [22,101].

TABLE 7.8

Function Table of a PRBS Generator

CLK Signal	O/P of XOR	Status of Shift Registers				PRBS O/P (y)
		FF3	FF2	FF1	FF3	
Rising edge of clock input	–	1	1	1	1	–
	0	0	1	1	1	1
	0	0	0	1	1	1
	0	0	0	0	1	1
	1	1	0	0	0	1
	0	0	1	0	0	0
	0	0	0	1	0	0
	1	1	0	0	1	0
	1	1	1	0	0	1
	0	0	0	1	0	0
	1	1	0	0	1	0
	1	1	1	0	0	1
	0	0	1	1	0	0
	1	1	0	1	1	0
	0	0	1	0	1	1
	1	1	0	1	0	1
	1	1	1	0	1	0
	1	1	1	1	0	1

DESIGN EXAMPLE 7.29

Develop a PRBS generator using LFSRs that are configured as "110111". Load the initial value of the LFSR from the external input whenever the control pin "load " is at logic 1. Display the status of the LFSR using the LEDs.

Solution

```
library ieee;
use ieee.std_logic_1164.all;
use ieee.std_logic_arith.all;
use ieee.std_logic_unsigned.all;
entity lfsr is
Port ( m_clk, load : in  STD_LOGIC:='0';
inti_val: in std_logic_vector(5 downto 0):="101011";
lfsr_st:out  std_logic_vector(5 downto 0):="000000";
prbs_out : out  STD_LOGIC:='0');
end lfsr;
architecture Behavioral of lfsr is
signal clk_sig, feed_back:std_logic:='0';
signal lfsr_reg:std_logic_vector(5 downto 0):="011101";
begin
lfsr_st<=lfsr_reg;
prbs_out<=lfsr_reg(0);
p0:process(m_clk)
variable clk_deci_cnt:integer:=0;
begin
if rising_edge(m_clk) then
```

```
if(clk_deci_cnt=2) then
clk_sig<=not(clk_sig);
clk_deci_cnt:=0;
else
clk_deci_cnt:=clk_deci_cnt+1;
end if;
end if;
end process;
p1:process(clk_sig,load)
begin
if (load='1') then
lfsr_reg<= inti_val;
else
if rising_edge(clk_sig) then
feed_back<=(lfsr_reg(1) xor lfsr_reg(2)xor lfsr_reg(4) xor lfsr_reg(5)
xor lfsr_reg(0));
end if;
if falling_edge(clk_sig) then
lfsr_reg<= (feed_back & lfsr_reg(5 downto 1));
end if;
end if;
end process;
end Behavioral;
```

7.8.2 Parallel Pseudorandom Binary Sequence Generator

Parallel PRBS generators and parallel detector circuits are commonly used to check the functionality and the data transmission/reception quality of newly designed transceiver and communication systems. The general schematic diagram of a parallel PRBS generator and receiver is shown in Figure 7.29. The parallel PRBS generator is constructed using an array (more than one) of serially connected shift registers. The functions of the LFSR and generation of parallel PRBSs are controlled by a common clock input. A serializer transfers the PRBS sequence generated in parallel one by one to the link channel. The PRBS generation and the serialization are appropriately synchronized by a state machine controller. The deserializer properly designates the sequence received from the channel. This arrangement helps to test and quantify the operating quality of the system, so it is called a device under test. A parallel PRBS generator is required whenever we need to pump a random binary sequence to examine the DUT in parallel, and, with the help of a serializer/deserializer, a serial DUT can also be examined as shown in Figure 7.29. An internal schematic of an n-bit parallel PRBS generator is shown in Figure 7.30. The VHDL logic models can be used to build this circuit in the FPGA. A fast parallel PRBS generator

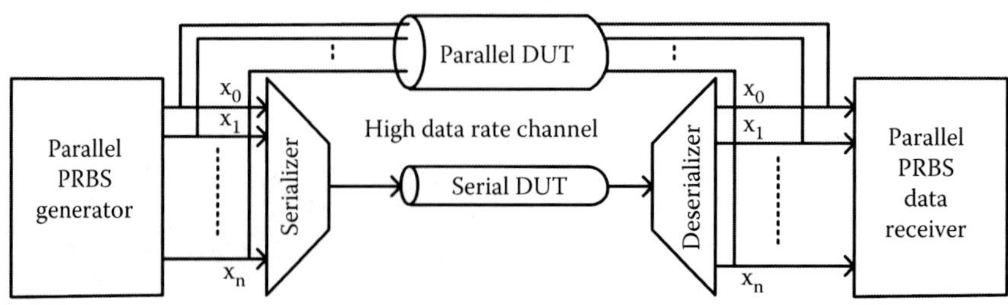

FIGURE 7.29
Illustration of application of a parallel PRBS generator.

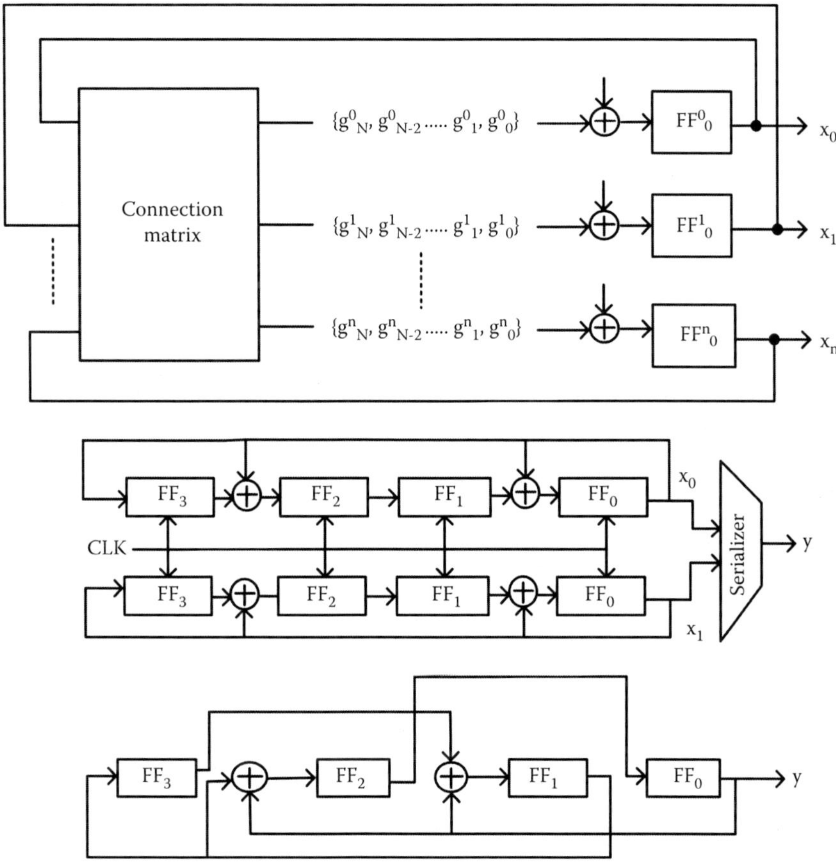

FIGURE 7.30
General and different designs of parallel PRBS generator.

can be implemented by using only N shift registers. There are n parallel outputs that are from the n N-order LFSRs [22,39,101].

The specific type of PRBSs, the so-called maximum length sequence, that are used for testing parallel channels are typically generated using an array of LFSRs. This is just a chain of flip-flops that is fed back on itself in a specific way to generate the PRBS of interest. To make this concept clear, let us consider a 4-bit two-array PRBS generator, as shown in Figure 7.30. Each array generates a repeating sequence of length 15. In the beginning, the LFSRs are loaded/initialized with different values. The nth signal tapping for the modulo 2 addition either be taken from (n − 1)th FFs and/or (n + 1, 2, 3 ...)th FFs along with the common feedback line, as shown in Figure 7.30. For every rising or falling edge of clock inputs (CLK), the input to the FFs will be generated based on the modulo 2 additions. However, two PRBS bits (x_0 and x_1), in Figure 7.30, enter the serializer through which we will send them one by one with appropriate header sequences, which are required for proper bit synchronization.

Now, an important point is that the FFs and modulo 2 adders have to be relatively slow, because the serializer has to convert multiple parallel inputs to a single output at a faster rate. The serializer can be done by a 4:1 MUX by taking every fourth element of the original PRBS sequence to get its serially shifted version. In fact, for longer sequences, one can take the shifted version of the original sequence for every repetition cycle of 2^{n-1}. The advantage of parallel PRBS is operating the entire LFSR array at a relatively lower rate to generate the multiple PRBS bits that are combined by a serializer for transferring at a faster rate. The internal IP core can be used for performing serialization and transferring at a faster rate.

DESIGN EXAMPLE 7.30

Develop a digital system to generate parallel PRBSs and transmit them serially to a LED. Initialize the four-stage LFSRs of width [5:0] by taking values from a suitable local counter with respect to a user input "load". Also, develop a system to realize the last design in Figure 7.30.

Solution

```
library ieee;
use ieee.std_logic_1164.all;
use ieee.std_logic_arith.all;
use ieee.std_logic_unsigned.all;
entity lfsr_par is
Port (m_clk,load : in STD_LOGIC:='0';
prbs_out1,prbs_out2 : out  STD_LOGIC:='0');
end lfsr_par;
architecture Behavioral of lfsr_par is
signal mux_cnt: std_logic_vector(1 downto 0):="00";
signal lfsr_clk: std_logic_vector(7 downto 0):="10000000";
signal mux_clk:  std_logic_vector(7 downto 0):="01010101";
signal lfsr_reg:std_logic_vector(5 downto 0):="011101";
signal lfsr_reg0:std_logic_vector(5 downto 0):="011101";
signal lfsr_reg1,temp:std_logic_vector(5 downto 0):="100101";
signal lfsr_reg2:std_logic_vector(5 downto 0):="011001";
signal lfsr_reg3:std_logic_vector(5 downto 0):="010001";
signal clk_mux:std_logic_vector(2 downto 0):="000";
signal lfsr,mux:std_logic:='0';
begin
prbs_out2<=lfsr_reg(0);
p0:process(m_clk)
variable cnt1:integer:=0;
begin
if rising_edge(m_clk) then
temp<=temp +1;
if(cnt1=2) then
clk_mux<=clk_mux +1;
cnt1:=0;
else
cnt1:=cnt1+1;
end if;
case clk_mux is
when "000" => lfsr<=lfsr_clk(7); mux<=mux_clk(7);
when "001" => lfsr<=lfsr_clk(6); mux<=mux_clk(6);
when "010" => lfsr<=lfsr_clk(5); mux<=mux_clk(5);
when "011" => lfsr<=lfsr_clk(4); mux<=mux_clk(4);
when "100" => lfsr<=lfsr_clk(3); mux<=mux_clk(3);
when "101" => lfsr<=lfsr_clk(2); mux<=mux_clk(2);
when "110" => lfsr<=lfsr_clk(1); mux<=mux_clk(1);
when "111" => lfsr<=lfsr_clk(0); mux<=mux_clk(0);
when others=> null;
```

```
end case;
end if;
end process;
p1:process(lfsr,load)
variable cnt: integer:=1;
begin
if rising_edge(lfsr) then
if (load='1') then
if (cnt=1) then
lfsr_reg0<= temp;
cnt:=cnt+1;
elsif (cnt=2) then
lfsr_reg1<= temp;
cnt:=cnt+1;
elsif (cnt=3) then
lfsr_reg2<= temp;
cnt:=cnt+1;
elsif (cnt=4) then
lfsr_reg3<= temp;
cnt:=5;
end if;
else
cnt:=1;
lfsr_reg0<=((lfsr_reg0(1) xor lfsr_reg0(2)xor lfsr_reg0(4) xor
lfsr_reg0(5) xor lfsr_reg0(0)) & lfsr_reg0(5 downto 1));
lfsr_reg1<=((lfsr_reg1(2) xor lfsr_reg1(3)xor lfsr_reg1(4) xor
lfsr_reg1(5) xor lfsr_reg1(0)) & lfsr_reg1(5 downto 1));
lfsr_reg2<=((lfsr_reg2(1) xor lfsr_reg2(1)xor lfsr_reg2(3) xor
lfsr_reg2(5) xor lfsr_reg2(0)) & lfsr_reg2(5 downto 1));
lfsr_reg3<=((lfsr_reg3(1) xor lfsr_reg3(3)xor lfsr_reg3(4) xor
lfsr_reg3(5) xor lfsr_reg3(0)) & lfsr_reg3(5 downto 1));
end if;
end if;
end process;
p2:process(load,mux, lfsr_reg0,lfsr_reg1,lfsr_reg2,lfsr_reg3)
begin
if (load='1') then
mux_cnt<="00";
elsif rising_edge(mux) then
mux_cnt<=mux_cnt+1;
case  mux_cnt is
when "00"=> prbs_out1<=lfsr_reg0(0);
when "01"=> prbs_out1<=lfsr_reg1(0);
when "10"=> prbs_out1<=lfsr_reg2(0);
when "11"=> prbs_out1<=lfsr_reg3(0);
when others=> null;
end case;
end if;
end process;
p3:process(mux)
begin
if rising_edge(mux) then
lfsr_reg<= ( lfsr_reg(3) & (lfsr_reg(1) xor lfsr_reg(2)) &
(lfsr_reg(3) xor
lfsr_reg(1)) & (lfsr_reg(5) xor lfsr_reg(4)) & (lfsr_reg(2) xor
lfsr_reg(3)) & (lfsr_reg(4) xor lfsr_reg(5)) );
end if;
end process;
end Behavioral;
```

7.8.3 Kasami Sequence Generator

Kasami sequences (Ks) are a set of sequences that have good cross-correlation properties. There are two sets of Kasami sequences, namely the (i) small set and (ii) large set. The large set contains all the sequences that are in the small set. Only the small set is optimal in the sense of matching Welch's lower bound for correlation functions. Kasami sequences have period $N = 2^n - 1$, where n is a nonnegative integer. Let u be a binary sequence of length N and let w be a sequence obtained by decimating u by $2^{n/2}+1$. The small set of Kasami sequences is defined by Equation 7.20 [102].

$$K_s(u, n, w) = \begin{cases} u & \text{if } m = -1 \\ u \oplus T^m w & \text{if } m = 0, \ldots, 2^{0.2n} - 2 \end{cases} \tag{7.20}$$

where T is the left shift operator, m is the shift parameter for w and $\oplus$ is the modulo 2 addition. Note that the small set contains $2^{n/2}$ sequences. Kasami sequences can be generated by multiplying the outputs of three different shift register (u, v and w) stages by appropriate feedback, as shown in Figure 7.31 [102]. Two shift registers (u and v) have equal length n, n = 6 in Figure 7.31, and form a preferred pair, while the third shift register (w) is of length 3. The first two shift registers create m sequences, while the last one is m/2 sequences. To obtain all possible Kasami codes, the shift registers should have different initial values with respect to each other. Two shift registers (u and w) play a major role in generating the Kasami sequence. In the Figure 7.31, the shift parameters are referred to as m′ and k′, as there is a translation step between the relative shift values and the tap values that are required to obtain those shifts.

The tapping at all stages, in Figure 7.31, is given by u(6,1), v(6,5,2,1) and w(2,1), which may also be given by u = "100001", v = "110011" and w = "011". The polynomial representations of the tapping at each shift register stage are given by $g_6x^6 + g_5x^5 + g_0x^0$, $g_6x^6 + g_5x^5 + g_4x^4 + g_1x^1 + g_0x^0$ and $g_1x^1 + g_0x^0$, respectively. Finally, the Kasami sequence of width (6, m, k) will be generated. Let us see an example to understand this concept more clearly. Assume that the u and v registers are initialized by "011101" and "100101", respectively. The initial value of the w register is "110". Once the LFSR shifting clocking is applied, then the three register statuses will be "101110", "010010" and "111". The statuses of the v and w registers have to be subjected to a bitwise AND operation with the values of m′ (110111) and k′(101). In this example, the value of the v and w stages after performing the AND

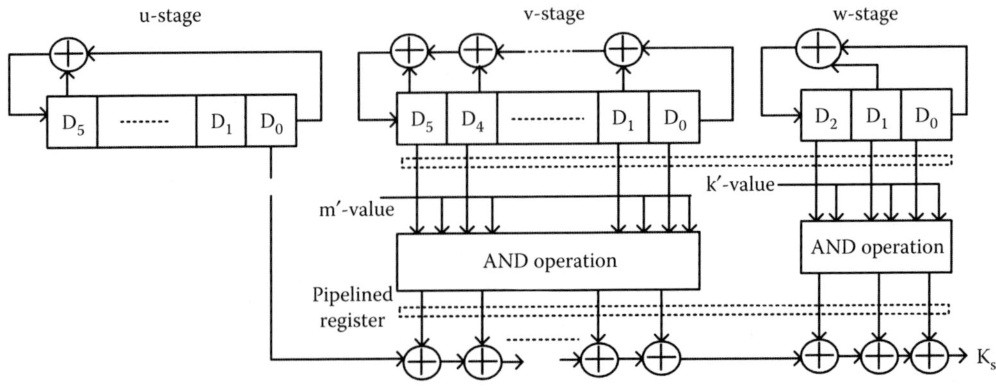

FIGURE 7.31
Schematic of a Kasami code generator.

operation are "010010" and "101". Then, the Kasami sequence is estimated from the LSB of the u register and v and u stage bits, as shown in Figure 7.31, which is '1' in this example. This computation is continued for every shifting of the registers.

DESIGN EXAMPLE 7.31

Develop a digital system to generate Kasami code. Assume that there are seven FFs in the u and v registers and four FFs in the w register. Perform the AND operation as required with pipelined registers, as shown in Figure 7.31. Keep a user input to reset the entire system when required.

Solution

```
library ieee;
use ieee.std_logic_1164.all;
use ieee.std_logic_arith.all;
use ieee.std_logic_unsigned.all;
use IEEE.NUMERIC_STD.ALL;
entity Kasami_code is
port (m_clk,reset:in  std_logic:='0';
Kasami_seq:out std_logic:='0');
end Kasami_code;
architecture Behavioral of Kasami_code is
signal clk_sig:std_logic:='0';
signal reg_u: std_logic_vector(6 downto 0):="1010110";
signal reg_v: std_logic_vector(6 downto 0):="1011110";
signal reg_w: std_logic_vector(3 downto 0):="1101";
signal mul_v: std_logic_vector(6 downto 0):="0000000";
signal mul_w: std_logic_vector(3 downto 0):="0000";
begin
p1:process(m_clk,reset)
variable cnt:integer:=1;
begin
if(reset='1') then
clk_sig<='0';
cnt:=0;
elsif rising_edge(m_clk) then
if(cnt=2)then
clk_sig<= not(clk_sig);
cnt:=1;
else
cnt:= cnt + 1;
end if;
end if;
end process;
p2:process(clk_sig)
begin
if rising_edge(clk_sig) then
reg_u<=( (reg_u(6) xor reg_u(0)) & reg_u(6 downto 1) );
reg_v<=( (reg_v(6) xor reg_v(5) xor reg_v(2) xor reg_v(0)) & reg_v(6
downto 1) );
reg_w<=( (reg_w(2) xor reg_w(0)) & reg_w(3 downto 1) );
mul_v<= (reg_v(6 downto 0) and "1101110");--m
mul_w<= (reg_w(3 downto 0) and "1101");--k
kasami_seq<=(reg_u(0) xor mul_v(6) xor mul_v(5) xor mul_v(4) xor
mul_v(3)xor mul_v(2) xor mul_v(1) xor mul_v(0)
xor mul_w(3) xor mul_w(2) xor mul_w(1) xor mul_w (0));
end if;
end process;
end Behavioral;
```

7.8.4 Analog Time Division Multiplexing and M-Array Pulse Amplitude Modulation

Time-division multiplexing (TDM) is a method of converting multiple parallel signals/data streams to a single signal by separating the signals into many segments, each having a very short duration. Each individual data stream is reassembled at the receiving end based on the timing. The circuit that combines signals at the source, the transmitting end, of a communications link is known as a multiplexer. It accepts input from each individual end user, breaks each signal into segments and assigns the segments to the composite signal as a repetitive access; hence, the composite signal contains data from multiple senders. At the other end of the long-distance cable, the individual signals are separated out by means of a circuit called a demultiplexer and routed to the proper end users, as shown in Figure 7.32. A two-way communication circuit requires a multiplexer/demultiplexer at each end of the long-distance high-bandwidth cable. If many signals must be sent along a single long-distance line, careful engineering is required to ensure that the system will perform properly and route to the appropriate receiver. The TDM scheme allows for variation in the number of signals being sent along the line and constantly adjusts the time intervals to make optimum use of the available bandwidth.

The concept of TDM can be demonstrated using either the IC4051, IC4052 or IC4053. The IC4051 is a single-pole octal-throw analog switch suitable for analog or digital 8:1 multiplexer/demultiplexer applications. The IC4051 features a digital enable input (E), three digital selection inputs (S_0, S_1 and S_2), eight independent inputs (CH0, CH1, … , CH7) and one output (y). Since the IC4051 has the capability of acting as a bidirectional switch, it is possible to feed one input (x) and get eight outputs (y_0, y_1, … , y_7) as well. When E is HIGH, the switches in IC4051 are turned off and all switches are in the high-impedance OFF state independent of S_0 to S_3. Inputs include clamp diodes that enable the use of current-limiting resistors to interface the inputs to voltages in excess of Vcc (+5 V). This device contains eight bidirectional analog switches with one side connected to input/output while the other side goes to output/input. With E at LOW, one of the eight switches can be selected by feeding the proper value to S_0, S_1 and S_3. The range of the power supply is 3 to 15 V [103].

The IC4052 is a two 4:1 analog/digital multiplexer/demultiplexer, while the IC4053 is a four 2:1 analog/digital multiplex/demultiplexer. Based on the intended applications, any one of these analog-switching ICs can be used. All these multiplexer/demultiplexer ICs are mainly used for (i) analog/digital multiplexing and demultiplexing, (ii) A/D and D/A conversion, (iii) signal gating, (iv) factory automation, (v) televisions, (vi) appliances, (vii) consumer audio, (viii) programmable logic circuits (PLCs), (ix) line coding, (x) sensor interfacing, and so on. We can design an eight-level PAM using a single IC4051, as shown

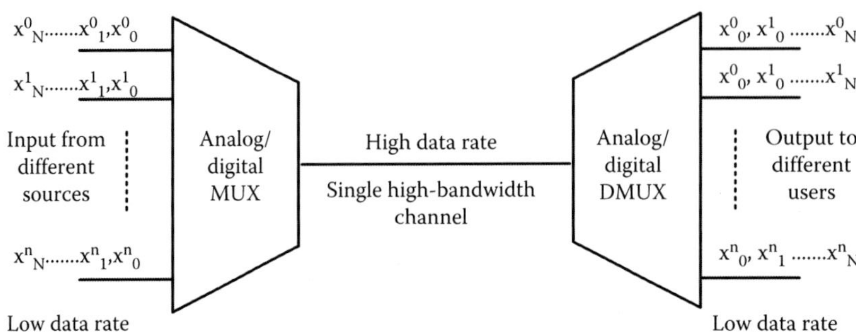

FIGURE 7.32
Schematic of a simplex TDM access.

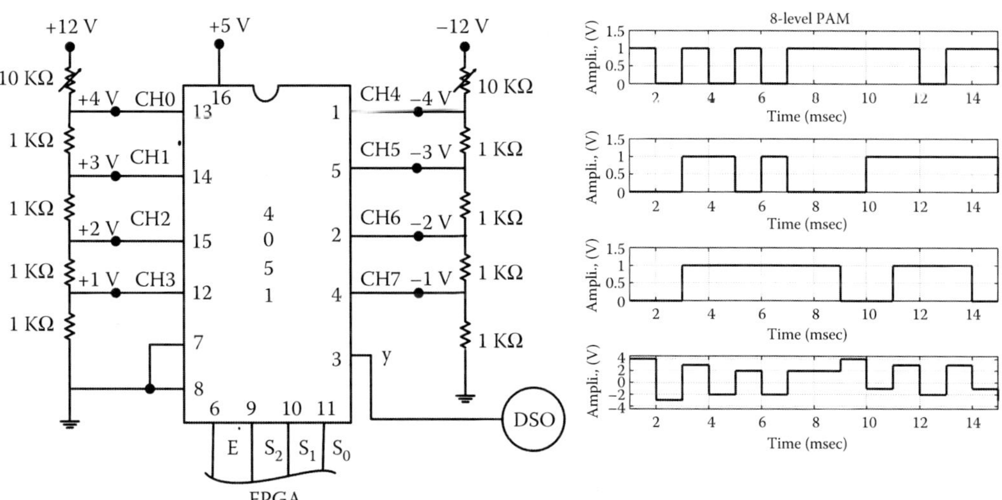

FIGURE 7.33
Circuit diagram and simulation result of a 8-level PAM.

in Figure 7.33. There, any one of the inputs will be selected at a time based on the value of the selection lines. The outputs for the different values of selection lines are also shown in Figure 7.33. The data of size [2:0], which are used at the selection line, would be properly decoded by a level-based decision device at the receiver end.

DESIGN EXAMPLE 7.32

Write MATLAB code for simulating the function of the circuit shown in Figure 7.33 and to get its simulation result as well.

Solution

```
clc;
clear all;
clear screen;
subplot(4,1,1)
d0=[1 0 1 0 1 0 1 1 1 1 1 0 1 1 0];
stairs(d0,'k-','LineWidth',1.5);
grid on;
ylim([0 1.5]);
xlim([1 15]);
xlabel('Time (msec)');
ylabel('Ampli., (V)');
title('8-Level PAM');
subplot(4,1,2)
d1=[0 0 1 1 0 1 0 0 0 1 1 1 1 1 0];
stairs(d1,'k-','LineWidth',1.5);
grid on;
ylim([0 1.5]);
xlim([1 15]);
xlabel('Time (msec)');
ylabel('Ampli., (V)');
subplot(4,1,3)
d2=[0 0 1 1 1 1 1 1 0 0 1 1 1 0 0];
stairs(d2,'k-','LineWidth',1.5);
grid on;
```

```
ylim([0 1.5]);
xlim([1 15]);
xlabel('Time (msec)');
ylabel('Ampli., (V)');
for i=1:1:length(d0),
x=(d2(i)*4 + d1(i)*2 + d0(i))
if x==0,
PAM(i)=-3;
elseif x==1,
PAM(i)=4;
elseif x==2,
PAM(i)=1;
elseif x==3,
PAM(i)=-1;
elseif x==4,
PAM(i)=-4;
elseif x==5,
PAM(i)=2;
elseif x==6,
PAM(i)=-2;
elseif x==7,
PAM(i)=3;
end
end
subplot(4,1,4);
stairs(PAM,'k-','LineWidth',1.5);
grid on;
ylim([-4.5 4.5]);
xlim([1 15]);
xlabel('Time (msec)');
ylabel('Ampli., (V)');
```

7.9 Signal Generator Design and Interfacing

In many applications, digital data that are generated from the FPGA cannot be directly applied to the subsequent system, for example, control, since it requires only analog signal, as they don't accept digital data. Thus, making use of the D/A converter, which converts digital data into equivalent analog voltage, becomes significant for many real-world applications. There are many types of D/A converters, as we saw in Section 7.2. In the following section, we discuss the interfacing techniques and application designs of the two most popular D/A converters.

7.9.1 Low Voltage Digital to Analog Conversion Using DAC0808

This section will show how to interface a D/A converter with the FPGA. Then, we will discuss how to generate a sine wave on the DSO using the D/A converter. The D/A converter is a device which is widely used to convert digital pulses to analog signals. Recall Section 7.1, where we discussed two methods of creating a D/A converter: binary weighted and R/2R ladder, out of which the second method is more commonly used in many integrated D/A converter circuits due to its ability to achieve a higher degree of precision. The main criteria for judging a DAC are (i) resolution, that is, number of binary bits; (ii) conversion speed, that is, frequency and (iii) polarity of output voltage, that is, whether the output is unipolar (+) or bipolar (±). The more common D/A converters have parallel digital input of 8, 10, and 12 bits. The number of input data bits (n) decides the resolution of the D/A converter, since the number of analog output levels is equal to 2^n, where n is the number of data bit inputs. Therefore, for an 8-bit input D/A converter, the maximum analog steps/levels are 256. Some D/A converter ICs, like DAC0808, give the equivalent current (I_{out}) as the output, while some other D/A converters, like DAC7728, give equivalent voltage (V_{out}) directly. However, I_{out} can be converted to V_{out} by using a suitable external current to voltage (I-V) converter circuit. In DAC0808, the total current provided by the I_{out} pin is a function of the binary data at the input D_0–D_7 and the reference current (I_{ref}) as follows [82].

$$I_{out} = I_{ref}\left(\frac{D_7}{2}+\frac{D_6}{4}+\frac{D_5}{8}+\frac{D_4}{16}+\frac{D_3}{32}+\frac{D_2}{64}+\frac{D_1}{128}+\frac{D_0}{256}\right) \tag{7.21}$$

where I_{ref} is the current being applied at pin 14 of the DAC, whose value is typically 2 mA and can be varied by adjusting the value of the resistor connected serially at that pin. Then, passing the output current (I_{out}) to an I-V converter, as shown in Figure 7.34, it is possible to get V_{out} as

$$V_{out} = I_{out}R \tag{7.22}$$

Suppose R is 5 KΩ and I_{ref} is 2 mA; then, the I_{out} for binary data of 99H, that is, $(10011001)_2$, is $2 \times 10^{-3}(1/2 + 1/16 + 1/32 + 1/256) = 2 \times 10^{-3}(153/256) = 1.195$ mA. Then, the V_{out} is 1.195 mA $5 \times 10^3 = 5.975$ V. Hence, the circuit shown in Figure 7.34 converts digital data into the equivalent analog current and then the equivalent analog voltage.

Now, we discuss how to generate a sine wave using a D/A converter. To generate a sine wave, first we need to find the values that have to be sent to the inputs of DAC0808 over time to display the sine wave on a scope. A cycle of a sine wave (y = 5sin(θ)) is shown in

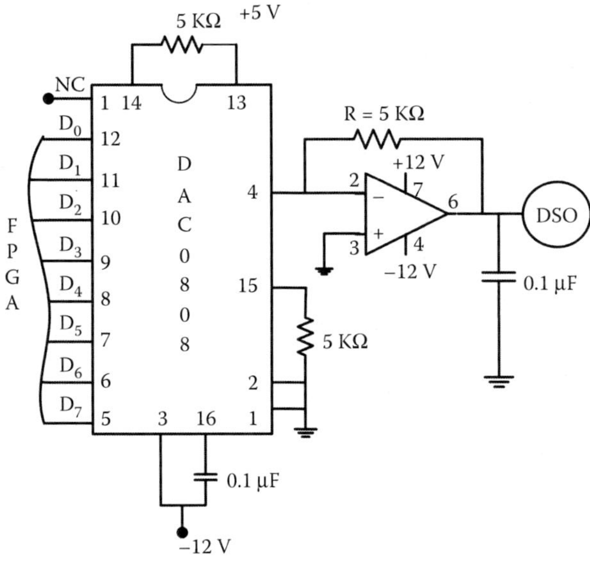

FIGURE 7.34
Circuit diagram of configuration of DAC0808 with FPGA.

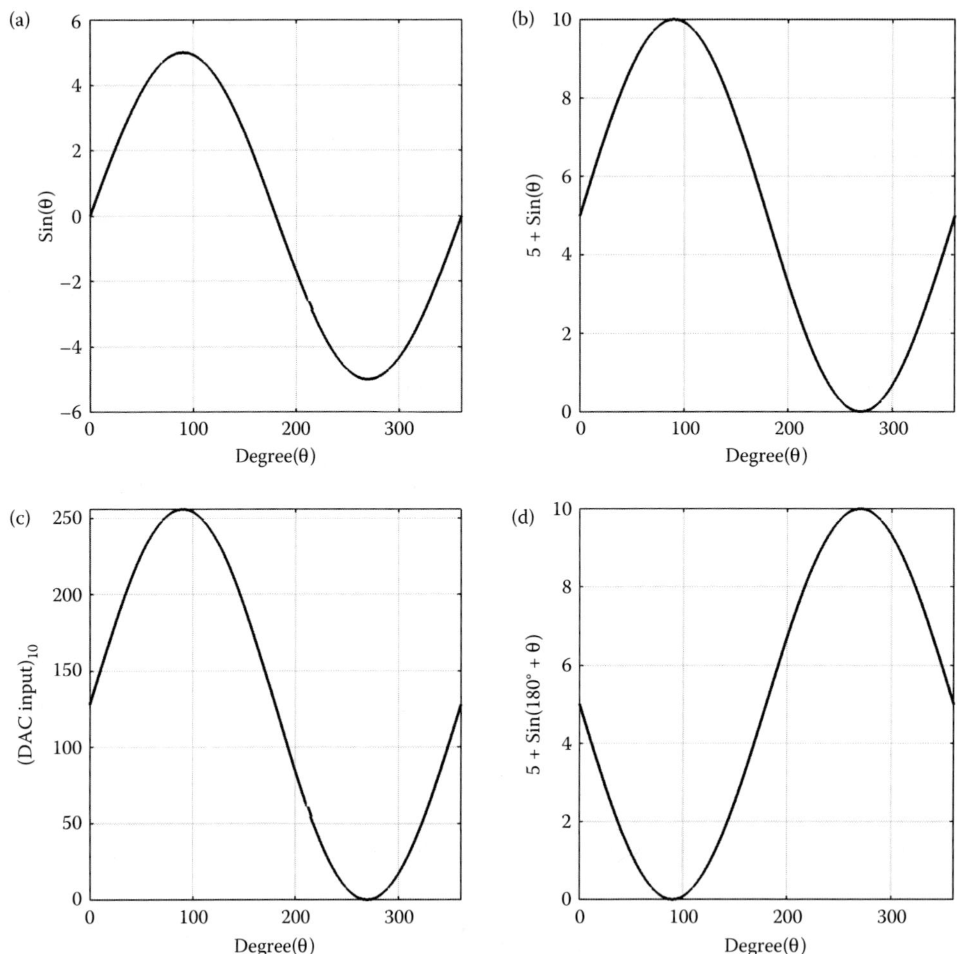

FIGURE 7.35
Sine wave with circuit diagram of configuration of DAC0808 with FPGA.

Figure 7.35a, where the amplitude varies from −5 V to +5 V for different values of the angle θ. The DAC0808 cannot produce negative current/voltage. Therefore, we need to provide a DC offset to linearly shift the sine wave up, that is, bipolar to unipolar sine wave. The DC offset 5 V is added to Figure 7.35a as $y = 5 + 5\sin(\theta)$. Then, the amplitude varies only from 0 to +10 V, as shown in Figure 7.35b, which now can be generated by this DAC. The necessary values to be input to the DAC0808 to generate a unipolar sine wave have to be found now. For example, θ is 30°, so the sin(θ) is 0.5, 5sin(θ) is 2.5 and $V_{out} = 5 + 5\sin(\theta)$ is 7.5. Then, its equivalent decimal value can be found as $(255/10)V_{out}$, which is ≈191, as shown in Figure 7.35c. Its binary equivalent value in 8 bit is "10111111", whose hexadecimal equivalent value is BFH. All these values can be found for the angle values from 0° to 360° at the angle resolution of 5° using the following MATLAB code.

```
clc;
clear all;
clear screen;
x = 0:5:360;
```

TABLE 7.9

A Portion of Calculated Values to be Sent to DAC0808 to Generate Sine Wave

Angle (θ) in Degree	Sin(θ) in Degree	V_{out} in V	Equivalent Decimal Value	Equivalent Binary Value	Equivalent Hexadecimal Value
0	0	5	128	10000000	80H
5	0.087	5.435	139	10001011	8BH
10	0.173	5.865	150	10010110	96H
15	0.258	6.29	160	10100000	A0H
.					
90	1	10	255	11111111	FFH
.					
180	0	5	128	10000000	80H
.					
270	−1	0	0	00000000	00H
.					
360	0	5	128	10000000	80H

```
y1=sind(x);
y = (5 + 5 * y1);
data=25.5 * y;
data1=uint8(data);
str = dec2bin(data1,8);
hex_str = dec2hex(bin2dec(str));
```

A portion of these values are given in Table 7.9. This method ensures that only integer numbers are the output of the FPGA and input to the DAC. As we increase the angle resolution, that is, reduce the angle step size, it is possible to increase to precision of the sine wave.

The amplitude of the sine wave can be further increased by varying the resistance value at the I-V converter as in Equation 7.22. The frequency of the sine wave is decided by the rate at which we send the above values (given in Table 7.9) to the DAC. The initial phase angle can be chosen at any value, as shown in Figure 7.35d, by appropriately sending the respective data and continuing the cycles from that value. In this way, any waveform can be generated. The DAC902, DAC1208, DAC1209, DAC1210, DAC1230, DAC1231 and DAC1232 are 12-bit D/A converter ICs, which can also be interfaced to the FPGA whenever more precision is needed.

DESIGN EXAMPLE 7.33

Design a digital system in the FPGA to interface with DAC0808 to produce a ramp signal. Display the ramp signal on a scope.

Solution

```
library IEEE;
use IEEE.STD_LOGIC_1164.ALL;
use IEEE.STD_LOGIC_ARITH.ALL;
use IEEE.STD_LOGIC_UNSIGNED.ALL;
entity ramp_sig is
Port ( clk : in  STD_LOGIC;
clk_test,clk_out : out  STD_LOGIC_vector(7 downto 0):="00000000");
```

```
end ramp_sig;
architecture Behavioral of ramp_sig is
signal clkout:STD_LOGIC_vector(7 downto 0):="00000000";
signal clk_sig:STD_LOGIC:='0';
begin
p0:process(clk)
variable cnt:integer;
begin
if rising_edge(clk) then
if(cnt=10)then
clk_sig<= not(clk_sig);
cnt:=0;
else
cnt:= cnt + 1;
end if;
end if;
end process;
p1:process(clk_sig)
begin
if rising_edge(clk_sig) then
clkout<= clkout+"00000001";
end if;
end process;
clk_out<=clkout;
clk_test<=clkout;
end Behavioral;
```

DESIGN EXAMPLE 7.34

Design a digital system in the FPGA to generate a sine wave using DAC0808. The sine wave peak-to-peak amplitude has to be 10 V and the DC offset has to be 5 V.

Solution

```
library IEEE;
use IEEE.STD_LOGIC_1164.ALL;
use IEEE.STD_LOGIC_ARITH.ALL;
use IEEE.STD_LOGIC_UNSIGNED.ALL;
entity clkd is
Port ( clk: in STD_LOGIC:='0';
sin: out STD_LOGIC_VECTOR(7 downto 0):="00000000");
end clkd;
architecture Behavioral of clkd is
signal clk_sig:STD_LOGIC:='0';
type freq_syn is array (0 to 72) of std_logic_vector(7 downto 0);
constant data : freq_syn:=(x"80",x"8B",x"96",x"A0",x"AB",x"B5",x"BF",
x"C9",x"D1",x"DA",x"E1",x"E8",x"EE",x"F3",x"F7",x"FB",x"FD", x"FE",x"FF"
,x"FE",x"FD",x"FB",x"F7",x"F3",x"EE",x"E8",x"E1",x"DA",x"D1",x"C9",x"BF"
,x"B5",x"AB",x"A0",x"96",x"8B",x"80",x"74",x"69",x"5F",x"54",x"4A",x"40"
,x"36",x"2E",x"25",x"1E",x"17",x"11",x"0C",x"08",x"04",x"02",x"01",x"00"
,x"01",x"02",x"04",x"08",x"0C",x"11",x"17",x"1E",x"25",x"2E",x"36",x"40"
,x"4A",x"54",x"5F",x"69",x"74",x"80");
begin
p0:process(clk)
variable cnt:integer:=0;
begin
if rising_edge(clk) then
if(cnt=1)then --Value has to be chosen based on the
clk_sig<= not(clk_sig);--on-board clock rate.
cnt:=0;
```

```
else
cnt:= cnt + 1;
end if;
end if;
end process;
p1:process(clk_sig)
variable cnt1:integer:=0;
begin
if rising_edge(clk_sig) then
if cnt1=72 then
cnt1:=0;
else
sin<=data(cnt1);
cnt1:=cnt1+1;
end if;
end if;
end process;
end Behavioral;
```

DESIGN EXAMPLE 7.35

Design a digital system in the FPGA to generate a QPSK output. Display the data and QPSK output in two different channels of a scope.

Solution

```
library IEEE;
use IEEE.STD_LOGIC_1164.ALL;
use IEEE.STD_LOGIC_ARITH.ALL;
use IEEE.STD_LOGIC_UNSIGNED.ALL;
entity clkd is
Port ( input: in  STD_LOGIC_vector(1 downto 0):="00";
clk: in STD_LOGIC:='0';
qpsk: out STD_LOGIC_vector(7 downto 0):="00000000");
end clkd;
architecture Behavioral of clkd is
signal clk_sig:STD_LOGIC:='0';
signal clkout1,clkout2,clkout3,clkout4:STD_LOGIC_VECTOR(7 downto
0):="00000000";
type freq_syn is array (0 to 72) of std_logic_vector(7 downto 0);
constant data : freq_syn:=(x"80",x"8B",x"96",x"A0",x"AB",
x"B5",x"BF",x"C9",x"D1",x"DA",x"E1",x"E8",x"EE",x"F3",x"F7",x"FB",x"FD",
x"FE",x"FF",x"FE",x"FD",x"FB",x"F7",x"F3",x"EE",x"E8",x"E1",x"DA",x"D1",
x"C9",x"BF",x"B5",x"AB",x"A0",x"96",x"8B",x"80",x"74",x"69",x"5F",x"54",
x"4A",x"40",x"36",x"2E",x"25",x"1E",x"17",x"11",x"0C",x"08",x"04",x"02",
x"01",x"00",x"01",x"02",x"04",x"08",x"0C",x"11",x"17",x"1E",x"25",x"2E",
x"36",x"40",x"4A",x"54",x"5F",x"69",x"74",x"80");
begin
p0:process(clk)
variable cnt:integer:=0;
begin
if rising_edge(clk) then
if(cnt=10)then
clk_sig<= not(clk_sig);
cnt:=0;
else
cnt:= cnt + 1;
end if;
end if;
```

```
end process;
p1:process(clk_sig,input)
variable cnt1:integer:=0;
variable cnt2:integer:=18;
variable cnt3:integer:=36;
variable cnt4:integer:=54;
begin
if rising_edge(clk_sig) then
if cnt1=72 then
cnt1:=0;
else
clkout1<=data(cnt1);
cnt1:=cnt1+1;
end if;
if cnt2=72 then
cnt2:=0;
else
clkout2<=data(cnt2);
cnt2:=cnt2+1;
end if;
if cnt3=72 then
cnt3:=0;
else
clkout3<=data(cnt3);
cnt3:=cnt3+1;
end if;
if cnt4=72 then
cnt4:=0;
else
clkout4<=data(cnt4);
cnt4:=cnt4+1;
end if;
case input is
when "00" => qpsk<= clkout1;
when "01"=> qpsk<= clkout2;
when "10"=> qpsk<= clkout3;
when "11"=> qpsk<= clkout4;
when others => null;
end case;
end if;
end process;
end Behavioral;
```

DESIGN EXAMPLE 7.36

Develop a digital system in the FPGA to produce a QPSK signal. The data that are required to modulate the carrier signal have to be generated from a parallel Kasami-sequence generator.

Solution

```
library ieee;
use ieee.std_logic_1164.all;
use ieee.std_logic_arith.all;
use ieee.std_logic_unsigned.all;
use IEEE.NUMERIC_STD.ALL;
entity Ks_qpsk is
port (m_clk,reset : in  std_logic:='0';
         qpsk_out  : out std_logic_vector(7 downto 0):="00000000");
end Ks_qpsk;
```

```
architecture Behavioral of Ks_qpsk is
type memory_type is array (0 to 72) of integer range 0 to 255;
signal sine : memory_type :=(128,139,150,160,171,181,191,201,209,218,225,232,
238,243,247,251,253,254,255,254,253,251,247,243,238,232,225,218,209,201,191,
182,171,160,150,139,128,116,105,95,84,74,64,54,46,37,30,23,17,12,8,4,2,1,0,
1,2, 4,8,12,17,23,30,37,46,54,
64,74,84,95,105,116,128);
signal count0:integer range 0 to 72:=0;
signal count1:integer range 0 to 72:=9;
signal count2:integer range 0 to 72:=18;
signal count3:integer range 0 to 72:=27;
signal count4:integer range 0 to 72:=36;
signal count5:integer range 0 to 72:=45;
signal count6:integer range 0 to 72:=54;
signal count7:integer range 0 to 72:=63;
signal clk_sig1,feed_u,feed_v,feed_w,out_v:std_logic:='0';
signal reg_u,reg_v: std_logic_vector(6 downto 0):="1000110" ;
signal reg_w: std_logic_vector(3 downto 0):="1101" ;
signal mul_v: std_logic_vector(5 downto 0):="000000" ;
signal kasami_seq,mul_w: std_logic_vector(2 downto 0):="000" ;
signal
dataout_temp0,dataout_temp1,dataout_temp2,dataout_temp3,dataout_temp4,
dataout_temp5,
dataout_temp6,dataout_temp7: std_logic_vector(7 downto 0):="00000000" ;
begin
p0:process(m_clk)
begin
if ( m_clk='1' and m_clk'event ) then
if (count0=72) then     count0<= 0 ; else     dataout_temp0 <=
conv_std_logic_vector (sine(count0),8); count0<=count0 + 1; end if;
if (count1=72) then     count1<= 0 ; else dataout_temp1 <=
conv_std_logic_vector (sine(count1),8); count1<=count1 + 1; end if;
if (count2=72) then     count2<= 0 ; else dataout_temp2 <=
conv_std_logic_vector (sine(count2),8); count2<=count2 + 1; end if;
if (count3=72) then     count3<= 0 ; else dataout_temp3 <=
conv_std_logic_vector (sine(count3),8); count3<=count3 + 1; end if;
if (count4=72) then     count4<= 0 ; else     dataout_temp4 <=
conv_std_logic_vector (sine(count4),8); count4<=count4 + 1; end if;
if (count5=72) then     count5<= 0 ; else dataout_temp5 <=
conv_std_logic_vector (sine(count5),8); count5<=count5 + 1; end if;
if (count6=72) then     count6<= 0 ; else dataout_temp6 <=
conv_std_logic_vector (sine(count6),8); count6<=count6 + 1; end if;
if (count7=72) then     count7<= 0 ; else dataout_temp7 <=
conv_std_logic_vector (sine(count7),8); count7<=count7 + 1; end if;
end if;
end process;
p1:process(m_clk,reset)
variable cnt:integer:=0;
begin
if(reset='1') then
clk_sig1<='0';
cnt:=0;
elsif rising_edge(m_clk) then
if(cnt=100)then
clk_sig1<= not(clk_sig1);
cnt:=0;
else
cnt:= cnt + 1;
end if;
end if;
end process;
```

```
p2:process(clk_sig1)
begin
if rising_edge(clk_sig1) then
feed_u<=(reg_u(6) xor reg_u(0));
feed_v<=(reg_v(6) xor reg_v(5) xor reg_v(2) xor reg_v(0));
feed_w<=(reg_w(2) xor reg_w(0));
end if;
if falling_edge(clk_sig1) then
reg_u<=(feed_u & reg_u(6 downto 1));
reg_v<=(feed_v & reg_v(6 downto 1));
reg_w<=(feed_w & reg_w(3 downto 1));
end if;
end process;
mul_v<= (reg_v(6 downto 1) and "101110");--m
mul_w<= (reg_w(3 downto 1) and "101");-----k
out_v<=(reg_u(1) xor mul_v(5) xor mul_v(4) xor mul_v(3)xor  mul_v(2)xor
mul_v(1) xor mul_v(0));
kasami_seq(0)<=(out_v xor mul_w(2) xor mul_w(1) xor mul_w(0));
kasami_seq(1)<=mul_w(1);
kasami_seq(2)<=mul_w(2);
p3:process(kasami_seq,dataout_temp0,dataout_temp1,dataout_temp2,
dataout_temp3,dataout_temp4,dataout_temp5,dataout_temp6,dataout_temp7)
begin
case kasami_seq is
when "000"=> qpsk_out<= dataout_temp0 ;
when "001"=> qpsk_out<= dataout_temp1 ;
when "010"=> qpsk_out<= dataout_temp2 ;
when "011"=> qpsk_out<= dataout_temp3 ;
when "100"=> qpsk_out<= dataout_temp4 ;
when "101"=> qpsk_out<= dataout_temp5 ;
when "110"=> qpsk_out<= dataout_temp6 ;
when "111"=> qpsk_out<= dataout_temp7 ;
when others=> null;
end case;
end process;
end Behavioral;
```

7.9.2 High-Voltage Digital to Analog Conversion Using DAC7728

The DAC7728 is a low-power, octal, 12-bit D/A converter. The output of the DAC7728 can be a bipolar ±15 V or unipolar 0 to +30 V when the operating supply voltage is ≥ ±15.5 and +30.5 V, respectively. In both cases, the reference voltage is +5 V. With a reference voltage of 5.5 V, the outputs ±16.5 and 0 to +33 V can be obtained when the operating power supply voltage is ≥ ±17 and ≥ +33.5 V, respectively. The DAC provides operation at low power with good linearity. The output range can be offset by using the DAC offset register. The DAC7728 features a 12-bit parallel interface that operates at up to 50 MHz and is 1.8, 3, and 5 V logic compatible to communicate with FPGAs. The eight D/A converters and the auxiliary registers are addressed with five address lines (A_0–A_4). The device also features double-buffered interface logic. An asynchronous load input (LDAC) transfers the data from the D/A converter "Data register" to the respective D/A converter latch, for example, DAC-0 to Latch-0. The asynchronous CLR input sets the output of all eight D/A converters to analog ground (AGND). The eight D/A converter channels and two offset D/A converters are arranged into two groups (A and B) with four channels and one offset D/A converter per group. Group A consists of DAC-0 through DAC-3 and offset DAC-A. Group B consists of DAC-4 through DAC-7 and offset DAC-B. Group A derives its reference voltage from REF-A, and Group B derives its reference voltage from REF-B. The analog monitor (V_{MON}) pin gives the output of individual analog outputs (DAC-0 through DAC-7), the offset of DAC (OFFSET-A and B) and the reference buffer outputs (Ref Buffer A and B) through a multiplexer (Mux). The power in each analog channel is 13.5 mW/Ch. The offsets and gain values can be controlled by the programs. The programmable gain multiplication factors are ×4 and ×6. The DAC7728 contains eight DAC channels and eight output amplifiers in a single package, as shown in Figure 7.36. Each channel consists of a resistor-string DAC followed by an output buffer amplifier. The resistor-string section is simply a string of resistors, each with a value of R, from REF to AGND. This type of architecture provides DAC monotonicity. The 12-bit binary digital code loaded to the DAC register determines at which node on the string the voltage is tapped off before being fed into the output amplifier. The output amplifier multiplies the DAC output voltage by a gain of ×4 or ×6. The output span is 9 V with a 1.5 V reference, 18 V with a 3 V reference and 30 V for a 5 V reference when using dual power supplies of ±16.5 V and a gain of 6 [104].

The analog outputs from all the DAC channels ($V_{out}0 - V_{out}7$) can be calculated as

$$V_{out} = \begin{cases} V_{Ref}Gain\left(\dfrac{\text{Input\_Code}}{4096}\right) - V_{Ref}(Gain - 1)\left(\dfrac{\text{OffsetDAC\_Code}}{4096}\right) & \text{if } SCE = 0 \\ V_{Ref}Gain\left(\dfrac{\text{DAC\_Data\_Code}}{4096}\right) - V_{Ref}(Gain - 1)\left(\dfrac{\text{OffsetDAC\_Code}}{4096}\right) & \text{if } SCE = 1 \end{cases} \tag{7.23}$$

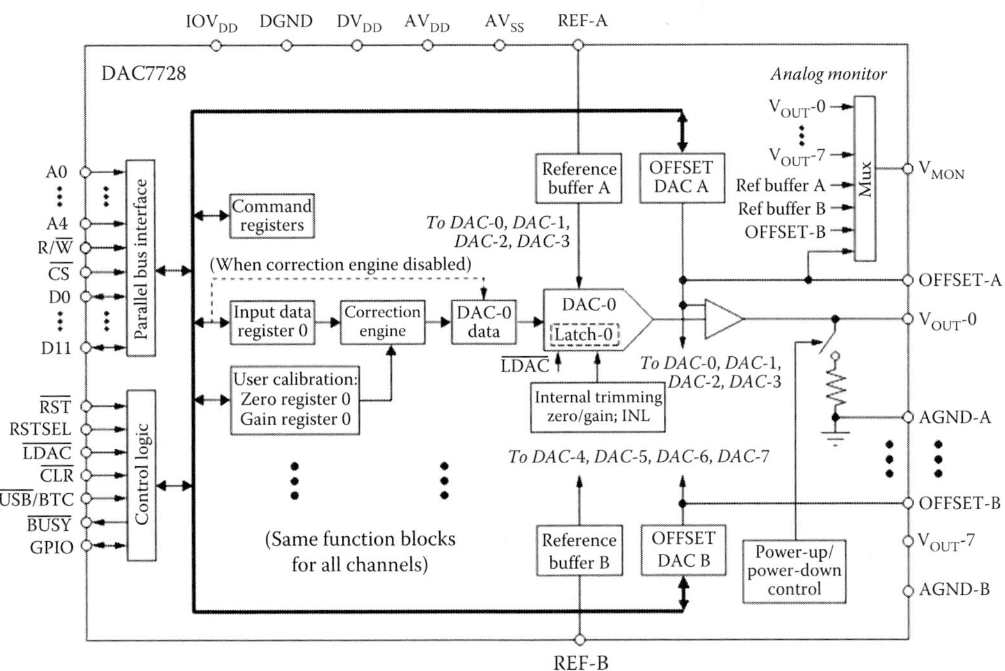

FIGURE 7.36
Functional block diagram of DAC7728.

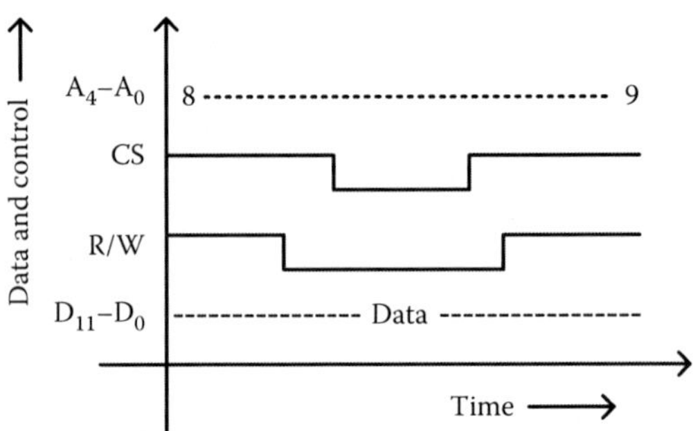

FIGURE 7.37
Timing diagram of write operation-1.

where SCE denotes the correction engine enable/disable bit, DAC_Data_Code is obtained by [(Input_code (User_gain+2^{11}))/2^{12}] + User_Zero, Gain is the DAC gain defined by the GAIN bit in the configuration register, Input_Code is the data written into the input data register, and Offset DAC_Code is the data written into the offset DAC register. In the estimation of DAC_Data_Code, User_gain is the code of the gain register and User_gain is the code of the zero register. The timing diagram of write operation 1, writing to the configuration register, offset register, monitor register, GPIO register, DAC input register, zero register and gain register in asynchronous mode, is shown in Figure 7.37. The address (A_0–A_4) values for loading 12-bit values into the DAC input registers (DAC-0 though DAC-7) are "01000" through "01111". All the address details and register maps can be found in [104]. The timing diagram shown in Figure 7.37 is the mode of writing data to the configuration register, offset register, monitor register, GPIO register, DAC input register, zero register, and gain register.

All the registers can be programmed by selecting their appropriate address values as "00000" for the configuration register, "00001" for the monitor register, "00010" for the GPIO register, "00011" for the offset DAC-A data register, "00100" for the offset DAC-B data register, and "10000", "10001", "10010", "10011", "10100", "10101", "10110", and "10111" for busy zero registers: 0, 1, 2, 3, 4, 5, 6, and 7, respectively. "11000", "11001", "11010", "11011", "11100", "11101", "11110" and "11111" are the addresses of the gain registers: 0, 1, 2, 3, 4, 5, 6, and 7, respectively. As discussed above, the DAC8728 (16-bit) and DAC8228 (14-bit) can also be interfaced to the FPGA whenever we are in need of high resolution. Discussion of many internal presets of the DAC7728 is out of the scope of this section. However, reference [17] gives more details about pin descriptions, read timing diagrams, timing diagrams of write operations 2 and 3, user calibration, gain errors, correction engines, output amplifiers, power-on/hardware reset, power-down mode, internal configuration register details and the physical layout of this high-voltage D/A converter.

DESIGN EXAMPLE 7.37

Develop a digital system to drive FSMs using DAC7728 to stabilize the optical beam at the centre of an OPD. There are two FSMs: one for x-plane steering and another for y-plane. The beam centroid displacement errors on the x and y planes are measured by 12-bit A/D converters. The error signals are directly mapped to DAC-0 and DAC-1 of DAC7728 to stabilize the beam. Assume that the DAC7728 has to be operated with its default settings. Further, all the necessary settings around the DAC7728 are done using the onboard jumpers.

Solution

```
library ieee;
use ieee.std_logic_1164.all;
use ieee.std_logic_arith.all;
use ieee.std_logic_unsigned.all;
use IEEE.NUMERIC_STD.ALL;
entity HV_DAC is
port (m_clk: in  std_logic:='0';
cs,rw:out std_logic:='1';
add:out std_logic_vector(4 downto 0):="00000";
x_error,y_error: in std_logic_vector(11 downto 0):="000000000000";
x_fsm,y_fsm: out std_logic_vector(11 downto 0):="000000000000");
end HV_DAC;
architecture Behavioral of HV_DAC is
signal clk_sig1:std_logic:='0';
begin
p0:process(m_clk)
variable cnt:integer:=0;
begin
if rising_edge(m_clk) then
if(cnt=1)then
clk_sig1<= not(clk_sig1);
cnt:=0;
else
cnt:= cnt + 1;
end if;
end if;
end process;
p1:process(clk_sig1)
variable cnt,cnt1,cnt2: integer:=1;
begin
if rising_edge(clk_sig1) then
if (cnt=1) then
cs<='1'; rw<='1';
cnt:=cnt+1;
elsif (cnt=2) then
if (cnt1=1) then
add<="01000";    cnt1:=cnt1+1;
elsif (cnt1=2) then
add<="01001"; cnt1:=1;
end if;
cnt:=cnt+1;
elsif (cnt=3) then
rw<='0'; cnt:=cnt+1;
elsif (cnt=4) then
if (cnt2=1) then
x_fsm<=x_error; cnt2:=cnt2+1;
elsif (cnt2=2) then
y_fsm<=y_error; cnt2:=1;
end if;
cnt:=cnt+1;
elsif (cnt=7) then
cs<='0'; cnt:=cnt+1;
elsif (cnt=8) then
cs<='1'; cnt:=cnt+1;
elsif (cnt=9) then
rw<='1'; cnt:=1;
else
cnt:=cnt+1;
end if;
```

```
end if;
end process;
end Behavioral;
```

LABORATORY EXERCISES

1. Modify Design Example 7.4 into implementable code.
2. Design an 8-bit binary A/D converter using a magnitude comparator (LM324). Assume that the common input is connected to the inverting inputs and the reference voltages are connected to the noninverting terminals.
3. Design a high degree of serial and parallel PRBS generator and realize different digital modulation schemes, for example ASK, FSK, PSK, M-FSK, and M-PSK, with necessary circuits.
4. Develop VHDL code to get different waveforms: clock pulse, square wave, triangular wave, staircase wave and exponential wave using a D/A converter.
5. Develop VHDL code for realizing an error-control coding technique using linear block code and the Hamming code algorithm.
6. Develop VHDL code and design a suitable circuit to get all line-coding data: UPRZ, PRZ, PNRZ, Manchester encoding, differential Manchester encoding, NRZ-inverted (differential encoding), NRZ-L, NRZ-M, RZ-AMI, and so on.
7. Develop a multichannel data acquisition system. All the data acquired from the 16 sensors (analog signals) have to be logged in an Excel file using MATLAB for the specified duration of time. All the data correspond to 16 channels, and the logging/updating rate is 1 second. Illustrate the data logging in MATLAB's command window in real time.
8. Design a digital system in the FPGA to realize a linear FM pulse (chirp) compression technique for a radar system. Display the linear sweep signal and FM signal in two different channels of a scope.
9. Develop a digital system to interface with an AD9858 (analog device) direct digital synthesizer and generate arbitrary wave forms based on user-selected inputs.
10. Develop a digital system in the FPGA to realize Barker codes, polyphase codes, and stretch processes for the generation of RADAR pulse compression wave forms at the based-band level.

Part III

Hardware Accelerated Designs

8

Real-Time Clock and Interface Protocol Programming

8.1 Introduction and Preview

In many real-world applications and peripheral interfacing, we need to follow certain protocols to successfully interface the devices with the FPGA (master) and establish command and data transactions between the master and peripherals without any conflict or missing of data. There are several protocols to interface devices with the FPGA, and they all are device dependent, which is decided by the device manufacturers. However, most of the devices follow some standard interfacing protocols. Thus, before beginning interface code/architecture design, we need to clearly understand what protocol the devices follow. This is most important, since writing code is simple, but knowing what we should write in the code is difficult until unless we are familiar with the interfacing protocol of the particular device with which we wish to interface. In this chapter, we will discuss several standard interfacing protocols with device pictures and pin details, application circuits, and design examples. This chapter is specifically devised to help readers master the required skills in the area of protocol-device interfacing with the FPGA.

This chapter begins with the pin details, interfacing circuit, and design examples of a popular RTC, DS12887. Register organization, initialization of registers, read/write methodologies, time and day data initialization, data storage, and control signal generation are explained, with appropriate tables and design examples. Another popular interfacing protocol, I^2C, is covered, with read/write clock cycles, start/stop sequences, address/data transactions, and design examples. Two real-world sensors, namely SHT11 and SCP1000-D01, that follow the TWI and SPI, respectively, are discussed. The significance of SCK, serial data (SDA), data ready (drdy), trigger (trig), tristate switch, master in slave out (MISO), and master out slave in (MOSI) of the SHT11 and SCP1000-D01 are explained, and their control state diagrams are given. GSM-based text sending/receiving, its associated attention (AT) comments and interfacing protocols are detailed, along with design examples.

The significance of GPS and its data receiving pattern are detailed. The method of extracting the necessary information like day details, location details, altitude, and so on is explained, with the necessary design examples. The interfacing protocol of the PS/2, data transaction codes, scan codes, make codes, and break codes are described, with examples. Displaying video frames; creating our own displays, like Radar A-scope and B-scope, and interfacing the VGA port with the FPGA are explained. The technique behind displaying an image on the VGA monitor is detailed, along with design examples. Finally, several laboratory exercises are given to help readers practice and improve their skills in this field.

8.2 Real-Time Clock (DS12887) Interface Programming

This section explains the interfacing and programming of the DS12887 RTC chip. It also describes the pin details, functionalities, programming of special registers, and interfacing digital architecture of the DS12887. The alarm and square wave features of the DS12287 are also touched upon. The RTC is a popular device that provides an accurate time and date, that is, calendar components, such as hour, minute, second, year, month, day, and leap-year compensation, for many applications. Many computers, such as the x86 IBM PC, come with such a chip on the motherboard. Although some processors come with an internal RTC embedded in the processor chip, we have to interface with the external RTC chip for many applications. The RTC chip uses an internal battery that keeps the time and date updated even when the main power is off. The DS12887 uses an internal lithium battery to keep operating for over 10 years in the absence of external power. All the above information is provided by the RTC chip in both binary (hex) and BCD formats. The pin details and a simple way of configuring the DS12887 to the FPGA are shown in Figure 8.1.

The DS12887 supports both 12-hr and 24-hr clock formats with AM and PM indication in the 12-hr form. The DS12887 has a total of 128 bytes of nonvolatile RAM, as shown in Figure 8.2. It uses 14 bytes of RAM for the clock/calendar and control registers, and the other 114 bytes of RAM are for general-purpose data storage. Before proceeding to discussing register programming, we need to understand some important pin details of the RTC chip as given below:

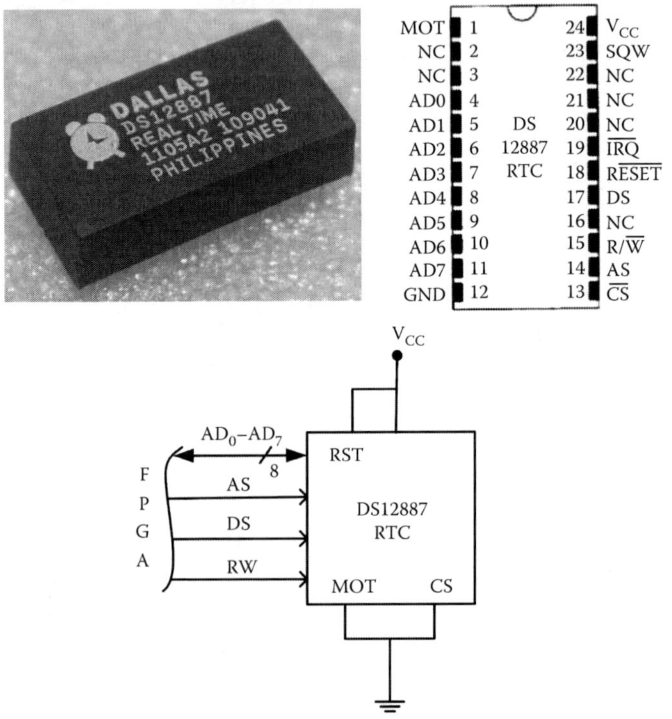

FIGURE 8.1
Pin details and application circuit of DS12887.

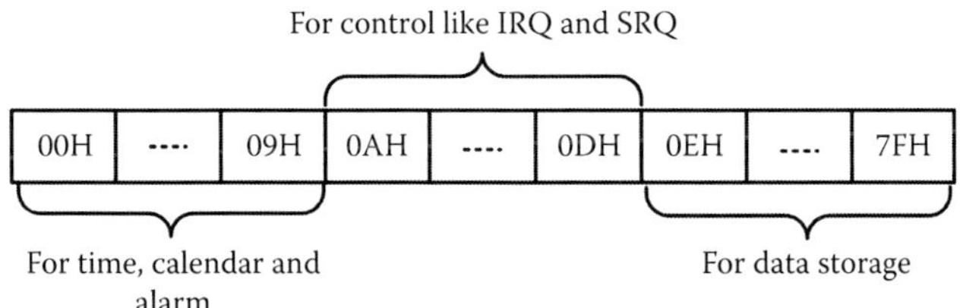

FIGURE 8.2
DS12887 address map.

Vcc: When the supply voltage (Vcc) falls below the 3V level, the external source is switched off and the internal lithium battery provides power to the RTC. This nonvolatile capability of the RTC prevents any loss of data or missing of update. According to the DS12887 data sheet, the RTC function continues to operate, and all of the RAM, time, calendar, and alarm memory locations remain nonvolatile regardless of the level of the Vcc input. However, in order to access the registers via a program, the Vcc must be supplied externally.

ADO-AD7: The multiplexed address/data pins provide both addresses and data to the chip. Addresses are latched into the DS12887 on the falling edge of the address strobe (AS) signal. The ADO–AD7 of the DS12887 can be directly connected to the FPGA without any pull-up resistors or latches, since the DS12887 provides the latch internally.

AS: The address strobe (AS) will cause the addresses to be latched into the DS12887 on the falling edge. The AS pin is used for demultiplexing the address and data.

MOT: This is an input pin that allows the choice between Motorola and Intel microcontroller bus timings. The MOT pin is always connected to Gnd for Intel timing.

DS: Data strobe (DS) is an input; it is called the read (RD) signal when the MOT is connected to the Gnd for Intel timing.

R/W: Read/write (R/W) is called the write (WR) signal when the MOT is connected to the Gnd for Intel timing.

CS: Chip select is an input pin and an active low signal. During the RD and WR cycles of Intel timing, the CS must be low in order to access the chip. It must be noted that the CS works only when the external Vcc is connected.

IRQ: The interrupt request is an output pin and active low signal. To use IRQ, the interrupt-enable bits in register B (refer to Figure 8.2), must be set high.

SQW: Square wave (SQW) is an output pin. We can program the DS12887 to provide up to 15 different square waves. The frequency of the square wave is set by programming register A; refer to Figure 8.2.

Reset: It is an input and is active low, and in most applications, the reset pin is connected to the Vcc pin. In applications where this pin is used, it has no effect on the clock, calendar or RAM if it is forced low. The low on this pin will cause the reset of the IRQ and clearing of the SQW pin.

With this information, we proceed to register programming. As we have seen above and as shown in Figure 8.2, the DS12887 has a total of 128 bytes of RAM with addresses 00H through 7FH. The first 10 locations, that is, 00H through 09H, are set aside for RTC

TABLE 8.1

Address Location for Time, Calendar, and Alarm IN DS12887

	Data Ranges			
Address	**Decimal**	**Binary (Hex)**	**BCD**	**Remarks**
00H	0-59	00H-3BH	00-59	Seconds
01H	0-59	00H-3BH	00-59	Second alarm
02H	0-59	00H-3BH	00-59	Minutes
03H	0-59	00H-3BH	00-59	Minutes alarm
04H	1-12	01H-0CH AM	01-12 AM	Hours (12-hr)
	1-12	81H-8CH PM	81-92 PM	Hours (12-hr)
	0-23	00H-17H	00-23	Hours (24-hr)
05H	1-12	01H-0CH AM	01-12 AM	Hours alarm (12-hr)
	1-12	81H-8CH PM	81-92 PM	Hours alarm (12-hr)
	0-23	00H-17H	00-23	Hours alarm (24-hr)
06H	1-7	01H-07H	01-07	Day of the week, Sun = 1
07H	1-31	01H-1FH	01-31	Day of the month
08H	1-12	01H-0CH	01-12	Month
09H	0-99	00H-63H	00-99	Year

values of time, calendar and alarm data, as marked in Figure 8.2. The next 4 bytes, that is, 0AH through 0DH, are used for the control and status registers. The next 114 bytes, from addresses OEH through 7FH, are available for data storage. The entire 128 bytes of RAM are accessible directly for read or write except registers A, C, and D, since they are read-only registers. All these register details are shown in Figure 8.2. The byte addresses 00H through 09H are set aside for the time, calendar, and alarm data. Table 8.1 gives the address locations, remarks (functions), and data mode ranges of these registers.

The DS12887 has an internal oscillator which is turned off at the factory in order to save the lithium battery and which we need to turn on before we use the time-keeping features of the DS12887. To do this, bits D_6 (DV2) through D_4 (DV0) of register A must be set to value "010," as given in Table 8.2, where UIP is update in progress, which is a read-only memory, and RS3 through RS0 are used for generating 13 different frequency square waves [105]. The value of RS3 through RS0 may be "0001" through "1111", which will generate the square wave at the frequency of 256 Hz, 128 Hz, 8.192 kHz, 4.096 kHz, 2.048 kHz, 1.024 kHz, 512 Hz, 256 Hz, 128 Hz, 64 Hz, 32 Hz, 16 Hz, 8 Hz, 4 Hz, and 2 Hz, respectively. The assignment of "0000" does not generate the square wave; hence, this assignment is invalid. More information about the UIP can be found in Ref. 105.

When we initialize the time or date, we need to set D_7 (SET) of register B to 1. This will prevent any update in the middle of initialization. After setting the time and date, we need to set D_7 (SET) of B register to 0 to make sure the clock and time are updated. The update occurs once per second. Refer to Table 8.3 for the more details about the B register.

In register B, SET = 0 indicates that the clock is counting once per second and time/dates are getting updated, and set = 1 indicates that the update is inhibited, that is, initialization is in progress. PIE is periodic interrupt enable. AIE is alarm interrupt enable, which is 1 when all the components of time, that is, yy:mm:dd, are the same as the alarm component to assert the IRQ. SQWE is square wave enable. DM is data mode, which is 0 for BCD and

TABLE 8.2

Register A and Its Contents

A Register								Remarks
UIP	**DV2**	**DV1**	**DV0**	**RS3**	**RS2**	**RS1**	**RS0**	
0	0	1	0	0	0	0	0	20H will turn on oscillator

TABLE 8.3

Register B and Its Contents

A Register								Remarks
SET	**PIE**	**AIE**	**UIE**	**SQW**	**DM**	**24/12**	**DSE**	
0	0	1	0	0	0	0	0	20H will turn on oscillator

1 for binary (Hex) mode. The value of 24/12 is 1 for 24-hr and 0 for 12-hr mode. DSE is daylight savings enable, which is 1 to enable daylight savings and 0 to disable it. More details about PIE, UIE, SQWE, individual bits of the registers, initializations, programming, and interrupt applications can be found in Ref. 105.

DESIGN EXAMPLE 8.1

Design and develop a digital system to interface the DS12887 RTC chip with the FPGA and get the time and day data in an 8-bit BCD number for each component. Use a decoder to convert these BCD data into seven-segment data and show the results using the required number of seven-segment displays.

Solution

The following code gives the time and day components in 8-bit BCD data form. Converting them into the suitable seven-segment format and then displaying them on a seven-segment display is not included in the following VHDL code and is left to the reader.

```
library ieee;
use ieee.std_logic_1164.all;
entity design0 is
port(clk,nrst:in std_logic;
irq:in std_logic;
sqw:in std_logic;
ad:inout std_logic_vector(7 downto 0);
as:out std_logic;
ds:out std_logic;
rw:out std_logic;
cs:out std_logic;
rst:out std_logic;
hrs:out std_logic_vector(7 downto 0);
min:out std_logic_vector(7 downto 0);
sec:out std_logic_vector(7 downto 0);
day:out std_logic_vector(7 downto 0);
mon:out std_logic_vector(7 downto 0);
year:out std_logic_vector(7 downto 0));
end entity;
architecture mark0 of design0 is
signal cnt200:integer range 0 to 5000000;
```

```
signal pos:integer range 0 to 40;
begin
process(nrst,clk)
begin
if(nrst='0') then
pos <= 0;
cnt200 <= 0;
ad <= x"00";
as <= '1';
ds <= '0';
rw <= '1';
cs <= '0';
rst <= '1';
else
if(falling_edge(clk)) then
case(pos) is
when 0=>
if(cnt200=5) then
pos <= 1;
cnt200 <= 0;
else
pos <= 0;
cnt200 <= cnt200 + 1;
end if;
when 1=>
ad <= x"0A";
as <= '0';
rw <= '1';
pos <= 2;
when 2=>
as <= '1';
pos <= 3;
when 3=>
ad <= x"20";
rw <= '0';
pos <= 4;
when 4=>
ad <= x"0B";
as <= '0';
rw <= '1';
pos <= 5;
when 5=>
as <= '1';
pos <= 6;
when 6=>
ad <= x"83";
rw <= '0';
pos <= 7;
when 7=>
ad <= x"00";
as <= '0';
rw <= '1';
pos <= 8;
when 8=>
as <= '1';
pos <= 9;
when 9=>
ad <= x"00";
rw <= '0';
pos <= 10;
when 10=>
```

```
ad <= x"02";
as <= '0';
rw <= '1';
pos <= 11;
when 11=>
as <= '1';
pos <= 12;
when 12=>
ad <= x"00";
rw <= '0';
pos <= 13;
when 13=>
ad <= x"04";
as <= '0';
rw <= '1';
pos <= 14;
when 14=>
as <= '1';
pos <= 15;
when 15=>
ad <= x"00";
rw <= '0';
pos <= 16;
when 16=>
ad <= x"07";
as <= '0';
rw <= '1';
pos <= 17;
when 17=>
as <= '1';
pos <= 18;
when 18=>
ad <= x"30";
rw <= '0';
pos <= 19;
when 19=>
ad <= x"08";
as <= '0';
rw <= '1';
pos <= 20;
when 20=>
as <= '1';
pos <= 21;
when 21=>
ad <= x"05";
rw <= '0';
pos <= 22;
when 22=>
ad <= x"08";
as <= '0';
rw <= '1';
pos <= 23;
when 23=>
as <= '1';
pos <= 24;
when 24=>
ad <= x"17";
rw <= '0';
pos <= 25;
when 25=>
ad <= x"0B";
```

```
as <= '0';
rw <= '1';
pos <= 26;
when 26=>
as <= '1';
pos <= 27;
when 27=>
ad <= x"03";
rw <= '0';
pos <= 28;
when 28=>
rw <= '1';
pos <= 29;
when 29=>
ad <= x"00";
as <= '0';
pos <= 30;
when 30=>
as <= '1';
sec <= ad;
pos <= 31;
when 31=>
ad <= x"02";
as <= '0';
pos <= 32;
when 32=>
as <= '1';
min <= ad;
pos <= 33;
when 33=>
ad <= x"04";
as <= '0';
pos <= 34;
when 34=>
as <= '1';
sec <= ad;
pos <= 35;
when 35=>
ad <= x"07";
as <= '0';
pos <= 36;
when 36=>
as <= '1';
sec <= ad;
pos <= 37;
when 37=>
ad <= x"08";
as <= '0';
pos <= 38;
when 38=>
as <= '1';
sec <= ad;
pos <= 39;
when 39=>
ad <= x"09";
as <= '0';
pos <= 40;
when 40=>
as <= '1';
sec <= ad;
pos <= 29;
```

```
end case;
end if;
end if;
end process;
end architecture;
```

8.3 Inter-Integrated Circuit Interface Programming

The I^2C interface is also called TWI because it is done by just two wires, namely serial clock (SCL) and SDA. The SCL is used to synchronize all data transfers over the I^2C bus, while the SDA actually transfers the data. The SCL and SDA lines are connected to all slave devices; thus, the I^2C cable is also called the I^2C bus. Other than these two wires, there are two more wires/connections for supply voltage (+5V) and common ground (Gnd). Both the SCL and SDA lines are open drain drivers; hence, they can drive output low, but cannot drive it high. Therefore, we need to use pull-up resistors, as shown in Figure 8.3a. We only need one set of pull-up resistors for all the slaves connected over the I^2C bus, not for each device. If the resistors are not used, then the SCL and SDA lines will always be logic 0 and the I^2C bus will not work; thus, there will be no proper correspondence between the master and slave devices. The master is always the device that initiates the communication and drives the SCL and SDA lines. The slaves are the devices that respond to the master in accordance with the commands of the master unit. A slave cannot initiate any data transfer or clock control over the I^2C bus. There can be multiple slaves on the I^2C bus; however, there is only one master. Both master and slave can transfer data over the I^2C bus, but that transfer is always controlled by the master.

When the master wishes to communicate to the slave device, it begins by issuing a start sequence on the I^2C bus, that is, the SDA goes logic 0, while the SCL is logic 1, as shown in Figure 8.3b. In the same way, the stop sequence is that the SDA goes logic 1 when the SCL is at logic 1, as shown in Figure 8.3b, which is the reverse of the start bit. The start and stop sequences are applied at the beginning and end of a transaction with the slave device. The data are transferred in sequences of 8 bits. The bits are placed on the SDA line starting with the MSB first, as shown in Figure 8.4. The SCL line is then pulsed high, then low. For every 8 bits transferred, the device receiving the data sends back an acknowledge bit, so that there are actually nine SCL clock pulses required to transfer each 8 bits of data, as shown in Figure 8.4a. If the receiving device sends back a low ACK bit, then it has received the data and is ready to accept another byte. If it sends back a high, then it is indicating it cannot accept any further data and the master should terminate the transfer by sending a stop

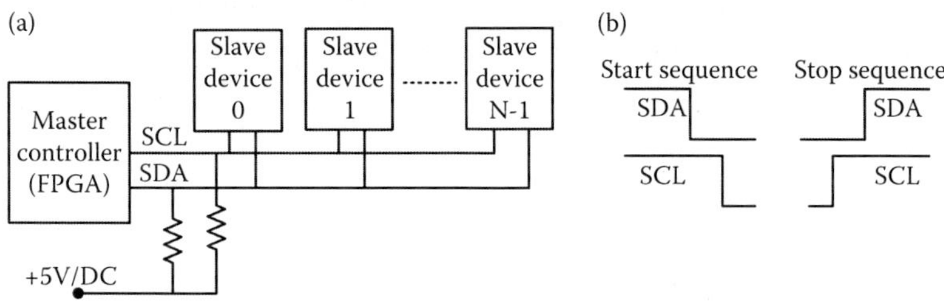

FIGURE 8.3
Configuration of master-slave over I^2C bus (a) and pattern of start and stop sequences of I^2C protocol (b).

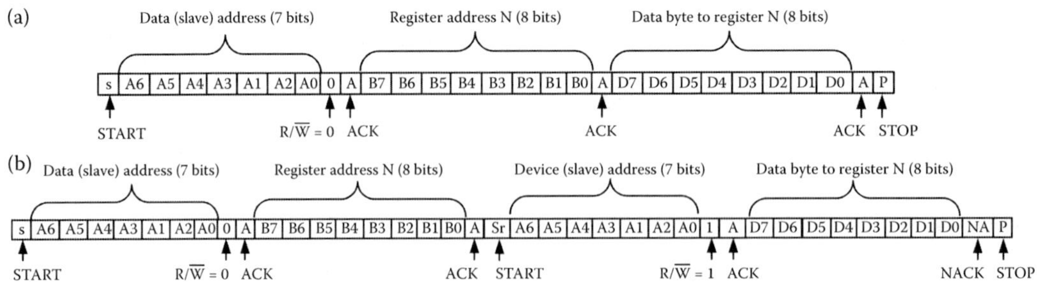

FIGURE 8.4
I²C bus address/data: write to slave device's register (a) and read from slave's register (b).

sequence [39,106]. Typically, the maximum speed of SCL over the I²C channel is 100 KHz; however, some devices, like Philips, have different speeds. For example, a faster speed is up to 400 KHz and high speed is up to 3.4 MHz. The address of the I²C devices used in this work is 7 bits; thus, we can connect up to 128 (2^7) devices on the I²C bus. Though the address bit is 7, we need to always send 8 bits, that is, 7 address bits and 1 extra bit at LSB, to indicate to the slave device that the address is for writing (logic 0) or reading (logic 1). The 7-bit address is placed in the upper 7 bits of the byte, that is, D_7 through D_1, and the R/W bit is at the LSB, that is, D_0. For example, to write to the address 21 = "00010101", we must send out 42 = "00101010"; to read from the same address, we must send 43 = "00101011". It is probably easier to think of the I²C bus addresses as 8-bit addresses, with even addresses, that is, LSB = 0, for writing and odd addresses, i.e., LSB = 1, for reading, as shown in Figure 8.4a.

The first process is that the master will send out a start sequence. This will alert all the slave devices configured on the I²C bus that a transaction is going to start and they should listen to the master. Next, the master will send out the device address. The slave that gets matched with this address will continue with the transaction and all others will ignore the rest of this transaction and wait for the next address. Having addressed the slave device, the master must now send out the internal location/register address of the slave to which the master wishes to write or read. This number is obviously dependent on what the slave actually is and how many internal registers it has [39,106]. Having sent the I²C address and the internal register address, the master can now send the data byte. In this way, any register content can be read or written. A sample serial clock, address, and data for writing and reading operation is shown in Figure 8.4a and b. To write on the I²C bus, the master will send a start condition on the bus with the slave's address, and the last bit, that is, the R/W bit, is set to logic 0, which signifies the write operation. After the slave sends the acknowledge bit, the master will then send the register address of the device to which it wishes to write. The slave will acknowledge again, letting the master know it is ready. After this, the master will start sending the register data to the slave. Until the master has sent all the data, the slave device keeps receiving, and after receiving the 8 bits, it sends acknowledgement to the master. Then, the master will terminate the transmission with a STOP condition, as shown in Figure 8.4a. In this way, any slave device can be selected and written any value by appropriately choosing the register.

Reading from the slave device is very similar to writing, but with some extra steps. In order to read from a slave device, the master must first instruct the slave whose register it wishes to read from. This is done by the master starting off the transmission in a similar fashion as the write, by sending the address with the R/W bit equal to 0 (signifying a write), followed by the register address it wishes to read from. Once the slave acknowledges this register address, the master will send a START condition again, followed by the slave address with the R/W bit set to 1 (signifying a read). This time, the slave will acknowledge the read request, and the master releases the SDA bus, but will continue supplying the clock to the slave. During this part of the transaction, the master will become the master-receiver and the slave will become the slave-transmitter. The master will continue sending out clock pulses, but will release the SDA line so that the slave can transmit data. At the end of every byte of data, the master will send an ACK to the slave, letting the slave know that it is ready for more data. Once the master has received the number of bytes it is expecting, it will send a NACK, signalling to the slave to halt communications and release the bus. The master will follow this up with a STOP condition [39,106]. In this way, reading operation can be performed over the I²C bus.

DESIGN EXAMPLE 8.2

Design and develop a digital system in the FPGA to realize an I²C data communication protocol. Design one master and two slave devices and realize the I²C protocol.

Solution

Master Unit

```
library ieee;
use ieee.std_logic_1164.all;
use ieee.std_logic_unsigned.all;
entity i2c is
port(clk,nrst:in std_logic;
scl:out std_logic;
sda:inout std_logic);
end entity;
architecture mark0 of i2c is
signal addr:std_logic_vector(7 downto 0);
signal ireg:std_logic_vector(7 downto 0);
signal val:std_logic_vector(7 downto 0);
signal cnt:integer range 0 to 9;
signal pos:integer range 0 to 15;
signal dclk:std_logic;
begin -- DCM --
dclk <= clkcnt(10);
clkgen:process(clk,nrst)
begin
if(nrst='0') then
clkcnt <= (others=>'0');
else
if(falling_edge(clk)) then
clkcnt <= clkcnt + 1;
end if;
end if;
end process;
process(dclk,nrst)
begin
if(nrst='0') then
addr <= x"FE";
ireg <= x"11";
val <= x"03";
scl <= '1';
sda <= '1';
pos <= 0;
else
```

```
if(falling_edge(dclk)) then
case(pos) is
when 0=>
sda <= '0';
pos <= 1;
cnt <= 1;
when 1=>
scl <= '0';
sda <= addr(7);
if(cnt=9) then
pos <= 3;
else
pos <= 2;
end if;
when 2=>
scl <= '1';
addr <= addr(6 downto 0)&'1';
pos <= 1;
cnt <= cnt + 1;
when 3=>
if(sda='0') then
pos <= 5;
else
scl <= '1';
pos <= 4;
sda <= '0';
end if;
when 4=>
sda <= '1';
pos <= 5;
when 5=>
sda <= '0';
pos <= 6;
cnt <= 1;
when 6=>
scl <= '0';
sda <= ireg(7);
if(cnt=9) then
pos <= 8;
else
pos <= 7;
end if;
when 7=>
scl <= '1';
ireg <= ireg(6 downto 0)&'1';
pos <= 6;
cnt <= cnt + 1;
when 8=>
if(sda='0') then
else
scl <= '1';
pos <= 9;
sda <= '0';
end if;
when 9=>
sda <= '1';
pos <= 10;
when 10=>
sda <= '0';
pos <= 11;
cnt <= 1;
```

```
when 11=>
scl <= '0';
sda <= val(7);
if(cnt=9) then
pos <= 13;
else
pos <= 12;
end if;
when 12=>
scl <= '1';
val <= val(6 downto 0)&'1';
pos <= 11;
cnt <= cnt + 1;
when 13=>
if(sda='0') then
pos <= 15;
else
scl <= '1';
pos <= 14;
sda <= '0';
end if;
when 14=>
sda <= '1';
pos <= 15;
when 15=>
sda <= '1';
scl <= '1';
pos <= 15;
end case;
end if;
end if;
end process;
end architecture;
```

Slave Device: The readers have to just append the following VHDL code for the above entity to get one slave device. The same VHDL code design can be used for another slave device just by changing the device address.

```
architecture mark1 of i2c is
signal addr:std_logic_vector(7 downto 0);
signal ireg:std_logic_vector(7 downto 0);
signal cnt:integer range 0 to 9;
signal pos:integer range 0 to 20;
signal rxreg:std_logic_vector(7 downto 0);
signal clkcnt:std_logic_vector(22 downto 0);
signal dclk:std_logic;
begin -- DCM --
dclk <= clkcnt(10);
clkgen:process(clk,nrst)
begin
if(nrst='0') then
clkcnt <= (others=>'0');
else
if(falling_edge(clk)) then
clkcnt <= clkcnt + 1;
end if;
end if;
end process;
-- BUS driver --
p1:process(dclk,nrst)
```

```
begin
if(nrst='0') then
addr <= x"FE";
ireg <= x"11";
scl <= '1';
sda <= '1';
pos <= 0;
else
if(falling_edge(dclk)) then
case(pos) is
when 0=>
sda <= '0';
pos <= 1;
cnt <= 1;
when 1=>
scl <= '0';
sda <= addr(7);
if(cnt=9) then
pos <= 3;
else
pos <= 2;
end if;
when 2=>
scl <= '1';
addr <= addr(6 downto 0)&'1';
pos <= 1;
cnt <= cnt + 1;
when 3=>
if(sda='0') then
pos <= 5;
else
scl <= '1';
pos <= 4;
sda <= '0';
end if;
when 4=>
sda <= '1';
pos <= 5;
when 5=>
sda <= '0';
pos <= 6;
cnt <= 1;
when 6=>
scl <= '0';
sda <= ireg(7);
if(cnt=9) then
pos <= 8;
else
pos <= 7;
end if;
when 7=>
scl <= '1';
ireg <= ireg(6 downto 0)&'1';
pos <= 6;
cnt <= cnt + 1;
when 8=>
if(sda='0') then
pos <= 10;
else
scl <= '1';
pos <= 9;
```

```
sda <= '0';
end if;
when 9=>
sda <= '1';
pos <= 10;
when 10=>
sda <= '0';
pos <= 11;
cnt <= 1;
addr <= x"FF";
when 11=>
scl <= '0';
sda <= addr(7);
if(cnt=9) then
pos <= 13;
else
pos <= 12;
end if;
when 12=>
scl <= '1';
addr <= addr(6 downto 0)&'1';
pos <= 13;
cnt <= cnt + 1;
when 13=>
if(sda='0') then
pos <= 15;
else
scl <= '1';
pos <= 14;
sda <= '0';
end if;
when 14=>
sda <= '1';
pos <= 15;
when 15=>
sda <= '0';
pos <= 16;
cnt <= 1;
rxreg <= x"00";
when 16=>
scl <= '0';
rxreg <= rxreg(6 downto 0)&sda;
if(cnt=9) then
pos <= 18;
else
pos <= 17;
end if;
when 17=>
scl <= '1';
pos <= 16;
cnt <= cnt + 1;
when 18=>
if(sda='0') then
pos <= 20;
else
scl <= '1';
pos <= 19;
sda <= '0';
end if;
when 19=>
sda <= '1';
```

```
pos <= 20;
when 20=>
pos <= 20;
end case;
end if;
end if;
end process;
end architecture;
```

8.4 Two-Wire Interface (SHT11 Sensor) Programming

A low-cost and low-power, surface-mountable, 8-pin SHT11 sensor is used to measure the temperature (T) in 14-bit resolution and relative humidity (RH) in 12-bit resolution. The sensor consists of a capacitive sensing element, a polymer, which absorbs and desorbs water molecules depending on the surrounding conditions and provides a fully calibrated digital output. The TWI and internal voltage regulation allow for fast system integration. The sensor's (SHT11) appearance, application circuit and the digital architecture required to interface it with the FPGA are shown in Figure 8.5. The digital architecture is developed to transfer the communication/control sequence to the sensor. The digital architecture logical design encompasses two distinct parts: (i) the data path processor unit performing data-processing operations consists of shift registers, multiplexers, counters, flip-flops, demultiplexers and tristate switches; and (ii) the control engine-finite state machine that sends commands to the data path processing unit to determine the sequence in which the various actions, like sending the sequence of connection reset, transmission start, address and command, are performed.

The SCK and data pins are accessed by the architecture and SHT11 sensor for internal register programming, measurement data acquisition, reception of acknowledgement, mode selection and so on. The SCK is used to synchronize communication between the architecture and sensor. Since the interface consists of fully static logic, there is no minimum SCK frequency. The data pin is used to transfer data in and out of the sensor. The output SCK and inout data are connected to the SHT11 and are used for sending the sequence and reading the measurement values as per the execution command from the control engine. The control engine state transition flow occurs from state 'a' through state 'k' for every T and RH measurement cycle. The signal `dmux1_ctrl` is used to control the Dmux1 to transfer the `rx_data` either to Dmux2 (`Mm_data`) or the control engine (ACK). This operation sequence is repeated to collect the T and RH data over a long period. Data shifting, circular rotations, enabling the sequence register and data loading at the desired accumulator are performed by the control engine, multiplexer and demultiplexer. The tristate switch output is inout, as discussed in Chapter 5; that is, it is driven by the input when enabled and otherwise driven by the SHT11. The timing diagram synchronisation among all the operations of the measurement is maintained by the clock manager. The control engine state transition flow for T and RH measurement is shown in Figure 8.6 and explained below:

State (a): The connection reset sequence is stored in a [25:0] circular shift register so as to toggle SCK nine times during data high, followed by toggling SCK one time at data low and forcing SCK and data high, then low again. This sequence resets the status register of the SHT11 with the default contents.

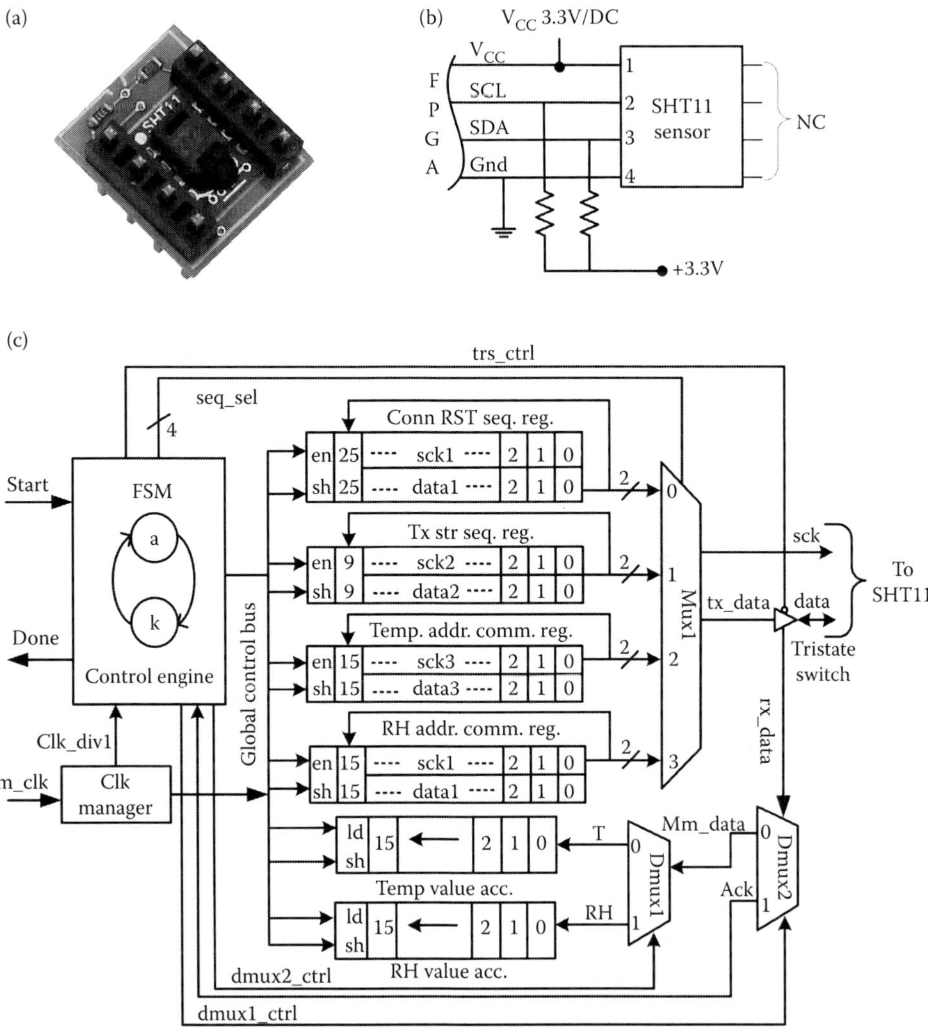

FIGURE 8.5
Picture of an SHT11 sensor (a), circuit configuration of SHT11 sensor (b), and digital architecture for temperature and relative humidity measurement (c).

State (b): The transmission start sequence is stored in a [9:0] circular shift register such as to lower the data while SCK is high, followed by a low pulse and rising data high while SCK is high.

State (c): The temperature measurement address and command (0 × 03H) is stored in a [15:0] circular shift register and transferred to the SHT11 for the rising edge of the SCK.

State (d): The SHT11 sensor indicates the proper reception of the address and command by pulling the data low after the trailing edge of the SCK (ACK1 Low).

State (e): Check whether data are zero.

State (f): Delay for measurement, approximately 320 ms. The completion of the measurement is signalled by pulling the data low.

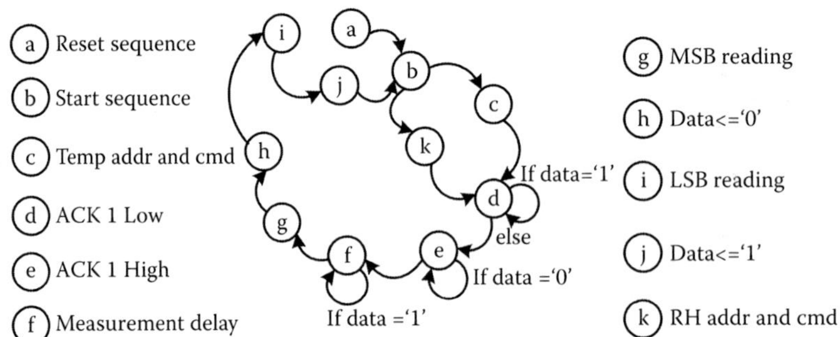

FIGURE 8.6
FSM control engine state transition flow for temperature and relative humidity measurement.

State (g): Read the MSB (D15–D8) of the measurement value bitwise and load into the [15: 8] of the T value accumulator for the rising edges of SCK.

State (h): An acknowledgment is passed to the SHT11 by forcing the data low after reading the MSB data, followed by a pulse on SCK.

State (i): Read the LSB (D7–D0) of the measurement value bitwise and load into the [7:0] of the T value accumulator.

State (j): Force the data high since the cyclic redundancy check (CRC) is not used, followed by the state(b) operation to read out the RH measurement data.

State (k): The RH measurement and command (0 × 05H) is stored in a [15:0] circular shift register and transferred to the SHT11 followed by state(d) to state(j) operations.

The signal `dmux1_ctrl` is used to control the Dmux1 to transfer the `rx_data` either to Dmux2 (`Mm_data`) or the control engine (`ACK`). In states (g) and (i), the RH value accumulator is enabled by `Dmux2_ctrl` while the relative humidity measurement is being carried out. This operation sequence is repeated to collect the `T` and `RH` data over a long period. Data shifting, circular rotations, enabling the sequence register, and data loading at the desired accumulator are performed by the control engine, multiplexer, and demultiplexer. The tristate switch output is inout, that is, it is driven by the input when enabled and otherwise driven by the SHT11. The timing synchronization among all the operations of the measurement is maintained by the clock manger.

The digital readout (T_{meas}) conversion formula to calculate the equivalent temperature in °C is [39,64,107]

$$\text{Temp} = T_{in^\circ C} = -40.1 + 0.01T_{meas} \tag{8.1}$$

The digital readout (Rh_{meas}) conversion second-order formula to calculate the true RH in percentage with the temperature compensation is [39,107]

$$\begin{aligned} RH_{in\%} &= (T_{in^\circ C} - 25)(0.01 + 0.00008Rh_{meas}) \\ &\quad + 0.0367RH_{meas} - 1.5955 \times 10^{-6}\, RH^2_{meas} - 2.0468 \end{aligned} \tag{8.2}$$

The dew point temperature (T_d) is calculated from `RH` and `T` readings with the following approximation with good accuracy [39,64,107]

$$T_d = 243.12\left(\frac{\ln\left(\frac{RH}{100}\right)+\left(\frac{17.72T}{243.12+T}\right)}{17.62-\ln\left(\frac{RH}{100}\right)-\left(\frac{17.62T}{243.12+T}\right)}\right) \tag{8.3}$$

The pseudocode of the measurement algorithm and simulation timing response for this interfacing and measurement data conversion can be found in References 39, 64, and 107. For example, if the value of the temperature is "01100100010100" = $(6420)_{10}$, it will be equal to 24.10°C according to Equation 8.1, and if the value RH is "010101111000" = $(1400)_{10}$, it will be equal to 46.09% according to Equation 8.2.

DESIGN EXAMPLE 8.3

Design and develop a digital system to interface a temperature and relative humidity sensor (SHT11) with the FPGA and display the measurement data using LEDs.

Solution

```
library IEEE;
use IEEE.STD_LOGIC_1164.ALL;
use IEEE.STD_LOGIC_ARITH.ALL;
use IEEE.STD_LOGIC_UNSIGNED.ALL;
entity SHT11_Inter is
port (clk,reset,data_r: in std_logic:='0';
sck:out std_logic:='0';
test_out:out std_logic_vector(3 downto 0):="0000";
data_disp: out std_logic_vector(15 downto 0):="0000000000000000";
data:out std_logic:='1');
end SHT11_Inter;
---------------------------------------------------------------
architecture RTL of SHT11_Inter is
signal clk_sig0: std_logic:='0';
signal data_temp_rh:std_logic_vector(43 downto 0) :="1111111111111111111
1110001100000000001100111";
signal sck_temp_rh :std_logic_vector(43 downto 0) :="0101010101010101010
1011011001010101010101010";
signal data_temp_t:std_logic_vector(29 downto 0) :="1111000001111000000
0001100111";
signal sck_temp_t :std_logic_vector(29 downto 0) :="0001110110000010101
01010101010";
begin
---------------------------------------------------------------
p1:process(clk,reset)
variable cnt:integer:=0;
begin
if(reset='0') then
clk_sig0<='0';
cnt:=0;
elsif rising_edge(clk) then
if(cnt=2000000)then
clk_sig0<= not(clk_sig0);
cnt:=0;
else
cnt:= cnt + 1;
end if;
end if;
end process;
```

```
p2:process(clk_sig0)
variable cnt1:integer:=0;
begin
if rising_edge(clk_sig0) then
if (cnt1<=43) then       data<=data_temp_rh(43);sck<=sck_temp_rh(43);
data_temp_rh<= (data_temp_rh(42 downto 0) & data_temp_rh(43));
sck_temp_rh <=(sck_temp_rh(42 downto 0)& sck_temp_rh(43));cnt1:=cnt1+1;
elsif (cnt1=44) then test_out<="0001";cnt1:=cnt1+1;
elsif (cnt1=45) then if (data_r='1') then sck<='0';cnt1:=cnt1; else
sck<='1';cnt1:=cnt1+1;end if;test_out<="0010";
elsif (cnt1=46) then sck<='0';cnt1:=cnt1+1;
elsif (cnt1=47) then if (data_r='0') then sck<='0';cnt1:=cnt1; else
data<='1';cnt1:=cnt1+1;end if;test_out<="0011";
elsif (cnt1=48) then if (data_r='1') then sck<='0';cnt1:=cnt1; else
sck<='0';cnt1:=cnt1+1;end if;test_out<="0100";
elsif (cnt1=49) then sck<='0';test_out<="0101";cnt1:=cnt1+1;
elsif (cnt1=50) then sck<='1';cnt1:=cnt1+1;
elsif (cnt1=51) then data_disp(15)<=data_r;cnt1:=cnt1+1;
elsif (cnt1=52) then sck<='0';cnt1:=cnt1+1;
elsif (cnt1=53) then sck<='1';cnt1:=cnt1+1;
elsif (cnt1=54) then data_disp(14)<=data_r;cnt1:=cnt1+1;
elsif (cnt1=55) then sck<='0';cnt1:=cnt1+1;
elsif (cnt1=56) then sck<='1';cnt1:=cnt1+1;
elsif (cnt1=57) then data_disp(13)<=data_r;cnt1:=cnt1+1;
elsif (cnt1=58) then sck<='0';cnt1:=cnt1+1;
elsif (cnt1=59) then sck<='1';cnt1:=cnt1+1;
elsif (cnt1=60) then data_disp(12)<=data_r;cnt1:=cnt1+1;
elsif (cnt1=61) then sck<='0';cnt1:=cnt1+1;
elsif (cnt1=62) then sck<='1';cnt1:=cnt1+1;
elsif (cnt1=63) then data_disp(11)<=data_r;cnt1:=cnt1+1;
elsif (cnt1=64) then sck<='0';cnt1:=cnt1+1;
elsif (cnt1=65) then sck<='1';cnt1:=cnt1+1;
elsif (cnt1=66) then data_disp(10)<=data_r;cnt1:=cnt1+1;
elsif (cnt1=67) then sck<='0';cnt1:=cnt1+1;
elsif (cnt1=68) then sck<='1';cnt1:=cnt1+1;
elsif (cnt1=69) then data_disp(9)<=data_r;cnt1:=cnt1+1;
elsif (cnt1=70) then sck<='0';cnt1:=cnt1+1;
elsif (cnt1=71) then sck<='1';cnt1:=cnt1+1;
elsif (cnt1=72) then data_disp(8)<=data_r;cnt1:=cnt1+1;
elsif (cnt1=73) then sck<='0';cnt1:=cnt1+1;
elsif (cnt1=74) then data<='0';sck<='0';cnt1:=cnt1+1;
elsif (cnt1=75) then sck<='1';cnt1:=cnt1+1;
elsif (cnt1=76) then sck<='0';cnt1:=cnt1+1;
elsif (cnt1=77) then sck<='1';cnt1:=cnt1+1;
elsif (cnt1=78) then data_disp(7)<=data_r;cnt1:=cnt1+1;
elsif (cnt1=79) then sck<='0';cnt1:=cnt1+1;
elsif (cnt1=80) then sck<='1';cnt1:=cnt1+1;
elsif (cnt1=81) then data_disp(6)<=data_r;cnt1:=cnt1+1;
elsif (cnt1=82) then sck<='0';cnt1:=cnt1+1;
elsif (cnt1=83) then sck<='1';cnt1:=cnt1+1;
elsif (cnt1=84) then data_disp(5)<=data_r;cnt1:=cnt1+1;
elsif (cnt1=85) then sck<='0';cnt1:=cnt1+1;
elsif (cnt1=86) then sck<='1';cnt1:=cnt1+1;
elsif (cnt1=87) then data_disp(4)<=data_r;cnt1:=cnt1+1;
elsif (cnt1=88) then sck<='0';cnt1:=cnt1+1;
elsif (cnt1=89) then sck<='1';cnt1:=cnt1+1;
elsif (cnt1=90) then data_disp(3)<=data_r;cnt1:=cnt1+1;
elsif (cnt1=91) then sck<='0';cnt1:=cnt1+1;
elsif (cnt1=92) then sck<='1';cnt1:=cnt1+1;
elsif (cnt1=93) then data_disp(2)<=data_r;cnt1:=cnt1+1;
```

```
elsif (cnt1=94) then sck<='0';cnt1:=cnt1+1;
elsif (cnt1=95) then sck<='1';cnt1:=cnt1+1;
elsif (cnt1=96) then data_disp(1)<=data_r;cnt1:=cnt1+1;
elsif (cnt1=97) then sck<='0';cnt1:=cnt1+1;
elsif (cnt1=98) then sck<='1';cnt1:=cnt1+1;
elsif (cnt1=99) then data_disp(0)<=data_r;test_out<="0000";cnt1:=cnt1+1;
elsif (cnt1=100) then sck<='0';cnt1:=cnt1+1;
elsif (cnt1=101) then data<='1';cnt1:=cnt1+1;
elsif (cnt1=102) then sck<='1';data<='1';cnt1:=cnt1+1;
elsif (cnt1=103) then sck<='0';data<='1';cnt1:=cnt1+1;
elsif (cnt1=108) then sck<='0';data<='1';cnt1:=0;
else
cnt1:=cnt1+1;
end if;
end if;
end process;
end RTL;
```

8.5 Serial Peripheral Interface (SCP1000D) Programming

The SCP1000-D01 sensor is used to measure the absolute pressure in 19-bit resolution. The sensor consists of a silicon bulk micromachined sensing element chip and a signal conditioning ASIC.

Overcoming the timing error, more accurate measurement can be carried out using SCP1000-D01 [39,64]. The pressure sensor element and the ASIC are mounted inside a plastic premoulded package and wire-bonded to appropriate contacts. The pressure output data are calibrated and compensated internally. The appearance and application interfacing circuits of the SCP1000-D01 sensor and the digital architecture of pressure measurement is shown in Figure 8.7. Pressure measurement digital architecture consists of two parts: one is the data processor unit and the other is the control engine. The communication protocol between the digital architecture and SCP1000-D01 is an SPI [39]. The SPI interface is a full duplex five-wire serial interface. The communication between the architecture and sensor is done with data ready (DRDY), trigger (TRIG), SCK, MOSI and MISO signals. The SPI communication frame consists of three 8-bit words. The first word defines the register address, followed by the type of access, that is, '0' for read, '1' for write and one '0' at the LSB, followed by the data words being read or written. The MSB of the words is sent first. Bits from the MOSI line are sampled for the rising edges of SCK, while bits to the MISO are latched out for the trailing edge of SCK. The register address and data sequence are stored in circular shift registers in the data processor unit. Finite state machine control engine pressure measurement state transition cycle occurs every second. The register address, content sequence shifting, enabling the register, loading the content, writing, and reading are performed at the data processor unit per the command sequence from control engine. The `SCK_ctrl` controls issuing `clk_div1` to the SCK. The timing synchronization among various operations of the whole measurement is maintained by the clock manager. The register address and data sequence are stored in the circular shift registers in the data processor unit. FSM control engine state transition flow for pressure measurement is shown in Figure 8.8 and explained below:

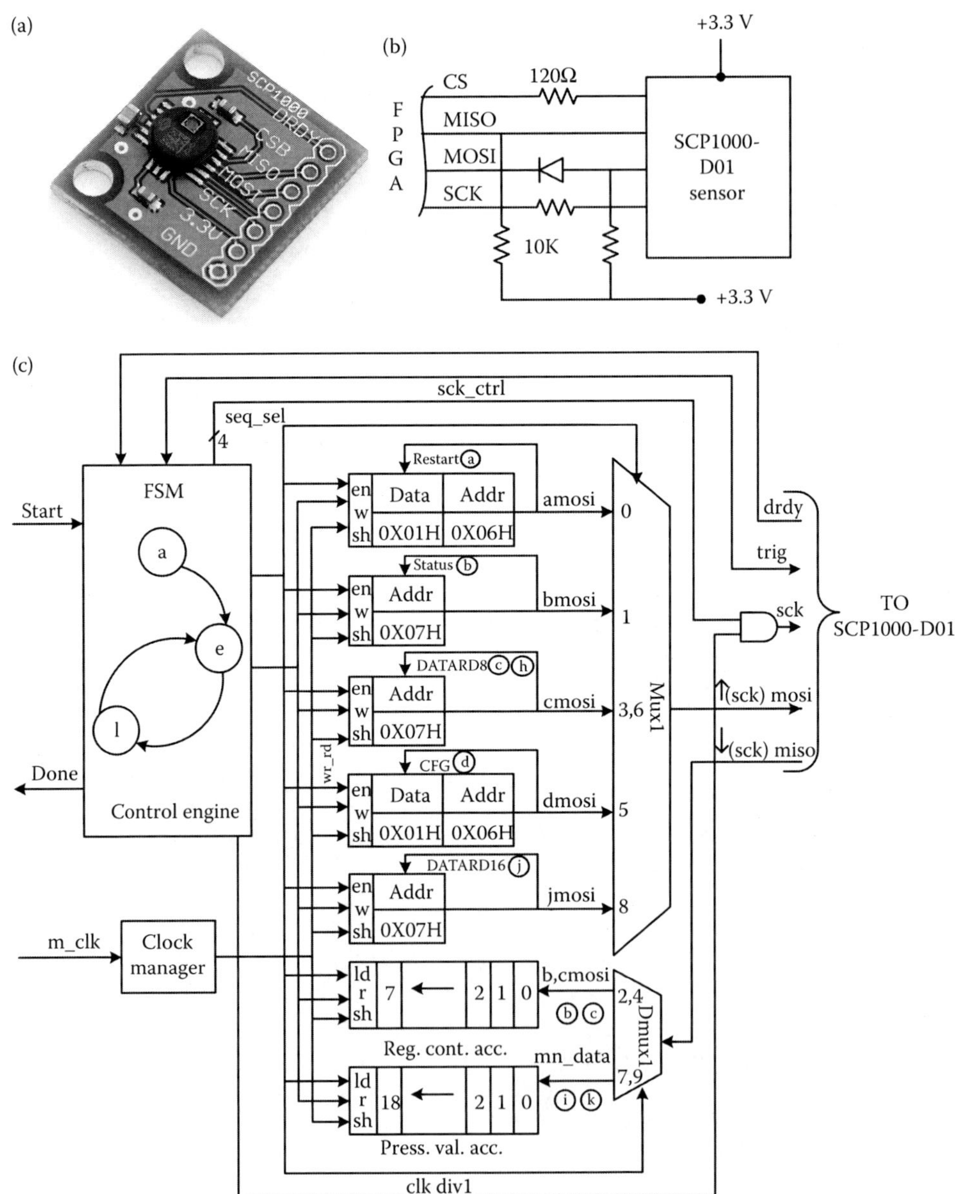

FIGURE 8.7
Picture (a), application circuit (b), and interface digital architecture of SCP1000-D01 sensor (c).

State (a): The restart register (Add: 0X06H) is written with restart sequence (data: `0X01H`).

State (b): Read the content of status register (Add: 0X07H) and load bitwise into the register content accumulator, followed by checking the LSB to verify that the startup procedure is finished.

State (c): Read the content of the datard8 register (Add: 0X1FH) and load bitwise into the register content accumulator, followed by checking the LSB to identify that the SCP100-D01 is in standby mode and waiting for measurement command.

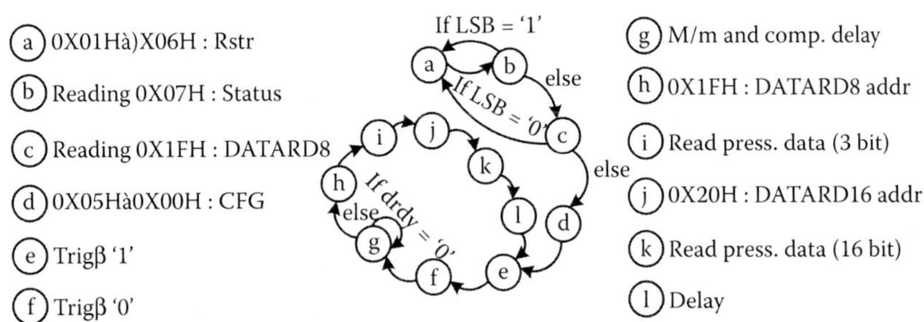

FIGURE 8.8
FSM control engine state transition flow for pressure measurement in triggered mode.

State (d): The configuration register (Add: 0X00H) is written with the 17-bit measurement sequence (data:0X05H).

State (e): Forcing the trigger signal high.

State (f): Forcing the trigger signal low.

State (g): Wait for measurement and computation delay until the data ready goes high.

State (h): Write the datard8 (Add:0X1FH) address sequence.

State (i): Read the content of the datard8 register and bitwise load into the pressure value accumulator.

State (j): Write the datard16 (0X20H) address sequence.

State (k): Read the content of the datard 16 register and bitwise load into the pressure value accumulator.

State (l): Wait for delay to start next measurement cycle.

The pseudocode of the measurement algorithm and simulation timing response of this pressure measurement can be found in [39,64]. The transmission goes to state (e) after finishing the first measurement cycle and continues collecting the pressure data over a long period.

The true pressure value Pr is calculated in kPa by [39,64]

$$\mathrm{Pr} = 0.25(\mathrm{P}_{\mathrm{meas}})_{10} \tag{8.4}$$

For example, when the measurement data is "1100010111000010000" = $(405008)_{10}$, then the actual pressure value will be equal to 101.252 kPa according to Equation 8.3. The measurement starts when the "Start" pin goes high and "Done" goes high and then low to signal the completion of every measurement cycle.

DESIGN EXAMPLE 8.4

Design and develop a digital system to interface a pressure sensor (SCP1000-D01) with the FPGA and display the measurement data using LEDs.

Solution

```
library IEEE;
use IEEE.STD_LOGIC_1164.ALL;
use IEEE.STD_LOGIC_ARITH.ALL;
```

```
use IEEE.STD_LOGIC_UNSIGNED.ALL;
entity SCP1000_D01 is
Port (clk,drdy,miso: in STD_LOGIC:='0';
csb : out std_logic:='1';
data_disp: out std_logic_vector(18 downto 0):="0000000000000000000";
mosi,sck : out std_logic:='0');
end SCP1000_D01;
architecture Behavioral of SCP1000_D01 is
signal mode_sel:std_logic_vector(15 downto
0):="0000111000001001";--0×03;0×09
signal temp_add:std_logic_vector(7 downto 0):="10000100";--0×21
signal press_add_msb:std_logic_vector(7 downto 0):="01111100";--0×1F
signal press_add_lsb:std_logic_vector(7 downto 0):="10000000";--0×20
signal sck_sig,sck_cnt,mux_sig: std_logic:='0';
signal mosi_cnt: std_logic:='1';
signal sck_temp: std_logic:='1';
signal mosi_mode_sel,mosi_temp_add,mosi_press_add_msb,mosi_press_add_
lsb: std_logic:='0';
signal wr_cnt: std_logic_vector(2 downto 0):="000";
signal temp_data: std_logic_vector(15 downto 0):="0000000000000000";
signal press_data_msb: std_logic_vector(7 downto 0):="00000000";
signal press_data_lsb: std_logic_vector(15 downto
0):="0000000000000000";
begin
sck_sig<=(sck_cnt and (not(sck_temp)));
sck<=sck_sig;
mosi<=((mosi_mode_sel or mosi_temp_add or mosi_press_add_msb or mosi_
press_add_lsb)and mosi_cnt);
-----Clk_Div-----
p0:process(clk)
variable cnt1:integer:=0;
begin
if rising_edge(clk) then
if(cnt1=40000)then
sck_temp<= not(sck_temp);
cnt1:=0;
else
cnt1:= cnt1 + 1;
end if;
end if;
end process;
---mosi_mode_sel----
p1:process(sck_temp)
begin
if (wr_cnt/="000")then
mosi_mode_sel<='0';
mode_sel<="0000111000001001";
elsif rising_edge(sck_temp) then
mosi_mode_sel<=mode_sel(15);
mode_sel<=(mode_sel(14 downto 0) & mode_sel(15));
end if;
end process;
---mosi_temp_add----
p2:process(sck_temp)
begin
if (wr_cnt/="001")then
mosi_temp_add<='0';
temp_add<="10000100";
elsif rising_edge(sck_temp) then
mosi_temp_add<=temp_add(7);
temp_add<=(temp_add(6 downto 0) & temp_add(7));
```

```
end if;
end process;
---miso_temp_data-------
p3:process(sck_sig)
begin
if (wr_cnt="010")then
if rising_edge(sck_sig) then
temp_data<=(temp_data(14 downto 0) & miso);
end if;
end if;
end process;
----mosi_press_add_msb------
p4:process(sck_temp)
begin
if (wr_cnt/="011")then
mosi_press_add_msb<='0';
elsif rising_edge(sck_temp) then
mosi_press_add_msb<=press_add_msb(7);
press_add_msb<=(press_add_msb(6 downto 0) & press_add_msb(7));
end if;
end process;
----miso_press_data_msb------
p5:process(sck_sig)
begin
if (wr_cnt="100")then
if rising_edge(sck_sig) then
press_data_msb<=(press_data_msb(6 downto 0) & miso);
end if;
end if;
end process;
-----mosi_press_add_lsb---
p6:process(sck_temp)
begin
if (wr_cnt/="101")then
mosi_press_add_lsb<='0';
press_add_lsb<="10000000";
elsif rising_edge(sck_temp) then
mosi_press_add_lsb<=press_add_lsb(7);
press_add_lsb<=(press_add_lsb(6 downto 0) & press_add_lsb(7));
end if;
end process;
----miso_press_data_lsb---
p7:process(sck_sig)
begin
if (wr_cnt="110")then
if rising_edge(sck_sig) then
press_data_lsb<=(press_data_lsb(14 downto 0) & miso);
end if;
end if;
end process;
----oe--------------
p8:process(mux_sig)
begin
case mux_sig is
when '0'=> data_disp<=("00000" & temp_data(13 downto 0));
when '1'=> data_disp<=(press_data_msb(2 downto 0) & press_data_lsb(15
downto 0));
when others=> null;
end case;
end process;
p9:process(sck_temp)
```

```
variable cnt3:integer:=0;
begin
if rising_edge(sck_temp)then
-----mosi_mode_sel---------
if (cnt3=0)then
csb<='0';
sck_cnt<='1';
mosi_cnt<='1';
mux_sig<='1';
wr_cnt<="000";
cnt3:=cnt3+1;
elsif (cnt3=16) then
csb<='0';
sck_cnt<='0';
mosi_cnt<='1';
mux_sig<='1';
wr_cnt<="111";
cnt3:=cnt3+1;
elsif (cnt3=17) then
csb<='1';
sck_cnt<='0';
mosi_cnt<='1';
mux_sig<='1';
wr_cnt<="111";
cnt3:=cnt3+1;
---drdy check--------
elsif (cnt3=18) then
if (drdy='0') then
csb<='1';sck_cnt<='0';
mosi_cnt<='1';
mux_sig<='1';
wr_cnt<="111";
cnt3:=cnt3;
else csb<='0';
sck_cnt<='0';
mosi_cnt<='1';
wr_cnt<="001";
mux_sig<='1';
cnt3:=cnt3+1;
end if;
--mosi_temp_add---miso_temp_data--
elsif (cnt3=19) then
csb<='0';sck_cnt<='1';mosi_cnt<='1';mux_sig<='1';wr_cnt<="001";
cnt3:=cnt3+1;
elsif (cnt3=26) then
csb<='0';sck_cnt<='1';mosi_cnt<='0';mux_sig<='1';wr_cnt<="001";
cnt3:=cnt3+1;
elsif (cnt3=27) then
csb<='0';sck_cnt<='1';mosi_cnt<='0';mux_sig<='1';wr_cnt<="010";
cnt3:=cnt3+1;
elsif (cnt3=43) then
csb<='0';sck_cnt<='0';mosi_cnt<='1';mux_sig<='1';wr_cnt<="111";
cnt3:=cnt3+1;
elsif (cnt3=44) then
csb<='1';sck_cnt<='0';mosi_cnt<='1';mux_sig<='0';wr_cnt<="111";
cnt3:=cnt3+1;
--mosi_press_add_msb-----miso_press_data_msb---
elsif (cnt3=45) then
csb<='0';sck_cnt<='0';mosi_cnt<='1';mux_sig<='0';wr_cnt<="011";
cnt3:=cnt3+1;
elsif (cnt3=46) then
```

```
csb<='0';sck_cnt<='1';mosi_cnt<='1';mux_sig<='0';wr_cnt<="011";
cnt3:=cnt3+1;
elsif (cnt3=54) then
csb<='0';sck_cnt<='1';mosi_cnt<='1';mux_sig<='0';wr_cnt<="100";
cnt3:=cnt3+1;
elsif (cnt3=62) then
csb<='0';sck_cnt<='0';mosi_cnt<='1';mux_sig<='0';wr_cnt<="111";
cnt3:=cnt3+1;
elsif (cnt3=63) then
csb<='1';sck_cnt<='0';mosi_cnt<='1';mux_sig<='0';wr_cnt<="111";
cnt3:=cnt3+1;
--mosi_press_add_lsb-----miso_press_data_lsb-----
elsif (cnt3=64) then
csb<='0';sck_cnt<='0';mosi_cnt<='1';mux_sig<='0';wr_cnt<="101";
cnt3:=cnt3+1;
elsif (cnt3=65) then
csb<='0';sck_cnt<='1';mosi_cnt<='1';mux_sig<='0';wr_cnt<="101";
cnt3:=cnt3+1;
elsif (cnt3=72) then
csb<='0';sck_cnt<='1';mosi_cnt<='0';mux_sig<='0';wr_cnt<="101";
cnt3:=cnt3+1;
elsif (cnt3=73) then
csb<='0';sck_cnt<='1';mosi_cnt<='0';mux_sig<='0';wr_cnt<="110";
cnt3:=cnt3+1;
elsif (cnt3=89) then csb<='0';sck_cnt<='0';
mosi_cnt<='1';mux_sig<='0';wr_cnt<="111";
cnt3:=cnt3+1;
elsif (cnt3=90) then csb<='1';sck_cnt<='0';
mosi_cnt<='1';mux_sig<='1';wr_cnt<="111";
cnt3:=18;
else
cnt3:=cnt3+1;
end if;
end if;
end process;
end Behavioral;
```

8.6 Global System for Mobile Communications Interface Programming

In this section, we discuss the GSM module interface with the FPGA and its application programming to send and receive text messages. There are different kinds of GSM modules available in the market. In this work, the one which has an RS232 serial interface is used. The GSM module is basically a modem, like a SIM900, which is mounted on a PCB with the provision to tap out different types of outputs from that PCB board, for example, TTL and RS232 output. In general, we need only three connections to make connections between the GSM module and the FPGA. The PCB board also has provisions to attach a microphone and speaker along with +5V and ground connections. Now, let's see how to connect a GSM module to the FPGA. The GSM module can be interfaced with any digital master unit in different modes. Here, we use serial data transmission mode. To wire for this mode, the transmitter (Tx) pin of the GSM module has to be connected to the receiver (Rx) pin of the UART built inside the FPGA. Similarly, the Tx pin of the UART has to be connected to the Rx pin of the GSM module. The grounds of both the FPGA and GSM module have to be connected with a common ground. It assumed here that the MAX233 is deployed between the FPGA and GSM modem to mutually convert the RS232 into TTL

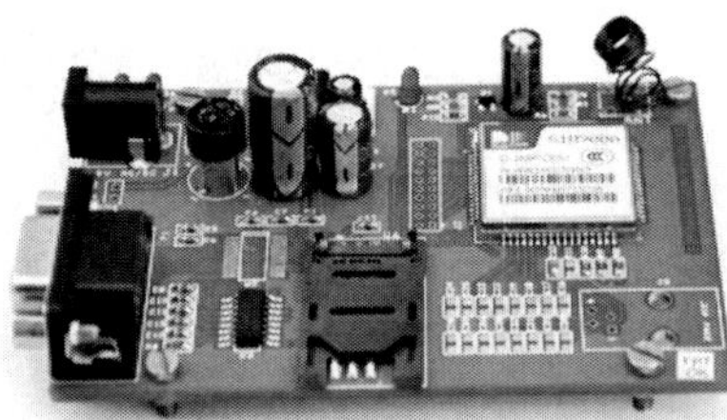

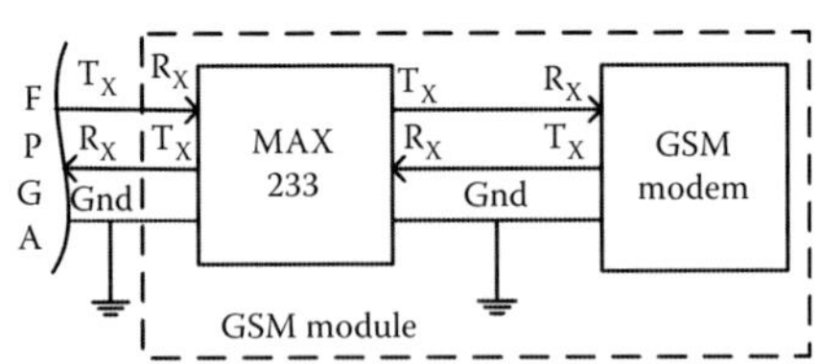

FIGURE 8.9
Picture and interface circuit of GSM module.

and vice versa, as shown in Figure 8.9. By properly communicating with the GSM module from the FPGA according to its standard protocol, it is possible to send and receive text messages via the GSM modem.

The first part of wiring the hardware components is over. Next, we should understand some special dedicative comments, known as Attention (AT) comments, which are required for establishing communication with the GSM module. There are different AT commands to perform different tasks through the GSM module. We can get a list of the complete AT commands and their use in Ref. 108. To send a Short Message Service (SMS) message using the GSM module, we have to set the GSM module to text mode first, which is achieved by sending the AT command `AT+CMGF=1`. After setting the GSM module to text mode, we have to send the mobile number to which we wish to send the SMS message. This is achieved with another AT command, `AT+CMGS=\"+XXxxxxxxxxxx\"\r`, where `X` denotes the country code and `x` denotes the mobile number. In the next step, we should send the actual content of the SMS message. The SMS content has to be completed with an identifier, that is, the CTRL + Z symbol, whose ASCII value is 26, so we need to send a "26" after the SMS content to the GSM module. Each and every AT command may require a 1-second delay for its execution; thus, we must allow some time for the GSM module to respond properly. Once these commands are sent to the GSM module, we will receive the SMS message at the mobile number we sent. Suppose we wish to receive an SMS message. The AT command to receive an SMS message is `AT + CNMI = 2,2,0,0,0`. We just need to send this command to the GSM module and then, after applying a 1-second delay, the GSM module keeps sending the received SMS data to the FPGA, which has to get it and display it on either an LCD or computer monitor. If we want to read all SMS messages stored in the SIM card, send the AT command `AT+CMGL = \"ALL\"\r` to the GSM module, which takes all the SMS messages stored in the SIM card of the GSM module and sends them to the FPGA one by one. The necessary AT commands to send and receive SMS messages are listed below:

AT commands to send/receive SMS messages using GSM module:

i. `AT+CMGF=1` // Set the GSM module in text mode.
ii. `AT+CMGS=\"+XXxxxxxxxxxx\"\r` // where `x` is the mobile number and `XX` is the country code.
iii. `"the message"` with stopping character `(char)26` // ASCII, Ctrl+z.
iv. `AT+CMGF=1` // Set the GSM module in text mode.
v. `AT+CNMI=2,2,0,0,0` // Receive live SMS messages.

The baud rate (from 1200 to 115,200) and frequency bands can also be set by AT commands. The GSM/GPRS modem has an internal TCP/IP stack to enable us to have connectivity with the internet via general packet radio service (GPRS). The SIM900A is an ultra-compact and reliable wireless module which has all these provisions [108]. Many other commands that are being used in mobile phones to control wired dial-up modems, such as dial (ATD), answer (ATA), hook control (ATH), and return to online data state (ATO), are also supported by GSM/GPRS modems. Some of the similar AT commands are listed below:

i. `AT + CPAS` – Mobile phone activity status
ii. `AT + CREG` – Mobile network registration status
iii. `AT + CSQ` – Radio signal strength
iv. `AT + CBC` – Battery charge level and battery charging status
v. `ATD, ATA` – Establish a data or voice connection to a remote modem
vi. `ATD, ATA, AT+F*` – Send and receive fax
vii. `AT + CMGS, AT+CMSS` – Send
viii. `AT + CMGR, AT+CMGL` – Read
ix. `AT + CMGW` – Write
x. `AT + CMGD` – Delete
xi. `AT + CNMI` – Receive SMS messages and obtain notifications of newly received SMS messages
xii. `AT + CPBR` – Read
xiii. `AT + CPBW` – Write
xiv. `AT + CPBF` – Search phonebook entries
xv. `AT + CLCK` – Perform security-related tasks, such as opening or closing facility locks
xvi. `AT + CLCK` – Check whether a facility is locked
xvii. `AT + CPWD` – Change passwords

In this way, a GSM module can be interfaced with the FPGA for any application.

DESIGN EXAMPLE 8.5

Design and develop a digital system to interface a GSM modem through UART architecture built inside the FPGA and display the received 4-byte text using an LCD.

Solution

```
library ieee;
use ieee.std_logic_1164.all;
use ieee.std_logic_unsigned.all;
entity gsminf is
port(clk,nrst:in std_logic;
sr:in std_logic;
gtx:out std_logic;
grx:in std_logic;
dout:out std_logic_vector(3 downto 0);
ctrl:out std_logic_vector(2 downto 0);
sf_ce0:out std_logic);
end entity;
architecture mark0 of gsminf is
component uart1 is
port(load:in std_logic;
```

```
clk,nrst:in std_logic;
din:in std_logic_vector(7 downto 0);
dout:out std_logic_vector(7 downto 0);
tx:out std_logic;
rx:in std_logic   );
end component;
signal entx,enrx,enx:std_logic;
signal tmsg,tmsg1,tmsg2:std_logic_vector(7 downto 0);
signal rmsg:std_logic_vector(7 downto 0);
signal tsig,rsig:std_logic;
signal spos:integer range 0 to 88;
signal rpos:integer range 0 to 29;
signal rreg:std_logic_vector(39 downto 0);
component lcdinf is
port(clk, nrst:in std_logic;
din:in std_logic_vector(31 downto 0);
set:in std_logic;
dout:out std_logic_vector(3 downto 0);
ctrl:out std_logic_vector(2 downto 0);
sf_ce0:out std_logic);
end component;
signal set:std_logic;-- clock signals --
signal clkcnt:std_logic_vector(15 downto 0);
signal dclk:std_logic;
begin
tmsg2 <= tmsg when(sr='1') else tmsg1;
enx <= entx when(sr='1') else enrx;
m1:uart1 port map(load=>enx,clk=>clk,nrst=>nrst,din=>tmsg2,dout=>rmsg,
tx=>tsig,rx=>rsig);
gtx <= tsig;
rsig <= grx;
m2:lcdinf port map(clk=>clk,nrst=>nrst, din=>rreg(39 downto8), set=>se
t,dout=>dout,ctrl=>ctrl,sf_ce0=>sf_ce0); -- DCM --
dclk <= clkcnt(14);
p_baud:process(clk,nrst)
begin
if(nrst='0') then
clkcnt <= (others=>'0');
else
if(falling_edge(clk)) then
clkcnt <= clkcnt + 1;
end if;
end if;
end process; -- send SMS --
psend:process(dclk,nrst,sr)
begin
if(nrst='0') then
spos <= 0;
entx <= '1';
else
if(sr='1') then
if(falling_edge(dclk)) then
case(spos) is
-- send 'AT' --
when 0=>
entx <= '0';
tmsg <= x"41";
spos <= 1;
when 1=>
entx <= '1';
spos <= 2;
```

```
when 2=>
entx <= '0';
tmsg <= x"54";
spos <= 3;
when 3=>
entx <= '1';
spos <= 4;
when 4=>
entx <= '0';
tmsg <= x"0D";
spos <= 5;
when 5=>
entx <= '1';
spos <= 6;
-- wait to receive 'OK' --
when 6=>
if(rmsg=x"4F") then
spos <= 7;
else
spos <= 6;
end if;
when 7=>
if(rmsg=x"4B") then
spos <= 8;
else
spos <= 7;
end if;
-- send 'AT+CMGF=1enter' --
when 8=>
entx <= '0';
tmsg <= x"41";
spos <= 9;
when 9=>
entx <= '1';
spos <= 10;
when 10=>
entx <= '0';
tmsg <= x"54";
spos <= 11;
when 11=>
entx <= '1';
spos <= 12;
when 12=>
entx <= '0';
tmsg <= x"2B";
spos <= 13;
when 13=>
entx <= '1';
spos <= 14;
when 14=>
entx <= '0';
tmsg <= x"43";
spos <= 15;
when 15=>
entx <= '1';
spos <= 16;
when 16=>
entx <= '0';
tmsg <= x"4D";
spos <= 17;
when 17=>
```

```
entx <= '1';
spos <= 18;
when 18=>
entx <= '0';
tmsg <= x"47";
spos <= 19;
when 19=>
entx <= '1';
spos <= 20;
when 20=>
entx <= '0';
tmsg <= x"46";
spos <= 21;
when 21=>
entx <= '1';
spos <= 22;
when 22=>
entx <= '0';
tmsg <= x"3D";
spos <= 23;
when 23=>
entx <= '1';
spos <= 24;
when 24=>
entx <= '0';
tmsg <= x"31";
spos <= 25;
when 25=>
entx <= '1';
spos <= 26;
when 26=>
entx <= '0';
tmsg <= x"0D";
spos <= 27;
when 27=>
entx <= '1';
spos <= 28;
-- wait to receive 'OK' --
when 28=>
if(rmsg=x"4F") then
spos <= 29;
else
spos <= 28;
end if;
when 29=>
if(rmsg=x"4B") then
spos <= 30;
else
spos <= 29;
end if;
-- send 'AT+CMGS="+918866124627"enter' --
when 30=>
entx <= '0';
tmsg <= x"41";
spos <= 31;
when 31=>
entx <= '1';
spos <= 32;
when 32=>
entx <= '0';
tmsg <= x"54";
```

```
spos <= 33;
when 33=>
entx <= '1';
spos <= 34;
when 34=>
entx <= '0';
tmsg <= x"2B";
spos <= 35;
when 35=>
entx <= '1';
spos <= 36;
when 36=>
entx <= '0';
tmsg <= x"43";
spos <= 37;
when 37=>
entx <= '1';
spos <= 38;
when 38=>
entx <= '0';
tmsg <= x"4D";
spos <= 39;
when 39=>
entx <= '1';
spos <= 40;
when 40=>
entx <= '0';
tmsg <= x"47";
spos <= 41;
when 41=>
entx <= '1';
spos <= 42;
when 42=>
entx <= '0';
tmsg <= x"53";
spos <= 43;
when 43=>
entx <= '1';
spos <= 44;
when 44=>
entx <= '0';
tmsg <= x"3D";
spos <= 45;
when 45=>
entx <= '1';
spos <= 46;
when 46=>
entx <= '0';
tmsg <= x"22";
spos <= 47;
when 47=>
entx <= '1';
spos <= 48;
when 48=>
entx <= '0';
tmsg <= x"2B";
spos <= 49;
when 49=>
entx <= '1';
spos <= 50;
when 50=>
```

```
entx <= '0';
tmsg <= x"39";
spos <= 51;
when 51=>
entx <= '1';
spos <= 52;
when 52=>
entx <= '0';
tmsg <= x"31";
spos <= 53;
when 53=>
entx <= '1';
spos <= 54;
when 54=>
entx <= '0';
tmsg <= x"38";
spos <= 55;
when 55=>
entx <= '1';
spos <= 56;
when 56=>
entx <= '0';
tmsg <= x"38";
spos <= 57;
when 57=>
entx <= '1';
spos <= 58;
when 58=>
entx <= '0';
tmsg <= x"36";
spos <= 59;
when 59=>
entx <= '1';
spos <= 60;
when 60=>
entx <= '0';
tmsg <= x"36";
spos <= 61;
when 61=>
entx <= '1';
spos <= 62;
when 62=>
entx <= '0';
tmsg <= x"31";
spos <= 63;
when 63=>
entx <= '1';
spos <= 64;
when 64=>
entx <= '0';
tmsg <= x"32";
spos <= 65;
when 65=>
entx <= '1';
spos <= 66;
when 66=>
entx <= '0';
tmsg <= x"34";
spos <= 67;
when 67=>
entx <= '1';
```

```
spos <= 68;
when 68=>
entx <= '0';
tmsg <= x"36";
spos <= 69;
when 69=>
entx <= '1';
spos <= 70;
when 70=>
entx <= '0';
tmsg <= x"32";
spos <= 71;
when 71=>
entx <= '1';
spos <= 72;
when 72=>
entx <= '0';
tmsg <= x"37";
spos <= 73;
when 73=>
entx <= '1';
spos <= 74;
when 74=>
entx <= '0';
tmsg <= x"22";
spos <= 75;
when 75=>
entx <= '1';
spos <= 76;
when 76=>
entx <= '0';
tmsg <= x"0D";
spos <= 77;
when 77=>
entx <= '1';
spos <= 78;
--------------------------
-- wait for '>' --
when 78=>
if(rmsg=x"3E") then
spos <= 79;
else
spos <= 78;
end if;
-- send message:DIAT --
when 79=>
entx <= '0';
tmsg <= x"44";
spos <= 80;
when 80=>
entx <= '1';
spos <= 81;
when 81=>
entx <= '0';
tmsg <= x"49";
spos <= 82;
when 82=>
entx <= '1';
spos <= 83;
when 83=>
entx <= '0';
```

```
tmsg <= x"41";
spos <= 84;
when 84=>
entx <= '1';
spos <= 85;
when 85=>
entx <= '0';
tmsg <= x"54";
spos <= 86;
when 86=>
entx <= '1';
spos <= 87;
when 87=>
entx <= '0';
tmsg <= x"1A";
spos <= 88;
when 88=>
entx <= '1';
spos <= 88;
end case;
end if;
end if;
end if;
end process; -- receive SMS(only 4 byte memory available) --
prec:process(dclk,nrst,sr)
begin
if(nrst='0') then
rpos <= 0;
enrx <= '1';
set <= '0';
else
if(sr='0') then
if(falling_edge(dclk)) then
case(rpos) is
-- send 'AT' --
when 0=>
enrx <= '0';
tmsg1 <= x"41";
rpos <= 1;
when 1=>
enrx <= '1';
rpos <= 2;
when 2=>
enrx <= '0';
tmsg1 <= x"54";
rpos <= 3;
when 3=>
enrx <= '1';
rpos <= 4;
when 4=>
enrx <= '0';
tmsg1 <= x"0D";
rpos <= 5;
when 5=>
enrx <= '1';
rpos <= 6;
when 6=>
if(rmsg=x"4F") then
rpos <= 7;
else
rpos <= 6;
```

```
end if;
when 7=>
if(rmsg=x"4B") then
rpos <= 8;
else
rpos <= 7;
end if;
when 8=>
enrx <= '0';
tmsg1 <= x"41";
rpos <= 9;
when 9=>
enrx <= '1';
rpos <= 10;
when 10=>
enrx <= '0';
tmsg1 <= x"54";
rpos <= 11;
when 11=>
enrx <= '1';
rpos <= 12;
when 12=>
enrx <= '0';
tmsg1 <= x"2B";
rpos <= 13;
when 13=>
enrx <= '1';
rpos <= 14;
when 14=>
enrx <= '0';
tmsg1 <= x"43";
rpos <= 15;
when 15=>
enrx <= '1';
rpos <= 16;
when 16=>
enrx <= '0';
tmsg1 <= x"4D";
rpos <= 17;
when 17=>
enrx <= '1';
rpos <= 18;
when 18=>
enrx <= '0';
tmsg1 <= x"47";
rpos <= 19;
when 19=>
enrx <= '1';
rpos <= 20;
when 20=>
enrx <= '0';
tmsg1 <= x"52";
rpos <= 21;
when 21=>
enrx <= '1';
rpos <= 22;
when 22=>
enrx <= '0';
tmsg1 <= x"3D";
rpos <= 23;
when 23=>
```

```
enrx <= '1';
rpos <= 24;
when 24=>
enrx <= '0';
tmsg1 <= x"31";
rpos <= 25;
when 25=>
enrx <= '1';
rpos <= 26;
when 26=>
enrx <= '0';
tmsg1 <= x"0D";
rpos <= 27;
when 27=>
enrx <= '1';
rpos <= 28; -- wait to receive 'OK' --
when 28=>
if(rmsg=x"4F") then
rpos <= 29;
else
rreg(7 downto 0) <= rmsg;
rreg(15 downto 8) <= rreg(7 downto 0);
rreg(23 downto 16) <= rreg(15 downto 8);
rreg(31 downto 24) <= rreg(23 downto 16);
rreg(39 downto 32) <= rreg(31 downto 24);
rpos <= 28;
end if;
when 29=>
if(rmsg=x"4B") then
rpos <= 0;
set <= '1';
else
rpos <= 29;
end if;
end case;
end if;
end if;
end if;
end process;
end architecture;
```

8.7 Global Positioning System Interface Programming

A GPS receiver is a device which has become an efficient tool today in the field of positioning, reliable navigation, surveillance, mapping, tracking, scientific use, and so on. The main purpose of the GPS is to find the location of a person or vehicle. A GPS receiver affords an exact location of an object in terms of longitude and latitude. GPS technology is now in everything ranging from wrist watches and cell phones to shipping containers, automatic teller machines (ATMs), and bulldozers. GPS increases productivity across a wide swathe of the economy, including construction, farming, mining, package delivery, surveying, banking systems, and financial markets. Some wireless communication services cannot operate without GPS technology. This system is used in fleet management, car navigation, and marine navigation. It is used for mapping and tracking devices. The GPS module works based on satellite navigation technology and provides information on time and location

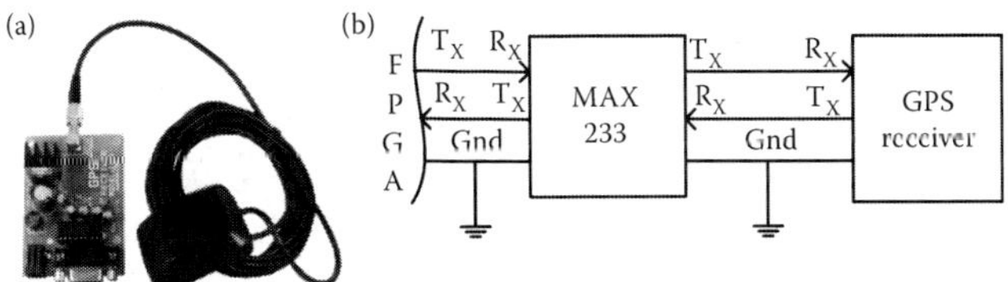

FIGURE 8.10
GPS module picture (a) and interface circuit with FPGA (b).

regardless of climatic conditions anywhere on earth. The GPS system mainly gets supports from 24–32 satellites to provide accurate data to the users. This system has become very important and popular worldwide because of its applications in navigation, tracking, surveillance, way/map marking, and so on. Before proceeding to understand its data receiving and processing protocols, let us discuss the basic GPS interfacing circuit with the FPGA, which is simple and straightforward. The block diagram of interfacing the GPS receiver with the FPGA is shown in Figure 8.10, which consists of GPS receiver modules, MAX233, FPGA, and LCD. The GPS module always transmits serial data in the form of sentences which have the longitude and latitude values, along with other information. To communicate to a GPS receiver over UART, we just need three basic connections, namely transmitter (Tx), receiver (Rx), and ground (Gnd). The received sentences have to be processed inside the FPGA to get the exact location of the GPS receiver in terms of longitude and latitude. The GPS module gives the output data in RS232 format. To convert RS232 format into TTL format, a line-converter MAX233 is used as discussed in Chapter 7. It is connected between the GPS module and the FPGA. The processed values of the location can be displayed on an LCD or any other display device and can also be used for any real-time application like tracking objects, trajectory plotting, object speed estimation/prediction and so on.

The UART and data processing units are developed inside the FPGA to communicate with the GPS receiver, receive the data from it and process the received data to extract the necessary information. The specifications of the UART for the GPS interfacing are baud rate: 4800, parity: none, data bit: 8 and stop bit: 1. Upon powering the circuit, the GPS receiver keeps sending the data in National Marine Electronics Association (NMEA) data format [109]. In the NMEA data format, the longitude and latitude values of exact locations are available. The data received from the GPS receiver are processed by the FPGA to take out the necessary information such as longitude and latitude. First, the six characters received from the GPS module are compared with the global positioning remote machine code (GPRMC) string. If the string is matched, then we need to wait until we get two commas. Then, next character specifies whether the GPS module is activated. If the next character is "A", then the GPS is activated; otherwise, it is inactive. If it is active, again, we have to wait until we get a comma. The next nine characters specify the latitude. Once again, wait until we get two commas, then the next 10 characters specify the longitude. The general format of an NMEA sentence is given below:

$GPGGA,080146.00,2342.9185,N,07452.7442,E,1,06,1.0,440.6M,−41.5, M, 0000*57

This data format has to be processed as follows to extract longitude, latitude, speed, time, altitude, and time information. The NMEA string always starts with a sign $. GPGGA indicates Global Positioning System Fix Data. A comma (,) specifies the separation between two values. 080146.00 is the GMT time as 08 hours: 01 minute: 46 seconds: 00 m seconds. 2342.9185, N indicates latitude 23 degrees: 42 minutes: 9185 seconds north. 07452.7442, E indicates longitude 074 degrees: 52 minutes: 7442 seconds east. 1 gives the fix quantity, where 0 = invalid data, 1 = valid data and 2 = differential global positioning system (DGPS) fix. 06 is the number of satellites currently viewed. 1.0 is horizontal dilution of precision (HDOP). 440.6, M indicates the altitude (height above sea level in meters). −41.5, M is the geoids height. 0000 is DGPS data and *57 is the checksum. Thus, the GPS receiver frequently sends information containing a number of data, such as global positioning system fixed data (GGA), geographic position—latitude/longitude (GLL), global navigation satellite system (GNSS) DOP and active satellites (GSA), GNSS satellites in view (GSV), recommended minimum specific GNSS data (RMC), course over ground and ground speed (VTG), date and time (ZDA), and datum reference (DTM). The GPS data start with "$GPXXX", where XXX is a three-letter identifier (GGA, GSA, etc.). We need to process this data format to extract the information we are interested in.

DESIGN EXAMPLE 8.6

Design and develop a digital system to interface a GPS receiver with the FPGA and extract all the information. Display the extracted information in the MATLAB® command widow and log it in an Excel file.

Solution

Since we covered UART design in the FPGA, serial communication with a computer, serial port data reading, displaying information in the MATLAB command window and storing serial port data in an Excel file in the preceding chapters, this simple design example is left to the readers as a laboratory exercise.

8.8 Personal System/2 Interface Programming

The IBM Personal System/2 (PS/2) is an interface for keyboards and mice to PC-compatible computer systems via a 6-pin mini-DIN connector which includes both clock and data, as shown in Figure 8.11. Both the mouse and keyboard drive the bus with identical signal timings and use 11-bit words that include start, stop, and odd parity bits. However, the data packets are organized differently for the mouse and keyboard. Furthermore, the keyboard interface allows bidirectional data transfers so that the master device can illuminate state LEDs on the keyboard. Interfacing of the PS/2 with the FPGA is very simple and straightforward. The keyboard is the most common way for humans to input information into a computer. This section describes the technique of interfacing a keyboard with the FPGA in order to understand what keys have been pressed and to enable the operations assigned to each key.

Both the PC keyboard and mouse use the two-wire interface that consists of data and a clock to communicate with a master device. When a key is pressed, released or held down, a

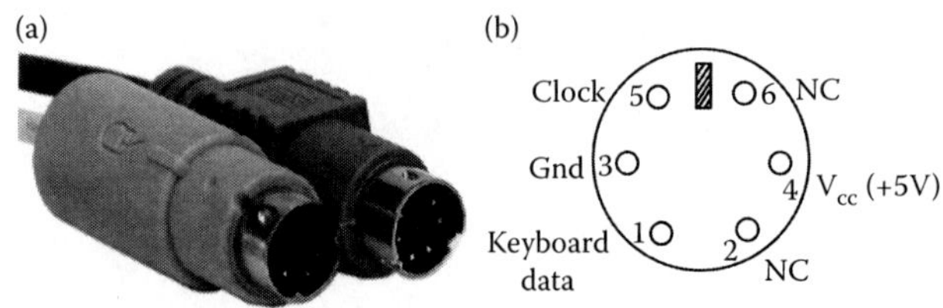

FIGURE 8.11
Picture of PS/2 connector (a) and its male pin out details (b).

packet of information known as a scan code is sent by the keyboard or mouse to the master device. Scan codes fall into the category of make codes or break codes. When a key is pressed or held down, a make code is sent, and when the key is released, a break code is sent. Every key has unique make and break codes. There are three standard scan code sets, namely set-1, set-2 and set-3. The default scan code set is set-2. In a situation where two keys are pressed, if one key is released, then the break code for that key will be sent, and then the make code of another pressed key continues to send at a periodic rate. If that key is also released, then the break code for that key also will be sent and no new scan codes will be sent until unless another key is pressed. Thus, it is possible to hold down any two keys. If too many keys are held down, then the scan code 0×00 is sent at a periodic rate until the keys over the maximum are released. The circuit of the PS/2 keyboard interface with the FPGA is shown in Figure 8.12a. As shown, both lines require pull-up resistors of 2kΩ and 120Ω series resistors to interface the 3.3V FPGA output to the 5V signals of the PS/2 keyboard. Once powered, the keyboard goes through a self-initialization sequence. Upon completion, it is ready to communicate the events over the PS/2 interface. Figure 8.12b illustrates the data and clock transaction formats. Both clock and data signals are logic level high when inactive. The clock has a frequency between 10 kHz and 16.7 kHz. The data begins with a start bit, that is, logic 0, followed by one byte of data (D_0-D_7), a parity bit and finally a stop bit, that is, logic 1, sending the LSB bit first. Each bit should be read only for the falling edge of the clock signal. Once complete, both the clock and data signals return to logic level high.

The data byte represents either a make code (key press) or a break code (key release) per scan code standard 2. The scan code for all the keys of the keyboard can be found in Ref. 110. A make code usually consists of one byte; however, for some keys, it is two or more bytes, as given in Table 8.4. If a make code uses two or more bytes, the first byte is x"E0." A given key's break code is typically the same as its make code, except that break codes start with an x"F0" byte, as given in Table 8.4. In the break code, x"F0" sits between two bytes if the make uses more than one byte, as given in Table 8.4 for page up and left arrow. Some of the keys are exceptions to the above, like "Prt Scr", as given in Table 8.4. However, we need to consider all these possibilities while developing VHDL code to interface the PS/2 with the FPGA. Sending data to the keyboard to change its settings is also possible, but it is mostly not applied.

The conceptual schematic diagram of PS/2 interface architecture and programming is shown in Figure 8.13; as shown there, the clock and data signals from the keyboard are first synchronized and then debounced. The resultant internal PS/2 data signal is then serially loaded into a shift register on the falling edges of the PS/2 clock. A bit receive counter keeps the transaction time and data information based on the PS/2 clock.

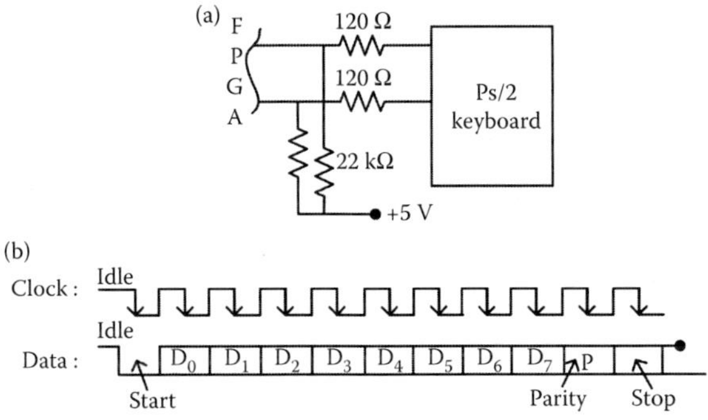

FIGURE 8.12
PS/2 interface circuit (a) and PS/2 keyboard transmission timing diagram (b).

TABLE 8.4

A Portion of Keyboard Scan Code

Key	Make	Break
A	1C	F01C
B	32	F032
C	21	F021
D	23	F023
Page Up	E07D	E0F07D
Left arrow	E06B	E0F06B
Prt Scr	E012E07C	E0F07CE0F012

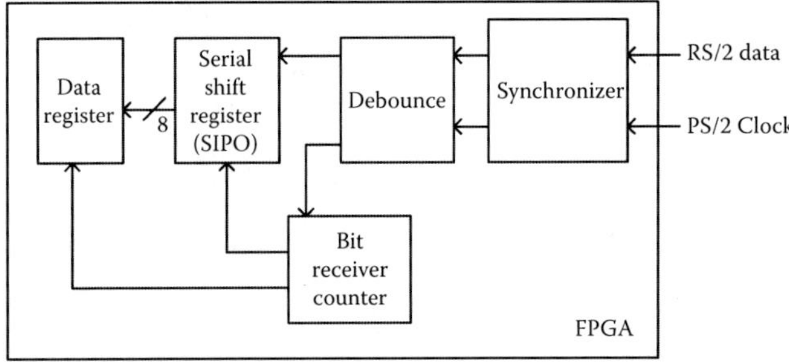

FIGURE 8.13
PS/2 keyboard interface logic architecture.

When the PS/2 port is idle and the received data are valid, then the bit counter signals the shift register to load the values in parallel to the data register. The key scan code can be used to perform subsequent operations from this register. The received scan code remains available at that register until another code is received. The PS/2 data and clock signal are high until another transaction begins, and the PS/2 clock signal clears the bit receive counter at the occurrence of the first clock. In this way, the PS/2 can be interfaced with the FPGA.

DESIGN EXAMPLE 8.7

Design and develop a digital system to receive and display the 1-byte scan code of a PS/2 keyboard. Show the received code using LEDs.

Solution

```
library ieee;
use ieee.std_logic_1164.all;
entity ps2 is
port(data:in std_logic;
kclk:in std_logic;
nrst:in std_logic;
dout:out std_logic_vector(7 downto 0));
end entity;
architecture mark0 of ps2 is
signal parity:std_logic;
signal dreg:std_logic_vector(7 downto 0);
signal pos:integer range 0 to 3;
signal cnt:integer range 0 to 9;
```

```
begin
p1:process(kclk,nrst)
begin
if(nrst='0') then
pos <= 0;
else
if(rising_edge(kclk)) then
case(pos) is
when 0=>
if(data='0') then
pos <= 1;
cnt <= 0;
parity <= '0';
else
pos <= 0;
end if;
when 1=>
dreg <= data&dreg(7 downto 1);
parity <= parity xor data;
if(cnt=8) then
pos <= 2;
else
cnt <= cnt + 1;
end if;
when 2=>
parity <= parity xor data;
pos <= 3;
when 3=>
if(parity='1') then
dout <= dreg;
end if;
pos <= 0;
end case;
end if;
end if;
end process;
end architecture;
```

DESIGN EXAMPLE 8.8

Design and develop a digital system to receive and display the all the scan code of a PS/2 keyboard. Show the received code using LEDs.

Solution

```
library ieee;
use ieee.std_logic_1164.all;
entity ps2_keyboard is
generic(clk_freq : integer := 50_000_000;
debounce_counter_size : integer := 8);
port(clk : in std_logic;
ps2_clk : in std_logic;
ps2_data : in std_logic;
ps2_code_new : out std_logic;
ps2_code : out std_logic_vector(7 downto 0));
end ps2_keyboard;
architecture logic of ps2_keyboard is
signal sync_ffs : std_logic_vector(1 downto 0);
signal ps2_clk_int : std_logic;
signal ps2_data_int : std_logic;
signal ps2_word : std_logic_vector(10 downto 0);
```

```
signal error : std_logic;
signal count_idle : integer range 0 to clk_freq/18_000;
component debounce is
generic(counter_size : integer);
port(clk : in std_logic;
button : in std_logic;
result : out std_logic);
end component;
begin
process(clk)
begin
if(clk'event and clk = '1') then
sync_ffs(0) <= ps2_clk;
sync_ffs(1) <= ps2_data;
end if;
end process;
debounce_ps2_clk: debounce
generic map(counter_size => debounce_counter_size)
port map(clk => clk, button => sync_ffs(0), result => ps2_clk_int);
debounce_ps2_data: debounce
generic map(counter_size => debounce_counter_size)
port map(clk => clk, button => sync_ffs(1), result => ps2_data_int);
process(ps2_clk_int)
begin
if(ps2_clk_int'event and ps2_clk_int = '0') then
ps2_word <= ps2_data_int & ps2_word(10 downto 1);
end if;
end process;
error <= not (not ps2_word(0) and ps2_word(10) and (ps2_word(9) xor
ps2_word(8) xor
ps2_word(7) xor ps2_word(6) xor ps2_word(5) xor ps2_word(4) xor ps2_
word(3) xor
ps2_word(2) xor ps2_word(1)));
process(clk)
begin
if(clk'event and clk = '1') then
if(ps2_clk_int = '0') then
count_idle <= 0;
elsif(count_idle /= clk_freq/18_000) then
count_idle <= count_idle + 1;
end if;
if(count_idle = clk_freq/18_000 and error = '0') then
ps2_code_new <= '1';
ps2_code <= ps2_word(8 downto 1);
else
ps2_code_new <= '0';
end if;
end if;
end process;
end logic;
```

8.9 Video Graphics Array Interface Programming

VGA is a video display standard developed by IBM and introduced in 1987 [111]. It is widely supported by PC graphic hardware and monitors. In the interface protocol of VGA, the image is divided into an array of small picture elements called pixels. Every pixel

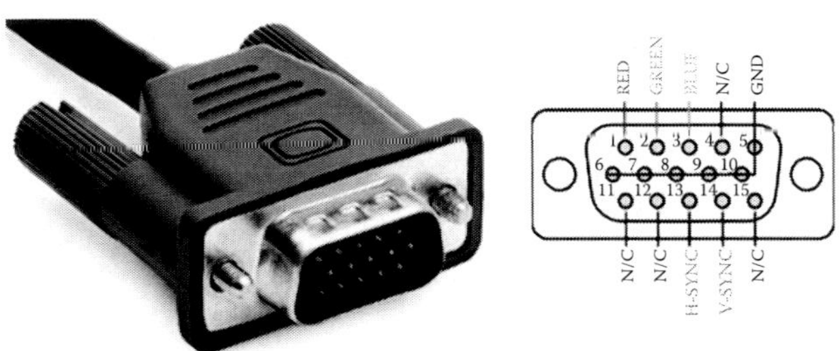

FIGURE 8.14
Picture and pin details of a VGA DB15 connector.

contains a sample/portion of the image. The display monitor continuously scans through the entire screen and rapidly controls individual pixels with appropriate values so as to color them. The VGA display port is a DB15 connector, as shown in Figure 8.14, which can be connected directly to the PC monitors or flat-panel LCD displays using a standard VGA cable. The main signals in the VGA port are red (1st pin), green (2nd pin), blue (3rd pin; these three pins are called RGB pins), horizontal synchronization (HS; 13th pin), vertical synchronization (VS; 14th pin), and the individual grounds (6th, 7th, 8th, 9th, 10th, and 5th pins), as shown in Figure 8.14. The RGB signals are called video signals. The standard VGA monitor consists of 640 × 480 pixel values. To display images on the LCD monitor, pixel values need to be continuously controlled at certain frame rate. The values of RGB for resulting in a single color of a pixel are "000" for black, "001" for blue, "010" for green, "011" for cyan, "100" for red, "101" for magenta, "110" for yellow, and "111" for white.

VGA interfacing with the FPGA is a very simple and straightforward configuration. There are three methods of configuration for interfacing the VGA with the FPGA, namely (i) direct configuration, (ii) 3-bit configuration, and (iii) DAC configuration. In direct configuration, the pins corresponding to RGB are connected to the FPGA through three 270 Ω resisters, as shown in Figure 8.15a. In the 3-bit configuration, 3 bits of the current divider network are used to generate analog voltage at the RGB pins, as shown in Figure 8.15b. The analog voltage can be varied just by feeding different values from "000" to "111" at the RGB pins. In this configuration, 0 to 511 possibilities of different colors can be produced. As marked in Figure 8.15b, the color inputs ($R_0R_1R_2G_0G_1G_2B_0B_1B_3$) can be varied as "000000000" to "111111111". In DAC configuration, the VGA connector is configured with the FPGA through a VGA driver DAC IC (ADV7123), as shown in Figure 8.15c. With the help of this driver's IC, it is possible to feed 0 to 255 different colors at every pin of RGB, which gives way to more accurately displaying images. The commercially available VGA driver module is shown in Figure 8.15d. Another two pins (HS and VS) of the VGA port are directly connected to the FPGA as shown in Figure 8.15a–c. Once we pass the standard code, that is, "000" through "111", to the RGB pins, we can design the images using only eight colors. If we wish to use multiple colors, we should use the current divider circuit or video DAC device (ADV7123). The HS and VS control the operations of the start and end of line (left to right) and frame (top to bottom) pixels, respectively. As given in Table 8.5, to get a pixel resolution of 640 × 480, the scanning frequency of the 25-MHz clock is required to perform VGA control operations.

While beginning the process of displaying the image, scanning has to be started from row 0, column 0, that is, the top left corner of the screen, and moved to the right until it reaches the last column, as shown in Figure 8.16, which is called horizontal scanning. When the scan reaches the last pixel of the row, it comes back to the beginning of the next

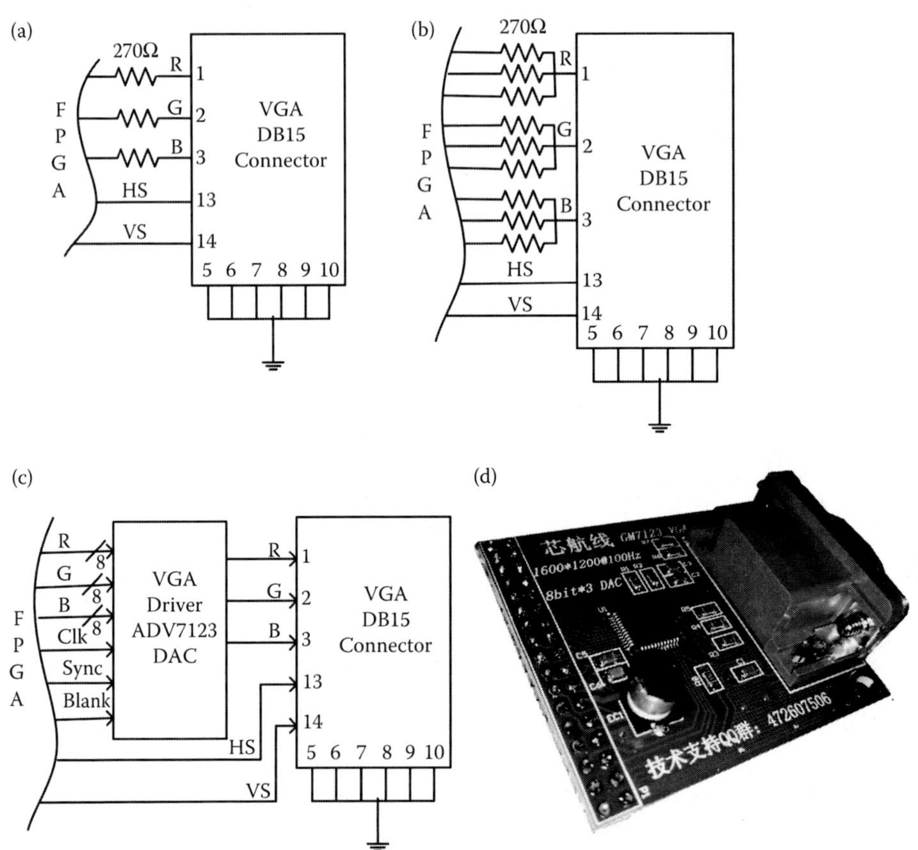

FIGURE 8.15
VGA DB15 port interfacing circuits: direct configuration (a), 3-bit configuration (b), video DAC–based configuration (c), and picture of a video driver (ADV7123) board (d).

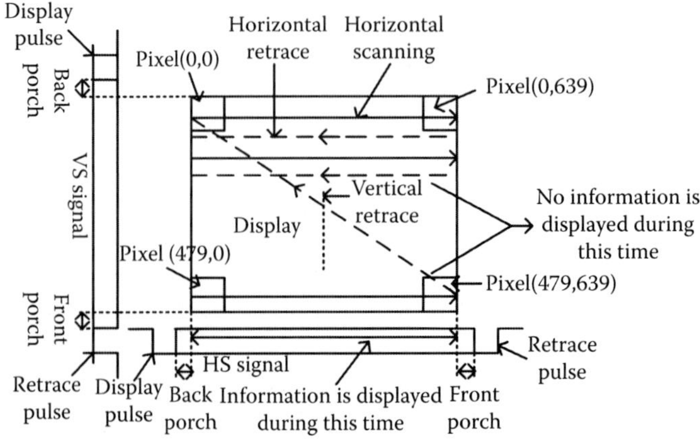

FIGURE 8.16
Mechanism of displaying image on the 640 × 480 screen.

row, as shown by dashed lines in Figure 8.16. Then, it continuous scanning on that row. The process of coming back to the starting point of the next row is called retracing. When it reaches the last pixel, that is, the pixel at the bottom-right corner of the screen, it retraces back to the top left corner pixel from where the scanning began and repeats the scanning process, as shown in Figure 8.16. The entire screen must be scanned (refreshed) not less than 60 times per second, which is called the refresh rate. The entire scanning process is controlled by the HS and VS signals. The combined video signals, that is, RGB, control the color of a pixel at a given location on the screen. They are analog signals with voltages ranging from 0.7 to 1 V. During the retrace period, the video color must be set to black. The blanking interval before applying the HS/VS retrace pulse is known as the front porch and forms the right/bottom border of the display region.

Similarly, the blanking interval after applying the HS/VS is called the back porch and forms the left/top border of the display region. In these blanking regions, the video signal should be disabled. All the necessary design parameters required to display an image on the screen are shown in Figure 8.16. After applying the HS retrace pulse, we must wait for a certain number of clocks before sending pixel values to the screen because the process of going back to the left and moving forward again on the next row requires some time. For example, we have to allow horizontal counting up to 640 pixels as the beam moves; after the 640th pixel, we have to wait for some amount of time, that is, the front porch time, and then assert the HS. The same is applicable for the VS control signal. We can design a circuit with two counters, one to keep track of horizontal location and another for vertical location. The horizontal counter should count up in 25-MHz clocks and the vertical counter should count once in every full horizontal scanning. Table 8.5 lists the timing values for several popular resolutions. In any design, the time synchronization module generates the HS/VS control signal for the VGA video display and row/column addresses to the image generation module.

TABLE 8.5

Control Signal Generation Details for Several Pixel Resolutions

Pixel Resolution	Scanning Clock (MHz)	Horizontal (in pixels)				Vertical (in lines)			
		Active Video	Front Porch	Sync Pulse	Back Porch	Active Video	Front Porch	Sync Pulse	Back Porch
640 × 480, 60 Hz	25	640	16	96	48	480	11	2	31
640 × 480, 72 Hz	31.5	640	24	40	128	480	9	3	28
640 × 480, 75 Hz	31.5	640	16	96	48	480	11	2	32
640 × 480, 85 Hz	36	640	32	48	112	480	1	3	25
800 × 600, 56 Hz	38	800	32	128	128	600	1	4	14
800 × 600, 60 Hz	40	800	40	126	88	600	1	4	23
800 × 600, 72 Hz	50	800	56	120	64	600	37	6	23
800 × 6000, 75 Hz	49.5	800	16	80	160	600	1	2	21
800 × 600, 85 Hz	56	800	32	64	152	600	1	3	27
1024 × 768, 60 Hz	65	1024	24	136	160	768	3	6	29
1024 × 768, 70 Hz	75	1024	24	136	144	768	3	6	29
1024 × 768, 75 Hz	78	1024	16	96	176	768	1	3	28
1024 × 768, 85 Hz	94.5	1024	48	96	208	768	1	3	36

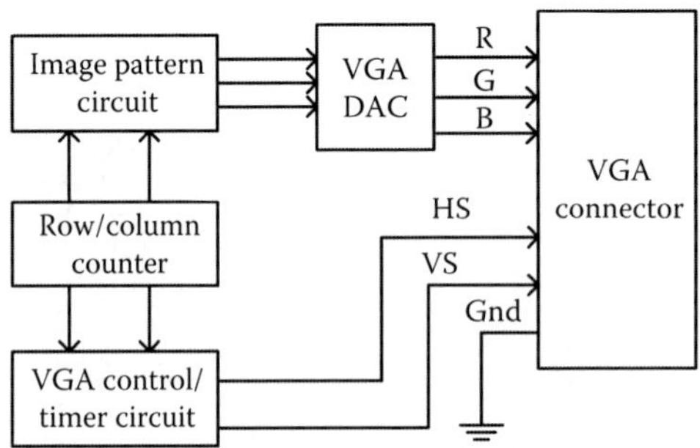

FIGURE 8.17
Schematic diagram of VGA interface.

Suppose the onboard clock frequency is 50 MHz. According to Table 8.5, we need only a 25-MHz clock signal in order to display an image at a resolution of 640 × 480 at a 60 Hz frame rate, which can be obtained by using a clock divider. We design two individual counters in the row/column circuit so as to count from 0 to 639 to meet horizontal resolution as well as to count from 0 to 479 to meet vertical resolution. The general application schematic diagram of the VGA interface is shown in Figure 8.17, where the control signals are generated from the VGA control/timer circuit and the video frames are given by the image pattern circuit.

The required image or video we wish to display on the screen is stored in the FPGA or external RAM. These data can be accessed by the image pattern circuit, and it passes them into the display screen through the VGA DAC in accordance with the row and column addresses given by the row/column circuit. The VGA control/timer circuit sends the HS and VS signal in accordance with the row and column data of the row/column circuit. In this way, the VGA can be interfaced with the FPGA and any pattern can be displayed on the computer screen.

DESIGN EXAMPLE 8.9

Design and develop a digital system to realize direct interfacing of VGA with the FPGA.

Solution

```
library ieee;
use ieee.std_logic_1164.all;
use ieee.std_logic_arith.all;
use ieee.std_logic_unsigned.all;
entity VGA_inte is
port(clk, reset : in std_logic;
red, green, blue : in std_logic;
r, g, b, hsync, vsync : out std_logic;
row : out std_logic_vector(8 downto 0);
column : out std_logic_vector(9 downto 0));
end VGA_inte;
architecture synt of VGA_inte is
signal videoon, videov, videoh : std_logic;
signal hcount, vcount : std_logic_vector(9 downto 0);
begin
hcounter: process (clk, reset)
begin if reset='1' then
hcount <= (others => '0');
else if (clk'event and clk='1') then
if hcount=799 then
hcount <= (others => '0');
```

```
else hcount <= hcount + 1;
end if;
end if;
end if;
end process;
process (hcount)
begin videoh <= '1';
column <= hcount;
if hcount>639 then
videoh <= '0';
column <= (others => '0');
end if;
end process;
vcounter: process (clk, reset)
begin
if reset='1' then
vcount <= (others => '0');
else if (clk'event and clk='1') then
if hcount=699 then
if vcount=524 then
vcount <= (others => '0');
else vcount <= vcount + 1;
end if;
end if;
end if;
end if;
end process;
process (vcount)
begin
videov <= '1';
row <= vcount(8 downto 0);
if vcount>479 then
videov <= '0';
row <= (others => '0');
end if;
end process;
sync: process (clk, reset)
begin
if reset='1' then
hsync <= '0';
vsync <= '0';
else if (clk'event and clk='1') then
if (hcount<=755 and hcount>=659) then
hsync <= '0';
else hsync <= '1';
end if;
if (vcount<=494 and vcount>=493) then
vsync <= '0';
else vsync <= '1';
end if;
end if;
end if;
end process;
videoon <= videoh and videov;
colors: process (clk, reset)
begin
if reset='1' then r <= '0';
g <= '0';
b <= '0';
elsif (clk'event and clk='1') then
r <= red and videoon;
```

```
g <= green and videoon;
b <= blue and videoon;
end if; end process;
end synt;
```

DESIGN EXAMPLE 8.10

Design and develop a digital system to realize direct interfacing of VGA with the FPGA. Design a separate digital system to generate the data corresponding to the pattern of an image.

Solution

```
library ieee;
use ieee.std_logic_1164.all;
entity vga_controller is
generic(h_pulse : integer := 208;
h_bp : integer := 336;
h_pixels : integer := 1920;
h_fp : integer := 128;
h_pol : std_logic := '0';
v_pulse : integer := 3;
v_bp : integer := 38;
v_pixels : integer := 1200;
v_fp : integer := 1;
v_pol : std_logic := '1');
port(pixel_clk : in std_logic;
reset_n : in std_logic;
h_sync : out std_logic;
v_sync : out std_logic;
disp_ena : out std_logic;
column : out integer;
row : out integer;
n_blank : out std_logic;
n_sync : out std_logic);
end vga_controller;
architecture behavior of vga_controller is
constant h_period : integer := h_pulse + h_bp + h_pixels + h_fp;
constant v_period : integer := v_pulse + v_bp + v_pixels + v_fp;
begin
n_blank <= '1';
n_sync <= '0';
process(pixel_clk, reset_n)
variable h_count : integer range 0 to h_period - 1 := 0;
variable v_count : integer range 0 to v_period - 1 := 0;
begin
if(reset_n = '0') then
h_count := 0;
v_count := 0;
h_sync <= not h_pol;
v_sync <= not v_pol;
disp_ena <= '0';
column <= 0;
row <= 0;
elsif(pixel_clk'event and pixel_clk = '1') then
if(h_count < h_period - 1) then
h_count := h_count + 1;
else
h_count := 0;
if(v_count < v_period - 1) then
v_count := v_count + 1;
```

```
else
v_count := 0;
end if;
end if;
if(h_count < h_pixels + h_fp or h_count > h_pixels + h_fp + h_pulse) then
h_sync <= not h_pol;
else
h_sync <= h_pol;
end if;
if(v_count < v_pixels + v_fp or v_count > v_pixels + v_fp + v_pulse) then
v_sync <= not v_pol;
else
v_sync <= v_pol;
end if;
if(h_count < h_pixels) then
column <= h_count;
end if;
if(v_count < v_pixels) then
row <= v_count;
end if;
if(h_count < h_pixels and v_count < v_pixels) then
disp_ena <= '1';
else
disp_ena <= '0';
end if;
end if;
end process;
end behavior;
```

Hardware Image Data Generation File

```
library ieee;
use ieee.std_logic_1164.all;
entity hw_image_generator is
generic(pixels_y : integer := 478;
pixels_x : integer := 600);
port(disp_ena : in std_logic;
row : in integer;
column : in integer;
red : out std_logic_vector(7 downto 0) := (others => '0');
green : out std_logic_vector(7 downto 0) := (others => '0');
blue : out std_logic_vector(7 downto 0) := (others => '0'));
end hw_image_generator;
architecture behavior of hw_image_generator is
begin
process(disp_ena, row, column)
begin
if(disp_ena = '1') then
if(row < pixels_y and column < pixels_x) then
red <= (others => '0');
green <= (others => '0');
blue <= (others => '1');
else
red <= (others => '1');
green <= (others => '1');
blue <= (others => '0');
end if;
else
red <= (others => '0');
green <= (others => '0');
blue <= (others => '0');
```

```
end if;
end process;
end behavior;
```

LABORATORY EXERCISES

1. Design and develop a digital system to display the real time and current date on a GLCD with automatic updates. Use the DS12887 for generating the current time and day details.
2. Design and develop a digital system to produce a time alarm and square waves of different frequencies using the RTC IC DS12887.
3. Design and develop a digital system to realize the I^2C data transfer protocol with one master unit and five slave devices. Connect all the slave devices with the master unit through the I^2C bus.
4. Design and develop a digital system to interface a SHT11 and SCP1000-D01 sensors and acquire the temperature, relative humidity and atmospheric pressure data in parallel. Log all these measurement data in an Excel file.
5. Design and develop a digital system to interface a GSM module that will control any five electrical appliances that are connected with a receiver mobile phone unit.
6. Design and develop a digital system to interface a GPS receiver with the FPGA and extract all the necessary possible information. Display the extracted information in the MATLAB command widow and log it in an Excel file.
7. Let us assume that we have an electronic system with a GPS receiver which receives GPS data and locally transmits them through a Zigbee transreceiver. Design and develop a digital system to interface with the Zigbee transreceiver; receive the GPS data; process them; send the processed data, that is, data relevant to the location, to a computer and plot the trajectory of the movement of the object that has the GPS module.
8. Design and develop a digital system to create radar displays like A-scope, B-scope, and PPI on the computer monitor.
9. Design and develop a digital system to read the keyboard (PS/2) data and indicate a marker point according to the keyboard input data (without shift is angle and with shift is range) on the PPI display designed on a computer monitor through the VGA interface.
10. Design and develop a digital system to interface the PS/2 and VGA display monitor to play a simple video game. The reader can choose any game.

9

Real-World Control Device Interfacing

9.1 Introduction and Preview

Almost all real-world applications are controlled by certain devices regardless of the algorithm used for the control action. A few of the most popular control devices are relay, solenoid valve, DC motor, servo motor, brush less direct current (BLDC) motor, stepper motor, SCR, Triac, Diac, and so on. The primary aim of this chapter is to give readers a clear idea of the working principles, control methodologies, limitations, and interfacing techniques of various real-world control devices. The chapter begins by introducing the types of relays, solenoid valves, opto-isolators, and DC motors and their working principles, control methodologies, applications, and interfacing methodologies. Several design examples are given to understand the interfacing methodologies and system design approaches. The significance of servomotors and BLDC motors are explained, along with their working principles and the generation of control signals to operate them either in clockwise or anti-clockwise directions. Types of stepper motors, rotation angle resolution, rotational speed, and control methodologies are explained, with application circuits and design examples. Liquid and fuel level measurement systems, their working principles, types, interfacing methodologies, and design examples are described. Differences among the AC/DC switches, digital switches, transistors, power transistors, and power electronics control devices are listed, and types of power electronics switches that are popularly used are explained. The working principles of SCR are briefed, with the necessary application circuits. The advantages and internal configuration of bidirectional power electronic switches over unidirectional are briefed. Internal configuration, working principles, control circuit design, and application circuit design are detailed, along with design examples. The design and modeling of a real-time closed-loop control system is also explained, with a description of the contribution of every element to the system. Finally, laboratory exercises are listed for readers' practice and to help improve their design skills.

9.2 Relay, Solenoid Valve, Opto-Isolator, and Direct Current Motor Interfacing and Control

In many real-world applications, control action becomes significant. Irrespective of the algorithm we use for performing the control action, we need to use real devices to accomplish tasks in real situations. The relay, solenoid valve, opto-isolator and DC motors are basic and popular control devices that cover a wide spectrum of requirements. In this section, we discuss their working principles, interfacing techniques, application circuits, and relevant design examples.

9.2.1 Relay Control

The relay is usually an electromechanical device which gets actuated by the electrical current. The current flow in one electrical circuit causes a mechanical opening or closing in another circuit. Relays are like remote control switches and are used in many applications, such as telephone exchanges, digital computers, measurement instruments, industrial process control, automation systems, and so on. Relay takes a relatively small amount of power to control even high-powered electrical devices like motors, heaters, lamps, AC circuits, and so on. Poles, also known as input terminals, are the number of separate switching circuits within the relay. The poles define how many separate circuits the relay can handle. The throws of a relay are the output terminals that define the number of different output connections the each pole can have. Based on the number of poles and throws, relays are classified as (i) single pole single throw (SPST), (ii) single pole double throw (SPDT), and (iii) double pole double throw (DPDT). The simplest relay is SPST. It has two terminals (one input and one output) like an ordinary electrical switch. As the name indicates, it can control only one circuit, that is, it can make one-to-one contact or disconnect. An SPDT relay routes one input to one of two outputs. This relay has three terminals (one input and two outputs), as shown in Figure 9.1a. A DPDT relay has two poles and each has two throws; hence, in all, it has six terminals (two inputs and four outputs), as shown in Figure 9.1b. It can be thought of as two SPDT switches that are always switched together by a single actuator. It is commonly used for phase or polarity reversal by crisscrossing the terminals [112]. The terminals in the relay are designated as normally open (NO) or "make contact" terminal, normally closed (NC) or "break contact" terminal and common terminal (CT), as indicated in Figure 9.1c. The NO is open when the coil is de-energized and closes when the coil is energized. The NC function is reversed, as it is closed in the de-energized position and opens when the coil is energized. The NO and NC terminals are tightly fixed and cannot move. The CT has contact with the NC terminal by default. When the relay is energized, the CT leaves NC and gets a new contact with the NO terminal. That means now the NC is opened and the NO is closed with the CT. To accomplish this moving action, we require a separate circuit. Therefore, the relay has two circuits; the first circuit is called the control circuit and the second is called the load circuit.

All the relays contain, at the control circuit, a sensing unit, that is, the electric coil, which is powered by AC or DC current, as shown in Figures 9.1b, c and 9.2a. When the applied current or voltage exceeds a threshold value, the coil activates the armature, which then

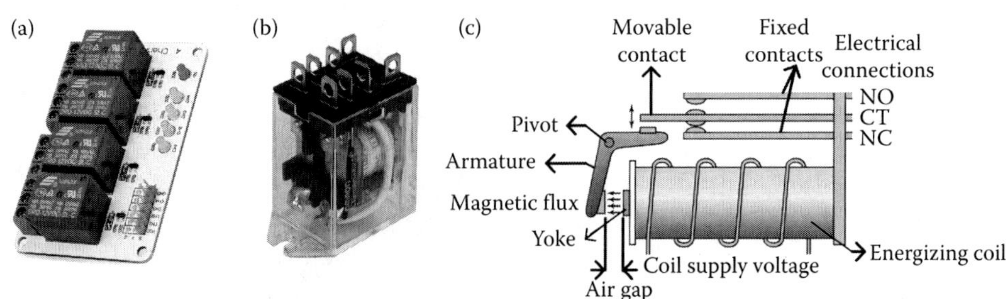

FIGURE 9.1
Photograph of four SPDT relays (a), photograph of DPDT relay (b), and schematic of SPDT relay (c).

becomes an electromagnet and attracts the opposite metals, that is, NC, NO, and CT. Since NC and NO are tightly fixed, CT moves towards the coil; hence, now the CT contacts the NO terminal. The magnetic force is, in effect, relaying the action from one control to another. The switch contacts in the relay remain in the NC position as long as no power is applied to the coil. When power is applied to the coil, the contacts move to a new position, that is, the NO position, and stay in that position as long as power is applied to the coil. Let us look at Figure 9.2a, which is the circuit of a DPDT relay, where an input (IP_1) is given at a common terminal (CT_1). This input can be routed to either NO_1 or NC_1, that is, OP_1 or OP_2. In other words, two inputs (IP_2 and IP_3) are connected to NO_2 and NC_2, respectively, and any one of these two inputs can be routed to the common terminal (CT_2) as output (OP_3). When the relay is not powered, IP_1 goes to OP_2 and IP_3 goes to OP_3, and after powering, IP_1 goes to OP_1 and IP_2 goes to OP_3. Thus, it possible to route one input to two outputs or two inputs to one output by using the relay.

Now we discuss the wiring of the relay. A typical relay circuit has the coil driven by a transistor switch, as shown in Figure 9.2b. When the base voltage of the transistor is zero (or negative), the transistor is cut off and acts as an electrically open circuit. In this condition, no collector current flows to the emitter, and the relay coil is de-energized because, being current devices, if no current flows into the base, then no current will flow through the relay coil. If a large enough positive current is now driven into the base to saturate the transistor, the current flowing from base to emitter controls the larger relay coil current

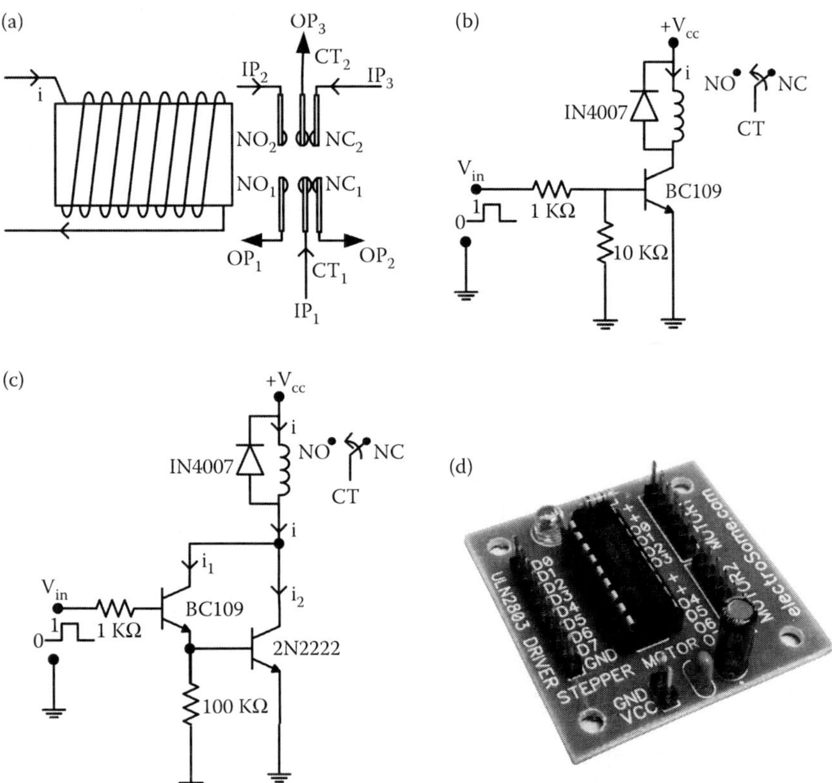

FIGURE 9.2
Circuit of DPDT relay: (a), typical relay control circuit (b), relay control circuit with Darlington pair of transistors (c), and 8-channel Darlington current amplifier ICULN2803A (d).

flowing through the transistor from the collector to the emitter. When the current is flowing through the coil, the relay gets energized and the corresponding action happens as discussed above. As we know, the relay coil is not only an electromagnet, it is also an inductor. When power is applied to the coil due to the switching action of the transistor, a maximum current will flow as a result of the DC resistance of the coil as defined by Ohm's Law ($I = V/R$). Some of this electrical energy is stored within the relay coil as the magnetic field. When the transistor turned off, the current flowing through the relay coil decreases and the magnetic field collapses. However, the stored energy within the magnetic field has to go somewhere and a reverse voltage is developed across the coil as it tries to maintain the current in the relay coil. This action produces a high voltage spike across the relay coil that can damage the switching operation of the transistor. Hence, in order to prevent damage to the transistor, a freewheeling diode is connected across the relay coil, as shown in Figure 9.2b. This flywheel diode clamps the reverse voltage across the coil to about 0.7 V, dissipating the stored energy and protecting the switching transistor [112].

The relay switch circuit, shown in Figure 9.2b, is ideal for switching low-power relay loads. However, sometimes it is required to switch larger relay coils or currents beyond the basic range of the transistor, for example, a stepper motor, and this can be achieved using the Darlington transistor [112]. The sensitivity and current gain of a relay switch circuit can be greatly increased by using a Darlington pair of transistors instead of a single switching transistor. Darlington transistor pairs can be made from two individually connected bipolar transistors, as shown in Figure 9.2c, or are available as one single device (IC), as shown in Figure 9.2d, with standard base, emitter and collector connecting leads. In the two-transistor connection, as shown in Figure 9.2c, the collector current of the first transistor becomes the base current of the second transistor. The application of a positive base current to the first transistor automatically turns on the second transistor, which leads to more current drawing from the source. Positive-negative-positive (PNP) transistors can also be used to design the relay control circuit. The PNP relay switching circuit is no different from the negative-positive-negative (NPN) relay switching circuit in terms of its ability to control the relay coil. However, it does require different polarities of operating voltages. For example, the collector-to-emitter voltage (V_{CE}) must be negative for the PNP type to cause current flow from the emitter to the collector. The PNP transistor circuit works in a way opposite to the NPN relay switching circuit. Load current flows from the emitter to the collector when the base is forward biased with a voltage that is more negative than that at the emitter. For the relay load current to flow through the emitter to the collector, both the base and the collector must be negative with respect to the emitter. In other words, when V_{in} is high, the PNP transistor is switched off and so too is the relay coil. When V_{in} is

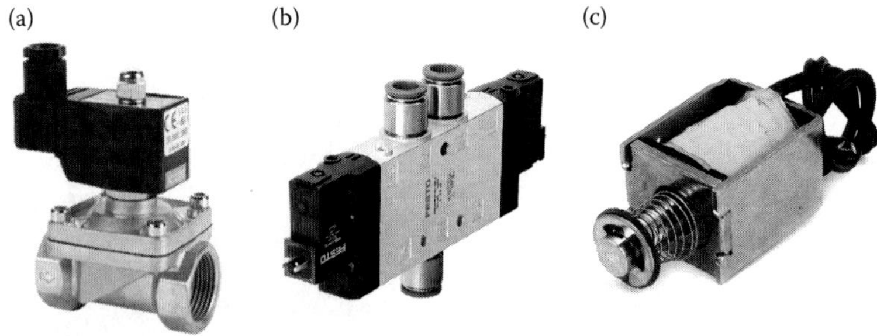

FIGURE 9.3
Solenoid outflow control valve (a), three-port solenoid valve, switching the outflow between two outlet ports (b), and solenoid plunger (c).

low, the base voltage is less than the emitter voltage and the PNP transistor turns on. The base resistor value sets the base current, which sets the collector current that drives the relay coil. PNP transistor switches can be used when the switching signal is the reverse of an NPN transistor.

DESIGN EXAMPLE 9.1

Design a digital system to control some of the parts of a robot. Rotate the left arm and head of the robot anti-clockwise while the right arm is being rotated clockwise. This rotation has to be changed after 5 seconds. Assume that the three motors that rotate the left arm, right arm and head are controlled by six separate SPST relays.

Solution

```
library IEEE;
use IEEE.STD_LOGIC_1164.ALL;
entityrobo is
port(start:instd_logic;
clk:instd_logic;
cout:outstd_logic;
relay1:outstd_logic;
relay2:outstd_logic;
relay3:outstd_logic;
relay4:outstd_logic;
relay5:outstd_logic;
relay6:outstd_logic);
endrobo;
architecture Behavioral of robo is
signalclk_sig:std_logic;
signalcounter:integer:=0;
begin
cout<= clk_sig;
p1:process(clk,start)
begin
if(start='0') then
clk_sig<='0';
elsif (rising_edge(clk)) then
if(counter=20000000) then
clk_sig<=not clk_sig;
counter<=0;
else
counter<=counter+1;
end if;
end if;
end process;
relay1<=start and clk_sig;
relay2<=start and clk_sig;
relay3<=start and (not clk_sig);
relay4<=start and (not clk_sig);
relay5<=start and (not clk_sig);
relay6<=start and (not clk_sig);
end Behavioral;
```

9.2.2 Solenoid Valve Control

Solenoids are simple electrical devices and have a vast impact on our daily life. A solenoid valve has two main parts: solenoid and valve. The solenoid converts electrical energy into mechanical energy which, in turn, opens or closes the valve mechanically. The term "solenoid" is derived from two Greek names, "Solen", which means a channel or pipe, and "Eidos", which means outlet [113]. The solenoid is used in a variety of applications, and numerous types of solenoids are commercially available. A few of them are shown in Figure 9.3a, b and c. Each has its own properties that make it useful in many precise/suitable applications. Design of solenoids can be done in different ways. Generally, solenoids work on general electrical principles, but the mechanical energy of this device is distributed in a different way in different designs. For example, the solenoid shown in Figure 9.3a can control the rate of liquid flow through a pipe. A three-port solenoid, as shown in Figure 9.3b, can switch the liquid outflow between two outlet ports.

A solenoid is a coil of wire that is used in electromagnets, inductors, antennas, relays, and so on. The application of a solenoid differs based on where it is installed, like medical equipment, computer printers, fuel injection gear, locking systems, industrial control, CNC machines, heating systems, washing machines, automotive systems, and so on. Mainly, solenoids are classified as AC-laminated solenoids, DC C–frame solenoids, DC D–frame solenoids, linear solenoids, rotary solenoids, tubular solenoid, and so on [113]. A solenoid includes a coil of wire around a core made out of a metal. When a current is applied to the solenoid, it has the effect of creating a consistent magnetic field. Electricity changes to magnetism, then to electricity; therefore, these two forces are united into one [113]. An attractive thing about the uniform field in a solenoid is that, if the solenoid has an immeasurable length, the magnetic field would be similar everywhere along the element. In a solenoid, this translates to very small electrical components being able to do the work. For example, a powerful solenoid can easily shut even a heavy pipe shutter. The solenoid plunge, shown in Figure 9.3c, is more popular for linear pushing/pulling applications. Solenoid plunges are capable of using a pulling or pushing force on a mechanical device and can be utilized for a variety of tasks. For example, a solenoid plunger is used in the starter device of a vehicle. Whenever the electrical current flows through a solenoid coil, a magnetic field is generated around the yoke and, due to this field, the center plunger moves away in a linear fashion to bring the two contacts together. A 2/2-way valve has two ports (inlet and outlet) and two positions (open or closed) which are in NC (closed in de-energized state) or NO (open in de-energized state). A 3/2-way valve has three ports and two positions and can therefore switch between two ports, NC, NO, diverting or universal. Solenoid valves with more ports and/or valves in a single device are also available.

Now, we discuss a very precise water flow control solenoid valve system. Let us assume a solenoid valve design as shown in Figure 9.4a. A spring is used to hold the valve open or closed while the valve is not activated. The liquid enters through an inlet port. The diaphragm is pushed down by the spring under normal conditions. The diaphragm has a pinhole through its center which allows a very small amount of liquid to flow through it. This liquid fills the diaphragm cavity on the upper side of the diaphragm so that pressure

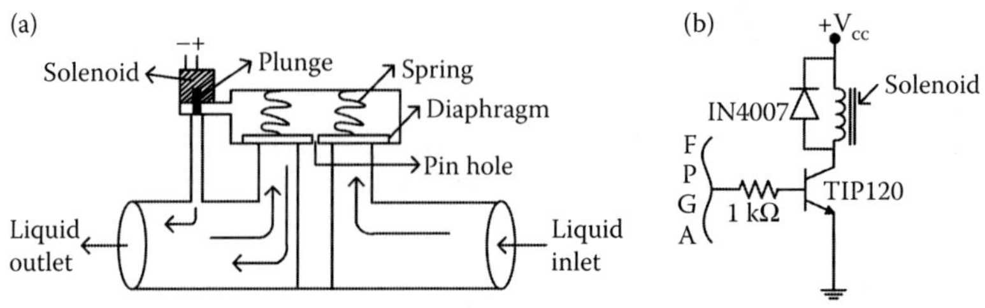

FIGURE 9.4
Solenoid valve for precise liquid flow control (a) and solenoid control circuit (b).

is equal on both sides of the diaphragm. While the pressure is the same on both sides of the diaphragm, the force is greater on the upper side, which forces the diaphragm shut against the incoming pressure. On the upper side, the pressure acts on the entire surface of the diaphragm, while on the lower side, it acts only on the incoming pipe. This results in the valve being securely shut to any flow, and the greater the input pressure, the greater the shutting force [113]. Until the solenoid coil is de-energized, the outlet pinhole is blocked by the solenoid plunger. When the solenoid is activated, the plunge moves up due to the magnetic force induced by the solenoid current; then, the liquid will flow through the outlet. The pressure in the diaphragm chamber will drop and the incoming pressure will lift the diaphragm, thus opening the main valve. Therefore, now liquid flows directly from the inlet to the outlet. When the solenoid is again deactivated, the previous control action will happen. The solenoid used in this system has an electric coil with a movable ferromagnetic core (plunger) in its center. In rest position, the plunger closes off a small orifice. An electric current through the coil creates a magnetic field. The magnetic field exerts a force on the plunger. As a result, the plunger is pulled towards the center of the coil so that the orifice opens. This is the basic principle that is used to open and close solenoid valves. The solenoid control circuit is shown in Figure 9.4b. The solenoid control circuit is the same as the relay control circuit. The control voltage applied at the base of the transistor allows current flow from collector to emitter; hence, the solenoid gets activated. A freewheeling diode is used to improve the response time of the solenoid as noted in the previous section. In the same way, any solenoid can be interfaced with and controlled.

DESIGN EXAMPLE 9.2

Let us assume that there is an electronic solenoid system that can pass a liquid in to two directions, right or left. The liquid is pumped into a valve control through a separate inlet point. A proximity sensor is configured with a solenoid control system that decides the direction of liquid based on the input of the receiver of the proximity sensor. Develop a digital system to read the sensor data and change the direction of the liquid flow and also display any of four different sentences when the sensor input is at logic 0.

Solution

```
LIBRARY ieee;
USE ieee.std_logic_1164.all;
USE ieee.numeric_std.all;
use ieee.std_logic_unsigned.all;
use IEEE.STD_LOGIC_ARITH.ALL;
entity sole_contr is
Port (m_clk,ir_rx: in STD_LOGIC:='0';
c:out std_logic_vector(2 downto 0):="000";
sol_cnt: out std_logic:='0';
d : out STD_LOGIC_vector(7 downto 0):="00000000");
end sole_contr;
architecture Behavioral of sole_contr is
-------------------------------------------------
signal order1: string (1 to 16):="Antony ,";
signal order2: string (1 to 16):="Jothy ,";
signal order3: string (1 to 16):="Bazil ,";
signal order4: string (1 to 16):="Rosita ,";
signal order5: string (1 to 16):="Feona ,";
signal order6: string (1 to 16):="---------------,";
signal sta,nsta:integer range 0 to 35:=0;
-----------------------------------------
```

```
begin
sol_cnt<=not ir_rx;
-----------------------------------------
p0:process(m_clk, ir_rx)
variable cnt: integer:=0;
begin
if (ir_rx='0') then
if rising_edge(m_clk) then
if (cnt=2) then --20000 for 4MHz on-board clock
sta<=nsta;
cnt:=0;
else
cnt:=cnt+1;
end if;
end if;
end if;
end process;
-----------------------------------------
p1:process(sta)
variable i: integer:=1;
begin
case sta is
when 0=> d<=x"38";c<="100"; nsta<=sta+1;--38
when 1|3|5|7|9|13|17|19|23|27|29|33=>c<="000";
nsta<=sta+1;
when 2=> d<=x"0C";c<="100"; nsta<=sta+1;--0C
when 4=> d<=x"01";c<="100"; nsta<=sta+1;--01
when 6=> d<=x"06";c<="100"; nsta<=sta+1;--06
when 8=> d<=x"80";c<="100"; nsta<=sta+1;--80
when 10=> d<=std_logic_vector(to_unsigned(character'pos(order1(i)),8));
c<="110";
nsta<=sta+1;
when 11=> if(order1(i)=',') then c<="010";i:=1;nsta<=12;
else c<="010";i:=i+1;nsta<=10;
end if;
when 12=> d<=x"C0";c<="100"; nsta<=sta+1;--C0
when 14=> d<=std_logic_vector(to_unsigned(character'pos(order2(i)),8));
c<="110";
nsta<=sta+1;
when 15=> if(order2(i)=',') then c<="010";i:=1;nsta<=16;
else c<="010";i:=i+1;nsta<=14;
end if;
when 16=> d<=x"0C";c<="100"; nsta<=sta+1;--0C
when 18=> d<=x"80";c<="100"; nsta<=sta+1;--80
when 20=> d<=std_logic_vector(to_unsigned(character'pos(order3(i)),8));
c<="110";
nsta<=sta+1;
when 21=> if(order3(i)=',') then
c<="010";i:=1;
nsta<=22;
else c<="010";i:=i+1;nsta<=20;
end if;
when 22=> d<=x"C0";c<="100"; nsta<=sta+1;--C0
when 24=> d<=std_logic_vector(to_unsigned(character'pos(order4(i)),8));
c<="110";
nsta<=sta+1;
when 25=> if(order4(i)=',') then
c<="010";i:=1;nsta<=26;
else c<="010";i:=i+1;nsta<=24;
end if;
when 26=> d<=x"0C";c<="100"; nsta<=sta+1;--0C
```

```
when 28=> d<=x"80";c<="100"; nsta<=sta+1;--80
when 30=> d<=std_logic_vector(to_unsigned(character'pos(order5(i)),8));
c<="110";
nsta<=sta+1;
when 31=> if(order5(i)=',') then c<="010";i:=1;nsta<=32;
else c<="010";i:=i+1;nsta<=30;
end if;
when 32=> d<=x"C0";c<="100"; nsta<=sta+1;--C0
when 34=> d<=std_logic_vector(to_unsigned(character'pos(order6(i)),8));
c<="110";
nsta<=sta+1;
when 35=> if(order6(i)=',') then c<="010";i:=1;nsta<=0;
else c<="010";i:=i+1;nsta<=34;
end if;
when others=> nsta<=0;
end case;
end process;
-----------------------------------------------------
end Behavioral;
```

9.2.3 Opto-Isolator Interfacing

In general, electrical signals are transmitted from one point to another using various components and media. For example, in the television remote control, the IR-Tx transmits the electrical signal from the remote circuit to the TV internal circuit, that is, IR-Rx. In the same way, RF, LED, acoustic signals, phonons, and lasers can also be used to transfer electrical signals. In this section, we discuss the working principles and applications of the opto-isolator circuit. As we know, transformers provide a step-down voltage as well as electrical isolation between the higher voltage on the primary side and the lower voltage on the secondary side. In other words, transformers isolate the primary input voltage from the secondary output voltage using electromagnetic coupling by means of a magnetic flux circulating within the iron-laminated core. We can also provide electrical isolation between an input source and an output load using just light by using an opto-isolator. An opto-isolator is also known as an optical isolator, photocoupler, or opto-coupler. It is one of the electrical/electronic components used for transferring electrical signals from one circuit to another with isolation using light [114]. In general, a combination of LEDs and phototransistors are used in a typical opto-isolator IC packed in a single opaque package, as shown in Figure 9.5a. The basic design of an opto-coupler consists of an LED that produces infra-red light and a semiconductor photosensitive device that is used to detect the emitted infrared beam. Both the LED and photosensitive device are enclosed in a light-tight body or package with metal legs for the external wiring.

Let us consider an opto-isolator circuit, as shown in Figure 9.5b, with an LED which emits light and a phototransistor (PT). This phototransistor is used to detect the incoming light and thereby directly generate electrical energy or modulate electrical energy of the external power supply connected to the phototransistor. Keeping another IR Tx/Rx pair in reverse inside the same package, it is possible to construct a symmetrical and bidirectional opto-isolator. If the light emitted from the LED falls on the phototransistor base, then the sensor is turned ON and the transistor starts conduction. If the current passing through the LED is set to zero or turned off, then the transistor stops conduction. This opto-isolator circuit is used generally for transferring digital signals, but can also be used to transfer analog signals with a few techniques. Current from the source signal passes through the input LED, which emits an infrared light whose intensity is proportional to the electrical signal. This emitted light falls upon the base of the phototransistor, causing it to switch ON and conduct

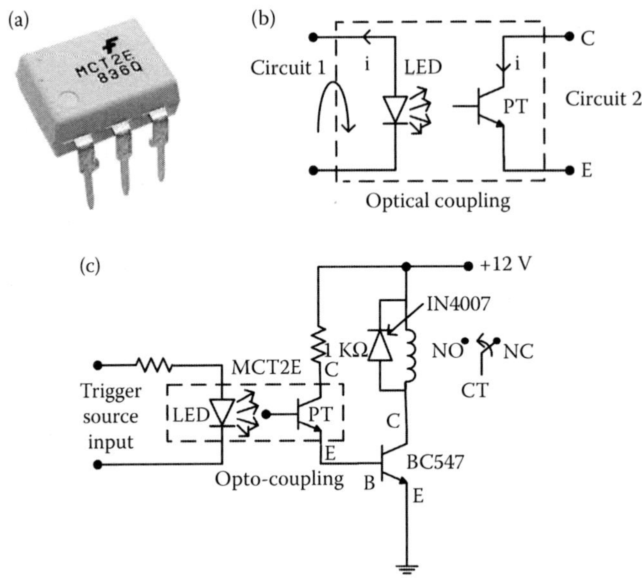

FIGURE 9.5
IC package of an opto-isolator (a), opto-isolator interfacing circuit (b), and application circuit of an opto-isolator (c).

in a similar way to a normal bipolar transistor. The base connection of the phototransistor can be left open for maximum sensitivity to the LED's infrared light energy or connected to ground via a suitable external high-value resistor to control the switching sensitivity, making it more stable and resistant to false triggering by external electrical noise or voltage transients. When the current flowing through the LED is interrupted, the infrared emitted light is cut off, causing the phototransistor to cease conducting. The phototransistor can be used to switch current in the output circuit. The spectral response of the LED and the photosensitive device are closely matched, being separated by a transparent medium such as glass, plastic or air. Since there is no direct electrical connection between the input and output of an opt-coupler, electrical isolation up to 10 kV is achieved.

Phototransistor and photo-Darlington devices are mainly for use in DC circuits, while photo-SCRs and photo-Triacs are for AC-powered circuit control. There are many other kinds of source–sensor combinations, such as LED-photodiode, LED-LASER, lamp-photoresistor pairs, and reflective and slotted opto-couplers. Opto-couplers and opto-isolators can be used on their own or to switch a range of other larger electronic devices such as transistors and Triacs provided the required electrical isolation between a lower voltage control signal and the higher voltage/current load signal. Common applications for opto-couplers include microprocessor input/output switching, DC and AC power control, PC communications, signal isolation, and power supply regulation, which suffer from current ground loops, and so on. The electrical signal being transmitted can be either analog (linear) or digital (pulses). An example application circuit to control two relays that are interfaced to the FPGA with an opto-isolator is shown in Figure 9.5c. The source which needs to be isolated is giving the trigger input to the anode of an LED via a limiting resistor (as we normally do with usual LEDs) and to switch the phototransistor in response to the applied input triggers. The above action illuminates the internal LED whose light is detected by the phototransistor, causing it to conduct across its relevant pin outs, that is, collector to emitter. Phototransistor output is normally used for driving the preceding isolated stage, for example, a relay driver stage. As shown in Figure 9.5c, the relay driver consists of an NPN transistor, which is trigged by the emitter of the phototransistor, quite like a Darlington pair configuration. Therefore, the triggering control and relay load circuits are perfectly isolated. The rapidly changing voltages or other electrical parameters on one side of the circuit may cause damage to components on the other side of the circuit if they are directly connected. Hence, an opto-isolator circuit can be used to isolate the high or rapidly changing voltages of one side of a circuit that causes damage to the components on the other side of the circuit. In general, for switching microprocessor input/output, AC or DC power control; computer communications and for the regulation of power supply, opto-isolator circuits are used. The opto-isolator circuit can be used to detect DC signals and data and to control AC-powered devices and mains lamps. In this way, the opto-isolator can be interfaced with the FPGA and used to control high power/load devices with proper isolation.

DESIGN EXAMPLE 9.3

Let us assume that there are three different process plants connected with a control unit through three separate opto-isolators and SPST relays. Design and develop a digital system to control them one by one at a predetermined period.

Solution

```
LIBRARY ieee;
USE icee.std_logic_1164.all;
USE ieee.numeric_std.all;
use ieee.std_logic_unsigned.all;
use IEEE.STD_LOGIC_ARITH.ALL;
entity pro_contr is
Port (m_clk,start: in STD_LOGIC:='0';
pro_cont:out std_logic_vector(2 downto 0):="000");
end pro_contr;
architecture Behavioral of pro_contr is
signal sta,nsta:integer range 0 to 2:=0;
begin
p0:process(m_clk,start)
variable cnt: integer:=0;
begin
if (start='1') then
if rising_edge(m_clk) then
if (cnt=2) then --20000 for 4MHz on-board clock
sta<=nsta;
cnt:=0;
else
cnt:=cnt+1;
end if;
end if;
end if;
end process;
p1:process(sta)
begin
case sta is
when 0=> pro_cont<="100";
nsta<=sta+1;
when 1=> pro_cont<="010";
nsta<=sta+1;
when 2=> pro_cont<="001";
nsta<=0;
when others=>
```

```
nsta<=0;
end case;
end process;
end Behavioral;
```

9.2.4 Direct Current Motor Control

The DC motor was one of the first electrical devices to convert electrical power into mechanical power. The working of the DC motor is based on the principle that when a current-carrying conductor is placed in a magnetic field, it experiences a mechanical force. The direction of that mechanical force is given by Fleming's left-hand rule, F = BIl Newton, where F is force in N, B is flux density in Wb/m^2, I is current in A and l is length of the conductor in m. One commercially available DC motor is shown in Figure 9.6a. The permanent magnet (PM) and DC convert the electrical energy into mechanical energy through the interaction of two fields. One field is produced by a PM assembly and the other by an electrical current flowing in the motor windings. These two fields result in a torque which tends to rotate the rotor, as shown in Figure 9.6b. As the rotor turns, the current in the windings is commutated to produce a continuous torque output. The stationary electromagnetic field of the motor can also be wire-wound like the armature or can be made up of PMs. In both models, that is, either wire-wound field or PM, the commutator acts as half of a mechanical switch and rotates with the armature as it turns. The commutator is composed of conductive segments which are usually made up of copper, and it gives termination of individual coils of wire distributed around the armature. The second half of the mechanical switch is completed by the brushes. These brushes typically remain stationary with the motor's housing but ride on the rotating commutator. As electrical energy is passed through the brushes and consequently through the armature, a torsional force is generated as a reaction between the motor's field and the armature, causing the motor's armature to turn. As the armature turns, the brushes switch to adjacent bars on the commutator. This switching action transfers the electrical energy to an adjacent winding on the armature which in turn perpetuates the torsional motion of the armature, and this action continues until it has applied the supply voltage.

PM DC motors are usually smaller in overall size and lighter for a given power rating. A simple PM DC motor is an essential electrical device in a variety of products such as toys, servo mechanisms, valve actuators, robots, automotive electronics, and so on. Since the motor's field created by the PM is constant in DC PM motors, the relationship between torque and speed is linear. When a DC motor is directly connected to a battery, it draws current as required when rotating. Further, a surge current is caused because the motor, when it is turning, acts as a generator. The generated voltage is directly proportional to

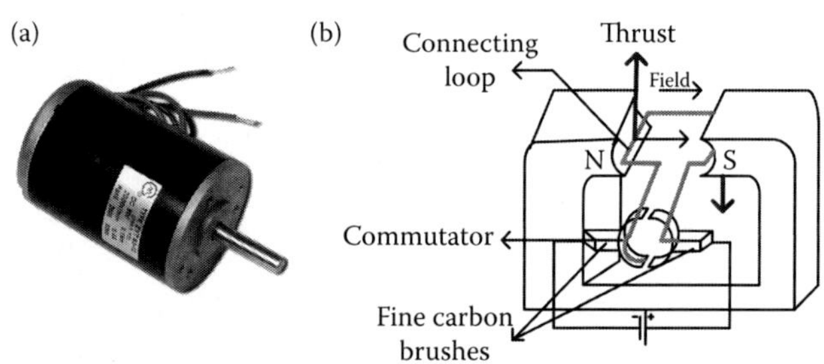

FIGURE 9.6
Picture of a DC motor (a) and constructional layout of a DC motor (b).

the speed of the motor [115,116]. The current through the motor is controlled by the difference between the battery voltage and the motor's generated voltage, which is known as back EMF. When the motor is first connected to a battery, there is no back EMF. Thus, the current is controlled only by the battery voltage, motor resistance/inductance, and battery leads. Without any back EMF, the motor, before it starts to turn, draws a larger amount of surge current. When the DC motor is operated through a controller, it varies the voltage fed to the motor. Initially, at zero speed, the controller will feed no voltage to the motor, so no current flows. As the controller increases the output voltage, the motor will start to turn. In the beginning, the voltage fed to the motor is small, so the current is also small, and as the controller output voltage increases, the back EMF is generated. The result is that the initial current surge is removed, and acceleration is smooth and fully under control [115,116].

The simplest method to control the rotation speed of a DC motor is to control its driving voltage, that is, the higher the supply voltage, the higher the speed of the motor. The PWM method is also used in many DC motor speed control applications. In the basic PWM method, the operating power to the motor is turned on and off to modulate the current to the motor, as shown in Figure 9.7a. The duty cycle, that is, ratio of on time to total time, determines the speed of the motor. Since the DC motor mainly consists of a large inductor, it is basically a low-pass device; hence, we need to take care when generating the PWM control signal. It is not capable of working at a high-frequency PWM control signal. The speed can be controlled just by varying the duty cycle, that is, ($T_{on}/(T_{on}+T_{off})$), between 0% and 100%, for example 90%, 50%, and 10%, as shown in Figure 9.7a. The PWM control signal can be generated from any standard timer circuit or from any programmable digital device. An application circuit for controlling the speed of a DC motor is shown in Figure 9.7b. The PWM control signal generated from FPGA is applied to the base of an NPN transistor, which results in current flow from collector to emitter through the DC motor. Since the internal resistance of the transistor becomes small, the voltage drop is across the motor and drives it. The average power available across the motor depends on the on period and frequency of the PWM input. The speed can be controlled by giving the appropriate PWM signal to the base of the transistor. A 0.1 μF capacitor is used to remove the repels/spikes in the voltage across the motor.

In many applications, for example, moving a robot backwards, the rotation direction of a DC motor needs to be changed. In a normal PM DC motor, the direction of rotation is changed just by changing the polarity of the operating voltage. Direction changing is typically implemented using relays or an H-bridge circuit. Further, the maximum current

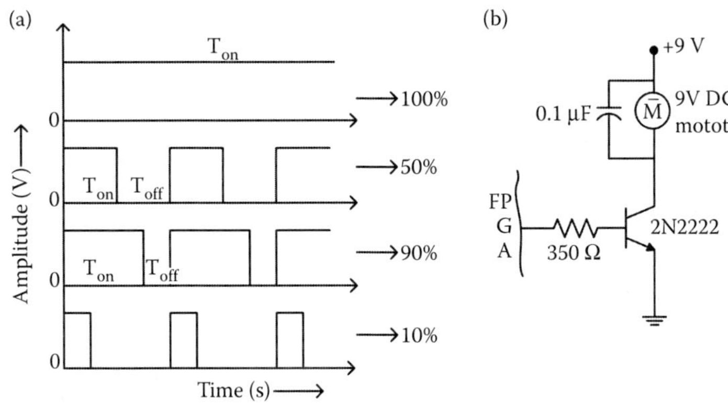

FIGURE 9.7
PWM control signals (a) and DC motor speed control circuit (b).

that can be sunk from the digital device is ≤15 mA at 5 V or 3.3 V. The DC motor needs much more current. It also needs more voltage, such as 6, 12, 24 V, and so on, depending upon the type of the DC motor we use. Another issue is that the back EMF produced by the motor may affect the proper functionality of the digital device. Due to these limitations, we should not connect a DC motor directly to a digital device like a microprocessor, microcontroller, and FPGA. To overcome the problems in their interfacing, we have to use either relays or a motor driver IC between the digital device and DC motor. Relay-based DC motor direction control is shown in Figure 9.8a and b. The DC power supply is connected to the motor through an SPST and two DPDT relays, as shown in Figure 9.8a. When the SPST relay is on, power goes to the DPDT relay and drives the motor with respect to the polarity of terminals A and B. For example, the motor rotes in a clockwise direction when terminals A and B are connected to the + and −, respectively, as shown in Figure 9.8a. In the same way, when the polarity across the motor is changed, as shown in Figure 9.8b, just by driving the DPDT relay, the motor starts rotating anti-clockwise. Speed control for a DC motor is very difficult using this method.

Another more comfortable method is using the H-bridge driver IC, which is shown in Figure 9.8c. The name H-bridge is derived from the shape of the switching circuit which controls the motion of the motor. The motor driver IC, that is, L293D, is a small-current amplifier. It takes a low current signal from the driving source and gives out a high current signal which can drive a DC motor. The L293D is a popular dual H-bridge motor driver through which we can interface two DC motors, and their speed can be controlled in both

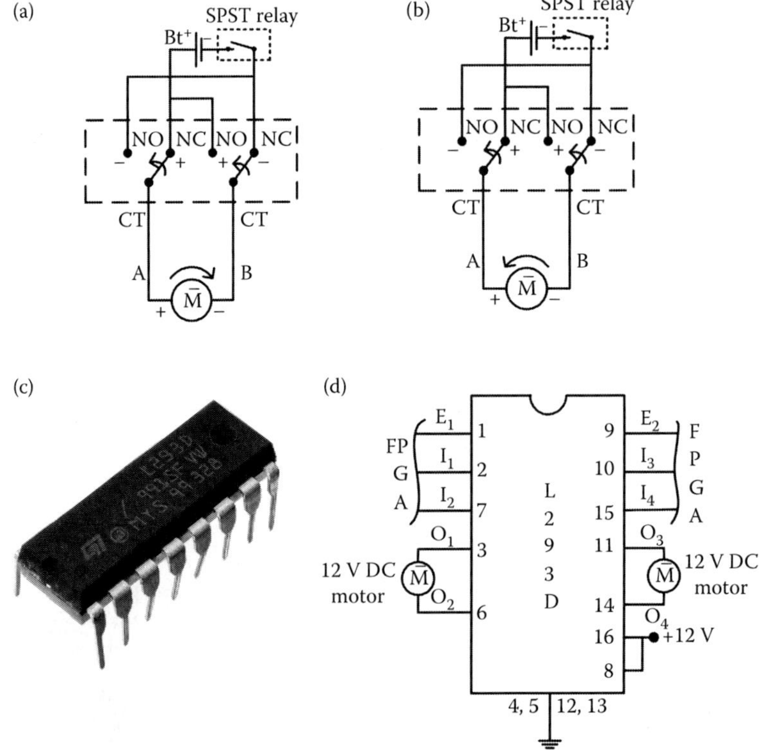

FIGURE 9.8
Relay-based DC motor driver circuit: clockwise direction (a) and anti-clockwise direction (b), picture of an L293D DC motor driver IC (c), and application circuit of L239D (d).

TABLE 9.1

Switching and Corresponding Action of a L293D

Left-Side Motor			Rotation Direction of Motors	Right-Side Motor		
I_1 (pin 2)	I_2 (pin 7)	E_1 (pin 1)		E_2 (pin 9)	I_3 (pin 10)	I_4 (pin 15)
H	L	H	Clockwise	H	L	H
L	H	H	Anti-clockwise	H	H	L
H/L	H/L	H	Motor stops or decelerates	H	H/L	H/L
–	–	L	Stops	L	–	–

the clockwise and counterclockwise directions by giving a PWM signal to the enable pin, L293D. The L293D has a peak output current of 1.2A per channel. Moreover, for protection of the circuit from back EMF output, diodes are included within the IC [115,116]. The output supply has a wide range, from 4.5V to 36V, which has made the L293D the best choice for driving DC motors.

Basically, in the H-bridge driver IC, there are four switching elements, as shown in Figure 9.8d, (i) input 1 (I_1), (ii) input 2 (I_2), (iii) input 3 (I_3), and (iv) input 4 (I_4), for motor switching/driving and two enables (E_1 and E_2) for motor controls. When these switches are turned on in pairs, as given in Table 9.1, the motor changes its direction accordingly. For example, if we switch on I_1 and switch off I_2, then motor will rotate in the clockwise direction as current from the power supply flows through I_1 and the motor coil and goes to ground through I_2. Similarly, when we switch on I_2 and switch off I_1, the current flows in the opposite direction and the motor rotates in the anti-clockwise direction. Enable 1 (E_1) has to be set to on to drive the motor. This pin can also be used to apply the enable signal as a PWM control signal to operate the motor at a decided speed. In the same way, either individually or in parallel, we can operate the right-side motor circuit as well. This is the basic working of an H-bridge. The complete function table of an H-bridge for driving two motors is given in Table 9.1. As shown in Figure 9.8d, three pins are needed for interfacing a DC motor (E_1, I_1, I_2, output 1 [O_1], output 2 [O_2], E_2, I_3, I_4, output 3 [O_3], and output 4 [O_4]). If we wish to control the speed of the DC motor, the corresponding enable pin has to be connected to the PWM output signal. Thus, using this simple switching driver IC, we can control the entire function of a DC motor.

DESIGN EXAMPLE 9.4

Let us assume that one DC motor is configured with a control system through a motor driver (L293D). Design and develop a digital system to operate the motor in the clockwise and anti-clockwise directions and stop it for the predetermined period.

Solution

```
LIBRARY ieee;
USE ieee.std_logic_1164.all;
USE ieee.numeric_std.all;
use ieee.std_logic_unsigned.all;
use IEEE.STD_LOGIC_ARITH.ALL;
entity dc_mot_contr is
Port (m_clk,start: in STD_LOGIC:='0';
mot_cont:out std_logic_vector(2 downto 0):="000");
end dc_mot_contr;
architecture Behavioral of dc_mot_contr is
signal sta,nsta:integer range 0 to 2:=0;
```

```
begin
p0:process(m_clk,start)
variable cnt: integer:=0;
begin
if (start='1') then
if rising_edge(m_clk) then
if (cnt=2) then --20000 for 4MHz on-board clock
sta<=nsta;
cnt:=0;
else
cnt:=cnt+1;
end if;
end if;
end if;
end process;
p1:process(sta)
begin
case sta is
when 0=> mot_cont<="101";
nsta<=sta+1;
when 1=> mot_cont<="011";
nsta<=sta+1;
when 2=> mot_cont<="000";
nsta<=0;
when others=>
nsta<=0;
end case;
end process;
end Behavioral;
```

9.3 Servo and BLDC Motor Interfacing and Control

In addition to DC motors, there are many types of commercially available motors, and they are used in a wide spectrum of applications. From the digital control point of view, servo and BLDC motors are very popular and are used in a wide spectrum of accurate positioning applications. In this section, we discuss the working principles, interfacing techniques, and control methodologies of servo and BLDC motors.

9.3.1 Servo Motor Control

A servo motor uses a servo mechanism, which is a closed-loop mechanism that uses position feedback to control the precise angular position of the shaft. The working principle on which a servo motor mainly depends is the Fleming left-hand rule. Basically, servo motors are adapted versions of DC motors, with a position sensor, gear reduction mechanism, and electronic circuit. In general, a normal DC motor powered by a DC source runs at high speed with low torque. Once we assemble a shaft and gear mechanism to that DC motor, then we can increase or decrease the motor speed gradually by increasing the torque. The position sensor senses the location of the shaft from its fixed position and sends the information to the control circuit. A picture of a servo motor with its parts labeled is shown in Figure 9.9. The control circuit decodes the signals accordingly from the position sensor and compares the actual location of the shaft with the reference position and accordingly

FIGURE 9.9
Picture of a servo motor with label for its important parts.

controls the direction of rotation of the DC motor to reach the decided position. Therefore, servo motors are self-contained mechanical devices that are used to control the shaft position with great precision. There are various kinds of motors, but servo motors are specially designed for specific angular position-related applications. They are found in many applications, like toys, robots, aeroplanes, elevators, rudders, radio-controlled cars, puppets, heavy industrial automation, and so on. Servo motor working principles depend on the PWM control pulse that we feed to the servo motor through a control wire. The advantage of using a servo motor is that the angular position of the motor can be controlled without any feedback mechanism, that is, in an open-loop control configuration. The reference position, that is, where we wish to move the shaft, has to be given from the external driver unit to the motor through its control wire. The servo motor's internal circuit takes care of comparing the actual and reference positions and then drives the motor shaft to the reference position.

Servo motor working principles and operation are very simple; it consists of three wires, where two of them (black and red) are used to provide power and the third is used to provide the control signal. The black wire is connected to ground, and the motor gets power from the red wire. The PWM signal can be used as a control signal to define the reference angular position by its width, that is, the duty cycle. We assume that we have a servo motor which has an angle of rotation of 0 through 180°. The control of the servo motor is connected to a pin of the FPGA through which we apply the PWM control signal. We can control the precise angular position by varying the pulse in a period from 1 msec through 2 msec, as shown in Figure 9.10. The shaft can be positioned to specific angular positions by sending a PWM coded signal. As long as the coded signal exists on the control wire, the servo will maintain the angular position of the shaft. As the coded signal changes, the angular position of the shaft changes. The servo motor has internal control circuits and

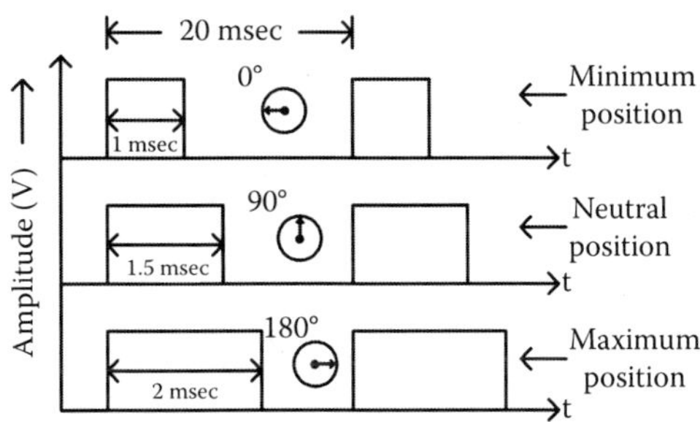

FIGURE 9.10
Duty cycles and corresponding positions of the shaft.

a potentiometer that is connected to the output shaft. This potentiometer allows the control circuitry to monitor the current angle of the servo motor. If the shaft is at the correct angle, then the motor shuts off. If the circuit finds that the angle is not correct, it will turn the motor to the correct direction until the angle is correct. Typically, the output shaft of the servo is capable of travelling up to max 180°. The maximum rotation angle varies with respect to the type we use [117].

A normal servo is mechanically not capable of turning by itself due to a mechanical stop built on the main output gear. The amount of power applied to the motor is proportional to the distance it needs to travel. If the shaft needs to turn a large distance, the motor will run at full speed. If it needs to turn only a small amount, the motor will run at a slower speed, which is called proportional control. The control wire is used to communicate the angle. The angle is determined by the duration of a pulse that we apply to the control wire. The servo expects to see a pulse every 20 msec, that is, 0.02 sec. The length of the pulse will determine how far the motor turns. For example, a 1.5 msec pulse will make the motor turn to the 90° position (often called the neutral position), as shown in Figure 9.10. If the pulse is shorter than 1.5 msec, then the motor will turn the shaft closer to 0°. If the pulse is longer than 1.5 ms, the shaft turns closer to 180°. In this way, we can control the servo motor to any decided position.

DESIGN EXAMPLE 9.5

A servomotor is connected to a control system. Design and develop a digital system to control the rotational angle of the servomotor at the minimum, natural and maximum positions in accordance with the external inputs. For example, if the switch input is "01", the angle is minimum; if the switch input is "10", the angle is natural and if the switch input is "11", the angle is maximum.

Solution

```
LIBRARY ieee;
USE ieee.std_logic_1164.all;
USE ieee.numeric_std.all;
use ieee.std_logic_unsigned.all;
use IEEE.STD_LOGIC_ARITH.ALL;
entity ser_mot_contr is
Port (m_clk,start: in STD_LOGIC:='0';
sel: in std_logic_vector (1 downto 0):="00";
mot_cont:out std_logic:='0');
end ser_mot_contr;
architecture Behavioral of ser_mot_contr is
signal ymin, ynut,ymax:std_logic:='0';
begin
p0:process(m_clk,start,sel)
variable cnt: integer:=0;
begin
if (start='1') then
if rising_edge(m_clk) then
if (cnt=2) then --Calculate based on on-board clock
ymin<='1';
ynut<='1';
ymax<='1';
cnt:=cnt+1;
elsif (cnt=6) then
ymin<='0';
ynut<='1';
ymax<='1';
```

```
cnt:=cnt+1;
elsif (cnt=15) then
ymin<='0';
ynut<='0';
ymax<='1';
cnt:=cnt+1;
elsif (cnt=20) then
ymin<='0';
ynut<='0';
ymax<='0';
cnt:=cnt+1;
elsif (cnt=22) then
ymin<='0';
ynut<='0';
ymax<='0';
cnt:=2;
else
cnt:=cnt+1;
end if;
end if;
end if;
case sel is
when "01" => mot_cont<=ymin;
when "10" => mot_cont<=ynut;
when "11" => mot_cont<=ymax;
when others=>null;
end case;
end process;
end Behavioral;
```

9.3.2 Brushless Direct Current Motor Control

BLDC motors have more satisfying results compared to brushed DC motors. The basic difference between them is that in a BLDC motor, the rotor itself contains permanent magnets, and the electromagnets move to the stator, which is the opposite of a brushed DC motor. A picture of a BLDC motor is shown in Figure 9.11a. BLDC motors are more precise and their speed can be factored into equations [118]. BLDC motors are more efficient as there is no sparking, less electrical noise and no brushes to wear out. They are superior to brushed DC motors in many ways, such as the ability to operate at high speeds, high efficiency, feedback-based control, constant torque and better heat dissipation. The major applications of BLDC motors are actuator drives, machine tools, electric propulsion, robotics, computer peripheral driving, and electrical power generation. With the development of sensorless technology in addition to digital control, these motors have become very effective in terms of total system cost, size and reliability. We can have more electromagnets (stator), as shown in Figure 9.11b–d, for more precise control. The only disadvantage of a BLDC motor is its higher initial cost, which can be recovered within a short period of time because of its more efficient operation. A BLDC motor is a permanent magnet synchronous electric motor which is driven by DC electricity and has an electronically controlled commutation system.

BLDC motors are also referred to as trapezoidal permanent magnet motors [118]. In BLDC motors, a permanent magnet rotates and current-carrying conductors are fixed. The armature coils are switched electronically by the specific sequence, as shown in Figure 9.12a–d, through transistors or silicon controlled rectifiers, as shown in Figure 9.13, at the correct rotor position in such a way that the armature field is in space quadrature with the rotor field poles [118]; hence, the force acting on the rotor causes it to rotate. Hall sensors, as

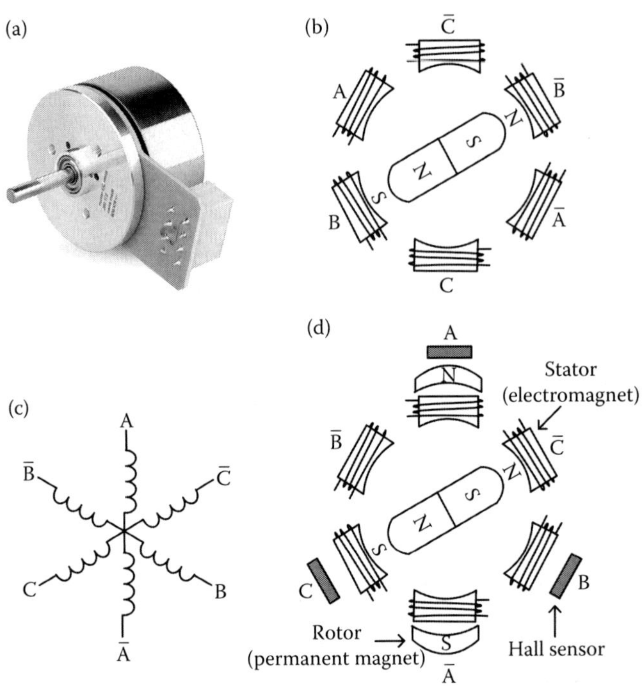

FIGURE 9.11
Picture and coil winding of a BLDC motor (a–d).

indicated in Figure 9.11d, or rotary encoders are positioned around the stator and used to sense the current position of the rotor. Rotor position feedback from the sensor helps to determine when (or from where) to switch the armature current. As shown in Figure 9.11b–d, the coil wiring consists of four connections: A, B, C and, obviously, a common, and every phase is divided into two equally separated windings. Six electrical commutations (A, A′, B, B′, C, C′) allow one full rotation of the BLDC motor. The very slow stepping mode of the BLDC motor does not deliver maximum torque to the motor shaft, and after a certain value, the speed can no longer be increased.

Now, we discuss the working principles of the BLDC motor. The rotor and stator of a BLDC motor are shown in Figure 9.11d. It is clearly shown that the rotor of the BLDC motor is a permanent magnet. The stator has a coil arrangement, as illustrated in the same figure. By applying DC power to the coil, the coil will energize and become an electromagnet. The operation of a BLDC is based on the simple force interaction between the permanent magnet and the electromagnet. In this condition, when the coil A is energized, the opposite poles of the rotor and stator are attracted to each other. As a result, the rotor poles move closer to the energized stator. For example, as in Figure 9.11b, if the rotor nears coil A, coil B is energized, and as the rotor nears coil B, coil C is energized. After that, coil A is energized with the opposite polarity with the switching sequence shown in Figure 9.12a. This process is repeated, and the rotor continues to rotate. The DC current, that is, switching sequence, required in each coil is shown in Figure 9.12a. If the coil arrangement is as shown in Figure 9.11c,d, that is, A, C′, B, A′, C, B′, A, then its corresponding switching sequence is slightly different, as shown in Figure 9.12b. However, in this mode of operation, at any instant, only one coil is individually energized. That means other coils are not contributing to rotating the rotor, which gives relatively low torque. This issue can be overcome by

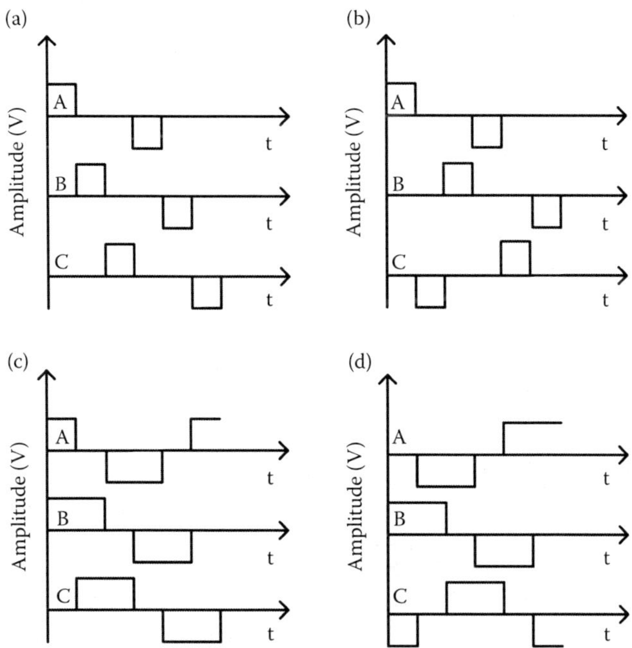

FIGURE 9.12
Patterns of BLDC motor switching sequence (a–d).

simultaneously energizing another appropriate coil. To do this, a small modification to the stator coil is required, as shown in Figure 9.11c,d, that is, simply connect the free ends of the respective coils together. When the rotor is in a position along with the first coil, which pulls the rotor, we can energize the coil behind it in such a way that it will push the rotor. The combined effect produces more torque and higher power output from the motor. The current (switching sequence) required for the BLDC configuration shown in Figure 9.11b and d to form a complete 360° rotation is shown in Figure 9.12c,d and given in Table 9.2. With this configuration, two coils can be energized separately to get more torque from the motor.

Though the source of supply is DC, switching generates an AC voltage waveform with a trapezoidal shape, as shown in Figure 9.12c,d. Consider the switching circuit shown in Figure 9.13, in which the motor stator is excited based on different switching inputs. With

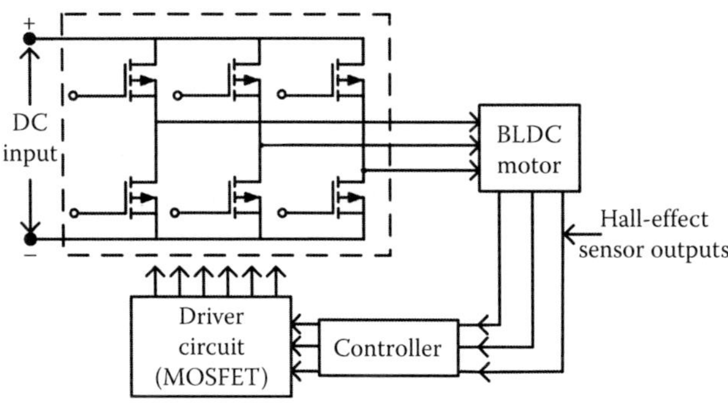

FIGURE 9.13
One of the BLDC motor switching circuits.

TABLE 9.2

BLDC Motor Switching Sequence

Coil Arrangement as in Figure 9.11b			Coil Arrangement as in Figure 9.11c		
A	B	C	A	B	C
+	+	0	0	+	–
0	+	+	–	+	0
–	0	+	–	0	+
–	–	0	0	–	+
0	–	–	+	–	0
+	0	–	+	0	–

the switching of windings as logic 1 and logic 0 signals, the corresponding winding is energized as north (N) and south (S) poles. The permanent magnet rotor with north and south poles aligns with the stator poles, causing the motor to rotate. Observe that the motor produces torque because of the development of attraction and repulsion forces. In this way, the motor moves in a clockwise direction.

This is because the motor's continuous rotation depends on the switching sequence around the coils. As discussed above, Hall sensors give shaft position feedback to the electronic controller unit, as shown in Figure 9.11d. Based on this signal from the sensor, the controller decides which particular coils to energize. Hall-effect sensors generate logic 0 and logic 1 signals whenever rotor poles pass near them. These signals determine the position of the motor shaft. The BLDC motor is driven as described above when the electronic controller circuit energizes appropriate motor winding by turning the transistor or other solid-state switches to rotate the motor continuously. Figure 9.13 shows a simple BLDC motor driver which consists of a MOSFET bridge circuit. The controller receives signals from the Hall sensors, processes them and sends the suitable control signals to the MOSFET driver circuit to drive the BLDC motor. In addition to the switching operation, the controller can change the motor speed based on applications for which it is designed or user inputs. The start switching sequence has to be identified by the controller based on the Hall sensor outputs.

DESIGN EXAMPLE 9.6

Let us assume that a BLDC motor is configured with the FPGA. Develop a digital system to operate that motor in a clockwise direction at a predetermined speed.

Solution

```
LIBRARY ieee;
USE ieee.std_logic_1164.all;
USE ieee.numeric_std.all;
use ieee.std_logic_unsigned.all;
use IEEE.STD_LOGIC_ARITH.ALL;
entity bldc_mot_contr is
Port (m_clk,start: in STD_LOGIC:='0';
mot_cont:out std_logic_vector(2 downto 0):="000");
end bldc_mot_contr;
architecture Behavioral of bldc_mot_contr is
signal sta,nsta:integer range 0 to 5:=0;
begin
p0:process(m_clk,start)
variable cnt: integer:=0;
```

```
begin
if (start='1') then
if rising_edge(m_clk) then
if (cnt=2) then --20000 for 4MHz on-board clock
sta<=nsta;
cnt:=0;
else
cnt:=cnt+1;
end if;
end if;
end if;
end process;
p1:process(sta)
begin
case sta is
when 0=> mot_cont<="110";
nsta<=sta+1;
when 1=> mot_cont<="011";
nsta<=sta+1;
when 2=> mot_cont<="001";
nsta<=sta+1;
when 3=> mot_cont<="000";
nsta<=sta+1;
when 4=> mot_cont<="000";
nsta<=sta+1;
when 5=> mot_cont<="100";
nsta<=0;
when others=>
nsta<=0;
end case;
end process;
end Behavioral;
```

9.4 Stepper Motor Control

A stepper motor is basically a BLDC motor whose rotor rotates through a fixed angular step in response to the input current pulse. That means the full rotation of the rotor is divided into an equal number of steps, and the rotor rotates through one step for each current pulse. Stepper motors are becoming very popular because they can be controlled directly by computers, microprocessors, microcontrollers, or plain programmable devices like FPGA. All electric motors run by electromagnetism; however, there are some other types of motors that utilize electrostatic forces or piezoelectric effects. Stepper motors are controlled electronically and do not require any costly feedback sensors or control circuits. A picture of a stepper motor is shown in Figure 9.14a. The shaft of the motor moves in distinct steps when electrical control pulses are applied. The current polarity and frequency of the applied pulses determine the direction and speed of the shaft movement [119]. One of the most significant advantages of a stepper motor is that its motion can be accurately controlled even in an open-loop configuration. A stepper motor is a good choice whenever controlled movement is required. They are recommended in applications where we need to control rotation angle, speed, position, and synchronism [119]. The movement created by each pulse is precise and repeatable, which is why stepper motors are so effective for positioning applications.

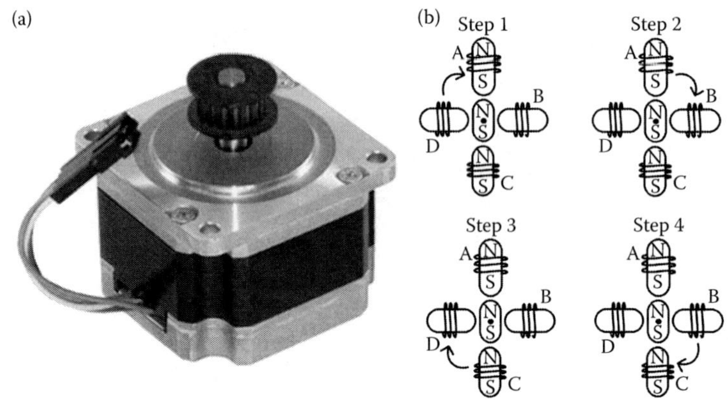

FIGURE 9.14
Picture of a stepper motor (a) and coil winding and shaft rotation of a stepper motor (b).

The magnetic field can be altered by sequentially energizing or stepping the stator coils, which generates rotary motion, as shown in Figure 9.14b, where, in step 1, the rotor is aligned to the stator A. In this condition, stator A is turned off and stator B is energized; hence, stator B becomes an electromagnet of the opposite pole of the rotor, so the rotor is attracted by stator B, that is, the rotor rotates 90° clockwise and is magnetically locked at that stator position. In step 2, stator B is turned off and C is energized; hence, the rotor rotates another 90° towards stator C. Repeating this sequence causes the rotor to rotate clockwise at an angle resolution of 90°. In this case, four steps are required to complete one full rotation, which is called one-phase-on rotation because we energize only one coil at a time. The stepping sequences are given in Table 9.3. A more common method of stepping is half-phase-on, where a maximum of two phases of the motor are energized at a time. In the two-phase-on method, stepping the rotor aligns it between two stators, that is, magnetic poles. For example, if the stators A and B are energized, the rotor will be attracted by A as well as B; hence, it will be aligned to a point between A and B. This means that the rotor rotated only 45°. In this condition, stator A will be turned off and B maintains an on state; hence, the rotor moves towards B, that is, another 45° rotation is achieved. In the same way, we can continue our rotation with a 45° step angle. Here, we need eight steps to complete one full rotation. The energizing sequence for this method is also given in Table 9.3. Rotation can also be achieved by continuously

TABLE 9.3
Data Sequence to Energize Stepper Motor Coils

360° Rotation at an Angle of 90°				360° Rotation at an Angle of 45°			
Yellow (A)	Red (B)	Blue (C)	White (D)	Yellow (A)	Red (B)	Blue (C)	White (D)
1	0	0	0	1	0	0	0
0	1	0	0	1	1	0	0
0	0	1	0	0	1	0	0
0	0	0	1	0	1	1	0
				0	0	1	0
				0	0	1	1
				0	0	0	1
				1	0	0	1

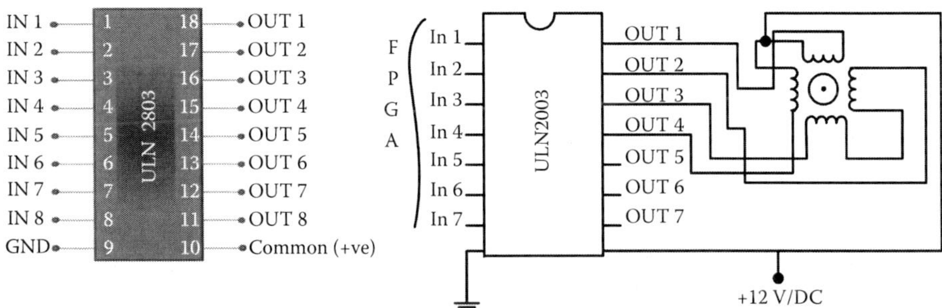

FIGURE 9.15
An application circuit for stepper motor interfacing with the FPGA.

turning on and off the subsequent coils, that is, AB on, BC on, CD, on and DA on, which is called the two-phase-on method and gives more torque than the above one-phase-on method. Since both coils are being energized in this method, it gives about 40% more torque than one-phase-on stepping. The half-stepping method typically results in a 20%–30% loss of torque depending on the step rate when compared to two-phase-on stepping [119]. Since one of the windings is not energized during each alternating half step, there is less electromagnetic force exerted on the rotor, resulting in a net loss of torque.

Stepper motors consist of a permanent magnetic rotating shaft called the rotor and electromagnets on the stationary portion called the stators, as shown in Figure 9.14b. A motor with a resolution of 5° would move its rotor 5° per step, thereby requiring 72 pulses (steps) to complete a full 360° rotation. You may double the resolution of some motors by a process known as half-phase-on. The unique feature of a stepper motor is that the shaft rotates in definite steps, one step being taken each time a command pulse is received. When a definite number of pulses is supplied, the shaft rotates through a definite known angle. The stator poles are magnetized in the appropriate manner by using the FPGA. In general, the stepper motor is driven by Darlington current driver IC ULN2003A due to the current-handling capability of the digital device. A stepper motor interfacing circuit through ULN2003 to the FPGA is shown in Figure 9.15. The stepper motor has six pins, two for power supply and four for control. The two power supply pins are connected together and connected to the power source (12V/DC), while the four control pins are connected to the ULN2003 driver.

The ULN2003A is used to drive the current of the stepper motor, as it requires more than 60 mA of current. It consists of seven pairs of Darlington arrays with common ground and supply provisions, as shown in Figure 9.15. The IC consists of 16 pins in which 7 are for inputs, 7 are for outputs and the remaining ones are V_{CC} and Ground. The first four input pins are connected to the FPGA. In the same way, four output pins are connected to the stepper motor. The stepper motor rotates according to the sequence given from the FPGA. The direction of rotation and rotating speed depend upon the control sequence shifting direction and sequence feed frequency, respectively.

DESIGN EXAMPLE 9.7

Develop a digital system to rotate a four-pole stepper motor. The degree of rotation has to be selected by the user using DIP switches. For example, "00" for 0°, 90°, 180°, 270°, and 360°; "01" for 45°, 135°, 225°, 315°, and 45° and "10" for 0°, 45°, 90°, 135°, 180°, 225°, 270°, 315°, and 360°.

Solution

```
library IEEE;
use IEEE.STD_LOGIC_1164.ALL;
```

```
use IEEE.STD_LOGIC_ARITH.ALL;
use IEEE.STD_LOGIC_UNSIGNED.ALL;
entity stepperc is
Port (m_clk: in STD_LOGIC:='0';
deg_sel : in STD_LOGIC_vector (1 downto 0):="00";
step_out : out STD_LOGIC_VECTOR (3 downto 0):="0000");
end stepperc;
architecture Behavioral of stepperc is
signal clk_sig:std_logic:='0';
signal step_out1: std_logic_vector(3 downto 0):="1000";
signal step_out2: std_logic_vector(3 downto 0):="1100";
signal step_out3: std_logic_vector(31 downto 0):="11000100011000100011000
0110011000";
begin
p0:process(m_clk)
variable cnt:integer :=0;
begin
if rising_edge(m_clk) then
if(cnt=2) then ----20000
clk_sig<=not clk_sig;
cnt:=0;
else
cnt:=cnt+1;
end if;
end if;
end process;
p1:process(clk_sig)
begin
if rising_edge(clk_sig) then
step_out1<=(step_out1(0) & step_out1(3 downto 1));
step_out2<=(step_out2(0) & step_out2(3 downto 1));
step_out3<=(step_out3(27 downto 0) & step_out3(31 downto 28));
if deg_sel="00" then
step_out<=step_out1;
elsif deg_sel="01" then
step_out<=step_out2;
elsif deg_sel="10" then
step_out<=step_out3(31 downto 28);
else
step_out<="0000";
end if;
end if;
end process;
end Behavioral;
```

9.5 Liquid/Fuel Level Control

The family of liquid/fuel level sensing and measurement systems is classified into different categories, such as liquids/solid level measurement, point/continuous level measurement, electromagnetic/electromechanical level measurement, and contacting/noncontacting level measurement. All these measurement techniques basically sense the density, viscosity, vapor mist, dust, chemical composition, interface/gradient, ambient temperature, humidity/moisture, process temperature, and pressure in the regulated or nonregulated environment. A few of the most popular measurement systems that sense liquid levels using float sensors, capacitance sensors, radar sensors, and conductivity probes are

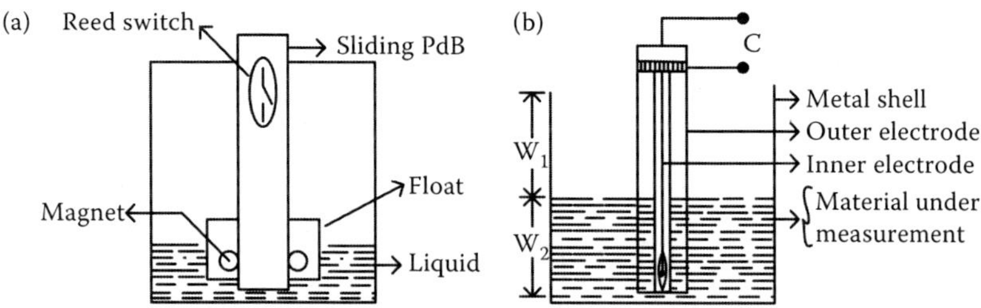

FIGURE 9.16
Liquid level monitoring: using float sensor (a) and capacitive sensor (b).

discussed in this section. A liquid level control system using a float sensor works on the principle of buoyancy [120]. The float sensor gets partially submerged on the liquid surface, as shown in Figure 9.16a, and covers the same distance the liquid level moves. A level measurement float system consists of a float, sensor stem, magnet, and reed switch. The float transfers on a mechanical arm, also called a sliding pole, and activates a reed switch when the level moves in the upward direction. Sometimes, the float itself contains a small magnet that varies the state of a switch when the liquid level moves up and moves into the original position. This type of level sensor comes with many advantages: it is very simple, highly accurate and suitable for various products. The external circuit begins the necessary control action once the reed switch is closed.

Capacitance-level sensors are made available for wide range of solids, aqueous solution, organic liquids, and slurries [120]. This technique is frequently described as radio-frequency signals applied to a capacitance circuit. The capacitive sensors are designed to sense material with dielectric constants as low as 1.1 for coke and fly ash, and as high as 88 for water or other liquids [120]. The principle of capacitive level measurement is based on the change of capacitance. There are two plates in a capacitive sensor: one acts as an insulated electrode and the other acts as a tank wall, as shown in Figure 9.16b. The capacitance depends on the liquid level. An empty tank has low capacitance, while a filled tank has higher capacitance. A simple capacitor consists of two electrode plates separated by a small distance of the insulation material such as solid, fluid, gas or vacuum. The value of capacitance (C) depends on the dielectric constant used, the area of the plate and the distance between the plates, $C = E(KA/d)$, where C is the capacitance in pF, E is a constant known as the absolute permittivity of free space, K is the relative dielectric constant of the insulating material, A is the effective area of the conductors, and d is the distance between the conductors. This change in capacitance can be measured by using an AC bridge. The measurement of the liquid level is done by applying an RF signal between the conductive probe and the tank wall. The RF signal results in a very low current which flows through the dielectric process material in the tank from the probe to the tank wall. If the liquid level in the tank drops, then the dielectric constant decreases, which leads to a drop in the capacitance reading as well as a minute drop in current flow. The external circuit detects these changes, processes them and translates them to the display and control data.

Radar level measurement systems are based on the principle of measuring the time required for a microwave pulse and its reflected echo to make a complete return trip between a noncontacting transducer and the sensed liquid level. Then, the transceiver converts this electrical signal into distance/level and presents it as an analog and/or digital signal. The liquid level or distance is $h_l = (ct/2)$, where h_l is height of the liquid, c is the velocity of the radar signal, and t is the round-trip time. Radar signals are transmitted

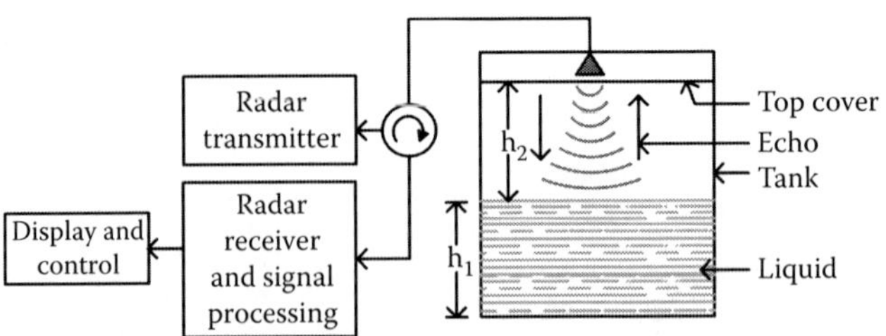

FIGURE 9.17
Liquid level monitoring: using radar sensor.

from an antenna placed at the top of the water tank, as shown in Figure 9.17. The radar signal is reflected by the liquid surface and the echo is carried out by the antenna. By varying the signal, the frequency varies during the time of the echo and the time of the signal transmission comparison. The difference in the arrival time delay is proportional to the distance of the liquid, and this statement is used to determine the accurate level of the liquid. Moreover, these sensors are very sensitive to buildup on the sensor surface. The liquid level sensor operates at a wide range of temperature, pressure and various process conditions. Instead of a radar signal, sonar signals and ultrasonic signals can also be used. The external circuit processes the echo signal and measures the liquid level accurately.

Now, we discuss an application circuit. The LM1830 is a monolithic integrated circuit that can be used for liquid level measurement and control systems. An AC signal is passed through the sensing probe immersed in the fluid/liquid. Usage of an AC signal for detection prevents electrolysis and makes the probes long lasting [120]. The LM1830 is capable of driving an LED, high impedance tweeter or low power relay at its output. A method for activating a relay when the liquid level exceeds a predetermined reference level is shown in Figure 9.18. The DC voltage is required for driving the relay. A capacitor connected from pin 9 to ground in LM1830 will keep the internal output transistor steadily on whenever the probe resistance goes higher than the reference resistor. The load that is the relay is connected at the collector of the external transistor. When the probe is not touching the water, it is equal to an open circuit situation, the probe resistance will be high and it is greater than R_{ref}, typically

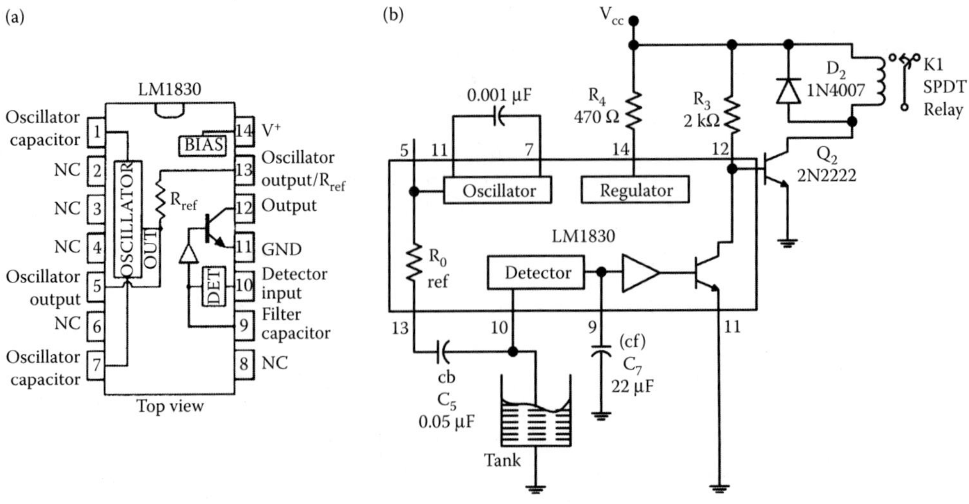

FIGURE 9.18
Pin details of LM1830 fluid lever sensor (a) and a liquid level control application circuit (b).

13 KΩ. The internal transistor will be switched on and the transistor whose base is connected to the collector of the internal transistor will be in the off condition, keeping the relay inactive. When the reverse scenario occurs, that is, the fluid level touches the probe, the internal transistor is switched off, and this in turn turns the transistor on, resulting in activation. The load connected through the relay, whether pump, lamp, alarm, solenoid valve or anything else, is driven. Resistor R_{cb} limits the collector current of the internal transistor, while resistor 470 Ω provides protection to the IC from transients. The probe used here can be any metal rod with size and shape as required. The tank must be made of metal and it should be properly grounded. For nonmetal, tanks fix a small metal contactor at the bottom level and ground it. The probe must be placed at the level you want to monitor. The collector of the external transistor can be connected to any control circuit to initiate the necessary actions once it is driven.

DESIGN EXAMPLE 9.8

Assume that there is a proximity laser sensor at the inner side of the roof of a water tank. The output of the optical receiver increases as the water rises inside the tank. Design and develop a digital system to digitize the output of the optical sensor and display the water level as Low (L), Center (C), and High (H) when the water level is below or equal to $(3)_{10}$, $(3)_{10}$ < water level <=$(200)_{10}$ and above $(200)_{10}$. Display the level-indicating letters L, C, and H on a seven-segment display.

Solution

```
LIBRARY ieee;
USE ieee.std_logic_1164.all;
use ieee.std_logic_unsigned.all;
USE ieee.numeric_std.all;
entity wat_tan is
Port (m_clk,start:in STD_LOGIC:='0';
int : in STD_LOGIC:='1';
rd : out STD_LOGIC:='1';
wr :out std_logic:='0';
adc_data_in:in std_logic_vector(7 downto 0):="00000000";
sev_seg_data_out : out STD_LOGIC_vector(7 downto 0):="00000000");
end wat_tan;
architecture Behavioral of wat_tan is
signal clk_sig1:std_logic:='0';
signal temp_magnitude:integer range 0 to 255:=0;
signal adc_data_temp: std_logic_vector(7 downto 0):="00000000";
begin
p0:process(m_clk,start)
variable cnt1:integer:=1;
begin
if (start='1') then
if rising_edge(m_clk) then
if(cnt1=2)then
clk_sig1<= not(clk_sig1);
cnt1:=1;
else
cnt1:= cnt1 + 1;
end if;
end if;
end if;
end process;
p1:process(clk_sig1,adc_data_in)
variable cnt:integer:=1;
begin
if rising_edge(clk_sig1) then
```

```
if (cnt=1) then
wr<='1';
cnt:=cnt+1;
elsif (cnt=2) then
if int='1' then
cnt:=2;
else
cnt:=cnt+1;
end if;
elsif (cnt=3) then rd<='0'; cnt:=cnt+1;
elsif (cnt=4) then adc_data_temp<=adc_data_in; cnt:=cnt+1;
elsif (cnt=5) then rd<='1'; cnt:=cnt+1;
elsif (cnt=6) then wr<='0'; cnt:=cnt+1;
elsif (cnt=7) then
temp_magnitude<=to_integer(unsigned(adc_data_temp));
cnt:=cnt+1;
if (temp_magnitude <= 3) then
sev_seg_data_out<="11100011";
elsif (temp_magnitude> 3 and temp_magnitude<=200) then
sev_seg_data_out<="01100011";
elsif (temp_magnitude>200) then
sev_seg_data_out<="10000001";
cnt:=cnt+1;
end if;
elsif (cnt=8) then cnt:=1;
end if;
end if;
end process;
end Behavioral;
```

9.6 Voltage and Current Measurement

Transformers are devices that change/transform the voltage of power supplied to meet the needs of consumers and/or industries. They use the principle of electromagnetic induction to change the voltage from one value to another, whether smaller or greater. A transformer is made up of a soft iron coil with two other coils wound around it, but not connected with one another, as shown in Figure 9.19. The iron coils can either be arranged on top of one another or be wound on separate copper core. The coil to which the alternating voltage is supplied is known as the primary winding/coil. The alternating current in the primary winding produces a changing magnetic field around it whenever an alternating potential is supplied. An alternating current is in turn produced by the changing field in the secondary coil, and the amount of current produced depends on the number of windings in the secondary coil. There are two types of transformers, step down and step up. Generally, the difference between them is the amount of voltage produced, depending on the number of secondary coils.

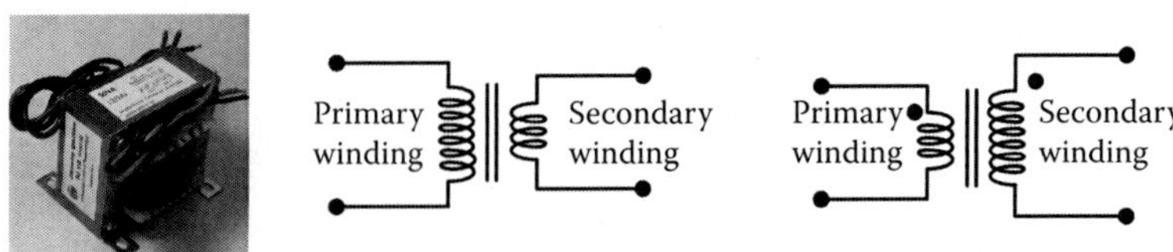

FIGURE 9.19
Simple illustration of step-up and step-down transformers.

The step-down transformer is one with fewer secondary windings (N_2) than primary windings (N_1). In other words, the transformer's secondary voltage (V_{out}) is less than the primary voltage (V_{in}). Therefore, the step-down transformer converts high-voltage, low-current power into a low-voltage, high-current power, and it is mainly used in domestic consumption [121,122]. A step-up transformer is the direct opposite of a step-down transformer. There are more turns on the secondary winding than in the primary winding in step-up transformers. Thus, the voltage in the secondary transformer is greater than that supplied across the primary winding. Because of the principle of conservation of energy, the step-up transformer converts low voltage, high current to high voltage, low current. In other words, the voltage has been stepped up. The type of metal winding used in the transformer is one of the considerations for determining the efficiency of transformers. Copper coils are more efficient than many other metal choices like aluminum. However, copper windings tend to cost more, but we can expect to save the initial cost over time, as the efficiency of the material will save on electrical cost [121,122]. The relationship between the input and output voltage and the number of turns in each coil is given by $(V_{out}/V_{in}) = (N_2/N_1)$. For example, if $V_{in} = 10$ V, then $V_{out} = 0.996$ V if the input is reduced by 10 times and the secondary current is approximately 10 times greater. That means the transformer steps the voltage down by a factor of 10 and steps down the current by a factor of 10. A turn ratio of 10:1 yields 10:1 primary:secondary voltage ratio and 1:10 primary:secondary current ratio.

This is a very useful device; indeed, we can easily multiply or divide the voltage and current in AC circuits. Transformers are also used in long-distance transmission of electric power, as stepping up the voltage and stepping down the current to reduce the wire's resistance power losses along power lines connecting generating stations with loads. At load end, voltage levels are reduced by transformers for safer operation and less expensive equipment. The step-down property of a transformer gives birth to designing a voltage measurement system, using it as shown in Figure 9.20. A half-wave rectifier diode and a capacitive AC filter are connected at the secondary terminal of the transformer. A rheostat and load resistance (R_L) are connected across the capacitor. The rheostat is adjusted so as to get 5 V DC for the maximum possible input at the primary terminals of the transformer. Therefore, the voltage across the R_L depends the input voltage, that is, V_{in}. If it is equal to the maximum value, the output will be 5 V; otherwise, the output will linearly go down. The output voltage is connected to an ADC, and the digital value will be processed and displayed. The accuracy and range of measurement depends upon the type of ADC we use.

Next, we discuss current measurement. A current transformer is a device that is used for the transformation of current from a higher value into a proportionate current to a lower value. It transforms a high-voltage current into a low-voltage current such that the heavy current flows through transmission lines and is safely monitored by an ammeter or other digital measurement system. A current transformer is used with an AC instrument,

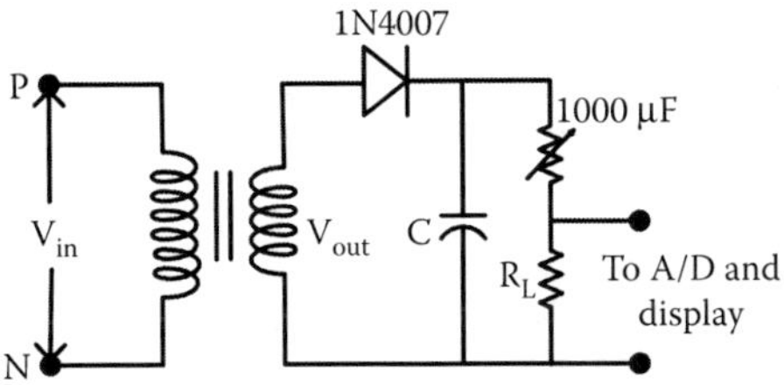

FIGURE 9.20
Simple voltage measurement circuit.

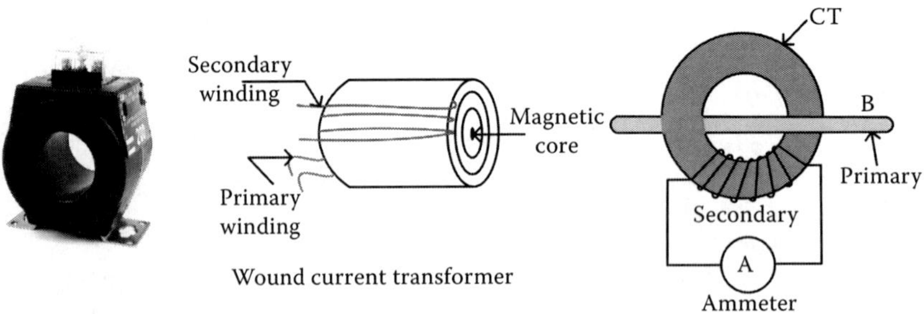

FIGURE 9.21
Illustration of current transformers.

meter or control apparatus when the current to be measured is of such magnitude that the meter or instrument coil cannot conveniently be made of sufficient current carrying capacity. A current transformer is shown in Figure 9.21. The primary and secondary currents of current transformers are proportional to each other. A current transformer is used for measuring high-voltage current because of the difficulty of inadequate insulation in the meter itself. The current transformer is used in meters for measuring current up to 100 A. The core of the current transformer is built up with lamination of silicon steel. For a high degree of accuracy, Permalloy or Mumetal is used for making cores [121,122]. The primary windings of current transformers carry the current to be measured, and it is connected to the main circuit. The secondary windings of the transformer carry current proportional to the current to be measured, and it is connected to the current windings of the meters or the instruments. The primary and secondary windings are insulated from the cores and each other. The primary winding is a single-turn winding and carries the full load current. The secondary winding of the transformer has a large number of turns.

The ratio of the primary current and the secondary current is known as the current transformer ratio of the circuit. The current ratio of the transformer is usually high. The secondary current ratings are of the order of 5, 1, and 0.1 A. The current primary ratings vary from 10 to 3000 A or more. A picture and different configurations of current transformers are shown in Figure 9.21. In a current transformer, when the current flows through the primary windings, it always flows through the secondary windings, and ampere-turns of each winding are subsequently equal and opposite. The secondary turns will be 1% and 2% less than the primary turns, and the difference is used in the magnetizing core. Thus, if the secondary winding is opened and the current flows through the primary windings, there will be no demagnetizing flux due to the secondary current. The secondary side of the current transformer may never be opened when the primary is carrying the current [121,122]. The current transformer is mainly classified into three types: (i) wound current transformer, (ii) toroidal current transformer, and (iii) bar-type transformer. We will briefly see them below:

i. Wound Transformer: In this transformer, the primary winding is inside the transformer. The primary winding has a single turn and is connected in series with the conductor that measures the current. The wound transformer is mainly used for measuring current from 1 to 100A.
ii. Toroidal Current Transformer: This transformer does not contain primary windings. The line through which the current flows in the network is attached through a hole or a window of the transformers. The major advantage of this transformer

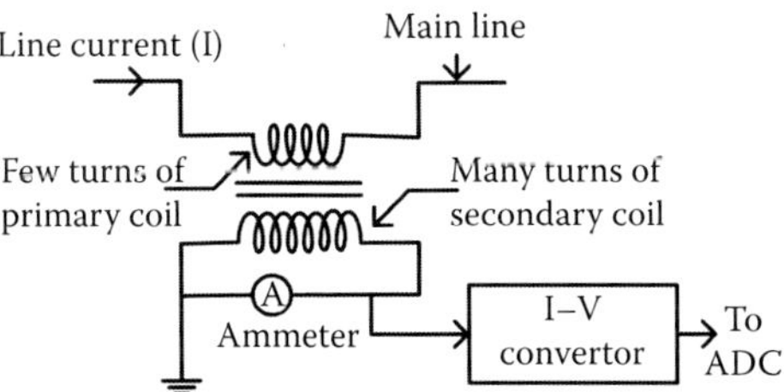

FIGURE 9.22
Simple current measurement circuit.

is that it has a symmetrical shape due to which it has a low leakage flux and thus less electromagnetic interference.

iii. Bar-Type Current Transformer: The bar-type transformer has only secondary windings. The conductor on which the transformer is mounted will act as the primary windings of the current transformer.

A simple application circuit to measure the current is shown in Figure 9.22, where the main load line is routed through the primary coil of the current transformer. Due to current flow in the primary coil, obviously, there is current flow in the secondary side but at a reduced rate because there are more turns in it. This current falls within the limit of the capability of the existing measurement instruments. Therefore, an ammeter of that scale can be directly connected across the secondary coil and the propionate current level can be measured. Instead of directly using an analog ammeter, the current can be converted to equivalent voltage and then processed to display the actual current, that is, line current (I), on a digital display. This type of low-cost measurement system is very useful and accurate for a single-phase power supply measurement which has to be slightly altered for three-phase power supply measurements.

DESIGN EXAMPLE 9.9

Let us assume that one voltage and one current transformer are connected across a load in parallel and serial, respectively, so that their outputs will represent the load voltage in voltage and load current, also in voltage. These two voltages are separately connected to an 8-bit voltage level comparator, for example, LM324. Design and develop a digital system to read the 8-bit comparator outputs and convert them to binary numbers so as to display the measurement values on two separate seven-segment displays.

Solution

```
LIBRARY ieee;
USE ieee.std_logic_1164.all;
use ieee.std_logic_unsigned.all;
USE ieee.numeric_std.all;
entity V_I_meas is
Port (start:in STD_LOGIC:='0';
V_in, I_in : in std_logic_vector(7 downto 0):="00000000";
V_seg,I_seg : out STD_LOGIC_vector(3 downto 0):="0000");
end V_I_meas;
architecture Behavioral of V_I_meas is
begin
p0:process(start,V_in,I_in)
begin
if (start='1') then
case V_in is
```

```
when "00000000"=> V_seg<="0000";
when "00000001"=> V_seg<="0001";
when "00000011"=> V_seg<="0010";
when "00000111"=> V_seg<="0011";
when "00001111"=> V_seg<="0100";
when "00011111"=> V_seg<="0101";
when "00111111"=> V_seg<="0110";
when "01111111"=> V_seg<="0111";
when "11111111"=> V_seg<="1000";
when others=> null;
end case;
case I_in is
when "00000000"=> I_seg<="0000";
when "00000001"=> I_seg<="0001";
when "00000011"=> I_seg<="0010";
when "00000111"=> I_seg<="0011";
when "00001111"=> I_seg<="0100";
when "00011111"=> I_seg<="0101";
when "00111111"=> I_seg<="0110";
when "01111111"=> I_seg<="0111";
when "11111111"=> I_seg<="1000";
when others=> null;
end case;
end if;
end process;
end Behavioral;
```

9.7 Power Electronic Device Interfacing and Control

Most homes, offices, factories, and other buildings are not powered by DC power supply or batteries. Wall outlets are supplied with AC of 40 or 50 Hz. We cannot drive DC motors directly with AC power; therefore, using a DC motor in this situation is almost impossible. If we want to run a motor from our household AC electricity supply, we need to do some modifications in the design approaches of the DC motor. Other than some low-power devices like radios, wall clocks, toys, small robots, flashlights, and so on, we are in need of driving larger loads directly with AC power, for example, AC motors, grinders, water heaters, drilling machines, and so on. Any technique of controlling AC voltage/current/phase is applicable for almost all appliances to operate them at the decided point. A picture of an AC motor is shown in Figure 9.23a. In the AC motor, a ring of electromagnets is used to produce a *rotating* magnetic field. Inside the stator, there is a solid metal axle, loop of wire, coil, squirrel cage, and interconnections which can conduct electricity. Unlike a DC motor, in an AC motor we send power to the outer coils that make up the *stator.* The coils are energized in pairs, in a sequence, and it produces a magnetic field that results in rotation force. The rotor, suspended inside the magnetic field is an electrical conductor. The magnetic field is constantly changing because of its rotation. According to the laws of electromagnetism, the magnetic field produces an electric current inside the rotor. If the conductor is a ring or a wire, then the current flow around it is also in a loop. The induced current produces its own magnetic field and, according to another law of electromagnetism, it tries to stop whatever it is that causes the rotating magnetic field. In synchronous AC motors, the rotor turns at exactly the same speed as the rotating magnetic field, whereas in an asynchronous AC motor, the rotor always turns at a lower speed than

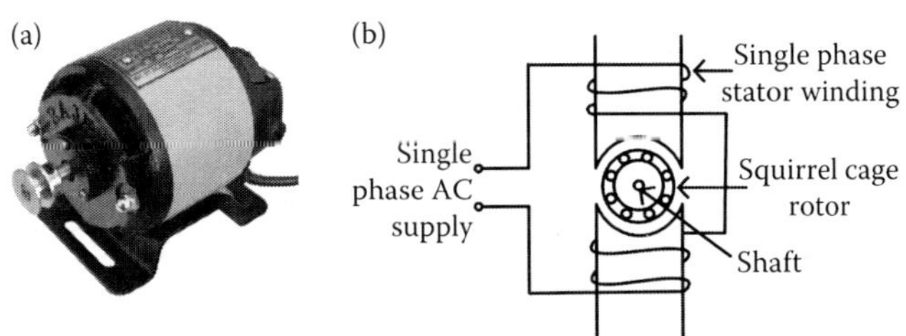

FIGURE 9.23
Picture of an AC motor (a) and simple design circuit of an AC motor (b).

the field [123]. The theoretical speed of the rotor in an induction motor depends on the frequency of the AC supply and the number of coils that make up the stator. With no load, the speed comes close to the speed of the rotating magnetic field. In practice, with load, the speed tends to relatively slow. A greater load increases the slip between the speed of the rotating magnetic field and the actual speed of the rotor.

AC motors designed for single-phase operation, as shown in Figure 9.23b, are used in a wide range of applications where only single-phase supply is the input. As the name suggests, these motors operate on single-phase AC supply. Single-phase induction motors are extensively used for low-power applications, especially in various domestic appliances. In many applications, we need to control the speed of the AC motor. Since this motor is being directly operated with a high-power AC supply, we need some power electronic control/switching devices to control the speed of the AC motors or any AC loads. There are many methods to control the speed of the AC motor. One way of controlling the speed of an AC motor is increasing/decreasing the voltage or frequency of the AC supply. When we adjust the speed of any AC motor, we actually operate a circuit that controls the voltage or frequency being delivered to the motor. A common AC motor speed control in closed-loop configuration is shown in Figure 9.24. The AC supply is applied to the AC motor through the power electronics controller or switching device. The speed of the motor is measured by a suitable measurement system, and it is applied as one of the inputs to a comparator, which compares it with the set or reference speed.

The difference, also called error, is the input to the control signal-generating algorithm, which is typically in computer, FPGA, microcontroller or ASIC chips. There are many algorithms, for example, proportional, integral, and derivative (PID), fuzzy logic controllers, neurocontrollers and model-based controllers [123]. We need to choose a suitable one or their combination to generate the appropriate control signal based on the value of the error signal so as to reduce the error toward zero. The control signal is applied to the controller or switching device, which makes the necessary changes in the supply voltage so as to increase or decrease the speed of the motor equal to the reference value. Since this control configuration is in a closed loop, the controller automatically controls the AC motor speed equal to the reference speed. The time required for the controller to bring the error

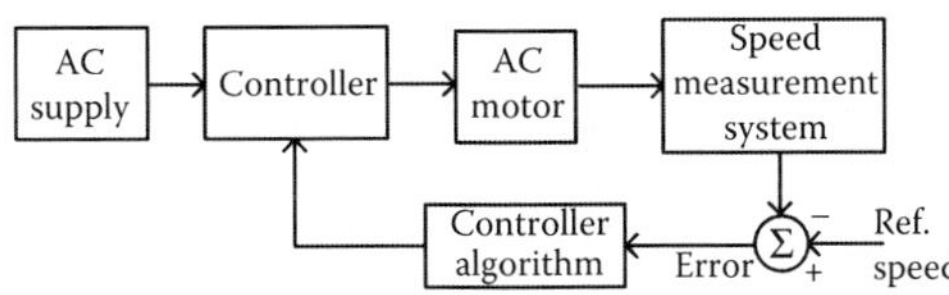

FIGURE 9.24
AC motor speed control in closed-loop control configuration.

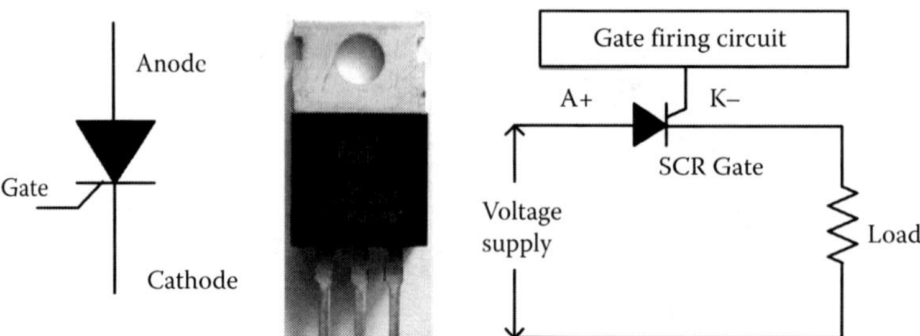

FIGURE 9.25
Electrical symbol, picture and simple application circuit of a SCR.

equal to zero depends upon the efficiency of the control algorithm we use. In this and the next sections, we discuss many power electronic control and switching devices, with the necessary application circuits and design examples.

A popular power electronic control or switching device is a thyristor , also known as an SCR. It is a special type of diode that only allows current to flow when a control voltage is applied to its gate terminal. In the presence of forward current, that is, after the thyristor is turned on by a suitable gate voltage, it will not turn off even after the gate voltage has been removed. The thyristor will only turn off when the forward current drops to zero. The electrical symbols and a picture of a SCR are shown in Figure 9.25. A thyristor is one of several controllable semiconductor devices that can act either like a switch, rectifier or voltage regulator. The thyristor is a solid-state analog of the thyratron vacuum tube.

The name thyristor is a combination of two words, thyratron and transistor. A thyristor functions a little like a transistor. It consists of three terminals, the gate, anode and cathode. The gate acts as the controlling terminal. When a small current flows into the gate, it allows a larger current to flow from the anode to the cathode. A thyristor can be switched from a blocking state, that is, high voltage, low current, to a conducting state, that is, low voltage, high current, by a suitable gate pulse. A simple SCR control circuit is also shown in Figure 9.25. Forward conduction is blocked until an external positive pulse is applied to the gate terminal. A thyristor cannot be turned off from the gate. The thyristor is a four-layer semiconductor device, consisting of alternating P-type and N-type materials (PNPN). The four layers act as bistable switches. As long as the voltage across the device has not reversed, that is, they are forward biased, the thyristor continues to conduct electric current. When the cathode is negatively charged relative to the anode, no current flows until a pulse is applied to the gate. Then, the SCR begins to conduct and continues to conduct until the voltage between the cathode and anode is reversed or reduced below a certain threshold value. Using this type of thyristor, large amounts of power can be switched or controlled using a small triggering current or voltage. The most common use of thyristors is in AC circuits. In an AC circuit, the forward current drops to zero during every cycle so there will always be a turn-off function. This operation means that the gate needs to be triggered every cycle just to turn it on again. Thyristors are also used in motor speed controls, light dimmers, pressure-control systems, and liquid-level regulators. Today, thyristors are manufactured and sold as modules to control up to 570 A [123].

Generally, the gate trigger pulse need to be only a few microseconds in duration, but the longer the gate pulse, the faster the internal avalanche breakdown that occurs in the thyristor, which means the faster the turn-on time. However, the maximum gate current must not be exceeded. In Figure 9.26a, the thyristor is forward biased and is triggered into conduction by briefly closing the push button (S_1), which connects the gate terminal to the

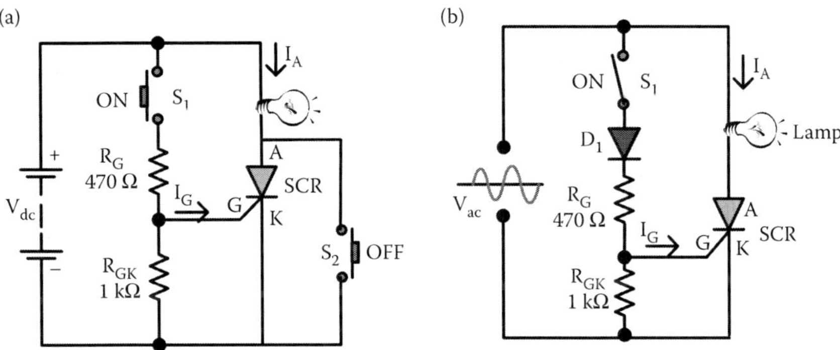

FIGURE 9.26
Simple SCR switching circuit: by DC source (a) and by AC source (b). The loads in both cases are DC and AC lamps respectively.

DC supply via the gate resistor (R_G), thus allowing current to flow into the gate. If the value of R_G is set too high with respect to the supply voltage, the thyristor may not trigger. Once the circuit has been turned on, then additional operations of S_1 will have no effect on the circuit state, as once latched, the gate loses all control. The thyristor is now fully turned on, allowing full load current to flow through the device in the forward direction and back to the battery supply. One of the main advantages of using a thyristor as a switch in a DC circuit is that it has a very high current gain [123]. The thyristor is a current-operated device because a small gate current can control a much larger anode current. The gate-cathode resistor (R_{GK}) is generally included to reduce the gate sensitivity and increase its dv/dt capability, thus preventing false triggering of the device [123]. As the thyristor has self-latched into the on state, the circuit can only be reset by interrupting the power supply and reducing the anode current to below the thyristor minimum holding current (I_H) value. Closing the normally-opened, that is, off, push button (S_2) breaks the circuit, reducing the circuit current flowing through the thyristor to zero and thus forcing it to turn off until again applying another gate signal. As in Figure 9.26b, when connected to an AC supply, the thyristor behaves differently from the previous DC-connected circuit. This is because AC power reverses polarity periodically and therefore any thyristor used in an AC circuit will automatically be reverse biased, causing it to turn off during each negative half-cycle.

Now, we will further discuss AC control operations with a thyristor. Phase control is the most common form of thyristor AC power control, and a basic circuit for phase control is shown in Figure 9.27. Here, a thyristor gate voltage is driven from the RC charging circuit via the trigger diode (D_1). During the positive half-cycle when the thyristor is forward

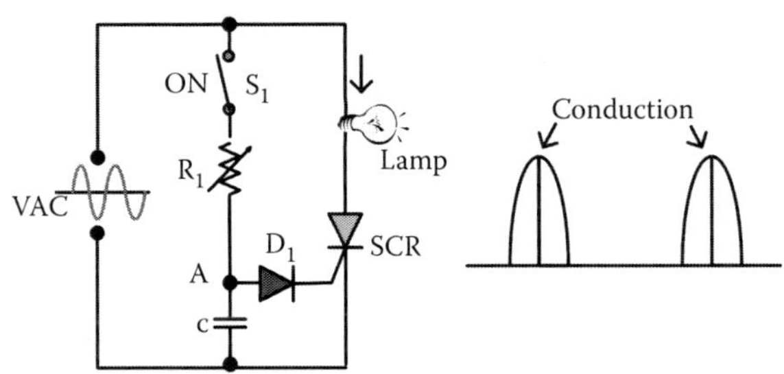

FIGURE 9.27
SCR phase control circuit.

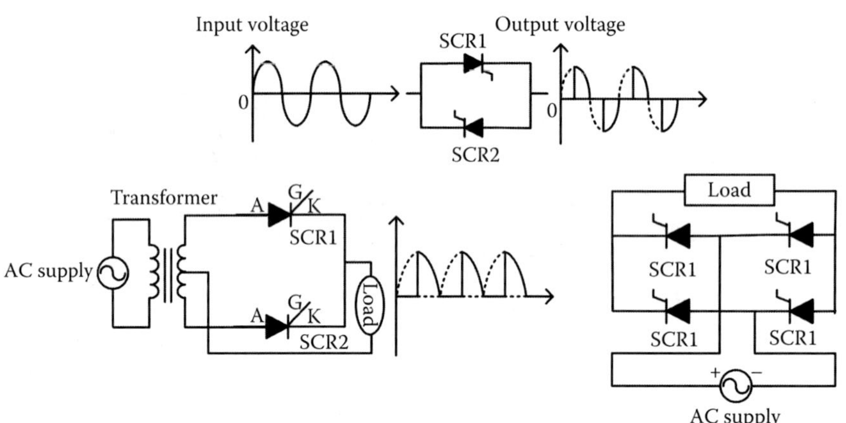

FIGURE 9.28
SCR phase control and its application circuits.

biased, the capacitor (C) charges up via resistor R_1 following the AC supply voltage. The gate is activated only when the voltage at point A has raised enough to trigger the D_1 to conduct and discharge the capacitor into the gate of the thyristor, which turns it on. The time duration in the positive half of the cycle at which conduction starts is controlled by the RC time constant, as shown in Figure 9.27. Increasing the value of R_1 has the effect of delaying the triggering voltage and current supplied to the gate of the thyristor, which in turn causes a lag in the device's conduction time. As a result, the fraction of the half-cycle over which the device conducts can be controlled between 0° and 180°, which means that the average power dissipated by the lamp can be adjusted. However, the thyristor is a unidirectional device; hence, only a maximum of 50% power can be supplied during each positive half-cycle.

SCRs are capable of controlling the power transmitted to the load. It is often required to vary the power delivered to the load depending on the load requirements such as motor speed control and light dimmers. Under such conditions, varying power with conventional adjustable potentiometers is not a reliable method due to the large power dissipation [123]. To reduce power dissipation in high-power circuits, SCRs are the best choice as power control devices. As we discussed previously, in AC circuits, the phase control is the most common form of SCR power control. In phase control, by varying the triggering angle (α) at the gate terminal, power control is obtained. Figure 9.28 shows a full-wave AC control circuit with the phase control method. Consider that the AC supply is given to two antiparallel SCRs, as in Figure 9.28. During the positive half cycle of the signal, SCR_1 conducts, while in the negative half cycle, SCR_2 conducts when proper gate pulses are applied to them. By varying the firing angle to the respective SCRs, the turn-on times are varied. This leads to varying the power consumed by the load. As in Figure 9.28, SCRs triggered at delayed pulses results in a decrease of the power delivered to the load. The main advantage of phase control is that the SCRs are turned off automatically at every current zero position of AC current. Hence, no commutation circuit is required to turn them on and off [123].

An SCR can also be connected as a full-wave rectifier, as shown in Figure 9.28, for high-power AC-to-DC conversion or generating a controlled full-wave rectified wave form. In that circuit, the simple PN junction diodes are replaced with SCRs. SCR_1 and SCR_2 alternatively work for the positive and negative cycles, respectively. By appropriately applying gate control pulses, it is possible to decide the output power. Instead of using a center-tapped transformer, as in Figure 9.28, it is also possible to use four SCRs in a bridge configuration to get a full-wave rectification, as also shown in Figure 9.28. During the positive half cycle of the input, SCR_1 and SCR_2 are in conduction, and during the negative half cycle, SCR_3 and SCR_4 are in conduction. The conduction angle of each thyristor is adjusted by varying the respective gate currents. Hence, the output voltage across the load is varied. Due to the fast switching action of the SCR, it can be employed as a protecting device. The circuit used for protection against overvoltages is referred to as a crowbar circuit [123,124]. As discussed above, SCRs can be used for a wide spectrum of AC current control applications.

DESIGN EXAMPLE 9.10

Design a digital system to control the speed of an AC motor using an SCR. Use a PWM signal to control the firing angle of the AC input signal to run the AC motor at the desired speed.

Solution

```
library ieee;
use ieee.std_logic_1164.all;
use ieee.std_logic_unsigned.all;
entity ctrlmotor is
port(clk,nrst:in std_logic;
setp:in std_logic_vector(7 downto 0); -- set point
fb:in std_logic_vector(7 downto 0); -- feedback
ctrlout:out std_logic); -- control pulse
end entity;
architecture mark0 of ctrlmotor is
-- PWM generator --
component pwmg is
port(clk,nrst:in std_logic;
dc:in std_logic_vector(7 downto 0); -- duty cycle
pwm:out std_logic); -- output pulses
end component;
signal err:std_logic_vector(8 downto 0); -- error signal
signal pulw:std_logic_vector(7 downto 0); -- pulsewidth control
begin
-- generate control pulses at output --
p0:pwmg port map(clk=>clk,nrst=>nrst,dc=>pulw,pwm=>ctrlout);
-- generate error signal --
err <= '0'&setp + (not ('0'&fb)) + '1'; -- error signal(2c subtraction)
process(nrst,err)
begin
if(nrst='0') then
pulw <= x"00";
else
case(err(8)) is
when '1'=> -- error is negative => speed is more => reduce pulse width
pulw <= pulw - 1;
when '0'=> -- error is positive => speed is less => increse pulse
width
if(err/=x"00") then -- if error is 0, stay constant
pulw <= pulw + 1;
end if;
when others=>
pulw <= pulw;
end case;
end if;
end process;
end architecture;
```

9.8 Power Electronics Bidirectional Switch Interfacing and Control

In this section, we discuss two very popular bidirectional power electronic control switches, the Triac and Diac. The pin details, working principles, application circuits, and design examples are explained. Unlike the SCR, these two devices have no specific terminals, which means anodes 1 and 2 can be interchangeable in wiring, since it has the ability of conducting current during both cycles; thus, the control signal can only turn on and off the bidirectional device.

9.8.1 Triac Control

A Triac is a three-terminal bidirectional semiconductor power electronic switching device which can control alternating current in both cycles. Triac is a word made from two words, Tri and AC, where Tri means three and AC means alternating current. It can conduct in both directions, that is, the positive half cycle and negative half cycle, for a specific time period. The symbol, picture and construction of a Triac are shown in Figure 9.29. As in that figure, a Triac is basically made by connecting two SCRs in inverse parallel combination with a common gate. A Triac is almost similar to an SCR, but the Triac is a bidirectional device, whereas the SCR is a unidirectional device like a diode. The internal layers and doping are done in such a way that the current can flow in both directions. The terminals are not called anode and cathode, as we know that a Triac can conduct with both terminals; the terminals are labeled Main Terminal 1 (MT1) or Anode 1 and Main Terminal 2 (MT2) or Anode 2, and the third terminal is the gate, which is common for both SCRs, but internally this gate is closer to MT1 [124].

To understand the working of a Triac, let us consider that we apply alternating current source across the terminals MT1 and MT2. Now, we connect a DC source to the gate through a switch (SW) that is kept open first. Now, when the gate supply switch is open, there is no current flowing through the gate; therefore, the maximum value of alternating current is less than the breakover voltage of the Triac. Therefore, the Triac will remain in the OFF state and not conduct. When we close the SW, the circuit becomes closed and current starts flowing through the gate. Now, the Triac breaks and starts conducting heavily. Thus, it works as a closed switch. However, we can control the firing angle and breakdown voltage of the Triac using gate current.

An application circuit with a Triac is shown in Figure 9.30, where the Triac is used as a simple static AC power switch to turn on and off the current flow to the load. When switch SW_1 is open, the Triac acts as an open switch and the lamp passes zero current. When SW_1 is closed, the Triac is gated on via current-limiting resistor R (100 Ω) and self-latches shortly

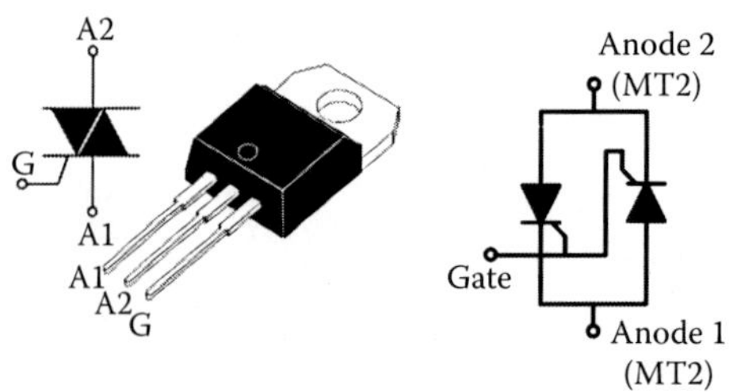

FIGURE 9.29
Symbol, picture and equivalent circuit of a Triac.

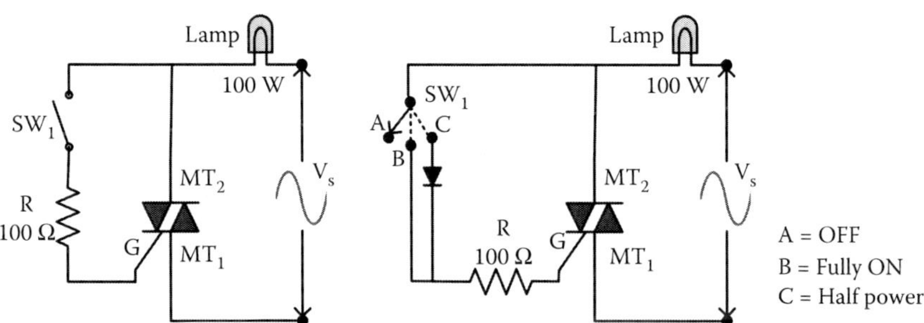

FIGURE 9.30
Application circuits of a Triac.

after the start of each half-cycle, thus switching full power to the lamp load. As the supply is sinusoidal AC, the Triac automatically unlatches at the end of each AC half cycle as the instantaneous supply voltage and thus the load current briefly fall to zero, but relatches again using the opposite thyristor half on the next half cycle as long as the switch remains closed. This type of switching control is generally called full-wave control due to the fact that both halves of the sine wave are being controlled. As the Triac is effectively two back-to-back connected SCRs, we can take the Triac switching circuit further by modifying the method of giving the control signal, as shown in Figure 9.30. As above, if switch SW_1 is open at position A, there is no gate current and the lamp is off. If the switch is moved to position B, gate current flows at every half cycle the same as before and full power is drawn by the lamp. However, this time when the switch is connected to position C, the diode will prevent the triggering of the gate when MT_2 is negative, as the diode is reverse biased [124]. Thus, the Triac only conducts on the positive half cycles and the lamp will light at half power. Then, depending upon the position of the switch, the load is off, at half power or fully on.

Another popular application of the Triac is as an optoisolation relay. An application circuit with the Triac providing the optoisolation is shown in Figure 9.31. These active semiconductor devices use light instead of magnetism to actuate a switch. The light comes from an LED. When control power is applied to the device's input, the light is turned on and shines across an open space. On the load side of this space, a part of the device senses the presence of the light and triggers a solid-state switch that either opens or closes the circuit under control. Often, solid-state relays are used where the circuit under control must protected from the introduction of electrical noises. Advantages of solid state relays include low electromagnetic interference (EMI), long life, no moving parts, no contact

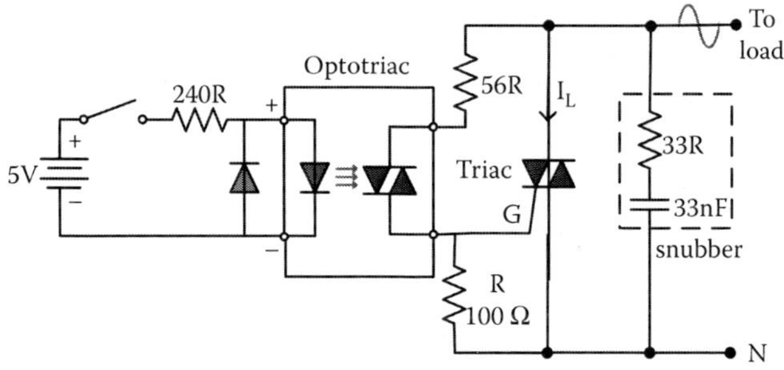

FIGURE 9.31
Triac opto-coupling relay.

bounce, and fast response. The drawback to using a solid-state relay is that it can only accomplish single pole switching. The quantity of electrical current that flows through the contacts directly influences the characteristic of the contacts.

SSRs are a faster alternative to electromechanical relays because the switching time is dependent on the time required to power the LED on and off, approximately 1 and 0.5 ms, respectively. Because there are no mechanical parts, their life expectancy is higher than electromechanical or reed relays. This also makes them less susceptible to physical vibrations. However, the downside is that contact resistance for SSRs is greater because the connection is made via a transistor instead of physical metal like in electromechanical relays. Although technology is continually improving the contact resistance of SSRs, it is still not uncommon to find them in production today with resistances of 100 Ohms or more. SSRs tend to be more expensive than other switches. They also dissipate more heat. SSRs are useful for high-voltage applications and are common on matrices and multiplexers. These optoisolator circuits are typically used for AC power control applications. The noise and voltage spikes (if any) of the AC power supply can be completely isolated using photo-SCRs or photo-Triacs. The thermal stress due to heavy inrush current caused by switching can be avoided by using the zero crossing detection technique. Optoisolator circuits provide high electrical isolation between input terminals and output terminals. Thus, small digitals are allowed for controlling high AC currents, voltages and power. Optocoupled Triacs such as the MOC 3020 have voltage ratings of about 400 volts, making them ideal for direct main connection and a maximum current of about 100 mA. For higher-powered loads, the opto-Triac may be used to provide a gate pulse to another larger Triac via a current-limiting resistor.

DESIGN EXAMPLE 9.11

Develop a digital system to control a heater using a Diac. Let us assume that a high-voltage heating element is powered through a Diac and has to be on until the temperature reaches the set point value, and once the temperature exceeds that value, the power to the heater has to be turned off. The same comparison and control operations have to be continued after some time. Also, log the temperature variation profile in the computer through the RS232 port. Use a single-channel ADC to read the temperature variations.

Solution

```
LIBRARY ieee;
USE ieee.std_logic_1164.all;
use ieee.std_logic_unsigned.all;
USE ieee.numeric_std.all;
entity triac_cont is
Port (m_clk,reset:in STD_LOGIC:='0';
int : in STD_LOGIC:='1';
rd : out STD_LOGIC:='1';
wr :out std_logic:='0';
txd:out std_logic:='0';
adc_data_in:in std_logic_vector(7 downto 0):="00000000";
adc_data_out : out STD_LOGIC_vector(7 downto 0):="00000000";
setpoint_sel: in std_logic_vector(1 downto 0):="00";
diac_ctrl:out std_logic:='0');
end triac_cont;
architecture Behavioral of triac_cont is
signal clk_sig1,clk_sig,txd_temp,mux_sig:std_logic:='0';
signal adc_data_temp: std_logic_vector(7 downto 0):="00000000";
signal sdata:std_logic_vector(10 downto 0):="11000000000";
signal setpoint:std_logic_vector(7 downto 0):="00000000";
```

```
signal temp_magnitude:integer range 0 to 400:=0;
begin
p0:process(m_clk)
variable cnt1:integer:=1;
begin
if rising_edge(m_clk) then
if(cnt1=2)then
clk_sig1<= not(clk_sig1);
cnt1:=1;
else
cnt1:= cnt1 + 1;
end if;
end if;
end process;
p1:process(clk_sig1,adc_data_in)
variable cnt:integer:=1;
begin
if rising_edge(clk_sig1) then
if (cnt=1) then wr<='1';cnt:=cnt+1;
elsif (cnt=2) then
if int='1' then cnt:=2;
else cnt:=cnt+1;
end if;
elsif (cnt=3) then rd<='0'; cnt:=cnt+1;
elsif (cnt=4) then adc_data_temp<=adc_data_in; cnt:=cnt+1;
elsif (cnt=5) then rd<='1'; cnt:=cnt+1;
elsif (cnt=6) then wr<='0'; cnt:=cnt+1;
elsif (cnt=7) then adc_data_out<=adc_data_temp;cnt:=cnt+1;
elsif (cnt=8) then
if (setpoint > adc_data_temp) then
diac_ctrl<='1';
else
diac_ctrl<='0';
end if;
cnt:=cnt+1;
elsif (cnt=20) then
cnt:=1;
else
cnt:=cnt+1;
end if;
end if;
end process;
p2:process(m_clk)
variable cnt1:integer:=1;
begin
if rising_edge(m_clk) then
if(cnt1=208)then
clk_sig<= not(clk_sig);
cnt1:=1;
else
cnt1:= cnt1 + 1;
end if;
end if;
end process;
p3:process(clk_sig,adc_data_temp)
variable cnt2:integer:=1;
begin
if rising_edge(clk_sig)then
if (cnt2>10) then
mux_sig<='1';
if (cnt2=11) then
```

```
sdata<=(sdata(0)& sdata(10 downto 1));
cnt2:=cnt2+1;
end if;
if (cnt2=16) then
sdata(8 downto 1)<=adc_data_temp;
mux_sig<='0';
cnt2:=1;
else
cnt2:=cnt2+1;
end if;
else
mux_sig<='0';
sdata<=(sdata(0)& sdata(10 downto 1));
cnt2:=cnt2+1;
end if;
case mux_sig is
when '0'=>txd_temp<=sdata(0);
when '1'=>txd_temp<='1';
when others=> null;
end case;
end if;
end process;
txd<=txd_temp;
p4:process(setpoint_sel)
begin
case setpoint_sel is
when "00"=> setpoint<=x"00";
when "01"=> setpoint<=x"78";
when "10"=> setpoint<=x"79";
when "11"=> setpoint<=x"C0";
when others=> null;
end case;
end process;
end Behavioral;
```

9.8.2 Diac Controller

The DIode Alternating Current (DIAC or Diac) is a full-wave and bidirectional semiconductor switch that can be turned on in both forward and reverse polarities. The Diac gains its name from the contraction of the words Diode and AC. The Diac is widely used to assist even triggering of a Triac when used in AC switches. Diacs are mainly used in dimmer applications and also in starter circuits for florescent lamps [125]. The Diac circuit symbol is generated from two triangles held between two lines, as shown in Figure 9.32. The two terminals of the device are designated as Anode 1 or MT_1 and Anode 2 or MT_2. The

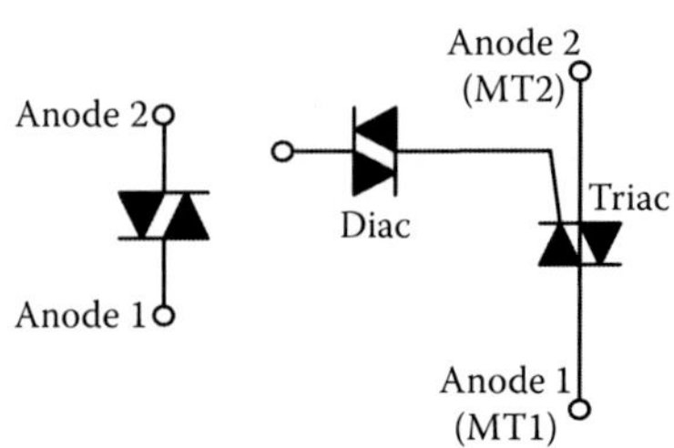

FIGURE 9.32
Symbol and general configuration of a Diac.

Diac is essentially a diode that conducts after a breakover voltage is exceeded. When the device exceeds this breakover voltage, it enters the region of negative dynamic resistance. This results in a decrease in the voltage drop across the diode with increasing voltage. Accordingly, there is a sharp increase in the level of current that is conducted by the device. The diode remains in its conduction state until the current through it drops below the holding current, and below the holding current, the Diac reverts to its high-resistance, that is, nonconducting, state. The Diac is a power electronic bidirectional switch, therefore, its operation occurs on both halves of an alternating cycle. Typically, the Diac is placed in series with the gate of a Triac because these devices do not fire symmetrically as a result of slight differences between the two halves of the device. This results in harmonics being generated, and the less symmetrically the device fires, the greater the level of harmonics produced. It is generally undesirable to have high levels of harmonics in a power system. To overcome this problem, a Diac is often placed in series with the gate [125]. This device helps to make the switching more even for both halves of the cycle. This results from the fact that its switching characteristic is far more even than that of the Triac. Since the Diac prevents any gate current flowing until the trigger voltage has reached a certain voltage in either direction, this makes the firing point of the Triac more even in both directions.

Now, we will discuss some of the application circuits. Wiring of Diacs as a trigger device to provide smooth control to the AC load is shown in Figure 9.33. We see the working principles of the heater control application. The load current (LC) combination across the Triac reduces the rate of rise of voltage during the turn-off of the Triac. The positive and negative half cycle of the input voltage to the heater is controlled by adjusting the resistance R_2. For all variable positions of R_2, a smooth control is ensured by placing resistance R_4 across the Diac. Diacs, because of their symmetrical bidirectional switching characteristics, are widely used as triggering devices in Triac phase control circuits. Although a Triac may be fired into the conducting state by a simple resistive triggering circuit, triggering devices are typically placed in series with the gates of SCRs and Triacs, as they give reliable/fast triggering.

The capacitor C_1 in series with choke L across the Triac slows up the voltage rise across the device during the off state. The resistor R_4 across the Diac ensures smooth control at all positions of potentiometer R_2. The Triac conduction angle is adjusted by adjusting the potentiometer R_2. The longer the Triac conducts, the larger the output will be from the heater. Thus, smooth control of the heat output from the heater is obtained. The Triac triggering voltage is derived from the VR_1–C_1 combination via the Diac. At the start of each cycle, C_1 charges up via the variable resistor (VR_1). This continues until the voltage across C_1 is sufficient to trigger the Diac into conduction, which in turn allows the capacitor (C_1) to discharge into the gate of the Triac, turning it on. Once the Triac is triggered into

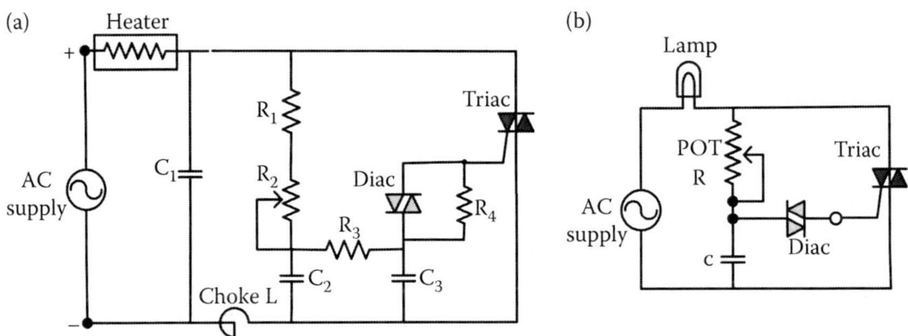

FIGURE 9.33
Application circuits with a Diac.

conduction and saturates, it effectively shorts out the gate, triggering the phase control circuit connected in parallel across it, and the Triac takes control for the remainder of the half cycle. As we have discussed above, the Triac turns off automatically at the end of the half cycle, and the VR_1–C_1 triggering process starts again on the next half cycle. However, because the Triac requires differing amounts of gate current in each switching mode of operation, a Triac is therefore asymmetrical, meaning that it may not trigger at the exact same point for each positive and negative half cycle. In the same way, AC lamp brightness can also be controlled using the circuit shown in Figure 9.33. The resistor (R) and capacitor (C) decide/discharge the charging time. A longer Triac conducting time increases the brightness of the lamp. The Diac acts like an open circuit until the voltage across the capacitor exceeds its breakover or switching voltage.

DESIGN EXAMPLE 9.12

Let us assume that several decoration lamps are connected in four separate wires and laid over a long wooden piece. The phase of each wire is connected through a separate Diac switch. Design and develop a digital system to operate all the lamps one row by one row in a predetermined sequence.

Solution

```
LIBRARY ieee;
USE ieee.std_logic_1164.all;
use ieee.std_logic_unsigned.all;
USE ieee.numeric_std.all;
entity diac_cntl is
Port (m_clk,start:in STD_LOGIC:='0';
diac_ctrl:inout STD_LOGIC_vector(3 downto 0):="1000");
end diac_cntl;
architecture Behavioral of diac_cntl is
begin
p0:process(start,m_clk,diac_ctrl)
variable cnt: integer:=0;
begin
if (start='1') then
if rising_edge(m_clk) then
if (cnt=2) then
diac_ctrl<=(diac_ctrl(0) & diac_ctrl(3 downto 1));
cnt:=0;
else
cnt:=cnt+1;
end if;
end if;
end if;
end process;
end Behavioral;
```

9.9 Real-Time Process Controller Design

Almost all automation and control systems are designed, developed and built using data computer software, microprocessors, DSPs, or any other advanced plain processing device like the FPGA. Real-time processing and fundamentals of real-time software design applying to automation and control systems are significantly growing up today. In many real-time applications, controllers play a vital role in performing certain tasks per the order we desire. We should have many prerequisites for designing an optimal controller to meet the necessary control actions at the plant. The design of controller requires knowledge of at least two basic components such as the plant, which will perform the task, and the output, what we expect. Real-time control systems are closed-loop control systems, where we have a tight time window to gather data, process that data, and update the system. If the time window is missed, then the stability of the system is degraded [126]. This reduced control can be catastrophic to some applications, such as power conversion, position control, computer numerical control (CNC) machine control, advanced motor speed control and so on. In general, the ADC reads information such as voltage and current and converts that information from the analog domain to the digital domain and then passes this information to the control engine. The ADC needs to have high resolution and needs to be able to convert an analog signal to a digital signal as fast as possible. One important feature is dual sample and hold, where both voltage and current can be sampled at the same time. In many products, such as inverters, dual sample and hold is important so that multiple voltages/currents can be sampled. Most high-power controls are accomplished with PWM control signals; hence, they have to be flexible and high resolution. PWMs in most cases control the output voltage or current in the embedded application. PWM flexibility is important in order to support a wide range of power topologies that enable different efficiencies and support for high resolution and to enable higher switching frequencies, which reduces the size/cost of the magnetic board [126].

A typical closed-loop control configuration is shown in Figure 9.34. The control system design consists of (i) controlled device, (ii) controlled element, (iii) controller, (iv) set point, (v) measuring element, and (vi) process or plant. As shown in Figure 9.34, the entire operation is in a closed loop; hence, the control system takes care of automatically adjusting the control parameters to get the expected or set point values. We briefly see the operation of this control scheme: the user has to set the reference values (set points) for the parameters based on which the controller has to work. Once the setpoint setting is over, the controller manipulates its outputs and passes into the controlled element, which activates a few parts of the system via controlled device. The response of the controlled device tunes the entire

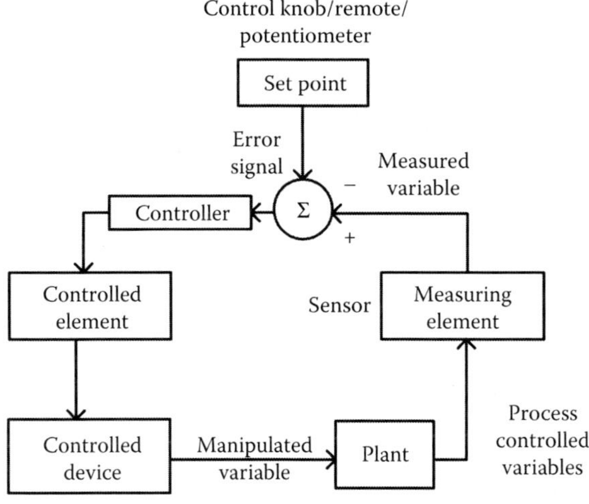

FIGURE 9.34
Schematic diagram of a closed-loop control system.

plant in accordance to the set points. This setting is measured at a regular discrete interval and fed back into the system. The controller reaches the steady state once the measurement results accurately or more closely match the set points. Typically, we need to control many physical parameters such as air, current, voltage, pressure, humidity, actuators, temperatures, stepper motor, AC/DC/servo/BLDC motor, and so on.

Therefore, designing a control chain becomes complex, plus each of the components involved has its own characteristics, and changing any single component will change the overall response of the control loop. The main variables more relevant to real-time control action are (i) controlled variables, (ii) measured variables, (iii) setpoints, (iv) error signals, (v) controller output, and (vi) manipulated variables, as shown in Figure 9.34, and all these variables are briefed below with a simple water tank level control example:

i. *Controlled variable*: The controlled variable is a process parameter being controlled. In the tank level control example, the water level is the controlled variable; however, it can be any process we wish to control. Controlling this variable is the primary function of process control.
ii. *Measured variable*: The measured variable is the electronic or pneumatic representation of the value of the controlled variable. The measured variable typically comes from a transmitter which measures the controlled variable and produces an output representative of it. In the tank level control example, the level transmitter measures the level in the tank (the controlled variable) and converts that level to a voltage or current signal.
iii. *Setpoint*: The measured variable is sent to the controller in the loop, where it is compared to a desired value called the setpoint. The setpoint and the measured variable are compared in order to produce an error signal. The setpoint is often manually entered by an operator, but it can also be automatically obtained from other systems. In the tank level control example, deciding the water level to be filled in the tank is the setpoint.
iv. *Error signal*: The error signal is the difference between the measured level and the setpoint. It can be either a negative or positive value. The error signal is then added to the base signal level of the controller to create the controller output. In the tank level control example, the error signal is the difference in between measured and setpoint values.
v. *Controller output*: The controller output is simply the total output of the controller. With the controller in automatic, the output is calculated by the controller itself. If the controller is placed in manual, the output can be manually adjusted to any desired position. In the tank level control example, the water supply rate is the controller output.
vi. *Manipulated variable*: The manipulated variable is the parameter that is adjusted to bring the process back to desired setpoint. The manipulated variable in this case is the water entering the tank. The water flow is manipulated in order to keep the level constant as the output demand changes.

The special paradigm of any controller design is closely related to many concepts involving sensors, signal acquisition, control set points and other aspects mostly related to process state variables, measurements, and control. Automation controllers are required to be real-time systems because they must control physical processes or plants that demand real-time control. For any physical plant to be governed, it requires a controller capable of

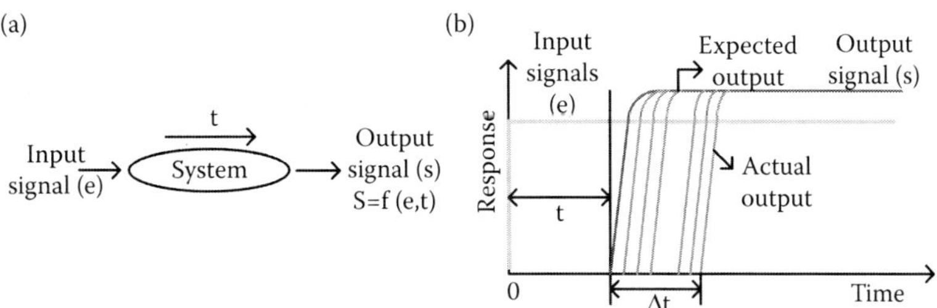

FIGURE 9.35
Illustration of: a simple control system (a) and its response over time (b).

acquiring some input signals and producing some specific output signals within a very specific time frame [126]. Suppose there is a very simple system with just one input and one output, as shown in Figure 9.35a. Such a system generates an output signal every time it receives an input signal. As in Figure 9.35b, the output signal is produced t seconds after the input signal is introduced. Now, suppose that we repeatedly, but not periodically, apply the same input signal to this system. We may expect that this system will generate the same output signal every time, at exactly t seconds after the input signal is applied. Unfortunately, this is not what happens in actual controllers [126]. The same output signal will be generated by the controller every time, but the time t that it takes the controller to produce each output may increase slightly, as shown in Figure 9.35b. If we repeat this experiment, we will find that the controller response times fall into a variation interval, say, Δt sec. It strongly depends on how the controller hardware was designed, which components were used and how the application software is running inside [126]. It is important to point out that no controller system can reach $\Delta t = 0$. Therefore, we need to carefully look all these aspects while designing any real-time controller.

DESIGN EXAMPLE 9.13

Let us assume a smoke chamber where we wish to perform various chemical processing at different levels of smoke. A smoke sensor is used in the chamber to sense the smoke level through a single-channel ADC. There are control input and output devices like read switch, door open/close control, solenoid valve, gas control valve and buzzer control circuit. All these controls getting input from the read switches, opening/closing the doors, applying the water, controlling the gas flow and buzzer control have to be performed at different instants of time as required. Log the profile of the smoke in a computer for the post-data-processing applications.

Solution

As stated in this design objective, we assume various operations at different instants of time (smoke level), as used in the below code. The scheduling of conditions and their corresponding operations can be deduced from the following code, which is left to the readers.

```
LIBRARY ieee;
USE ieee.std_logic_1164.all;
USE ieee.numeric_std.all;
entity proc_ctrl is
port (m_clk,int,read_sw_open,read_sw_clos:in STD_LOGIC:='1';
reset:in STD_LOGIC:='0';
cs,rd,wr :out std_logic:='1';
range_data: in std_logic_vector(7 downto 0):="00000000";
door_open,door_close,main_gas_cont,water_soli_cont,buz_cont,tx:out
std_logic:='0');
```

```
end proc_ctrl;
architecture Behavioral of proc_ctrl is
type statetype is (s0,s1,s2,s3,s4,s5,s6,s7,s8,s9,s10);
signal state,nextstate: statetype;
signal smoke_value:std_logic_vector(7 downto 0):="00000000";
signal clk_sig,txd_temp,mux_sig:std_logic:='0';
signal sdata:std_logic_vector(10 downto 0):="11000000000";
signal smoke_magnitude:integer range 0 to 255:=0;
begin
p0:process(m_clk,reset)
begin
if (reset='1') then
state<=s0;
elsif rising_edge(m_clk) then
state<=nextstate;
end if;
end process;
p1:process(state,int)
begin
case state is
when s0=> cs<='0'; nextstate<=s1;
when s1=> wr<='0'; nextstate<=s2;
when s2=> wr<='1'; nextstate<=s3;
when s3=> cs<='1'; nextstate<=s4;
when s4=> if int='1' then nextstate<=s4;
else nextstate<=s5;
end if;
when s5=> cs<='0'; nextstate<=s6;
when s6=> rd<='0'; nextstate<=s7;
when s7=> smoke_value<= range_data;     nextstate<=s8;
when s8=> rd<='1'; nextstate<=s9;
when s9=> cs<='1'; nextstate<=s10;
when s10=>nextstate<=s0;
when others=> nextstate<=s0;
end case;
end process;
p2:process(m_clk)
variable cnt1:integer:=1;
begin
if (reset='1') then
clk_sig<='0';
elsif rising_edge(m_clk) then
if(cnt1=208)then
clk_sig<= not(clk_sig);
cnt1:=1;
else
cnt1:= cnt1 + 1;
end if;
end if;
end process;
p3:process(clk_sig,range_data)
variable cnt2:integer:=1;
begin
if rising_edge(clk_sig)then
if (cnt2>10) then
mux_sig<='1';
if (cnt2=11) then
sdata<=(sdata(0)& sdata(10 downto 1));
cnt2:=cnt2+1;
end if;
if (cnt2=16) then
```

```
sdata(8 downto 1)<=range_data;
smoke_magnitude<=to_integer(unsigned(smoke_value));
mux_sig<='0';
cnt2:=1;
else
cnt2:=cnt2+1;
end if;
else
mux_sig<='0';
sdata<=(sdata(0)& sdata(10 downto 1));
cnt2:=cnt2+1;
end if;
case mux_sig is
when '0'=>txd_temp<=sdata(0);
when '1'=>txd_temp<='1';
when others=> null;
end case;
end if;
end process;
tx<=txd_temp;
p4:process(smoke_magnitude,read_sw_open,read_sw_clos)
begin
if (smoke_magnitude >= 0 and smoke_magnitude<=20) then
if (read_sw_clos='0') then
door_close<='1';
door_open<='0';
water_soli_cont<='0';
main_gas_cont<='0';
buz_cont<='0';
else
door_close<='0';
door_open<='0';
water_soli_cont<='0';
main_gas_cont<='0';
buz_cont<='0';
end if;
elsif (smoke_magnitude >= 21 and smoke_magnitude<=80) then
if (read_sw_open='0') then
door_open<='1';
door_close<='0';
water_soli_cont<='0';
main_gas_cont<='0';
buz_cont<='0';
else
door_close<='0';
door_open<='0';
water_soli_cont<='0';
main_gas_cont<='0';
buz_cont<='0';
end if;
elsif (smoke_magnitude >= 81 and smoke_magnitude<=150) then
if (read_sw_open='0') then
door_open<='1';
door_close<='0';
water_soli_cont<='1';
main_gas_cont<='0';
buz_cont<='0';
else
door_close<='0';
door_open<='0';
water_soli_cont<='1';
```

```
main_gas_cont<='0';
buz_cont<='0';
end if;
elsif (smoke_magnitude >= 151 and smoke_magnitude<=200) then
if (read_sw_open='0') then
door_open<='1';
door_close<='0';
water_soli_cont<='0';
main_gas_cont<='1';
buz_cont<='0';
else
door_close<='0';
door_open<='0';
water_soli_cont<='0';
main_gas_cont<='1';
buz_cont<='0';
end if;
elsif (smoke_magnitude >= 201 and smoke_magnitude<=255) then
if (read_sw_open='0') then
door_open<='1';
door_close<='0';
water_soli_cont<='0';
main_gas_cont<='0';
buz_cont<='1';
else
door_close<='0';
door_open<='0';
water_soli_cont<='0';
main_gas_cont<='0';
buz_cont<='1';
end if;
end if;
end process;
end Behavioral;
```

LABORATORY EXERCISES

1. Construct an experimental setup and study the magnitude and frequency response of an opto-electronic device.
2. Develop a digital system to drive six different loads in different orders as selected by the user. For example, if the user input is "01", the order of driving the load has to be load 1-on, load 2-off, load 3-off, load 4-on, load 5-on, and load 6-on. In the same way, make a lookup table to map the user input to a decided order. Assume that six different loads are connected with relays.
3. Develop a digital system to draw simple shapes using two servo motors. Assume that one servo motor is for the x-axis and the other for the y-axis. The user has to select the shapes using a three DIP switches. Assume that a paintbrush is connected to one of the servo motors.
4. Develop a digital system with a lookup table to operate an AC motor at different specified controlled speeds.
5. Develop a simple system using ultrasonic sensors to demonstrate the principle of radar. Interface the target detection with the computer and display the range and coordinates in it.
6. Develop a digital system and RF unit to measure the height of different objects using the principles of a radar system.

7. Develop a digital system to measure the voltage and current for different power loads and display them on a 16 × 4 LCD. Calibrate the developed measurement system against a standard master unit.
8. Develop a digital system to measure the liquid level in a chemical tank using an optical sensor and display it as a one-bin histogram on a GLCD. The height of the histogram has to be directly proportional to the liquid level.
9. Develop a digital system to control the speed of an AC motor using a Diac and Triac. The speed control has to be based on the phase variations. Use a lookup table to relate the used selection to the desired speed. Display the speed (in rpm) of the motor on an LCD.
10. Develop a digital architecture to control a modern water irrigation system for cultivating land. Assume and use sensors for soil humidity measurement and solenoid valves to control the water flow.

10

Floating-Point Computations with Very-High-Speed Integrated Circuit Hardware Description Language and Xilinx System Generator (SysGen) Tools

10.1 Introduction and Preview

With advancements in the field of sensor technology, it is now possible to measure and monitor a large number of parameters in a number of fields such as telecommunications, signal processing, medical, defense, and commercial applications. Real-time sensor-based applications require a system which can read, store and process the sensor data. Figure 10.1 represents a real-time signal processing system based on the FPGA. Such a system comprises sensors, a signal conditioning unit, A/D converters, a digital processor and a memory device. It is important for an FPGA processor to store the real-time sensor values to an external memory device for signal processing using a suitable algorithm for the intended application. For example, analyzing a speech signal in real time requires recording of sound signals and processing them in a floating-point format to maintain a specific accuracy and resolution.

In this chapter, we discuss the different types of floating-point formats (single, double, customized, and extended precision), floating-point arithmetic (addition, subtraction, multiplication, division) and floating-point computation (DSP system design). Before proceeding to that section, we need to recall unsigned and signed number systems. The unsigned and signed representation of all possible 4-bit data is given in Table 10.1. The unsigned decimal values and their binary equivalents are given in the first two columns, respectively. The third and fourth columns show the same representation for signed numbers of the same length, that is, 4 bits. Since now, the fourth bit is used to denote the sign of the number, the remaining 3 bits are only available to represent the magnitude, that is, the integer part. Hence, the magnitude varies in a 4-bit representation from –8 through +7, whose binary equivalent is "1000" through "0111". This is the 2's complement representation, as discussed in Chapter 4. The floating-point number system, in comparison with the binary number system, has a better dynamic range and would handle underflow/overflow situations during mathematical calculations. Therefore, storing a sensor value in floating-point format and processing it is the basis of accurate signal processing. Floating-point number system-based arithmetic and signal processing gives better accuracy than dealing only with the integer part.

Developing a digital system using VHDL code to perform floating-point computations is a classical way of doing digital signal processing. While using this method, much real-time signal processing would be a more challenging design. Xilinx provides

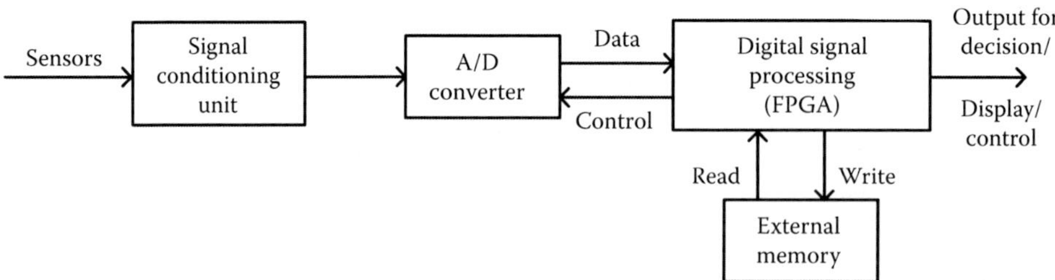

FIGURE 10.1
FPGA-based signal-processing system.

MATLAB®-assisted SysGen tools as a set of Simulink® blocks for several hardware operations that would be implemented on various Xilinx FPGA platforms. The SysGen tools run within the Simulink simulation environment, which is part of the MATLAB mathematical package. The SysGen blocks can be used to simulate the functionality of a hardware system using the MATLAB Simulink environment. The nature of most DSP applications requires floating-point format for data representation and processing. While this is easy to implement on computer systems, it is more challenging in the hardware world due to the complexity of the implementation of floating-point arithmetic. This is one of the important applications of designing the DSP system using SysGen tools. This chapter explains the design procedure of many DSP system designs and hardware co-simulations on the SysGen platform. The co-simulation integrates Simulink simulation capabilities with a hardware implementation to verify the functionality of the system [127,128].

TABLE 10.1
Signed/Unsigned Representation of 4-Bit Binary Word

Unsigned		Signed (i.e., 2's Complement)	
Decimal	**Bit Pattern**	**Decimal**	**Bit Pattern**
15	1111	7	0111
14	1110	6	0110
13	1101	5	0101
12	1100	4	0100
11	1011	3	0011
10	1010	2	0010
9	1001	1	0001
8	1000	0	0000
7	0111	−1	1111
6	0110	−2	1110
5	0101	−3	1101
4	0100	−4	1100
3	0011	−5	1011
2	0010	−6	1010
1	0001	−7	1001
0	0000	−8	1000

10.2 Representation of Fixed and Floating-Point Binary Numbers

Numerical values and numbers can be represented with integer and fractional parts, for example, 10 V, 25.345 mA, 0.0234, and 13.234 μF. The range of a numerical value is limited by the integer part, while the accuracy is limited by the fractional part; for example, assume two and four digits are allotted for the integer and fractional parts, respectively. Then, the integer can vary from 00 through 99, while the fraction can vary from 0000 through 9999 for every value of the integer part. Therefore, discussing computation techniques with different formats of floating-point numbers becomes significant to implement various real-time applications. This section discusses fixed-point, customized floating-point, single-precision, and double-precision number formats and representations. In all these representations, the first bit is always reserved for the sign of the number, that is, the sign bit.

10.2.1 Fixed-Point Number System

In any digital system, unsigned/singed numbers are stored in binary form, which is a certain-length sequence of bits: 1s and 0s. How hardware or software functions interpret this sequence of 1s and 0s is defined by the data type [127,129]. Binary numbers are represented as either fixed-point or floating-point data types. In order to implement a digital signal processing algorithm, the data type has to be a fixed-point number and has to be described with VHDL. In the VHDL coding process, it is necessary to indicate the size of the variables and registers that process fixed-point numbers. In the other cases, when an algorithm is implemented in floating-point numbers, all of the variables have a large number of bits; hence, all of the operations have to be done with a large number of bits. Then, obviously, we require a large number of flip-flops, larger area and more power consumption. However, selection of either the fixed-point or floating-point number system depends on the processing accuracy we except in the intended application. Fixed-point numbers are a simple way to express fractional numbers using a fixed number of bits. In this section, we discuss fixed-point representation and probe floating-point representation in subsequent sections. Normally, fixed-point numbers are denoted by (S,W,F) if it is a signed number, or simply (W,F) if it is a unsigned number, that is, only a positive number, where S is a sign bit, W is the total sequence length and F is the length of the fractional part. For example, (1,8,6) or (8,6) means that there is a bit for the sign, 2 bits for the integer part, and 6 bits for the fractional part. This example clearly illustrates that the W does not contain the sign bit. In some systems, the length W is equal to sign bit + integer bits + fractional bits. It extends our finite word length from a finite set of integers to a finite set of rational real numbers.

It is obvious that larger W and F results in better performance and lower resolution error, but the design needs significant hardware resources. On the other hand, a smaller W and F result in a larger resolution error but less implementation area. Therefore, the designer should choose suitable values of (W,F) for each parameter in the algorithm based on the accuracy we expect. There is a need of binary point in the fixed-point data type to disguise the decimal and fractional parts. The position of the binary point (also known as the decimal point) is the means by which fixed-point values are scaled and interpreted. For example, a binary representation of a generalized signed fixed-point number is shown in Figure 10.2, where b_i is the ith binary bit, b_0 is the LSB, b_{W-1} is the MSB, s is the sign bit, and '.' is binary point. The binary point is shown three places to the left of the LSB. The number is said to have one sign bit, W-1-3 bits for the integer part and three bits for the fractional part.

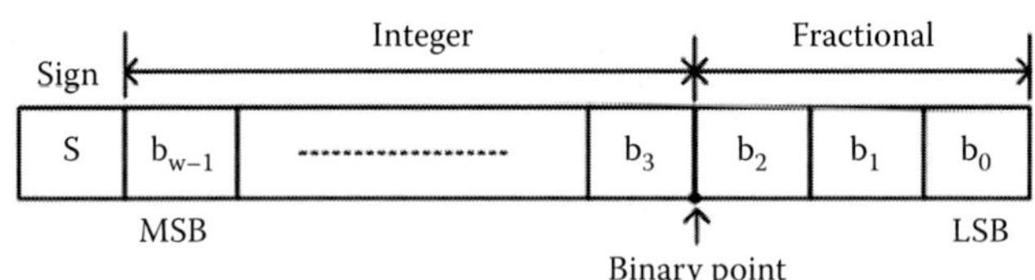

FIGURE 10.2
General representation of a fixed-point number.

We see an example to understand the above concept, the conversion of a decimal number to the equivalent fixed-point binary number and conversion back to the decimal number, below. Consider a positive fixed-point number of size (10,6) as "01011.111001." The size says that there are 6 bits for the fractional part; therefore, there are only 4 bits for the integer part and a default bit for the sign. There is no symbol to indicate the binary point in the digital system; however, according to the data type, that is, (W,F), the number of bits will be taken by the digital circuit to process the algorithm. As in this example, the integer part can be processed with the weighted binary values, that is, 2^0, 2^1, 2^2..., to the left side from the binary point while processing the fractional part with 2^{-1}, 2^{-2}, 2^{-3}... to the right side of the binary point, as given in Table 10.2. Note that the appropriate binary weighting is assigned according to the size of the fixed-point number, that is, (10,6).

This example also shows that the integer part varies from "0000" through "1111", while the fraction part varies from "000000" through "111111", as given in Table 10.2. Further, there are 64 levels in all at the fractional part for every increment of the integer part. That means if the integer value is initially "0000", it gets incremented to "0001" once the fractional part reaches "111111" from "000000". The fractional part increment by 1 bit, that is, "000000" to "000001", is equal to $(1/2^6) = (0.015625)_{10}$. Therefore, in this example, the range (integer) varies from 0 through 15 with a precision of 0.015625. The range of a fixed-point number is the difference between the minimum and maximum integer number possible. The precision of a fixed-point number is the total number of bits allotted for the fractional part of that number.

The range and precision of a fixed-point number depend on where we fix the binary point and total width of the number. Note that, therefore, we can have more precision with less range, more range with less precision, or both, more or less. When designing a fixed-point system, it is mandatory to have the range and resolution requirements. With this knowledge, we continue our discussion to understand the procedure required to convert a fixed-point number into the equivalent decimal value and vice-versa. For example, consider a signed fixed-point number of size (8,3), "111011111". In this example, the sign bit is 1; hence, it is a negative number. The integer value is "11011", whose value, as discussed in Chapter 4, is 27, and the fractional part is "111", whose value has to be calculated as

TABLE 10.2
Binary Weighing of a Fixed-Point Number

Sign	**Integer**				**Fraction**					
S	2^3	2^2	2^1	2^0	2^{-1}	2^{-2}	2^{-3}	2^{-4}	2^{-5}	2^{-6}
0	1	0	1	1	1	1	1	0	0	1
	Range				**Precision or Resolution**					
	"0000" to "1111"				"000000" to "111111"					
	0 to 15				0 to 63					

TABLE 10.3

Fixed-Point Fractional Part Conversion

Binary Position	Fraction Value	Binary Bits
$2^{-1} = 1/2$	0.5	.1
$2^{-2} = 1/4$	0.25	.01
$2^{-3} = 1/8$	0.125	.001
$2^{-4} = 1/16$	0.0625	.0001
$2^{-5} = 1/32$	0.03125	.00001
$2^{-6} = 1/64$	0.015625	.000001
$2^{-7} = 1/128$	0.0078125	.0000001

given in Table 10.3. The values of the fractional part have to be multiplied by the appropriate binary weight values. The first bit, next to the binary point, has to be multiplied by $2^{-1} = (1/2) = 0.5$, the next bit by $2^{-2} = (1/4) = 0.25$, and so on.

The fractional decimal values for certain values of binary positions are given in Table 10.3. As the fractional part in this example is "111", we need to add $0.5 + 0.25 + 0.125 = 0.875$. Therefore, the equivalent decimal number is –27.875. Consider another example of size (8,7) as $(001011001)_2$. Apply the values given in Table 10.3 as $(1/2 + 1/8 + 1/16 + 1/128) = (0.5 + 0.1 25 + 0.0625 + 0.0078125) = (0.6953125)_{10}$. Now, we discuss the vice-versa case, that is, obtaining the fixed-point binary number of any fixed-point size from the given decimal number. Assume two decimal numbers, 13.15625 and 26.9. We represent these numbers with a fixed-point number of size (9,5) and (12,7), respectively. The decimal part of both numbers can be directly calculated by the binary weighting method, and they are "1101" and "11000", respectively. The binary value of the fractional part can be calculated in a slightly different way as given in Table 10.4. There, the first column shows the steps of getting the binary value for 0.15625. The equivalent value is "00101"; here, we need to note some important approaches: (i) at the fifth iteration, the result is 1, so we need to stop the iteration, which means 0.15625 = "00101"; (ii) writing the binary value is started, keeping the resultant integer part of the first iteration as the MSB, that is, "00101"; and (iii) the length of this result is equal to the length of bits allotted for the fractional part. The above value can be verified with Table 10.3 as well. Therefore, the binary value in a fixed-point number of size (9,5) for 13.15625 is

TABLE 10.4

Conversion of Decimal to Fixed-Point Binary Value

Convert the Numbers Below into Equivalent Binary Numbers			
0.15625	**0.9**	**0.9624**	**0.6875**
$0.15625 \times 2 = 0.3125$ $0.3125 \times 2 = 0.625$ $0.625 \times 2 = 1.25$ $0.25 \times 2 = 0.5$ $0.5 \times 2 = 1$	$0.9 \times 2 = 1.8$ $0.8 \times 2 = 1.6$ $0.6 \times 2 = 1.2$ $0.2 \times 2 = 0.4$ $0.4 \times 2 = 0.8$ $0.8 \times 2 = 1.6$ $0.6 \times 2 = 1.2$ . . .	$0.9624 \times 2 = 1.9248$ $0.9248 \times 2 = 1.8496$ $0.8496 \times 2 = 1.6992$ $0.6992 \times 2 = 1.3984$ $0.3984 \times 2 = 0.7968$ $0.7968 \times 2 = 1.5936$ $0.5936 \times 2 = 1.1872$ . . .	$0.6875 \times 2 = 1.375$ $0.375 \times 2 = 0.75$ $0.75 \times 2 = 1.5$ $0.5 \times 2 = 1$
$(0.15625)_{10} = (00101)_2$	$(0.9)_{10} = (1110011\ldots)_2$	$(0.9624)_{10} = (1111011\ldots)_2$	$(0.6875)_{10} = (1011)_2$

"110100101". Take the fractional part of another example, 0.9. We apply the same procedure as given in the second column of Table 10.4. Note that the multiplication does not reach the 1 even at the seventh iteration. Further, it started roll-off from the sixth iteration onwards, and we do not have bits to represent the result of the eighth and further iterations since only 7 bits are allotted for the fractional part. Hence, we need to stop the iteration at the seventh level, and the equivalent binary value for 0.9 is "1110011". Therefore, 26.9 ≈ "110001110011". Refer to the third and fourth columns of Table 10.4. We use them in subsequent sections.

DESIGN EXAMPLE 10.1

Write MATLAB code to get the value of W,F and a signed decimal number and convert it to the given fixed-point (W,F) binary number.

Solution

```
clc;
clear all;
W=input('Enter your W value:=');
F=input('Enter your F value:=');
x=input('Enter your signed decimal value:=');
n=W-F;
if x>0
     y=0;
     y1=floor(x);
     y2=x-y1;
     y3=de2bi(y1,n);
     m=n;
     for i=1:1:n,
          y33(i)=y3(m);
          m=m-1;
     end
     y5=zeros(1,F);
     for i=1:F,
        y4=(y2*2);
        y5(i)=floor(y4);
        if y4>0
           y4=y4-y5(i);
           y2=y4;
        else
           y2=y4;
        end
      end
     fix_bin_nu=[y y33 y5]
else
     y=1;
     y1=(floor(x)+1)*-1;
     y2=(x+y1)*-1;
     y3=de2bi(y1,n);
     m=n;
     for i=1:1:n,
          y33(i)=y3(m);
          m=m-1;
     end
   y5=zeros(1,F);
     for i=1:F,
       y4=(y2*2);
       y5(i)=floor(y4);
       if y4>0
          y4=y4-y5(i);
```

```
            y2=y4;
         else
            y2=y4;
         end
          end
      fix_bin_nu=[y y33 y5]
end
```

Upon running this MATLAB code with the inputs W = 12, F = 8 and x = –14.2345, the result will be "1111000111100".

10.2.2 IEEE754 Single-Precision Floating-Point Number System

Apart from the fixed-point number system, there is the floating-point number system in practice, in which the binary point gets floated to the left or right side based on logic we discuss below. There are a few important terms in this number system: sign (S), exponent (E), mantissa (M), and bias (B). We discuss and explain the single-precision floating-point number system in this section. Let us start with an example for decimal to IEEE754 single-precision floating-point number and floating-point number to decimal conversion, which will make the concepts and notations of floating-point numbers much clearer. Assume a decimal number, 286.75. We represent it in IEEE754 single-precision floating-point format, that is, 32 bits, as 1 sign bit, 8 exponent bits, and 23 mantissa bits as given below [127–130]:

S	Exponent (8 bits)								Mantissa (23 bits)																						

We need to find the values for the sign, exponent, and mantissa bits as explained step-by-step below:

1. The given number is a positive number; hence, the value of the sign bit is 0.
2. Represent the decimal number 286.75 as the equivalent binary number as explained in the previous section as "100011110.11". Note that the conversion bit length for the integer as well as the fractional part of the decimal number has to be decided by keeping in mind that only 8 bits for the exponent and 23 bits for the mantissa are available.
3. We need to normalize the binary number to get the value of the mantissa. To normalize it, shift the binary point so as to get a 1 at the left side of the binary point, and increment the power of 2, that is, the normalized power (Np). In this example, we need to shift the binary point to the left side to get one 1 at the left of it; after shifting, the binary value becomes $(1.0001111011) \times 2^8$. Therefore, the value of the mantissa is "0001111011". The 1, which is the left side of the binary point, has to be omitted and we will consider this while finding the equivalent decimal value.
4. Now, we need to find the bias (B) value, $B = 2^{N-1}-1$, where N is the number of bits allotted for the exponent (E). In the single-precision floating-point number system, the value of N is 8. Hence, the value of B is $2^{8-1}-1 = 2^7-1 = 128-1 = B = 127$.
5. In this example, the normalized power (Np) obtained from normalization (step 3) is 8. Now, we need to find the value for the exponent as E = bias + Np = 127 + 8 = E = 1 35. The binary value of 135 is "10000111"; therefore, this is the value of the exponent.

6. We need to concatenate the zeros to fill the allotted bit positions. In this example, we have only 10 bits for the mantissa part; hence, we need to add zeros from the remaining positions. Therefore, the IEEE754 single-precision floating-point binary number for 286.75 is "01000011100011110110000000000000", which is appropriately filled in the respective positions below:

s	Exponent (8 bits)								Mantissa (23 bits)																						
0	1	0	0	0	0	1	1	1	0	0	0	1	1	1	1	0	1	1	0	0	0	0	0	0	0	0	0	0	0	0	0

7. The decimal value of a normalized single-precision floating-point number in the IEEE754 standard is represented by decimal value $= (-1)^s \times 1.M \times 2^{Np}$

From the above equation, it is understood that we need to append a 1 to the mantissa of a floating-point word while the conversion takes place to get the decimal value. The 1 is hidden in the representation of the IEEE754 floating-point word since it takes up an extra bit location; hence, it was avoided. Now, we discuss an example to understand the procedure of getting a decimal value from the single-precision floating-point number below. For example, assume the following value

s	Exponent (8 bits)								Mantissa (23 bits)																						
0	1	0	0	0	0	0	0	1	0	1	0	1	0	0	0	0	0	0	0	0	0	0	0	0	0	0	0	0	0	0	0

1. The sign bit is 0; hence, it is a positive decimal number.
2. From the above example, the value of the mantissa is "01010000000000000000000"; actually, it has to be (1.M), so the actual mantissa is "1.01010000000000000000000".
3. The value of the normalized bias (Np) is E–B. In this example, the value of E is "10000001", which is equal to 129. We know that the value of the bias for the single-precision floating-point number is 127, so Np = 129–127 = Np=2.
4. Substitute all these results in the equation given in step 7. Then, the decimal value $= (-1)^0 \times (1.0101000\ldots00) \times 2^2 = 1 \times (101.01000\ldots00) =$ "101.01000…00".
5. Then, find the equivalent decimal value, as discussed in Chapter 4 and the previous section, for the above value as $(1 \times 2^2 + 0 + 1 \times 2^0 . 0 \times 2^{-1} + 1 \times 2^{-2} + 0 + 0 + \cdots + 0) = (4 + 1 + 0.25) = 5.25$.

DESIGN EXAMPLE 10.2

Find the (i) single-precision floating-point number for $(-639.6875)_{10}$ and (ii) decimal value of a single-precision floating-point number "11000011100011110110000000000000".

Solution

The above two problems are solved one by one below:

1. The given number is −639.6875, so the sign bit is 1.
2. The binary form of this number is "1001111111.1011".
3. Normalize and find the mantissa and Np. $(1.0011111111011) \times 2^9$; here, M = "0011111111011" and Np = 9.

4. We know that the bias is 127, so the value of E is E = Np + B, E = 9 + 127. E = 136, whose binary form is "10001000".
5. Then, the floating-point value is "11000100000111111110110000000000".

Now, we solve the second problem,

1. The given floating-point number is "11000011100011110110000000000000".
2. The sign bit is 1; hence, the decimal number will be a negative number.
3. The mantissa part with hidden one is "1. 00011110110000000000000".
4. The exponent value is "10000111", which is equal to 135, so the Np is 135–127, Np = 8.
5. Then, the decimal value is $(-1)^1 \times (1.00011110110000000000000) \times 2^8$, which is equal to $-1 \times (100011110.1100000\ldots0000) = -286.75$

DESIGN EXAMPLE 10.3

Develop VHDL code to convert a fixed-point binary number (W,F) into a single-precision floating-point number. Take n to represent the length of the integer part, that is, W-F, and m to represent the length of the fractional part, that is, F. Use an enable control (pulse) to read the signed fixed-point binary number.

Solution

```
library IEEE;
use IEEE.STD_LOGIC_1164.ALL;
use IEEE.STD_LOGIC_arith.ALL;
use IEEE.STD_LOGIC_unsigned.ALL;
entity ieee_754_sp is
generic(n:integer:=8;--integer part
    m:integer:=6);--fraction part
port(clk:in std_logic;
enable:in std_logic;
data_in:in std_logic_vector(n+m downto 0);
data_out:out std_logic_vector(31 downto 0));
end ieee_754_sp;
architecture Behavioral of ieee_754_sp is
type state_type is(rst_state,s1,s2,s3);
signal state:state_type:=rst_state;
signal latch_data:std_logic_vector((n+m) downto 0):=(others=>'0');
signal data_temp:std_logic_vector(31 downto 0):=(others=>'0');
signal count,exponent:std_logic_vector(7 downto 0):=(others=>'0');
signal mantissa:std_logic_vector((n+m-2) downto 0);
signal done:std_logic:='0';
signal count1:integer:=0;
begin
process(clk)
begin
if(clk='1' and clk'event)then
case state is
when rst_state=>done<='0';
if(enable='1')then
state<=s1;
data_temp<=(others=>'0');
latch_data<=data_in;
else
state<=rst_state;
end if;
when s1=>
data_temp(31)<=latch_data(n+m);
```

```
if(latch_data(n+m-1)='0')then
latch_data((n+m-1) downto 0)<=latch_data((n+m-2) downto 0) & '0';
count<=count+"00000001";
state<=s1;
else
mantissa<=latch_data((n+m-2) downto 0);
exponent<=conv_std_logic_vector((n-1),8) - count+"01111111";
state<=s2;
end if;
when s2=>
data_temp(30 downto 23)<=exponent;
data_temp(22 downto (22-(n+m-2)))<=latch_data((n+m-2) downto 0);
done<='1';
state<=rst_state;
when others=>null;
end case;
end if;
end process;
data_out<=data_temp;
end Behavioral;
```

10.2.3 IEEE754 Double-Precision Floating-Point Number System

The double-precision floating-point number system provides more digits to the right side of the binary point than a single-precision number. The term double-precision is something of a misnomer because the precision is not really double; however, the double-precision number system uses twice as many bits as the single-precision floating-point number system. For example, the single-precision floating-point number system requires 32 bits and its double, that is, 64 bits, is required for the double-precision floating-point number system. The additional 32 bits increase not only the precision but also the range of magnitudes that can be covered. Double-precision floating-point format is a computer number format that occupies 8 bytes, that is, 64 bits, in computer memory and represents a wide and dynamic range of values by using a floating point [127–131].

The IEEE754 standard specifies a double-precision floating-point number with 64 bits as 1 bit for the sign, 11 bits for the exponent, and 52 bits for the mantissa, as shown in Figure 10.3. Now, we solve one example to understand the conversion of a decimal number to a double-precision floating-point number below. We have to apply the same steps as applied for the single-precision floating-point number, but with appropriate exponent, mantissa, and bias values. Assume that the decimal number is 891.9624.

1. The given number is a positive number; hence, the sign (S) bit is 0.
2. The given decimal number is 891.9624, whose equivalent binary number is "11011 11011.1111011001011111110110001010110110101011101".

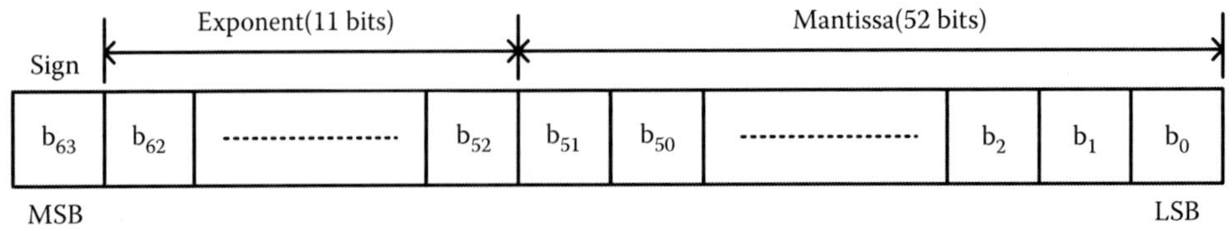

FIGURE 10.3
Length of double-precision floating-point format.

3. Now, we need to normalize the above binary number by incrementing the normalized power (Np). After normalizing, the above binary number becomes (1.1011 110111111011001011111110110001010110110101011101) $\times$ 2^9. We need to leave the hidden one. Therefore, the value of the mantissa and normalized power Np becomes M = "1011110111111011001011111110110001010110110101011101" and Np = 9.
4. We know how to find the bias (B) value, that is, $B = 2^{N-1}-1$, where N is the number of bins allotted for the exponent part. In the double-precision floating-point number system, 11 bits are allotted for the exponent; hence, N = 11, then $B = 2^{11-1}-1$, $B = 2^{10}-1$, B = 1023.
5. Then, the value of the exponent (E) is E = Np + B, E = 9 + 1023, E = 132; E = "10000001000".
6. Now, we can formulate the double-precision floating-point number as

$$X = (0100000010001011110111111011001011111110110001010110110101011101)_2$$

This double-precision binary floating-point value can be rearranged as below to easily get its equivalent hexadecimal value.

$$X = (0100\ 0000\ 1000\ 1011\ 1101\ 1111\ 1011\ 0010\ 1111\ 1110\ 1100\ 0101\ 0110\ 1101\ 0101\ 1101)_2.$$

Please note that the sign bit is included as the MSB while finding the equivalent hexadecimal value. Now we get the hexadecimal value as $X = (408BDFB2FEC56D5D)_{16}$ or simply$\times$ = "408BDFB2FEC56D5DH".

As explained in Design Example 10.2, we can convert the double-precision floating-point number into the equivalent decimal value with the appropriate bias value. This practice is left to the reader as an exercise.

DESIGN EXAMPLE 10.4

Develop a digital system to convert a fixed-point floating-point number (W,F) into the equivalent double-precision floating-point number. Take n to represent the length of the integer part, that is, (W-F), and m to represent the length of the fractional part, that is, F. Use an enable control (pulse) to read the signed fixed-point binary number.

Solution

```
library IEEE;
use IEEE.STD_LOGIC_1164.ALL;
use IEEE.STD_LOGIC_arith.ALL;
use IEEE.STD_LOGIC_unsigned.ALL;
entity ieee_754_dp is
generic(n:integer:=8;--integer part
    m:integer:=6);--fraction part
port(clk:in std_logic;
enable:in std_logic;
data_in:in std_logic_vector(n+m downto 0);
data_out:out std_logic_vector(63 downto 0));
end ieee_754_dp;
architecture Behavioral of ieee_754_dp is
type state_type is(rst_state,s1,s2,s3);
signal state:state_type:=rst_state;
signal latch_data:std_logic_vector((n+m) downto 0):=(others=>'0');
signal data_temp:std_logic_vector(63 downto 0):=(others=>'0');
```

```
signal count,exponent:std_logic_vector(10 downto 0):=(others=>'0');
signal mantissa:std_logic_vector((n+m-2) downto 0);
signal done:std_logic:='0';
signal count1:integer:=0;
begin
process(clk)
begin
if(clk='1' and clk'event)then
case state is
when rst_state=>
done<='0';
if(enable='1')then
state<=s1;
data_temp<=(others=>'0');
latch_data<=data_in;
else
state<=rst_state;
end if;
when s1=>
data_temp(63)<=latch_data(n+m);
if(latch_data(n+m-1)='0')then
latch_data((n+m-1) downto 0)<=latch_data((n+m-2) downto 0) & '0';
count<=count+"00000000001";
state<=s1;
else
mantissa<=latch_data((n+m-2) downto 0);
exponent<=conv_std_logic_vector((n-1),11) - count+"01111111111";
state<=s2;
end if;
when s2=>
data_temp(62 downto 52)<=exponent;
data_temp(51 downto (51-(n+m-2)))<=latch_data((n+m-2) downto 0);
done<='1';
state<=rst_state;
when others=>null;
end case;
end if;
end process;
data_out<=data_temp;
end Behavioral;
```

The limitations and features of single- and double-precision floating-point formats are compared, and the comparison results are given in Table 10.5. Based on the number of bits allotted for the exponent, mantissa, representation of underflow or ± 0, not a number (NaN), overflow ($\pm\infty$), and the value of the bias, the following conditions are derived for these two number systems.

The IBM introduced a one number system known as the extended-precision or quadruple-precision floating-point number system. This number system has 128-bit-long representation. It was introduced in the IBM series S/360, S/360−85, and −195 models and others by special request for operating system software. The extended precision mantissa field is wider, and the extended-precision number is stored as two double words, that is, 16 bytes.

10.2.4 Customized Floating-Point Number System

In this section, we discuss the representation of custom-precision floating-point numbers. Floating-point operations are difficult to implement in FPGAs because of their complexity in calculations as well as their hardware utilization for such calculations. The custom-precision

TABLE 10.5

Feature Comparison of Single- and Double-Precision Floating-Point Numbers

Parameter	IEEE754	
	Single Precision	**Double Precision**
Word length	32	64
Mantissa	23	52
Exponent	8	11
Sign bit	1	1
Bias	127	1023
NaN	If the value of E is 255 and the value of the mantissa is $\neq 0$	If the value of E is 2047 and the value of the mantissa is $\neq 0$
$\pm\infty$	If the value of E is 255 and the value of the mantissa is $= 0$	If the value of E is 2047 and the value of the mantissa is $= 0$
Renormalized number	If the value of E is 0 and the value of the mantissa is $\neq 0$	If the value of E is 0 and the value of the mantissa is $\neq 0$
± 0	If the value of E is 0 and the value of the mantissa is $=0$	If the value of E is 0 and the value of the mantissa is $=0$
± 1	If the value of E is 127 and the value of the mantissa is $=0$	If the value of E is 1023 and the value of the mantissa is $=0$
Valid number	If value of E is between 0 and 255	If the value of E is between 0 and 2047
E_{min} and E_{max}	-126 and 127	-1022 and 1023
Range	0 to $>3.8 \times 10^{38}$	0 to $>1.8 \times 10^{308}$
Precision	1.4×10^{-45}, i.e., LSB of mantissa is only 1	$\approx 4.95 \times 10^{-324}$, i.e., LSB of mantissa is only 1

floating-point format is developed to provide high resolution over a large dynamic range. Floating-point systems can often provide a solution when a fixed-point system, with its limited dynamic range, does not meet the required resolution. Customized precision floating-point systems, however, bring a speed and complexity penalty if it is longer than single- or double-precision floating-point formats. Most microprocessor floating-point systems comply with the single or double-precision IEEE754 floating-point standard, while FPGA-based systems often employ custom formats. As we discussed in the previous sections, a standard floating-point format consists of a sign bit, exponent bits and normalized mantissa bits, as shown in Figure 10.4, where 1 bit is allotted for the sign (S) of the number, 6 bits are allotted for the exponent (E), and 10 bits are allotted for the mantissa (M). Therefore, a total of 17 bits is allotted in the custom floating-point number representation [131,132].

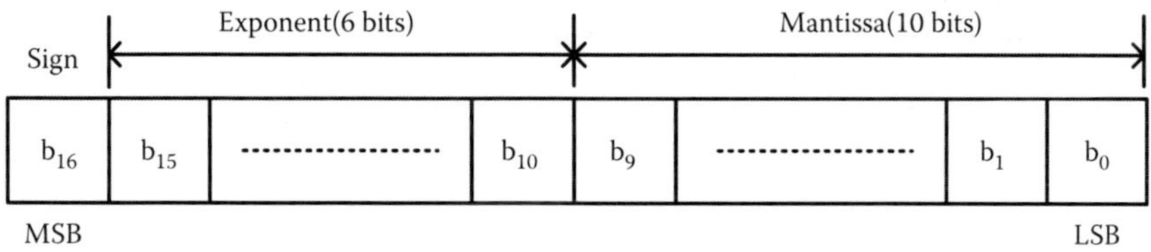

FIGURE 10.4
Customized floating-point format of size (1,6,10).

We discuss examples to understand the representation of customized precision floating-point number formats. We recall the general steps we followed in the previous sections in general as given below to discuss the customized precision floating-point number system.

1. Decide the sign bit based on the polarity of the given decimal number.
2. Convert the given decimal number into the equivalent binary number, as discussed in Section 10.1.
3. Normalize the binary value by keeping one 1 at the left side of the binary point and increment the normalized power (Np) value accordingly. After the normalization, the binary number will be in the form of $1.M \times 2^{Np}$.
4. Take the mantissa (M) value from the above step, append the required number of zeros, if required, and devise the value for the mantissa of allotted length.
5. Calculate the bias value. The bias has to be calculated as bias (B) $= 2^{N-1} - 1$.
6. Find the value for exponent (E) as E = Np + bias.

We apply the above steps on an example. Consider a decimal number –13.9 and represent it in a floating-point format of size (1,4,3). From the given size, it is understood that 1 bit is for the sign, 4 bits for the exponent and 3 bits for the mantissa. We apply the above steps one by one:

1. The given number is a negative number; hence, the sign bit is 1.
2. The binary equivalent of $(-13.9)_{10}$ is $(1101.11100)_2$.
3. Normalize the binary value and get the value for Np. In this example, after normalization, the number will be $(1.10111100) \times 2^3$, where M = "10111100" and Np = 3. The hidden 1 is omitted. Only 3 bits are allotted for the mantissa; therefore, this value has to be truncated as M = "101". This is an important step in a customized precision floating-point number system.
4. The value of N is 4; hence, $B = 2^{4-1}-1$, $B = 2^3 - 1$, $B = 7$.
5. Then, the value of the exponent is E = Np + B, E = 3 + 7, E = 10, E = "1010".
6. The customized precision floating-point binary number of size (1,4,3) for –13.9 is $(11010101)_2$. The first bit is the sign bit, the second 4 bits are the exponent bits and the last 3 bits are the mantissa bits.

In the same way, a customized floating-point number can also be converted back into the decimal fractional number. The necessary steps are as below:

1. Find the sign bit, $(-1)^S$. Suppose S = 0; then, $(-1)^0$ is 1, so the number is a positive number; otherwise, the value will be –1, meaning the number is negative.
2. Form the mantissa part by considering the hidden 1; here, it can be given by 1.M.
3. Compute the Np value from the value of the exponent and bias as Np = E-bias.
4. Now, concatenate all the above findings as $(-1)^S \times 1.M \times 2^{Np}$.
5. Simplify the values, and the final result will be the equal to the signed decimal fractional number.

We discuss one example now: assume a custom floating-point number of size (1,3,4), "00100010". Here, the sign bit is 0, the exponent is "010" and the mantissa is "0010". We apply the above steps one by one as below:

1. The sign value is 0, so $(-1)^0$ is 1; therefore, the given number is a positive number.
2. The hidden 1 and mantissa part is "1.0010".
3. The exponent value is E = "010", E = 2.
4. In this example, 3 bits are allotted for the exponent; hence, the value of the bias $= 2^{3-1}-1$, $B = 2^2-1$, $B = 4-1$, $B = 3$. Then, $Np = 2-3$, $E = -1$.
5. Then, the final form is $(-1)^S \times 1.M \times 2^{Np} = 1 \times (1.0010) \times 2^{-1}$.
6. Simplify the above values as $1 \times (1.0010) \times 0.5 = 1 \times (1.125) \times 0.5 = 0.5625$. The final value is $(0.5625)_{10}$. The 1.0010 is converted to its equivalent decimal value as discussed in Chapter 4 and Section 10.1.

DESIGN EXAMPLE 10.5

(i) Convert $(9.25)_{10}$ and $(0.0625)_{10}$ to a customized precision floating-point representation of size (1,6,5) and (1,6,10), respectively.

Solution

$(9.25)_{10}$ *to* (1,6,5) *format:*

1. The sign bit for the given number is 0.
2. The binary equivalent is $(9.25)_{10} = (1001.01)_2$.
3. Normalize the binary value: 1.00101×2^3.
4. Omit the hidden 1, then 00101×2^3. Five bits are allotted for the mantissa; therefore, m = 00101 and Np = 3.
5. Here, N = 6, so bias $B = 2^{6-1}-1$, $B = 2^5-1$, $B = 31$.
6. Then, the value of the exponent E = 3 + B, E = 3 + 31, E = 34, E = "100010".
7. The floating-point representation of $(9.25)_{10}$ is $(010001000101)_2$.

$(0.0625)_{10}$ *to* (1,6,10) *format:*

1. The sign bit for the given number is 0.
2. The binary equivalent is $(0.0001)_2$.
3. Normalized value is 1.0×2^{-4}.
4. Omit the hidden 1, and the mantissa value is "0000000000" since 10 bits are allotted for the mantissa. The value of Np is −4.
5. Bias $= 2^{6-1} -1$, $b=2^5-1$, $b = 31$.
6. Then, the value of the exponent is E = −4 + 31, E = 27, E = "011011".
7. The floating-point representation of $(0.0625)_{10}$ is $(00110110000000000)_2$.

10.3 Floating-Point Arithmetic

Floating-point arithmetic such as addition, subtraction, multiplication, and division are slightly different from the classical binary arithmetic operations. In this section, we discuss the specific algorithms that need to be followed to perform arithmetic operations with fixed, single-precision, double-precision, and customized floating-point numbers.

10.3.1 Floating-Point Addition

In floating-point addition, the binary points need to be aligned before beginning the addition. We discuss fixed-point addition first and then proceed to floating-point addition. For example, take two decimal numbers, $X = (6.5625)_{10}$ and $Y = (4.25)_{10}$. Their equivalent binary form in the fixed-point size of (1,6,4) is $X = (0000110.1001)_2$ and $Y = (0000100.0100)_2$. The first bit indicates that X and Y are positive numbers, and then the sign bit can be omitted while adding, then

$$\begin{aligned} X+Y &= (000110.1001+000100.0100) \\ &= (001010.1101) \end{aligned}$$

Convert it back to decimal value, then

$$X + Y = (10.8125)_{10}$$

As above, the bits at the decimal and integer parts can be directly added. However, when adding two numbers, an additional bit is required for storing the full result, that is, sum and final carry. Fixed-point representation is convenient and useful when dealing with fixed-point signal processing techniques and implementations. Floating-point addition has so many bits to work with. Suppose we want to add two floating-point numbers X and Y. The addition process has to be carried out as given below [133,134]:

$$\begin{aligned} X+Y &= 1.M_x \times 2^{Np} + 1.M_Y \times 2^{Np} \\ X+Y &= (1.M_x + 1.M_y) \times 2^{Np} \end{aligned} \tag{10.1}$$

where M_x and M_y are the actual mantissa parts of X and Y, respectively, and Np is common (similar) value of power of 2. We realize Equation 10.1 step by step, with an example below. Assume two customized precision floating-point numbers of size (1,4,3), X = "01001110" and Y = "00111000". Since the length of the exponent is 4, the bias value is obtained by $2^{4-1}-1$, B = 7. In X, the sign is +, exponent is "1001" = $(9)_{10}$, mantissa is "110" = $(6)_{10}$, and Np = 9–7, Np = 2; and in Y, the sign is +, exponent is "0111" = $(7)_{10}$, mantissa is "000" = $(0)_{10}$, and Np = 0. First, convert these two representations to ordinary scientific notation as in Equation 10.1; thus, we need to use the hidden 1. Then the X and Y are given by

$$X = 1.110 \times 2^2 \text{ and } Y = 1.000 \times 2^0$$

In the above X and Y, the powers of 2 are not equal. In order to add, we need to keep these two numbers the same. In this example, we do this by rewriting the Y. Then, the result of Y will not be normalized; however, the value of the power of 2 will be equal to the power of 2 of X. To meet this requirement, shift the decimal point of Y left to compensate for the difference of power of 2. Then the Y will be

$$Y = 0.0100 \times 2^2.$$

Now, we can add as

$$X + Y = (1.110 \times 2^2 + 0.0100 \times 2^2)$$

$$X + Y = (1.110 + 0.0100) \times 2^2$$

Next, add the two mantissas of X and Y together, that is, add 1.110 with 0.010, then

$$X + Y = (10.000) \times 2^2.$$

We need to normalize the above result to represent the result in floating-point format again. The normalized form of the above result is

$$X + Y = 1.000 \times 2^3$$

Apply the steps discussed in the previous section and convert again into a floating-point number of size (1,4,3), and the result will be

$$X + Y = \text{"01010000"}.$$

In the following sections, the term "customized precision" will not be used; however, the size of the floating-point format will be given, so the readers will understand that the given number is a single-, double- or customized precision number.

DESIGN EXAMPLE 10.6

(i) Two floating-point numbers of size (1,4,3) are X = "01001110" and Y = "00110110". Add these two numbers and get the result in the given format. (ii) Represent X = 9.75 and Y = 0.5625 in single-precision floating-point format, add them and show the result in single-precision floating-point format.

Solution

Addition of X = "01001110" *and* Y = "00110110" – (1,4,3):

The given numbers are X = "01001110" and Y = "00110110" with sizes of (1,4,3), The addition procedures are given below step by step, as explained in the previous section.

$$X = 1.110 \times 2^2 \text{ and } Y = 1.110 \times 2^{-1}$$

$$X = 1.110 \times 2^2 \text{ and } Y = 0.00111 \times 2^2$$

$$X + Y = (1.110 + 0.00111) \times 2^2$$

$$X + Y = (1.11000 + 0.00111) \times 2^2$$

$$X + Y = (1.11111) \times 2^2$$

In this example, the sum 1.11111 has only one 1 at the left side of the binary point; hence, the sum itself is normalized. We do not need to adjust anything. Therefore, just convert this sum value into a floating point of size (1,4,3). Here, we need to truncate the mantissa to "111", since only 3 bits are allotted for it. Then, the result will be

$$X + Y = \text{"01001111"}$$

Addition of X = 9.75 *and* Y = 0.5625 – *(single-precision floating-point):*

The size of the single-precision floating-point format is (1,8,23); hence, the bias value is 127. Then, convert the above X and Y into single-precision floating-point format as discussed in the preceding sections

$$X = (01000001000111000000000000000000) \text{ and}$$

$$Y = (00111111000100000000000000000000)$$

The sign bit for both numbers is 0, so the result will also be a positive number. Now, the sign bits can be omitted. Then, the X and Y are X = (1000001000111000000000000000000) and Y = (0111111000100000000000000000000). Since the size of this floating-point format is (1,8,23), the values of Ex and Ey are (10000010) = 130 and (01111110) = 126, respectively, and the Mx and My are (00111000000000000000000) and (00100000000000000000000), respectively. From the above values, we can take the Mx and My as they are, and we need to include the hidden 1. The Ex and Ey are not the actual Npx and Npy; hence, we need to calculate them as Npx = Ex − 127, Npx = 130 − 127, Npx = 3 and Npy = Ey − 127, Npy = 126 − 127, Npy = −1.

Now, X and Y become X = (1.00111000…0) $\times 2^3$ and Y = (1.001000000…000) $\times 2^{-1}$.

Now, we need to adjust the smaller Np to be equal to the larger Np, that is, −1–3 by left-shifting the decimal point in Y. Then, X and Y become X = (1.00111000000000…000) $\times 2^3$ and Y = (0.00010010000000…000) $\times 2^3$. Now, we can add these two values. The result is X + Y = (1.010010100000…000) $\times 2^3$, which is equal to $(10.3125)_{10}$.

We need to represent this result in the single-precision floating-point number format. Apply the standard procedure as discussed in the preceding sections, and the result will be $(0100000100100101000\ldots0000)_2$.

DESIGN EXAMPLE 10.7

Develop a digital system in the FPGA to perform the addition operation of two floating-point numbers and store the result in the given format.

Solution

```
library IEEE;
use IEEE.STD_LOGIC_1164.ALL;
use IEEE.STD_LOGIC_arith.ALL;
use IEEE.STD_LOGIC_unsigned.ALL;
entity ieee_754_dp is
generic(n:integer:=8;--integer part
    m:integer:=6);--fraction part
port(clk:in std_logic;
enable:in std_logic;
data_in:in std_logic_vector(n+m downto 0);
data_out:out std_logic_vector(63 downto 0));
end ieee_754_dp;
architecture Behavioral of ieee_754_dp is
type state_type is(rst_state,s1,s2,s3);
signal state:state_type:=rst_state;
signal latch_data:std_logic_vector((n+m) downto 0):=(others=>'0');
signal data_temp:std_logic_vector(63 downto 0):=(others=>'0');
signal count,exponent:std_logic_vector(10 downto 0):=(others=>'0');
signal mantissa:std_logic_vector((n+m-2) downto 0);
signal done:std_logic:='0';
signal count1:integer:=0;
begin
process(clk)
begin
if(clk='1' and clk'event)then
case state is
when rst_state=>
done<='0';
if(enable='1')then
state<=s1;
data_temp<=(others=>'0');
latch_data<=data_in;
else
state<=rst_state;
```

```
end if;
when s1=>
data_temp(63)<=latch_data(n+m);
if(latch_data(n+m-1)='0')then
latch_data((n+m-1) downto 0)<=latch_data((n+m-2) downto 0) & '0';
count<=count+"00000000001";
state<=s1;
else
mantissa<=latch_data((n+m-2) downto 0);
exponent<=conv_std_logic_vector((n-1),11) - count+"01111111111";
state<=s2;
end if;
when s2=>
data_temp(62 downto 52)<=exponent;
data_temp(51 downto (51-(n+m-2)))<=latch_data((n+m-2) downto 0);
done<='1';
state<=rst_state;
when others=>null;
end case;
end if;
end process;
data_out<=data_temp;
end Behavioral;
```

10.3.2 Floating-Point Subtraction

Floating-point subtraction is like addition, but with a minus sign for the second number, that is, X + (−Y). Therefore, the standard 2's complement method for subtraction has to be followed. We see an example step by step to understand floating-point subtraction below [135].

Suppose we assume two decimal numbers, X = 0.5 and Y = −0.4375. Now, we find the subtraction result.

The binary representations of X and Y are

$$X = 0.1 \times 2^0 \text{ and } Y = -0.0111 \times 2^0$$

Here, the power of 2 is common, so we can take out 2^0, then subtract as

$$X + (-Y) = (0.1000 - 0.0111) \times 2^0$$

Now, we need to take the 2's complement of the second number and then add to the first number.

$$X - Y = (0.1000 - 1.0000) \times 2^0 \text{ One's complement is taken}$$

$$X - Y = (0.1000 - 1.0001) \times 2^0 \text{ Two's complement is taken, then add them}$$

$$X - Y = (10.0001) \times 2^0$$

Note the MSB bit, that is, the EoC bit, is very important. If EoC is 1, then we can omit it and the result will be a positive number, that is, the sign bit is 0; otherwise, we have to again take the 2's complement of the result, and the result will be a negative number, i.e., the sign bit is 1. In this example, the EoC bit is 1; therefore, we can omit it, so the result is a positive number.

$$X - Y = (0.0001) \times 2^0$$

Now, we have to check that either the power of 2 is overflow/underflow or within the range, that is, $(-126 \leq E \leq 127)$. Here, the power of 2 is within the range, and the actual value of this result is

$$X - Y = 0.0001 = 0.0625, \text{ which can be verified as } (0.5 - 0.4375) = 0.0625.$$

DESIGN EXAMPLE 10.8

Perform the subtraction operation (X − Y) on X = 9.75 and Y = 0.5625 in single-precision floating-point number format and represent the final result in the same format.

Solution

We need to subtract Y from X, that is, X − Y = 9.75 − 0.5625. We discuss this operation step by step below.

Convert the given decimal numbers into the equivalent single-precision floating-point format, as discussed in the preceding sections. The value will be

$$X = (01000001000111000000000000000000) \text{ and}$$
$$Y = (10111111000100000000000000000000)$$

The sign bit can be omitted; then, X and Y are

$$X = (1000001000111000000000000000000) \text{ and}$$
$$Y = (0111111000100000000000000000000)$$

Reformulate the above X and Y with actual mantissas (M_x and M_y), hidden 1s, and respective exponents (E_x and E_y), then

$$X = (1.0011100000\ldots000) \times 2^3 \text{ and}$$

$$Y = (1.0010000000\ldots000) \times 2^{-1}$$

Adjust the power of 2 to have a common value as

$$X = (1.00111000000000\ldots000) \times 2^3 \text{ and}$$

$$Y = (0.00010010000000\ldots000) \times 2^3.$$

To subtract Y from X, we can keep X as it is and take the 2's complement of Y as $Y = (1.11101101111\ldots11) \times 2^3$. The 1's complement is taken. Add 1 at the LSB of this result to get the 2's complement value; so the Y is

$$Y = (1.11101110000\ldots00) \times 2^3 \text{ 2's complement is taken.}$$

Now, we can add X to Y, which is equal to X − Y, then

$$X - Y = [(1.00111000000\ldots000) + (1.11101110000\ldots000)] \times 2^3$$

$$X - Y = (1.00100110000\ldots000) \times 2^3.$$

The EoC can be omitted since it is 1. The final value is $X - Y = (1001.001100000\ldots0000)_2$, which can be verified by (9.75 − 0.5625) = 9.1875. The final result has to be represented in a single-precision floating-point number, so apply the standard procedure to get the result, which will be X − Y = $(010000010001001100\ldots00)_2$.

DESIGN EXAMPLE 10.9

Develop a digital system to realize a floating-point subtraction.

Solution

```
library IEEE;
use IEEE.STD_LOGIC_1164.ALL;
use IEEE.STD_LOGIC_arith.ALL;
use IEEE.STD_LOGIC_unsigned.ALL;
entity ieee_754_dp is
generic(n:integer:=8;--integer part
    m:integer:=6);--fraction part
port(clk:in std_logic;
enable:in std_logic;
data_in:in std_logic_vector(n+m downto 0);
data_out:out std_logic_vector(63 downto 0));
end ieee_754_dp;
architecture Behavioral of ieee_754_dp is
type state_type is(rst_state,s1,s2,s3);
signal state:state_type:=rst_state;
signal latch_data:std_logic_vector((n+m) downto 0):=(others=>'0');
signal data_temp:std_logic_vector(63 downto 0):=(others=>'0');
signal count,exponent:std_logic_vector(10 downto 0):=(others=>'0');
signal mantissa:std_logic_vector((n+m-2) downto 0);
signal done:std_logic:='0';
signal count1:integer:=0;
begin
process(clk)
begin
if(clk='1' and clk'event)then
case state is
when rst_state=>
done<='0';
if(enable='1')then
state<=s1;
data_temp<=(others=>'0');
latch_data<=data_in;
else
state<=rst_state;
end if;
when s1=>
data_temp(63)<=latch_data(n+m);
if(latch_data(n+m-1)='0')then
latch_data((n+m-1) downto 0)<=latch_data((n+m-2) downto 0) & '0';
count<=count+"00000000001";
state<=s1;
else
mantissa<=latch_data((n+m-2) downto 0);
exponent<=conv_std_logic_vector((n-1),11) - count+"01111111111";
state<=s2;
end if;
when s2=>
data_temp(62 downto 52)<=exponent;
data_temp(51 downto (51-(n+m-2)))<=latch_data((n+m-2) downto 0);
done<='1';
state<=rst_state;
when others=>null;
```

```
end case;
end if;
end process;
data_out<=data_temp;
end Behavioral;
```

10.3.3 Floating-Point Multiplication

With the knowledge we gained in the preceding sections, we proceed to floating-point multiplication. In this section, we first discuss fixed-point multiplication and then move to floating-point multiplication. The main difference between them is floating the binary point on both numbers: multiplier and multiplicand. Fixed-point multiplication is the same as 2's compliment multiplication [25,133–136] but requires the position of the decimal point to be determined after the multiplication to interpret the correct result. We can understand floating-point multiplication with examples. Assume two decimal numbers, $(6.5627)_{10}$ and $(4.25)_{10}$, and represent them in fixed-point binary numbers of size (1,3,4) and (1,5,2), respectively. Then, their binary equivalent values in the given size are $(0110.1001)_2$ and $(000100.01)_2$. In this example, both numbers are positive numbers, so the sign bit can be omitted. Now, we multiply them as

$$
\begin{aligned}
110.1001 \times 000100.01 &= 0011011111001 \\
&= 0000011011.111001 \\
&= 27.890625
\end{aligned}
$$

Suppose we take another example as (+7 × −5). This can be represented in binary as (0111 × 1101), where the first bit is used to represent the sign of the number. Here, one number is positive and the other is negative; hence, we need to put 1 at the sign bit after multiplication. Then, the result will be (1100011), which is equal to −35. With this knowledge, we enter floating-point multiplication. A brief overview of the floating-point multiplication algorithm is given below. Assume there are two numbers, X and Y. Then, the multiplication result can be represented as

$$
X \times Y = (Sx \text{ XOR } Sy)\,(1.Mx \times 1.My) \times 2^{(Npx + Npy + Np + B)} \tag{10.2}
$$

where Sx and Sy are the sign bits of X and Y, respectively; Mx and My are the mantissa bits of X and Y, respectively and Npx and Npy are the normalized powers of Ex and Ey, respectively. These values can be found by Npx = Ex−B and Npy = Ey−B, and Np is the normalized power of the multiplication of Mx and My, that is, (1.Mx × 1.My). Upon having all the values, sign, exponent, and mantissa, we can write the product value in a single-precision floating-point number.

We can understand the above algorithm with an example. Let us consider two decimal numbers, X and Y, as X = $(125.125)_{10}$, and Y = $(12.0625)_{10}$. The single-precision, that is, (1,8,23), floating-point binary values for these two numbers are X = "01000010111110100100000000000000" and Y = "01000001010000010000000000000000". As explained above, find the sign bit by performing the XOR operation with Sx and Sy, that is, in this example, the sign bit (S) = (0 XOR 0), S = 0; hence, the sign of the final result is positive. We can bring the values of the mantissa from X and Y as Mx = "11110100100…00" and My = "10000010000…00". Then, multiply the mantissa values as given in Equation 10.2, that is, including the hidden 1. The resultant product of the 24 bits (1 hidden bit and 23 mantissa bits) will be in 48 bits.

In this example, $1.Mx_x \times 1.M_y$ = (1.11110100100…00) × (1.10000010000…00), 1.Mx × 1. M_y = (10.1111001010101001000000000000000000000000000000).

If only 1 is at the left side of the binary point, then we can take the product in the normalized form; otherwise, we need to normalize it by floating the binary point to either the right or left side as required by varying the power of 2. In this example, there are 2 bits at the left side of the binary point, so float the point one position to the left and increase the power of 2 to 1 to normalize the result as 1.Mx × 1.My = (1.01111001010101001 00…00) × 2^1. Because of this normalization, the hidden 1 is omitted. Truncate the final mantissa value to 23 bits. Here, the normalized power (Np) is 1.

Now, we need to find the value for the exponent. Ex and Ey can be obtained from the single-precision floating-point representation of the given decimal number. Therefore, Ex = "10000101", Ex = 133, then Npx = 133−127, Npx = 6 and Ey = "10000010", Ey = 130, then Npy = 130−127, Npy = 3. The value for the exponent part of the product can be found by E = Npx + Npy + Np, E = 6 + 3 + 1 + 127, E = 137, and E = "10001001". Now, we have the values for the sign, mantissa, and exponent bits, so the final result in single-precision floating-point format is M1 × M2 = "01000100101111001010101001000000", which is equal to $(1509.3203125)_{10}$. This can also be verified by X × Y = (125.125 × 12.0625) = 1509.3203125.

DESIGN EXAMPLE 10.10

Let us assume two signed decimal numbers, +197.625 and −31.8634, are stored in variables X and Y, respectively, in single-precision floating-point format. Multiply these two numbers and store the final results in the same format.

Solution

The given numbers are X = +197.625 and Y = −31.8634. The single-precision floating-point representations of these numbers are X = "01000011010001011010000000000000" and Y = "11000001111111011101000000111110". The sign bit of the final multiplication result is obtained by S = (Sx XOR Sy), S = (0 XOR 1), S = 1, so the final result is a negative number. Now, we formulate the mantissa parts with the hidden 1, as given in Equation 10.2, as (1.Mx × 1.My). Then, the value will be (1.Mx × 1.My) = (1.10001011010000000000 000) × (1.11111011101000000111110). After multiplication, the result will be (11.0001001100 100000001000110111001100000000000000) × 2^0. Now, we need to normalize this result to get the value for Np. Then, (1.10001001100100000001000110111001100000000000000) × 2^1. The value of Np = 1. The hidden 1 can be omitted; however, in this result, more than 23 bits are present, but only 23 bits are allotted for the mantissa part in the single-precision floating-point number system, so we need to truncate to 23 bits. The truncated result, i.e., the mantissa, will be M = "10001001100100000001001".

Now, we need to find the value for the exponent, for which we require Npx and Npy. Npx = Ex−B, Npx = "10000110"−B, Npx = 134−127, Npx = 7 and Npy = "10000011"−B, Npy = 131−127, Npy = 4. Already, we found the value of Np = 1; then, E = Npx + Npy + Np + B, E = 7 + 4 + 1 + 127, E = 139, E = "10001011".

Now, we have all the values, and the multiplication result in single-precision floating-point format is XxY = (11000101110001001100100000001001).

DESIGN EXAMPLE 10.11

Develop a digital system to perform the multiplication operation on two fixed-point numbers of size (1,m,n), where m is the number of bits for the integer and n is the number of bits for the fractional part. Finally, display the result in a single-precision floating-point format. Use structural modeling to develop the digital architecture to do this operation.

Solution

Since we need to follow structural modeling, we have to develop two parts, namely (i) main and (ii) component parts. In this way, the code is written and given below.

Main Part

```
library IEEE;
use IEEE.STD_LOGIC_1164.ALL;
use IEEE.STD_LOGIC_arith.ALL;
use IEEE.STD_LOGIC_unsigned.ALL;
entity ieee_754_mul is
generic(n:integer:=4;--integer part
    m:integer:=5);--fraction part
port(clk:in std_logic;
enable:in std_logic;
data_in_1:in std_logic_vector(n+m downto 0);
data_in_2:in std_logic_vector(n+m downto 0);
data_out:out std_logic_vector(31 downto 0));
end ieee_754_mul;
architecture Behavioral of ieee_754_mul is
type state_type is(rst_state,s1,s2,s3,s4,s5,s6,s7,s8,s9);
signal state:state_type:=rst_state;
signal latch_data_1,latch_data_2:std_logic_vector((n+m) downto 0):=
(others=>'0');
signal data_temp_1,data_temp_2,data_temp_3,data_temp,datatemp:std_logic_
vector(31 downto 0):=(others=>'0');
signal
count1,exponent1,count2,exponent2,exponent3,exponent,exponent_1,
exponent_f:std_logic_vector(7 downto 0):=(others=>'0');
signal mantissa1,mantissa2,mantissa:std_logic_vector((n+m-1) downto 0);
signal mantissa_1,mantissa3:std_logic_vector((m+m+n+n-1) downto 0);
signal done:std_logic:='0';
component ieee754 is
generic(n:integer:=4;--integer part
    m:integer:=5);--fraction part
port(clk:in std_logic;
enable:in std_logic;
data_in:in std_logic_vector(n+m downto 0);
data_out:out std_logic_vector(31 downto 0);
done:out std_logic;
mantissa_o:out std_logic_vector((n+m-1) downto 0);
exponent_o:out std_logic_vector(7 downto 0));
end component;
signal conv_start,conv_start_1:std_logic:='0';
signal conv_done,conv_done_1:std_logic:='0';
signal mantissa_mul:std_logic_vector((n+n+m+m-1) downto 0);
signal data_in1:std_logic_vector((n+n+m+m) downto 0);
signal data_in:std_logic_vector(n+m downto 0);
signal m1,n1:integer;
signal sign_bit:std_logic:='0';
begin
a1:ieee754
generic map(n=>n,m=>m)
port map(clk=>clk,
enable=>conv_start,
data_in=>data_in,
data_out=>data_temp,
done=>conv_done,
mantissa_o=>mantissa,
exponent_o=>exponent);
a2:ieee754
```

```
generic map(n=>2,m=>(2*m)+(2*n)-2)
port map(clk=>clk,
enable=>conv_start_1,
data_in=>data_in1,
data_out=>datatemp,
done=>conv_done_1,
mantissa_o=>mantissa_1,
exponent_o=>exponent_1);
process(clk)
begin
if(clk='1' and clk'event)then
case state is
when rst_state=>
done<='0';
conv_start<='0';
n1<=n;
m1<=m;
data_out<=(others=>'0');
if(enable='1')then
state<=s1;
data_temp_1<=(others=>'0');
data_temp_2<=(others=>'0');
latch_data_1<=data_in_1;
latch_data_2<=data_in_2;
else
state<=rst_state;
end if;
when s1=>
sign_bit<=latch_data_1(n+m) xor latch_data_2(n+m);
data_in<=latch_data_1;
conv_start<='1';
state<=s2;
when s2=>
conv_start<='0';
if(conv_done='1')then
data_temp_1<=data_temp;
mantissa1<=mantissa;
exponent1<=exponent;
state<=s3;
else
state<=s2;
end if;
when s3=>
n1<=2;
m1<=2*m;
data_in<=latch_data_2;
conv_start<='1';
state<=s4;
when s4=>
conv_start<='0';
if(conv_done='1')then
data_temp_2<=data_temp;
mantissa2<=mantissa;
exponent2<=exponent;
state<=s5;
else
state<=s4;
end if;
when s5=>
mantissa_mul<=mantissa1 * mantissa2;
state<=s6;
```

```
when s6=>
data_in1<=sign bit & mantissa_mul;
conv_start_1<='1';
state<=s7;
when s7=>
conv_start_1<='0';
if(conv_done_1='1')then
data_temp_3<=datatemp;
mantissa3<=mantissa_1;
exponent3<=exponent_1;
state<=s8;
else
state<=s7;
end if;
when s8=>
exponent_f<=exponent1+exponent2+exponent3-"11111110";
state<=s9;
when s9=>
data_out(31)<=sign_bit;
data_out(30 downto 23)<=exponent_f;
data_out(22 downto (22-(m+m+n+n-2)))<=mantissa3((m+m+n+n-2) downto 0);
end case;
end if;
end process;
end Behavioral;
```

Component Part

```
library IEEE;
use IEEE.STD_LOGIC_1164.ALL;
use IEEE.STD_LOGIC_arith.ALL;
use IEEE.STD_LOGIC_unsigned.ALL;
entity ieee754 is
generic(n:integer:=4;--integer part
    m:integer:=5);--fraction part
port(clk:in std_logic;
enable:in std_logic;
data_in:in std_logic_vector(n+m downto 0);
data_out:out std_logic_vector(31 downto 0);
done:out std_logic;
mantissa_o:out std_logic_vector((n+m-1) downto 0);
exponent_o:out std_logic_vector(7 downto 0));
end ieee754;
architecture Behavioral of ieee754 is
type state_type is(rst_state,s1,s2,s3);
signal state:state_type:=rst_state;
signal latch_data:std_logic_vector((n+m) downto 0):=(others=>'0');
signal data_temp:std_logic_vector(31 downto 0):=(others=>'0');
signal count,exponent:std_logic_vector(7 downto 0):=(others=>'0');
signal mantissa:std_logic_vector((n+m-2) downto 0);
--signal done:std_logic:='0';
signal count1:integer:=0;
--signal n:integer:=9;
begin
process(clk)
begin
if(clk='1' and clk'event)then
case state is
when rst_state=>
done<='0';
count<=(others=>'0');
```

```
mantissa_o<=(others=>'0');
exponent_o<=(others=>'0');
--data_temp<=(others=>'0');
if(enable='1')then
state<=s1;
data_temp<=(others=>'0');
latch_data<=data_in;
else
state<=rst_state;
end if;
when s1=>
--data_temp(31)<=latch_data(9);
data_temp(31)<=latch_data(n+m);
if(latch_data(n+m-1)='0')then
latch_data((n+m-1) downto 0)<=latch_data((n+m-2) downto 0) & '0';
count<=count+"00000001";
state<=s1;
else
mantissa<=latch_data((n+m-2) downto 0);
exponent<=conv_std_logic_vector((n-1),8) - count+"01111111";
--exponent<="00000011" - count+"01111111";
state<=s2;
end if;
when s2=>
data_temp(30 downto 23)<=exponent;
data_temp(22 downto (22-(n+m-2)))<=latch_data((n+m-2) downto 0);
done<='1';
mantissa_o<=latch_data((n+m-1) downto 0);
exponent_o<=exponent;
state<=rst_state;
when others=>null;
end case;
end if;
end process;
data_out<=data_temp;
end Behavioral;
```

10.3.4 Floating-Point Division

The floating-point division algorithm is explained below. Division of two floating-point numbers can be done by dividing the mantissas and subtracting the normalized power of 2, that is, Npx and Npy. Assume two floating-point numbers, X and Y. Their division is given by [25,133–137]

$$X/Y = \frac{\left[(-1)^{Sx}(1.Mx \times 2^{Npx})\right]}{\left[(-1)^{Sy}(1.My \times 2^{Npy})\right]}$$

$$X/Y = S\left(\frac{1.Mx}{1.My}\right)2^{(Npx-Npy)}$$

As per the procedure, we need to add the appropriate bias (B) to the power of 2; therefore, the above equation becomes

$$X/Y = S\left(\frac{1.Mx}{1.My}\right)2^{(Npx-Npy)+B} \tag{10.3}$$

If divisor Y is 0, then set the result to infinity; if both X and Y are 0, set it to NaN.

We continue with an example: assume two numbers, X = 127.03125 and Y = 16.9375. The equivalent single-precision floating-point numbers of the above two numbers are X = (010 0001011111110000100000…00) and Y = (010000011000011110000000…0000). Here, the values of X and Y are not 0; therefore, the dividing procedure can be started to get the result. The sign bit(S) is obtained by (Sx XOR Sy) as S = (0 XOR 0) = 0; therefore, the result is a positive number and the values of (1.Mx and 1.My) are 1.Mx = (1.11111100001000000000000) and 1.My = (1.00001111000000000000000). Now, we need to find 1.Mx/1.My. The result for this division will be 1.Mx/1.My = (1.1110000000000000000). Normalize this result if required. This result is already in normalized form; hence, Np = 0.

Now, we have to find the value of the exponent, as given in Equation 10.3, as E = (Npx–Npy) + B. In this example, Npx = Ex–B, Npx = "10000101" – 127, Npx = 133–127, Npx = 6 and Npy = Ey–B, Npy = "10000011" –127, Npy = 131–127, Npy = 4. Then, E = (Npx–Npy) + 127, E = (6–4) + 127, E = 129, E = "10000001". Check for overflow/underflow if $E > E_{max}$ returns overflow, i.e. infinity, and if $E < E_{min}$ returns underflow, i.e. 0. Now, we have found all the values, so the result in single-precision floating-point binary format is X/Y = (01000000111100000000000000000000), which is equal to 7.5.

DESIGN EXAMPLE 10.12

Perform the arithmetic operation X/Y when X = 91.34375 and Y = 0.14453125. Represent the final result in single-precision floating-point format.

Solution

The given decimal numbers are X = 91.34375 and Y = 0.14453125. These numbers are not in any form; therefore, we represent them in binary form as X = "1011011.01011" and Y = "0.00100101". We need to normalize them to get 1.Mx, 1.My, Npx and Npy. The normalized forms of X and Y are X = (1.01101101011) × 2^6 and Y = (1.00101) × 2^{-3}. Now, the value of 1.Mx is "1.01101101011", 1.My is "1.00101", Npx = 6, and Npy = –3.

Now, we have to find 1.Mx/1.My, which is "1.001111"; therefore, the value of the mantissa is "001111". Then, the value of the exponent can be found, as given in Equation 10.3, as E = (Npx–Npy) + B, E = (6–(–3)) + 127, E =136, E = 10001000". Given the two numbers are positive numbers, the sign bit is 0. Then, the single-precision floating-point representation of the result is X/Y = "01000100000111100…0", which is equal to 632.

DESIGN EXAMPLE 10.13

Design and develop a digital system to realize a floating-point division.

Solution

```
library IEEE;
use IEEE.STD_LOGIC_1164.ALL;
use IEEE.STD_LOGIC_arith.ALL;
use IEEE.STD_LOGIC_unsigned.ALL;
entity ieee_754_dp is
generic(n:integer:=8;--integer part
    m:integer:=6);--fraction part
port(clk:in std_logic;
enable:in std_logic;
data_in:in std_logic_vector(n+m downto 0);
data_out:out std_logic_vector(63 downto 0));
end ieee_754_dp;
architecture Behavioral of ieee_754_dp is
type state_type is(rst_state,s1,s2,s3);
```

```
signal state:state_type:=rst_state;
signal latch_data:std_logic_vector((n+m) downto 0):=(others=>'0');
signal data_temp:std_logic_vector(63 downto 0):=(others=>'0');
signal count,exponent:std_logic_vector(10 downto 0):=(others=>'0');
signal mantissa:std_logic_vector((n+m-2) downto 0);
signal done:std_logic:='0';
signal count1:integer:=0;
begin
process(clk)
begin
if(clk='1' and clk'event)then
case state is
when rst_state=>
done<='0';
if(enable='1')then
state<=s1;
data_temp<=(others=>'0');
latch_data<=data_in;
else
state<=rst_state;
end if;
when s1=>
data_temp(63)<=latch_data(n+m);
if(latch_data(n+m-1)='0')then
latch_data((n+m-1) downto 0)<=latch_data((n+m-2) downto 0) & '0';
count<=count+"00000000001";
state<=s1;
else
mantissa<=latch_data((n+m-2) downto 0);
exponent<=conv_std_logic_vector((n-1),11) - count+"01111111111";
state<=s2;
end if;
when s2=>
data_temp(62 downto 52)<=exponent;
data_temp(51 downto (51-(n+m-2)))<=latch_data((n+m-2) downto 0);
done<='1';
state<=rst_state;
when others=>null;
end case;
end if;
end process;
data_out<=data_temp;
end Behavioral;
```

10.4 Xilinx System Generator (SysGen) Tools

In the previous sections, we discussed some basic arithmetic operations with fixed- and floating-point numbers. We have seen some VHDL code design examples for the implementation of all these number-based arithmetic systems. Over and above developing any high-level signal processing applications/algorithms, for example, an FIR filter, it is difficult, or the design becomes highly complicated, once we prefer to use only VHDL code. Therefore, Xilinx-MATLAB provides a hardware simulation environment with many digital block sets through which it is possible to simply build any digital system for FPGAs. The hardware simulation environment supports the various arithmetic, trigonometric, digital,

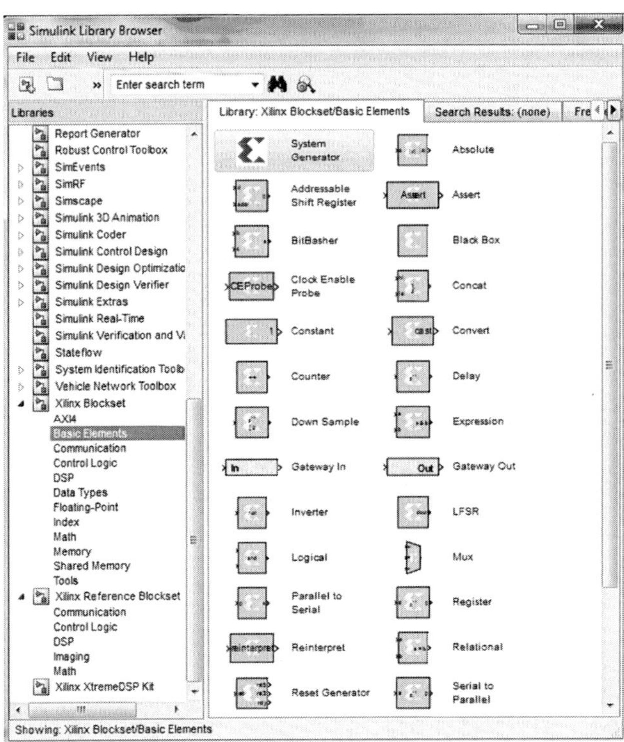

FIGURE 10.5
MATLAB-Xilinx simulation environment window.

communication, control, and other operations, with many data types. To start Xilinx System Generator, select Start → All Programs → Xilinx design tools → System Generator → System Generator. This will start MATLAB and the Simulink simulation environment, as shown in Figure 10.5. The Simulink library browser shows a list of all the different toolboxes installed within MATLAB. Xilinx System Generator components will also appear under the Xilinx blockset that contains all the basic blocks, communication, control logic, data types, DSP, math, indexes and so on. Block sets, as shown in Figure 10.6, are required for various applications. The MATLAB Simulink provides several blocks that can be used for giving input and displaying the outputs to/from the model generated using the Xilinx system generator block sets. Therefore, the developed Xilinx SysGen model can be verified in the Simulink environment. The System Generator supports four data types, namely unsigned numbers, signed numbers (in two's complement form), Boolean (1-bit) and floating-point numbers. Each block will typically have quantization parameters. The initial quantization is defined by the Gateway In blocks. The System Generator fixed-point data type is defined by specifying the total number of bits, then specifying the location of the binary point. The difference which represents the number of bits to the left of the binary point is the integer part. Xilinx FPGAs do not require that fixed-point numbers fall in predefined 8-bit boundaries like DSP processors [42,39,138]. The logic can grow bit by bit to accommodate the required fixed-point precision in FPGA. The Xilinx System Generator Blockset provides several blocks for floating-point arithmetic operations.

There are many advanced block sets available in the recent release of the ISE design suits, namely addressable shift register, assert, output types, bitbasher, CIC compiler 2.0, clock

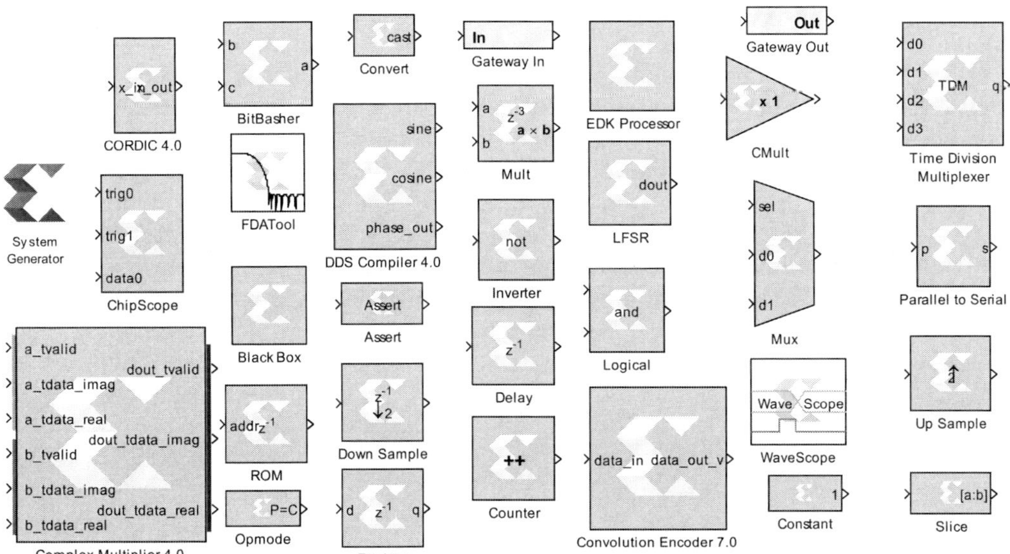

FIGURE 10.6
Some of the Xilinx System generator blocks.

enable port, complex multiplier, convolution encoder, distributive arithmetic finite impulse response (DAFIR) filter, direct digital synthesizer (DDS) complier, depuncture, down-sample, dual-port RAM, FFT, FIFO, from FIFO, from register, indeterminate probe, interleaver/deinterleaver, MicroBlaze processor, negate, network-based Ethernet co-simulation, opmode, parallel-to-serial shift register, pause simulation, picoBlaze Microcontroller, Reed-Solomon decoder, register, reinterpret, relational, reset generator, resource estimator, ROM, sample time, scale, shared memory read/write, shift, simulation multiplexer, single-port RAM, single-step simulation, slice, threshold, time division demultiplexer, time division multiplexer, to register, toolbar, up-sampler, 2-channel decimate by 2 MAC FIR filter, BPSK AWGN channel, CIC filter, CORDIC sincos, CORDIC sqrt, dual-port memory interpolation MAC filter and multiplier. All these block sets have separate specifications based on their operations. Most of the block operations are more dynamic; hence, the operation can be specified by the designer by assigning the appropriate values to the block internal parameters.

10.4.1 Use and Interfacing Methods of Some Blocksets

We discuss the use and interfacing methods of some block sets briefly in this section. Later in this section, following the brief explanation of some SysGen blocks, many more designs are discussed with different block sets so as to expose the readers to gain confidence and the necessary practice to build a SysGen tools-based system for any new application they have. The appropriate explanation of the blocks used in the running text or design examples is given here and there.

BitBasher: The Xilinx BitBasher block performs slicing, concatenation, and augmentation of inputs attached to the block. The block may have up to four output ports. The number of output ports is equal to the number of expressions. Multiple expressions (limited to a maximum of four) can be specified using a new line as a separator between expressions.

CIC Compiler 2.0: The Xilinx Cascaded Integrator-Comb (CIC) Compiler provides the ability to design and implement filters for a variety of Xilinx FPGA devices. The CIC filters, also known as Hogenauer filters, are multirate filters often used for implementing large sample rate changes in digital systems. They are typically employed in applications that have a large excess sample rate. Implementations of CIC filters have structures that use only adders, subtractors, and delay elements. These structures make CIC filters appealing for their hardware-efficient implementations of multirate filtering.

Clock Enable Probe: The Xilinx Clock Enable (CE) Probe provides a mechanism for extracting derived clock enable signals from Xilinx signals in System Generator models. The probe accepts any Xilinx signal type as input and produces a Bool output signal. The Bool output can be used at any point in the design where Bools are acceptable. The probe output is a cyclical pulse that mimics the behavior of an ideal clock enable signal used in the hardware implementation of a multirate circuit. The frequency of the pulse is derived from the input signal's sample period. The probe outputs are run to output gateways and then to the scope for analysis.

Convolution Encoder 7.0: The Xilinx Convolution Encoder block implements an encoder for convolution codes. Ordinarily used in tandem with a Viterbi decoder, this block performs forward error correction (FEC) in digital communication systems. Values are encoded using a linear feed-forward shift register which computes modulo 2 sums over a sliding window of input data. The length of the shift register is specified by the constraint length. The convolution codes specify which bits in the data window contribute to the modulo 2 adder/sum. Resetting the block will set the shift register to zero. The encoder rate is the ratio of input to output bit length.

Down-Sample: The Xilinx Down-Sample block reduces the sample rate at the point where the block is placed in your design. The input signal is sampled at even intervals at either the beginning (first value) or end (last value) of a frame. The sampled value is presented on the output port and held until the next sample is taken. A Down-Sample frame consists of one input sample, where l is the sampling rate. This is an essential block for designing multirate filters.

Dual Port RAM: The Xilinx Dual Port RAM block implements RAM. Dual ports enable simultaneous access to the memory space at different sample rates using multiple data widths. The block has two independent sets of ports for simultaneous reading and writing. Independent address, data, and write enable ports allow shared access to a single memory space. By default, each port set has one output port and three input ports for address, input data and write enable. Optionally, we can also add a port enable and synchronous reset signal to each input port set.

Fast Fourier Transform 7.1: The Xilinx Fast Fourier Transform 7.1 block implements an efficient algorithm for computing the DFT. The N-point, where N = 2m, m = 3 – 16, inverse DFT (IDFT) is computed on a vector of N complex values represented using data widths from 8 through 34. The transform computation uses the Cooley-Tukey decimate-in-time algorithm for Burst I/O architectures and decimation in frequency for pipelined and streaming I/O architectures. Some other advanced blocks like Fast Fourier Transform 8.0 are also available in Xilinx SysGen tools.

FIFO: The Xilinx FIFO block implements a FIFO memory queue. Values presented at the module's data-input port are written to the next available empty memory location when the write-enable input is 1. By asserting the read-enable input port, data can be read out of the FIFO via the data output port (`dout`) in the order in which they were written. The FIFO can be implemented using block or distributed RAM. The full output port is asserted to 1 when no unused locations remain in the module's internal memory. The `percent_full` output port indicates the percentage of the FIFO that is full, represented with user-specified precision. When the empty output port is asserted, the FIFO is empty.

Indeterminate Probe: The output of the Xilinx Indeterminate Probe indicates whether the input data is indeterminate (MATLAB value NaN). An indeterminate data value corresponds to a VHDL indeterminate logic data value of "X". The probe accepts any Xilinx signal as input and produces a double signal as output. Indeterminate data on the probe input will result in an assertion of the output signal indicated by a value 1; otherwise, the probe output is 0.

EDK Processor Block: The EDK Processor block has to be used as an alternative to the MicroBlaze processor block. The Xilinx MicroBlaze processor block provides a way to design and simulate peripherals created to target the EDK MicroBlaze Processor. We can connect customized IP to the MicroBlaze processor via Fast Simplex Links (FSLs) that are available on the processor. The MicroBlaze processor can include a maximum of 8 input and 8 output FSLs (a total of 16). Both synchronous and asynchronous FIFOs can be used. We can create System Generator peripherals that run synchronously or asynchronously to the MicroBlaze processor. Simulation of such systems may be done via hardware co-simulation. Creation, management and configuration of the simulation model is accessed via the block parameter mask.

Network-Based Ethernet Co-Simulation: The Xilinx Network-Based Ethernet Co-Simulation block provides an interface to perform hardware co-simulation through an Ethernet connection over the IPv4 network infrastructure. The port interface of the co-simulation block varies. When a model is implemented for network-based Ethernet hardware co-simulation, a new library is created that contains a custom network-based Ethernet co-simulation block with ports that match the gateway names or port names from the original model. The co-simulation block interacts with the FPGA hardware platform during a Simulink simulation. Simulation data that are written to the input ports of the block are passed to the hardware by the block. Conversely, when data is read from the co-simulation block's output ports, the block reads the appropriate values from the hardware and drives them on the output ports so they can be interpreted in Simulink.

Opmode: The Xilinx Opmode block generates a constant that is a DSP48 or DSP48E instruction. The instruction is an 11-bit value for the DSP48 or a 15-bit value for the DSP48E. The instruction consists of the opmode, carry-in, carry-in select and either the subtract or alumode bits depending upon the selection of DSP48 or DSP48E. The Opmode block is useful for generating DSP48 or DSP48E control sequences. The opmode blocks supply the desired instructions to a multiplexer that selects each instruction in the desired sequence.

Pause Simulation: The Xilinx Pause Simulation block pauses the simulation when the input is nonzero. The block accepts any Xilinx signal type as input. When the simulation is paused, it can be restarted by selecting the Start button on the model toolbar.

PicoBlaze Microcontroller: The Xilinx PicoBlaze Microcontroller block implements an embedded 8-bit microcontroller using the PicoBlaze macro. The block provides access to two versions of PicoBlaze. PicoBlaze 2 only supports Virtex-II and is superseded by PicoBlaze 3, which supports Spartan-3, Spartan-6, Virtex-4, Virtex-5, and Virtex-6. The PicoBlaze 2 macro provides 49 instructions, 32 8-bit general-purpose registers, 256 directly and indirectly addressable ports, and a maskable interrupt. By comparison, the PicoBlaze 3 provides 53 instructions, 16 8-bit general purpose registers, 256 directly and indirectly addressable ports, and a maskable interrupt, as well as 64 bytes of internal scratch pad memory accessible using the STORE and FETCH instructions.

Reed-Solomon Decoder 7.1: The Reed-Solomon (RS) codes are block-based error-correcting codes with a wide range of applications in digital communications and storage. They are used to correct errors in many systems such as digital storage devices, wireless/mobile communications and digital video broadcasting. The RS decoder processes blocks generated by a RS encoder, attempting to correct errors and recover information symbols. The number and type of errors that can be corrected depend on the characteristics of the code. RS codes are Bose-Chaudhuri-Hocquenghem (BCH) codes, which in turn are linear block codes. An (n,k) linear block code is a k-dimensional subspace of an n-dimensional vector space over a finite field. Elements of the field are called symbols. For an RS code, n is ordinarily 2^{s-1}, where s is the width in bits of each symbol. When the decoder processes a block, there are three possibilities: (i) the information symbols are recovered. This is the case provided $2p + r < n-k$, where p is the number of errors and r is the number of erasures; (ii) the decoder reports it is unable to recover the information symbols; or (iii) the decoder fails to recover the information symbols, but does not report an error. The probability of each possibility depends on the code and the nature of the communications channel. Simulink provides excellent tools for modeling channels and estimating these probabilities.

Register: The Xilinx Register block models a D flip-flop-based register, having latency of one sample period. The block has one input port for the data and an optional input reset port. The initial output value is specified by the user in the block parameters dialog box. Data presented at the input will appear at the output after one sample period. Upon reset, the register assumes the initial value specified in the parameters dialog box. The Register block differs from the Xilinx Delay block by providing an optional reset port and a user-specifiable initial value.

Relational: The Xilinx Relational block implements a comparator. The supported comparisons are the following: equal-to (a = b), not-equal-to (a != b), less-than (a < b), greater-than (a > b), less-than-or-equal-to (a <= b), and greater-than-or-equal-to (a >= b). The output of the block is a Boolean. The block parameters dialog box can be invoked by double-clicking the icon in the Simulink model. The only parameter specific to the Relational block is Comparison, which specifies the comparison operation to be computed by the block.

Reset Generator: The Reset Generator block captures the user's reset signal that is running at the system sample rate and produces one or more down-sampled reset signal(s) running at the rates specified on the block. The down-sampled reset signals are synchronized in the same way as they are during startup. The DRDY output signal indicates when the down-sampled resets are no longer asserted after the input reset is detected.

ROM: The Xilinx ROM block is a single port ROM. Values are stored by word and all words have the same arithmetic type: width and binary point position. Each word is associated with exactly one address. An address can be any unsigned fixed-point integer from 0 to d-1, where d denotes the ROM depth (number of words). The memory contents are specified through a block parameter. The block has one input port for the memory address and one output port for data out. The address port must be an unsigned fixed-point integer.

Scale: The Xilinx Scale block scales its input by a power of two. The power can be either a positive or negative integer. The block has one input and one output. The scale operation has the effect of moving the binary point without changing the bits in the container. The only parameter that is specific to the Scale block is Scale factor s. The output of the block is `i*2^k`, where `I` is the input value and `k` is the scale factor. The effect of scaling is to move the binary point, which in hardware has no cost.

Shared Memory: The Xilinx Shared Memory block implements a RAM that can be shared among multiple designs or sections of a design. Instances of Shared Memory, whether within the same model or in different models, or even different instances of MATLAB, will share the same memory space. The System Generator's hardware co-simulation interfaces allow shared memory blocks to be compiled and co-simulated in FPGA hardware. These interfaces make it possible for hardware-based shared memory resources to map transparently to common address spaces on a host PC. When used in System Generator co-simulation hardware, shared memories facilitate high-speed data transfers between the host PC and FPGA, and bolster the tool's real-time hardware co-simulation capabilities.

Shift: The Xilinx Shift block performs a left or right shift on the input signal. The result will have the same fixed-point container as that of the input. Shift direction specifies a direction of shift: left or right. The right shift moves the input towards the least significant bit within its container with appropriate sign extension. The left shift moves the input towards the most significant bit within its container with zero padding of the least significant bits. Bits shifted out of the container are discarded. Number of bits specifies how many bits are shifted. If the number is negative, the direction selected with shift direction is reversed.

Slice: The Xilinx Slice block allows you to slice off a sequence of bits from your input data and create a new data value. This value is presented as the output from the block. The output data type is unsigned, with its binary point at zero. The block provides several mechanisms by which the sequence of bits can be specified. If the input type is known at the time of parameterization, the various mechanisms do not offer any gain in functionality. If, however, a Slice block is used in a design where the input data width or binary point position are subject to change, the variety of mechanisms becomes useful. The block can be configured, for example, always to extract only the top bit of the input, only the integral bits or only the first three fractional bits.

Up-Sample: The Xilinx Up-Sample block increases the sample rate at the point where the block is placed in your design. The output sample period is l/n, where l is the input sample period and n is the sampling rate. The input signal is up-sampled so that within an input sample frame, an input sample is either presented at the output n times if samples are copied, or presented once with (n − 1) zeroes interspersed if zero padding is used. In hardware, the Up-Sample block has two possible implementations. If the Copy Samples option is selected on the block parameters dialog box, the Din port is connected directly to Dout and no hardware is expended. Alternatively, if zero padding is selected, a mux is used to switch between the input sample and inserted zeros.

WaveScope: The System Generator WaveScope block provides a powerful and easy-to-use waveform viewer for analyzing and debugging System Generator designs. The viewer allows you to observe the time-changing values of any wires in the design after the conclusion of the simulation. The signals may be formatted in a logic or analog format and may be viewed in binary, hex, or decimal radices.

Channel Decimate by 2 MAC FIR Filter: The Xilinx n-tap 2 Channel Decimate by 2 MAC FIR Filter reference block implements a multiply-accumulate-based FIR filter. One dedicated multiplier and one Dual Port Block RAM are used in the n-tap filter. The same MAC engine is used to process both channels that are TDMed together. Completely different coefficient sets can be specified for each channel as long as they have the same number of coefficients. The filter also provides a fixed decimation by 2 using a polyphase filter technique. The filter configuration helps illustrate techniques for storing multiple coefficient sets and data samples in filter design. The Virtex FPGA family (and Virtex family derivatives) provide dedicated circuitry for building fast, compact adders, multipliers and flexible memory architectures. The filter design takes advantage of these silicon features by implementing a design that is compact and resource efficient. Implementation details are provided in the filter design subsystems.

BPSK AWGN Channel: The Xilinx BPSK AWGN Channel reference block adds scaled white Gaussian noise to an input signal. The noise is created by the White Gaussian Noise Generator reference block. The noise is scaled based on the SNR to achieve the desired noise variance. The SNR is defined as (E_b/N_o) in dB for uncoded BPSK with unit symbol energy (Es = 1). The SNR input is `UFix8_4` and the valid range is from 0.0 to 15.9375 in steps of 0.0625dB. To use the AWGN in a system with coding and/or to use the core with different modulation formats, it is necessary to adjust the SNR value to accommodate the difference in spectral efficiency. If we have BPSK modulation with rate 1/2 coding and keep Es = 1 and no constant, then $E_b = 2$ and E_b/N_o = SNR + 3 dB. If we have uncoded QPSK modulation with I = +/−1 and Q = +/−1 and add independent noise sequences, then each channel looks like an independent BPSK channel and E_b/N_o = SNR. If we then add rate 1/2 coding to the QPSK case, we have E_b/N_o = SNR + 3 dB. The overall latency of the AWGN Channel is 15 clock cycles. Channel output is a 17-bit signed number with 11 bits after the binary point. The input port SNR can be any type. The reset port must be Boolean, and the input port din must be of unsigned 1-bit type with binary point position at zero.

CIC Filter: Cascaded integrator-comb (CIC) filters are multirate filters used for realizing large sample rate changes in digital systems. Both decimation and interpolation structures are supported. CIC filters contain no multipliers and consist only of adders, subtractors, and registers. They are typically employed in applications that have a large excess sample rate; that is, the system sample rate is much larger than the bandwidth occupied by the signal. CIC filters are frequently used in digital down-converters and digital up-converters. To read the annotations, place the block in a model, then right-click on the block and select Explore from the popup menu. Double-click on one of the sub-blocks to open the subblock model and read the annotations.

CORDIC SINCOS: The Xilinx CORDIC SINCOS reference block implements Sine and Cosine generator circuits using a fully parallel CORDIC algorithm in circular rotation mode. That is, given input angle z, it computes the output cosine (z) and sine (z). The CORDIC processor is implemented using building blocks from the Xilinx blockset. The CORDIC sine cosine algorithm is implemented in the three steps. (i) Coarse angle rotation: the algorithm converges only for angles between –pi/2 and pi/2. If z > pi/2, the input angle is reflected to the first quadrant by subtracting pi/2 from the input angle. When z < –pi/2, the input angle is reflected back to the third quadrant by adding pi/2 to the input angle. The sine cosine circuit has been designed to converge for all values of z, except for the most negative value. (ii) Fine angle rotation: by setting x equal to 1/1.646760 and y equal to 0, the rotational mode CORDIC processor yields the cosine and sine of the input angle z. (iii) Coordinate correction: if there was a reflection applied in step (i), this step applies the appropriate correction. For z > pi/2, using z = t + pi/2, then sin(z) = sin(t).cos(pi/2) + cos(t).sin(pi/2), which is also equal to cos(t)cos(z) = cos(t).cos(pi/2)−sin(t). sin(pi/2), which is equal to −sin(t). For z < pi/2, using z = t − pi/2, then sin (z) = sin(t).cos(−pi/2) + cos(t).sin(−pi/2) = −cos(t) cos (z) = cos(t).cos(−pi/2)−sin(t).sin(−pi/2) = sin(t).

CORDIC SQRT: The Xilinx CORDIC SQRT reference block implements a square root circuit using a fully parallel CORDIC algorithm in hyperbolic vectoring mode. It computes the output sqrt (x) for the given x. The CORDIC processor is implemented using building blocks from the Xilinx blockset. The square root is calculated indirectly by the CORDIC algorithm by applying the identity listed as follows. sqrt (w) = sqrt $\{(w + 0.25)^2 (w - 0.25)^2\}$. The CORDIC square root algorithm is implemented in the following four steps. (i) Coordinate rotation: the CORDIC algorithm converges only for positive values of x. If x < zero, the input data is converted to a nonnegative number. If x = 0, a zero detect flag is passed to the coordinate correction stage. The square root circuit has been designed to converge for all values of x, except for the most negative value. (ii) Normalization: the CORDIC algorithm converges only for x through 0.25 and 1. During normalization, the input x is shifted to the left until it has a 1 in the most significant nonsigned bit. If the left shift results in an odd number of shift values, a right shift is performed, resulting in an even number of left shifts. The shift value is divided by 2 and passed on to the coordinate correction stage. The square root is derived using the identity sqrt(w) = sqrt{(w + $0.25)^2$ – (w – $0.25)^2$}. Based on this identity, the input x gets mapped to, X = x + 0.25 and Y = x – 0.25. (iii) Hyperbolic rotations: for sqrt ($X^2 - Y^2$) calculation, the resulting vector is rotated through progressively smaller angles, such that Y goes to zero. (iv) Coordinate correction: if the input was negative and a left shift was applied to x, this step assigns the appropriate sign to the output and multiplies it with 2-shift. If the input was zero, the zero detect flag is used to set the output to 0.

Multiplier: There are two basic properties for deciding the output precision in the multiplier block; one is full and another is user-defined. In the full-precision option, the multiplier block uses the input fixed-point format to determine the format of the output. In this case, the full precision requires 32 bits with the binary point at bit position 24. In the user-defined precision mode, the designer can specify a different format. In this case, the designer needs to specify a rounding method for excess data values. Also, in the implementation menu, the user has the option of optimizing for speed/area and also using the dedicated multipliers embedded in the FPGA or just the LUTs instead.

More explanations of the SysGen blocks briefed above and other blocks can be found in References 39, 138, and 139. There are plenty of resources are available on the internet to understand the basic applications of all the block sets available, even in the recent Xilinx ISE or Vivado Design Suites.

10.4.2 System Design and Implementation Using SysGen Tools

We will become familiar with the Xilinx SysGen tools using the example-based approach below. Given this aim, now we discuss some simple Xilinx System-Generator-based model designs in the following sections. First, we discuss the design of a Simulink model for evaluating Equation 10.4. In this example, we discuss a step-by-step approach to design and understand the Xilinx Simulink model.

$$y(a,b) = 5a + 3b \tag{10.4}$$

The main arithmetic operations required to implement this equation are: (i) two multiplications, (ii) one addition, (iii) two memory elements to store the constants 5 and 3, and (iv) two input terminals for getting inputs a and b. Each of these operations can be performed using the existing System Generator Block sets. Before starting any design, the first block that needs to be used for any System Generator model is the System Generator Block. This block can be found in the Simulink category by navigating to Xilinx Blockset → Basic Elements → System Generator. Place the System Generator block in the new model window, as shown in Figure 10.7, either by dragging and dropping or right-clicking on the blockset and choosing the file name. All blocksets can be brought to the simulation window by the same way. The blocksets used in Figure 10.7 to realize Equation 10.4 are explained below.

System Generator: The System Generator block provides control of system and simulation parameters and is used to invoke the code generator. The System Generator block is also referred to as the System Generator "token" because of its unique role in the design. Every Simulink model containing any element from the Xilinx blockset must contain at least one System Generator block (token). Once a System Generator block is added to a

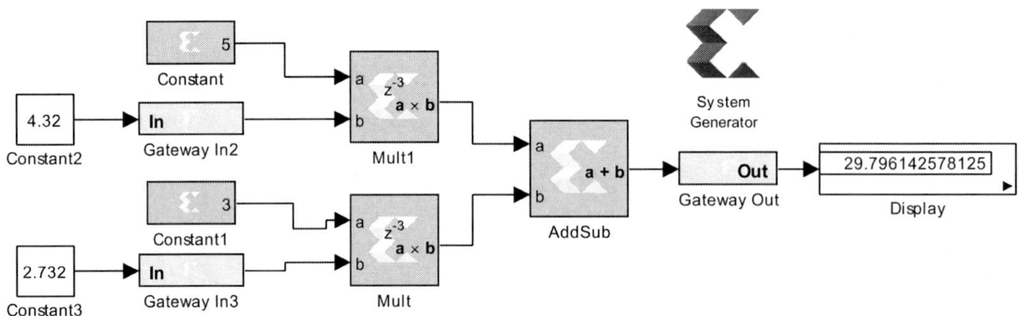

FIGURE 10.7
A simple Xilinx model for realizing Equation 10.4.

model, it is possible to specify how code generation and simulation should be handled. The System Generator token differs from other blocks in one significant manner, as multiple sets of parameters are stored for an instance of a System Generator block. The different sets of parameters stored correspond to different compilation targets available to the System Generator block. The `compilation` parameter is the switch used to toggle between different compilation targets stored in the System Generator block. In order to get or set parameters associated with a particular compilation type, it is necessary to first use `xlsetparam` to change the `compilation` parameter to the correct compilation target before getting or setting further values.

Constant: The Xilinx Constant block generates a constant that can be a fixed-point value, a Boolean value or a DSP48 instruction. This block is similar to the Simulink constant block, but can be used to directly drive the inputs on Xilinx blocks, as shown in Figure 10.7, or DSP48 instruction mode. The constant block, when set to create a DSP48 instruction, is useful for generating DSP48 control sequences. The example implements a 35 × 35-bit multiplier using a sequence of four instructions in a DSP48 block and can be found in [138–141]. The constant blocks supply the desired instructions to a multiplexer that selects each instruction in the desired sequence. The parameter settings for this block are type: signed (2's comp), constant value: 5, number of bits: 7, and binary value: 3.

Gateway In: The Xilinx Gateway In blocks are the inputs into the Xilinx portion of our Simulink design. These blocks convert Simulink integer, double, and fixed-point data types into the System Generator fixed-point type. Each block defines a top-level input port in the HDL design generated by System Generator. While converting a double type to a System Generator fixed-point type, the Gateway In uses the selected overflow and quantization options. For overflow, the options are to saturate to the largest positive/smallest negative value, to wrap, that is, to discard bits to the left of the most significant representable bit, or to flag an overflow as a Simulink error during simulation. For quantization, the options are to round to the nearest representable value or to the value furthest from zero if there are two equidistant nearest representable values, or to truncate, that is, to discard bits to the right of the least significant representable bit. It is important to realize that overflow and quantization do not take place in hardware; they take place in the block software itself before entering the hardware phase. The parameters settings for this block are output type: signed (2's comp), number of bits: 7, binary point: 3, quantization: truncate, overflow: saturate and sample period: 1.

Mult: The Xilinx Mult block implements a multiplier. It computes the product of the data on its two input ports, producing the result on its output port. Parameters specific to the Basic tab are as follows. Latency: this defines the number of sample periods by which the block's output is delayed. When saturation or rounding is selected on the user data type of a multiplier, latency is also distributed so as to pipeline the saturation/rounding logic first, and then additional registers are added to the core. For example, if a latency of three is selected and rounding/saturation is selected, then the first register will be placed after the rounding or saturation logic and two registers will be placed to pipeline the core. Registers will be added to the core until optimum pipelining is reached, then further registers will be placed after the rounding/saturation logic. However, if the data type we select does not require additional saturation/rounding logic, then all the registers will be used to pipeline the core. The parameters setting for this block are: under head Basic; precision: full, latency: 3 and under head implementation; optimize for: speed and use embedded multiplier has to be ticked.

AddSub: The Xilinx AddSub block implements an adder/subtractor. The operation can be fixed (+ or −) or changed dynamically under control of the submode signal that specifies the block operation to be Addition, Subtraction, or Addition/Subtraction. When Addition/Subtraction is selected, the block operation is determined by the sub input port, which must be driven by a Boolean signal. When the sub input is 1, the block performs subtraction. Otherwise, it performs addition. Provide carry-in Port, when selected, allows access to the carry-in port, that is, Cin. The carry-in port is available only when user-defined precision is selected and the binary point of the inputs is set to zero. Provide carry-out Port, when selected, allows access to the carry-out port, that is, Cout. The carry-out port is available only when user-defined precision is selected, the inputs and output are unsigned and the number of output integer bits equals x, where x = max (integer bits a, integer bits b). The pout parameter settings for this block are: under basic; operation: addition, latency: 0, under output type, precision: full and under implementation, implement using: Fabric.

Gateway Out: Xilinx Gateway Out blocks are the outputs from the Xilinx portion of our Simulink design. This block converts the System Generator fixed-point data type into Simulink Double. According to its configuration, the Gateway Out block can either define an output port for the top level of the HDL design generated by System Generator or be used simply as a test point that will be trimmed from the hardware representation. The Xilinx Gateway Out block is used to provide a number of functions: converting data from SysGen fixed-point type to Simulink double, defining I/O ports for the top level of the HDL design generated by System Generator, defining test-bench result vectors when the System Generator Create Test-bench box is checked and naming the corresponding output port on the top-level HDL entity. The parameter settings for this block are translate into output port has to be ticked and input/output buffer (IoB) timing constraint: None.

In Figure 10.7, the Constant 2, Constant 3 and Display blocks are taken from the MATLAB Simulink library. The System Generator, Constant, Gateway In, Mult, Addsub, and Gateway out are taken from the SysGen bock sets. Double-click on the blocksets to change their properties. We can change the block name by clicking on the existing name. The system we are building needs to performs two multiplications between the two inputs (a and b) and two constant/factors (5 and 3). The inputs can be given/varied through the MATLAB constants while the constant values for factors are being given by Xilinx constants. However, in a typical DSP application, we may need to use memory elements to give the values for the constant. Double-click on the Xilinx constant blocksets to display the constant properties, change the constant values to 5 and 3 and directly connect them to the multipliers, as shown in Figure 10.7. To make connections between blocks, click on the output terminal of one block with the mouse button and drag the connection link down to the input terminal of another block. Connect MATLAB constant blocks into multipliers through Gateway In blocks, as shown in Figure 10.7. Connect the outputs of the two multipliers to the inputs of the AddSub block.

Connect the output of the AddSub block into the input of Gateway Out and then, in the same way, into Display as well. The construction of Xilinx hardware blocks is over for the intended application. Now, we will use the MATLAB Simulink environment to verify its functionality. Simulink offers a very flexible simulation environment which allows the building of different testing/analysis scenarios. We will build a testing scenario for our system by applying a constant input at the variable and displaying the result on the single-value display. The simulation process can be started by clicking the Start Simulation button in the toolbar of the Xilinx model simulation window. The execution of the model can be performed using several methods that we discuss one by one in the forthcoming sections. Now, we use MATLAB Simulink to perform functional simulation. For this purpose, no change is required in the present form of the design. After starting the simulation process, System Generator starts to process each block in the design/model and generate a simulation model according to the specific configurations of each block, as shown in Figure 10.7. Upon successful operation of the design, the corresponding result will be displayed in the Display block. For example, if $a = 4.32$ and $b = 2.732$, as shown in Figure 10.7, then $y = 4.32 \times 5 + 2.732 \times 3$, $y = 29.796$.

The SysGen tool will also create project files (HDL) that we can synthesize, simulate, and implement in the FPGA with Xilinx ISE Project Navigator or Vivado design suite. One of the advantages of Xilinx System Generator is the capability of generating HDL code from our designs. By following the steps given below, we can generate the VHDL code and analyze the implemented design using reports generated in ISE foundation. Once the simulation is over at the expected accuracy, then we need to enter the IoB pad locations of the target device for the implementation of our design in FPGA. In the case where the accuracy of the final result of our design is not up to the expected level, then we need to adjust the total number of bits and number of bits allotted to the binary point at each stage and/or the main stage in the design. By increasing them, it is possible to achieve the expected accuracy. However, we need to encounter the number of bits available for input and output of the external devices that connected to the FPGA. We need to note that the largest fixed-point precision declared at the inputs (i.e., Gateway in) is copied for the output (i.e., Gateway out) by the software itself. To implement our design in the FPGA, we need to enter the pin locations for the inputs and outputs of the design. Just by double-clicking on Gateway in and Gateway out, we will see a window, as shown in Figure 10.8, through which we need to give the pin details. Once the window, as shown in Figure 10.8, pops up, select the implementation and enter the pin details in the given location under the IoB pad locations. The entry has to be MSB through LSB, such as {`'p34','p35','p36',...'p24'`}. Finally, to conform the pin entry, click apply and then OK. We need to repeat this type of pin detail entry for all the external inputs and outputs. We can generate the new bit file by editing the pin assignment netlist before fusing/programming into the targeted FPGA.

Now our design has logic, that is, input-output mapping and input-output pin details. Next, we need to choose the target device of interest to implement the design in the FPGA. To do this and other subsequent tasks, we need to double-click on the System Generator token, which will then pop up a window, as shown in Figure 10.9. There are three different heads at the top of the window, "Compilation", "Clocking" and "General". Under the main head "Compilation", there are many subheads; we will see them one by one below. The first is "compilation", through which we can select HDL Netlist, NGC Netlist, Bitstream, EDK Export Tool, Hardware Co-simulation and timing and power analysis. Here, we need to select the HDL Netlist since we are interested in generating the HDL

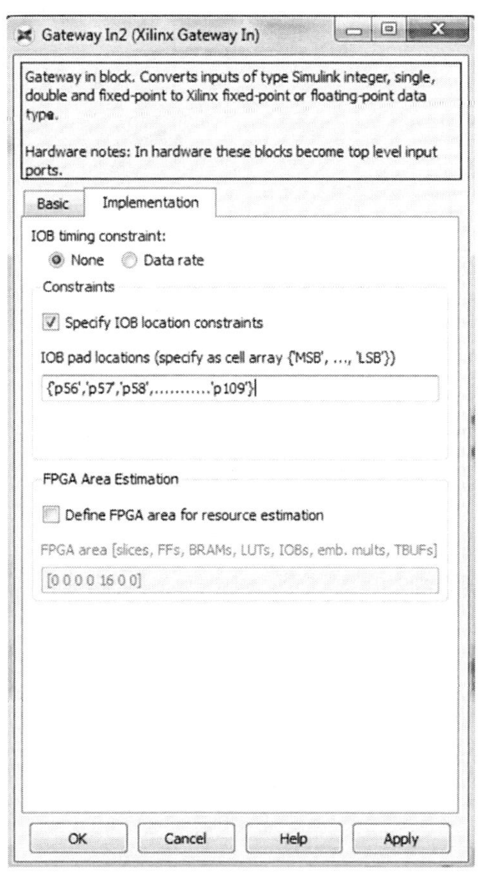

FIGURE 10.8
Input pin configuration for implementation.

code for our design to implement it in the FPGA. The remaining options are discussed in Section 9 and subsequent chapters. The second one is "part", which helps us choose all the details associated with the target device. The third one is "Synthesis tools" and "Hardware Description Language", whose selections have to be "XST" and "VHDL", respectively. The fourth one is "Target directory"; this is the place where we need to select the location we wish to have the HDL code. The next option is "Project Type", where we need to select the Project Navigator. The settings in this example are Compilation: HDL Netlist, Part: Virtex6 xc6vsx315t–3ff1156, Synthesis Tool: XST, Hardware Description Language: VHDL, Target Directory: .../ise (select the directory where we wish to save the VHDL project), Project Type: Project Navigator and the other options, namely "Create interface document" and "Create Testbench", have to be unchecked.

Now, select the second head, that is, "Clocking", where we need to give the FPGA onboard clock period in terms of nanoseconds (ns) and clock pin location on the board. For example, FPGA clock period (ns): 100 (if the clock frequency is 10 MHz) and Clock pin location: p149 (enter clock pin number). This information can be found in the user's manual of the FPGA development board. In the third head, that is, "General", the Default has to be set under "Block icon display". Once all these settings are done, then our design is ready to generate the HDL code. Click "OK" and then "Generate" at the bottom of the same window to generate the VHDL code and ISE Project files. Once code generation is complete, the

FIGURE 10.9
Configurations of target device details.

VHDL file will be ready in the selected directory. To open the generated project, that is, the VHDL code, go to selected directory and open the <…>.xise, that is, ISE file. The file will be opened in the Xilinx environment from where we can begin our classical methods of checking syntax, synthesizing XST, simulation, and implementation. Generated VHDL code can be simulated using any simulator like ISim, Model Simulator, or Test Bench Waveform.

The simulation result of Figure 10.7 is shown in Figure 10.10. Just for explanation purposes, a fixed-point of size (7,3) is used in all the input blocks, that is, constant, Gateway in2, constant1, and Gateway in3. Therefore, the resolution of all the inputs will be 4 bits for the

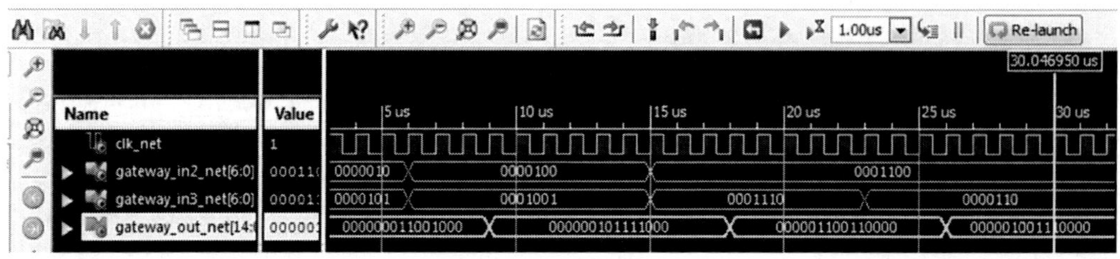

FIGURE 10.10
SysGen VHDL code simulation window (time versus data).

magnitude and 3 bits for the fractional part. Since the type: signed (2's comp) is selected, there is a bit to represent the sign of the inputs. As long as the MSB bit is '0', the input will be a positive number; otherwise, it is a negative number. Therefore, the result will be in 15 bits with 1 bit for the sign, 8 bits for the integer and 6 bits for the fractional part. As we have seen already in Chapter 4, $N \times N$ requires 2N bits to store the final product. In Figure 10.10, four different simulation results are given. In the first example, two external inputs 0.25, that is, "000010", and 0.625, that is, "0000101", are given to the system. The output will be $y = 5 \times 0.25 + 3 \times 0.625$, $y = 1.25 + 1.875$, $y = 3.125$, which is equal to, in 15-bit form, "000000011001000", as shown in Figure 10.10. Here, the first bit is for the sign, another 8 bits are for the integer and the last 6 bits are for the fraction. In the same way, the second result is $y = 5 \times 0.5 + 3 \times 1.125$, $y = 2.5 + 3.375$; $y = 5.875$, which is equal to "000000101111000". The same analysis can be applied for another result as well. Then, the classical method of bit file generation and implementation can be followed to download the code into the selected FPGA. In the same way, VHDL code can be generated and simulated for any Xilinx SysGen-based design, and, finally, the design can be implemented into the FPGA for real-time application/verification. In the following sections, we will discuss only Xilinx SysGen based designs for different applications/logics, and generating/simulating VHDL code and its implementation in the FPGA for onboard verification are left to the readers.

DESIGN EXAMPLE 10.14

Design a MAC unit using the Xilinx SysGen tools. Assume that one input to the MAC unit is a constant, that is, $(14)_{10}$, while another input is from a 4-bit unsigned binary counter. Accumulate the product values and display the constant, counter value and accumulated results in a scope.

Solution

Before starting the model design as given in this design example, we need to understand basic operation of the necessary blocks. In this example, we are in need of constant, counter, multiplier, accumulator, and scope blocks to complete our design task. We have discussed the operation of the constant block already in this section. We now discuss the counter and accumulator blocks below.

Counter: The Xilinx Counter block implements a free-running or count-limited type of an up, down, or up/down counter. The counter output can be specified as a signed or unsigned fixed-point number. The free-running up, down, or up/down counter can also be configured to load the output of the counter with a value on the input din port by selecting the Provide Load Pin (PLP) option in the block's parameters. For the free-running up/down counter, the counter performs addition when the input up port is 1 or subtraction when the input up port is 0. A count-limited counter is implemented by combining a free-running counter with a comparator, which is limited to only 64 bits of output precision. Count-limited types of a counter can be configured to step between the initial and ending values provided the step value evenly divides the difference between the initial and ending values. The count-limited down counter calculation replaces addition with subtraction. For the count-limited up/down counter, the counter performs addition when the input up port is 1 or subtraction when the input up port is 0. The parameters setting for this block are: under basic, type: unsigned, constant value: 14, number of bits: 4 and binary input: 0.

Accumulator: The Xilinx Accumulator block implements an adder- or subtractor-based scaling accumulator. The block's current input is accumulated by a scaled current stored value. The scale factor is a block parameter. The block has an input b and an output q. The output must have the same width as the input data. The output will have the same arithmetic type and binary point position as the input. The output q is calculated as follows: A subtractor-based accumulator replaces addition of the current input b(n) with subtraction. When reset is selected, the output of the accumulator is reset to the data on input port b; otherwise, the output of the accumulator is reset to zero. This option is available only when the block has a reset port. Using this option has clock speed implications if the accumulator is in a multirate system. In this case, the accumulator is forced to run at the system rate since the Clock Enable (CE) signal driving the accumulator runs at the system rate and the reset-to-input operation is a function of the CE signal. The parameters setting for this block are: under basic, operation: add, number of bits: 13, over flow: wrap, feedback scaling: 1 and latency: 0.

The parameters settings for other blocks are given below:

1. Constant: under basic, output type: fixed-point, arithmetic type: unsigned, number of bits: 4 and binary point:0.
2. Mult: under basic, precision: full and latency: 0. Under implementation, optimized for: speed and tick the use embedded multipliers.
3. Gateway out: tick "translate into output port" and IoB timing constraint: data rate.
4. Scope: under general, number of axis: 3, time range: auto, tick label: all and sampling decimation: 1.

With this information about the blocks, we start our design as required for this example, as shown in Figure 10.11, where a constant 14 and a 4-bit counter are connected to a multiplier. The output of the multiplier is connected to the accumulator. Finally, all the outputs are passed to a scope through Gateway out blocks. Once the design is complete, set the simulation time to 650 and run the design. Once the design has run, open the scope, which will display the simulation response of the design, as shown in Figure 10.12, where the upper plot is a constant, that is, $(14)_{10}$, and the middle plot corresponds to the counter output. Since the a 4-bit counter is used in this example, it runs from "0000" through "1111", that is, $(0)_{10}$ through $(15)_{10}$. The lower plot corresponds to the value of the accumulator at every sampling instant. Figure 10.12 is generated from the scope plots using the MATLAB code given in Appendix A.

DESIGN EXAMPLE 10.15

Design a digital system using the Xilinx SysGen tools to add two sine waves. Assume that the two sine waves are in the same amplitude, frequency and phase. Display the

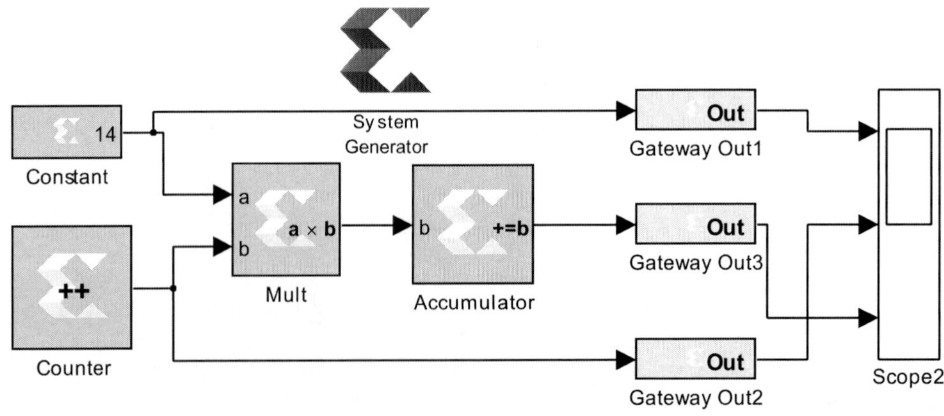

FIGURE 10.11
A MAC unit design.

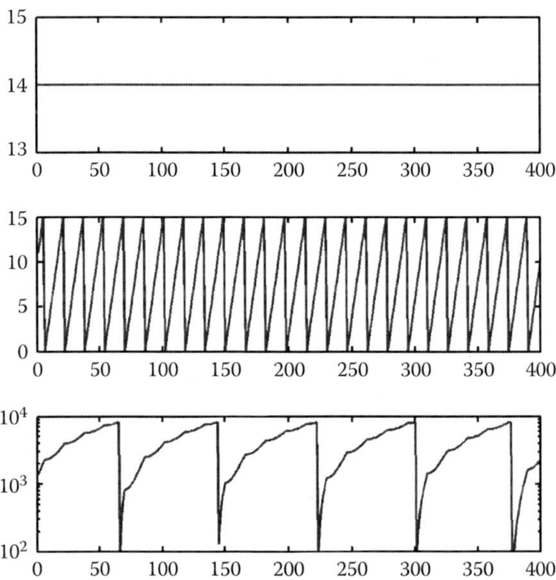

FIGURE 10.12
Performance simulation of a MAC unit (magnitude versus time).

sine waves that are being generated by MATLAB sine wave sources, their digital forms, that is, Gateway in outputs, and finally the sum result in a scope.

Solution

We have discussed all the blocks that are needed for designing the digital system as given in this example. Hence, we can directly start the design process, which will appear as shown in Figure 10.13 upon completing the design.

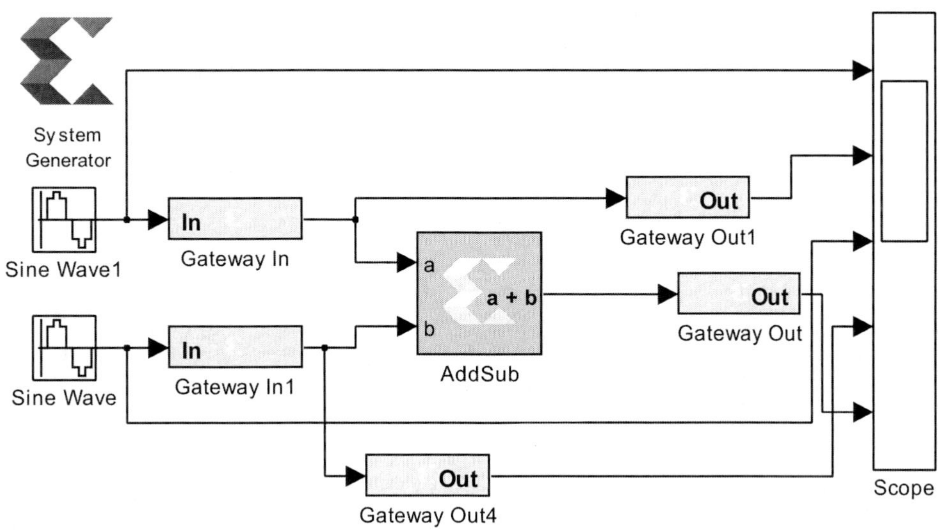

FIGURE 10.13
A design to add and display two sine waves.

The design parameter settings in this design are given below:

1. Sine Wave: sine type: time based, time (t): user estimation time, amplitude: 1, bias: 0, frequency (rad/sec): 0.25, phase (rad): 0 sample time: 0.01 and tick "interpret vector parameters as 1-D".
2. Gateway in: under head base, output type: fixed-point, arithmetic type: signed (2's comp), number of bits: 20, binary point: 16, quantization: truncate, overflow: wrap and sample period: 1.
3. AddSub: under basic, operation: addition, latency: 0 and under output, precision: full.
4. Gateway out: tick "truncate into output port" and IoB timing constraint: none.
5. Scope: under general, number of axis: 3, time range: auto, tick label: all and sampling decimation: 1.

Once the design processes and parameter setting are over, then set the simulation time to 10000 and then simulate the design. The simulation results in the display scope will be as shown in Figure 10.14, where the first and third plots correspond to the sine waves generated by the MATLAB blocks, while the second and fourth plots correspond to their digital forms, i.e., respective Gateway In blocks. The last plot is the sum of the second and fourth plots. Since the frequency and phase are same for the both input signals, only the amplitude variation ±2 V is obtained.

DESIGN EXAMPLE 10.16

Develop a digital system to sample a sine wave with a 1-bit PRBS sequence using the natural sampling technique. Display all the signals in a scope.

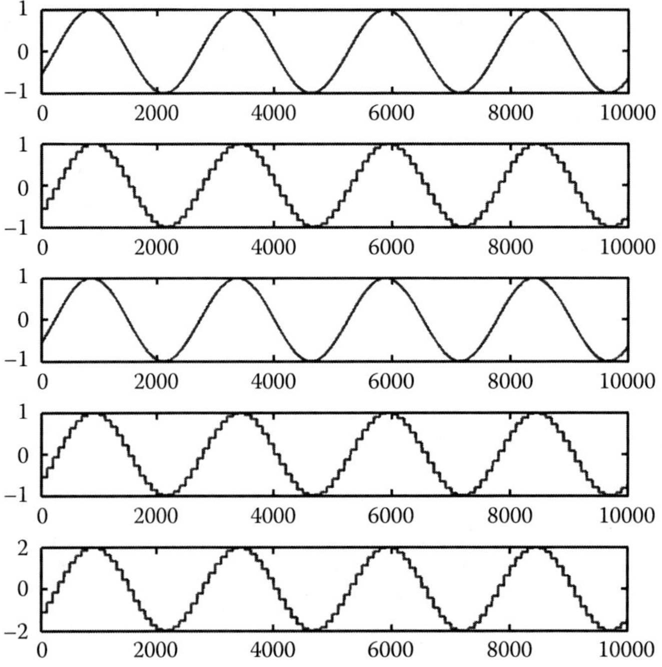

FIGURE 10.14
Simulation result of a design shown in Figure 10.13 (magnitude versus time).

Solution

The SysGen model can be constructed to perform the given task with the blocks we discussed in the preceding sections. However, we are in need of a new block called the PRBS generator block. In Xilinx SysGen tools, an LFSR would perform the operation of a PRBS generator; hence, we need to understand the operation of an LFSR block first as below.

LFSR: The Xilinx LFSR block implements an LFSR. This block supports both the Galois and Fibonacci structures using either the XOR or XNOR gate and allows a reloadable input to change the current value of the register at any time. The LFSR output and reloadable input can be configured as either serial or parallel. The parameter settings for this block are: under basic, type: galois, gate type: XNOR, number of bits in LFSR: 16, Feedback polynomial (enter hex value enclosed with ticks): '1011' and initial value (enter hex values enclosed with ticks): '03AD'. Under advanced, explicit sample period: 1.

Now, we start designing the system module with all the necessary blocks, and it will appear as shown in Figure 10.15.

The design parameters setting of other blocks in this design are given below:

1. Sine Wave: sine type: time based, time (t): user estimation time, amplitude: 2, bias: 0, frequency (rad/sec): 0.025, phase (rad): 0 sample time: 0.01 and tick the interpret vector parameters as 1-D.
2. Gateway in: under head base, output type: fixed-point, arithmetic type: signed (2's comp), number of bits: 8, binary point: 3, quantization: round, overflow: saturate and sample period: 1.
3. Mult: under basic, precision: full and latency: 0. Under implementation, optimized for: speed and tick the "use embedded multipliers".
4. Gateway out: tick "truncate into output port" and IoB timing constraint: none.
5. Scope: under general, number of axis: 4, time range: 5, tick label: all and sampling decimation: 1.

The scope output is shown in Figure 10.16, where the first two figures correspond to the sine wave input of MATLAB and Gateway in, respectively. The third plot is the output of the LFSR. Finally, the fourth plot is the sampled version of second plot, that is, Gateway in output. The output of the LFSR is used as the sampling signal, and the sampled version of the sine wave is generated. The final output is also like a PAM output of the second pot.

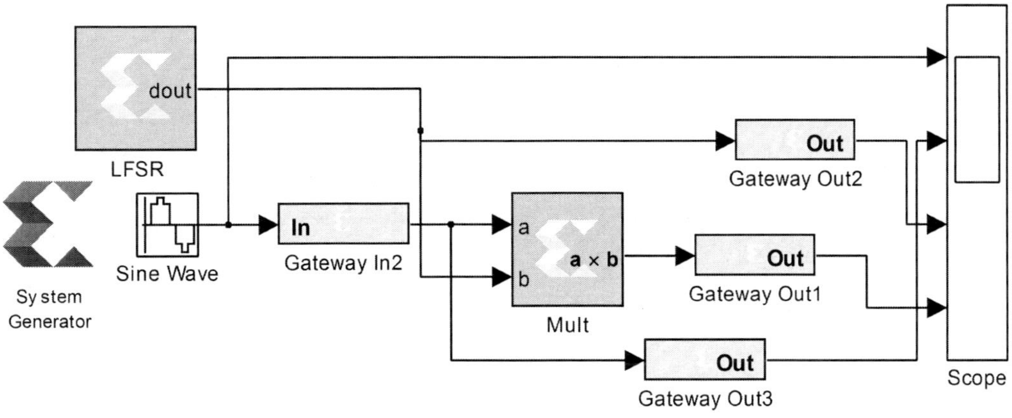

FIGURE 10.15
A design to sample a sine wave with a PRBS sequence.

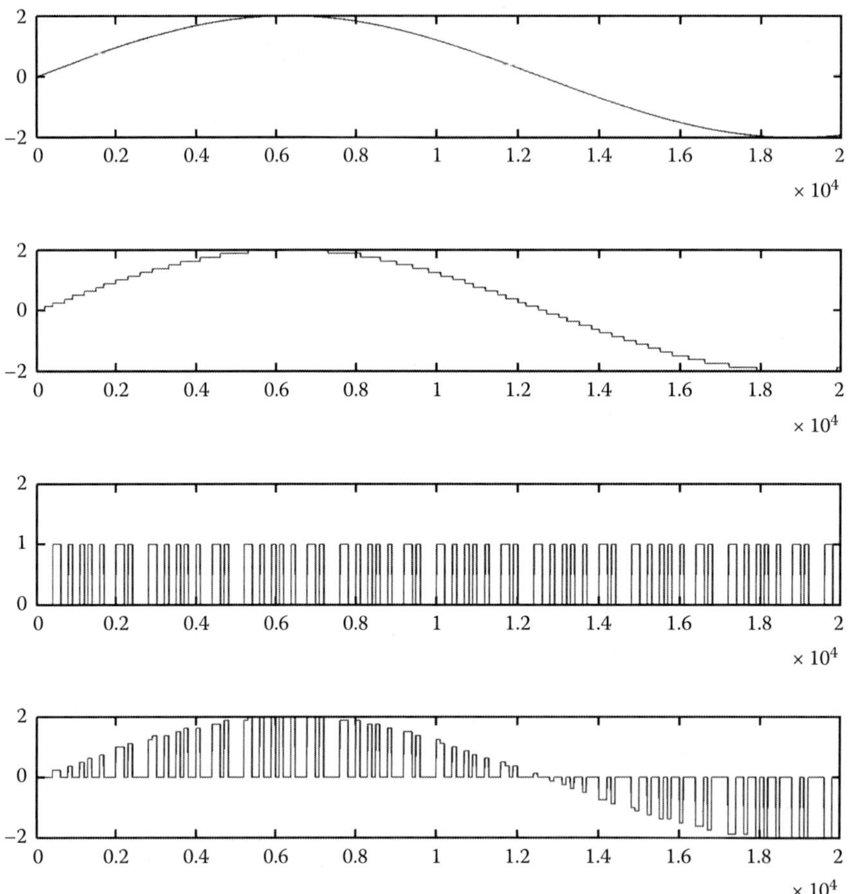

FIGURE 10.16
Simulation result of Figure 10.15 (magnitude versus time).

10.5 Fractional-Point Computation Using SysGen Tools

In the previous section, we introduced some basic designs using the Xilinx SysGen tools. In this section, we discuss some signed floating-point computation and complex multiplications along with time division multiplexing and demultiplexing, serial-to-parallel conversion, simple FIR filters and delay elements. Consider Figure 10.17, which corresponds to signed fractional-point computation with two signed inputs and one signed output. The new block used in this design is CMult, whose operations are seen below before analyzing the computation flow of the design.

CMult: The Xilinx CMult block implements a gain operator with output equal to the product of its input by a constant value. This value can be a MATLAB expression that evaluates to a constant. The block parameter dialog box can be invoked by double-clicking the icon in the Simulink model. The basic tab parameters are: constant value, which may be a constant or an expression. If the constant cannot be expressed exactly in the specified fixed-point type, its value can be rounded and saturated as needed. A positive value is implemented as an unsigned number, while a negative value is as signed. The constant

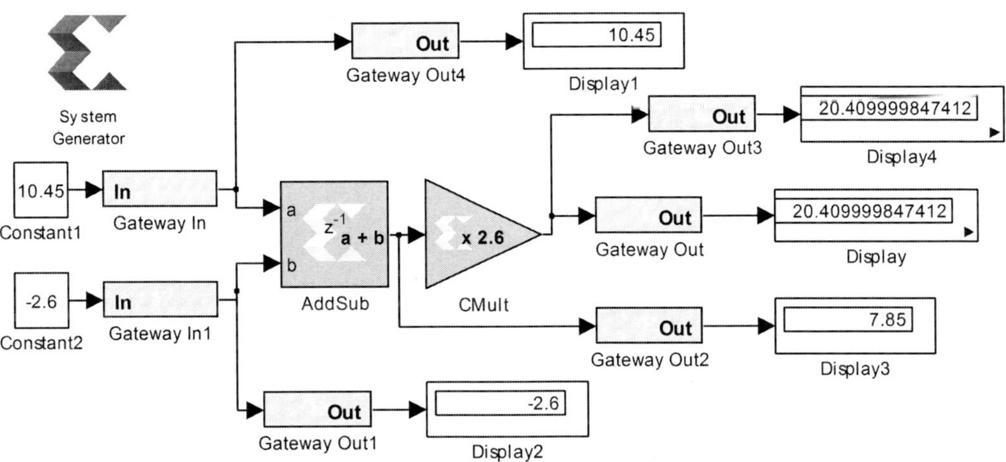

FIGURE 10.17
Fractional-point calculations with stagewise-result display.

number of bits specifies the bit location of the binary point of the constant where bit zero is the least significant bit. The parameters on the output type tab define the precision of the output of the CMult block [142]. The parameter settings of this block are: under basic, constant value: 2.6, constant-type: fixed-point, number of bits: 32, binary point: 23, latency: 0. Under output, precision: user defined, arithmetic type: signed (2's comp), number of bits: 32, binary point: 23, quantization: truncate, overflow: wrap. Under implementation: implement using: distributed RAM.

The design parameter settings of other blocks in this design are given below:

1. Constant: under main, constant value: 10.45, tick "interpret vector parameters as 1-D", sample time: 1. Under signal attributes: output data type: double.
2. Gateway in: under head base, output type: fixed-point, arithmetic type: unsigned, number of bits: 32, binary point: 23, quantization: truncate, overflow: wrap and sample period: 1.
3. Gateway in1: under head base, output type: fixed-point, arithmetic type: signed (2's comp), number of bits: 32, binary point: 23, quantization: truncate, overflow: wrap and sample period: 1.
4. AddSub: under basic, operation: addition, latency: 1 and under output, precision: user defined, arithmetic type: signed (2's comp), number of bits: 32, binary point: 23, quantization: truncate, overflow: wrap.
5. Gateway out: tick "truncate into output port" and IoB timing constraint: none.
6. Display: format: long and decimation: 1.

With this information, we proceed to analyze the design. The equation corresponding to the fractional-point computation in Figure 10.17 is $y = (a + b)2.5$, where y is a fractional-point output and a, b are the fractional-point inputs. The values of the two input constants are 10.45 and -2.6. These two inputs are applied into the Gateway In and Gateway In1, respectively. Since the gateway in is set by (1,32,23) bit fixed-point resolution, its outputs are at a greater resolution than MATLAB constant inputs, 10.45 and -2.6, which are displayed in Display1 and Display2, respectively. The outputs of Gateway In are the input to the adder block, whose output is 7.85, as displayed in Display3. Since the bit resolution of the adder unit is (1,32,23), this block keeps the same resolution as the Gateway In blocks. The output of the adder block is applied into the CMult block, whose final output is 20.409999847412 in long display format. The bit resolution of Gateway out blocks is like the Gateway In block; hence, the same accuracy is maintained throughout the computation.

DESIGN EXAMPLE 10.17

Design a Xilinx SysGen–based system to realize a complex multiplier and show the result in Display blocks. Make another design to realize the function of time division multiplexing and demultiplexing with a serial-to-parallel counter. Store some of the intermediate results in the MATLAB workspace. Also, estimate the resources utilized for these designs using the Resource Estimator block.

Solution

Before beginning the design, we need to understand many new SysGen blocks that are needed for it. All those blocks' operations and interfacing principles are discussed below one by one:

Manual Switch: This is a MATLAB Simulink block which acts as a manual switch. This switch passes either one of its two inputs based on the present knob position. The first input will be the output once the knob is connected to the upper terminal; otherwise, the output will be equal to the input of the lower terminal. The position change is accomplished by the user; that is, the switch position is manually controlled.

Threshold: The Xilinx Threshold block tests the sign of the input number. If the input number is negative, the output of the block is –1; otherwise, the output is 1. The output is a signed fixed-point integer that is 2 bits long. The block has one input and one output. The block parameters dialog box can be invoked by double-clicking the icon in your Simulink model. The block parameters do not control the output data type because the output is always a signed fixed-point integer that is 2 bits long.

Complex Multiplier 3.1: The Xilinx Complex Multiplier block multiplies two complex numbers. All operands and the results are represented in signed two's complement format. The operand widths and the result width are parameterisable. Either a 3-real-multiplier structure or a 4-real-multiplier structure can be used. Always use the 4-real multiplier structure to allow the best frequency performance to be achieved. Output product range selects the required MSB and LSB of the output product. The values are automatically set to provide a full-precision product when the A and B operand widths are set. The output is sign-extended if required. If rounding is required, set the output LSB to a value greater than zero to enable the rounding options.

With this information about some of the new blocks, we proceed to design the system as stated in Design Example 10.17, which will look like Figure 10.18, where two designs are made. The first is for a complex multiplier, while the second is for multiplexing and demultiplexing. In the first design, the manual switch routes either 3 or −6 with respect to its knob position. The output of the switch is given to the threshold block through a Gateway In block. The output of this device is given as one input, that is, `ar`, into the complex multiplier. Other inputs of the complex multiplier are `ai` = −2.25, `br` = 8.75, and `bi` = 4.5, where, `r` denotes the real part and `i` denotes the imaginary part. The actual computation is `p = (pr + pi) = (ar + ai) × (br + bi)`, where `pr` is the product's real part and `pi` is the product's imaginary part. Therefore, the block "complex multiplier

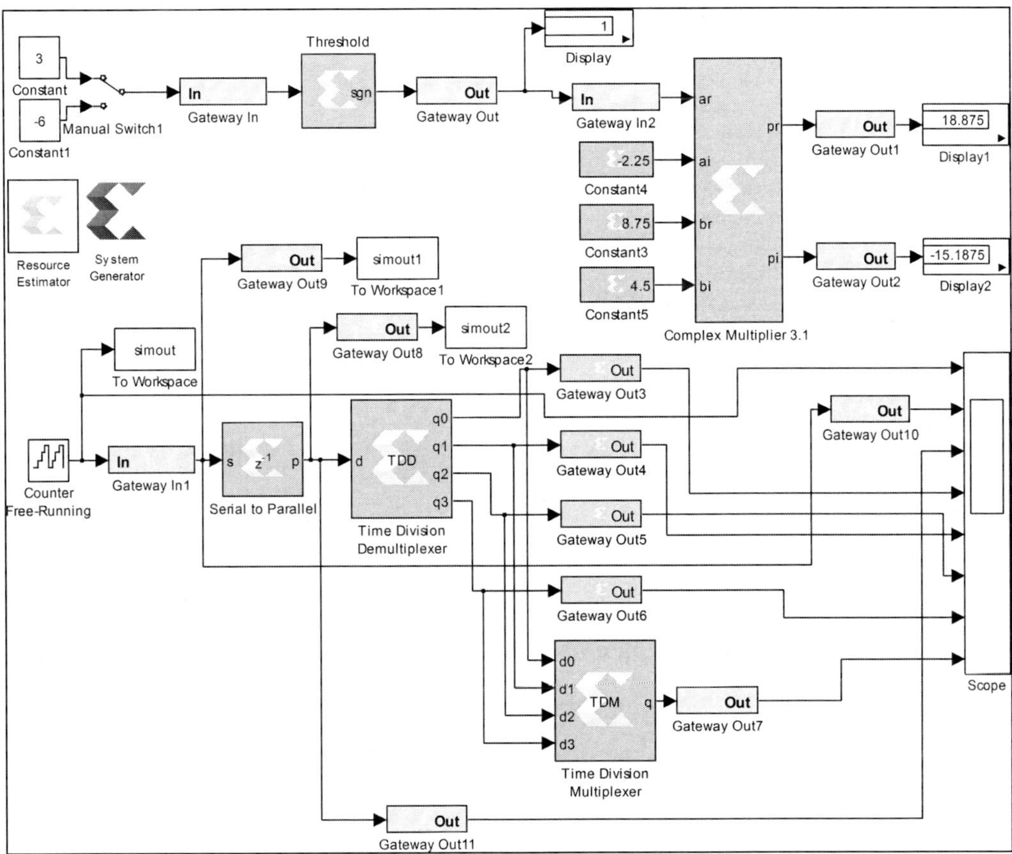

FIGURE 10.18
Xilinx SysGen-based system as per design Example 10.17.

3.1" multiplies two complex numbers and gives the complex results at `pr` and `pi`, as shown in Figure 10.18. The final result in the algebraic form is p = (18.875 – 15.1875i), as shown in Figure 10.18. The parameter settings in this design are given below:

1. Constanl: under head main, constant value: 3, tick "interpret vector parameters as 1 – D" and sample time = 1. The same setting for constantl, but the value is −6.
2. Gateway In: under head base, output type: fixed-point, arithmetic type: signed (2's complement), number of bits: 10, binary point: 3, quantization: truncate, overflow: wrap and sample period: 1.
3. Threshold: latency: 0.
4. Gateway Out: tick "truncate into output port" and IoB timing constraint: none.
5. Display: format: long and decimation: 1.
6. Constant: under basic, constant value:–2.2, output type: fixed-point, arithmetic type: signed (2's comp), number of bits: 10, binary point: 3 and tick "sampled constant".
7. Complex Multiplier 3.1: under page 1, multiplier construction options: use_mults, optimization goal: performance, output MSB: 20 and output LSB: 0. Under page 2, select-1 to pipeline the core for maximum performance: 0.

Suppose the manual switch is connected to the second input, that is, –6. Then, the first input to the complex multiplier is ar = –1, and the product is p = (–1–2.25i) × (8.75 + 4.5i) = (1.375–24.1875i).

Now, we proceed to the second design. Here, we also need to discuss some new blocks, namely, counter free-running, simout, serial to parallel, time division demultiplexer, and time division multiplexer.

Counter free-running: This block is a free-running counter that overflows back to zero after it has reached the maximum value possible for the specified number of bits. The counter is always initialized to zero. The output is normally an unsigned integer with the specified number of bits.

Simout: Simout is a MATLAB Simulink block. This is used to load the simulated result into the MATLAB workspace. It writes the result into the specified timeseries, array or structure in a workspace. For menu-based simulation, data are written in the MATLAB base workspace. Data will not be available until the simulation is stopped or paused.

Serial-to-Parallel: The serial-to-parallel block takes a serial input and groups some of it into parallel samples. This block gives the parallel output, that is, more than one bit at a time, once the predefined number of bits is received at the input.

Time Division Demultiplexer: The Xilinx time division demultiplexer block accepts input serially and presents it to multiple outputs at a slower rate. This block has one data input port and a user-configurable number of data outputs, ranging from 1 to 32. The data output ports have the same arithmetic type and precision as the input data port. The time division demultiplexer block also has an optional input-valid port and output-valid port. Both valid ports are of type Boolean. The block has two possible implementations, single or multiple channel. For single-channel implementation, the time division demultiplexer block has one data input and output port.

Time Division Multiplexer: The Xilinx time division multiplexer block multiplexes values presented at input ports into a single faster rate output stream. This block has 2 to 32 input ports and 1 output port. All input ports must have the same arithmetic type, precision and rate. The output port has the same arithmetic type and precision as the inputs. The output rate is NR, where N is the number of input ports and R is their common rate. This block also has optional ports and v_{out} that specify when input and output, respectively, are valid. Both valid ports are of type Boolean. A parameter, number of inputs, specifies the number of inputs (2 to 32).

With this information, we start designing the system, which will appear as Figure 10.18, where the output of a counter (free-running) is connected to the Gateway In and simout. The simout block writes the values of the counter in the workspace. The Gateway In block converts the input to the declared format and gives it to the serial-to-parallel block. The output of the serial-to-parallel block is given to simout2 as well as the time division demultiplexer. The demultiplexer's outputs are given to the scope as well as to a time division multiplexer, whose output is connected to the last channel of the scope. The parameter settings in this design are given below:

1. Counter free-running: number of bits: 4 and ample time: 1.
2. Simout: variable name: simout, limit data points to last: inf, decimation: 1 and sample time: –1.
3. Gateway In: under head base, output type: fixed-point, arithmetic type: unsigned, number of bits: 4, binary point: 1, quantization: truncate, overflow: wrap and sample period: 1.
4. Serial to Parallel: under basic, input order: most significant word first, arithmetic type: unsigned, number of bits: 4, binary point: 0 and latency: 1.
5. Gateway Out: tick "truncate into output port" and IoB timing constraint: none.
6. Time division demultiplexer: under basic, frame sampling pattern: [1 1 1 1] and implementation: Multiple channel.
7. Time division multiplexer: under basic, number of inputs: 4.

8. Scope: under general, number of axex: 8, time range: 100, tick labels: bottom axis only and decimation: 1. Under history: tick "limit data points to last" with the value of 5000, tick "save data to workspace", variable name: ScopeData1 and format: structure with time.

Upon completing the design and simulating it, the scope will show all the results, as shown in Figure 10.19. The first plot shows that the free-running counter is running from "0000" through "1111". Once it reaches the maximum value, that is, $(15)_{10,}$ it falls back to $(0)_{10}$ and then continuously counts up to $(15)_{10}$. That is why it is called the free-running counter. The result of the free-running counter is given to the serial-to-parallel converter block through a Gateway In 1. Since the data width in the Gateway In 1 is (3,1), that is, 3 bits for the integer, 1 bit for the fraction and no bit for the sign because the input is unsigned number, the output of the Gateway In 1 varies from "000.0" through "111.0", that is, $(0)_{10}$ through $(7)_{10}$, because the counter was never given a fractional number; hence, the fractional part is always zero. Therefore, the output of Gateway In 1 completes two cycles within the duration of one cycle of the free-running counter. This result is shown in second plot of Figure 10.19. The input of the serial-to-parallel converter is 4 bits in which the first bit (LSB) is always 0 and the remaining 3 bits will be varied. However, all together, the parallel output is "0000", "0010", "0100", "0110",… "1110", "0000", "0010", that is, $(0)_{10}$, $(2)_{10,}$ $(4)_{10,}$ $(6)_{10}$,… $(8)_{10,}$ $(10)_{10,}$ $(12)_{10,}$ $(14)_{10,}$ $(0)_{10,}$ $(2)_{10}$, and so

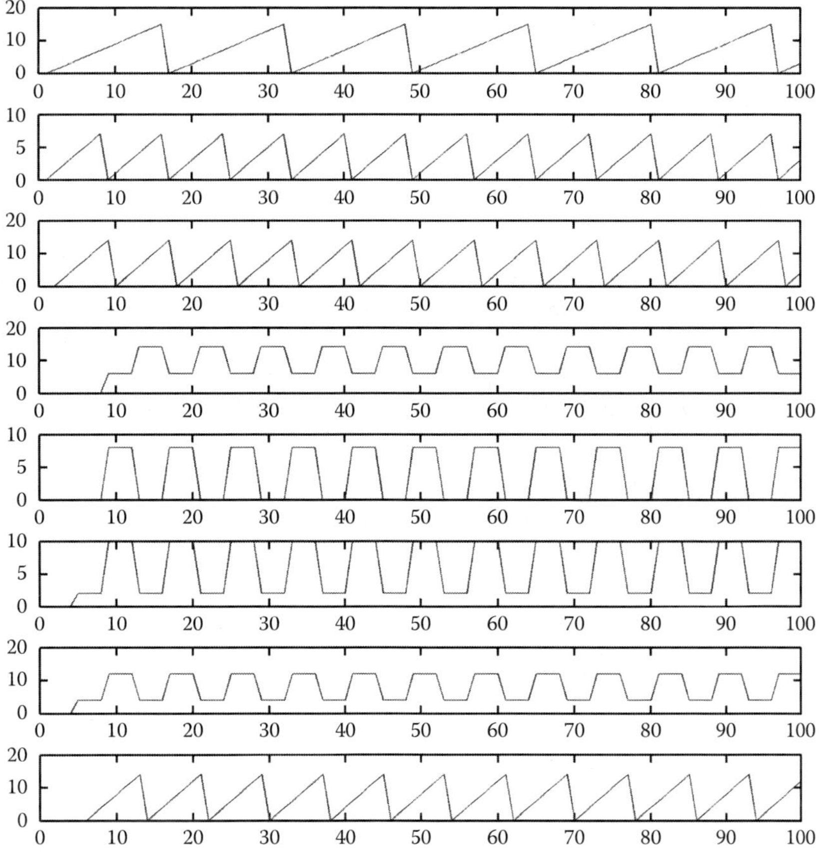

FIGURE 10.19
Simulation result of second example of Figure 10.18.

on. This result is shown in the third plot of Figure 10.19. Since the latency of the serial-to-parallel block is 1, there is one sample time delay.

The output of the serial-to-parallel converter is given to the time division demultiplexer, which routes the input sample into the appropriate output channel. The output of the time division demultiplexer is (–, 0, 2, 4) in the first cycle, (6,8,10,12) in the second cycle, (14,0,2,4) in the third cycle and so on. Due to latency, there is one unit of time delay, which is indicated as "–" above. These results are shown in the fourth to seventh plots of Figure 10.19. These values are also given to a time division multiplexer, which routes the input one by one to output. The result of the time division demultiplexer is shown in the eighth plot of Figure 10.19, which proves that the input of the time division demultiplexer is equal to the output of the time division multiplexer with some amount of latency.

As a third part of this design example, we need to estimate the resources required for this design. This estimation would be accomplished using the Resource Estimator token, which is shown in Figure 10.18. The use and interfacing of Resource Estimator is given below:

Resource Estimator: The Xilinx Resource Estimator block provides fast estimates of the FPGA resources required to implement a System Generator subsystem or model. These estimates are computed by invoking block-specific estimators for Xilinx blocks and summing these values to obtain aggregated estimates of lookup LUTs, FFs, block memories (block random access memory [BRAM]), 18 × 18 multipliers, tristate buffers, and GP I/Os. Every Xilinx block that requires FPGA resources has a mask parameter that stores a vector containing its resource requirements. The Resource Estimator block can invoke underlying functions to populate these vectors or aggregate previously computed values that have been stored in the vectors. An estimator block can be placed in any subsystem of a model.

To make use of the Resource Estimator, just double-click on the token. It pops up a window, where we need to choose one option out of estimate, quick, post mao and read mrp, to perform the required estimation. In this example, we choose quick and estimate the resources. After closing the Resource Estimator window, simulate the design model once again and then again open the window, which now will have complete details about the resource requirements, as shown in Figure 10.20. Based on this result, the design may be altered so as to get the optimum design and resource utilization.

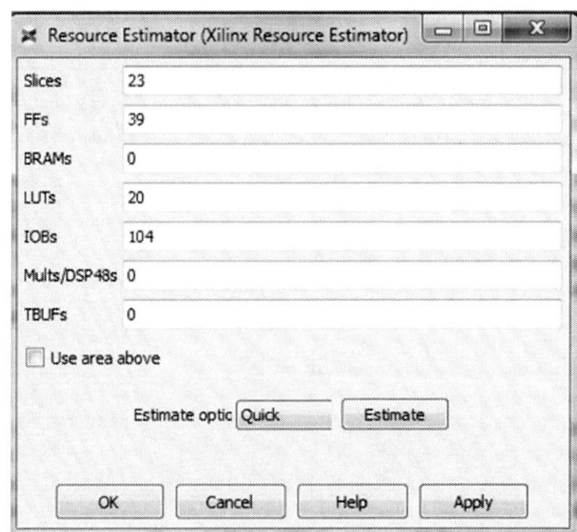

FIGURE 10.20
Design resource estimation.

DESIGN EXAMPLE 10.18

Write MATLAB code to realize a simple FIR filter using y[n] = x[n] + x[n−1]. Finally, implement the same filter using Xilinx SysGen tools.

Solution

As per this design example, we need to feed the noisy signal, that is, the multifrequency signal, into a simple FIR filter to show that the filter removes many unwanted frequency components and its output is maximum only at a single frequency. To accomplish this task, we generate two sinusoidal signals of low and high frequency and add them to get a multifrequency signal. This will be the input to the filter whose output will be maximum only for one frequency. We plot all these results and analyze them to understand how a simple FIR filter removes many frequency components. The MATLAB code for application is given below.

```
clc;
x=0:0.01:1*pi;
y=sin(2*pi*2*x);
subplot(2,2,1);
plot(x,y);
grid on;
xlim([0 3.0]);
xlabel('time');
ylabel('Amplitude');
y1=sin(2*pi*47*x);
subplot(2,2,2);
plot(x,y1);
grid on;
xlim([0 3.0]);
xlabel('time');
ylabel('Amplitude');
subplot(2,2,3);
y2=y+y1;
plot(x,y2);
ylim([-3 3]);
hold on;
y3=y2(2:1:length(y2));
y4=y2(1:1:length(y2)-1);
y5=y3 + y4;
plot(x(1:1:length(x)-1),y5,'k-','LineWidth',1.5);
grid on;
xlim([0 3.0]);
xlabel('time');
ylabel('Amplitude');
subplot(2,2,4);
plot(x(1:1:length(x)-1),y5,'k-','LineWidth',1.5);
grid on;
xlim([0 3.0]);
xlabel('time');
ylabel('Amplitude');
```

The time-domain response of the filter is shown in Figure 10.21, where the first plot correspond to low frequency (with uniform amplitude) signal and the second plot corresponds to the high frequency (with un-uniform amplitude) signal. The sums of these two signals are shown in the third plot. The filter response is shown in the fourth plot, and also shown in the third plot just to demonstrate the correlation of the input and output of the filter. The fourth plot clearly shows that the signal has the maximum response for only one frequency.

Now, we proceed to implementing the same FIR filter using the Xilinx SysGen tools. In this implementation, we need new blocks, namely sum to add two sine signals and

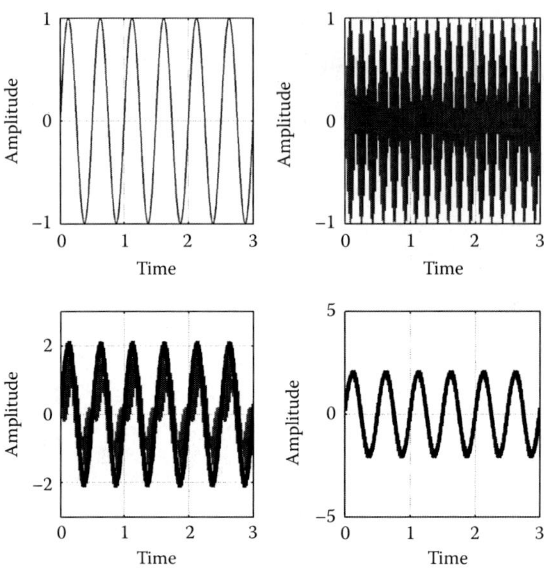

FIGURE 10.21
Time-domain input-output response of a FIR filter.

delay to get x[n−1] while x[n] is the present input sample. The operation and interfacing techniques of the sum and delay blocks are given below.

Sum: The Sum is a MATLAB Simulink block. It does either add or subtract operations. Specify one of the following: (i) a string containing + or − for each input port, | for spacer between ports (e.g., ++|−|++) and (ii) scalar, >= 1, which specifies the number of input ports to be summed. When there is only one input port, add or subtract elements over all dimensions or one specified dimension.

Delay: The Xilinx Delay block implements a fixed delay of L cycles. The delay value is displayed on the block in the form z^{-L}, which is the Z transform of the block's transfer function. Any data provided to the input of the block will appear at the output after L cycles. The rate and type of the data of the output will be inherited from the input. This block is used mainly for matching pipeline delays in other portions of the circuit. The delay block differs from the register block in that the register allows a latency of only one cycle and contains an initial value parameter. The delay block supports a specified latency but no initial value other than zeros.

With this information, we begin the design of the Xilinx SysGen model for the given equation, which will look like Figure 10.22. There, two sinusoidal signals are summed by an adder and then applied to the delay element through a Gateway In. The multiplier block gets delayed input, that is, x[n−1], from the delay element and a coefficient, that is, 1, from the constant block. The multiplier output is x[n−1], which goes into the addsub block that gets one input, that is, x[n], directly from the Gateway In. The intermediate results and final output are displayed in a scope.

The parameter settings in this design are given below:

1. Sine wave 1: sine type: time based, time (t): use simulation time, amplitude: 1, bias: 0, frequency (rad/sec): 5, phase (rad): 0. Sample time: 0.01 and tick "interpret vector parameters as 1-D".
2. Sine wave: sine type: time based, time (t): use simulation time, amplitude: 1, bias: 0, frequency (rad/sec): 45, phase (rad): 0. Sample time: 0.01 and tick "interpret vector parameters as 1-D".
3. Sum: under main, Icon shape: round, list of signs: |++ and sample time: 0.1.

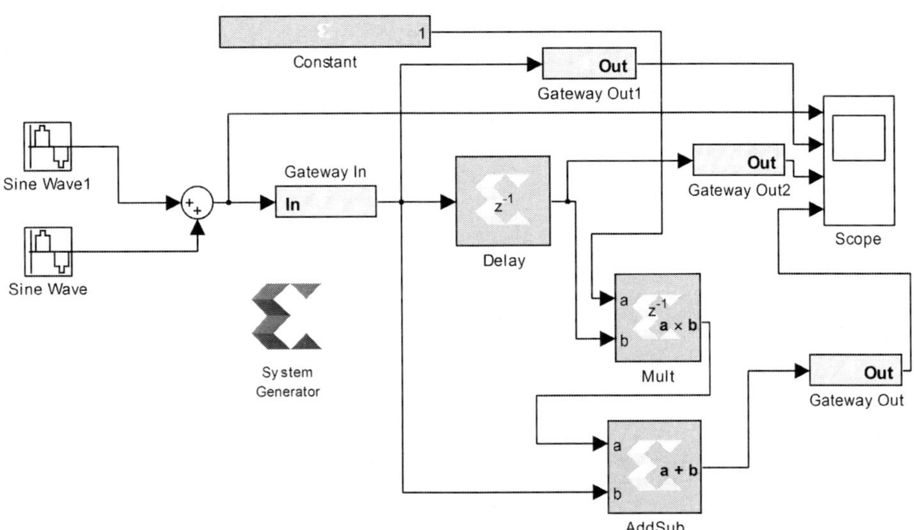

FIGURE 10.22
A design for FIR filter y[n] = x[n] + x[n − 1].

4. Gateway In: under head base, output type: fixed-point, arithmetic type: signed (2's complement), number of bits: 8, binary point: 4, quantization: truncate, overflow: wrap and sample period: 1.
5. Delay: under basic, latency: 1.
6. Constant: under head base, output type: fixed-point, arithmetic type: signed (2's complement), number of bits: 8, binary point: 6, tick the "sampled constant" and sample period: 1.
7. Mult: under basic, precision: used defined, arithmetic type: signed (2's comp), number of bits: 16, binary point: 12, quantization: truncate, overflow: wrap and latency: 1.
8. Addsub: under basic, operation: addition, latency: 0. Under output, precision: user defined, arithmetic type: signed (2's comp), number if bits: 16, binary point: 12, quantization: truncate and overflow: wrap.
9. Gateway Out: tick "truncate into output port" and IoB timing constraint: none.
10. Scope: under general: number of axes: 4, time range: 50, labels: all and sample time: 0. Under history: tick "save data to workspace", variable name: ScopeData1 and format: structure with time.

In Figure 10.23, the first plot corresponds to the sum of two sinusoidal signals. The second plot is the digital version of that sinusoidal signal. The third plot is the delayed version of the second plot. The fourth plot is the filter output. The appearance of the display of the signals in Figure 10.23 is unlike Figure 10.21 because of the deviations in the minimum sample time of the Xilinx blocks. In real time, that is, after implementing in the FPGA, the sample time is obviously very high; hence, all the signals will be more accurate than the MATLAB results.

10.6 System Engine Model Using Xilinx Simulink Block Sets

In this section, we will discuss designing system engine models using logical expressions, DSP intellectual property (IP) cores, and Xilinx SysGen subsystems. All these system

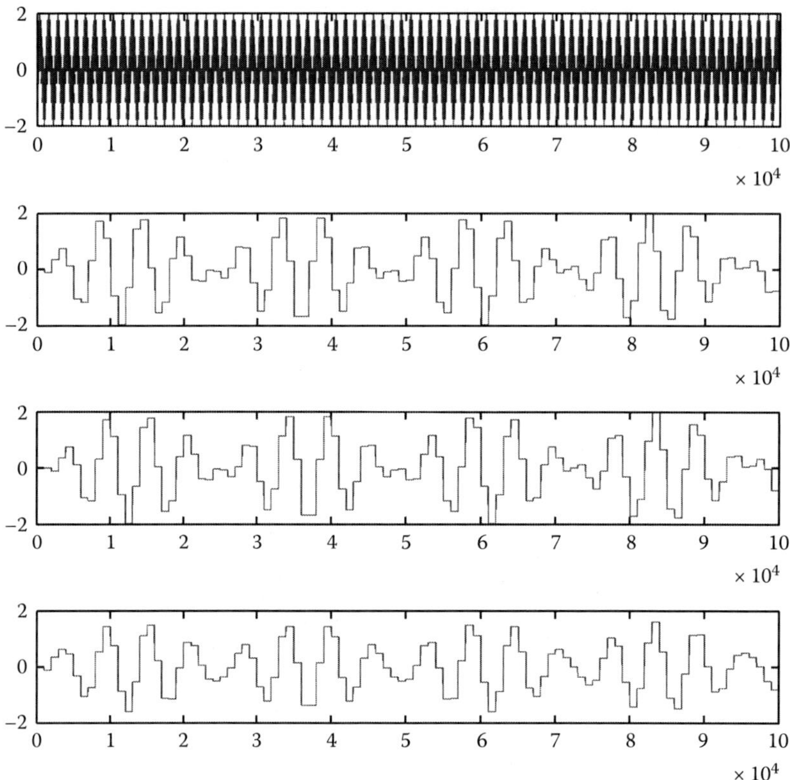

FIGURE 10.23
Xilinx SysGen simulation result of Figure 10.22 (time versus amplitude).

engine design models are briefed below. The Xilinx SysGen model provides features to build block sets with our own logical expressions whenever needed. That means if the available Xilinx block sets do not fulfil our design requirements in an optimized way, we can make our own block with the required logical relations. IP refers to preconfigured logic functions that can be used in our design. Xilinx provides a wide selection of IP that is optimized for Xilinx FPGAs. These can include functions delivered through the Xilinx CORE generator software, through the Xilinx Architecture Wizard, as standalone archives, from third parties, or through Xilinx Platform Studio (XPS) or System Generator. Xilinx and its partner companies produce IP ranging in complexity from simple arithmetic operators and delay elements to complex system-level building blocks such as DSP filters, multiplexers, transformers, memory, fixed-point FFT, quadrature conversion, advanced polyphase filters, power spectrum, and 64-bit external memory interface (EMIF) interface to Texas instrument (TI) DSPs, software defined radios (SDR), data logging, image processing, motion pictures expert group (MPEG), and joint photographic expert group (JPEG) compression and so on. As in MATLAB, the Xilinx SysGen tool also has the ability to create custom blocks that have the same appearance as the built-in blocks. This is called masking. Masks are placed on top of subsystems to create a masked subsystem. Subsystems are the mechanism for adding hierarchy to a model. This is achieved by grouping blocks together and encapsulating them into a single block. The blocks that need to be encapsulated into the subsystem need to be selected. Next, select all of the blocks in your model (Ctrl-G) and select Create subsystem from the Edit menu at the top of the model window. It helps reduce the number of blocks displayed in the model window. Functionally related blocks can be kept together [139–142]. This permits the establishment of a hierarchical block diagram wherein a subsystem block can be built.

In the following design examples, we discuss creating a system model using logical expressions, designing models using IP cores and building Xilinx SysGen subsystem modules.

DESIGN EXAMPLE 10.19

Develop a digital system using some combinational and sequential circuits and logical expressions to understand system engine modeling.

Solution

We need to discuss some new block sets that we plan to use in this design. For example, we use the NOT gate, AND gate, concat block, counter block, mux block, expression block, serial-to-parallel block, and so on. The operation and interfacing principles of all the new blocks are discussed one by one below.

NOT: In Xilinx block sets, this is also called Inverter. The Xilinx Inverter block calculates the bitwise logical complement of a fixed-point number. The block is implemented as a synthesizable VHDL module. The block parameters dialog box can be invoked by double-clicking the icon in our Simulink model.

AND: In Xilinx block sets, it is also called logical. The Xilinx Logical block performs bitwise logical operations on 2, 3, or 4 fixed-point numbers. Operands are zero padded and sign extended as necessary to make binary point positions coincide; then, the logical operation is performed and the result is delivered at the output port. In hardware, this block is implemented as synthesizable VHDL. If we build a tree of logical gates, this synthesizable implementation is best, as it facilitates logic collapsing in synthesis and mapping.

Concat: This block is listed in the following Xilinx Blockset libraries: Basic Elements, Data Types, and Index. The Xilinx Concat block performs a concatenation of n bit vectors represented by unsigned integer numbers, that is, n unsigned numbers with binary points at position zero. The Xilinx Reinterpret block provides capabilities that can extend the functionality of the Concat block. The block has n input ports, where n is some value between 2 and 1024, inclusively, and one output port. The first and last input ports are labeled hi and low, respectively. Input ports between these two ports are not labeled. The input to the hi port will occupy the most significant bits of the output and the input to the lo port will occupy the least significant bits of the output.

Mux: The Xilinx Mux block implements a multiplexer. The block has one select input (type unsigned) and a user-configurable number of data bus inputs, ranging from 2 to 1024, that is, 2^{10}.

Expression: The Xilinx Expression block performs a bitwise logical expression. The expression is specified with operators as "~" for NOT gate, "&" for AND gate, "|" for OR gate, and "^" for XOR gate. The number of input ports is inferred from the expression. The input port labels are identified from the expression, and the block is subsequently labeled accordingly. For example, Equation 10.5 can be implemented as y = ~ ((a1|a2) & (b1^b2)), resulting in 4 input ports labeled 'a1', 'a2', 'b1', and 'b2'. The expression will be parsed and an equivalent statement will be written in VHDL. A logical expression for implementation in the expression block is

$$y = \overline{(a_1 + a_2)\cdot(b_1 \oplus b_2)} \tag{10.5}$$

where a_1, a_2, a_3, a_4 are the inputs and y is the output.

Serial to Parallel: The serial-to-parallel block takes a series of inputs of any size and creates a single output of a specified multiple of that size. The input series can be ordered either with the most significant word first or the least significant word first.

With this information, we can begin the system modeling, which will look like Figure 10.24, where all these SysGen blocks are employed. The parameter settings in this design are given below:

1. Constant: under head base, output type: fixed-point, arithmetic type: unsigned, number of bits: 1, binary point: 0, and sample period: 1.
2. Inverter: under basic, latency: 0.
3. Logical: under basic, logical function: AND, number of inputs: 2, latency: 0. Under output type, precision: user defined, output type: unsigned, number of bits: 1, binary point: 0 and tick "align binary point".
4. Concat: number of inputs: 3.
5. Counter: under basic, counter type: free running, count direction: up, initial value: 0, step: 1, output type: unsigned, number of bits: 1, binary point: 0 and explicit period: 1.
6. Mux: under basic, number of inputs: 2, latency: 0. Under output, precision: user defined, arithmetic type: unsigned, number of bits: 3, binary point: 0, quantization: truncate and overflow: wrap.
7. Expression: under basic, expression: (~(a^b) & (c|d)), latency:0. Under output, precision: user defined, arithmetic type: unsigned, number of bits: 3 and binary point: 0.
8. Serial to parallel: under basic, input order: least significant word first, arithmetic type: unsigned, number of bits: 3, binary point: 0 and latency: 1.
9. Gateway Out: tick "truncate into output port" and IoB timing constraint: none.
10. Scope: under general: number of axes: 4, time range: 50, labels: all and sample time: 0. Under history: tick "save data to workspace", variable name: ScopeData1 and format: structure with time.

As in Figure 10.24, a constant "1" goes into the NOT gate, whose output will be "0". This "0" and another constant "1" go into the AND gate, whose output will always be "0". This "0" and another two constants (both "1") go into a concat block, which concatenates all these inputs as "011", that is, $(3)_{10}$, as shown in Figure 10.24. The output of the concat block goes into an inverter, whose output will be "100", that is, $(4)_{10}$, as shown in Figure 10.24. The "011" and "100" are the first and second inputs, respectively, of a mux block. The selective line of the multiplexer is taken from the single-bit binary counter. The counter output will either be "0" or "1"; when "0", "011" is the output of mux block, and when "1", the output will be "100". The expression block is written by an expression as given in Equation 10.6

$$y = \overline{(a \oplus b)}(c + d) \quad (10.6)$$

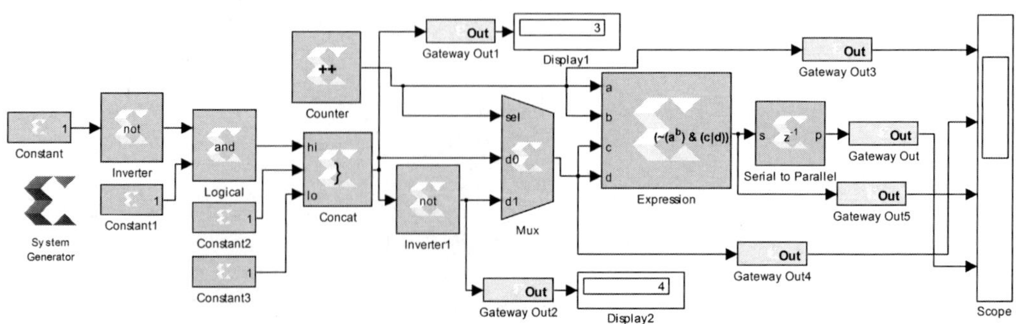

FIGURE 10.24
System modeling using expression block.

where, a, b, c, d are the inputs and y is the output. According to this logical expression, that is, Equation 10.6 and Figure 10.24, the inputs a and b will either be "0" or "1". However, in both cases, the result of not(a ⊕ b) = 1; hence, the output now depends only c and d, and c and d will be either "011" or "100". Hence, as in Equation 10.6, (c + d) will either be "011" or "100". Therefore, the output of the expression block is either "011" or "100". This result goes into a serial-to-parallel block whose block is also the same, with a one-unit time delay. The simulation result of this design example is shown in Figure 10.25, where the first plot corresponds to the counter output, the second plot is the mux block output, the third plot is the output of the expression block and the fourth plot is the output of the serial-to-parallel converter. Figure 10.25 clearly illustrates the signal flow in this design example.

DESIGN EXAMPLE 10.20

Develop a DSP system using an IP core DSP48 macro 2.0 block; make it a subsystem and realize its operations.

Solution

First, we discuss some of the required new blocks before proceeding to designing the DSP system. The new MATLAB and Xilinx blocks are discussed one by one below.

Random Number: This is a MATLAB Simulink block. Its output is a normally (Gaussian) distributed random signal. Output is repeatable for a given seed.

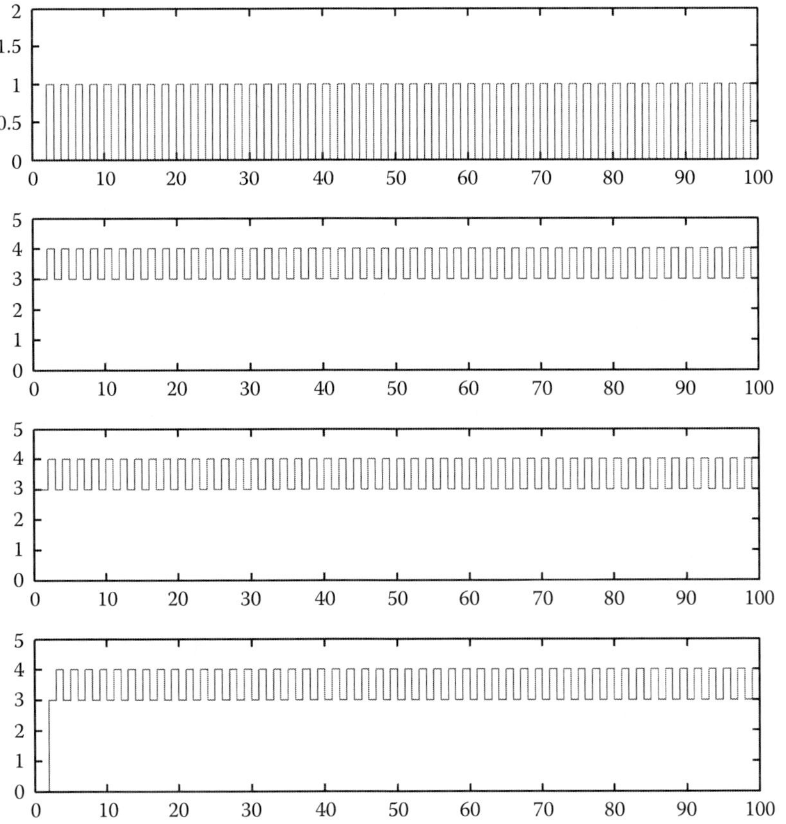

FIGURE 10.25
Simulation result of Figure 10.24 (time versus amplitude).

DSP48 Macro: The System Generator DSP48 Macro block provides a device-independent abstraction of the blocks DSP48, DSP48A and DSP48E. Using this block, instead of using a technology-specific DSP slice, helps make the design more portable between Xilinx technologies. Depending on the target technology specified at compile time, the block wraps one DSP48/DSP48E/DSP48A block along with reinterpret and convert blocks for data type alignment, multiplexers to handle multiple opmodes and inputs and registers. The DSP48 Macro block has a variable number of inputs and outputs determined from user-specified parameter values. The input data ports are determined by the opmodes entered in the Instructions field of the DSP48 Macro. Input port Sel appears if more than one opmode is specified in the Instructions field. Port P, an output data port, is the only port appearing in all configurations of the DSP48 Macro. Output ports PCOUT, BCOUT, ACOUT, CARRYOUT, and CARRYCASCOUT appear depending on the user selections.

With this information, as stated in this design example, we need to use the DSP 48 macro 2.0 core. Drag and drop this block on the simulation environment and design a system as shown in Figure 10.26. There are three real-world inputs (`in1`, `in2` and `in3`) which are given to the DSP48 macro 2.0 through the Gateway In blocks. The output of the DSP48 macro 2.0 is taken out as `out1` through a Gateway Out block. The actual inputs and outputs of the DSP48 macro 2.0 core are `a`, `b` and `c` and `p`, respectively. Now, we need to enter the expression (instruction) to the DSP48 macro 2.0 core based on which it will work and produce the output. To do this task, double-click on DSP48 macro 2.0, which brings up the properties dialog box editor window. Under head instruction, enter the instruction as "$a * b + c$" and leave all other available instruction (implementation) options at their default values. Then, move to head pipeline options, where we need to select the pipelining we expect. In this design example, click the A, B and C boxes in tier 1, the M and C boxes in tier 5 and the P box in tier 6. Unclick all other boxes and click OK. Now, a module using the DSP48 macro 2.0 core is ready. We are interested in making the subsystem using all these blocks. Hence, select all the Xilinx blocks (components) shown in Figure 10.26, and click "Ctrl + G". Then the subsystem will be ready with all the selected components, and it appears as a rectangular box as shown at the right of the Xilinx System Generator Token in Figure 10.26. Now, this subsystem is ready for our use in any design.

We proceed with our DSP system design using this subsystem, as shown in Figure 10.27. Note that, as shown in Figure 10.27, the MATLAB block outputs are directly connected to the subsystems since there is a Gateway In the subsystems, as shown in Figure 10.26. Also note that System Generator Tokens are used in the subsystem and represent

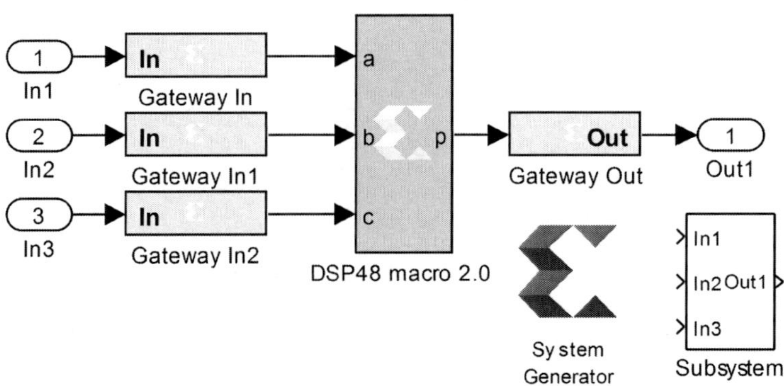

FIGURE 10.26
Subsystem design with DSP48 macro 2.0 IP core.

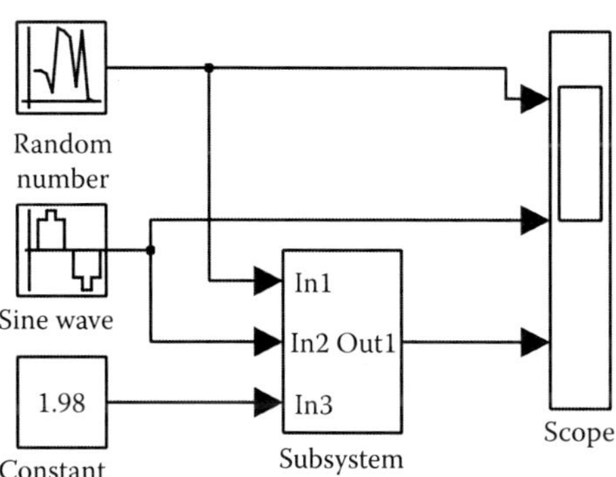

FIGURE 10.27
DSP system design with subsystems.

two blocks of a single FPGA within a larger DSP system. Each system generator token creates a top-level entity from the subsystem with which it is associated [139–142]. The parameter settings in this design are given below:

1. Random number: mean: 0.5, variance: 2, seed: 1, sample time: 0.1 and tick "interpret vector parameters as 1−D".

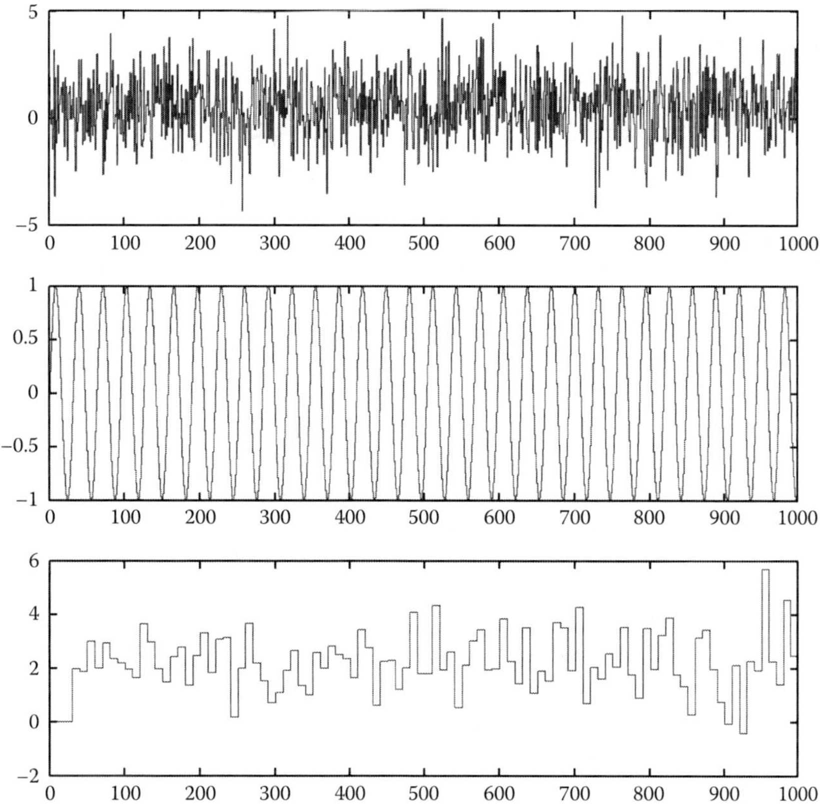

FIGURE 10.28
Simulation result of Figure 10.27 (time versus amplitude).

2. Sine wave: sine type: time based, time (t): user simulation time, amplitude: 1, bias: 0, frequency (rad/sec): 2, phase (rad): 0, sample time: 0.1 and tick "interpret vector parameters as 1–D".
3. Constant: constant value: 1.98, tick "interpret vector parameters as 1-D" and sample time: 0.1.
4. Gateway In: under head base, output type: fixed-point, arithmetic type: signed (2's complement), number of bits: 16, binary point: 8, quantization: truncate, overflow: wrap and sample period: 1. The Gateway In2 has to be loaded with number of bits: 24 and binary point: 16.
5. DSP48 macro 2.0: as discussed above.
6. Gateway Out: tick "truncate into output port" and IoB timing constraint: none.
7. Scope: under general: number of axes: 3, time range: 1000, labels: all and sample time: 0. Under history: tick "save data to workspace", variable name: ScopeData1 and format: structure with time.

Upon completing the design and simulation, the scope will show the results, as shown in Figure 10.28. The first plot corresponds to a random number (signal), the second plot is the sine wave, and the last plot is the output of the DSP system we designed.

10.7 MATLAB® Code Interfacing with SysGen Tools

Xilinx SysGen tools provide many features to design systems for any application. One of them is interfacing MATLAB code with the SysGen environment, through which it is possible to directly make use of the design we have in MATLAB. This section explains writing MATLAB code for SysGen–based applications and interfacing it with the system generator blocks. There is an unique SysGen block called the MCode block through which we can link MATLAB functions to the Xilinx domain. Before starting an example design, we discuss some new SysGen blocks we plan to use now.

Convert: The Xilinx Convert block converts each input sample to a number of a desired arithmetic type. For example, a number can be converted to a signed (two's complement) or unsigned value. Quantization errors occur when the number of fractional bits is insufficient to represent the fractional portion of a value. The options are truncate or round. This is similar to the MATLAB `round()` function. This method rounds the value to the nearest desired bit away from zero, and when there is a value at the midpoint between two possible rounded values, the one with the larger magnitude is selected. For example, to round 01.0110 to a Fix_4_2, this yields 01.10, since 01.0110 is exactly between 01.01 and 01.10 and the latter is further from zero. Round is also known as "Convergent Round" or "Unbiased Rounding". Symmetric rounding is biased because it rounds all ambiguous midpoints away from zero, which means the average magnitude of the rounded results is larger than the average magnitude of the raw results. That is, midpoints are rounded towards the nearest even number. For example, to round 01.0110 to a Fix_4_2, this yields 01.10, since 01.0110 is exactly between 01.01 and 01.10 and the latter is even.

MCode: The Xilinx MCode block is a container for executing a user-supplied MATLAB function within Simulink. A parameter on the block specifies the M-function name. The block executes the M-code to calculate block outputs during a Simulink simulation. The same code is translated in a straightforward way into equivalent behavioral VHDL when hardware is generated. The block's Simulink interface is derived from the MATLAB function signature and from block mask parameters. There is one input port for each parameter to the function and one output port for each value the function returns. Port names and ordering correspond to the names and ordering of parameters and return values. The MCode block supports a limited subset of the MATLAB language that is useful for implementing arithmetic functions, finite state machines and control logic. The MCode block has the following primary coding guidelines that must be followed [143]:

1. All block inputs and outputs must be of Xilinx fixed-point type.
2. The block must have at least one output port.
3. The code for the block must exist on the MATLAB path or in the same directory as the directory as the model that uses the block.

With this information, we begin our design. Now, we take a simple example for adding two inputs, a and b, and producing an output c as c = (3ab) + (2a) − (5b). First, we will write a MATLAB function for this requirement in the MATLAB environment. The MATLAB code for this computation is given below.

```
function c=add_m(a,b)
c=3*a*b + 2*a -5*b;
```

From this MATLAB code, we can deduce that the file name is "add_m", the number of inputs is 2, the number of outputs is 1 and the number of constants or coefficients is 3, 2, and 5. After completing the MATLAB code (as a function), start designing the system in the SysGen environment, as shown in Figure 10.29. Once the design is done, then double-click on the MCode block and interface the block with the MATLAB function either by directly entering the file name or through browsing. Upon correctly interfacing the MCode block with the MATLAB function, the file name appears on the MCode block. Now, our design is ready for simulation.

The parameter settings in this design are given below:

1. Constant: constant value: 9, tick "interpret vector parameters as 1-D" and sample time: 0.1.
2. Gateway In: under head base, output type: fixed-point, arithmetic type: unsigned, number of bits: 8, binary point: 0, quantization: truncate, overflow: wrap and sample period: 1.
3. Convert: output type: fixed-point, arithmetic type: unsigned, number of bits: 4, binary point: 0, quantisation: truncate, over flow: wrap and latency: 0.

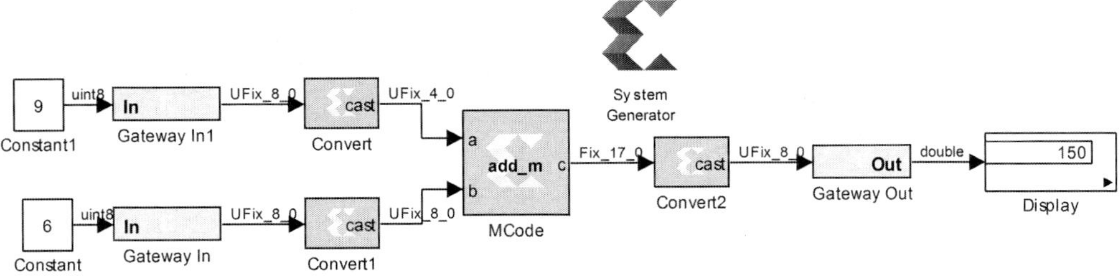

FIGURE 10.29
Design for Xilinx SysGen tool and MATLAB code interfacing: Data types are displayed at all the stages.

4. MCode: as explained above, and MATLAB function: add_m.
5. Gateway Out: tick "truncate into output port" and IoB timing constraint: none.
6. Display: format: long and decimation: 1.

The port data type at all the stages can be displayed by enabling it as Display → Signals & Ports → Port Data Type. All the data types are shown in Figure 10.29. According to the MATLAB function code in this example, the value of c when $a = 9$ and $b = 6$ is, $c = 3 \times 9 \times 6 + 2 \times 9 - 5 \times 6$, $c = (162 + 18 - 30)$, $c = 150$, which is the result in Figure 10.29. Thus, this section has explained the way to interface MATLAB code with the Xilinx SysGen environment.

DESIGN EXAMPLE 10.21

Develop a system using SysGen tools with an interface to a MATLAB function to recognize a sequence "1011". The Mealy model has to be used for designing the FSM.

Solution

It is understood from the question that the sequence of interest is "1011" and the Mealy model has to be used for devising the FSM. We need to recall all the discussions we had about FSM design using the Mealy model and designing state diagrams from Chapter 6. Based on the knowledge we gained in Chapter 6, we can generate a suitable Mealy FSM state diagram to detect the sequence "1011". The corresponding MATLAB code is given below.

```
function dout = FSM(din)
persistent state,state=xl_state(0,{xlUnsigned,3,0});
switch state;
case 0
   if din==1
      state=1;
   else
      state=0;
   end
      dout=0;
case 1
   if din==1
      state=1;
   else
      state=2;
   end
      dout=0;
case 2
   if din==1
      state=3;
   else
      state=0;
   end
      dout=0;
case 3
   if din==1
      state=4;
   else
      state=2;
   end
      dout=0;
```

```
case 4
   if din==1
      state=1;
   else
      state=0;
   end
      dout=1;
otherwise
   state=0;dout=0;
end
```

Once the MATLAB code is complete, we can start designing the SysGen model, as shown in Figure 10.30. This design consists of a repeating sequence stair to generate the random binary sequence, Gateway In, MCode block, Gateway Out and scope. The new block in this design is the repeating sequence stair, whose function is discussed below.

Repeating Sequence Stair: This is a MATLAB Simulink block. This will be used to repetitively generate a predefined sequence of certain length. For example, if the initial sequence is [1 0 1 1 0 0 0 0 1 1 0 1 1 0 1], then this block repetitively generates this sequence over the time.

Then, use the same approach we discussed above to interface the MATLAB function with the Xilinx SysGen environment. The parameter settings in this design are given below:

1. Repeating sequence stair: vector of output values: [1 0 1 1 1 0 0 1 1 0 1 0 1 1] and sample time: 1.
2. Gateway In: under head base, output type: fixed-point, arithmetic type: unsigned, number of bits: 1, binary point: 0, quantization: truncate, overflow: wrap and sample period: 1.
3. MCode: MATLAB function: FSM.
4. Gateway Out: tick "truncate into output port" and IoB timing constraint: none.
5. Scope: under general: number of axes: 2, time range: 100, labels: all and sample time: 0. Under history: tick "save data to workspace", variable name: ScopeData1 and format: structure with time.

The SysGen model can now be simulated. After completing the simulation, the result will be as shown in Figure 10.31. The first plot corresponds to the random binary sequence and the second is the sequence detector output. Whenever the binary serial input arrives at the Mealy machine as "1011", the output goes high to indicate the detection of a valid sequence.

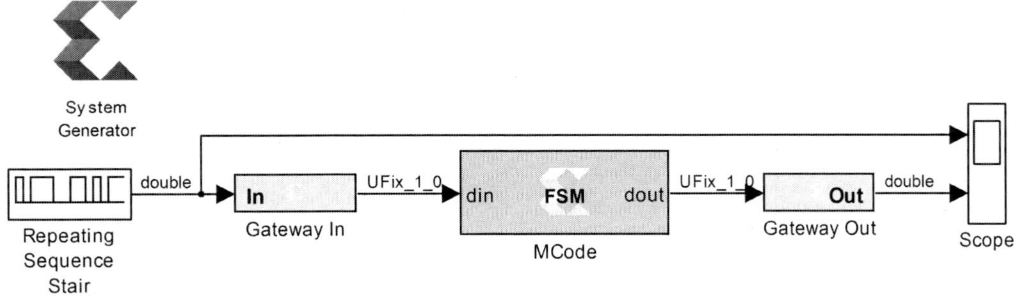

FIGURE 10.30
Mealy model sequence detector using SysGen tools and MATLAB function.

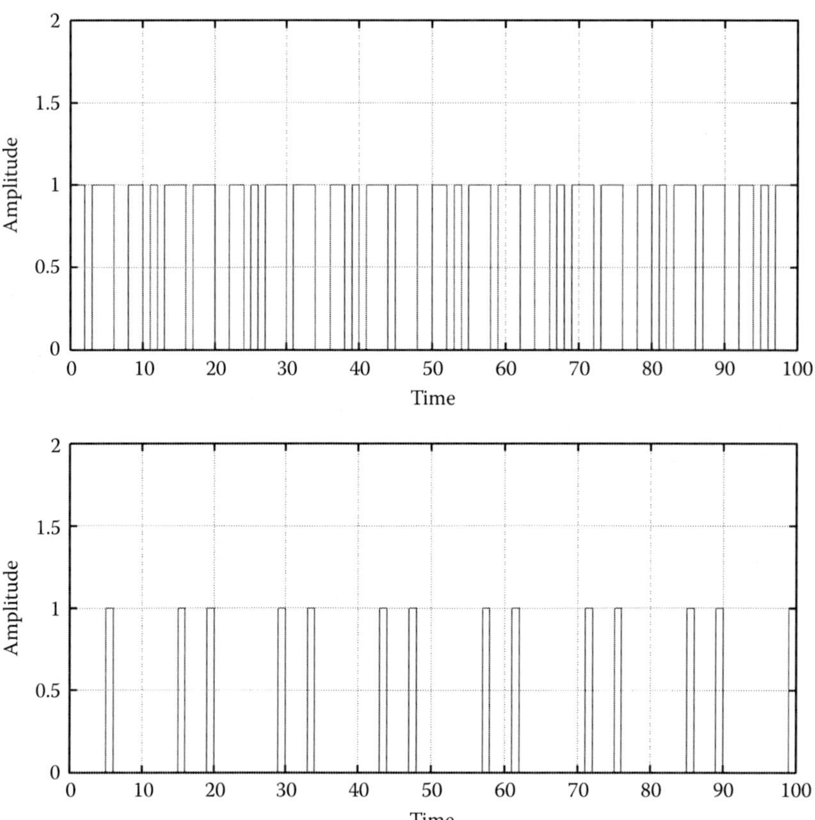

FIGURE 10.31
Simulation result of Figure 10.29 (time versus amplitude).

10.8 Very-High-Speed Integrated Circuit Hardware Description Language Code Interfacing with SysGen Tools

As we discussed in the previous section, the Xilinx SysGen tools provide provisions to interface/call VHDL code in the SysGen environment. In this section, we discuss making a VHDL Black Box with our own code and using it in a SysGen-based system design. Before starting the design, we discuss the Black Box below.

Black Box: The system generator Black Box block provides a way to incorporate HDL models into the system generator. The block is used to specify both the simulation behavior in Simulink and the implementation files to be used during code generation with system generator. A Black Box's ports produce and consume the same sorts of signals as other system generator blocks. When a Black Box is translated into hardware, the associated HDL entity is automatically incorporated and wired to other blocks in the resulting design. The Black Box can be used to incorporate VHDL into a Simulink model. In addition to incorporating HDL into a system generator model, the Black Box can be used to define the implementation associated with an external simulation model. The general primary requirements and conventions are: (i) the entity name must not collide with any entity name that is reserved by system generator and (ii) bidirectional ports are supported in

HDL Black Boxes; however, they will not be displayed in the system generator as ports; they will only appear in the generated HDL after netlisting [140–143].

With this information, we begin the design. The first step is completing the VHDL code part. Now, we take an example of implementing a straight line equation $y = mx + c$. We develop the VHDL code to plot a line using this equation. The project we create for this equation has to be saved in the directory where the SysGen tools are installed. The developed VHDL code is given below. There, the inputs `m` and `c` are of size [2:0], the master clock `m_clock` is a bit and y is output of size [6:0].

```
library IEEE;
use IEEE.STD_LOGIC_1164.ALL;
use IEEE.STD_LOGIC_ARITH.ALL;
use IEEE.STD_LOGIC_UNSIGNED.ALL;
entity HDL_Sys_Gen is
   Port (m,c : in std_logic_vector(2 downto 0):="000";
      y : out std_logic_vector(6 downto 0):="0000000";
      m_clk : in std_logic:='0');
end HDL_Sys_Gen;
architecture Exa of HDL_Sys_Gen is
begin
process(m,c,m_clk)
variable x: std_logic_vector(3 downto 0):="0000";
begin
if rising_edge(m_clk) then
x:=x+1;
y <=(m*x)+ c;
end if;
end process;
end Exa;
```

Once the code development, check syntax and simulations are over, open the Xilinx SysGen simulation window. There, we design the circuit as we expect according to the VHDL code written. Once the Black Box is dragged to the simulation window, a small window will automatically pop up and give an instruction. Upon clicking OK, all the VHDL files available in that directory will be made available for selection. By selecting and choosing the appropriate VHDL file, the design will be completed. The design for `y = mx + c` is shown in Figure 10.32. Once the VHDL code is properly interfaced with the Black Box, then the input and output we considered in the VHDL code will appear on the Black Box icon. Now, it is necessary to configure Black Box simulation mode from "Inactive" to "ISE Simulator". For this, double-click on Black Box; the property window will open, and there select "ISE Simulator" instead of the default mode "Inactive" under "simulation mode".

The parameter settings in this design are given below:

1. Constant: constant value: 2, tick "interpret vector parameters as 1–D" and sample time: 0.1.
2. Pulse Generator: pulse type: time-based, time (t): use estimation time: amplitude: 1, period (secs): 2, pulse width (% of period): 50, phase delay (secs): 0 and tick "interpret vector parameters as 1–D".
3. Gateway In: under head base, output type: Boolean and sample period: 1.

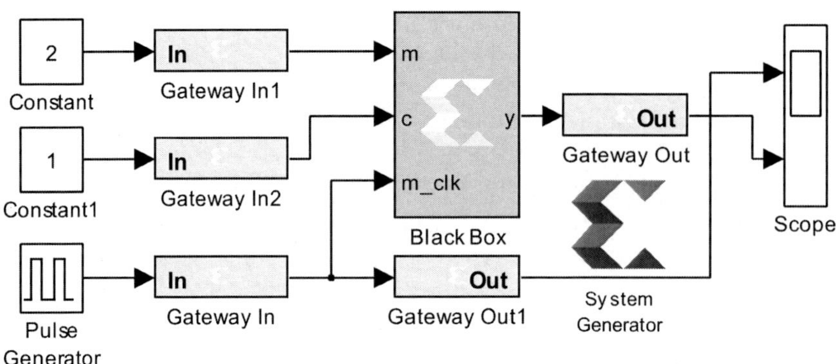

FIGURE 10.32
VHDL code interfacing with SysGen tools.

4. Gateway In1 and 2: under head base, output type: unsigned, number of bits: 3, binary point: 0, quantization: truncate, overflow: wrap and sample period: 1.
5. Black Box: under basic, block configuration m-function: <CHOSEN file> and simulation mode: ISE Simulator.
6. Gateway Out: tick "truncate into output port" and IoB timing constraint: none.
7. Scope: under general: number of axes: 2, time range: 100, labels: all and sample time: 0. Under history: tick "save data to workspace", variable name: ScopeData1 and format: structure with time.

Now the model is ready for simulation. Click the start simulation button. Once simulation is over, double-click on scope, and it shows the result, as shown in Figure 10.33. The first plot corresponds to the clock input to the Black Box and the second is the output of the Black Box, that is, `y = mx + c`. In the VHDL code, the `x` is taken as a variable to increment from "0000" through "1111" for every `rising_edge(m_clk)`; hence, `x` is incremented every time. Therefore, `y` will also be incremented every time. The values of given constants are 2 and 1 for `m` and `c`, respectively; hence, the equation becomes `y = 2x + 1`. The value of `x` varies from $(0)_{10}$ through $(15)_{10}$, and accordingly, the value of `y` varies from initially $(1)_{10}$ through $(31)_{10}$. This is clearly shown in Figure 10.33. We discuss one more example below.

DESIGN EXAMPLE 10.22

Assume that there is an analog input to a 14-bit IP core ADC whose output data have to be input to the VHDL code, that is, the Black Box. Compare the first 8 bits of ADC output with 02H and display $(110)_{10}$ if it exceeds or is equal; otherwise, the display data is $(28)_{10}$. Also, the system has to decrease and increase the value of an 8-bit counter with respect to a signal value; that is, if signal is '0', decrement, otherwise increment. Develop a digital system to do this task using the Black Box.

Solution

Before starting the VHDL code design as well as the SysGen design, we need to understand how the IP core ADC would be interfaced with the SysGen tools. The principles and interfacing methods of ADC are discussed below.

XtremeDSP Analog to Digital Converter: The Xilinx XtremeDSP ADC block allows system generator components to connect to the analog input channels on the Nallatech BenAdda board when a model is prepared. A separate ADC block (ADC1) is provided

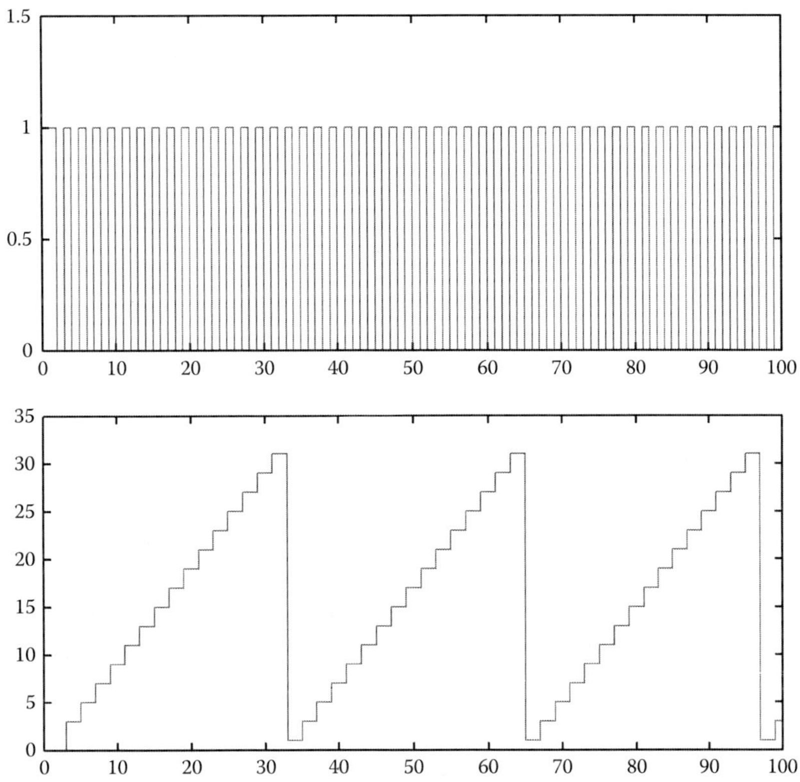

FIGURE 10.33
Simulation result of Figure 10.32 (time versus amplitude).

for the analog input channel. In Simulink, the ADC block is modeled using an input gateway that drives a register. The ADC block accepts an analog signal as input and produces a signed 14-bit Xilinx fixed-point signal as output. In hardware, a component that is driven by the ADC block output will be driven by one of the two 14-bit AD6644 ADC devices on the BenAdda board. When a System Generator model that uses an ADC block is translated into hardware, the ADC block is translated into a top-level input port on the model HDL. The appropriate pin location constraints are added in the BenAdda constraints file, thereby ensuring the port is driven appropriately by the ADC component.

With this information, first we develop the VHDL code. The developed VHDL code to accomplish the above task is given below.

```
LIBRARY ieee;
USE ieee.std_logic_1164.all;
USE ieee.numeric_std.all;
use ieee.std_logic_unsigned.all;
use IEEE.STD_LOGIC_ARITH.ALL;
entity vhdl_sys_gen_adc_dac is
Port (data_in:in STD_LOGIC_vector(7 downto 0):="00000000";
data_seg: out std_logic_vector(7 downto 0):="00000000";
data_out:out STD_LOGIC_vector(7 downto 0):="00000000";
m_clk:in std_logic:='0');
end vhdl_sys_gen_adc_dac;
architecture Behavioral of vhdl_sys_gen_adc_dac is
signal sel:std_logic:='0';
signal data_out_temp:STD_LOGIC_vector(7 downto 0):="00000000";
```

```
begin
p0:process(m_clk)
variable cnt:integer:=0;
begin
if rising_edge(m_clk) then
if cnt=255 then
sel<=not sel;
cnt:=0;
else
cnt:=cnt+1;
end if;
case sel is
when '0'=> data_out_temp<=data_out_temp-'1';
when '1'=> data_out_temp<=data_out_temp+'1';
when others=> data_out<=x"FF";
end case;
data_out <= data_out_temp;
end if;
end process;
p1: process(data_in)
begin
if data_in >= x"02" then
data_seg<="01101110";
else
data_seg<="00011100";
end if;
end process;
end Behavioral;
```

We follow the same steps that we followed in this section above. The SysGen design will look like Figure 10.34. Since a Gateway In already exists in ADC1, the MATLAB sine wave is directly connected to the ADC1. The pulse generator applies the driving clock to the Black Box, and the two outputs, counter output and comparator input, are connected to the Scope. The ADC output follows the sine wave; hence, the comparator output follows the sine wave. The counter output depends only on the signal status. The parameter settings in this design are given below:

1. Sine wave: sine type: time-based, time (t): use simulation time, amplitude: 5, bias: 5, frequency (rad/sec): 1, phase (rad): 0, sample time: 0.001and tick "interpret vector parameters as 1–D".

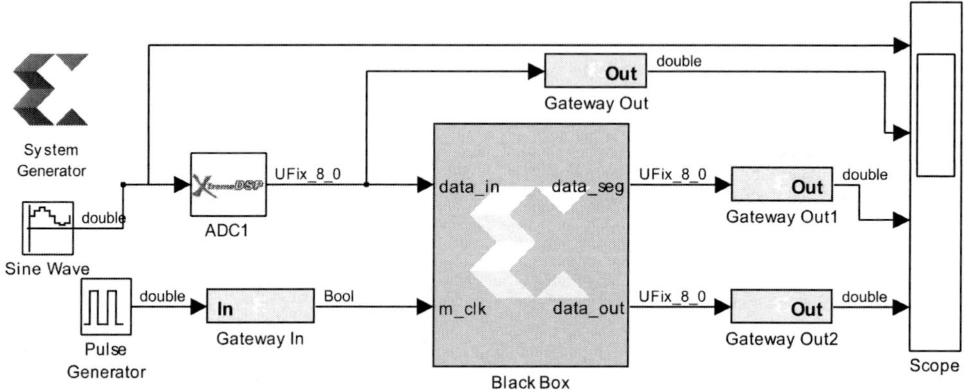

FIGURE 10.34
IP core ADC interfacing with Black Box.

2. Pulse Generator: pulse type: time-based, time (t): use estimation time: amplitude: 1, period (secs): 2, pulse width (% of period): 50, phase delay (secs): 0 and tick "interpret vector parameters as 1–D".
3. Gateway In: under head base, output type: unsigned, number of bits: 3, binary point: 0, quantization: truncate, overflow: wrap and sample period: 1.
4. ADC1: sample period: 1.
5. Black Box: under basic, block configuration m-function: <CHOSEN file> and simulation mode: ISE Simulator.
6. Gateway Out: tick "truncate into output port" and IoB timing constraint: none.
7. Scope: under general: number of axes: 4, time range: 100, labels: all and sample time: 0. Under history: tick "save data to workspace", variable name: ScopeData1 and format: structure with time.

After completing all the parameter settings, simulate the design, and the result will look like Figure 10.35. The first plot corresponds to the sine wave, the second is the ADC output, the third is the comparator output and the last is the counter output. Since the signal is complemented at an instant when the internal integer count reaches 255 in the VHDL code, the counter increments and decrements alternatively, as shown in Figure 10.35. In this way, any VHDL code can be interfaced with the Xilinx SysGen tools.

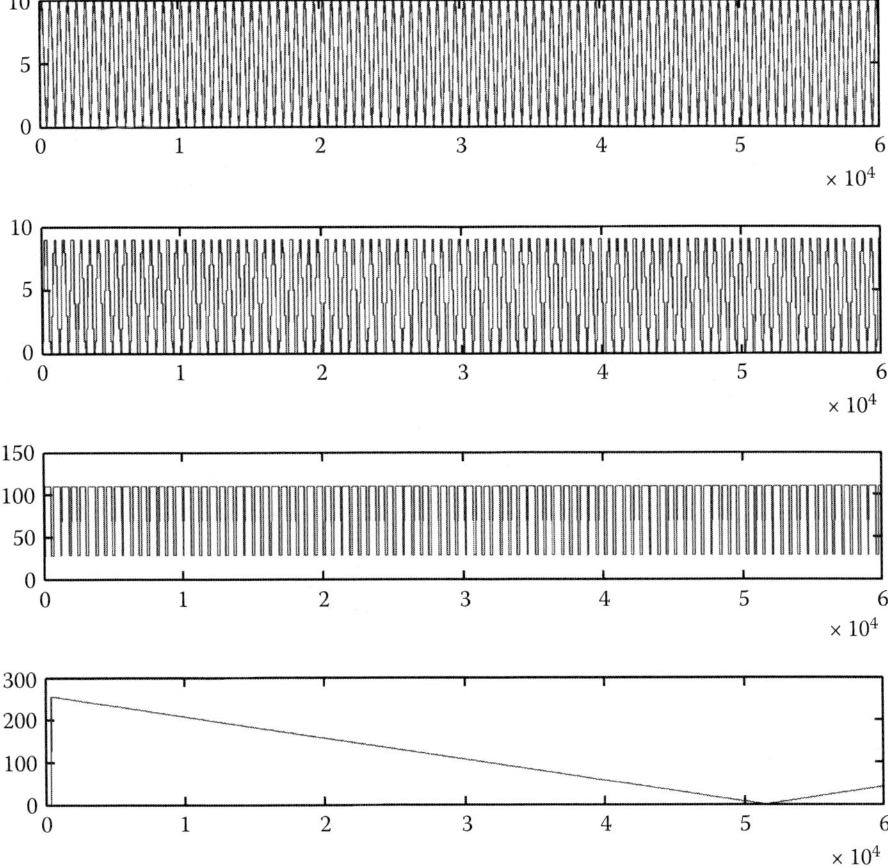

FIGURE 10.35
Simulation result of Figure 10.34 (time versus amplitude).

10.9 Real-Time Verification and Reconfigurable Architecture Design

As we discussed above, the Xilinx system generator tools have become increasingly important because of time-to-market constraints [140]. Apart from that, another advantage is the design's real-time function verification. This section presents a methodology for verifying the real-time operation of the design we built inside the FPGA using Xilinx system generator tools. This section also addresses the significance of reconfigurable architecture designs.

10.9.1 Design Flow for Hardware Co-Simulation

The Xilinx joint text action group (JTAG) co-simulation block allows us to perform hardware co-simulation using either JTAG, a parallel cable or a platform USB. The JTAG hardware co-simulation block interfaces the system generator to the FPGA platforms and keeps it in the simulation loop. When a model is implemented for JTAG hardware co-simulation, a new library is created that contains a custom JTAG co-simulation block with ports that match the gateway names from the original model. The co-simulation block interacts with the FPGA hardware platform during the simulation [140,144]. Simulation data that are written to the input ports of the block is passed to the hardware by the block. Conversely, when data are read from the co-simulation block's output ports, the block reads the appropriate values from the hardware and drives them on the output ports; hence, they can be interpreted in Simulink. In addition, the block automatically opens, configures, steps and closes the platform. Hardware/software co-simulation enables building a hardware version of the model and using the flexible simulation environment of Simulink. Therefore, we can perform several tests to verify the functionality of the system in hardware.

To generate the co-simulation module, double-click on the system generator token corresponding to the design for which we wish to perform the hardware co-simulation. A dialog box will pop up, as shown in Figure 10.36. This dialog box allows us to select the type of hardware we plan to use to perform the hardware co-simulation. There are three main icons, compilation, clocking and general, along with few heads below them in Figure 10.36. We have already seen all of them except the head compilation in Section 10.4 above. To select the hardware for the co-simulation, see Figure 10.36 under the sub-head "compilation", for example, for the XtremeDSP Development Kit, > → Hardware co-simulation → XtremeDSP Development Kit → JTAG. In this way, we can add any supporting device for the hardware co-simulation. Now, these details will appear in the respective box below the subhead "compilation". If the board is not listed therein, we can add it by following the procedures given in References 140, 144, and 145. Once the device selection is appropriately over, click Generate to build the hardware system. This step will generate a bit-stream that will later be used to configure the FPGA. ISE flow is used by the system generator to build this bit-stream.

When the compilation is complete, a new library is created, including one block, as shown in Figure 10.37. The library name will be "<FILE name>_hwcosim_lib" and the block name should be "Co_simuhwcosimu". The block has two inputs and one output as required by the DSP system. This block includes all the functionality required for the system to be executed on the FPGA. Now, we are ready to perform hardware co-simulation with the device we selected for the design we built using SysGen tools. The new block has to be linked to a bit-stream that will be downloaded into the FPGA for co-simulation. In this section, we need to modify the SysGen design with this new block, as shown in Figure

FIGURE 10.36
System generator token – window.

10.37. Then, connect the FPGA USB prog. cable and UART cable to the USB port and power it on. Double-click the "Co_simuhwcosimu" block. To download the design, select device and click OK. Now the design is ready for co-simulation. Click the Start Simulation button in the model window toolbar to start the co-simulation. The system generator will first download the bit-stream associated with the block. When the download completes, the system generator reads the inputs from the Simulink simulation environment and sends them to the design on the board using the JTAG connection. The system generator then reads the output back from JTAG and sends it to Simulink for display.

After the simulation is completed, the results will be displayed as per the design. We can verify the results by comparing the simulation output to the expected output. Another

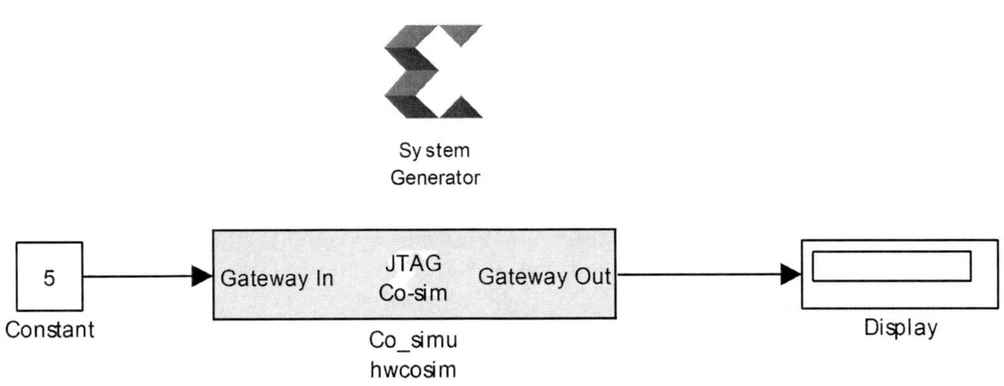

FIGURE 10.37
Hardware co-simulation window.

option to do runtime debugging is ChipScope Pro. The Xilinx ChipScope block enables runtime debugging and verification of signals within the FPGA. Deep capture memory and multiple trigger options are provided. Data is captured based on user-defined trigger conditions and stored in internal block memory. The Xilinx ChipScope block can be accessed at runtime using the ChipScope Pro analyzer software. The analyzer is used to configure the FPGA, set up trigger conditions and view the captured data at runtime. All control and data transfer are done via the JTAG port, eliminating the need to drive data off-chip using I/O pins. Data can be exported from the analyzer and read back into the MATLAB workspace. More details on hardware co-simulation and ChipScope Pro can be found in References 140 and 145.

10.9.2 Reconfigurable Architecture Design

A fundamental aspect of digital signal processing is filtering. Filtering is a selective system which passes a certain range of frequencies by attenuating the other frequencies. A digital filter is a system that performs mathematical operations on the sampled or discrete time-variant signal to shrink or enhance certain aspects of that signal. In general, hardware programmability in FPGA introduces design overhead that results in 21 times more area, 3 times longer delay and 10 times more dynamic power consumption compared to ASICs [140,145,146]. This section briefs the architecture designs and optimization techniques aimed at bridging the FPGA-ASIC gaps. The explorations are based on the advanced FPGA reconfiguration model, called temporal logic folding, that partitions applications into a sequence of stages to temporally share the same hardware resources. The distributed high-performance low-power memory blocks enable cycle-by-cycle runtime reconfiguration without a large power overhead. The area usage is significantly reduced, which also improves the interconnect performance and power consumption. Next, observe that logic folding reduces area significantly and most interconnects are localized.

Reconfigurable computing is becoming increasingly attractive for many DSP applications. There are three aspects of reconfigurable computing: (i) architecture, (ii) design method, and (iii) application. There are many applications for reconfigurable technology, such as data encryption, video processing, network security, and image processing. It has been shown that reconfigurable computing designs are capable of achieving up to 500 times speedup and 70% energy savings over microprocessor implementations for specific applications [146,147]. Reconfigurable computing involves the use of reconfigurable devices such as FPGAs for computing purposes. Sheer speed is not the only strength of reconfigurable computing; other compelling advantages are reduced energy and power consumption. In a reconfigurable system, the circuitry is optimized for the application such that the power consumption will tend to be much lower than that for a general-purpose processor. Other advantages of reconfigurable computing include a reduction in size and component count, improved time-to-market and improved flexibility and upgradability. These advantages are especially important for embedded applications. Indeed, there is evidence that embedded systems developers show a growing interest in reconfigurable computing systems, especially with the introduction of soft cores, which can contain one or more instruction processors.

We discuss an example for temporal and spatial computations. The computation flow of an operation $y = ax^2 + bx + c$ using software is given below:

1. Reading x @ t_1
2. Multiplying x by a @ t_2
3. Adding b to xa @ t_3

4. Multiplying (xa + b) by x @ t_4
5. Adding c to $ax^2 + bx$ @ t_5
6. Output $y = ax^2 + bx + c$ @ t_5

The same computation can also be performed in spatial domain as

1. Simultaneously read x and b and multiply x by x @ t_1
2. Multiply a by x^2 and add c to bx @ t_2
3. Add ax^2 to bx + c @ t_3
4. Output $y = ax^2 + bx + c$ @ t_3

The above computation flow is shown in Figure 10.38, which clearly shows that the temporal processor divides the computation across time and the dedicated hardware divides it across space.

Therefore, we need to understand when we should use software custom hardware or reconfigurable hardware. Software-based systems are simple to implement, but there is a huge performance gap between software and hardware systems, as listed below [146,147].

- There is often a 100:1 ratio of performance speed and area.
- Custom hardware systems, for e.g., ASIC-core based systems, are not so good for general computing. A major design effort is a barrier to implementation.
- Reconfigurable systems, for e.g., FPGA, offer the best runtime programmability.
- No instruction set for performing any operations as we need in microcontroller/processor.
- Multipliers, dividers and instruction memory are in temporal processors that need lots of memory to hold the instructions to make up a program and hold intermediate results.
- In general, processors have a fixed bit width and all computations are performed on that number of bits only.

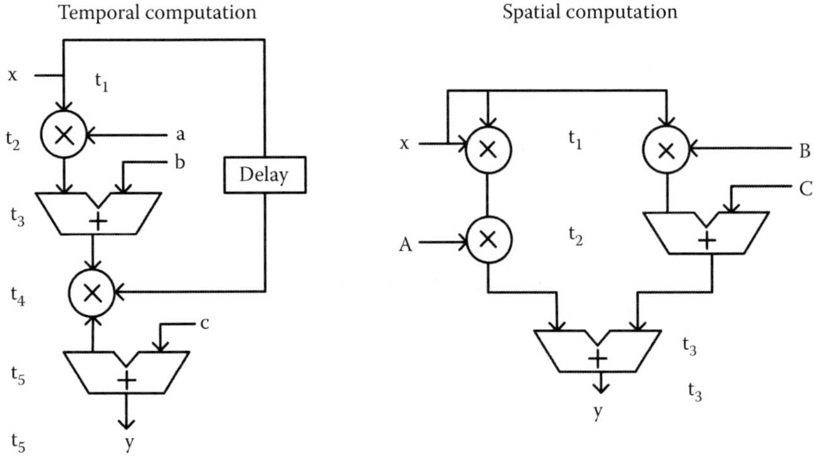

FIGURE 10.38
Illustration of temporal and spatial computation architecture.

- Execute the reconfigurable computations on many independent data elements and pipeline computations through the hardware with small and/or varying bit widths; take the advantage of the ability to customize the size of operators.
- Low-volume applications require rapid design turnaround time and hardware-like speeds, for example, DSP filters, radar, genomics (DNA sequence matching), processor emulation, and neural network applications.

However, dedicated hardware takes a long time for its design and development. Therefore, there is a strong need for a design approach which has performance comparable to dedicated hardware with an ease of programmability comparable to software. More details about reconfigurable architecture design can be found in Ref. 147.

LABORATORY EXERCISES

1. Develop VHDL code to build a digital system in the FPGA to convert a signed fixed-point precision binary number to a signed customized precision floating-point number.
2. Construct a GUI using the Xilinx System Generator model for converting decimal values into fixed-point, single-precision, double-precision, extended precision and customized precision floating-point numbers without using the available conversion Xilinx black sets.
3. Develop an Xilinx System Generator model for performing addition, subtraction, multiplication, and division with signed/unsigned fixed-point and single/double/extended/customized precision floating-point numbers without using the available Xilinx black sets.
4. Develop a digital system in FPGA to read the pixel values of an image from the computer, normalize the values to 100, and send them back to the computer display. Use MATLAB code to write and read the image values to/from the FPGA board through the RS232 port.
5. Develop an image-filtering algorithm using the MedianFilter in MATLAB and interface the code with the Xilinx SysGen environment. Display the original and filtered images.
6. Realize some operations like changing the sample rate, serial to parallel, parallel to serial, memory usage and reading/writing data from/to external onboard memory devices using VHDL code and the Xilinx SysGen block Black Box.
7. Develop a digital system using the Xilinx System Generator tools to generate the ECG signal. Assume a 12-bit bidirectional DAC is connected to the FPGA for converting the digital data to the equivalent analog signal.
8. Develop a digital circuit using the Xilinx System Generator tools to get the ASK, FSK, and PSK modulated signals. Assume a 12-bit bidirectional DAC is connected to the FPGA for converting the digital data to the equivalent analog signal.
9. Develop a digital system using the SysGen CORDIC block set to generate an LFM signal for a radar pulse compression application.
10. Develop a pipewrench reconfigurable architecture in FPGA and compile/demonstrate the reconfigurability while the system is working in real time.

Part IV

Miscellaneous Design and Applications

11

Digital Signal Processing with Field-Programmable Gate Array

11.1 Introduction and Preview

DSP is an essential section in many real-world applications such as speech processing, communication systems, measurements, radar, sonar, navigation systems, satellite communication, image processing, weather forecasting, and so on. Due to the vast advantages of the FPGA technology, all the real-time analog signals have to be converted to digital data to process them in the digital domain and then back to analog for further operations. Therefore, DSP techniques play a major role in almost all modern electronic applications, and the need for them significantly increases day by day to fulfil the potential demands. This chapter begins with a detailed description of the DFT and its implementations. The significance of correlations, measurement of frequency and representation of analog signals in the frequency domain (spectrum) are explained, with a design example.

The computation of real and imaginary components, power spectrum estimation, magnitude and phase response are also detailed. Construction of a simple FIR filter and IIR filter are explained, with necessary theoretical proofs. Different forms of design of the IIR filter are also detailed, and many signed/unsigned filter design examples are given with VHDL code. The necessities of operating a digital system at different sampling/processing rates are explained under multirate rate signal processing. The concepts of decimation and interpolation are explained, with design examples. Spectral shift due to decimation and interpolation are illustrated with MATLAB® code. A simple multirate filter design and its functions are given, with operation tables.

The advantages of Chinese remainder theorem (CRT), modulo arithmetic, and RNA systems are explained, with examples. Signed and unsigned addition, subtraction and multiplication using RNA systems are explained, with different numerical examples. VHDL code for a FIR filter design using RNA is detailed. Complete theoretical proofs for unsigned and signed DA number systems and their computations are described. Numerical examples, VHDL code and filter design using the DA system are explained. Algorithms and step-by-step numerical examples are given to enable the readers to clearly understand the Booth multiplier. The contribution of adaptive filters/equalizers to many applications is given. Theoretical proof for the least mean square (LMS) algorithm and a simple adaptive equalizer design using LMS are explained, with some design examples. Finally, laboratory exercises are listed.

11.2 Discrete Fourier Transform

Every real-time measurement signal contains one or more frequency, amplitude and phase components, for example, a sound wave on which any N samples have a set of frequencies, amplitudes, and phases that describe that wave. According to mathematician Joseph Fourier, we can take a set of sine waves of different amplitudes and frequencies and sum them together to get any wave form [148]. A plot of frequency versus amplitude (magnitude) on a 2D plane of these sine wave components is called a frequency spectrum, which is essential for digital signal processing. In this section, we discuss the principles and applications of the DFT technique. There are several ways of understanding the FT; in this section, we see it in terms of correlation between a signal and various frequency signals. The discrete samples of a signal can be represented by $x[n] = \{x_0, x_1, x_2, \ldots, x_N\}$, where n is the time index and N is the total number of samples produced during the observation period. The elements in x[n] are the actual amplitude of the samples, for example, $x[n] = \{-1, 0.34, -3.24, 0.98, 9.12, 13.24, \ldots, 12.23\}$. To determine the frequency content of x[n], we need to decompose it into simple parts. The time-domain to frequency domain or reverse operation can be done using the DFT and inverse DFT (IDFT), respectively, as shown in Figure 11.1.

Any signal can be described as a sum of many individual frequency components. The DFT is used to determine how many frequencies a complicated signal is composed of (see Reference 148). Correlation is a popular concept in signal processing which helps to measure how similar two signals are as

$$\sum_{i=0}^{N} x[i]y[i] \tag{11.1}$$

where x[i] and y[i] are the signals of interest to calculate the correlation/similarity between them; that means correlation is nothing but comparison. If the correlation is high, then the signals are similar, and if the correlation is near zero the signals are not very similar. Suppose a signal x[i] is very similar to a signal y[i]. Then, the sign of x[i] and y[i] will be equal, that is, positive or negative. If we multiply the positive part of these two signals, we get another positive number. If we multiply the negative part, we get another positive number [148]. If x[i] and y[i] are opposite or opposite in sign, then the final sum will have a few positive numbers and a few negative numbers, which results in a number near zero.

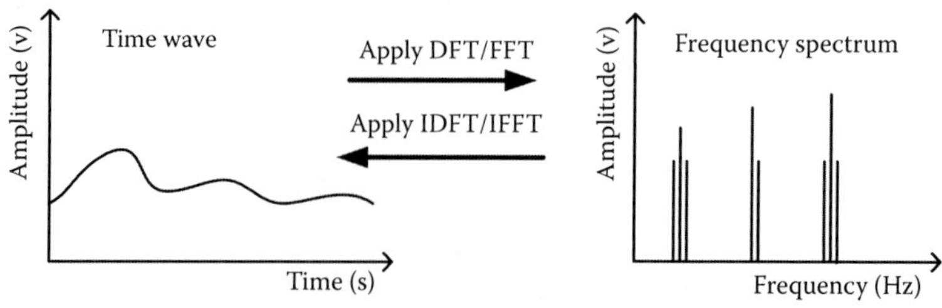

FIGURE 11.1
Illustration of time wave and frequency spectrum.

Therefore, very similar signals will yield a high correlation, while very different signals will yield a low correlation, and signals that are a little bit similar will get a correlation score somewhere in between.

A large negative correlation shows the signals are very similar, but one is inverted with respect to the other [148,149]. Now, we discuss how correlation helps us obtain the frequency spectrum of a signal. Consider the following DFT equation

$$X[k] = \sum_{n=0}^{N-1} x[n]e^{-ink2\pi/N} \tag{11.2}$$

where the sweep factor k can be from 0 to N − 1 to calculate all the DFT coefficients, that is, X[k] as {X[0], X[1], X[2], … , X[N − 1]}; X[0] is the first coefficient, X[1] the second, and so on. Equation 11.2 is very similar to Equation 11.1 because it calculates the correlation between a signal x[n] and a function exp(−i2πkn/N). It means that if we break up that exponential part into sine and cosine components as exp(−iθ) = cos θ − isin θ, then Equation 11.2 becomes

$$\begin{aligned} X[k] &= \sum_{n=0}^{N-1} x[n]\left(\cos\left(\frac{2\pi kn}{N}\right) - i\sin\left(\frac{2k\pi n}{N}\right)\right) \\ X[k] &= \sum_{n=0}^{N-1} x[n]\cos\left(\frac{2\pi kn}{N}\right) - i\sum_{n=0}^{N-1} x[n]\sin\left(\frac{2\pi kn}{N}\right) \end{aligned} \tag{11.3}$$

This equation clearly shows that the first term computes the correlation of x[n] with the cosine function and second term uses the sine function. Therefore, the correlation result, that is, X[k], is stored in a DFT bin of size k. First, we calculate the correlation between x[n] and a cosine function of a certain frequency and put the value into the real part of X[k]. Then, we do the same computation with a sine function and put the value into the imaginary part of X[k]. Therefore, the DFT bin has two values (real and imaginary) for every value of k, as shown in Figure 11.2. Now, we discuss the computation of correlation of x[n] with a sequence of cosine and sine signals of increasing frequency. Equation 11.3 can be used to calculate the coefficients X[k] for various values of k. When k = 0, Equation 11.3 becomes

$$\begin{aligned} X[0] &= \sum_{n=0}^{N-1} x[n]\cos\left(\frac{2n0\pi}{N}\right) - i\sum_{n=0}^{N-1} x[n]\sin\left(\frac{2n0\pi}{N}\right) \\ X[0] &= \sum_{n=0}^{N-1} x[n] - i\sum_{n=0}^{N-1} 0 \\ X[0] &= \sum_{n=0}^{N-1} x[n] \end{aligned} \tag{11.4}$$

From Equation 11.4, we can calculate only the direct/fluctuating DC components. This, in turn, is equal to the sum of the unmodified signal, that is, just all the samples. When k = 1, Equation 11.3 becomes

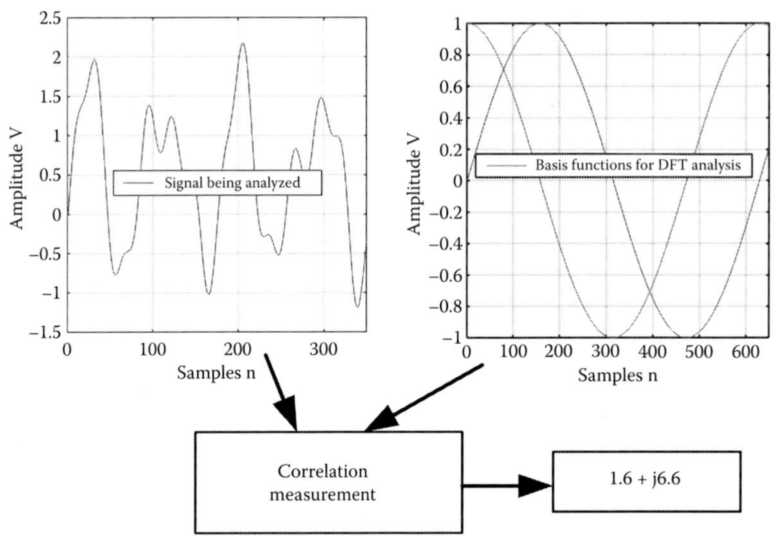

FIGURE 11.2
Illustration of signal correlation measurement. Cosine and sine basis functions with 1 cycle over N samples.

$$X[1] = \sum_{n=0}^{N-1} x[n]\cos\left(\frac{2\pi 1n}{N}\right) - i\sum_{n=0}^{N-1} x[n]\sin\left(\frac{2n1\pi}{N}\right)$$
$$X[1] = \sum_{n=0}^{N-1} x[n]\cos\left(\frac{2\pi n}{N}\right) - i\sum_{n=0}^{N-1} x[n]\sin\left(\frac{2\pi n}{N}\right) \quad (11.5)$$

From Equation 11.5, we can calculate the correlation of the signal with 1 Hz of cos(.) and sin(.) functions. In the same way, the correlation of the signal with 2 Hz, 3 Hz, 4 Hz, … kHz of cos(.) and sin(.) functions can be calculated. Finally, we will get many coefficient values which will give the magnitude for a wide spectrum of frequencies through which we would be able to plot the spectral characteristics of the signal x[n]. The magnitude and phase spectrum can be obtained by using

$$\text{Magnitude} = \sqrt{X^2_{\text{Real}} + X^2_{\text{Ima}}}$$
$$\text{Phase} = \tan^{-1}\left(\frac{X_{\text{Ima}}}{X_{\text{Real}}}\right) \quad (11.6)$$

We will see an example to generate a sensor signal; plot the basis functions; plot the correlation at each value of k and plot the frequency, magnitude, and phase spectrum to understand this concept more clearly below. However, calculating a DFT is sometimes too slow because of the number of multiplications required. An FFT is an algorithm that speeds up the calculation of a DFT. An inverse Fourier transform (IFT) converts the frequency domain components back into the original time wave, as shown in Figure 11.1.

$$\text{frequency (f)} = \left(\frac{kf_s}{N}\right) \quad (11.7)$$

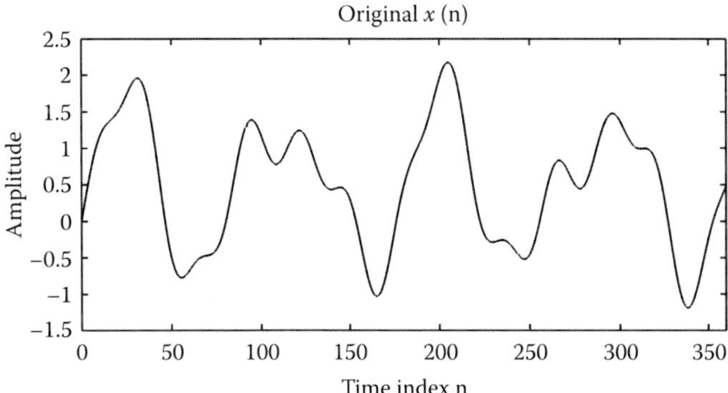

FIGURE 11.3
Random (sensor) signal with 350 samples.

The actual frequency can be computed using Equation 11.7 upon knowing the original sampling frequency (f_s) and number of samples considered for frequency analysis, that is, N.

DESIGN EXAMPLE 11.1

Generate a random signal by combining various sine and cosine frequency components and assume it as a real-time sensor signal as shown in Figure 11.3. Measure its frequency spectrum with 200 samples for each measurement and 11 basis sine and cosine functions using the DFT without MATLAB functions. Plot the frequency, magnitude and phase spectrum as well.

Solution

```
clc;
clear all;
clear screen;
figure(1);
subplot(2,2,3:4);
t=0:1:360;
x_t= 0.5 + 0.5*sind(2*pi*t-90)+ 0.3*sind(4*pi*t)+ sind(4*t);
plot(t,x_t,'k-','LineWidth', 1.5);
xlim([0 max(t)]);
ylabel('Amplitude');
xlabel('Time index n');
title('Original x[n]');
set(gca, 'FontName','Timenewromain','FontSize',10, 'LineWidth', 1.5);
The sensor output samples are shown in Figure 11.3.
figure(2);
N=200;
n=0:1:N-1;
subplot(4,2,1);
x_n=x_t(1:length(n));
plot(n,x_n,'k-','LineWidth', 1.5);
hold on;
dc_val=ones(1,length(n));
plot(n,dc_val,'r-','LineWidth', 1.5);
xlim([0 length(n)]);
ylabel('Amplitude');
xlabel('Time index (n)');
%For X(0)
hold off;
Real_part=x_n.*1;
```

```
Imag_part=zeros(1,length(n));
title('k=0 i.e., DC');
set(gca, 'FontName','Timenewromain','FontSize',10, 'LineWidth', 1.5);
subplot(4,2,3);
plot(n,Real_part,'b-','LineWidth', 1.5);
hold on;
plot(n,Imag_part,'g-','LineWidth', 1.5);
hold off;
ylabel('Amplitude');
xlabel('Time index n');
xlim([0 length(n)]);
X_0_real=sum(Real_part);
X_0_imag=sum(Imag_part);
X_0=sum(Real_part) -1i*sum(Imag_part);
set(gca, 'FontName','Timenewromain','FontSize',10, 'LineWidth', 1.5);
%for X(1)
subplot(4,2,2);
plot(n,x_n,'k-','LineWidth', 1.5);
hold on;
plot(n,cos((2*pi*1*n)/length(n)),'b-','LineWidth', 1.5);
hold on;
plot(n,sin((2*pi*1*n)/length(n)),'g-','LineWidth', 1.5);
hold off;
xlim([0 length(n)]);
Real_part=x_n.*(cos((2*pi*1*n)/length(n)));
Imag_part=x_n.*(sin((2*pi*1*n)/length(n)));
title('k=1');
set(gca, 'FontName','Timenewromain','FontSize',10, 'LineWidth', 1.5);
subplot(4,2,4);
plot(n,Real_part,'b-','LineWidth', 1.5);
hold on;
plot(n,Imag_part,'g-','LineWidth', 1.5);
hold off;
ylabel('Amplitude');
xlabel('Time index n');
xlim([0 length(n)]);
X_1_real=sum(Real_part);
X_1_imag=sum(Imag_part);
X_1=sum(Real_part) -1i*sum(Imag_part);
set(gca, 'FontName','Timenewromain','FontSize',10, 'LineWidth', 1.5);
%For X(2)
subplot(4,2,5);
plot(n,x_n,'k-','LineWidth', 1.5);
hold on;
plot(n,cos((2.*pi.*2.*n)./length(n)),'b-','LineWidth', 1.5);
hold on;
plot(n,sin((2.*pi.*2.*n)./length(n)),'g-','LineWidth', 1.5);
hold off;
xlim([0 length(n)]);
Real_part=x_n.*(cos((2*pi*2*n)/length(n)));
Imag_part=x_n.*(sin((2*pi*2*n)/length(n)));
title('k=2');
set(gca, 'FontName','Timenewromain','FontSize',10, 'LineWidth', 1.5);
subplot(4,2,7);
plot(n,Real_part,'b-','LineWidth', 1.5);
hold on;
plot(n,Imag_part,'g-','LineWidth', 1.5);
hold off;
ylabel('Amplitude');
xlabel('Time index n');
xlim([0 length(n)]);
```

```
X_2_real=sum(Real_part);
X_2_imag=sum(Imag_part);
X_2=sum(Real_part) -1i*sum(Imag_part);
set(gca, 'FontName','Timenewromain','FontSize',10, 'LineWidth', 1.5);
%for X(3)
subplot(4,2,6);
plot(n,x_n,'k-','LineWidth', 1.5);
hold on;
plot(n,cos((2*pi*3*n)/length(n)),'b-','LineWidth', 1.5);
hold on;
plot(n,sin((2*pi*3*n)/length(n)),'g-','LineWidth', 1.5);
hold off;
xlim([0 length(n)]);
Real_part=x_n.*(cos((2*pi*3*n)/length(n)));
Imag_part=x_n.*(sin((2*pi*3*n)/length(n)));
title('k=3');
set(gca, 'FontName','Timenewromain','FontSize',10, 'LineWidth', 1.5);
subplot(4,2,8);
plot(n,Real_part,'b-','LineWidth', 1.5);
hold on;
plot(n,Imag_part,'g-','LineWidth', 1.5);
ylabel('Amplitude');
xlabel('Time index n');
xlim([0 length(n)]);
hold off;
X_3_real=sum(Real_part);
X_3_imag=sum(Imag_part);
X_3=sum(Real_part) -1i*sum(Imag_part);
set(gca, 'FontName','Timenewromain','FontSize',10, 'LineWidth', 1.5);
```

The original sensor output, sine and cosine functions and correlation results for values of k from 0 to 3 are shown in Figure 11.4.

```
figure(3);
%for X(4)
subplot(4,2,1);
plot(n,x_n,'k-','LineWidth', 1.5);
hold on;
plot(n,cos((2*pi*4*n)/length(n)),'b-','LineWidth', 1.5);
hold on;
plot(n,sin((2*pi*4*n)/length(n)),'g-','LineWidth', 1.5);
hold off;
xlim([0 length(n)]);
ylabel('Amplitude');
xlabel('Time index n');
Real_part=x_n.*(cos((2*pi*4*n)/length(n)));
Imag_part=x_n.*(sin((2*pi*4*n)/length(n)));
title('k=4');
set(gca, 'FontName','Timenewromain','FontSize',10, 'LineWidth', 1.5);
subplot(4,2,3);
plot(n,Real_part,'b-','LineWidth', 1.5);
hold on;
plot(n,Imag_part,'g-','LineWidth', 1.5);
hold off;
ylabel('Amplitude');
xlabel('Time index n');
xlim([0 length(n)]);
X_4_real=sum(Real_part);
X_4_imag=sum(Imag_part);
X_4=sum(Real_part) -1i*sum(Imag_part);
set(gca, 'FontName','Timenewromain','FontSize',10, 'LineWidth', 1.5);
```

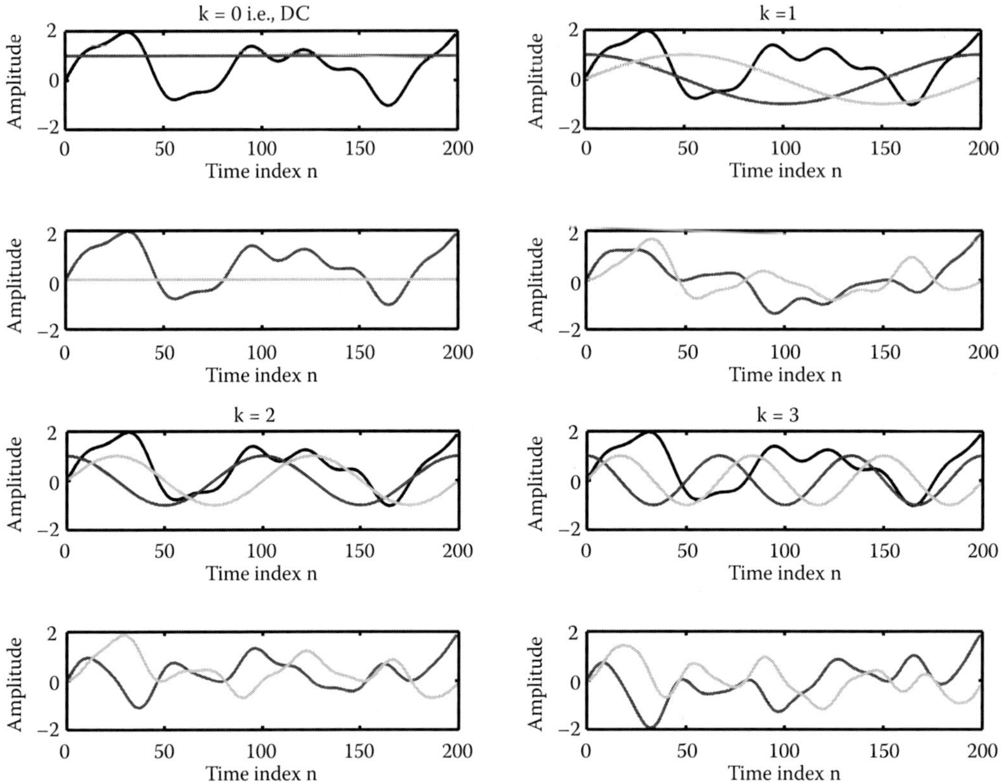

FIGURE 11.4
Basis function correlation with k = 0 to 3 over 200 samples of x[n]. In Figure 11.4, black, red, green, and blue represent the x[n], DC, sine and cosine functions, respectively. The correlation results are shown below the plots corresponding to k.

```
%for X(5)
subplot(4,2,2);
plot(n,x_n,'k-','LineWidth', 1.5);
hold on;
plot(n,cos((2*pi*5*n)/length(n)),'b','LineWidth', 1.5);
hold on;
plot(n,sin((2*pi*5*n)/length(n)),'g','LineWidth', 1.5);
hold off;
xlim([0 length(n)]);
ylabel('Amplitude');
xlabel('Time index n');
Real_part=x_n.*(cos((2*pi*5*n)/length(n)));
Imag_part=x_n.*(sin((2*pi*5*n)/length(n)));
title('k=5');
set(gca, 'FontName','Timenewromain','FontSize',10, 'LineWidth', 1.5);
subplot(4,2,4);
plot(n,Real_part,'b-','LineWidth', 1.5);
hold on;
plot(n,Imag_part,'g-','LineWidth', 1.5);
hold off;
ylabel('Amplitude');
```

```
xlabel('Time index n');
xlim([0 length(n)]);
X_5_real=sum(Real_part);
X_5_imag=sum(Imag_part);
X_5=sum(Real_part) -1i*sum(Imag_part);
set(gca, 'FontName','Timenewromain','FontSize',10, 'LineWidth', 1.5);
%for X(6)
subplot(4,2,5);
plot(n,x_n,'k-','LineWidth', 1.5);
hold on;
plot(n,cos((2*pi*6*n)/length(n)),'b-','LineWidth', 1.5);
hold on;
plot(n,sin((2*pi*6*n)/length(n)),'g-','LineWidth', 1.5);
hold off;
xlim([0 length(n)]);
ylabel('Amplitude');
xlabel('Time index n');
Real_part=x_n.*(cos((2*pi*6*n)/length(n)));
Imag_part=x_n.*(sin((2*pi*6*n)/length(n)));
title('k=6');
set(gca, 'FontName','Timenewromain','FontSize',10, 'LineWidth', 1.5);
subplot(4,2,7);
plot(n,Real_part,'b-','LineWidth', 1.5);
hold on;
plot(n,Imag_part,'g-','LineWidth', 1.5);
hold off;
ylabel('Amplitude');
xlabel('Time index n');
xlim([0 length(n)]);
X_6_real=sum(Real_part);
X_6_imag=sum(Imag_part);
X_6=sum(Real_part) -1i*sum(Imag_part);
set(gca, 'FontName','Timenewromain','FontSize',10, 'LineWidth', 1.5);
%for X(7)
subplot(4,2,6);
plot(n,x_n,'k-','LineWidth', 1.5);
hold on;
plot(n,cos((2*pi*7*n)/length(n)),'b-','LineWidth', 1.5);
hold on;
plot(n,sin((2*pi*7*n)/length(n)),'g-','LineWidth', 1.5);
hold off;
xlim([0 length(n)]);
ylabel('Amplitude');
xlabel('Time index n');
Real_part=x_n.*(cos((2*pi*7*n)/length(n)));
Imag_part=x_n.*(sin((2*pi*7*n)/length(n)));
title('k=7');
set(gca, 'FontName','Timenewromain','FontSize',10, 'LineWidth', 1.5);
subplot(4,2,8);
plot(n,Real_part,'b-','LineWidth', 1.5);
hold on;
plot(n,Imag_part,'g-','LineWidth', 1.5);
hold off;
ylabel('Amplitude');
xlabel('Time index n');
xlim([0 length(n)]);
X_7_real=sum(Real_part);
X_7_imag=sum(Imag_part);
X_7=sum(Real_part) -1i*sum(Imag_part);
set(gca, 'FontName','Timenewromain','FontSize',10, 'LineWidth', 1.5);
```

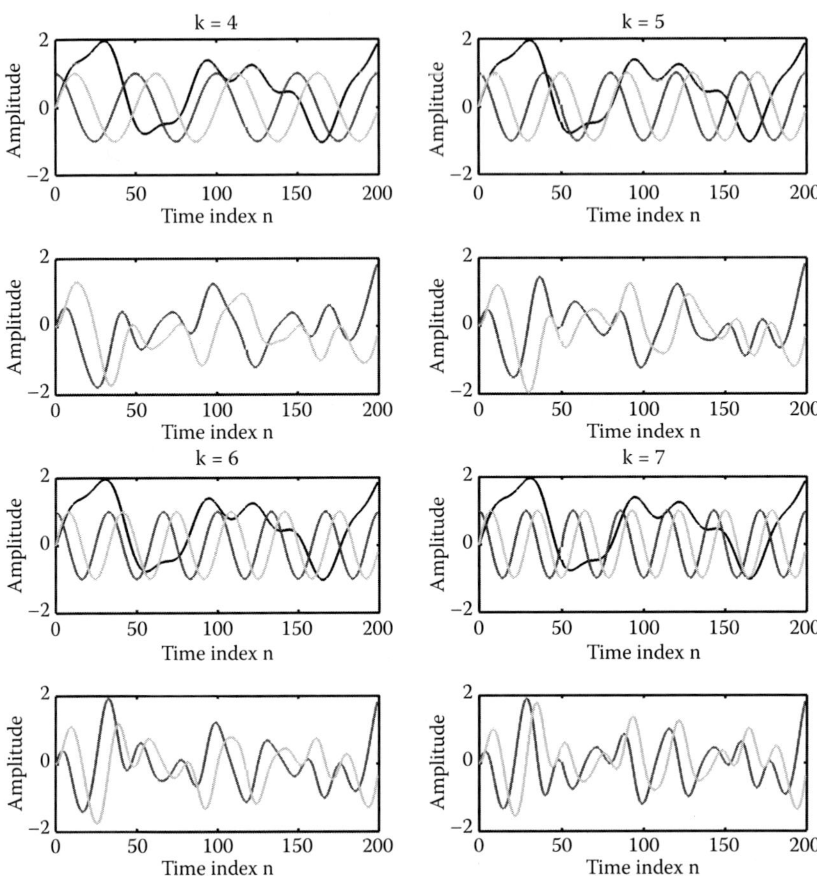

FIGURE 11.5
Same as Figure 11.4 with k = 4 to 7.

The original sensor output, sine and cosine functions and correlation results for values of k from 4 to 7 are shown in Figure 11.5.

```
figure(4);
%for X(8)
subplot(4,2,1);
plot(n,x_n,'k-','LineWidth', 1.5);
hold on;
plot(n,cos((2*pi*8*n)/length(n)),'b-','LineWidth', 1.5);
hold on;
plot(n,sin((2*pi*8*n)/length(n)),'g-','LineWidth', 1.5);
hold off;
xlim([0 length(n)]);
ylabel('Amplitude');
xlabel('Time index n');
Real_part=x_n.*(cos((2*pi*8*n)/length(n)));
Imag_part=x_n.*(sin((2*pi*8*n)/length(n)));
title('k=8');
set(gca, 'FontName','Timenewromain','FontSize',10, 'LineWidth', 1.5);
subplot(4,2,3);
plot(n,Real_part,'b-','LineWidth', 1.5);
```

```
hold on;
plot(n,Imag_part,'g-','LineWidth', 1.5);
hold off;
ylabel('Amplitude');
xlabel('Time index n');
xlim([0 length(n)]);
X_8_real=sum(Real_part);
X_8_imag=sum(Imag_part);
X_8=sum(Real_part) -1i*sum(Imag_part);
set(gca, 'FontName','Timenewromain','FontSize',10, 'LineWidth', 1.5);
%for X(9)
subplot(4,2,2);
plot(n,x_n,'k-','LineWidth', 1.5);
hold on;
plot(n,cos((2*pi*9*n)/length(n)),'b-','LineWidth', 1.5);
hold on;
plot(n,sin((2*pi*9*n)/length(n)),'g-','LineWidth', 1.5);
hold off;
xlim([0 length(n)]);
ylabel('Amplitude');
xlabel('Time index n');
Real_part=x_n.*(cos((2*pi*9*n)/length(n)));
Imag_part=x_n.*(sin((2*pi*9*n)/length(n)));
title('k=9');
set(gca, 'FontName','Timenewromain','FontSize',10, 'LineWidth', 1.5);
subplot(4,2,4);
plot(n,Real_part,'b-','LineWidth', 1.5);
hold on;
plot(n,Imag_part,'g-','LineWidth', 1.5);
hold off;
ylabel('Amplitude');
xlabel('Time index n');
xlim([0 length(n)]);
X_9_real=sum(Real_part);
X_9_imag=sum(Imag_part);
X_9=sum(Real_part) -1i*sum(Imag_part);
set(gca, 'FontName','Timenewromain','FontSize',10, 'LineWidth', 1.5);
%for X(10)
subplot(4,2,5);
plot(n,x_n,'k-','LineWidth', 1.5);
hold on;
plot(n,cos((2*pi*10*n)/length(n)),'b-','LineWidth', 1.5);
hold on;
plot(n,sin((2*pi*10*n)/length(n)),'g-','LineWidth', 1.5);
hold off;
xlim([0 length(n)]);
ylabel('Amplitude');
xlabel('Time index n');
Real_part=x_n.*(cos((2*pi*10*n)/length(n)));
Imag_part=x_n.*(sin((2*pi*10*n)/length(n)));
title('k=10');
set(gca, 'FontName','Timenewromain','FontSize',10, 'LineWidth', 1.5);
subplot(4,2,7);
plot(n,Real_part,'b-','LineWidth', 1.5);
hold on;
plot(n,Imag_part,'g-','LineWidth', 1.5);
hold off;
ylabel('Amplitude');
xlabel('Time index n');
xlim([0 length(n)]);
X_10_real=sum(Real_part);
```

```
X_10_imag=sum(Imag_part);
X_10=sum(Real_part) -1i*sum(Imag_part);
set(gca, 'FontName','Timenewromain','FontSize',10, 'LineWidth', 1.5);
subplot(4,2,6);
X_real=[X_0_real X_1_real X_2_real X_3_real X_4_real X_5_real X_6_real
X_7_real X_8_real X_9_real X_10_real];
stem(0:length(X_real)-1,X_real,'LineWidth', 1.5);
xlabel('Index k');
ylabel('Real Magnitude');
set(gca, 'FontName','Timenewromain','FontSize',10, 'LineWidth', 1.5);
subplot(4,2,8);
X_imag=[X_0_imag X_1_imag X_2_imag X_3_imag X_4_imag X_5_imag X_6_imag
X_7_imag X_8_imag X_9_imag X_10_imag];
stem(0:length(X_imag)-1,X_imag,'LineWidth', 1.5);
xlabel('Index k');
ylabel('Imag. Magnitude');
set(gca, 'FontName','Timenewromain','FontSize',10, 'LineWidth', 1.5);
X_k=[X_0 X_1 X_2 X_3 X_4 X_5 X_6 X_7 X_8 X_9 X_10]
```

The original sensor output, sine and cosine functions, correlation results for values of k from 8 to 10 and real and imaginary part spectra are shown in Figure 11.6.

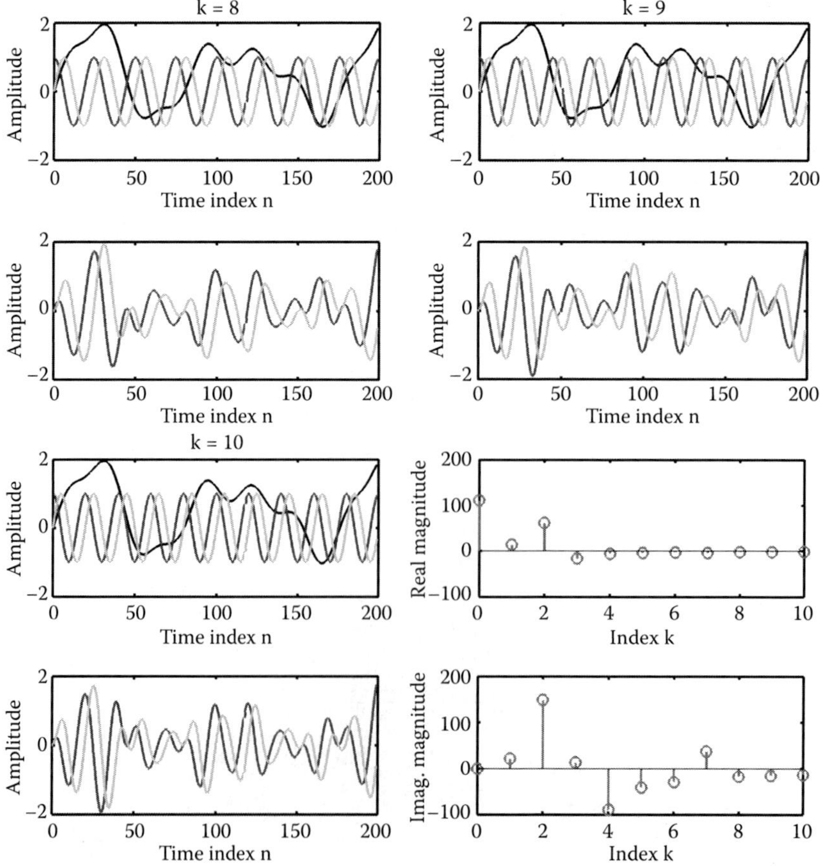

FIGURE 11.6
Same as Figure 11.4 with k = 8 to 10 and real and imaginary magnitude spectrum.

```
figure(5);
subplot(2,2,1:2);
PS= (X_real).^2 + (X_imag).^2;
stem(0:length(PS)-1,PS,'LineWidth', 1.5);
xlabel('Index k');
ylabel('Estimated Power');
set(gca, 'FontName','Timenewromain','FontSize',10, 'LineWidth', 1.5);
subplot(2,2,3);
magni=sqrt(PS)
stem(0:length(PS)-1,magni,'LineWidth', 1.5);
xlabel('Index k');
ylabel('Magnitude');
subplot(2,2,4);
pha=atand(X_imag./X_real)
stem(0:length(PS)-1,pha,'LineWidth', 1.5);
xlabel('Index k');
ylabel('Phase');
```

The estimated power, magnitude and phase spectrum for values of k from 0 to 10 is shown in Figure 11.7.

DESIGN EXAMPLE 11.2

Develop a digital system to read out the coefficient values stored in two different LUTs based on the input values. Use a multiplexer to obtain the output from the LUTs.

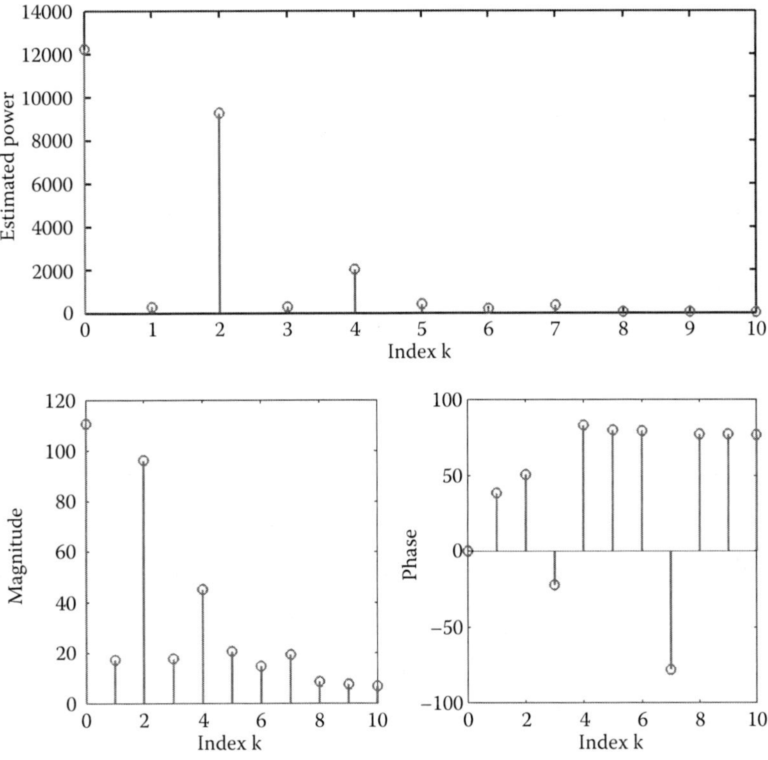

FIGURE 11.7
Power, magnitude, and phase spectrum over 200 samples.

Solution

```
library ieee;
use ieee.std_logic_1164.all;
use ieee.std_logic_arith.all;
entity case5p is
port ( clk   : in  std_logic;
table_in  : in  std_logic_vector(4 downto 0);
table_out : out integer range 0 to 25);
end case5p;
architecture les of case5p is
signal lsbs : std_logic_vector(3 downto 0);
signal msbs0 : std_logic_vector(1 downto 0);
signal table0out00, table0out01 : integer range 0 to 25;
begin
process
begin
wait until clk = '1';
lsbs(0) <= table_in(0);
lsbs(1) <= table_in(1);
lsbs(2) <= table_in(2);
lsbs(3) <= table_in(3);
msbs0(0) <= table_in(4);
msbs0(1) <= msbs0(0);
end process;
process
begin
wait until clk = '1';
case msbs0(1) is
when  '0' =>    table_out <=  table0out00;
when  '1' =>    table_out <=  table0out01;
when   others  =>    table_out <=  0;
end case;
end process;
process
begin
wait until clk = '1';
case lsbs is
when  "0000" =>    table0out00 <=  0;
when  "0001" =>    table0out00 <=  1;
when  "0010" =>    table0out00 <=  3;
when  "0011" =>    table0out00 <=  4;
when  "0100" =>    table0out00 <=  5;
when  "0101" =>    table0out00 <=  6;
when  "0110" =>    table0out00 <=  8;
when  "0111" =>    table0out00 <=  9;
when  "1000" =>    table0out00 <=  7;
when  "1001" =>    table0out00 <=  8;
when  "1010" =>    table0out00 <=  10;
when  "1011" =>    table0out00 <=  11;
when  "1100" =>    table0out00 <=  12;
when  "1101" =>    table0out00 <=  13;
when  "1110" =>    table0out00 <=  15;
when  "1111" =>    table0out00 <=  16;
when   others =>   table0out00 <=  0;
end case;
end process;
process
begin
wait until clk = '1';
case lsbs is
```

```
when  "0000" =>  table0out01 <=  9;
when  "0001" =>  table0out01 <=  10;
when  "0010" =>  table0out01 <=  12;
when  "0011" =>  table0out01 <=  13;
when  "0100" =>  table0out01 <=  14;
when  "0101" =>  table0out01 <=  15;
when  "0110" =>  table0out01 <=  17;
when  "0111" =>  table0out01 <=  18;
when  "1000" =>  table0out01 <=  16;
when  "1001" =>  table0out01 <=  17;
when  "1010" =>  table0out01 <=  19;
when  "1011" =>  table0out01 <=  20;
when  "1100" =>  table0out01 <=  21;
when  "1101" =>  table0out01 <=  22;
when  "1110" =>  table0out01 <=  24;
when  "1111" =>  table0out01 <=  25;
when  others =>  table0out01 <=  0;
end case;
end process;
end les;
```

11.3 Digital Finite Impulse Response Filter Design

Digital filters can be divided into two categories, namely, FIR filters and IIR filters. FIR filters, in general, require higher taps than IIR filters to obtain similar frequency characteristics. We discuss FIR filters in this section and IIR filters in the next section. FIR filters are widely used because they have linear phase characteristics, guarantee stability, and are easy to implement with multipliers, adders, and delay elements [25,149]. The number of taps in a FIR filter varies according to the application and decides the order of the filter. Today, commercial filter chips are available with a fixed number of taps and provisions to load the required coefficient values through the registers to perform the necessary computations. A standard structure of a FIR filter is shown in Figure 11.8, where the input is x[n], the filter coefficient or impulse response is h[n] and the filter output is y[n]. The output signal of a FIR filter can be described by a simple convolution operation, as expressed in Equation 11.8.

Thus, the operation of a FIR filter is just a convolution of the input signal x[n] with the filter's impulse response h[n]. The use of the transversal structure allows relatively straightforward implementation of a filter, as in Figure 11.8, which also can be implemented in a software environment. The primary function of each block of the FIR filter is accepting the input signal, blocking prespecified frequency components, and passing the original signal minus those components to the output. For example, a typical phone line

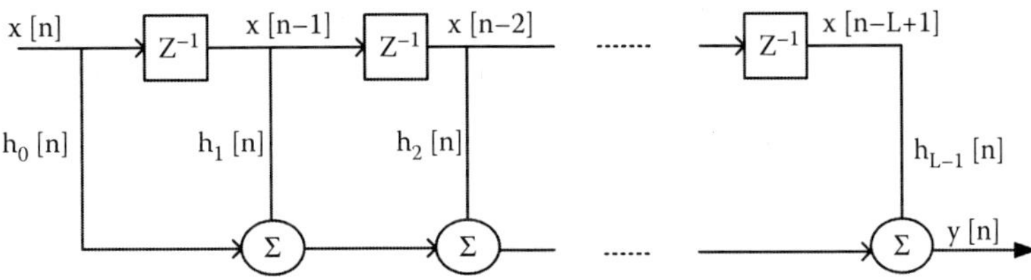

FIGURE 11.8
Standard structure of a FIR filter.

acts as a filter that limits frequencies to a range considerably smaller than the range of frequencies human beings can hear [25,149]. That is why listening to CD-quality music over the phone is not as pleasing to the ear as listening to it directly [149].

$$y[n] = h_0x[n] + h_1x[n-1] + h_2x[n-2] + \cdots + h_Mx[n-M)$$
$$= \sum_{m=0}^{M} h_m x[n-m]$$

In general,

$$y[n] = \sum_{m=-\infty}^{\infty} h[m]x[n-m]$$
$$y[n] = h(n) * x(n) \tag{11.8}$$

where M is "L – 1", L is the total number of tapping elements and * is the convolution operator. A digital filter consists of digital components, takes a digital input and gives a digital output. FIR filters are usually defined by their responses to the individual frequency components that constitute the input signal. The responses to different frequencies are characterized as pass-band and stop-band. The pass-band response is the filter's effect on frequency components that are passed through it. Frequencies within a filter's stop-band are highly attenuated. The transition band represents frequencies in the middle which may receive some attenuation but are not removed completely from the output signal. These filters can be designed software packages with required equations and coefficients for implementation on a DSP platform. A FIR filter is usually implemented by using a series of delays, multipliers and adders to create the filter's output. The values of `h[n]` are used for multiplication so that the output at time `n` is the summation of all the delayed samples multiplied by the appropriate coefficients. The longer the filter, that is, the more taps, the more finely the response can be tuned.

DESIGN EXAMPLE 11.3

Develop a digital system in the FPGA to realize a FIR filter. The impulse response `h[n]` of the filter is {–2, –1, 3, 4}. Get signed input samples of width 8 from the eight GPIO pins. Display the filter response using 16 LEDs [25].

Solution

```
library IEEE;
use IEEE.STD_LOGIC_1164.ALL;
use IEEE.NUMERIC_STD.ALL;
entity fir_4tap is
port(   Clk : in std_logic:='0';
Xin : in signed(7 downto 0):="11001111";
Yout : out signed(15 downto 0):="0000000000000000");
end fir_4tap;
architecture Behavioral of fir_4tap is
component DFF is
port( Q : out signed(15 downto 0);
Clk :in std_logic;
D :in  signed(15 downto 0));
end component;
signal H0,H1,H2,H3 : signed(7 downto 0)  := (others => '0');
```

```
signal MCM0,MCM1,MCM2,MCM3,add_out1,add_out2,add_out3 : signed(15 downto
0) := (others => '0');
signal Q1,Q2,Q3 : signed(15 downto 0) := (others => '0');
begin
H0 <= to_signed(-2,8);
H1 <= to_signed(-1,8);
H2 <= to_signed(3,8);
H3 <= to_signed(4,8);
MCM3 <= H3*Xin;
MCM2 <= H2*Xin;
MCM1 <= H1*Xin;
MCM0 <= H0*Xin;
add_out1 <= Q1 + MCM2;
add_out2 <= Q2 + MCM1;
add_out3 <= Q3 + MCM0;
dff1 : DFF port map(Q1,Clk,MCM3);
dff2 : DFF port map(Q2,Clk,add_out1);
dff3 : DFF port map(Q3,Clk,add_out2);
p0:process(Clk)
begin
if(rising_edge(Clk)) then
Yout <= add_out3;
end if;
end process;
end Behavioral;
--Components
library IEEE;
use IEEE.STD_LOGIC_1164.ALL;
use IEEE.NUMERIC_STD.ALL;
entity DFF is
port(Q : out signed(15 downto 0);
Clk :in std_logic;
D :in  signed(15 downto 0));
end DFF;
architecture Behavioral of DFF is
signal qt : signed(15 downto 0) := (others => '0');
begin
Q <= qt;
p0:process(Clk)
begin
if ( rising_edge(Clk) ) then
qt <= D;
end if;
end process;
end Behavioral;
```

DESIGN EXAMPLE 11.4

Develop a digital system in the FPGA to realize a FIR filter whose coefficients are h[n]={–1, 3.75, 3.75, –1} and obtain input from eight GPIO pins. Display the filter response using eight LEDs [25].

Solution

```
package eight_bit_int is
subtype byte is integer range -128 to 127;
type array_byte is array (0 to 3) of byte;
end eight_bit_int;
library work;
use work.eight_bit_int.all;
library ieee;
```

```
use ieee.std_logic_1164.all;
use ieee.std_logic_arith.all;
entity fir_srg is
port (clk   :   in  std_logic;
x     :   in  byte;
y     :   out byte);
end fir_srg;
architecture flex of fir_srg is
signal tap : array_byte := (0,0,0,0);
begin
p1: process
begin
wait until clk = '1';
y <= 2 * tap(1) + tap(1) + tap(1) / 2 + tap(1) / 4
+ 2 * tap(2) + tap(2) + tap(2) / 2 + tap(2) / 4
- tap(3) - tap(0);
for i in 3 downto 1 loop
tap(i) <= tap(i-1);
end loop;
tap(0) <= x;
end process;
end flex;
```

11.4 Digital Infinite Impulse Response Filter Design

IIR filters are also an efficient type of filter for implementing DSP algorithms. The amount of processing that is required to compute an IIR filter is relatively small, which enables the design of low-cost miniaturized DSP processors/products. A standard lattice structure of an IIR filter is shown in Figure 11.9. This lattice structure consists of only the feed-forward signal flow from the input `x[n]` to output `y[n]`. The delay elements are used only in the bottom section of the IIR filter; however, they can be used any where in the filter signal flow path based on the application for which the IIR filter has to serve. The cross interconnections with the respective weights h_N, the so-called filter coefficients, are also the part of IIR filter designs. IIR filters are mostly represented by difference equations, as in Equation 11.9. Therefore, IIR filter design using Equation 11.9 includes both feedback and feed-forward terms. The system transfer function for this filter is the rational function, where both the zeros and the poles are at nonzero locations in the Z-plane [25,150].

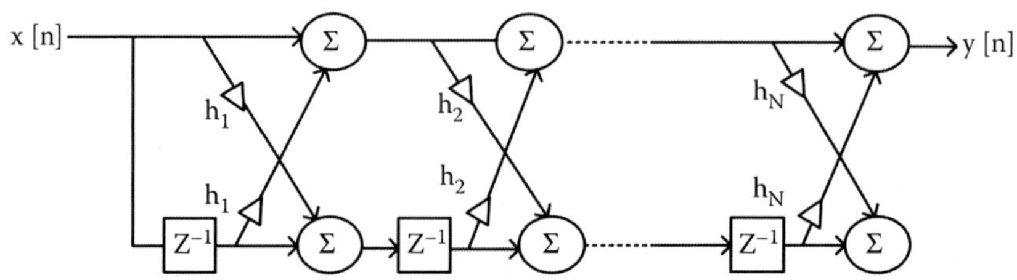

FIGURE 11.9
General lattice structure of an IIR filter.

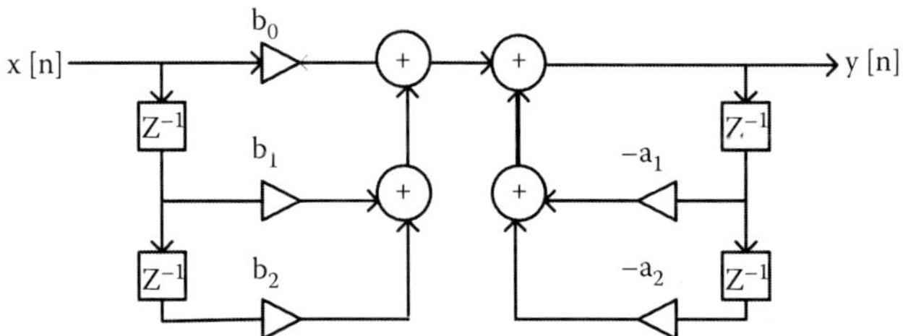

FIGURE 11.10
Direct form-I of an IIR filter.

The general IIR filter difference equation in the direct form-I for the output function y[n] is given by

$$y[n] = \sum_{l=0}^{N} a_l y[n-1] + \sum_{k=0}^{M} b_k x[n-k] \tag{11.9}$$

where y[n] is the output, x[n] is the input, M is the order of the filter at the input section, N is the order of the filter at the output section and a_l and b_k are the filter coefficients at the output and input, respectively. A direct form-I structure of an IIR filter of N = M = 3 is shown in Figure 11.10. In the above equation, a_0 is always taken as 1, since it is the path connecting the adder result to the output y[n] as well as to the feedback loop, as shown in Figure 11.10. The coefficient count is obtained by N + M + 1, through which we can calculate how many multipliers are needed to compute each new output from the difference equation. IIR filters can be implemented on digital platform as a direct form-II, cascade form or lattice form. Assume that N = 1 and M = 0; then, the output y[n] will, as per Equation 11.9, be $y[n] = a_1y[n - 1] + b_0x[n]$, which is called an IIR time-domain first-order filter.

The impulse response of IIR filters can be analyzed using the difference equation. For example, consider the first-order IIR system. The impulse response can be obtained by setting the input impulse input δ[n] at the input x[n] while the system is initially at rest. The rest means that the input and output of the system prior to the start time are '0'. Assume that the δ[n] is applied to the IIR system when n = 0. Then, the first-order difference equation becomes

$$\begin{aligned}
y[n] &= a_1 y[n-1] + b_0 x[n] \\
&\quad n = 0 \quad \text{and} \quad x[n] = \delta[n], \quad \text{then} \\
y[0] &= a_1 y[-1] + b_0 \delta[0] = b_0 \\
&\quad n = 1 \\
y[1] &= a_1 y[0] + b_0 \delta[1] = a_1 b_0 \\
&\quad n = 2 \\
y[2] &= a_1 y[1] + b_0 \delta[2] = a_1^2 b_0 \\
&\quad \ldots \\
y[n] &= a_1^n b_0 \quad \text{for } n \geq 0
\end{aligned}$$

The above results show that the impulse response of the IIR system is $h[n] = b_0a_1{}^nu[n]$. The δ[n] is chosen to ensure that the system response is '0' other than the time n = 0.

In the same way, the other characteristics, for example, linearity, time invariance and the step response of the IIR filter, can also be obtained [25,150]. From the Z-transform, the convolution in the time domain is equal to multiplication in the z-domain:

$$y[n] = x[n] * h[n] \overset{z}{\longleftrightarrow} X(z)H(z) = Y(z)$$

In the IIR filter, `H(z)` will be a fully rational function. By using the z-transform on both sides of the general IIR difference equation, using the delay property is given by

$$\begin{aligned}
Y(z) &= \sum_{l=1}^{N} a_l z^{-l} Y(z) + \sum_{k=0}^{M} b_k z^{-k} X(z) \\
Y(z) - \sum_{l=1}^{N} a_l z^{-l} Y(z) &= \sum_{k=0}^{M} b_k z^{-k} X(z) \\
Y(z)\left(1 - \sum_{l=1}^{N} a_l z^{-l}\right) &= \sum_{k=0}^{M} b_k z^{-k} X(z) \\
\frac{Y(z)}{X(z)} = H(z) &= \frac{\sum_{k=0}^{M} b_k z^{-k}}{1 - \sum_{l=1}^{N} a_l z^{-l}} = \frac{b_0 + b_1 z^{-1} + \cdots + bMz - M}{1 - a_1 z^{-1} - \cdots - a_N z^{-N}}
\end{aligned} \tag{11.10}$$

The coefficients in the numerator represent the feed-forward terms, while the denominator polynomial represents the feedback terms in the difference equation.

DESIGN EXAMPLE 11.5

Develop a digital system to realize a first-order IIR filter as a lossy integrator whose output `y[n + 1]` is given by `y[n + 1] = (3/4)y[n] + x[n]`. Prove that this integrator can be used to suppress the effect of the noise [25].

Solution

```
--package
package n_bit_int is
subtype bits15 is integer range -2**14 to 2**14-1;
end n_bit_int;
--Main Code
library work;
use work.n_bit_int.all;
library ieee;
use ieee.std_logic_1164.all;
use ieee.std_logic_arith.all;
entity iir is
port (x_in   : in  bits15;
y_out  : out bits15;
clk    : in  std_logic);
end iir;
architecture fpga of iir is
signal x, y : bits15 := 0;
begin
process
```

```
begin
wait until clk = '1';
x  <= x_in;
y  <= x + y / 4 + y / 2;
end process;
y_out <= y;
end fpga;
```

DESIGN EXAMPLE 11.6

Develop a digital system in the FPGA to realize an IIR filter with lossy integrator II whose output is given by `y[n] = (9/16)y[n – 2] + (3/4)x[n – 2] + x[n – 1]` using pipelined stage registers in look-ahead form [25].

Solution

```
package n_bit_int is
subtype bits15 is integer range -2**14 to 2**14-1;
end n_bit_int;
library work;
use work.n_bit_int.all;
library ieee;
use ieee.std_logic_1164.all;
use ieee.std_logic_arith.all;
entity iir_pipe is
port ( x_in  : in   bits15;
y_out : out  bits15;
clk   : in   std_logic);
end iir_pipe;
architecture fpga of iir_pipe is
signal  x, x3, sx, y, y9 : bits15 := 0;
begin
process
begin
wait until clk = '1';
x   <= x_in;
x3  <= x / 2 + x / 4;
sx <=  x + x3;
y9  <= y / 2 + y / 16;
y   <= sx + y9;
end process;
y_out <= y ;
end fpga;
```

DESIGN EXAMPLE 11.7

Develop a digital system in the FPGA to realize an IIR filter with parallel lossy integrator III. Assume that the even discrete time samples in `x[n]` are applied to the upper part of the filter chain, while the odd discrete time samples are applied to the lower part. Tape-out the output `y[n]` separately from the upper and lower chains. The even and odd outputs of the filter are given by $y_{even}[n + 1] = (3/4)x_{even}[n - 3] + x_{odd}[n - 2] + (9/16)y_{even}[n]$ and $y_{odd}[n + 1] = (3/4)x_{odd}[n - 3] + x_{even}[n - 3] + (9/16)y_{odd}n]$, respectively [25].

Solution

```
package n_bit_int is
subtype bits15 is integer range -2**14 to 2**14-1;
```

```
end n_bit_int;
library work;
use work.n_bit_int.all;
library ieee;
use ieee.std_logic_1164.all;
use ieee.std_logic_arith.all;
entity iir_par is
port ( clk, reset : in  std_logic;
x_in      : in  bits15;
x_e, x_o, y_e, y_o : out bits15;
clk2      : out std_logic;
y_out     : out bits15);
end iir_par;
architecture fpga of iir_par is
type state_type is (even, odd);
signal  state                      : state_type;
signal  x_even, xd_even            : bits15 := 0;
signal  x_odd, xd_odd, x_wait      : bits15 := 0;
signal  y_even, y_odd, y_wait, y   : bits15 := 0;
signal  sum_x_even, sum_x_odd      : bits15 := 0;
signal  clk_div2                   : std_logic;
begin
multiplex: process (reset, clk)
begin
if reset = '1' then
state <= even;
elsif rising_edge(clk) then
case state is
when even =>
x_even <= x_in;
x_odd <= x_wait;
clk_div2 <= '1';
y <= y_wait;
state <= odd;
when odd =>
x_wait <= x_in;
y <= y_odd;
y_wait <= y_even;
clk_div2 <= '0';
state <= even;
end case;
end if;
end process multiplex;
y_out <= y;
clk2  <= clk_div2;
x_e <= x_even;
x_o <= x_odd;
y_e <= y_even;
y_o <= y_odd;
arithmetic: process
begin
wait until clk_div2 = '0';
xd_even <= x_even;
sum_x_even <= (xd_even * 2 + xd_even) /4 + x_odd;
y_even <= (y_even * 8 + y_even )/16 + sum_x_even;
xd_odd <= x_odd;
sum_x_odd <= (xd_odd * 2 + xd_odd) /4 + xd_even;
y_odd  <= (y_odd * 8 + y_odd) / 16 + sum_x_odd;
end process arithmetic;
end fpga;
```

DESIGN EXAMPLE 11.8

Design a digital system to realize a lattice Daubechies filter with a length (L) of 4 [25].

Solution

```
package n_bits_int is
subtype bits8 is integer range -128 to 127;
subtype bits9 is integer range -2**8 to 2**8-1;
subtype bits17 is integer range -2**16 to 2**16-1;
type array_bits17_4 is array (0 to 3) of bits17;
end n_bits_int;
library work;
use work.n_bits_int.all;
library ieee;
use ieee.std_logic_1164.all;
use ieee.std_logic_arith.all;
use ieee.std_logic_unsigned.all;
entity db4latti is
port (clk, reset : in  std_logic;
clk2        : out std_logic;
x_in        : in  bits8;
x_e, x_o    : out bits17;
g, h        : out bits9);
end db4latti;
architecture fpga of db4latti is
type state_type is (even, odd);
signal state                  : state_type;
signal sx_up, sx_low, x_wait  : bits17 := 0;
signal clk_div2               : std_logic;
signal sxa0_up, sxa0_low      : bits17 := 0;
signal up0, up1, low0, low1   : bits17 := 0;
begin
multiplex: process (reset, clk)
begin
if reset = '1' then
state <= even;
elsif rising_edge(clk) then
case state is
when even =>
sx_up   <= 4 * (32 *   x_in - x_in);
sx_low  <= 4 * (32 * x_wait - x_wait);
clk_div2 <= '1';
state <= odd;
when odd =>
x_wait <= x_in;
clk_div2 <= '0';
state <= even;
end case;
end if;
end process;
sxa0_up  <= (2*sx_up  - sx_up /4)
- (sx_up /64 + sx_up/256);
sxa0_low <= (2*sx_low - sx_low/4)
- (sx_low/64 + sx_low/256);
up0  <= sxa0_low + sx_up;
lowertreeff: process
begin
wait until clk = '1';
if clk_div2 = '1' then
```

```
low0 <= sx_low - sxa0_up;
end if;
end process;
up1  <= (up0 - low0/4) - (low0/64 + low0/256);
low1 <= (low0 + up0/4) + (up0/64  +  up0/256);
x_e  <= sx_up;
x_o  <= sx_low;
clk2 <= clk_div2;
outputscale: process
begin
wait until clk = '1';
if clk_div2 = '1' then
g <=  up1 / 256;
h <= low1 / 256;
end if;
end process;
end fpga;
```

11.5 Multirate Signal Processing

When we try to send/receive information using electronic systems like telegraphs, telephones, televisions, radar, Identify Friend or Foe (IFF), and so on, there is a realization that the input signals may be affected by the same system while acquiring, transmitting and processing the signal. That means these systems introduce noise, distortion, and other artefacts [25]. Understanding these effects, the system has to be designed in such a way as to correct them at the signal-processing level. In many modern DSP applications, we need to change the sampling/processing rate of the input signal. When two devices that operate at different rates are to be interconnected, it is necessary to use a rate changer between them. For example, the typical frequency for broadcasting is 32 KHz, a digital CD is 44.1 KHz, a digital audio tape is 48 KHz, video analog signal is 3.5 MHz, and a telephone is 8 KHz. Once a DSP system needs to process more than one of these, a frequency changer is mandatory. Therefore, instead of choosing a single rate-processing technique in the DSP system that needs to deal with many kind of signals, a multirate processing technique is preferred. In many applications, filter banks and wavelet transforms depend on multirate methods. Multirate signal processing plays a major role in various standard DSP techniques, and its applications are increasing.

Multirate systems are building blocks that are commonly used in DSP applications. Their function is to alter the rate of the discrete-time signals, which is achieved by deleting or adding a portion of the signal samples, as shown in Figure 11.11, the so-called decimation and interpolation, respectively. In many systems, multirate DSP increases processing efficiency and reduces the DSP hardware requirements. Whenever a signal at one rate has to be used by a system that expects a different rate, the rate has to be increased or decreased, and some processing is required to accomplish this kind of operation. Using these techniques, design engineers can gain an added degree of freedom that could improve the overall performance of a system architecture. At other times, multirate processing is used to reduce the computational overhead of a system. For example, an algorithm requires k operations to be completed per cycle. By reducing the sample rate of a signal or system by a factor of M, the arithmetic bandwidth requirements are reduced from kf_s operations to kf_s/M operations per second. These include down-conversion or

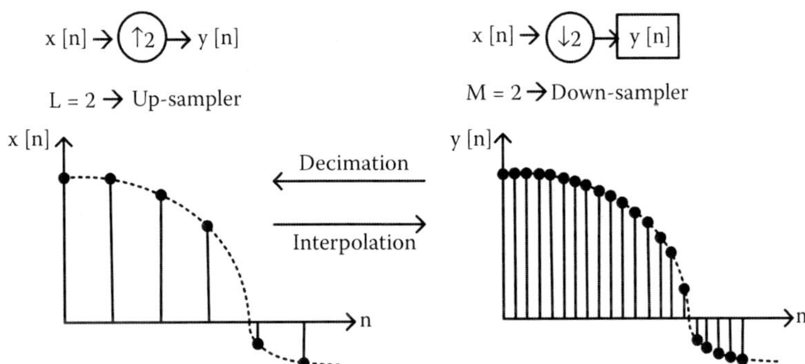

FIGURE 11.11
Representation and illustration of decimation and interpolations.

decimation by a factor M, up-conversion by a factor L and sampling rate conversion by a ratio of M to L.

We discuss some examples below to understand these concepts clearly. Suppose `x[n] = {... 3, 5, 2, 9, 6, ...}`; then, its up-sampling version `y[n]` can be obtained as `y[n] = {x[n/2]` for even n; 0 for odd n}. That means `y[n] = {... 0, 3, 0, 5, 0, 2, 0, 9, 0, 6, 0, ...}`; the up-sampling can be done by inserting zeros between samples. Interpolation is another way of up-sampling where the inserted (new) sample is obtained by taking the average of adjacent samples, that is, the Nth new sample is obtained from the $(N-1)^{th}$ and $(N+1)^{th}$ samples as `(x[N - 1] + x[N + 1])/2`. In the same way, suppose `x[n] = {.... 7, 3, 5, 2, 9, 6, 4, ...}`. The down-sampling of `x[n]` can be obtained by keeping only every second sample: `y[n] = {... 7, 5, 9, 4, ...}`. This operation is also called decimation. With this information, we discuss the operation of a multirate signal processor constructed as shown in Figure 11.12, where the input is `x[n]` and output is `y[n]`. The upper signal-processing chain consists of one down-sampler (by a factor of 2), one up-sampler (by a factor of 2) and one delay element (Z^{-1}). The input and output of the upper chain are represented by {$x[n]$, $x^1[n]$, $x^2[n]$ and $x^3[n]$}.

This means that $x^1[n]$ is the 2-sample downed version of $x[n]$, $x^2[n]$ is the 2-sample upped version of $x^1[n]$ and $x^3[n]$ is just one clock-delayed version of $x^2[n]$. In a similar way, the lower processing chain is also constructed by keeping the delay element as a first block, and the input-output is represented by {$x[n]$, $x^4[n]$, $x^5[n]$ and $x^6[n]$}. Finally, the output of the multirate signal processor is $y[n] = x^3[n] + x^6[n]$. The complete operation of this multirate filter is given in Table 11.1. There, the first column represents the input/output of each block used in this multirate processor, and the remaining columns represent the actual samples that are being processed by the this processor. Assume that

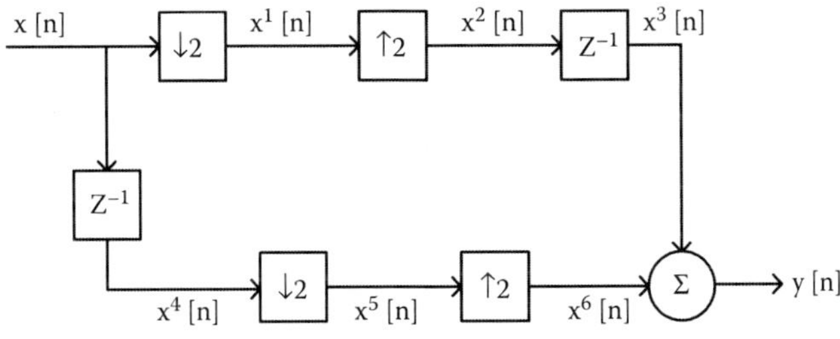

FIGURE 11.12
Structure of a simple multirate filter.

TABLE 11.1

Operation of a Multirate Signal Processor

Sample Sequence	Actual Samples That Are Being Processed							
x[n]	x[0]	x[1]	x[2]	.	.	.	...	x[10]
x^1[n]	x[0]	x[2]	x[4]	.	.	.	...	x[20]
x^2[n]	x[0]	0	x[2]	0	x[4]	0	...	x[10]
x^3[n]	0	x[0]	0	x[2]	.	.	...	0
x^4[n]	x[−1]	x[0]	x[1]	x[2]	.	.	...	x[9]
x^5[n]	x[−1]	x[1]	x[3]	.	.	.	...	x[19]
x^6[n]	x[−1]	x[0]	x[1]	0	x[3]	0	...	x[9]
y[n]	x[−1]	x[0]	x[1]	x[2]	.	.	...	x[9]

x[n]={... x[−1], x[0], x[1], x[2], ... x[25]}; then, x^1[n] is the down-sampled version of x[n]; therefore, x^1[n] is {x[0], x[2], x[4], x[6], ...}, which is given in the third row of Table 11.1. Similarly, all the rows have been filled with the appropriate samples. The output clearly shows that multirate signal processing is a linear and time-variant function. A more common multirate architecture is cascade integrator comb (CIC), which is out of the scope of this book; hence, the readers are directed to refer to [25] for more information on CIC. Some design examples using the CIC architecture are given below.

DESIGN EXAMPLE 11.9

Generate MATLAB code to illustrate down-sampling and up-sampling by a factor of M with N samples. Generate the samples from the sine functions. Illustrate the samples and their analog versions as well.

Solution

```
clc;
clear all;
N = input('Enter the total No. of samples:=');
M = input('Enter the value of Up/Down factor:=');
n = 0:1:N;
x=sin(2*pi*0.1*n) + sin(2*pi*0.3*n);
figure (1)
subplot(2,2,1);
plot(x,'b-')
hold on;
stem(x,'K')
xlabel('Time index n');
ylabel('x[n]');
xlim([0 length(x)]);
y = downsample(x,M);
subplot(2,2,2);
plot(y,'b-');
hold on;
stem(y,'k');
xlabel('Time index n')
ylabel('y[n]');
title('Down-sampling by factor 2');
xlim([0 length(y)]);
x1 = sin(2*pi*0.15*n);
```

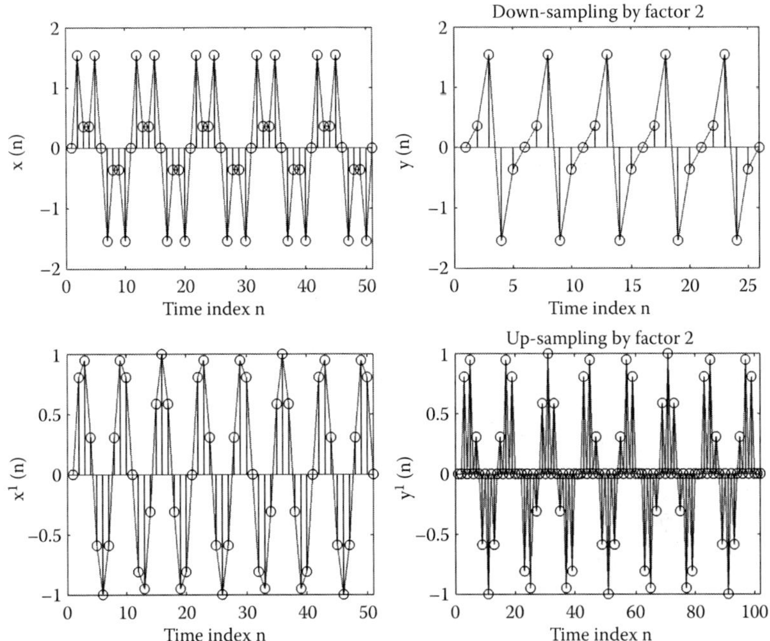

FIGURE 11.13
Down/up-sampling result for the input values of N = 50 and M = 2.

```
subplot(2,2,3);
plot(x1,'b-');
hold on;
stem(x1,'k');
ylabel('x1[n]');
xlabel('Time index n');
xlim([0 length(x1)]);
subplot(2,2,4);
y1 = upsample(x1,M);
plot(y1,'b-');
hold on;
stem(y1,'k');
ylabel('y1[n]');
xlabel('Time index n');
xlim([0 length(y1)]);
title('Up-sampling by factor 2');
```

The down and up-sampled version of two different sinusoidal functions x[n] and x1[n] are shown in Figure 11.13.

DESIGN EXAMPLE 11.10

Generate MATLAB code to illustrate the spectral characteristics of the samples obtained from up-sampling and the interpolation process. Generate the samples using the sine function and obtain the spectrum using the MATLAB function `fft`.

Solution

```
clc;
clear all;
close all;
```

```
N = input('Enter the total No. of samples:=');
M = input('Enter the value of Up/Down factor:=');
n= 0:1:N-1;
f1 = 0.05;
f2 = 0.09;
x1=sin(2*pi*f1*n)+0.9.*sin(2*pi*f2*n);
s=randn(1,N);
x=x1+0.8.*s;
figure (1);
subplot(3,2,1);
stem(0:31,x(1:32),'k');
ylabel('x[n]');
xlabel('Time index n');
legend('Original signal');
X = fft(x,1024);
f = 0:1/512:(512-1)/512;
subplot(3,2,2);
plot(f,abs(X(1:512)),'k');
ylabel('|X(e^j^\omega)|');
xlabel('\omega/\pi');
legend('Original signal');
x1 = upsample(x,M);
subplot(3,2,3);
stem(M-1:M*32-1,x1(M-1:M*32-1),'k')
ylabel('x1[n]');
xlabel('Time index n');
legend('Upsampled signal',4)
axis([M-1,M*32-1,-5,5])
X1=fft(x1,1024);
subplot(3,2,4);
plot(f,abs(X1(1:512)),'k')
ylabel('|X1(e^j^\omega)|');
xlabel('\omega/\pi');
legend('Upsampled signal');
subplot(3,2,5);
yi = interp(X1,M);
stem(M-1:M*32-1,yi(M-1:M*32-1),'k')
ylabel('y[m]');
xlabel('Time index n');
legend('Interpolated signal',4);
Yi = fft(yi,1024);
subplot(3,2,6);
plot(f,abs(Yi(1:512)),'k');
ylabel('|Y(e^j^\omega)|');
xlabel('\omega/\pi');
legend('Interpolated signal');
```

A sinusoidal sample x[n], up-sampled version of x[n], interpolated version of x[n] and the frequency spectrum of all these signals are shown in Figure 11.14.

DESIGN EXAMPLE 11.11

Develop a digital system in the FPGA to realize a first-order CIC filter [25] with Decimator-I.

Solution

```
library ieee;
use ieee.std_logic_1164.all;
use ieee.std_logic_arith.all;
use ieee.std_logic_signed.all;
```

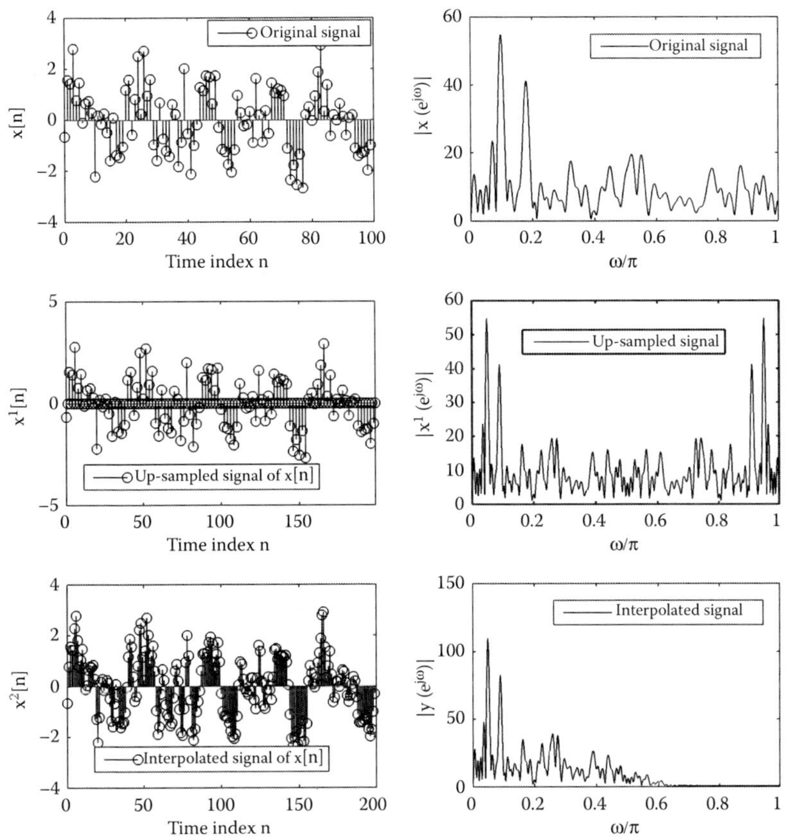

FIGURE 11.14
Spectral characteristics of up-sampled and interpolated samples for the input values of N = 100 and M = 2.

```
entity cic3r32 is
port ( clk, reset : in  std_logic;
x_in        : in  std_logic_vector(7 downto 0);
clk2        : out std_logic;
y_out       : out std_logic_vector(9 downto 0));
end cic3r32;
architecture fpga of cic3r32 is
subtype word26 is std_logic_vector(25 downto 0);
type    state_type is (hold, sample);
signal  state    : state_type ;
signal  count     : integer range 0 to 31;
signal  x : std_logic_vector(7 downto 0) :=(others => '0');
signal  sxtx : std_logic_vector(25 downto 0);
signal  i0, i1 , i2 : word26 := (others=>'0');
signal  i2d1, i2d2, c1, c0 : word26 := (others=>'0');
signal  c1d1, c1d2, c2 : word26 := (others=>'0');
signal  c2d1, c2d2, c3 : word26 := (others=>'0');
begin
fsm: process (reset, clk)
begin
if reset = '1' then
state <= hold;
count <= 0;
```

```
clk2  <= '0';
elsif rising_edge(clk) then
if count = 31 then
count <= 0;
state <= sample;
clk2  <= '1';
else
count <= count + 1;
state <= hold;
clk2  <= '0';
end if;
end if;
end process fsm;
sxt: process (x)
begin
sxtx(7 downto 0) <= x;
for k in 25 downto 8 loop
sxtx(k) <= x(x'high);
end loop;
end process sxt;
int: process
begin
wait until clk = '1';
x    <= x_in;
i0   <= i0 + sxtx;
i1   <= i1 + i0 ;
i2   <= i2 + i1 ;
end process int;
comb: process
begin
wait until clk = '1';
if state = sample then
c0   <= i2;
i2d1 <= c0;
i2d2 <= i2d1;
c1   <= c0-i2d2;
c1d1 <= c1;
c1d2 <= c1d1;
c2   <= c1  - c1d2;
c2d1 <= c2;
c2d2 <= c2d1;
c3   <= c2-c2d2;
end if;
end process comb;
y_out <= c3(25 downto 16);
end fpga;
```

DESIGN EXAMPLE 11.12

Develop a digital system in the FPGA to realize a three-stage CIC filter with Decimator-III. The filter has to be designed with pruning [25].

Solution

```
library ieee;
use ieee.std_logic_1164.all;
use ieee.std_logic_arith.all;
use ieee.std_logic_signed.all;
entity cic3s32 is
port ( clk, reset  : in std_logic;
x_in       : in std_logic_vector(7 downto 0);
```

```
clk2        : out std_logic;
y_out        : out std_logic_vector(9 downto 0));
end cic3s32;
architecture fpga of cic3s32 is
subtype word26 is std_logic_vector(25 downto 0);
subtype word21 is std_logic_vector(20 downto 0);
subtype word16 is std_logic_vector(15 downto 0);
subtype word14 is std_logic_vector(13 downto 0);
subtype word13 is std_logic_vector(12 downto 0);
subtype word12 is std_logic_vector(11 downto 0);
type    state_type is (hold, sample);
signal  state       : state_type ;
signal  count       : integer range 0 to 31;
signal  x           : std_logic_vector(7 downto 0):= (others => '0');
signal  sxtx  : std_logic_vector(25 downto 0);
signal  i0  : word26 := (others => '0');
signal  i1  : word21 := (others => '0');
signal  i2  : word16 := (others => '0');
signal  i2d1, i2d2, c1, c0 : word14 := (others => '0');
signal  c1d1, c1d2, c2 : word13 := (others=>'0');
signal  c2d1, c2d2, c3 : word12 := (others=>'0');
begin
fsm: process (reset, clk)
begin
if reset = '1' then
state <= hold;
count <= 0;
clk2  <= '0';
elsif rising_edge(clk) then
if count = 31 then
count <= 0;
state <= sample;
clk2  <= '1';
else
count <= count + 1;
state <= hold;
clk2  <= '0';
end if;
end if;
end process fsm;
sxt : process (x)
begin
sxtx(7 downto 0) <= x;
for k in 25 downto 8 loop
sxtx(k) <= x(x'high);
end loop;
end process sxt;
int: process
begin
wait
until clk = '1';
x   <= x_in;
i0  <= i0 + x;
i1  <= i1 + i0(25 downto 5);
i2  <= i2 + i1(20 downto 5);
end process int;
comb: process
begin
wait until clk = '1';
if state = sample then
c0   <= i2(15 downto 2);
```

```
i2d1 <= c0;
i2d2 <= i2d1;
c1   <= c0 - i2d2;
c1d1 <= c1(13 downto 1);
c1d2 <= c1d1;
c2   <= c1(13 downto 1) - c1d2;
c2d1 <= c2(12 downto 1);
c2d2 <= c2d1;
c3   <= c2(12 downto 1) - c2d2;
end if;
end process comb;
y_out <= c3(11 downto 2);
end fpga;
```

DESIGN EXAMPLE 11.13

Develop a digital system to realize an arbitrary fractional sampling rate converter. Assume that the value of the rate changer R is equal to (3/4) = 0.75 and realize the changer with three sinc filters [25].

Solution

```
package n_bits_int is
subtype bits8 is integer range -128 to 127;
subtype bits9 is integer range -2**8 to 2**8-1;
subtype bits17 is integer range -2**16 to 2**16-1;
type array_bits8_11 is array (0 to 10) of bits8;
type array_bits9_11 is array (0 to 10) of bits9;
type array_bits8_3 is array (0 to 2) of bits8;
type array_bits8_4 is array (0 to 3) of bits8;
type array_bits17_11 is array (0 to 10) of bits17;
end n_bits_int;
library work;
use work.n_bits_int.all;
library ieee;
use ieee.std_logic_1164.all;
use ieee.std_logic_arith.all;
use ieee.std_logic_signed.all;
entity rc_sinc is
generic (ol : integer := 2;
il : integer := 3;
l  : integer := 10);
port (clk                 : in  std_logic;
x_in                      : in  bits8;
reset                     : in  std_logic;
count_o                   : out integer range 0 to 12;
ena_in_o, ena_out_o,ena_io_o : out boolean;
f0_o, f1_o, f2_o          : out bits9;
y_out                     : out bits9);
end rc_sinc;
architecture fpga of rc_sinc is
signal count  : integer range 0 to 12;
signal ena_in, ena_out, ena_io : boolean;
constant c0 : array_bits9_11:= (-19,26,-42,106,212,-53,29,-21,16,-13,11);
constant c2 : array_bits9_11:= (11,-13,16,-21,29,-53,212,106,-42,26,-19);
signal x : array_bits8_11 := (0,0,0,0,0,0,0,0,0,0,0);
signal ibuf : array_bits8_4 := (0,0,0,0);
```

```
signal obuf : array_bits8_3 := (0,0,0);
signal f0, f1, f2     : bits9 := 0;
begin
fsm: process (reset, clk)
begin
if reset = '1' then
count <= 0;
elsif rising_edge(clk) then
if count = 11 then
count <= 0;
else
count <= count + 1;
end if;
case count is
when 2 | 5 | 8 | 11 =>
ena_in <= true;
when others =>
ena_in <= false;
end case;
case count is
when 4 | 8 =>
ena_out <= true;
when others =>
ena_out <= false;
end case;
if count = 0 then
ena_io <= true;
else
ena_io <= false;
end if;
end if;
end process fsm;
inputmux: process
begin
wait until clk = '1';
if ena_in then
for i in il downto 1 loop
ibuf(i) <= ibuf(i-1);
end loop;
ibuf(0) <= x_in;
end if;
end process;
ouputmux: process
begin
wait until clk = '1';
if ena_io then
obuf(0) <= f0 ;
obuf(1) <= f1;
obuf(2) <= f2 ;
elsif ena_out then
for i in ol downto 1 loop
obuf(i) <= obuf(i-1);
end loop;
end if;
```

```
end process;
tap: process
begin
wait until clk = '1';
if ena_io then
for i in 0 to 3 loop
x(i) <= ibuf(i);
end loop;
for i in 4 to 10 loop
x(i) <= x(i-4);
end loop;
end if;
end process;
sop0: process (clk, x)
variable sum : bits17;
variable p : array_bits17_11;
begin
for i in 0 to 1 loop
p(i) := c0(i) * x(i);
end loop;
sum := p(0);
for i in 1 to 1  loop
sum := sum + p(i);
end loop;
if clk'event and clk = '1' then
f0 <= sum /256;
end if;
end process sop0;
sop1: process (clk, x)
begin
if clk'event and clk = '1' then
f1 <= x(5);
end if;
end process sop1;
sop2: process (clk, x)
variable sum : bits17;
variable p : array_bits17_11;
begin
for i in 0 to 1 loop
p(i) := c2(i) * x(i);
end loop;
sum := p(0);
for i in 1 to 1  loop
sum := sum + p(i);
end loop;
if clk'event and clk = '1' then
f2 <= sum /256;
end if;
end process sop2;
f0_o <= f0;
f1_o <= f1;
f2_o <= f2;
count_o <= count;
ena_in_o <= ena_in;
```

```
ena _ out _ o <= ena _ out;
ena _ io _ o <= ena _ io;
y _ out <= obuf(ol);
end fpga;
```

DESIGN EXAMPLE 11.14

Develop a digital system in the FPGA to realize a Farrow design–based rate changer IV designed using a Lagrange polynomial of order 3. Assume that the value of the rate changer is (3/4) = 0.75 [25].

Solution

```
package n_bits_int is
subtype bits8 is integer range -128 to 127;
subtype bits9 is integer range -2**8 to 2**8-1;
subtype bits17 is integer range -2**16 to 2**16-1;
type array_bits8_4 is array (0 to 3) of bits8;
end n_bits_int;
library work;
use work.n_bits_int.all;
library ieee;
use ieee.std_logic_1164.all;
use ieee.std_logic_arith.all;
use ieee.std_logic_signed.all;
entity farrow is
generic (il : integer := 3);
port (clk                      : in  std_logic;
x_in                       : in  bits8;
reset                      : in  std_logic;
count_o                    : out integer range 0 to 12;
ena_in_o, ena_out_o        : out boolean;
c0_o, c1_o, c2_o, c3_o : out bits9;
d_out, y_out               : out bits9);
end farrow;
architecture fpga of farrow is
signal count  : integer range 0 to 12;
constant delta : integer := 85;
signal ena_in, ena_out : boolean;
signal x, ibuf : array_bits8_4 := (0,0,0,0);
signal  d : bits9 := 0;
signal c0, c1, c2, c3     : bits9 := 0;
begin
fsm: process (reset, clk)
variable dnew : bits9 := 0;
begin
if reset = '1' then
count <= 0;
d <= delta;
elsif rising_edge(clk) then
if count = 11 then
count <= 0;
else
count <= count + 1;
end if;
case count is
when 2 | 5 | 8 | 11 =>
ena_in <= true;
when others =>
ena_in <= false;
end case;
```

```
case count is
when 3 | 7 | 11 =>
ena_out <= true;
when others =>
ena_out <= false;
end case;
if ena_out then
dnew := d + delta;
if dnew >= 255 then
d <= 0;
else
d <= dnew;
end if;
end if;
end if;
end process fsm;
tap: process
begin
wait until clk = '1';
if ena_in then
for i in 1 to il loop
ibuf(i-1) <= ibuf(i);
end loop;
ibuf(il) <= x_in;
end if;
end process;
get: process
begin
wait until clk = '1';
if ena_out then
for i in 0 to il loop
x(i) <= ibuf(i);
end loop;
end if;
end process;
sop: process (clk, x, d, c0, c1, c2, c3, ena_out)
variable y : bits9;
begin
if ena_out then
if clk'event and clk = '1' then
c0 <= x(1);
c1 <= -85 * x(0)/256 - x(1)/2 + x(2) - 43 * x(3)/256;
c2 <= (x(0) + x(2)) /2 - x(1) ;
c3 <= (x(1) - x(2))/2 + 43 * (x(3) - x(0))/256;
end if;
y := c2 + (c3 * d) / 256;
y := (y * d) / 256 + c1;
y := (y * d) / 256 + c0;
if clk'event and clk = '1' then
y_out <= y;
end if;
end if;
end process sop;
c0_o <= c0;
c1_o <= c1;
c2_o <= c2;
c3_o <= c3;
count_o <= count;
ena_in_o <= ena_in;
```

```
ena_out_o <= ena_out;
d_out <= d;
end fpga;
```

11.6 Modulo Adder and Residual Number Arithmetic Systems

Modulo arithmetic is essential for many modern computing systems. This type of computation enables the results to wrap around the modulo number and give only the remainder portion, through which further calculations can be carried out with significantly reduced complexity. One of the main applications of modulo arithmetic is RNA systems. RNA is a carry-independent arithmetic, or so-called residue arithmetic, which is applicable for many DSP applications with some limitations. In this section, we discuss the concept of modulo and RNA systems with some design examples.

11.6.1 Modulo 2^n and 2^n–1 Adder Design

In mathematics, modulo or modular arithmetic is a system of arithmetic for integers where the numbers wrap around upon reaching a certain value. The modern approach to modulo arithmetic was developed by Carl Friedrich Gauss in 1801 [25,151]. A good example for modulo arithmetic is the 12-hour clock in which the day is divided into two 12-hour periods. If the time is 7 o'clock now, then 8 hours later, it will be 3 o'clock. Usual addition would suggest that the later time should be 7 + 8 = 15 o'clock, but this is not the answer because clock time wraps around every 12 hours and in 12-hour time, there is no 15 o'clock [25,151]. Likewise, if the clock starts at 12 o'clock, after 21 hours elapse, the time will be 9 o'clock the next day rather than 33 o'clock. Because the hour number starts over after it reaches 12, this is modulo 12 arithmetic calculation. In the same way, when we divide two integers A and B as A/B, we will have the quotient (Q) and remainder (R). Sometimes, we are only interested in the remainder; for these cases, there is an operator called the modulo operator denoted "mod". For example, if we add two numbers A and B in modulo C arithmetic, it can be written as (A + B) mod C. Suppose A = 10, B = 15, and C = 8; then, (10 + 15) mod 8 = 25 mod 8 = 1. This means that C = 8 wrapped around the 25 three times and ends with a remainder of 1. The above expression can also be written as (A mod C + B mod C) mod C. Substituting the same values, (10 mod 8 + 15 mod 8) mod 8 = (2 + 7) mod 8 = 9 mod 8 = 1. A very similar proof can be given for modular subtraction; that is, (A – B) mod C = (A mod C – B mod C) mod C as well. The digital implementation of modulo 2^n and modulo 2^n–1 is shown in Figure 11.15. The inputs `A` and `B` of `n`-bit width are applied to the first parallel `n + 1` bit adder. The EoC is discarded, which means the carry bit will not be considered in the second adder that adds the sum to the '1'. The outputs of both adders are connected to a 2:1 multiplexer. The user input, that is, the selective line (sel) of the multiplexer, decides the end results of Figure 11.15 either as a modulo 2^n or modulo 2^n–1 adder. The output is equal to `(A + B) mod` 2^n when the 'sel' input is '0'; otherwise, the output will be `(A + B) mod` 2^n–1. We discuss some examples to understand the operation of modulo additions in view of design and implementation. Assume `n` = 4. That means Figure 11.15 is a 4-bit modulo adder. Assume that the values of `A` and `B` are "1100" and "0110", respectively. Then, the summation is "10010".

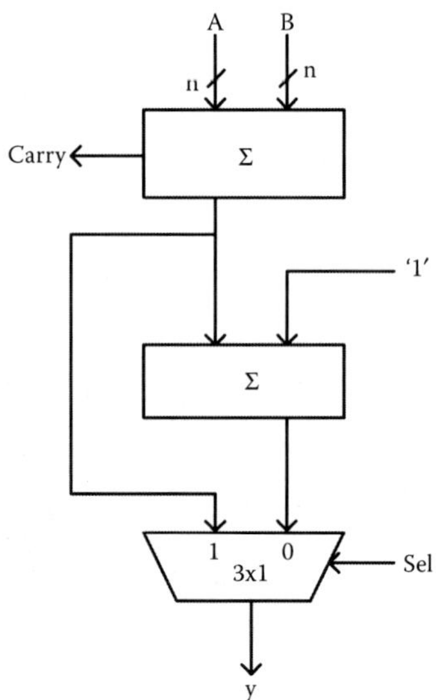

FIGURE 11.15
Schematic diagram of 2^n and 2^n–1 modulo adders.

The EoC is discarded and "0010" is applied to the second adder to add to '1', and the sum is "0011". The values "0010" and "0011" are available at the inputs of the multiplexer. Based on the input sel, the multiplexer routes "0010" if the sel is '0'; otherwise, it routes "0011" to the output. In this case, (A + B) has to be greater than 2^n, that is, $2^4 = 16$. Assume A = "1000" and B = "1000". The first sum is "10000", which is equal to 2^4; therefore, the multiplexer one input is "0000", which is correct for the (A + B) mod 2^n, and another input is "0001", which is correct for (A + B) mod 2^n–1. As another example, suppose A = "0001" and B = "1000". Then the sum of the first adder is "01001", which is smaller than 2^n as well as 2^n–1; hence, mod 2^n and mod 2^n–1 would not be performed, and the output would have to be just the sum of the first adder. In this way, a digital system can be designed for any modulo adder.

DESIGN EXAMPLE 11.15

Develop a modulo adder to perform the following functions: (i) the output has to be (A + B) mod 2^n when a selective line input is '0' and (ii) the output has to be (A + B) mod 2^n–1 when a selective line input is '1'. Assume that the inputs A and B are of width 'n' and the selective line is a bit.

Solution

```
library IEEE;
use IEEE.STD_LOGIC_1164.ALL;
use ieee.std_logic_arith.all;
use ieee.std_logic_unsigned.all;
entity test1 is
generic(n:integer:=4);
Port ( sel: in std_logic;
a,b : in  STD_LOGIC_vector(n-1 downto 0);
y : out  STD_LOGIC_vector(n-1 downto 0));
end test1;
architecture Behavioral of test1 is
```

```
signal sum1: std_logic_vector(n downto 0);
signal cnt:integer range 0 to 256;
begin
sum1<=('0'& a) + ('0' & b);
cnt<=2**n;
p0:process(sel,sum1)
begin
if sel='0' then
y<=sum1 (n-1 downto 0);
elsif sel='1' then
if sum1>=cnt-1 then
y<=(sum1  (n-1 downto 0) + '1');
else
y<=sum1 (n-1 downto 0);
end if;
end if;
end process;
end Behavioral;
```

11.6.2 Residue Number System

The residue numeral system (RNS) represents a large integer using a set of smaller integers to perform computations more efficiently [25,151,152]. The basic principle of the RNA is CRT and modular arithmetic, as discussed in the previous section. RNA-based computations and their operations are the mathematical ideas of Sun Tsu Suan-Ching, who lived in the fourth century. By applying this idea, a large integer can be decomposed into a set of smaller integers. The larger digit calculation can be performed as a series of smaller digit calculations that can be performed independently as well as in parallel. We discuss some examples to understand that how the RNS supports improving the speed of arithmetic operations. In most arithmetic systems, the speed is limited by the nature of arithmetic building blocks that make the logic decisions. For example, in the addition operation, a lower-order/intermediate carry can have a ripple effect on the subsequent/total sum. In the residue number system, the positional bases are relatively prime to each other; for example, the bases may be 2, 3 and 5, that is, prime numbers, a number that is divisible only by itself and 1, for example, 2, 3, 5, 7, 11, and so on. With these prime numbers, the maximum decimal number possible to perform RNS representation is computed by $2 \times 3 \times 5 = 30$. This means that the decimal numbers 0 through 29 can be uniquely represented by the RNS with the bases of 2, 3 and 5. For example, in a (5,3,2) RNS, $(3)_{10}$ is represented by (3 mod 5) = 3, (3 mod 3) = 0 and (3 mod 2) = 1; therefore, the RNS representation for $(3)_{10}$ is (3,0,1). Suppose a decimal number is $(28)_{10}$. Then, (28 mod 5) = 3, (28 mod 3) = 1 and (28 mod 2) = 0, and the RNS of $(28)_{10}$ is (3,1,0).

However, using three RNS bases, we can uniquely represent a number from 0 through 29, and we cannot represent any number greater than 29 using only (2,3,5) as the bases. For example, take $(35)_{10}$, whose RNS using these bases is (0,2,1), which is equal to $(5)_{10}$; hence, the uniqueness is lost. Therefore, we have to use sufficient RNS bases to represent a large range of decimal numbers. The residues in the RNS uniquely identify/represent a decimal number. The main advantage of the RNS is the absence of carries between columns in addition and multiplication. Arithmetic can be done completely within each residue position. Therefore, it is possible to perform addition and multiplication on long numbers at the same speed as on short numbers. For example, RNS addition of (9 + 16) and multiplication of (7×4) with the bases of (5,3,2) is given in Table 11.2 under the respective heads. As given in that table, the addition and multiplication are performed to the base irrespective of any interposition carries.

TABLE 11.2

RNS-Based Addition and Multiplication

Addition			Multiplication		
Decimal	RNS in (5,3,2)	Remarks	Decimal	RNS in (5,3,2)	Remarks
9+	(4,0,1)	(4 + 1 mod 5, 0 + 1 mod 3, 1 + 0 mod 2) = (0,1,1)	7×	(2,1,1)	(2 × 4 mod 5, 1 × 1 mod 3, 1 × 0 mod 2) = (3,1,0)
16=	(1,1,0)		4=	(4,1,0)	
25	(0,1,1)		28	(3,1,0)	

TABLE 11.3

RNA-Based Subtraction

Subtraction			
Decimal	RNS in (5,3,2)	Complement	Remarks
8−	(3,2,0)	–	Not needed
9=	(4,0,1)	9^c = (1,0,1)	(5-4 mod 5, 3-0 mod 3, 2-1 mod 2) = (1,0,1)
−1	−1 is equal to 29 in signed RNS, i.e., (4,2,1)	–	Add 8 to 9^c i.e., (3,2,0) + (1,0,1) = (4,2,1)
16−	(1,1,0)	–	Not needed
12	(2,0,0)	12^c = (3,0,0)	(5-2 mod 5, 3-0 mod 3, 2-0 mod 2) = (3,0,0)
4	4 is equal to 4 only in signed RNS, i.e., (4,1,0)	–	Add 16 to 12^c i.e., (1,1,0) + (3,0,0) = (4,1,0)

Based on this information, we can understand the very popular CRT as "Given a set of relatively prime moduli ($m_1, m_2, \ldots, m_i, \ldots, m_n$), then for any $X < M$, the set of residues ($X \bmod m_i \mid 1 \le i \le n$) is unique, in which the M is computed by Equation 11.11 [25,152]"

$$M = \prod_{i=1}^{n} m_i \tag{11.11}$$

Now, we discuss creating a signed RNS by designating numbers $X < M/2$ as positive and numbers $X \ge M/2$ as negative for all $X < M$. Therefore, when the bases are (5,3,2), 0 through 14 are positive numbers 0,1,2,3, … 14 and 15 through 29 are negative numbers −15, −14, −13, … −1. Based on these positive and negative representations, it is possible to perform RNS subtractions. We need to understand a term complement, such as 1's and 2's complements in the binary system. For example, the RNS for $(8)_{10}$ is (3,2,0), and its complement is (5-3) mod 5, (3-2) mod 3, (2-0) mod 2, that is, (2,1,0). A subtraction example is given in Table 11.3. There are two examples to illustrate negative and positive results. The RNS of the first number can be taken as it is and has to be added to the complement of the second number to get the results of the subtraction of the second number from the first. Additional reading on operations with general moduli, conversion to/from residue representation, limits of fast arithmetic and modeling of ROM speed in gate delays can be found in [25,151,152].

DESIGN EXAMPLE 11.16

Develop a digital system to perform a five-tape FIR time invariant filter. The coefficients of the filter are h[n] = {3,2,1,3,1}. The samples are arriving at the FPGA through 15-bit

GPIOs. Obtain the filter estimation using the RNA system. Write explicit VHDL code to illustrate the RNA-based parallel computation. Assume that the initial samples arriving at the FPGA are x[n] = {4,2,1,3,3}. Display the filter output for each clock (shift) using five LEDs.

Solution

```
LIBRARY ieee;
USE ieee.std_logic_1164.all;
USE ieee.numeric_std.all;
use ieee.std_logic_unsigned.all;
use IEEE.STD_LOGIC_ARITH.ALL;
entity RNA_pro is
port (start,m_clk:in std_logic:='0';
xn:in std_logic_vector(0 to 14):="100010001011011";
RNA_out: out std_logic_vector(4 downto 0):="00000");
end RNA_pro;
architecture Behavioral of RNA_pro is
signal xn_temp,data: std_logic_vector(0 to 14):="000000000000000";
type memory_type0 is array (0 to 4) of integer range 0 to 7;
signal coef : memory_type0 :=(3,2,1,3,1);
type memory_type1 is array (0 to 4) of integer range 0 to 7;
signal sample : memory_type1 :=(0,0,0,0,0);
signal RNA_clk,mul_clk:std_logic:='0';
type memory_type2 is array (0 to 14) of integer range 0 to 20;
signal x,y,z,w : memory_type2 :=(0,0,0,0,0,0,0,0,0,0,0,0,0,0,0);
type memory_type3 is array (0 to 2) of integer range 0 to 30;
signal mul1,mul2,v : memory_type3 :=(0,0,0);
signal a1,b1,c1: std_logic_vector(2 downto 0):="000";
begin
P0:process(m_clk)
variable cnt1,cnt2:integer:=0;
begin
if rising_edge(m_clk) then
if  (cnt1=1)then
if (start='1') then
xn_temp<=xn;
cnt1:=cnt1+1;
end if;
elsif (cnt1=2)then
data<=(xn_temp(0 to 2)& data(3 to 14));
cnt1:=cnt1+1;
elsif (cnt1=3)then
sample(0)<=conv_integer(data(0 to 2));
sample(1)<=conv_integer(data(3 to 5));
sample(2)<=conv_integer(data(6 to 8));
sample(3)<=conv_integer(data(9 to 11));
sample(4)<=conv_integer(data(12 to 14));
cnt1:=cnt1+1;
elsif (cnt1=4)then
RNA_clk<='1';
cnt1:=cnt1+1;
elsif (cnt1=5)then
RNA_clk<='0';
cnt1:=cnt1+1;
elsif (cnt1=6)then
z(0)<=(x(0)*y(0));
z(1)<=(x(1)*y(1));
z(2)<=(x(2)*y(2));
z(3)<=(x(3)*y(3));
z(4)<=(x(4)*y(4));
```

```
z(5)<=(x(5)*y(5));
z(6)<=(x(6)*y(6));
z(7)<=(x(7)*y(7));
z(8)<=(x(8)*y(8));
z(9)<=(x(9)*y(9));
z(10)<=(x(10)*y(10));
z(11)<=(x(11)*y(11));
z(12)<=(x(12)*y(12));
z(13)<=(x(13)*y(13));
z(14)<=(x(14)*y(14));
cnt1:=cnt1+1;
elsif (cnt1=7)then
mul1(0)<=z(0);
mul1(1)<=z(1);
mul1(2)<=z(2);
cnt1:=cnt1+1;
elsif (cnt1=8)then
mul_clk<='1';
cnt1:=cnt1+1;
elsif (cnt1=9)then
mul_clk<='0';
w(0)<=mul2(0);
w(1)<=mul2(1);
w(2)<=mul2(2);
cnt1:=cnt1+1;
elsif (cnt1=10)then
mul1(0)<=z(3);
mul1(1)<=z(4);
mul1(2)<=z(5);
cnt1:=cnt1+1;
elsif (cnt1=11)then
mul_clk<='1';
cnt1:=cnt1+1;
elsif (cnt1=12)then
mul_clk<='0';
w(3)<=mul2(0);
w(4)<=mul2(1);
w(5)<=mul2(2);
cnt1:=cnt1+1;
elsif (cnt1=13)then
mul1(0)<=z(6);
mul1(1)<=z(7);
mul1(2)<=z(8);
cnt1:=cnt1+1;
elsif (cnt1=14)then
mul_clk<='1';
cnt1:=cnt1+1;
elsif (cnt1=15)then
mul_clk<='0';
w(6)<=mul2(0);
w(7)<=mul2(1);
w(8)<=mul2(2);
cnt1:=cnt1+1;
elsif (cnt1=16)then
mul1(0)<=z(9);
mul1(1)<=z(10);
mul1(2)<=z(11);
cnt1:=cnt1+1;
elsif (cnt1=17)then
mul_clk<='1';
cnt1:=cnt1+1;
```

```
elsif (cnt1=18)then
mul_clk<='0';
w(9)<=mul2(0);
w(10)<=mul2(1);
w(11)<=mul2(2);
cnt1:=cnt1+1;
elsif (cnt1=19)then
mul1(0)<=z(12);
mul1(1)<=z(13);
mul1(2)<=z(14);
cnt1:=cnt1+1;
elsif (cnt1=20)then
mul_clk<='1';
cnt1:=cnt1+1;
elsif (cnt1=21)then
mul_clk<='0';
w(12)<=mul2(0);
w(13)<=mul2(1);
w(14)<=mul2(2);
cnt1:=cnt1+1;
elsif (cnt1=22)then
mul1(0)<=(w(0)+w(3)+w(6)+w(9)+w(12));
mul1(1)<=(w(1)+w(4)+w(7)+w(10)+w(13));
mul1(2)<=(w(2)+w(5)+w(8)+w(11)+w(14));
cnt1:=cnt1+1;
elsif (cnt1=23)then
mul_clk<='1';
cnt1:=cnt1+1;
elsif (cnt1=24)then
mul_clk<='0';
v(0)<=mul2(0);
v(1)<=mul2(1);
v(2)<=mul2(2);
cnt1:=cnt1+1;
elsif (cnt1=25)then
a1<=conv_std_logic_vector(v(0),3);
b1<=conv_std_logic_vector(v(1),3);
c1<=conv_std_logic_vector(v(2),3);
cnt1:=cnt1+1;
elsif (cnt1=26)then
case (a1 & b1 & c1) is
when "000000000" =>RNA_out<="00000";
when "001001001" =>RNA_out<="00001";
when "010010000" =>RNA_out<="00010";
when "011000001" =>RNA_out<="00011";
when "100001000" =>RNA_out<="00100";
when "000010001" =>RNA_out<="00101";
when "001000000" =>RNA_out<="00110";
when "010001001" =>RNA_out<="00111";
when "011010000" =>RNA_out<="01000";
when "100000001" =>RNA_out<="01001";
when "000001000" =>RNA_out<="01010";
when "001010001" =>RNA_out<="01011";
when "010000000" =>RNA_out<="01100";
when "011001001" =>RNA_out<="01101";
when "100010000" =>RNA_out<="01110";
when "000000001" =>RNA_out<="01111";
when "001001000" =>RNA_out<="10000";
when "010010001" =>RNA_out<="10001";
when "011000000" =>RNA_out<="10010";
when "100001001" =>RNA_out<="10011";
```

```
when "000010000" =>RNA_out<="10100";
when "001000001" =>RNA_out<="10101";
when "010001000" =>RNA_out<="10110";
when "011010001" =>RNA_out<="10111";
when "100000000" =>RNA_out<="11000";
when "000001001" =>RNA_out<="11001";
when "001010000" =>RNA_out<="11010";
when "010000001" =>RNA_out<="11011";
when "011001000" =>RNA_out<="11100";
when "100010001" =>RNA_out<="11101";
when others =>RNA_out<="00000";
end case;
cnt1:=cnt1+1;
elsif (cnt1=27)then
if (cnt2=4)then
cnt1:=1; cnt2:=0;
xn_temp<="000000000000000";
data<="000000000000000";
else
xn_temp<=(xn_temp(3 to 14)& xn_temp(0 to 2));
data<=("000"& data(0 to 11));
cnt2:=cnt2+1;
cnt1:=2;
end if;
else
cnt1:=cnt1+1;
end if;
end if;
end process;
P1:process(RNA_clk,sample,coef)
begin
if rising_edge(RNA_clk) then
case sample(0) is
when 0=>x(0)<=0;x(1)<=0;x(2)<=0;
when 1=>x(0)<=1;x(1)<=1;x(2)<=1;
when 2=>x(0)<=2;x(1)<=2;x(2)<=0;
when 3=>x(0)<=3;x(1)<=0;x(2)<=1;
when 4=>x(0)<=4;x(1)<=1;x(2)<=0;
when 5=>x(0)<=0;x(1)<=2;x(2)<=1;
when 6=>x(0)<=1;x(1)<=0;x(2)<=0;
when 7=>x(0)<=2;x(1)<=1;x(2)<=1;
when others => null;
end case;
case sample(1) is
when 0=>x(3)<=0;x(4)<=0;x(5)<=0;
when 1=>x(3)<=1;x(4)<=1;x(5)<=1;
when 2=>x(3)<=2;x(4)<=2;x(5)<=0;
when 3=>x(3)<=3;x(4)<=0;x(5)<=1;
when 4=>x(3)<=4;x(4)<=1;x(5)<=0;
when 5=>x(3)<=0;x(4)<=2;x(5)<=1;
when 6=>x(3)<=1;x(4)<=0;x(5)<=0;
when 7=>x(3)<=2;x(4)<=1;x(5)<=1;
when others => null;
end case;
case sample(2) is
when 0=>x(6)<=0;x(7)<=0;x(8)<=0;
when 1=>x(6)<=1;x(7)<=1;x(8)<=1;
when 2=>x(6)<=2;x(7)<=2;x(8)<=0;
when 3=>x(6)<=3;x(7)<=0;x(8)<=1;
when 4=>x(6)<=4;x(7)<=1;x(8)<=0;
when 5=>x(6)<=0;x(7)<=2;x(8)<=1;
```

```
when 6=>x(6)<=1;x(7)<=0;x(8)<=0;
when 7=>x(6)<=2;x(7)<=1;x(8)<=1;
when others => null;
end case;
case sample(3) is
when 0=>x(9)<=0;x(10)<=0;x(11)<=0;
when 1=>x(9)<=1;x(10)<=1;x(11)<=1;
when 2=>x(9)<=2;x(10)<=2;x(11)<=0;
when 3=>x(9)<=3;x(10)<=0;x(11)<=1;
when 4=>x(9)<=4;x(10)<=1;x(11)<=0;
when 5=>x(9)<=0;x(10)<=2;x(11)<=1;
when 6=>x(9)<=1;x(10)<=0;x(11)<=0;
when 7=>x(9)<=2;x(10)<=1;x(11)<=1;
when others => null;
end case;
case sample(4) is
when 0=>x(12)<=0;x(13)<=0;x(14)<=0;
when 1=>x(12)<=1;x(13)<=1;x(14)<=1;
when 2=>x(12)<=2;x(13)<=2;x(14)<=0;
when 3=>x(12)<=3;x(13)<=0;x(14)<=1;
when 4=>x(12)<=4;x(13)<=1;x(14)<=0;
when 5=>x(12)<=0;x(13)<=2;x(14)<=1;
when 6=>x(12)<=1;x(13)<=0;x(14)<=0;
when 7=>x(12)<=2;x(13)<=1;x(14)<=1;
when others => null;
end case;
case coef(0) is
when 0=>y(0)<=0;y(1)<=0;y(2)<=0;
when 1=>y(0)<=1;y(1)<=1;y(2)<=1;
when 2=>y(0)<=2;y(1)<=2;y(2)<=0;
when 3=>y(0)<=3;y(1)<=0;y(2)<=1;
when 4=>y(0)<=4;y(1)<=1;y(2)<=0;
when 5=>y(0)<=0;y(1)<=2;y(2)<=1;
when 6=>y(0)<=1;y(1)<=0;y(2)<=0;
when 7=>y(0)<=2;y(1)<=1;y(2)<=1;
when others => null;
end case;
case coef(1) is
when 0=>y(3)<=0;y(4)<=0;y(5)<=0;
when 1=>y(3)<=1;y(4)<=1;y(5)<=1;
when 2=>y(3)<=2;y(4)<=2;y(5)<=0;
when 3=>y(3)<=3;y(4)<=0;y(5)<=1;
when 4=>y(3)<=4;y(4)<=1;y(5)<=0;
when 5=>y(3)<=0;y(4)<=2;y(5)<=1;
when 6=>y(3)<=1;y(4)<=0;y(5)<=0;
when 7=>y(3)<=2;y(4)<=1;y(5)<=1;
when others => null;
end case;
case coef(2) is
when 0=>y(6)<=0;y(7)<=0;y(8)<=0;
when 1=>y(6)<=1;y(7)<=1;y(8)<=1;
when 2=>y(6)<=2;y(7)<=2;y(8)<=0;
when 3=>y(6)<=3;y(7)<=0;y(8)<=1;
when 4=>y(6)<=4;y(7)<=1;y(8)<=0;
when 5=>y(6)<=0;y(7)<=2;y(8)<=1;
when 6=>y(6)<=1;y(7)<=0;y(8)<=0;
when 7=>y(6)<=2;y(7)<=1;y(8)<=1;
when others => null;
end case;
case coef(3) is
when 0=>y(9)<=0;y(10)<=0;y(11)<=0;
```

```
when 1=>y(9)<=1;y(10)<=1;y(11)<=1;
when 2=>y(9)<=2;y(10)<=2;y(11)<=0;
when 3=>y(9)<=3;y(10)<=0;y(11)<=1;
when 4=>y(9)<=4;y(10)<=1;y(11)<=0;
when 5=>y(9)<=0;y(10)<=2;y(11)<=1;
when 6=>y(9)<=1;y(10)<=0;y(11)<=0;
when 7=>y(9)<=2;y(10)<=1;y(11)<=1;
when others => null;
end case;
case coef(4) is
when 0=>y(12)<=0;y(13)<=0;y(14)<=0;
when 1=>y(12)<=1;y(13)<=1;y(14)<=1;
when 2=>y(12)<=2;y(13)<=2;y(14)<=0;
when 3=>y(12)<=3;y(13)<=0;y(14)<=1;
when 4=>y(12)<=4;y(13)<=1;y(14)<=0;
when 5=>y(12)<=0;y(13)<=2;y(14)<=1;
when 6=>y(12)<=1;y(13)<=0;y(14)<=0;
when 7=>y(12)<=2;y(13)<=1;y(14)<=1;
when others => null;
end case;
end if;
end process;
P2:process(mul_clk,mul1)
begin
if rising_edge(mul_clk) then
case mul1(0) is
when 0|5|10|15|20|25=>mul2(0)<=0;
when 1|6|11|16|21|26=>mul2(0)<=1;
when 2|7|12|17|22|27=>mul2(0)<=2;
when 3|8|13|18|23|28=>mul2(0)<=3;
when 4|9|14|19|24|29=>mul2(0)<=4;
when others => null;
end case;
case mul1(1) is
when 0|3|6|9|12|15|18|21|24|27=>mul2(1)<=0;
when 1|4|7|10|13|16|19|22|25|28=>mul2(1)<=1;
when 2|5|8|11|14|17|20|23|26|29=>mul2(1)<=2;
when others => null;
end case;
case mul1(2) is
when 0|2|4|6|8|10|12|14|16|18|20|22|24|26|28=>mul2(2)<=0;
when 1|3|5|7|9|11|13|15|17|19|21|23|25|27|29=>mul2(2)<=1;
when others => null;
end case;
end if;
end process;
end Behavioral;
```

11.7 Distributed Arithmetic-Based Computations

DA-based computation plays a key role in implementing DSP functions in FPGA devices. This section discusses the theoretical background of DA algorithms and some design examples to understand the effectiveness of DA-based computations for gate-efficient implementations. DA is a computation algorithm that performs multiplications with LUTs as the SOP expressions that cover many of the important DSP operations. Due to the potential of the FPGA LUT architecture, the DA algorithm can be implemented to design a very efficient filter/equalizer. The derivation of the DA algorithm is simple, but its applications are extremely broad [25,152]. Now, we discuss the theoretical proofs for the DA-based unsigned/signed computations below. In general, the arithmetic SOP for a linear and time-invariant network is given by

$$y = c[0]x[0] + c[1]x[1] + c[2]x[2] + \cdots + c[N-1]x[N-1]$$

Which also can also be written as

$$y = \sum_{n=0}^{N-1} c[n]x[n] \tag{11.12}$$

where y is the response of the network at a given time, N is the number of required consecutive multiplications, N – 1 is the number of required additions per sample, c[n] is the network coefficients, that is, constants, which can also be denoted by h[n], and x[n] is the network input, that is, variable [25,152]. Since in this network design, the filter coefficients c[n] are known already, then the multiplication in Equation 11.12 becomes an SOP expression of a variable x[n] with a constant c[n]. In this situation, the DA-based computation becomes possible as well as more efficient. In an unsigned binary number system, the data can be represented with a coding formula as

$$x[n] = \sum_{b=0}^{B-1} x_b[n]2^b \tag{11.13}$$

where B is the total number of binary digits used to represent the data, $x_b[n]$ is the binary bit at the bth position, $x_b[n] \in [0,1]$, and 2^b is the weighted value of the binary bit position; $2^0 = 1, 2^1 = 2, 2^2 = 4 \ldots 2^{B-1}$. Substituting Equation 11.13 in Equation 11.12

$$y = \sum_{n=0}^{N-1} c[n] \sum_{b=0}^{B-1} x_b[n]2^b$$

This equation indicates that we need to distribute the second summation for each element of summation of the first summation. Therefore, this equation can be expanded as

$$\begin{aligned} y = {} & c[0]\left(x_{B-1}[0]2^{B-1} + x_{B-2}[0]2^{B-2} + \cdots + x_0[0]2^0\right) \\ & + c[1]\left(x_{B-1}[1]2^{B-1} + x_{B-2}[1]2^{B-2} + \cdots + x_0[1]2^0\right) \\ & \vdots \\ & + c[N-1]\left(x_{B-1}[N-1]2^{B-1} + x_{B-2}[N-1]2^{B-2} + \cdots + x_0[N-1]2^0\right) \end{aligned}$$

Distribute the coefficients inside to get an SOP equation as

$$\begin{aligned} y = {} & \left(c[0]x_{B-1}[0]2^{B-1} + c[0]x_{B-2}[0]2^{B-2} + \cdots + c[0]x_0[0]2^0\right) \\ & + \left(c[1]x_{B-1}[1]2^{B-1} + c[1]x_{B-2}[1]2^{B-2} + \cdots + c[1]x_0[1]2^0\right) \\ & \vdots \\ & + \left(c[N-1]x_{B-1}[N-1]2^{B-1} + c[N-1]x_{B-2}[N-1]2^{B-2} + \cdots + c[N-1]x_0[N-1]2^0\right) \end{aligned}$$

Upon adding all the binary positions with respect to their binary weights, the above equation becomes

$$\begin{aligned} y = {} & [c[0]x_{B-1}[0] + c[1]x_{B-1}[1] + \cdots + c[N-1]x_{B-1}[N-1]]2^{B-1} \\ & + [c[0]x_{B-2}[0] + c[1]x_{B-2}[1] + \cdots + c[N-1]x_{B-2}[N-1]]2^{B-2} \\ & \vdots \\ & + [c[0]x_0[0] + c[1]x_0[1] + \cdots + c[N-1]x_0[N-1]]2^0 \end{aligned}$$

This equation can also be written as

$$y = \sum_{b=0}^{B-1} 2^b \sum_{n=0}^{N-1} c[n]\, x_b[n] \tag{11.14}$$

The above equation clearly shows that the second summation can be treated as an SOP expression and has N multiplications and N – 1 additions for every value of b. This multiplications can be carried out using the LUT. In the same way, the theoretical proof for a signed DA system can also be derived as discussed below. In the signed number system, the last bit, that is, the Bth bit, is used to denote the sign of the number, while the remaining bits, that is, B – 1 through 0, represent the magnitude. The negative number is always represented in its equivalent 2's complement form. For example, a signed number "1101" is –5 in radix-10 whose 2's complement value is "1011", and "01011" is +11 in radix-10. Therefore, in general, the coding formula for the signed number is

$$x[n] = (-2^B)x_B[n] + \sum_{b=0}^{B-1} x_b[n]2^b \tag{11.15}$$

In Equation 11.15, the first part contributes to get the 2's complement of x[n] if it is a negative number. For example, assume a negative number "111001", which is –25 in radix-10. Here, B = 5, $x_B[n] = 1$ and b varies from 0 to 4, whose value is 25 in radix-10. Substituting these values in Equation 11.15, then $x[n] = -2^5 \times 1 + 25$, which is equal to –7, which is equal to –25 in radix-10. This can be verified using the classical approach of converting the binary to the 2's complement by taking the 1's complement for "11001", that is, "00110", then adding '1' at the LSB, that is, "00111", which is equal to 7 in 2's complement form, and the sign bit is '1'; hence, the final value is –7. Suppose the number is positive. Then, $x_B[n]$ will be '0'; hence, the first part in Equation 11.15 becomes '0' and represents only the actual value in radix-10 as well as in 2's complement form. Therefore, we can substitute Equation 11.15 in Equation 11.12 to derive the theoretical proof for the signed DA system, so Equation 11.12 becomes

$$y = \sum_{n=0}^{N-1} c[n]\left((-2^B)x_B[n] + \sum_{b=0}^{B-1} x_b[n]2^b\right)$$

$$y = \sum_{n=0}^{N-1} c[n]\left(\sum_{b=0}^{B-1} x_b[n]2^b + (-2^B)x_B[n]\right)$$

The summations can be distributed for all the values of 'n' and 'b'. Then, the above equation can be given, in general, by

$$
\begin{aligned}
y = {} & c[0]\left(x_{B-1}[0]2^{B-1} + x_{B-2}[0]2^{B-2} + \cdots + x_0[0]2^0 + (-2^B)x_B[0]\right) \\
& + c[1]\left(x_{B-1}[1]2^{B-1} + x_{B-2}[1]2^{B-2} + \cdots + x_0[1]2^0 + (-2^B)x_B[1]\right) \\
& \vdots \\
& + c[N-1]\left(x_{B-1}[N-1]2^{B-1} + x_{B-2}[N-1]2^{B-2} + \cdots + x_0[N-1]2^0 + (-2^B)x_B[N-1]\right)
\end{aligned}
$$

Distribute the filter coefficients inside the multiplications

$$
\begin{aligned}
y = {} & c[0]x_{B-1}[0]2^{B-1} + c[0]x_{B-2}[0]2^{B-2} + \cdots + c[0]x_0[0]2^0 + (-2^B)c[0]x_B[0] \\
& + c[1]x_{B-1}[1]2^{B-1} + c[1]x_{B-2}[1]2^{B-2} + \cdots + c[1]x_0[1]2^0 + (-2^B)c[1]x_B[1] \\
& \vdots \\
& c[N-1]x_{B-1}[N-1]2^{B-1} + c[N-1]x_{B-2}[N-1]2^{B-2} \\
& + \cdots + c[N-1]x_0[N-1]2^0 + (-2^B)c[N-1]x_B[N-1]
\end{aligned}
$$

Adding all possible powers with respect to their positions, the above equation becomes

$$
\begin{aligned}
y = {} & \left(c[0]x_{B-1}[0] + c[1]x_{B-1}[1] + \cdots + c[N-1]x_{B-1}[N-1]\right)2^{B-1} \\
& + \left(c[0]x_{B-2}[0] + c[1]x_{B-2}[1] + \cdots + c[N-1]x_{B-2}[N-1]\right)2^{B-2} \\
& \vdots \\
& + (c[0]x_0[0] + c[1]x_0[1] + \cdots + c[N-1]x_0[N-1])2^0 + (-2^B) \\
& + (c[0]x_B[0] + c[1]x_B[1] + \cdots + c[N-1]x_B[N-1])
\end{aligned}
\tag{11.16}
$$

In Equation 11.16, the first part corresponds to the magnitude, which is an SOP; hence, the DA can be employed. The second part corresponds to the sign bit; however, it also a, SOP, so we can use the DA technique along with -2^B. Therefore, this equation can be written as

$$
\begin{aligned}
y &= \sum_{b=0}^{B-1} 2^b \sum_{n=0}^{N-1} c[n]x_b[n] + \sum_{n=0}^{N-1} c[n]x_B[n](-2^B) \\
y &= \sum_{b=0}^{B-1} 2^b \sum_{n=0}^{N-1} c[n]x_b[n] + (-2^B)\sum_{n=0}^{N-1} c[n]x_B[n]
\end{aligned}
\tag{11.17}
$$

This equation clearly shows that the second and third summation are SOP expressions, which can be dealt with using the DA computation technique.

DESIGN EXAMPLE 11.17

Design a general DA-LUT for 3-bit unsigned samples and compute the response of a FIR filter when the samples and coefficients are `x[n] = {4,3,7}` and `c[n] = {7,5,6}` using the DA computation technique.

TABLE 11.4

DA-LUT for 3-Bit Samples

$X_b[n]$			$\sum_{n=0}^{N-1} c[n]xb[n]; \quad b \in \{0 \text{ to } -1\}$			
$x_b[2]$	$x_b[1]$	$x_b[0]$	$C[2]x_b[2]$	$C[1]x_b[1]$	$C[0]x_b[0]$	LUT Value
0	0	0	$6 \times 0 +$	$5 \times 0 +$	$7 \times 0 +$	0
0	0	1	$6 \times 0 +$	$5 \times 0 +$	$7 \times 1 +$	7
0	1	0	$6 \times 0 +$	$5 \times 1 +$	$7 \times 0 +$	5
0	1	1	$6 \times 0 +$	$5 \times 1 +$	$7 \times 1 +$	12
1	0	0	$6 \times 1 +$	$5 \times 0 +$	$7 \times 0 +$	6
1	0	1	$6 \times 1 +$	$5 \times 0 +$	$7 \times 1 +$	13
1	1	0	$6 \times 1 +$	$5 \times 1 +$	$7 \times 0 +$	11
1	1	1	$6 \times 1 +$	$5 \times 1 +$	$7 \times 1 +$	18

Solution

The samples and filter coefficients are `x[n] = [4,3,7]` and `c[n] = [7,5,6]`, respectively; hence, `x[0] = 4`, `x[1] = 3` and `x[2] = 7`; `c[0] = 7`, `c[1] = 5` and `c[2] = 6`. The sample values and filter coefficients given in the above design example are the positive numbers. Therefore, we can use the following equation to find the response of the filter using the DA technique.

$$y = \sum_{b=0}^{B-1} 2^b \sum_{n=0}^{N-1} c[n]x_b[n]$$

Now, we form the general DA-LUT as required using the value of the constant `c[n] = {7,5,6}` for all the possible values of `x[n]`. Here, there are only 3-bit samples, so we need an 8 × 3 LUT as given in Table 11.4.

The given samples for `x[n]` are {4,3,7}, and these samples will be loaded to the respective registers as given in Table 11.5, and the bits (from LSB onward) will come out for

TABLE 11.5

Filter Response Computation Using DA Technique

Register Part					
CLK-->		t = 2	t = 1	t = 0	
Register values	**x[0]**	1	0	0	
	x[1]	0	1	1	
	x[2]	1	1	1	
Computation Part					
t	**$x_t[2]$**	**$x_t[1]$**	**$x_t[0]$**	**$2^b \times$ LUT value + Acc[t − 1]**	**Acc[t]**
0	1	1	0	$2^0 \times 11 + 0$	11
1	1	1	0	$2^1 \times 11 + 11$	33
2	1	0	1	$2^2 \times 13 + 33$	85

every CLK(t) and enter to the computation process, where the multiplication of 2^b and addition takes place as in Equation 11.14.

Initially, the value of the accumulator (Acc) is taken as '0'. Finally, the filter response is `y = 85`, which can be verified by the classical method as `y = 7 × 4 + 5 × 3 + 6 × 7 = 85`.

DESIGN EXAMPLE 11.18

Design a general DA-LUT for 4-bit signed samples and compute the response of a FIR filter when the samples and coefficients are `x[n] = {−6, −4, 5}` and `c[n] = {−7, −4, 3}` using the DA computation technique.

Solution

The coefficients of the filter are `c[0] = −7`, `c[1] = −4` and `c[2] = 3`. With these values, we first construct the DA-LUT as given in Table 11.6. Since the samples are negative numbers, we need to use the following equation to find the response of the filter.

$$y = \sum_{b=0}^{B-1} 2^b \sum_{n=0}^{N-1} c[n]x_b[n] + (-2^B)\sum_{n=0}^{N-1} c[n]x_B[n]$$

The given samples are `x[0] = −6`, `x[1] = −4` and `x[2] = 5`. Here, two samples are negative numbers; therefore, we need to represent them in the 2's complement form as $x[0] = (1110)_2 = (1010)_{2c}$, $x[1] = (1100)_2 = (1100)_{2c}$ and $x[2] = (0101)_2$. These values will be loaded to the respective registers as given in Table 11.7 and entered bitwise for further computations.

The filter response is `y = 73`, which can be verified by the classical method as `y = −7 × −6 + −4 × −4 + 3 × 5 = 73`.

DESIGN EXAMPLE 11.19

Develop a digital system in the FPGA to compute the response of a 3-tap FIR filter whose coefficients are `c[n] = {32421, 11234, 32056, 13268}`. Assume that the input

TABLE 11.6

DA_LUT for Signed Samples and Coefficients

$X_b[n]$			$\sum_{n=0}^{N-1} c[n]xb[n];\ b \in \{0 \text{ to } -1\}$			
$x_b[2]$	$x_b[1]$	$x_b[0]$	$C[2]x_b[2]$	$C[1]x_b[1]$	$C[0]x_b[0]$	LUT Value
0	0	0	3 × 0 +	−4 × 0 +	−7 × 0 +	0
0	0	1	3 × 0 +	−4 × 0 +	−7 × 1 +	-7
0	1	0	3 × 0 +	−4 × 1 +	−7 × 0 +	−4
0	1	1	3 × 0 +	−4 × 1 +	−7 × 1 +	−11
1	0	0	3 × 1 +	−4 × 0 +	−7 × 0 +	3
1	0	1	3 × 1 +	−4 × 0 +	−7 × 1 +	−4
1	1	0	3 × 1 +	−4 × 1 +	−7 × 0 +	-1
1	1	1	3 × 1 +	−4 × 1 +	−7 × 1 +	-8

TABLE 11.7

Filter Response Computation for Signed Samples/Coefficients Using DA Techniques

Register Part					
CLK-->		sign	t = 2	t = 1	t = 0
Register values	x[0]	1	0	1	0
	x[1]	1	1	0	0
	x[2]	0	1	0	1
Computation Part					
t	**$x_t[2]$**	**$x_t[1]$**	**$x_t[0]$**	**2^b × LUT value + Acc[t − 1]**	**Acc[t]**
0	1	0	0	$2^0 \times 3 + 0$	3
1	0	0	1	$2^1 \times -7 + 3$	−11
2	1	1	0	$2^2 \times -1 + (-11)$	−15
Sign Part Computation					
3	0	1	1	$(-2^3) \times (-11) + (-15)$	73

samples are 4 bits applied using 16 dip switches (four switches per sample) and are unsigned numbers. Display the end results in a 128-bit sequence.

Solution

```
library IEEE;
use IEEE.STD_LOGIC_1164.ALL;
use IEEE.STD_LOGIC_ARITH.ALL;
use IEEE.STD_LOGIC_UNSIGNED.ALL;
use IEEE.NUMERIC_BIT.ALL;
use IEEE.NUMERIC_STD.ALL;
entity DA_fir is
Port ( m_clk: in  STD_LOGIC:='0';
sample_in: in STD_LOGIC_vector(15 downto 0):="0100001100100001";
mul_vlu:out std_logic_vector(127 downto 0):="000000000000000000000000000
0000000000000000000000000000000000000000000000000000000000000000000000
00000000000000000000000000000");
end DA_fir;
architecture Behavioral of DA_fir is
signal clk_sig,lut_enable:std_logic:='0';
signal xn_temp: STD_LOGIC_vector(15 downto 0):="0000000000000000";
signal sampt0,sampt1,sampt2,sampt3,sampt_temp:STD_LOGIC_vector(3 downto
0):="0000";
signal lut_val:integer range 0 to 1577458;
type memory_type1 is array (0 to 3) of integer range 0 to 1577458;
signal mul:memory_type1 :=(0,0,0,0);
type memory_type2 is array (0 to 2) of integer range 0 to 1577458;
signal fi_mul:memory_type2 :=(0,0,0);
begin
p0:process(m_clk)
variable cnt:integer:=0;
begin
if rising_edge(m_clk) then
if(cnt=1) then
clk_sig<=not(clk_sig);
```

```
cnt:=0;
else
cnt:=cnt+1;
end if;
end if;
end process;
p1:process(clk_sig,sample_in)
variable cnt1:integer:=0;
begin
if rising_edge(clk_sig) then
if (cnt1=1) then
xn_temp<=sample_in;
cnt1:=cnt1+1;
elsif (cnt1=2) then
sampt0<=(xn_temp(0) & xn_temp(4) & xn_temp(8) & xn_temp(12));
sampt1<=(xn_temp(1) & xn_temp(5) & xn_temp(9) & xn_temp(13));
sampt2<=(xn_temp(2) & xn_temp(6) & xn_temp(10) & xn_temp(14));
sampt3<=(xn_temp(3) & xn_temp(7) & xn_temp(11) & xn_temp(15));
cnt1:=cnt1+1;
elsif (cnt1=3) then
sampt_temp<=sampt0;
cnt1:=cnt1+1;
elsif (cnt1=4) then
lut_enable<='1';
cnt1:=cnt1+1;
elsif (cnt1=5) then
lut_enable<='0';
cnt1:=cnt1+1;
elsif (cnt1=6) then
mul(0)<=lut_val;
cnt1:=cnt1+1;
elsif (cnt1=7) then
sampt_temp<=sampt1;
cnt1:=cnt1+1;
elsif (cnt1=8) then
lut_enable<='1';
cnt1:=cnt1+1;
elsif (cnt1=9) then
lut_enable<='0';
cnt1:=cnt1+1;
elsif (cnt1=10) then
mul(1)<=lut_val;
cnt1:=cnt1+1;
elsif (cnt1=11) then
sampt_temp<=sampt2;
cnt1:=cnt1+1;
elsif (cnt1=12) then
lut_enable<='1';
cnt1:=cnt1+1;
elsif (cnt1=13) then
lut_enable<='0';
cnt1:=cnt1+1;
elsif (cnt1=14) then
mul(2)<=lut_val;
cnt1:=cnt1+1;
elsif (cnt1=15) then
sampt_temp<=sampt3;
cnt1:=cnt1+1;
elsif (cnt1=16) then
lut_enable<='1';
cnt1:=cnt1+1;
```

```
elsif (cnt1=17) then
lut_enable<='0';
cnt1:=cnt1+1;
elsif (cnt1=18) then
mul(3)<=lut_val;
cnt1:=cnt1+1;
elsif (cnt1=19) then
fi_mul(0)<=( mul(0) + (mul(1)* 2));
cnt1:=cnt1+1;
elsif (cnt1=20) then
fi_mul(1)<=( fi_mul(0) + (mul(2)* 4));
cnt1:=cnt1+1;
elsif (cnt1=21) then
fi_mul(2)<=( fi_mul(1) + (mul(3)* 8));
cnt1:=cnt1+1;
elsif (cnt1=22) then
mul_vlu<=conv_std_logic_vector(fi_mul(2),128);
cnt1:=cnt1+1;
elsif (cnt1=23) then
cnt1:=0;
else
cnt1:=cnt1+1;
end if;
end if;
end process;
p2:process(lut_enable,sampt_temp)
begin
if rising_edge(lut_enable) then
case  sampt_temp is
when "0000"=>lut_val<=0;
when "0001"=>lut_val<=32421;
when "0010"=>lut_val<=11234;
when "0011"=>lut_val<=43655;
when "0100"=>lut_val<=32056;
when "0101"=>lut_val<=64477;
when "0110"=>lut_val<=43290;
when "0111"=>lut_val<=75711;
when "1000"=>lut_val<=13268;
when "1001"=>lut_val<=45689;
when "1010"=>lut_val<=24502;
when "1011"=>lut_val<=56923;
when "1100"=>lut_val<=45324;
when "1101"=>lut_val<=77745;
when "1110"=>lut_val<=56558;
when "1111"=>lut_val<=88979;
when others=>lut_val<=0;
end case;
end if;
end process;
end Behavioral;
```

DESIGN EXAMPLE 11.20

Develop a digital system in the FPGA to estimate the response of a FIR filter. Assume that the inputs are of 3-bit positive numbers and the filter coefficients are `c[n] = {2,3,1}` [25].

Solution

```
library ieee;
use ieee.std_logic_1164.all;
```

```
use ieee.std_logic_arith.all;
entity dafsm is
port (clk, reset : in std_logic:='0';
x0_in, x1_in, x2_in :in  std_logic_vector(2 downto 0):="000";
lut   : out integer range 0 to 7:=0;
y     : out integer range 0 to 63:=0);
end dafsm;
architecture fpga of dafsm is
component case3
port ( table_in   : in   std_logic_vector(2 downto 0);
table_out  : out  integer range 0 to 6);
end component;
type state_type is (s0, s1);
signal state     : state_type;
signal x0, x1, x2, table_in: std_logic_vector(2 downto 0);
signal table_out : integer range 0 to 7;
begin
table_in(0) <= x0(0);
table_in(1) <= x1(0);
table_in(2) <= x2(0);
p0:process (reset, clk)
variable p     : integer range 0 to 63;
variable count : integer range 0 to 3;
begin
if reset = '1' then
state <= s0;
elsif rising_edge(clk) then
case state is
when s0 =>
state <= s1;
count := 0;
p := 0;
x0 <= x0_in;
x1 <= x1_in;
x2 <= x2_in;
when s1 =>
if count = 3 then
y <= p;
state <= s0;
else
p := p / 2 + table_out * 4;
x0(0) <= x0(1);
x0(1) <= x0(2);
x1(0) <= x1(1);
x1(1) <= x1(2);
x2(0) <= x2(1);
x2(1) <= x2(2);
count := count + 1;
state <= s1;
end if;
end case;
end if;
end process;
lc_table0: case3
port map(table_in => table_in, table_out => table_out);
lut <= table_out;
end fpga;
--components
library ieee;
use ieee.std_logic_1164.all;
use ieee.std_logic_arith.all;
```

```
entity case3 is
port ( table_in  : in  std_logic_vector(2 downto 0);
table_out : out integer range 0 to 6);
end case3;
architecture les of case3 is
begin
p0:process (table_in)
begin
case table_in is
when  "000" =>   table_out <=  0;
when  "001" =>   table_out <=  2;
when  "010" =>   table_out <=  3;
when  "011" =>   table_out <=  5;
when  "100" =>   table_out <=  1;
when  "101" =>   table_out <=  3;
when  "110" =>   table_out <=  4;
when  "111" =>   table_out <=  6;
when   others  =>   table_out <=  0;
end case;
end process;
end les;
```

DESIGN EXAMPLE 11.21

Develop a digital system in the FPGA to compute the response of a FIR filter. The inputs are of 4-bit words and the filter's coefficients are `c[n] = {−2,3,1}` [25].

Solution

```
library ieee;
use ieee.std_logic_1164.all;
use ieee.std_logic_arith.all;
entity dasign is
port (clk, reset : in std_logic;
x_in0, x_in1, x_in2
: in  std_logic_vector(3 downto 0);
lut  : out integer range -2 to 4;
y    : out integer range -64 to 63);
end dasign;
architecture fpga of dasign is
component case3s
port ( table_in : in  std_logic_vector(2 downto 0);
table_out : out integer range -2 to 4);
end component;
type state_type is (s0, s1);
signal state      : state_type;
signal table_in   : std_logic_vector(2 downto 0);
signal x0, x1, x2 : std_logic_vector(3 downto 0);
signal table_out  : integer range -2 to 4;
begin
table_in(0) <= x0(0);
table_in(1) <= x1(0);
table_in(2) <= x2(0);
process (reset, clk)
variable  p : integer range -64 to 63:= 0;
variable count : integer range 0 to 4;
begin
if reset = '1' then
state <= s0;
```

```
elsif rising_edge(clk) then
case state is
when s0 =>
state <= s1;
count := 0;
p := 0;
x0 <= x_in0;
x1 <= x_in1;
x2 <= x_in2;
when s1 =>
if count = 4 then
y <= p;
state <= s0;
else
if count = 3 then
p := p / 2 - table_out * 8;
else
p := p / 2 + table_out * 8;
end if;
for k in 0 to 2 loop
x0(k) <= x0(k+1);
x1(k) <= x1(k+1);
x2(k) <= x2(k+1);
end loop;
count := count + 1;
state <= s1;
end if;
end case;
end if;
end process;
lc_table0: case3s
port map(table_in => table_in, table_out => table_out);
lut <= table_out;
end fpga;
--component
library ieee;
use ieee.std_logic_1164.all;
use ieee.std_logic_arith.all;
entity case3s is
port ( table_in  : in   std_logic_vector(2 downto 0);
table_out : out  integer range -2 to 4);
end case3s;
architecture les of case3s is
begin
process (table_in)
begin
case table_in is
when  "000" =>    table_out <=  0;
when  "001" =>    table_out <=  -2;
when  "010" =>    table_out <=  3;
when  "011" =>    table_out <=  1;
when  "100" =>    table_out <=  1;
when  "101" =>    table_out <=  -1;
when  "110" =>    table_out <=  4;
when  "111" =>    table_out <=  2;
when   others  =>    table_out <=  0;
end case;
end process;
end les;
```

11.8 Booth Multiplication Algorithm and Design

Signed or unsigned multipliers are the key components of many high-performance systems such as filters, equalizers, controllers, and digital signal processors. Generally, system performance is determined by the performance of the multiplier because it is generally the slowest element in any digital system and occupies more area in the hardware. Hence, improving the computation speed at a reasonable implementation area becomes significant. However, area and speed are usually conflicting constraints so that improving speed results mostly in larger areas [153]. Different algorithms are being used for the multiplication of numbers; however, Booth's algorithm is an often-used technique due to its simplicity and efficient implementation in the FPGA. Booth's algorithm, which is also called fixed-point 2's complement multiplication, is an algorithm that multiplies two signed/unsigned binary numbers on a two's complement platform. The algorithm was invented by Andrew Donald Booth in 1950 while doing research on crystallography at Birkbeck College in Bloomsbury, London [153]. Booth used desk calculators that were faster at shifting than adding and created the algorithm to increase their speed. There are two important terms we should know before attempting Booth's algorithm: (i) right shift circulate (RSC) and (ii) right shift arithmetic (RSA). RSC is simply shifting the binary string to the right by 1 bit position, taking the last bit (LSB) in the string and appending it to the beginning of the string. For example, assume x = "1100100." After performing RSC, x becomes "0110010". In RSA, shift all bits right by 1 bit position and put the first bit (MSB) of the result at the beginning of the new string. For example, assume x = "1001110". After one RSA, x becomes "11001110", which means one bit is increased in the width of register x. With this information, we discuss Booth's multiplication algorithm step by step below:

Step 1: Convert two operands into equivalent signed 2's complement binary numbers, then identify which operand has less transition and set it as the multiplier (Mr) while the other operand is set to multiplicand (Md). For example, in a multiplication of 4 × –6, the signed binary equivalent is "0100" × "1110" and the 2's complement binary equivalent is "0100" × "1010"; here, "0100" has two transitions and "1010" has three transitions. Therefore, the Mr is "0100" and the Md is "1010". Then, –Md is the 2's complement of the full Md; that is, –Md = "0110"; this is required to subtract Md from Mr via adding –Md to Mr in future steps.

Step 2: Now, prepare a table with six columns, as in Table 11.8. The count value is taken from the number of bits required to represent the largest magnitude of the operands. The operands are "0100" and "1010", so the value of count is 3. The register Q is formed by concatenating a temporary (temp) register of length Mr and actual Mr. The temp is concatenated so as to accommodate the final product within temp and Mr. In the above example, the total length of the operands is 4 bits; hence, we need to chose a temp register with "0000" and take the value of Mr as it is to form the Q register. In this way, the Q register is formed and Q_0 is equal

TABLE 11.8

Initialization Process of Booth's Multiplication

	Q				
Count	**Temp**	**Mr**	Q_{-1}	Q_0Q_{-1}	**Operation**
3	0000	0100	0	00	Initialization

TABLE 11.9

Booth's Algorithm – Multiplication Example

Count	Q		Q_{-1}	Q_0Q_{-1}	Operation
	temp	Mr			
3	0000	0100	0	00	Initialization
3	0000	0100	0	00	RSA
2	0000	0010	0	00	RSA
1	0000 0110	0001	0	10	Subtract Md from temp, i.e., temp+ (−Md)
	0110	0001			RSA
0	0011 1010	0000	1	01	Add Md to temp, i.e., temp + Md
	1101	0000			RSA
Below 0	1110	1000	0	00	**Complete**

to Mr[0]. Initially, the Q_{-1} is '0', and Q_0Q_{-1} is "00". This process is called initialization, and the respective values are given in the second row of Table 11.8.

Step 3: All our analysis henceforth will be based on the value of Q_0Q_{-1}, that is, the fifth-column values. We need to subtract Md from temp if the value of Q_0Q_{-1} is "10", add Md to temp if the value is "01" and perform no operation if the value is "00"or "11". The subtraction can be done in 2's complement form; that is, adding –Md to temp is equal to subtraction of Md from the temp register. The final carry has to be omitted while adding the temp to Md or –Md. In every operation, we need to finally perform the RSA in Mr if the value of temp is '0'; otherwise, in the Q register that is, [temp Mr] register as given in Table 11.9. In the above example, initially, the value of Q_0Q_{-1} is "00", so there is no operation other than RSA, as indicated in the third row of Table 11.9. The values after performing RSA are given in the fourth row of Table 11.9. Continue these operations until the counter value goes below '0'.

Step 4: In the value of the Q register, when the counter value is below '0', the MSB represents the sign, while the other bits (MSB-1 to LSB) represent the magnitude. In the above example, the value of the Q register is "11101000". This result is in 2's complement; hence, convert it back to binary by again taking the 2's complement. Then, the result is "0010111", which is equal to 24 in radix-10. Since the sign (the MSB of Q) bit is '1', the final result is –24, which is equal to 4 × –6.

DESIGN EXAMPLE 11.22

Apply Booth's algorithm and multiply 14 by –5. Illustrate all the steps involved in the multiplication process.

Solution

The binary representations of 14 and –5 are "01110" and "10101", respectively. The –5 has to be represented in 2's complement as "11011". The bit transition in both operands is 2; therefore, we can chose any operand as Mr. Choose "11011" as Mr and "01110" as Md; hence, –Md is "10010". The step-by-step results as per Booth's algorithm are given in Table 11.10.

TABLE 11.10

Multiplication of 14 × –5 Using Booth's Algorithm

Count	Q		Q_{-1}	Q_0Q_{-1}	Operation
	Temp	Mr			
4	00000	11011	0	10	Initialization
4	00000 10010	11011	0	10	Subtract: temp + (–Md)
	10010	11011			RSA
3	11001	01101	1	11	RSA
2	11100 01110	10110	1	01	Add: temp + Md
	01010	10110			RSA
1	00101 10010	01011	0	10	Subtract: temp + (–Md)
	10111	01011			RSA
0	11011	10101	1	11	RSA
	11101	11010	1	01	Complete

The final result is "1110111010". Here, the sign bit is '1'; therefore, the value is a negative number. The 2's complement of the magnitude part "110111010" is "001000110", which is equal to –70 in radix-10 format.

DESIGN EXAMPLE 11.23

Design a digital system to multiply two signed 4-bit words using Booth's multiplication algorithm.

Solution

```
library IEEE;
use IEEE.STD_LOGIC_1164.ALL;
use IEEE.STD_LOGIC_arith.ALL;
use IEEE.STD_LOGIC_unsigned.ALL;
entity booth_mul is
port(
clk,enable:in std_logic;
data1,data2:in std_logic_vector(4 downto 0);
c:out std_logic_vector(9 downto 0));
end booth_mul;
architecture Behavioral of booth_mul is
type state_type is(s1,s2,s3,s4,s5,s6,rsa_state,s7,s8,result);
signal state,state1:state_type:=s1;
signal a,b,tmp1,tmp2,mr,md,md_p,md_n,latch1,latch2:std_logic_vector(4
downto 0);
signal count,count1,count2,counter,transition1,transition2,cnt:inte
ger:=0;
signal x,y:std_logic;
signal latch:std_logic_vector(10 downto 0);
begin
process(clk)
begin
if(clk='1' and clk'event)then
```

```
case state is
when s1=>
if(enable='1')then
state<=s2;
a<=data1,
b<=data2;
else
state<=s1;
end if;
when s2=>
--find 2's complement
if(a(4)='1')then
tmp1<=(('0' & a(3 downto 0)) xor "11111")+"00001";
else
tmp1<=a;
end if;
if(b(4)='1')then
tmp2<=(('0' & b(3 downto 0)) xor "11111")+"00001";
else
tmp2<=b;
end if;
state<=s3;
when s3=>
---Find no of transition
x<=tmp1(0);
y<=tmp2(0);
count<=1;
state<=s4;
when s4=>
if(count=5)then
count<=0;
state<=s5;
else
if(x=tmp1(count))then
transition1<=transition1;
else
x<=tmp1(count);
transition1<=transition1+1;
end if;
if(y=tmp2(count))then
transition2<=transition2;
else
y<=tmp2(count);
transition2<=transition2+1;
end if;
count<=count+1;
state<=s4;
end if;
when s5=>
if(transition1>transition2)then
mr<=tmp1;
md<=tmp2;
else
mr<=tmp2;
md<=tmp1;
end if;
state<=s6;
when s6=>
latch<="00000" & mr & '0';
md_p<=md;
md_n<=(md xor "11111")+"00001" ;
```

```
state<=s7;
when s7=>
if(cnt=counter+1)then
state<=s8;
cnt<=0;
else
if(latch(1 downto 0)="01")then
latch<=latch+(md_p & "000000");
state<=rsa_state;
elsif(latch(1 downto 0)="10")then
latch<=latch+(md_n & "000000");
state<=rsa_state;
else
state<=rsa_state;
end if;
cnt<=cnt+1;
end if;
when rsa_state=>
if(latch(10)='1')then
latch<='1' & latch(10 downto 1);
else
latch<='0' & latch(10 downto 1);
end if;
state<=s7;
when s8=>
if(latch(10)='1')then
latch(10 downto 1)<=(latch(10 downto 1) xor "1111111111")+"0000000001";
else
latch<=latch;
end if;
state<=result;
when result=>
c(9)<=data1(4) xor data2(4);
c(8 downto 0)<=latch(9 downto 1);
end case;
end if;
end process;
---find count value
process(clk,enable)
begin
if(clk='1' and clk'event)then
case state1 is
when s1=>
if(enable='1')then
latch1<=data1;
latch2<=data2;
state1<=s2;
else
state1<=s1;
end if;
when s2=>
if(latch1(3)='0')then
latch1<=latch1(3 downto 0) & '0';
count1<=count1+1;
state1<=s2;
else
state1<=s3;
count1<=4 -count1;
end if;
when s3=>
if(latch2(3)='0')then
```

```
latch2<=latch2(3 downto 0) & '0';
count2<=count2+1;
statel<=s3;
else
statel<=s4;
count2<=4 -count2;
end if;
when s4=>
if(count1>count2)then
counter<=count1;
else
counter<=count2;
end if;
statel<=s4;
when others=>null;
end case;
end if;
end process;
end Behavioral;
```

DESIGN EXAMPLE 11.24

Design a digital system to perform the multiplication of two signed inputs, each of 16 bits, using Booth's multiplication algorithm. Assume that the digital system has to consider the input(s) in 2's complement in case they are negative numbers, and the result also has to be in 2's complement format. Implement the algorithm using the `for` loop.

Solution

```
library ieee;
use ieee.std_logic_1164.all;
use ieee.numeric_std.all;
use ieee.std_logic_unsigned.all;
entity booth_mul is
port(x,y: in std_logic_vector(15 downto 0):="0000000000001111";
mul: out std_logic_vector(31 downto 0):="0000000000000000000000000000000
0");
end booth_mul;
architecture Behavioral of booth_mul is
begin
p0:process(x,y)
variable a: std_logic_vector(32 downto 0);
variable s,p : std_logic_vector(15 downto 0);
variable i:integer;
begin
a := "000000000000000000000000000000000";
s := y;
a(16 downto 1) := x;
for i in 0 to 15 loop
if(a(1) = '1' and a(0) = '0') then
p := (a(32 downto 17));
a(32 downto 17) := (p - s);
elsif(a(1) = '0' and a(0) = '1') then
p := (a(32 downto 17));
a(32 downto 17) := (p + s);
end if;
a(31 downto 0) := a(32 downto 1);
end loop;
mul(31 downto 0) <= a(32 downto 1);
end process;
end Behavioral;
```

11.9 Adaptive Filter/Equalizer Design

Adaptive filters and equalizers are widely used in system/channel/plant identification, echo cancellation in long-distance transmission, acoustic (mechanical longitudinal wave) echo cancellation, adaptive noise cancelling, channel equalization, interference cancelling, inverse plant modeling, signal prediction, adaptive array processing, linear predictive coding, and adaptive line enhancement [154]. In the last few decades, the field of adaptive digital signal processing has developed enormously due to the growing availability of VLSI technology for the implementation of modern algorithms. Adaptive filters may be called self-modifying/self-tuning filters since they automatically adjust their coefficients in order to minimize an error/cost function, that is, the error between the reference/desired signal 'd(k)' and the output of the adaptive filter 'y(k)'. The basic configuration of an adaptive filter, operating in the discrete-time domain k, is shown in Figure 11.16, where the input signal is denoted by `x(k)` and the error signal is defined as `e(k) = d(k) - y(k)`. The input–output that is, `x(k)` and `y(k)`, relation in the adaptive filter depends on its transfer function, which is in general the tapped-delay lines associated with the standard FIR filter, also called a transversal filter.

Another popular filter structures is the FIR lattice, that is, the IIR filter. However, this structure greatly increases the computational complexity of the adaptive algorithm and the overall speed of the adaptation process, as previously discussed in Sections 11.3 and 11.4 in this chapter. The rate of convergence, misadjustment tracking and coefficients of an adaptive filter can be made to converge to the optimum solution using a more suitable adaptation technique. In general, a faster convergence yields a higher misadjustment, and slow convergence requires more processing time [154].

The input-output relationship of an adaptive transversal filter is given by

$$\begin{aligned} y[k] &= w_0x[k] + w_1x[k-1] + w_2x[k-2] + \cdots + W_Nx[k-N] \\ &= \sum_{i=0}^{N} w_ix[k-i] \\ y(k) &= W^Tx[k] \end{aligned} \tag{11.18}$$

where N is the filter order and w and x[k] are the vector components composed of filter coefficients and input signal samples, which may be represented by

$$x[k] = [x[k] \quad x[k-1] \quad x[k-2] \ldots x[k-N]]^T$$
$$W = [w_0 \quad w_1 \quad w_2 \ldots w_N]^T$$

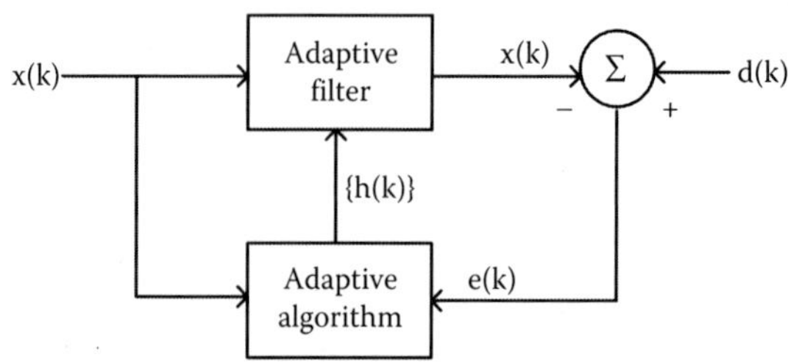

FIGURE 11.16
Simple structure of an adaptive filter.

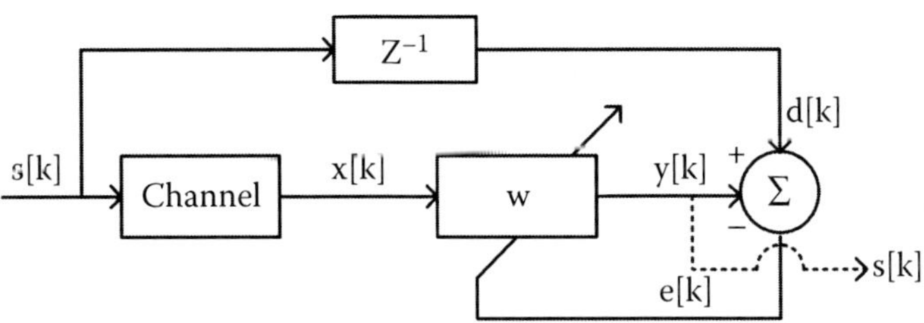

FIGURE 11.17
Signal flow graph of an adaptive channel equalizer.

A more precise computation drives the attention of designers on numerical stability, numerical accuracy and convergence robustness during the adaptation process. The error signal is used by the adaptation algorithm to update the adaptive filter coefficient vector w[k] according to some performance criterion. In general, the whole adaptation process aims at minimizing some metric of the error signal e[k] by forcing the adaptive filter output signal y[k] to approximate the reference signal d[k] in the statistical sense. This is the main principle for all adaptive filter designs. The same can be employed for adaptive equalization as well, as shown in Figure 11.17. In practical communication systems, a transmitted signal may be heavily distorted by the channel. One way to attempt to recover the original signal is employing an adaptive equalizer to mitigate/suppress the channel effects. In such a scheme, a training sequence x[k], known by the receiver, is sent via the channel generating the distorted signal.

The same sequence x[k], after a proper time shift to compensate for transmission delays, is used as a reference signal d[k] at the receiver for the adaptation process. When the error function e[k] approximates zero, the output signal y[k] resembles the transmitted signal x[k], indicating that the adaptive equalizer is compensating the channel distortions. After this training process, the desired information can be sent through the channel, which will be properly recovered back by this adaptive equalizer. Different metrics yield different adaptation processes with quite distinct characteristics. The commonly employed adaptive algorithms and their implementations can be found in detail in Ref. 154. One of the more popular adaptation algorithms is the LMS, which is discussed below. In general, the mean square error is given by

$$\begin{aligned}\xi[k] &= E[e^2[k]]\\ &= E[|d[k]-y[k]|^2]\end{aligned}$$

From Equcation 11.18, the above equation becomes

$$\begin{aligned}\xi[k] &= E[|d[k]-W^T x[k]|^2]\\ &= E[d^2[k]+(W^T x[k])^2 - 2d[k]\cdot W^T x[k]]\\ &= E[d^2[k]+(W^T)^2 x[k]^2 - 2d[k]\cdot W^T x[k]]\\ &= E[d^2[k]+W^T\cdot W\cdot x[k]^T x[k] - 2d[k]\cdot W^T x[k]]\\ &= E[d^2[k]] - 2W^T E[d[k]x[k]] + W^T E[x[k]x^T[k]]W\\ \xi[k] &= E[d^2[k] - 2W^T p + W^T R W]\end{aligned} \tag{11.19}$$

where R and p are the input signal correlation matrix and the crosscorrelation vector between the reference signal and the input signal, respectively. Those values can be represented by

$$R = E[x[k]xT[k]] \text{ and } p = E[d[k]xT[k]]$$

R and p are not represented as a function of the iteration k and are not time-varying, due to the assumed stationarity of the input and reference signal. From Equation 11.19, the gradient vector of the MSE function with respect to the adaptive filter coefficient is given by

$$\nabla_w \xi[k] = -2p + 2RW$$

This is called the Wiener solution W_0 that minimizes the MSE cost function, which is obtained by equating the gradient vector in Equation 11.19 to zero. Assuming that R is nonsingular,

$$W_0 = R^{-1}p$$

One can estimate the Wiener solution in a computationally efficient manner, iteratively adjusting the coefficient vector W at each time instant k, in such a manner that the resulting sequence W[k] converges to the desired W_0 solution, possibly in a sufficiently small number of iterations. The steepest-descent scheme searches the minimum of a given function following the opposite direction of the associated gradient vector. A factor μ/2, where μ is the convergence factor, adjusts the step size between consecutive coefficient vector estimates, yielding the following updating procedure.

$$W[k] = W[k-1] - \frac{\mu}{2}\nabla_W \xi[k] \tag{11.20}$$

The Wiener solution requires knowledge of the autocorrelation matrix R and the crosscorrelation vector p. To do that, one must have access to the complete second-order statistics of signals x[k] and d[k], which makes Equation 11.19 unsuitable for most practical applications. Naturally, the Wiener solution can be approximated by a proper estimation of R and p based on sufficiently long time intervals. A rather simpler approach is to approximate the MSE by the integral square error (ISE) function, using the gradient vector of the latter, given in Equation 11.19, to adjust the coefficient vector in Equation 11.20. The resulting algorithm is an LMS algorithm that is characterized by

$$W[k] = W[k-1] - \frac{\mu}{2}\nabla_w \xi[k]$$
$$W[k] = W[k-1] + \mu e[k]x[k]$$

In this equation, the e[k] can be estimated from ($d[k] - W^T[k-1]x[k]$). The step size parameter μ plays an important role in the convergence characteristics of the LMS algorithm as well as in its stability condition. The pseudocode for the LMS algorithm is given below, where the ∗ denotes the complex-conjugate.

```
initialize μ
for each k
  {
```

```
    E[k]=d[k]-W^-1[k-1]X[k];
    W[k]=W[k-1]+μe*[k]X[k];
}
```

An approximation for the upper bound of this parameter is given in the technical literature and may be stated as $0 < \mu < (2/tr(R))$, where tr(.) is the trace operator of a matrix. The LMS algorithm is very popular and has been widely used due to its extreme simplicity. Its convergence speed, however, is highly dependent on the condition number ρ of the input-signal autocorrelation matrix, defined as the ratio between the maximum and minimum eigenvalues of this matrix. More details about the many other types of adaptation techniques can be found in Ref. 154.

DESIGN EXAMPLE 11.25

Design MATLAB code to realize the LMS algorithm to recover a line generated by $y = \sin(2\pi ft + \theta)$ from its distorted version. Assume that a linear function is distorted by additive white Gaussian (AWG) noise. Plot the linear function, noise signal, distortion signal, error/weight profiles, and recovered signals.

Solution

```
clc;
clear all;
N=input('length of sequence N is:= ');
mu=input('Value of step function is:= ');
t=[1:N];
d=sin(2*pi*0.001*[1:N]+0.1);
subplot(2,2,1);
plot(t,d,'k','LineWidth',1.5);
xlabel('Samples');
ylabel('Signal (V)');
axis([0 max(t) min(d) max(d)]);
n=randn(1,N)*0.5;
subplot(2,2,2);
plot(t,n,'k','LineWidth',1.5);
xlabel('Samples');
ylabel('Noise (V)');
axis([0 max(t) min(n) max(n)]);
x=d+n;
subplot(2,2,3);
plot(t,x,'k','LineWidth',1.5);
xlabel('Samples');
ylabel('Noisy Signal (V)');
axis([0 max(t) min(x) max(x)]);
w=zeros(1,N);
for i=1:N,
   e(i)=d(i)-w(i)*x(i);
   w(i+1)=w(i)+mu*e(i)*x(i);
end
subplot(2,2,4),
plot(t,e,'k','LineWidth',1.5);
xlabel('Samples');
ylabel('Error');
axis([0 max(t) min(e) max(e)]);
figure(2)
subplot(2,2,1);
plot(t,w(2:length(w)),'k','LineWidth',1.5);
xlabel('Samples');
```

```
ylabel('Weight');
axis([0 max(t) min(w) max(w)]);
for i=1:N,
y(i)=w(i)*x(i);
end
subplot(2,2,2);
plot(t,y,'k','LineWidth',1.5);
xlabel('Samples');
ylabel('Correction Voltage');
axis([0 max(t) min(y) max(y)]);
subplot(2,2,3);
plot(t,(e+y),'k','LineWidth',1.5);
xlabel('Samples');
ylabel('O/P Voltage');
axis([0 max(t) min(e+y) max(e+y)]);
subplot(2,2,4);
plot(t,(d-(e+y)),'k','LineWidth',1.5);
xlabel('Samples');
ylabel({'Expected - O/P';'Voltage'});
axis([0 max(t) min(d-(e+y)) max(d-(e+y))]);
```

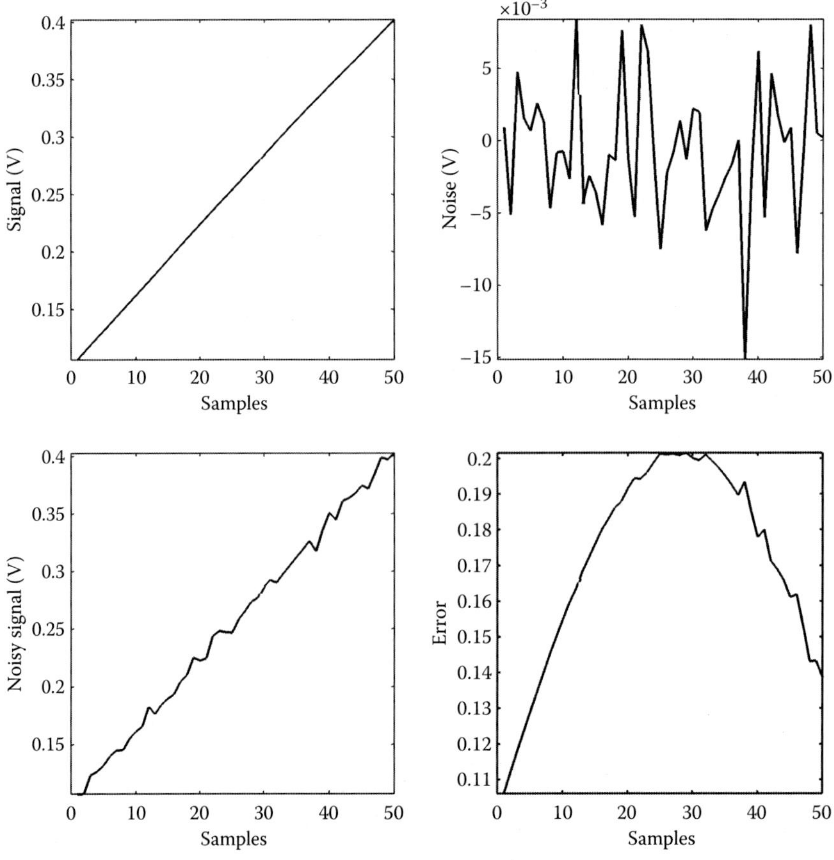

FIGURE 11.18
Actual and noise signal with error profile for the input values of 50 and 0.3.

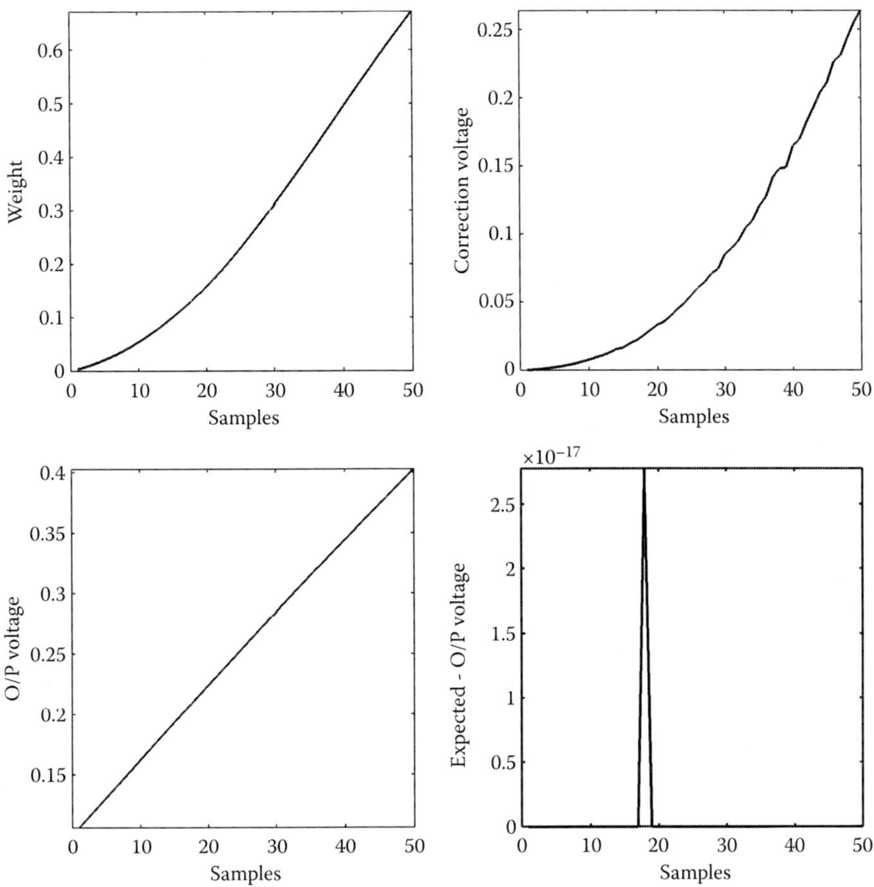

FIGURE 11.19
Profile of weight variations, correction voltage and recovered signal for the input values of 50 and 0.3.

A linear function, random noise used to corrupt the linear function, noise linear function and deviation errors are shown in Figure 11.18.

The profile of the weight function, correction values, equalized output signal and equalization accuracy in terms of error are shown in Figure 11.19.

DESIGN EXAMPLE 11.26

Write MATLAB code to realize a simple adaptive FIR filter whose output `y[n]` is `x[n]` W_0 + `x[n − 1]`W_1. Represent the actual signal by `d[n]` and the noise input signal by `x[n]`. Converge the output of the adaptive FIR filter close to the actual/desired signal using the LMS algorithm. Illustrate all the intermediate results along with the histogram of the actual and noise signals.

Solution

```
clc;
clear all;
N=input('length of sequence N is:= ');
mu=input('Value of step function is:= ');
figure(1);
```

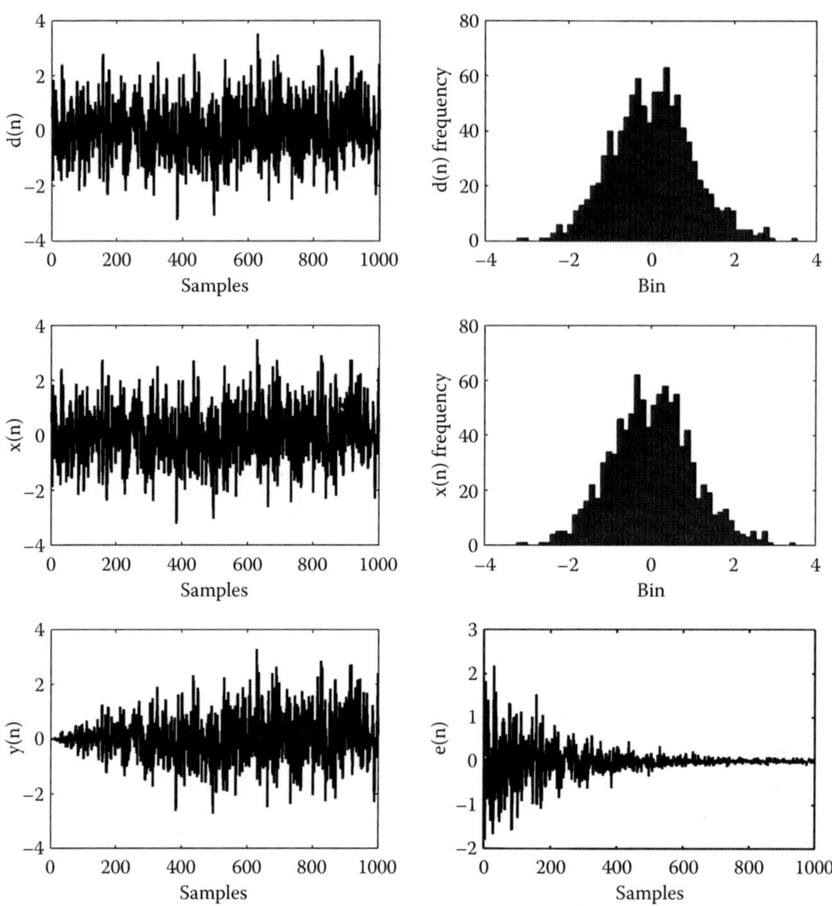

FIGURE 11.20
Desired and noise signal profile along with filter output and error function for the input values of 1000 and 0.005.

```
subplot(3,2,1);
d = 1.*randn(1,N);
plot(d,'k','LineWidth',1.5);
xlabel('Samples');
ylabel('d(n)');
xlim([0 N]);
subplot(3,2,2);
hist(d,50);
xlabel('Bin');
ylabel('d(n) Frequency');
subplot(3,2,3);
x = d + 0.04.*randn(1,N);
plot(x,'k','LineWidth',1.5);
xlabel('Samples');
ylabel('x(n)');
xlim([0 N]);
subplot(3,2,4);
hist(x,50);
xlabel('Bin');
```

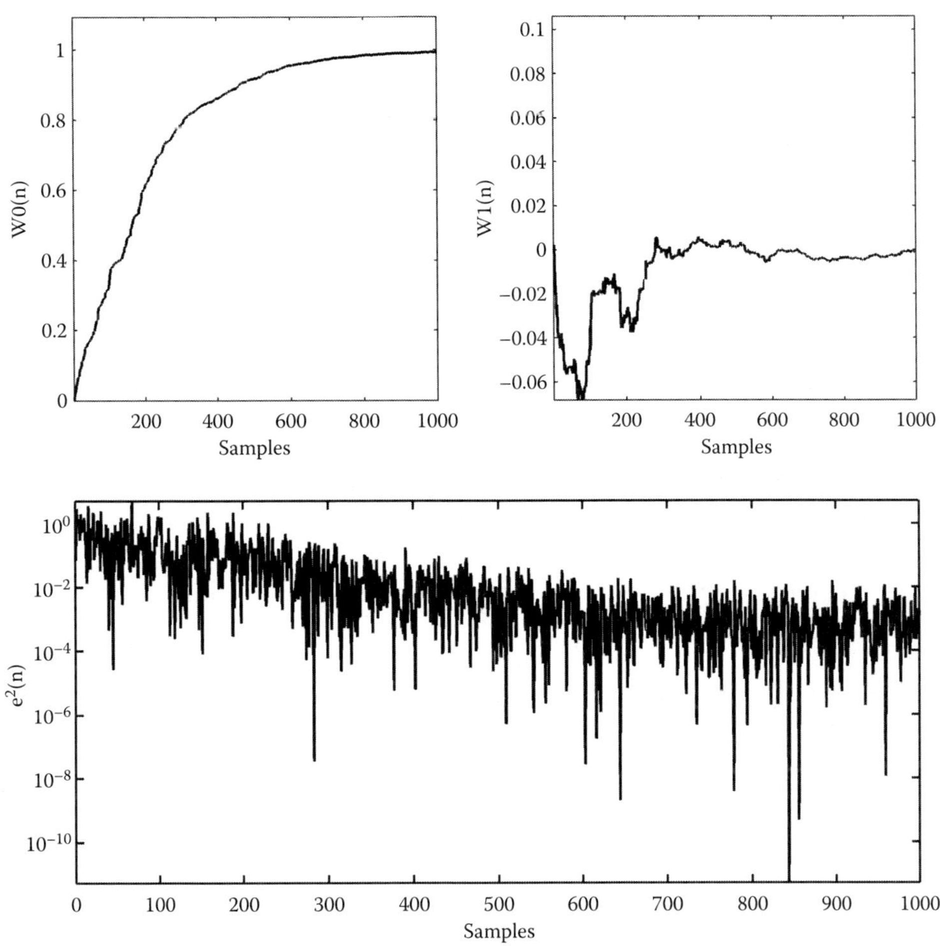

FIGURE 11.21
Weights (W0 and W1) and error-square profile for the input values of 1000 and 0.005.

```
ylabel('x(n) Frequency');
w0(1) = 0;
w1(1) = 0;
y(1) = w0(1)*x(1);
e(1) = d(1) - y(1);
w0(2) = w0(1) + mu*e(1)*x(1);
w1(2) = w1(1);
for n=2:N,
y(n) = w0(n)*x(n) + w1(n)*x(n-1);
e(n) = d(n) - y(n);
w0(n+1) = w0(n) + mu*e(n)*x(n);
w1(n+1) = w1(n) + mu*e(n)*x(n-1);
end
n = 1:N+1;
subplot(3,2,5)
plot(y,'k','LineWidth',1.5);
xlabel('Samples');
ylabel('y(n)');
xlim([0 N]);
```

```
subplot(3,2,6);
plot(e,'k','LineWidth',1.5);
xlabel('Samples');
ylabel('e(n)');
xlim([0 N]);
figure(2)
subplot(2,2,1)
plot(w0,'k','LineWidth',1.5);
xlabel('Samples');
ylabel('W0(n)');
axis([1 N min(w0) max(w0)+0.1]);
subplot(2,2,2)
plot(w1,'k','LineWidth',1.5);
xlabel('Samples');
ylabel('W1(n)');
axis([1 N min(w1) max(w1)+0.1])
subplot(2,2,3:4);
semilogy(e.^2,'k','LineWidth',1.5);
xlabel('Samples');
ylabel('e^2(n)');
ylim([min(e.^2) max(e.^2)]);
```

The simulation results of Design Example 11.26 are shown in Figures 11.20 and 11.21.

LABORATORY EXERCISES

1. Develop a digital system to realize FIR and IIR filters using different windowing techniques. Design MATLAB code to obtain the spectral characteristics of a real-time signal using DFT and FFT and get back the time domain signal using IDFT and IFFT, respectively.
2. Develop a digital system in the FPGA to realize a simple compressed sensing technique to read a sensor output analog signal sample and correlate it with a sufficient number of reference (correlator) signals of different frequency sufficient to the bandwidth of the sensor signal. Integrate the results at every sampling instant and store the results in the frequency domain.
3. Develop a RAM in the FPGA and store the digital values of 12 bits corresponding to the sine function. Generate a sinusoidal signal using a 12-bit bipolar DAC at different resolutions using a suitable multirate architecture. The frequency, phase and resolution of the sine function have to be selected using GPIOs.
4. Develop a digital system in the FPGA to perform modulo $(2^n \pm 1)$ addition.
5. Develop a digital system to perform a 10-tape FIR time-invariant filter. The coefficients of the filter are h[n] = {3, −2, 1, −3, −1}. The samples arrive at the FPGA through 15-bit GPIOs. Obtain the filter estimation using the RNA system. Write explicit VHDL code to illustrate the RNA-based parallel computation. Assume that the initial samples arriving at the FPGA are x[n] = {−4,−2,1,3,−3}. Display the filter output for each clock (shift) using five LEDs.
6. Develop a FIR filter using the DA computation technique for signed numbers x = {−12, −9, −5, 13} and c[n] = {−15,4,19, 6}.
7. Develop a digital system to implement an equation $y = (w_1a_1 + w_2a_2 + w_3a_3 + w_4a_4)/(w_5a_5 + w_6a_6 + w_7a_7)$. w and a are 8-bit data arriving at the FPGA from external inputs. Use Booth's multiplication algorithm to realize the above equation.

8. Develop a digital system in the FPGA to realize Figure 11.12 for the input sequence of `x[n] = {-3,-8,9,12,14,-4,-6,-9,-10}`. Realize these operations using different decimation and interpolation rate values and show the final result `y[n]` using a sufficient number of LEDs. Select the values for the decimation and interpolation using external input and an internal LUT.
9. Develop MATLAB code to study the equalization effect of an adaptive LMS-based equalizer for speech echo cancellation.
10. Design a digital architecture using the Xilinx System Generator tools to implement an LMS-based adaptive noise cancellation system.

12

Advanced SysGen-Based System Designs

12.1 Introduction and Preview

MATLAB®-Xilinx SysGen has several toolboxes for specific advanced designs and applications. For example, if the DSP toolbox is installed, we can see the FIR and IIR filter design blocks, which may be used to design lowpass, highpass, bandpass, or bandstop filters using the SysGen GUI window. To design a filter, one has to specify the passband with the maximum attenuation that can be allowed in that frequency band and the stopband with the minimum attenuation required for those frequencies. In this chapter, we discuss many advanced system designs using the MATLAB-Xilinx SysGen blocks. The advantages of FFTs are discussed, and a digital system design for computing the FFT is presented. FIR and IIR filter working principles and design using the standard design SysGen tools are detailed, along with design examples. Filter coefficient generation, simulation results, and application of the DA technique for FIR filter design are detailed. The design simplicities of the MAC FIR filter are explained, and design examples are presented.

Different design approaches to the CIC filter and associated expressions are explained, along with design examples and time-domain analysis. Realisation of the CORDIC architecture using SysGen blocks and conversion of polar to rectangular using the CORDIC architecture are described. Image processing using the DWT technique is also presented, with a design example. Different advanced design debugging techniques such as ChipScope Pro Analyzer, VHDL test-bench design and data/text file reading are detailed, with many design examples. Finally, design examples to interface with the 2×16 LCD in 4-bit mode are given. Most of the system designs presented in this chapter consist of SysGen blocks along with MATLAB Simulink tools. The function and interface details of the SysGen blocks used in the design examples are not explicitly explained in this chapter since they are easily available in many of the relevant online sources. However, the readers can find them in the System Generator user guide by Xilinx.

12.2 Fast Fourier Transform Computation Using SysGen Design

The DFT is today a widespread analysis tool used in all the branches of science and engineering. However, before the emergence of computers, the DFT was seldom used, as the DFT calculation of 32 samples requires 1024 operations of complex multiplication and summing. The function expansion algorithm in a trigonometric series, where the periodicity properties of trigonometric functions are used for accelerated analysis, used the DFT calculation. The first program realization of the FFT algorithm was carried out in the early 1960s of the 20th century by James Cooley in the IBM center under the supervision of John Tukey, and in 1965, they published an article devoted to the FFT algorithm [25,155]. From that point, the real FFT mania began. FFT has become the main tool of spectrum signal analysis. As seen in Chapter 11, consider the equation for the DFT:

$$S(k) = \sum_{n=0}^{N-1} s(n) e^{-j\frac{2\pi}{N}nk} \quad k = 0,1,\ldots, N-1 \tag{12.1}$$

The DFT associates N complex spectrum samples S(k), k = 0, 1, … , N − 1 with N signal samples s(n), n = 0, 1, … , N − 1. To calculate one spectrum sample, N operations of complex multiplication and summing are required. Thus, the computational complexity of the DFT algorithm consists of N^2 complex multiplication and summing. As the algorithm complexity grows in a quadratic correlation concerning the size of an input signal, it is possible to reach essential calculation acceleration if we manage to reduce the N-point DFT calculation to a two (N/2)-point DFT, as shown in Figure 12.1.

Replacement of one N-point DFT with two (N/2)-point DFTs will lead to halving the number of operations, but the operations of halving the sequence and association of two (N/2)-point DFTs into one N-point are additionally required. At the same time, each of the (N/2)-point DFTs can also be calculated by replacing the (N/2)-point DFT with two (N/4)-point DFTs, which in their turn can be calculated by means of the (N/8)-point DFT. This recursion can be continued; so far it is possible to divide the input sequence into two. In general, if $N = 2^L$, where L is a positive integer, then we can halve the sequence L times; for example, if $N = 8 = 2^3$, then L = 3 and three stages of dividing are possible. The FFT algorithms which use selections with length $N = 2^L$ are called radix-2 FFT algorithms. These algorithms have gained the greatest distribution because of their efficiency and the relative simplicity of the program realisation. We will consider two ways of dividing (./.) and conquering (+), that is, decimation in time and decimation in frequency, while reducing the points. The efficient direct FFT calculation algorithm can be used for the inverse transform, too. The complex exponents in the equations for the direct and inverse DFTs are complex conjugated

$$e^{j\frac{2\pi}{N}nk} = e^{-j\frac{2\pi}{N}nk} \tag{12.2}$$

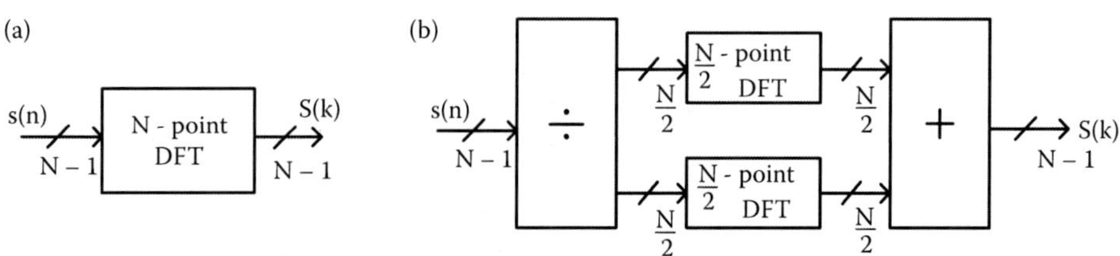

FIGURE 12.1
Representation of N- (a) and N/2- (b) point DFT.

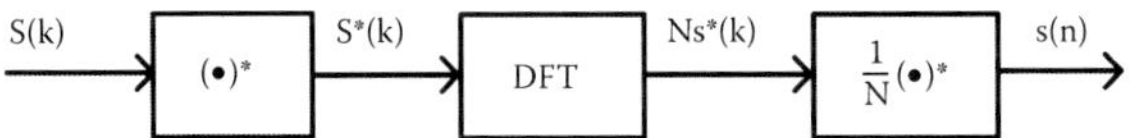

FIGURE 12.2
Representation of calculation of IFFT.

where (·)* is a complex conjugate operator. Further, in connection with the IDFT equation, it is possible to write:

$$s(n) = \frac{1}{N}\left(\sum_{k=0}^{N-1} S^{*}(k)e^{-j\frac{2\pi}{N}nk}\right) \quad (12.3)$$

Thus, take the complex conjugated spectrum $S^{*}(k)$. The direct DFT is carried out and the result is exposed to the complex conjugate. The IDFT calculation when using the DFT is shown in Figure 12.2.

If, instead of the DFT, we use the FFT, we will get the IFFT. Using it, to carry out the complex conjugate, it is necessary only to change a sign before an imaginary spectrum part before the FFT function call and the result after the FFT. Thus, we have considered an acceleration method of calculations upon calculating the DFT, and also converted the IFFT to the direct one.

DESIGN EXAMPLE 12.1

Design and develop a digital system to compute the FFT of signals that are generated using DDS 5.0. Use FFT 8.0 blocks and other SysGen blocks as required to design the digital system [143,155,156].

Solution

The main design and two submodule designs are shown in Figures 12.3 through 12.5.

Main Design

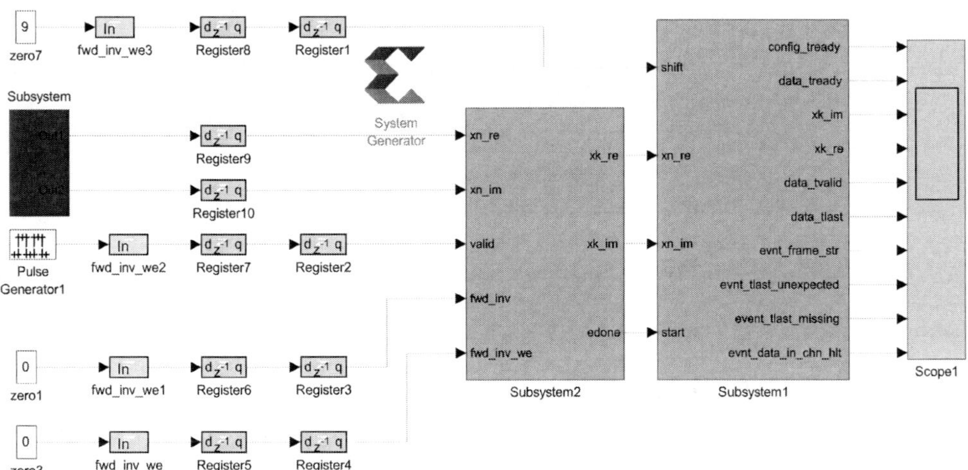

FIGURE 12.3
Main design module with two FFT blocks.

Subsystem2 Design

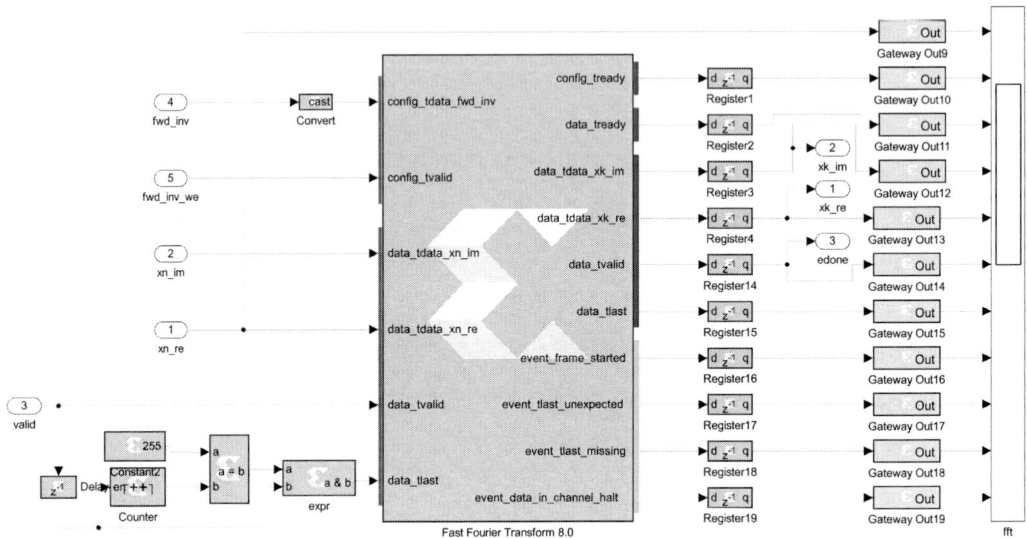

FIGURE 12.4
First subsystem design with one FFT 8.0 core.

Subsystem1 Design

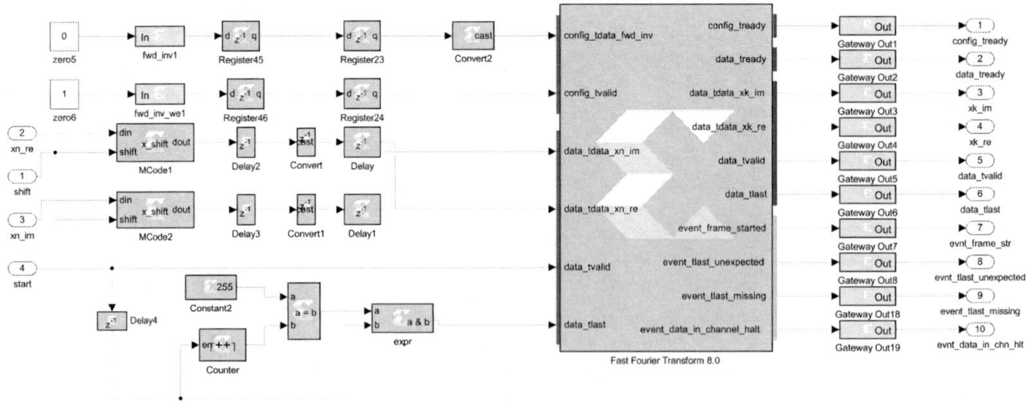

FIGURE 12.5
Second subsystem design with one FFT 8.0 core.

12.3 Finite Impulse Response Digital Filter Design

The Xilinx FDATool block provides an interface to the MATLAB signal-processing toolbox. The block does not function properly and should not be used if the MATLAB signal-processing toolbox is not installed. This block provides a means of defining an

FDATool object and storing it as a part of a SysGen model, that is, an .mdl, file. The FDATool provides a powerful means of defining digital filters with a graphical user interface. Double-clicking the block icon, which is shown in Figure 12.6, opens up an FDATool session and graphical user interface, which is shown in Figure 12.7. The filter is stored in a data structure internal to the FDATool interface block, and the coefficients can be extracted using MATLAB helper functions provided as part of SysGen design. The function call `>>xlfda_numerator('FDATool')` returns the numerator of the transfer function, that is, the filter's coefficients. Similarly, the helper function `>>xlfda_denominator('FDATool')` retrieves the denominator for a non-FIR filter. The Xilinx FIR Compiler 5.0 block implements a MAC- or DA-based FIR filter. It accepts a stream of input data and computes filtered output with a fixed delay based on the filter configuration.

Now, we will discuss designing a FIR filter using this tool. Open the System Generator under the Simulink environment in MATLAB to implement a FIR digital filter, that is, a type-I FIR lowpass filter of the sixth order. We can use multiple MAC units to implement the filter design. A separate MAC unit has to be used for every filter coefficient multiplication and accumulation. The filter coefficients can be obtained using the FDA Xilinx tool. Assume that in our filter design, two sine wave sources, shown in Figure 12.6, are sampled at 1 kHz; the frequencies are low and high, 5Hz and 300 Hz, and it has 16-bit input: signed data (2's complement) binary point 13 and sampling period of 0.001 sec. The multiplier block latency is 1 with full precision. The simulation parameters are: Stop time 0.5 seconds. Configure the inputs by double-clicking the Gateway In blocks to open their parameter dialog boxes, set the number of bits to 16 and the binary point to 14, set the overflow to saturate, and the sample period to 0.001 sec. Upon completing the design for six stages, as shown in Figure 12.6, we can click the FDATool block. Design the six-order lowpass filter by double-clicking on the FDATool block. For response type, choose lowpass; set the design method to FIR and from the bottom-down menu, select least-squares. Set the filter order to 6. For frequency specifications, select normalized for the units and set the wpass to 0.1 and the wstop to 0.25. Click on the design filter button. Click on the file menu → export, and the export window appears. Choose export to workspace, set the variable names to filter_coeff and click on the export button.

In the MATLAB workspace, double-click the filter_coeff variable. Click the view menu → numerical array format → long in MATLAB. Double-click the Xilinx constant block and set the number of bits to 16, the binary point to 14 and the signed (2's complement) type of data. In the same block, edit the constant value to the values of the filter_coeff variable according to their position in the array; we have to do this for the other five coefficients. Do not change the order of the coefficients, since each is associated with a specific delay value, and keep in mind that MATLAB indexes start at 1, which corresponds to the first coefficient value which has no delay, that is, c_0. Double-click the System Generator block and set the Simulink system period to 1/1000. Save the design in our work directory as lowpass.mdl and run the simulation by clicking on the start simulation button. The graphs of the input and output should look like Figures 12.6 and 12.7.

DESIGN EXAMPLE 12.2

Design and develop a digital system to realize a lowpass FIR filter using the Xilinx SysGen tools.

Solution

Main Design

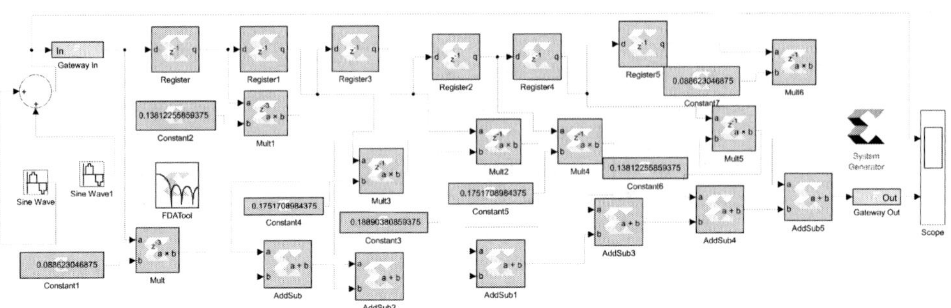

FIGURE 12.6
Filter design using FDA core.

FDA Tool Design

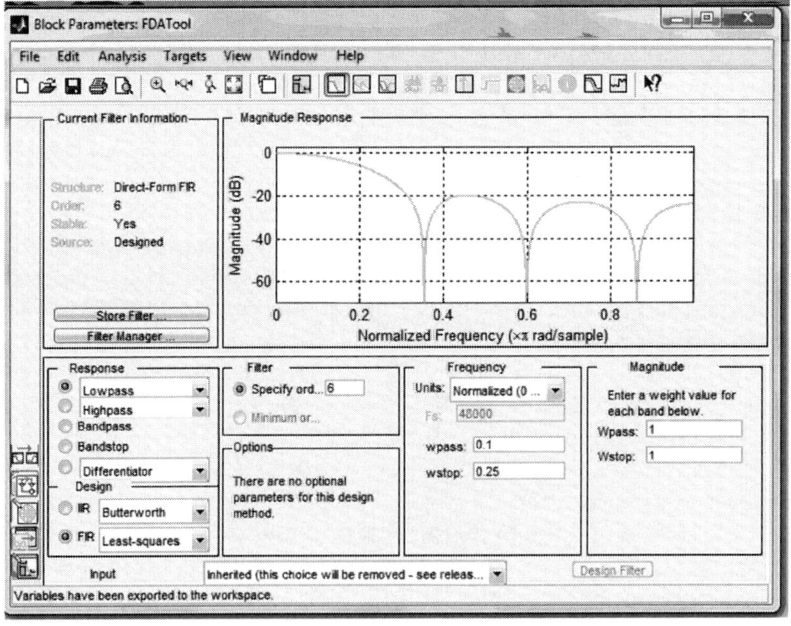

FIGURE 12.7
Filter design using FDA core.

Upon running the FDA tool shown in Figure 12.7, the filter coefficients, given in Table 12.1, can be obtained.

TABLE 12.1
Filter Coefficients

Coefficient (C_0)	Coefficient (C_1)	Coefficient (C_2)
0.0886435639221770	0.138113513306075	0.175192865474914
Coefficient (C_3)	**Coefficient (C_4)**	**Coefficient (C_5)**
0.138113513306075	0.188946582810158	0.175192865474914

Filter Simulation Result

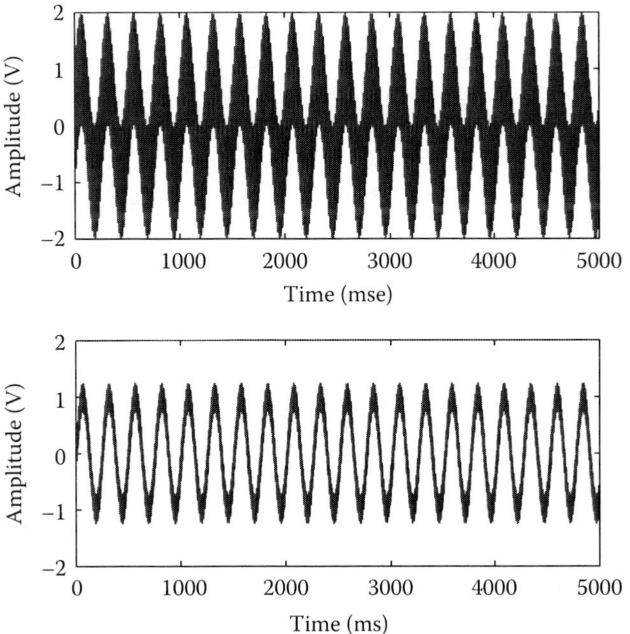

FIGURE 12.8
Time-domain result of the lowpass filter.

The simulation result of Design Example 12.2 is shown in Figure 12.8.

DESIGN EXAMPLE 12.3

Design and develop a digital system to realize a FIR bandpass filter. Choose the pass-band 300 KHz to 450 KHz. Show the FDA tool design and response of the designed filter.

Solution

Main Design

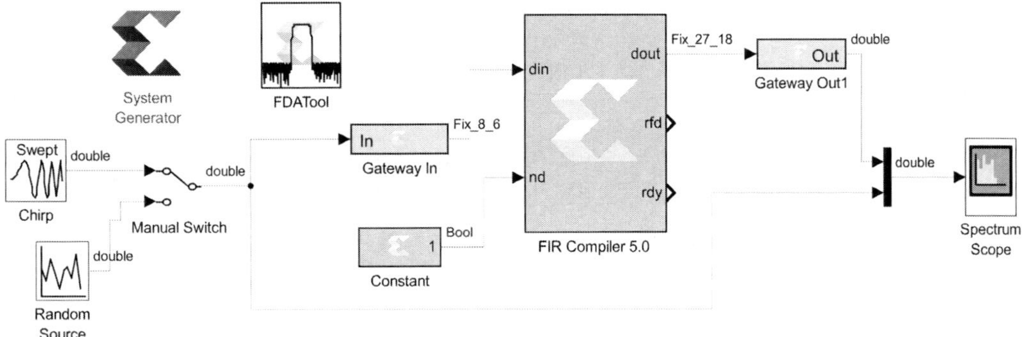

FIGURE 12.9
Filter design using FDA core.

FDA Tool Design

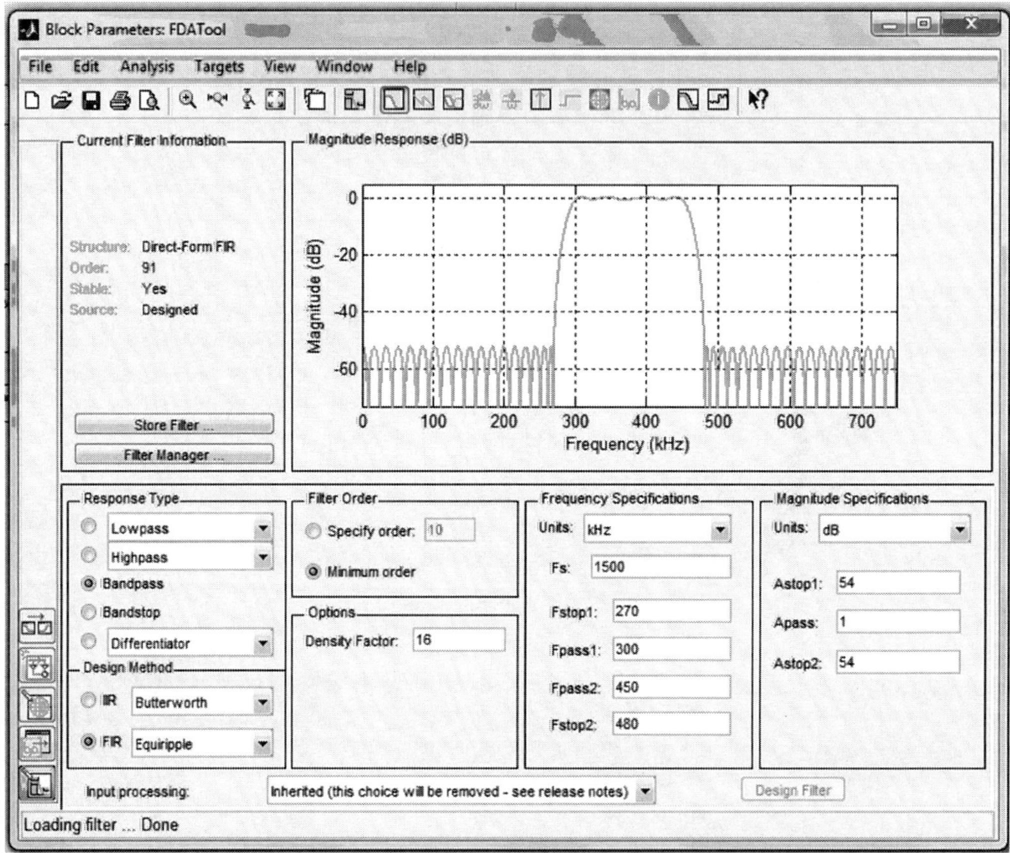

FIGURE 12.10
Filter design using FDA core.

Upon simulating the design shown in Figure 12.9, the following filter coefficients can be obtained.

The filter coefficients (C_0 through C_{91}) for this design are [−0.0014 0.0025 0.0035 −0.0035 −0.0048 0.0043 0.0050 −0.0031 −0.0028 −0.0005 −0.0017 0.0060 0.0073 −0.0113 −0.0114 0.0138 0.0116 −0.0115 −0.0066 0.0041 −0.0025 0.0056 0.0118 −0.0133 −0.0164 0.0142 0.0127 −0.0059 −0.0001 −0.0096 −0.0170 0.0258 0.0304 −0.0337 −0.0309 0.0250, 0.0124 0.0036 0.0251 −0.0484 −0.0741 0.0986 0.1218 −0.1404 −0.1541 0.1610 0.1610 −0.1541 −0.1404 0.1218 0.0986 −0.0741 −0.0484 0.0251 0.0036 0.0124 0.0250 −0.0309 −0.0337 0.0304 0.0258 −0.0170 −0.0096 −0.0001 −0.0059 0.0127 0.0142 −0.0164 −0.0133 0.0118 0.0056 −0.0025 0.0041 −0.0066 −0.0115 0.0116 0.0138 −0.0114 −0.0113 0.0073 0.0060 −0.0017 −0.0005 −0.0028 −0.0031 0.0050 0.0043 −0.0048 −0.0035 0.0035 0.0025 −0.0014]

Filter Simulation Result

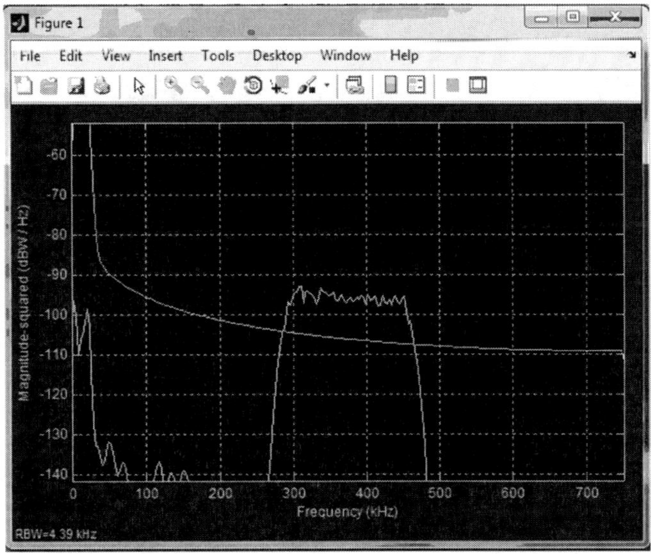

FIGURE 12.11
Time-domain simulation result of the designed bandpass FIR filter.

The simulation results of Design Example 12.3 are shown in Figures 12.9 through 12.11.

DESIGN EXAMPLE 12.4

Design a second-order FIR filter using distributive arithmetic logic. The coefficient of the filter is `c[n]={7,5,6}`. Assume that 3 bits are required to represent the magnitude of the samples `x[n]`.

Solution

```
library ieee;
use ieee.std_logic_1164.all;
use ieee.numeric_std.all;
use ieee.std_logic_unsigned.all;
entity dist_fir is
port(x:in std_logic_vector(2 downto 0);
nrst,clk:in std_logic;
y:out std_logic_vector(6 downto 0));
end entity;
architecture mark0 of dist_fir is
type lut is array (0 to 7) of integer range 0 to 50;
type reg is array (2 downto 0) of std_logic_vector(2 downto 0);
signal filt:lut;
signal xreg:reg;
signal m0,m1,m2:std_logic_vector(2 downto 0);
signal a0,a1,a2:integer range 0 to 7;
signal l0,l1,l2:std_logic_vector(5 downto 0);
signal s0,s1,s2:std_logic_vector(5 downto 0);
begin
-- controlling of x[n] --
p0:process(clk,nrst)
begin
if(nrst='0') then
```

```
xreg <= (others=>"000"); -- initialize with 0
filt <= (0,7,5,12,6,13,11,18);  -- generate LUT c[n] = {7,5,6}
else
if(falling_edge(clk)) then              -- shift of x register
xreg(2) <= x;
xreg(1) <= xreg(2);
xreg(0) <= xreg(1);
end if;
end if;
end process; -- FIR filter --
m0 <= xreg(0)(0) & xreg(1)(0) & xreg(2)(0);    -- take bitwise values
m1 <= xreg(0)(1) & xreg(1)(1) & xreg(2)(1);
m2 <= xreg(0)(2) & xreg(1)(2) & xreg(2)(2);
a0 <= to_integer(unsigned(m0));                -- convert to address
a1 <= to_integer(unsigned(m1));
a2 <= to_integer(unsigned(m2));
l0 <= std_logic_vector(to_unsigned(filt(a0),6));      -- resolve the
values
l1 <= std_logic_vector(to_unsigned(filt(a1),6));
l2 <= std_logic_vector(to_unsigned(filt(a2),6));
s0 <= l0;                               -- multiply with 2⊥b
s1 <= l1(4 downto 0) & '0';
s2 <= l2(3 downto 0) & "00";
y <= ('0'&s0) + ('0'&s1) + ('0'&s2);   -- final filter output
end architecture;
```

12.4 Infinite Impulse Response Digital Filter Design

One type of digital filter is the IIR filter, which is generally used at lower sample rates, that is, <200 kHz. The IIR uses feedback in order to compute outputs; hence, it is known as a recursive filter. The advantages of the IIR filter are (i) better magnitude response, (ii) fewer coefficients, (iii) less storage required for storing variables, (iv) a lower delay, and (v) it is closer to analog models. Two popular types of different classical IIR filters are Butterworth and Bessel filters. The Butterworth filter provides the best approximation of the ideal low-pass filter response at analog frequencies. Passband and stopband response is maximally flat. Analog Bessel lowpass filters have a maximum flat group delay at zero frequency and retain nearly constant group delay across the entire passband. Filtered signals therefore maintain their wave shapes in the passband frequency range [155,156]. Frequency-mapped and digital Bessel filters, however, do not have this maximum flat property. Bessel filters generally require a higher filter order than other filters for satisfactory stopband attenuation. In the IIR filter, the output is fed back to make a contribution in generating the output.

The expression for the IIR filter is given in Equation 12.4; note that a delayed version of the output, that is, `y(n)`, plays a part in generating the output

$$y(n) = \sum_{i=0}^{n} b(i)x(n-i) - \sum_{i=0}^{n} a(i)y(m-i) \tag{12.4}$$

where `a(i)` and `b(i)` are the coefficients of the IIR filter. Another way to express an IIR filter is as a transfer function with numerator coefficients (b_i) and denominator coefficients (a_i) as

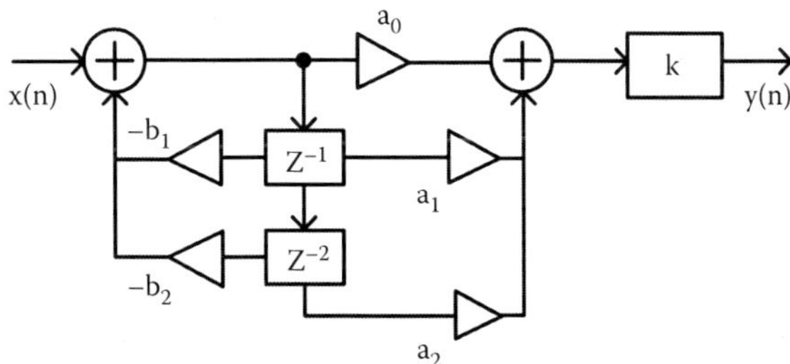

FIGURE 12.12
Direct Form-II representation of a biquad IIR filter.

$$H(z)=\frac{\sum_{i=0}^{n} b_i z^{-i}}{1+\sum_{i=0}^{n} a_i z^{-i}} \tag{12.5}$$

In the direct form-I architecture, the forward and reverse FIR filter stages can be swapped, which creates a center consisting of two columns of delay elements. From this, one column can be formed; hence, this type of structure is also known as canonical, meaning it requires the minimum amount of storage (Figure 12.12).

The Biquad filter structure is that of a direct form-II, but it includes a second-order numerator and denominator coefficient; that is, it is simply two poles and two zeros. This structure is used in FPGA/DSP implementations, because it is not terribly sensitive to quantization effects. The Butterworth biquad filter's transfer function is given by

$$H(z)=\frac{1+2z^{-1}+z^{-2}}{1+b_1z^{-1}+b_2z^{-2}} \tag{12.6}$$

The coefficients in the denominator simplify the calculations such that no multiplier is needed and the entire denominator can be calculated using shift and accumulators. As we have seen in the previous chapters, multiplications are expensive operations in FPGA/DSP implementations. Several issues must be examined in detail to ensure satisfactory fixed-point operation of the IIR filter. The main issues are coefficient/internal quantization, wraparound/saturation and scaling. In order to examine the effect of quantization, it is useful to look at the pole/zero plot [155,156]. This shows how the zeros, that is, depths in the frequency response plot, and poles, that is, peaks in the frequency response plot, are positioned. In fact, an issue with IIR stability relates to the denominator coefficients and their positions as poles on the pole/zero plots.

The poles for the floating-point version of the plot are within the unit circle, that is, the values of the coefficients are less than 1. Once the coefficients are quantized, these poles move, which affects the frequency response. If they move onto the unit circle, that is, the poles approach 1, we will potentially have an oscillator. A fixed-point implementation has a certain bit width and hence has a range. Calculations may cause the filter to exceed its maximum/minimum ranges. For example, let us consider a 2's complement value of "01111000", that is, +120 + "00001001", that is, +9 = "10000001" = –127. The large positive number becomes a large negative number; this is known as wraparound and it can cause large errors. There are two methods of dealing with overflows; if scaling is used, values can never overflow. DSP processors tend to use different kinds of scaling in order to fit within their fixed structure and use saturation logic. The main artefacts of IIR filtering are the quantization noise and the limit cycle oscillations. The truncation or rounding of the IIR accumulator at the output of filter creates quantization noise. This noise is fed into the filter recursive path. The noise is multiplied by the IIR recursive path transfer function. The impact of this noise source is very significant in the lower cut-off frequency filters of the second order, since the recursive path gain is proportional to the second power of the F_c/F_s ratio. The filter stages with high Q can also suffer from this effect because the gain is proportional to Q [155,156].

DESIGN EXAMPLE 12.5

Design and develop a digital system to realize an IIR filter using the SysGen core blocks (Figure 12.13).

Solution

Main Design

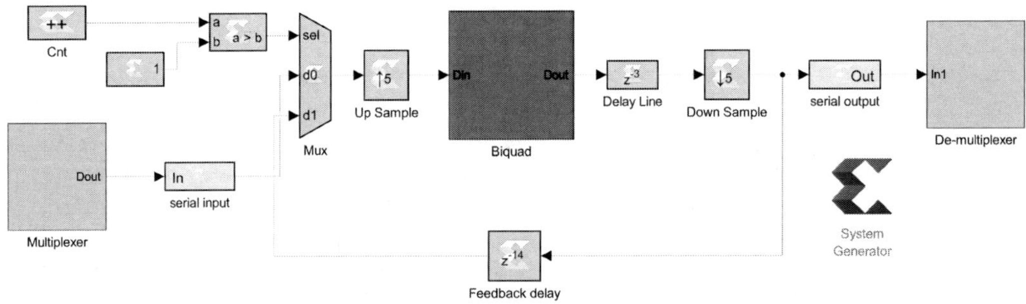

FIGURE 12.13
Top-level design of an IIR filter.

The subsystems of the multiplexer, biquad, demultiplexer and operation and interface protocols can be found in [156].

12.5 Multiply Accumulation Finite Impulse Response Filter Using SysGen Design

FIR filters are widely used in most DSP applications. Many high-sampling-rate FIR filters are employed in different applications. However, in some applications, filter circuits with sampling rates using low power are desired. Either to increase the efficient speed or to reduce the power consumption of the original digital filter, parallel processing techniques are applied to FIR filters. For many years, parallel processing has been applied to an FIR filter that uses the units of hardware with replications that exist in the original filter. The choice of the multiplier circuit also affects the resultant power consumption. If the multiplier is chosen wisely with fewer calculations, both speed and power can be optimized. Multipliers consume considerable power, occupy large areas and take long latency. The MAC unit plays a vital role in different DSP applications, including converters, removing undesired components, inner products, some nonlinear function implementations such as discrete cosine transform (DCT), and DWT and so on. They are mainly proficient in multiplication and addition applications. The reason is due to the enhanced proficiency of circular execution of arithmetic operations, like multiplication and addition, by improving the performance in terms of aggregate execution rate. The MAC unit consists of a multiplier unit, along with an accumulator unit to sum the previous successive products. The input of a multiply and accumulate unit is retrieved from the memory location and fed to the multiplier block. In a MAC unit, as soon as the input value, in 16-bit format, is given, the multiplier initiates the computation from the 16-bit information and generates 32-bit information as output. This output is fed as an input to the carry skip adder to perform fast addition operation. Actually, the carry bypass adder generates 33-bit information, of which 32 bits are the sum and 1 bit is carry. These data are given to the accumulator as input and are loaded in the accumulator register using PIPO shifting. Because the carry skip adder uses PIPO, it produces all the possible values simultaneously as output. The output of the accumulator register is fed back or received as an input to the carry skip adder. The basic architecture of a MAC unit is presented in Figure 10.12.

System Generator is a high-level design tool well suited to creating custom DSP data paths in FPGAs. While providing a high level abstraction of an FPGA circuit, it can be used to build designs comparable to hand-crafted implementations in terms of area and performance [155,156]. In this section, we discuss a MAC-based FIR filter design to understand the interplay between mathematical abstraction and hardware-centric considerations enabled by System Generator tools. We demonstrate how an algorithm can be efficiently mapped onto FPGA resources and present the hardware results of several System Generator FIR filter implementations. System Generator is a high-level design tool for Xilinx FPGAs that extends the capabilities of MATLAB Simulink to include bit- and cycle-accurate modeling of FPGA circuits and generation of an FPGA circuit from it.

Virtex family FPGAs provide dedicated circuitry for building fast, compact adders, multipliers and flexible memory architectures in the logic fabric. The Addressable Shift Register (ASR) block abstracts the SRL16 memory configuration with the capability of running delay and address inputs at different rates. System Generator multiplier blocks provide options for combinational and pipelined structures built in the FPGA fabric and embedded multipliers. A versatile FIR filter architecture that maps well onto FPGA resources employs MAC engines. In general, N-MAC operations are required to compute output samples for an N-tap FIR filter. A System Generator model of a single-MAC FIR filter is shown in Figure 12.14. The tapped delay line is implemented using an ASR, which

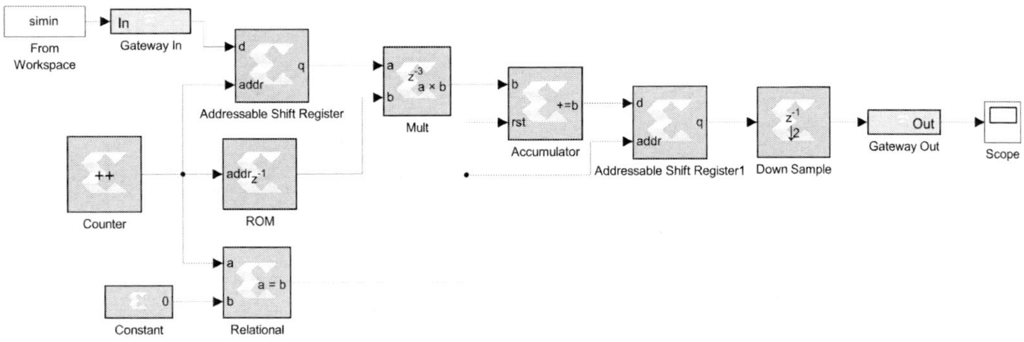

FIGURE 12.14
Design of a MAC FIR filter.

provides both a compact design and simple addressing requirements. The delay line runs at the data rate, but the rest of the filter, including the address port of the ASR, runs at N times this rate. The multiplier of the MAC computes the product of a filter tap and a sample from the data buffer, and the accumulator computes a running sum of these products. The accumulator is configured to reinitialize upon reset to its current input value to avoid a one-clock cycle stall at the end of each SOP computation. A register captures the output of the MAC engine before it is reset. The capture register output is down-sampled so the filter output rate matches its input rate. The filter coefficient ROM is initialized from a MATLAB array bound to the model from the MATLAB workspace. The implications of this ability in System Generator to exploit the MATLAB interpreter during model customization should not be underestimated.

A single counter, configured to count from 0 to N – 1 repeatedly, generates addresses for both the coefficient ROM and input data ASR block. The counter's output sample period defines the rates of the downstream data and coefficient buffers using Simulink's sample time propagation rules. Since the filter requires addresses to change at N times the input data rate, the sample period is 1. For every new input sample, the accumulator block is reset to its current input and the capture register latches the MAC engine output. This occurs when the address is zero; a relational block detects the condition. A single MAC architecture has the drawback that throughput is inversely proportional to the number of filter taps. Throughput can be increased dramatically by exploiting parallelism. The tapped delay line of the single-MAC architecture can be partitioned into cascaded sections, each serviced by a separate MAC engine, whose outputs are combined in an adder tree to compute the filter output [155,156].

DESIGN EXAMPLE 12.6

Design and develop a digital system to realize a MAC unit–based FIR filter (Figure 12.15).

Solution

Main Design

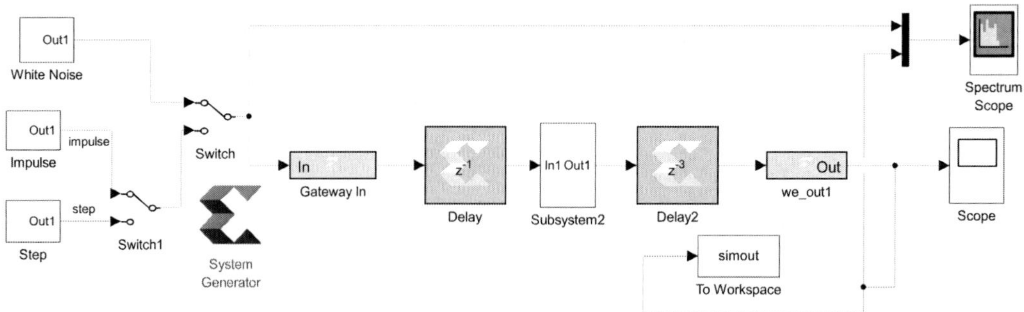

FIGURE 12.15
Top-level design of a MAC-FIR filter.

The internal design of Subsystem2, results of spectrum analysis, scope, simout and operation and interface techniques of all the SysGen blocks can be found in References 155 and 156.

12.6 Cascaded Integrator Comb Filter Design

The previously obscure CIC filter is now vital to many high-volume wireless communication tasks and equipment. Using CIC filters, we can cut costs, improve reliability and get improved performance. CIC digital filters are computationally efficient implementations of narrow-band lowpass filters and are often embedded in hardware implementations of decimation and interpolation in modern communications systems. CIC filters were introduced to the signal-processing community by Eugene Hogenauer more than two decades ago, but their application possibilities have grown in recent years [155–157]. Improvements in chip technology, the increased use of polyphase filtering techniques, advances in delta-sigma converter implementations, and the significant growth in wireless communications have all spurred much interest in CIC filters. While the behavior and implementation of these filters is not complicated, their coverage has been scarce in the literature of embedded and VLSI systems. CIC filters are well-suited for anti-aliasing filtering prior to decimation, that is, sample-rate reduction, and for anti-imaging filtering for interpolated signals, that is, sample-rate increase.

The frequency response of this filter looks like a (sin(x)/x) function; thus, CIC filters are typically followed by higher-performance linear-phase low-pass tapped-delay-line FIR filters whose tasks are to compensate for the CIC filter's nonflat passband. We can greatly reduce the computational complexity of narrow-band lowpass filtering using decimation relative to a single lowpass FIR filter. In addition, the filter operates at reduced clock rates, minimizing power consumption in high-speed hardware applications. A very important advantage in using CIC filters is that they require no multiplication. The arithmetic needed to implement these digital filters is strictly addition and subtraction only.

The CIC filter originates from the notion of a recursive running-sum filter, shown in Figure 12.16a, which itself is an efficient form of a nonrecursive moving averager, as

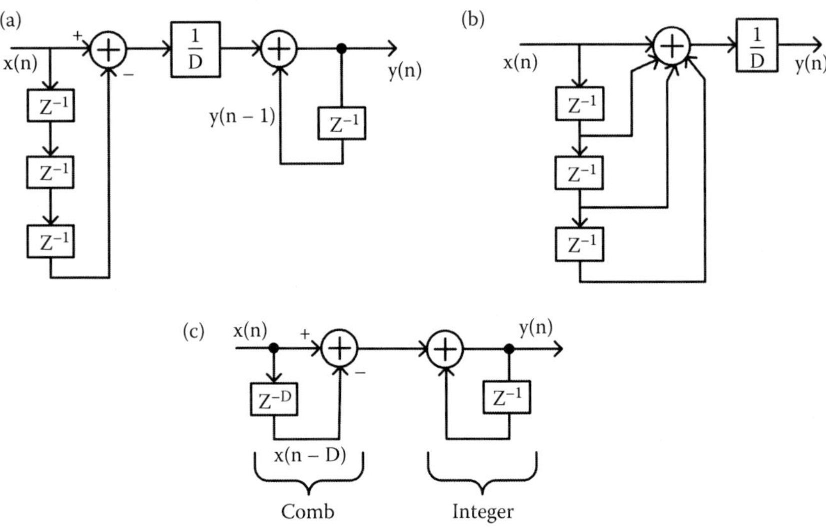

FIGURE 12.16
Structure of moving average (a), recursive running sum (b), and CIC (c) filters.

shown in Figure 12.16b. The D-point moving-average filter's output in the time domain is expressed as

$$y[n] = \frac{1}{D}[x[n] + x[n-1] + x[n-2] + \cdots + x[n-D+1]] \tag{12.7}$$

where n is our time-domain index; then, as we know, the z-domain expression for this moving averager is

$$Y(z) = \frac{1}{D}[X(z) + X(z)z^{-1} + X(z)z^{-2} + \cdots + X(z)z^{-D+1}] \tag{12.8}$$

While its z-domain transfer function (H(z)) is

$$H(z) = \frac{Y(z)}{x(z)} = \frac{1}{D}[1 + z^{-1} + z^{-2} + \cdots + z^{-D+1}] = \frac{1}{D}\sum_{n=0}^{D-1} z^{-n} \tag{12.9}$$

Equation 12.7 shows how to build a moving averager, and Equation 12.9 is in the form used by commercial signal-processing software to model the frequency-domain behavior of the moving averager. Next, we examine the CIC filter in the form of the moving averager, as given in Equations 12.7, 12.8, and 12.9. There, we see that the current input sample `x(n)` is added and the oldest input sample `x(n - D)` is subtracted from the previous output average `y(n - 1)`. It's called recursive because it has feedback. Each filter output sample is retained and used to compute the next output value. The recursive running-sum filter's difference equation is given by

$$y(n) = \frac{1}{D}[x(n) - x(n-D) + y(n-1)) \tag{12.10}$$

with a z-domain, H(z) is

$$H(z) = \frac{1}{D}\frac{1-z^{-D}}{1-z^{-1}} \tag{12.11}$$

We use the same H(z) variable for the transfer functions of the moving-average filter and the recursive running-sum filter because their transfer functions are equal to each other. Equation 12.9 is the nonrecursive expression and Equation 12.11 is the recursive expression for a D-point averager. The standard moving averager, as in Figure 12.16, must perform D – 1 additions per output sample. The recursive running-sum filter has the significant advantage that only one addition and one subtraction are required per output sample regardless of the delay length D. This computational efficiency makes the recursive running-sum filter attractive in many applications seeking noise reduction through averaging. If we condense the delay-line representation and ignore the 1/D scaling in Figure 12.16, we obtain the classic form of a first-order CIC filter, whose cascade structure is shown in Figure 12.16c. The feed-forward portion of the CIC filter is called the comb section, whose differential delay is D, while the feedback section is typically called an integrator. The comb stage subtracts a delayed input sample from the current input

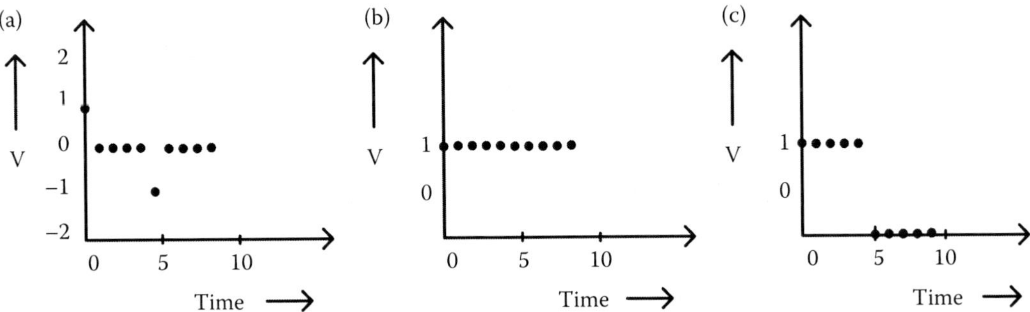

FIGURE 12.17
Time-domain impulse responses of single-stage CIC filter: comb (a), integrator (b), and CIC (c).

sample, and the integrator is simply an accumulator. The CIC filter's difference equation is given by

$$y(n) = x(n) - x(n-D) + y(n-1) \tag{12.12}$$

and its z-domain transfer function is given by

$$H_{CIC}(z) = \frac{1-z^{-D}}{1-z^{-1}} \tag{12.13}$$

Now, we examine the time-domain behavior of a CIC filter for D = 5, as shown in Figure 12.17. If a unit-impulse-sequence followed by many zero-valued samples were applied to the comb stage, the output will be as shown in Figure 12.17a. The initial positive impulse from the comb filter starts the integrator's all-ones output as in Figure 12.17b. After D samples, the negative impulse from the comb stage arrives at the integrator to zero all further CIC filter output samples, as in Figure 12.17c.

The key issue is that the combined unit-impulse response of the CIC filter, being a rectangular sequence, is identical to the unit-impulse responses of a moving-average filter and the recursive running-sum filter. The frequency magnitude and linear-phase response of a D = 5 CIC can be found in [155–157]. We can obtain an expression for the CIC filter's frequency response by evaluating the transfer function, as in Equation 12.14, on the z-plane's unit circle by setting $z = e^{j2\pi f}$

$$H_{CIC}(e^{j2\pi f}) = \frac{1-z^{-j2\pi fD}}{1-e^{-j2\pi f}} = e^{-j2\pi f(D-1)/2}\frac{\sin(\pi fD)}{\sin(\pi f)} \tag{12.14}$$

If we ignore the phase factor $(e^{-j2\pi f(D-1)/2})$ in Equation 12.14, then the ratio of sin(.) terms can be approximated by a sin(x)/x function. This means the CIC filter's frequency magnitude response is approximately equal to a sin(x)/x function centered at 0Hz. CIC filters are primarily used for anti-aliasing filtering prior to decimation and for anti-imaging filtering for interpolated signals. With these things in mind, we swap the order of Figure 12.16c's comb and integrator and include decimation by a sample rate change factor R as in Figure 12.18. In most CIC filter applications; the rate change R is equal to the comb's differential delay D.

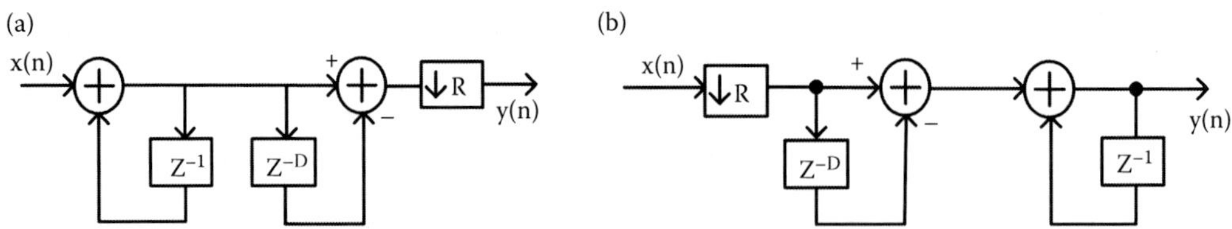

FIGURE 12.18
Structure of single-stage CIC filters with decimation (a) and interpolation (b).

The decimation operation ↓R means to discard all but every Rth sample, resulting in an output sample rate of f_s ($f_{s,in}$/R). Figure 12.18a shows the frequency magnitude response of a CIC filter when D = 8 prior to decimation. The spectral band of width B centered at 0Hz is the desired passband of the filter. A key aspect of CIC filters is the spectral folding that takes place due to decimation. Those B-width shaded spectral bands centered about multiples of $f_{s,in}$/R, as in Figure 12.18a, will alias directly into our desired passband after decimation by R = 8. Notice how the largest aliased spectral component in this example is roughly 16dB below the peak of the band of interest. Of course, the aliased power levels depend on the bandwidth B, as the smaller B is, the lower the aliased energy after decimation. Figure 12.18b shows a CIC filter used for interpolation where the ↑R symbol means insert R^{-1} zeros between each `x(n)` sample, yielding a `y(n)` output sample rate of f_s ($Rf_{s,in}$). In this CIC filter, interpolation is defined as zero-insertion followed by filtering. The most common method to improve CIC filter anti-aliasing and image-reject attenuation is by increasing the order D of the CIC filter using multiple stages. More details and further analysis, implementation and applications of CIC filters can be found in References 155–157.

DESIGN EXAMPLE 12.7

Design and develop a simple CIC filter using the SysGen tools (Figure 12.19).

Solution

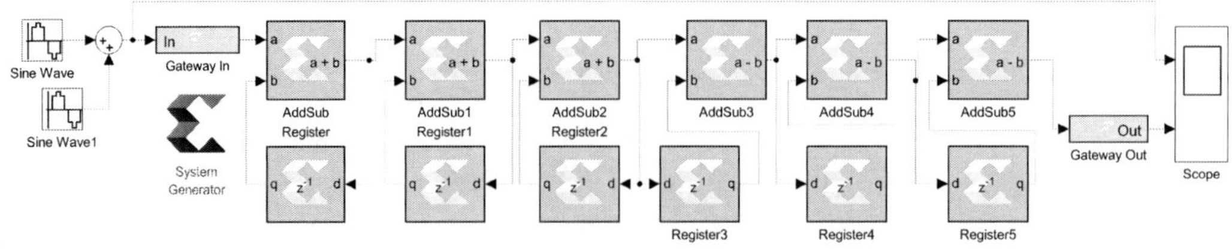

FIGURE 12.19
Structure of a simple CIC filter.

The internal design of Subsystem2, results of spectrum analysis, scope, simout and operation and interface techniques of all the SysGen blocks can be found in References 155 to 157.

12.7 COordinate Rotation DIgital Computer Design Using SysGen Tools

The Xilinx block CORDIC 4.0 implements a generalized CORDIC algorithm, which we already discussed in Chapter 4. The CORDIC core implements rectangular-to-polar and polar-to-rectangular conversion and trigonometric, hyperbolic, and square root functions. Two architectural configurations are available for the CORDIC core: a fully parallel configuration with single-cycle data throughput at the expense of silicon area and a word serial implementation with multiple-cycle throughput. A coarse rotation is performed to rotate the input sample from the full circle into the first quadrant. The coarse rotation stage is required, as the CORDIC algorithm is only valid over the first quadrant; please refer to Chapter 4 for more clarity on CORDIC rotation. An inverse coarse rotation stage rotates the output sample into the correct quadrant. The CORDIC algorithm introduces a scale factor to the amplitude of the result, and the CORDIC core provides the option of automatically compensating for the CORDIC scale factor. Now, we discuss some of the coordinate conversions and later implement them using the SysGen tools.

Next, we see an example for the conversion of polar to rectangular coordinates and vice versa. The rectangular coordinates (x, y) and polar coordinates (R, θ) are related as follows: $y = R\sin(\theta)$ and $x = R\cos(\theta)$; $R^2 = x^2 + y^2$ and $\tan(\theta) = y/x$. All these notations are shown in Figure 12.20.

To find the polar angle θ, we have to take into account the signs of x and y, which gives us the quadrant. Angle θ is in the range (0 through 2π radians or 0° through 360°). Convert the polar coordinates (5, 2.01) and (0.2, 53°) to rectangular coordinates to three decimal places. For the first point (5, 2.01), R = 5 and θ = 2.01°; then

$$x = R\cos(\theta) = 5\cos(2.01) = -2.126 \quad \text{and} \quad y = R\sin(\theta) = 5\sin(2.01) = 4.525.$$

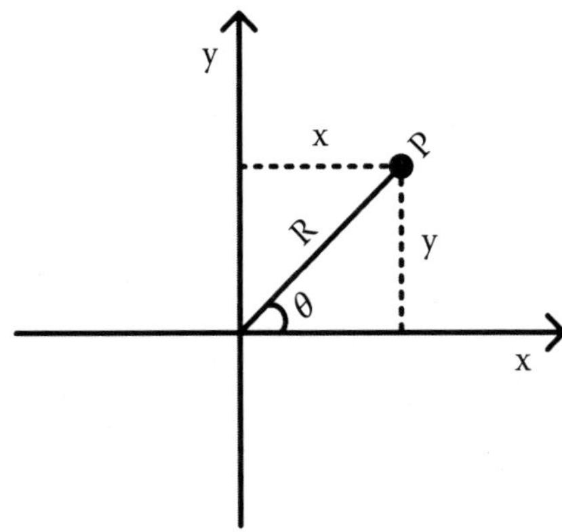

FIGURE 12.20
Representation of polar and rectangular coordinate variables.

For the second point (0.2, 53°), R = 0.2 and θ = 53° and it is in degrees. Set our calculator to degrees and use the above formulas for x and y in terms of R and θ to obtain

$$x = R\cos(\theta) = 0.2\cos(53) = 0.120 \quad \text{and} \quad y = R\sin(\theta) = 0.2\sin(53) = 0.160.$$

Convert the rectangular coordinates (1, 1) and (−2, −4) to polar coordinates to three decimal places. Express the polar angle θ in degrees and radians. We first find R using the formula R = sqrt [$x^2 + y^2$] for the point (1, 1). R = sqrt [$x^2 + y^2$] = sqrt [1 + 1] = sqrt (2). We now find tan (θ) using the formula tan(θ) = y/x. Tan(θ) = 1/1. Using the arctan function of the calculator, we obtain t = Pi/4 or t = 45°. Point (1,1) in rectangular coordinates may be written in polar for as follows: sqrt (2), Pi/4) or (sqrt (2), 45°). Let us find the R for the point (−2, −4). R = sqrt [$x^2 + y^2$] = sqrt [4 + 16] = sqrt (20) = 2sqrt (5). We now find tan(θ). tan t = −4/−2 = 2. Using the arctan function of the calculator, we obtain. θ = 1.107 or θ = 63.435°. However, since the rectangular coordinates x and y are both negative, the point is in quadrant III and we need to add pi or 180° to the value of θ given by the calculator. Hence, the polar angle θ is given by θ = 4.249 or θ = 243.435°. Point (−2,−4) in rectangular coordinates may be written in polar as follows: (2sqrt (5), 4.249) or (2 sqrt(5), 243.435°). More information on polar coordinates and trigonometry topics can be found in References 25, 44, 45, and 158.

DESIGN EXAMPLE 12.8

Design and develop a digital system to convert a rectangular value to a polar value using the CORDIC Xilinx SysGen core [156–158] (Figures 12.21 and 12.22).

Solution

Main Design

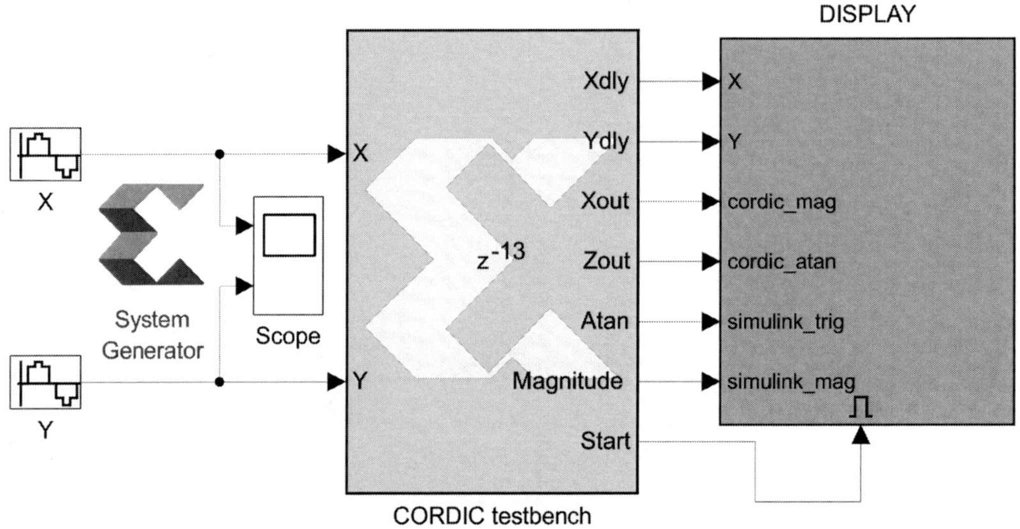

FIGURE 12.21
CORDIC-based SysGen design for rectangular-to-polar conversion.

Display Subsystem Design

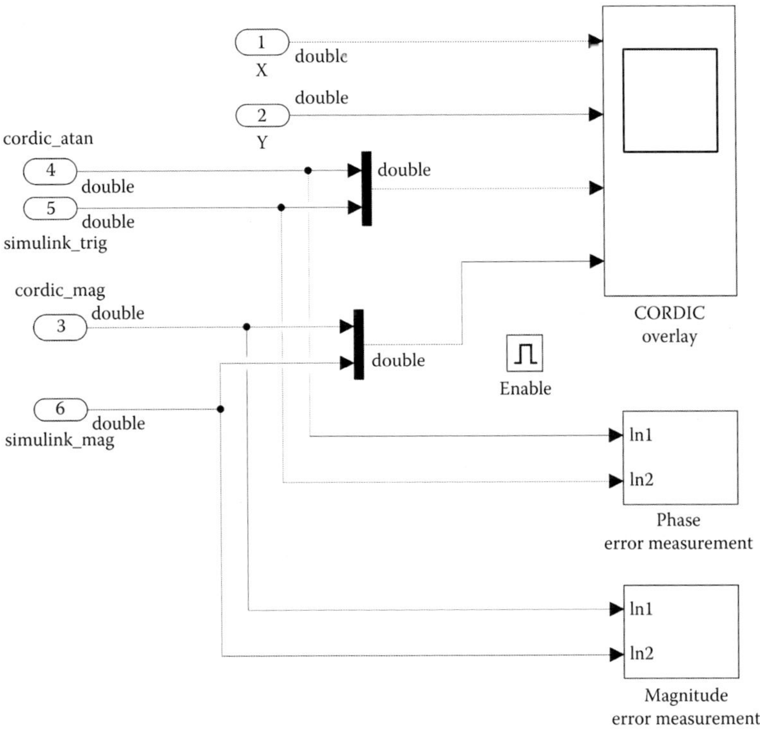

FIGURE 12.22
Design of display subsystem.

Phase/Magnitude Error Measurement Subsystem Design

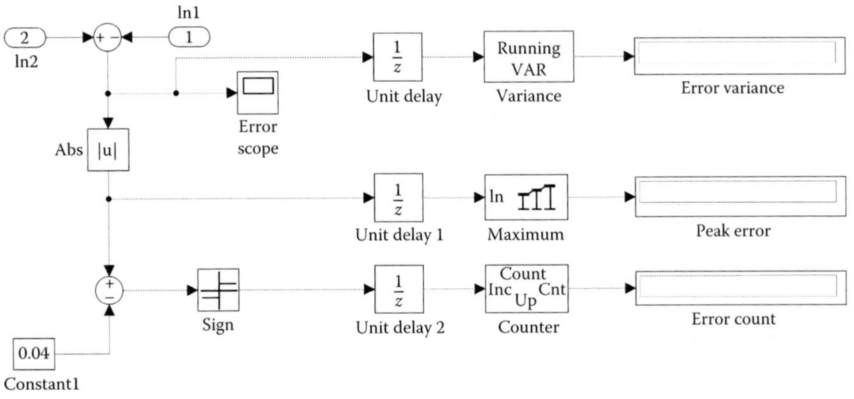

FIGURE 12.23
Design of phase error measurement subsystem.

The design shown in Figure 12.23 can be followed with appropriate inputs for the magnitude error measurement. The same can be found in References 25, 156, and 158.

12.8 Image Processing Using Discrete Wavelet Transform

When the data of any kind is recorded, it hardly reveals its true nature from the point where you are standing. A basic transform in our physical world is a transform from the time domain to the frequency domain and vice versa. Our ears are made to find the frequencies and phases of a sound signal. The frequencies are used to understand the message where the phases carry information, which is mostly used for spatial hearing. It is clear that such transforms have a high analytical value. The idea of having a basis function is that any signal can be described as a weighted sum of a family of functions. These functions are the basis functions of the transform. Transforms are often just changes from one basis function to another, although other transform types exist. For example, in cryptology, it is not practical to have a codebook; rather, it uses other mathematical methods for performing the transform. Some possible basis function conditions, depending on the features of the transform, the basis functions that will have to fulfil different requirements. These conditions allow for a number of nice mathematical tricks. From an analysis point of view, it is practical that a set of functions is orthogonal. For a set of functions $f_1 \ldots f_n$, orthogonality exists if and only if, on a given interval $a \leq x \leq b$, the correlation between different functions is zero. The δ denotes the Kronecker delta-function as [159]

$$\int_a^b f_n(x) f_m(x) dx = \delta(m - n) \tag{12.15}$$

Not all transforms aim to analyze the signal afterwards. If we want to protect the data against errors, it might be rational to add redundancy. In these cases, the choice of basis functions leads to a redundant set, where only certain basis functions exist and others are forbidden. In case of an error, the basis function can be corrected to the nearest legal one according to the distance metric of the function space. Compact support means that wherever the function is not defined, it will have a value of zero, which is an advantage in numerical calculation. Compact support is an important requirement in wavelet decompositions, because it also tells us about the locality of the wavelet in the time domain. The Fourier transform is, by all means, the most well known frequency domain analysis tool. There still exist certain problems that cannot be solved by using Fourier transformations [159].

As we have seen in the previous chapters, the DFT uses sines and cosines as basis functions, which limits its practical use to periodic and stationary signals. There are also boundary effects, as the transform is originally designed for infinite signals. The biggest problem is that frequency components in a signal cannot be localized in time. The original DFT was slow, but with the internal redundancy taken away, the FFT is a nice tool for signal analysis, which is implemented by exploiting many times the fact that one DFT transform can be separated into two transforms of half the size. This halves the calculation in the 1D transform. The mathematical representation associated with this point can be found in References 25 and 159. The short-term Fourier transform (STFT) was designed to overcome the problem by localizing the frequency components in time. The idea is to use a window to get a part of the signal. Heisenberg's uncertainty relation states the relation of the window width to the frequency localization capability. This basically means that the narrower the window is, the more inaccurate the results will be, that is, low frequencies cannot be detected with small windows. A very big problem, however, is that there exists no inverse transform for the STFT and therefore it is suitable for only gathering data and so on.

The wavelets automatically use optimal window sizes for different wavelet widths, and they are localized in both the time and frequency domains. As they model only small portions of the signal, the signal does not have to be stationary. A huge difference to the STFT is that wavelets can be used for decomposition and exact reconstruction of the signal. The CWT forms a family whose basic form is called the mother wavelet. All the daughter wavelets are derived from this wavelet according to Equation 12.16:

$$\Psi_{s,\tau}(t) = \frac{1}{\sqrt{s}}\Psi\left(\frac{t-\tau}{s}\right) \tag{12.16}$$

where two variables s and τ are the scale and translation of the daughter wavelet, respectively. The term $s^{-1/2}$ normalizes the energy for different scales, whereas the other terms define the width and translation of the wavelet. The CWT is defined as

$$\gamma(s,\tau) = \int f(t)\Psi^{*}_{s,\tau}(t)dt \tag{12.17}$$

The inverse transform, on the other hand, is

$$f(t) = \iint \gamma(s,t)\Psi_{s,\tau}(t)d\tau ds \tag{12.18}$$

A very important fact about wavelets is that the transform itself poses no restrictions whatsoever on the form of the mother wavelet. This is an important difference from other transforms. All square integratable functions that satisfy the admissibility [25,156,159] condition can be used to analyze and reconstruct any signal. The function must have a value of zero at zero frequency. Therefore, the wavelet itself has an average value of zero. This means that the wavelet describes an oscillatory signal, where the positive and negative values cancel each other. This gives the "wave" in wavelet. However, the CWT has three major problems: redundancy, infinite solution space and efficiency make continuous wavelets hard to implement for solving any real problem. First, the continuous variables s and τ are discretized so that the mother wavelet can only be scaled and translated in discrete steps. The discretized values of τ are made dependent on s in such a manner that low-frequency components are sampled less often. As a result of the discretization, we get a new equation for deriving daughter wavelets:

$$\Psi_{j,k}(t) = \frac{1}{\sqrt{s_0^j}}\Psi\left(\frac{t-k\tau_0 s_0^j}{s_0^j}\right) \tag{12.19}$$

S_0 is usually assigned a value of 2 and t_0 a value of one, which results in dyadic sampling. As we already stated, the wavelets can be used to implement a filter bank. The actual implementation is quite straightforward. The filtering is done iteratively so that the upper half of the spectrum is analyzed by the wavelet and the lower part continues on to the next round. Our wavelet itself acts as a half-band high-pass filter. As the result is bound to

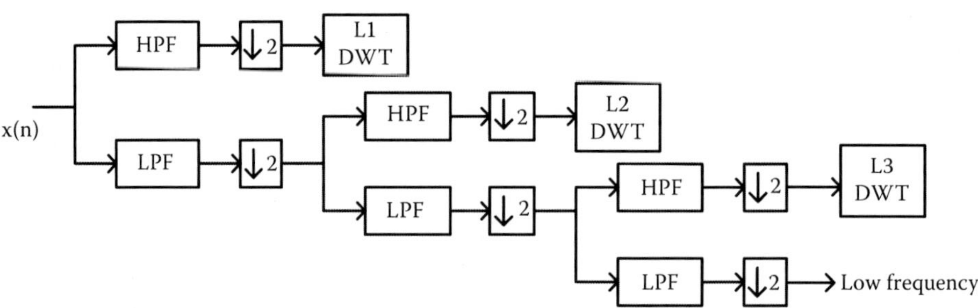

FIGURE 12.24
Representation of subband coding scheme.

be redundant, we can discard every second value after the filtering by subsampling. The lower half of the spectrum is filtered with the scaling function. The wavelet and scaling function form an orthonormal basis. Because the scaling function is a half-band low-pass filter, we get a similar effect as we do with the high-pass filter. The number of samples becomes twice the size of the spectrum. We can discard half the samples simply by subsampling by two. This algorithm can be carried out until we are left with a single value for the low-pass spectrum. The final result is a spectrum of equal size of that of the signal. The high-pass filtering is placed in the result vector so that the remaining single value (average) goes to the first "slot" in the vector. This algorithm is called subband coding, which is shown in Figure 12.24.

The DWT refers to wavelet transforms for which the wavelets are discretely sampled. The transform localizes a function both in space and scaling. It has some desirable properties compared to the Fourier transform. The transform is based on a wavelet matrix, which can be computed more quickly than the analogous Fourier matrix. Most notably, the DWT is used for signal coding, where the properties of the transform are exploited to represent a discrete signal in a more redundant form, often as a precondition for data compression. The discrete wavelet transform has a huge number of applications in science, engineering, mathematics, and computer science. Wavelet compression is a form of data compression well suited for image compression, sometimes also video compression and audio compression. The goal is to store image data in as little space as possible in a file. Using a wavelet transform, the wavelet compression methods are better at representing transients, such as percussion sounds in audio or high-frequency components in 2D images. Signal can be represented by a smaller amount of information than would be the case if some other transform, such as the more widespread DCT, had been used. This produces as many coefficients as there are pixels in the image, that is, there is no compression yet since it is only a transform. These coefficients can then be compressed more easily because the information is statistically concentrated in just a few coefficients. The Haar wavelet transformation is composed of a sequence of low-pass and high-pass filters, known as a filter bank. These filter sequences can be applied in the same way as a discrete FIR filter in the DSP using the MAC processing command, except as multiple successive FIR filters. The low-pass filter performs an averaging/blurring operation and is expressed as

$$H = \frac{1}{\sqrt{2}}(1, -1) \tag{12.20}$$

and the high-pass filter performs a differencing operation which can be expressed as

$$G = \frac{1}{\sqrt{2}}(1, -1) \tag{12.21}$$

On any adjacent pixel pair, the complete wavelet transform can be represented in matrix format by the second half, that is, applying 1D transformation to columns of the image:

$$T = W_N A W_N^T \tag{12.22}$$

where A is the matrix representing the 2D image pixels, T is the Haar wavelet transformation of the image and, in the case of transforming a 4 × 4 pixel image, $W_N = (H/G)$, where

$$H = \left[\frac{1}{\sqrt{2}}\frac{1}{\sqrt{2}}000000;00\frac{1}{\sqrt{2}}\frac{1}{\sqrt{2}}0000;0000\frac{1}{\sqrt{2}}\frac{1}{\sqrt{2}}00;000000\frac{1}{\sqrt{2}}\frac{1}{\sqrt{2}}\right]$$

and

$$G = \left[-\frac{1}{\sqrt{2}}\frac{1}{\sqrt{2}}000000;00-\frac{1}{\sqrt{2}}\frac{1}{\sqrt{2}}0000;0000-\frac{1}{\sqrt{2}}\frac{1}{\sqrt{2}}00;000000-\frac{1}{\sqrt{2}}\frac{1}{\sqrt{2}}\right].$$

The result of the complete transformation (T) is composed of four new subimages, which correspond to the blurred image and the vertical, diagonal, and horizontal differences between the original image and the blurred image. The blurred representation of the image removes the details of the high-frequency component, which are represented separately in the other three images, in a manner that produces a sparser representation overall, making it easier to store and transmit. Figure 12.25 is an example of a wavelet-transformed image portraying the four subimages as explained above.

The inverse transformation can be applied to T, resulting in lossless compression. Lossy compression can be implemented by manually setting to zero the elements below a certain threshold in T. The equation of the inverse transformation is

$$\tilde{A} = W_N^{-1} T W_N^{-T} \tag{12.23}$$

$\tilde{A} = A$ after lossless compression. The wavelet transformation employs the use of a technique called averaging and differencing. Below is a simple example of the technique performed on a single row of the matrix of pixels that represent an image. Imagine extracting one row from a 16 × 16 black-and-white, 256-colour image in its matrix form. For example, an original sequence may be [45,45,46,46,47,48,53,101,104,105,106,106,107,106,106,106]. Every element in this 16-element vector has a value between 0 and 255 since the image is greyscale, that is, 256-color. The magnitude of the element represents the brightness of its corresponding pixel. Hence, the darker it is, the closer it is to black (0). Inside most of the pixel sequence, we notice that most of the elements have values very close to the values of their neighbors this indicates a gradual shift in color. Sudden jumps in values occur at the outlines (edges) of objects in the image (53 to 101), which are sudden transitions in brightness. We also know that any two numbers can be completely characterized by their average and their difference. In the sequence above, the averages and differences of successive pairs are [45,46,47.5,77,105,106,107,106] and [0,0,-1,-48,-1,0,1,0], respectively.

Now, we observe that the sequence of differences is quite small; recall those neighbor values that are very similar, implying small differences. This sequence is therefore very easy

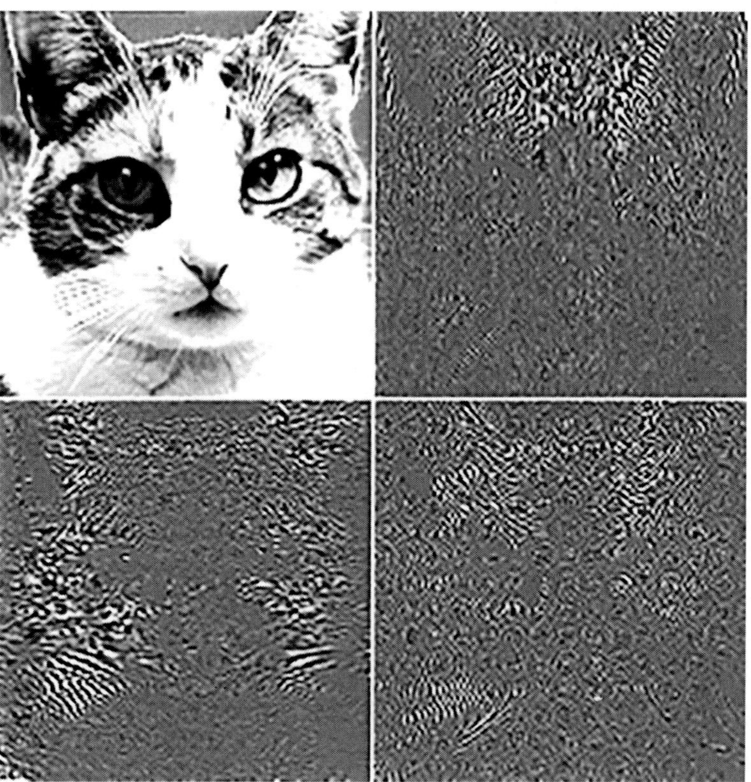

FIGURE 12.25
Wavelet-transformed images.

to compress. We perform the compression by making all the values of small magnitude 0. Hence, we have a very sparse sequence, that is, differences (with threshold) [0,0,0,-48,0,0,0,0]. This sequence is highly compressible. Using this new sequence of differences and the same sequence of averages, we can easily recompute the original sequence that defined a section of the image as [45,45,46,46,47,47,53,101,104,104,106,106,106,106,106,106]. This is a very good approximation/recovery of the original sequence. To perform the transformation in DSP/FPGA, we load the pixel values for the image into DSP/FPGA memory, process the image in the DSP assembly program, and output the transformed image or transformed-compressed image to a memory location below the original image. We can successfully perform the Haar wavelet transform for image compression in MATLAB and also on DSP and FPGA.

DESIGN EXAMPLE 12.9

Design and develop a digital system to realize image processing using the 2D-DWT technique (Figures 12.26 and 12.27).

Solution

Main Design

Simulation Results

The design of internal subsystems and associated MATLAB codes can be found in Ref. 159.

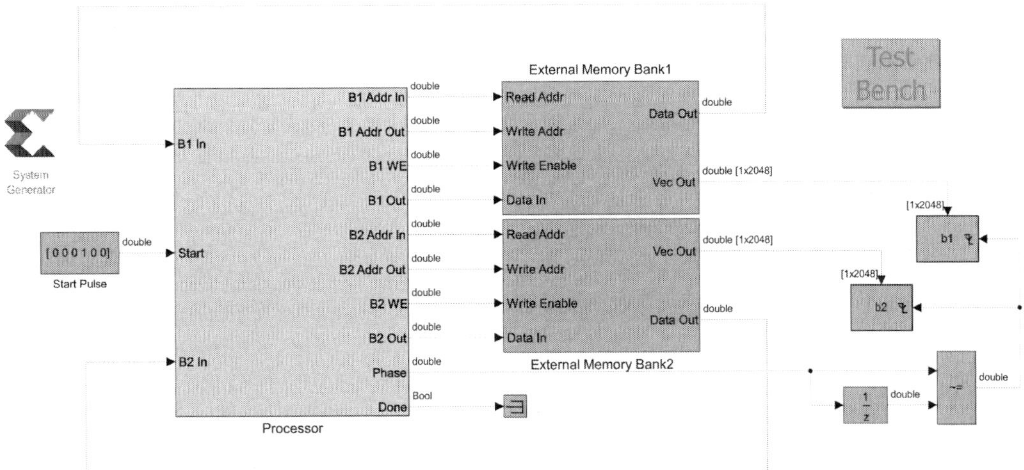

FIGURE 12.26
Design of image-processing system using DWT.

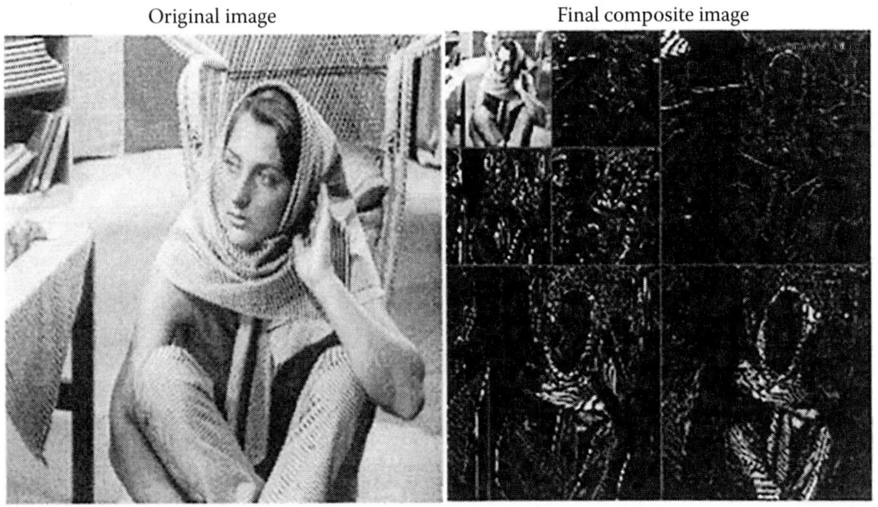

FIGURE 12.27
DWT-applied subband coded images.

12.9 Very-High-Speed Integrated Circuit Hardware Description Language Design Debugging Techniques

In this section, we discuss three different popular techniques of debugging our VHDL designs. Design verification can be done either with the hardware, that is, after implementation in FPGA, or within the design environment, that is, on the VHDL design itself. The ChipScope Pro tool is used to verify the design with hardware, while the test-bench and data/text reading/writing methods are used with VHDL design. We briefly discuss these techniques in the following sections, with some design examples.

12.9.1 ChipScope Pro Analyzer

This section shows how to take advantage of enhanced ChipScope Pro Analyzer features in the PlanAhead design environment that make the debug process faster and simpler. This tool facilitates the debugging of some common problems in FPGA logic designs. ChipScope is a logic analyzer implemented in the FPGA together with the designed hardware to test, that is, the DUT, in real time. Both the DUT and ChipScope use the onboard system clock, so ChipScope is used generally for state analysis. ChipScope can be added to a design and configured with the core inserter. Just by double-clicking the .cdc file, we can open the core inserter. In the first window, just click next. The Integrated CONtroller (ICON) is the main module interfacing with the PC through the JTAG cable and the Integrated Logic Analyzer (ILA) modules. Using a Spartan-3 FPGA, ICON has no configuration parameters and only one ICON module can be used; thus, click next. Sixteen ILAs can be connected to one ICON, but during the testing, normally, we will need only one. The ILA configuration defines the parameters of the logic analyzer. In the main window, we can see the currently used resources, that is, the LUT, FF, and BRAM count, and the configuration options on the right side. Now, we can define a maximum of 16 trigger ports and each of them can use different trigger signals.

The trigger conditions can be defined by means of match units. The match units can count the fulfilled trigger conditions, so it is possible to trigger, for example, on the 42nd rising edge of a signal. First and last, the number of trigger ports must be given, and you have to define the trigger bit width, number of match units and counter width. The match type must be either basic, in which the trigger condition can be defined with 0, 1 or X (don't care) bit values; basic edges, where, besides the signal levels, edges can be specified in the trigger: rising edge, falling edge, both edges and no transition; or range, which is extended and range check. We can select whether the data bits to sample are the same as the trigger port bits. The sampling occurs on the rising or falling edge of the system clock. Data depth gives the size of the state memory in words.

The connections between the DUT and ChipScope can be defined after clicking on modify connections. The postsynthesis signals of the DUT can be assigned to ChipScope channels. Generally, in a proper design register, names will be kept after synthesis, but combinational logics can be simplified. The base type gives the module type of the signal source: BUFGP, the clock buffer, which must be used as the ChipScope clock signal; IBUF, that is, the input buffer, connected to an FPGA pin, which we have to use as input; OBUF, the output buffer, connected to an FPGA pin, which we have to use as output; IOBUF, a bidirectional buffer; GND/VCC, constant logical 0/1; FD: some version of D flip-flop; and LUTx, which stands for a look-up table with x used inputs. The field can help search for the desired signals. After assigning all ChipScope channels, we need to return to the project navigator. The design with the CDC file must be resynthesized. Normally, the synthesis of the analyzer takes quite a long time and must be repeated every time you change the chip scope configuration. So double-check the settings if you want to analyze a design. Download the new bit file and start the ChipScope Analyzer GUI.

Select the JTAG chain/Xilinx platform USB cable in the upper menu, and click OK in the pop-up window to connect to ChipScope module. Select file/import to import the signal names and define busses based on the CDC file we made before. For match units, we can set up a trigger condition. The possible settings depend on the match unit type. Don't care bits are allowed only if the function is equal or not equal; in all other cases, all bits of a match unit must be defined. The trigger condition can be defined as a logical function or a sequence of match unit conditions. Click on "trigger condition equation" to set it. The sampling can be set up using the capture panel. In window mode, we define how many trigger conditions should be stored in the state memory during the session. In N samples mode, the number of words to be stored after one trigger is defined. By setting the (trigger) position to greater than 0, the samples before the trigger event can be examined. We can assign the logical combination of match units not only to a trigger condition, but also to a storage qualification enable signal.

After the appropriate setup, the analysis itself can be started using the toolbar of ChipScope. In "Single" trigger run mode, after pressing the "Play" button, the analyzer waits for a trigger; after the trigger, it fills the state memory and, if it is full, the result can be examined in the waveform window of the Logic analyzer. We can stop the analyzer any time, but the waveform is refreshed only if the state memory is totally filled. When "Repetitive" trigger run mode is selected, the analyzer waits automatically for the next trigger event after the state memory is filled.

12.9.2 Very-High-Speed Integrated Circuit Hardware Description Language Test-Bench Design

To simulate our design, we need both the DUT and the stimulus provided by the test bench. A test bench is the HDL code that allows us to provide a documented repeatable set of stimuli that is portable across different simulators. A test bench can be as simple as a file with clock and input data or a more complicated file that includes error checking, file input/output, and conditional testing. We can create the test bench using either of the following methods. The first method is a text editor, which is the recommended method for verifying complex designs. This method allows us to use all the features available in the HDL and gives us flexibility in verifying the design. Although this method may be more challenging in that we must create the code, the advantage is that it may produce more precise and accurate results. To get the test-bench, we have to create a template that lays out the initial framework, including the instantiation of the DUT and initialization of the stimulus for the intended design. Create a template as required selecting VHDL test-bench as the source type.

The second method is the Xilinx test-bench waveform editor, which is a recommended method for verifying less complicated simulation tasks and is apt if we are new to HDL simulation [160]. This method allows us to graphically enter the test-bench to drive the stimulus to the design. Make the test-bench the top level of the code which instantiates the DUT, and the stimulus is applied from the top-level test-bench to the lower-level design or portion of the design being tested. Use the instance name DUT for the instantiated design under test. We can use the same test-bench for both functional and timing simulation. Initialize all input ports at simulation time zero, but do not drive the expected stimulus until after 100 ns simulation time.

During timing simulation, a global set/reset signal is automatically pulsed for the first 100 ns of simulation. To keep the test-bench consistent for both timing and functional simulation, it is recommended that we hold off the input stimulus until the global set/reset has completed. We can still run the clocks during the first 100 ns of simulation. There is no need to provide data to the input ports at the same time as the clock. For nontiming simulation, this can cause some signals to be applied before the clock and some after the clock. Apply data only after the clock is applied to the input ports. This makes it easier to keep track of which clock edge data changes are being applied. If output checking is performed in the test-bench, apply data just before the next clock cycle. For timing simulation, it could take up to an entire clock cycle for the data to propagate through the logic and settle to a known value. Checking data too early in the clock cycle may yield incorrect results. We see some design examples in the following parts of this section.

DESIGN EXAMPLE 12.10

Design a digital system with a combinational circuit to realize a VHDL test-bench–based simulation.

Solution

Main Design

```
library IEEE;
use IEEE.std_logic_1164.all;
use IEEE.std_logic_unsigned.all;
entity test_bench is
port(a,b,c: in std_logic:='0';
d: out std_logic);
end test_bench;
architecture RTL of test_bench is
begin
d<=((a and b) xor c);
end RTL;
```

As in the above simple main design, create a test-bench file named "test_bench_simu" and then proceed by clicking "next", "next", and "finish" as required. Then a test-bench window will be opened. Edit the same window as given below.

Test-Bench Design

```
LIBRARY ieee;
USE ieee.std_logic_1164.ALL;
-- Uncomment the following library declaration if using
-- arithmetic functions with Signed or Unsigned values
--USE ieee.numeric_std.ALL;
ENTITY test_bench_simu IS
END test_bench_simu;
ARCHITECTURE behavior OF test_bench_simu IS
-- Component Declaration for the Unit Under Test (UUT)
COMPONENT test_bench
PORT(
a : IN  std_logic;
b : IN  std_logic;
c : IN  std_logic;
d : OUT  std_logic);
END COMPONENT;
--Inputs
signal a : std_logic := '0';
signal b : std_logic := '0';
signal c : std_logic := '0';
--Outputs
signal d : std_logic;
-- No clocks detected in port list. Replace <clock> below with
-- appropriate port name
--constant <clock>_period : time := 10 ns;
BEGIN
-- Instantiate the Unit Under Test (UUT)
uut: test_bench PORT MAP (
a => a,
b => b,
c => c,
d => d);
--   Clock process definitions
---<clock>_process :process
--    begin
-- <clock> <= '0';
```

```
--wait for <clock>_period/2;
--<clock> <= '1';
--wait for <clock>_period/2;
--end process;
--Stimulus process
stim_proc1: process
begin
wait for 100 ns;
a<= not a;
end process;
stim_proc2: process
begin
wait for 50 ns;
b<= not b;
end process;
stim_proc3: process
begin
wait for 10 ns;
c<= not c;
end process;
END;
```

Select behavioral simulation and the test_bench_simu at the hierarchy window and open the modelSim simulator window. Just keep clicking the run button; the inputs will be applied as given in the test_bench code and appropriate output will be generated.

DESIGN EXAMPLE 12.11

Design a digital system with a sequential circuit to realize a test-bench–based simulation.

Solution

Main Design

```
library IEEE;
use IEEE.std_logic_1164.all;
use IEEE.std_logic_unsigned.all;
entity test_bench2 is
port(a: inout std_logic_vector(7 downto 0):="01101010";
reset,m_clk:in std_logic:='0';
d: out std_logic:='0');
end test_bench2;
architecture RTL of test_bench2 is
begin
process(m_clk,a)
begin
if reset='1' then
if rising_edge(m_clk) then
a<=(a(0) & a(7 downto 1));
d<=a(0);
end if;
end if;
end process;
end RTL;
```

As in the above simple main design, create a test-bench file named "test_bench_simu" and then proceed by clicking "next", "next", and "finish" as required. Then a test-bench window will be opened. Edit the same window as given below. Please note that we should not load any value for the input a, because it is an inout, that is, as far as the test-bench is concerned, this inout is a bidirectional variable; thus, it will take a value from assignment given at the main design.

Test-bench Design

```
LIBRARY ieee;
USE ieee.std_logic_1164.ALL;
-- Uncomment the following library declaration if using
-- arithmetic functions with Signed or Unsigned values
--USE ieee.numeric_std.ALL;
ENTITY test_bench_simu IS
END test_bench_simu;
ARCHITECTURE behavior OF test_bench_simu IS
-- Component Declaration for the Unit Under Test (UUT)
COMPONENT test_bench2
PORT(
a : INOUT  std_logic_vector(7 downto 0);
reset : IN  std_logic;
m_clk : IN  std_logic;
d : OUT  std_logic);
END COMPONENT;
--Inputs
signal reset : std_logic := '0';
signal m_clk : std_logic := '0';
--BiDirs
signal a : std_logic_vector(7 downto 0);
--Outputs
signal d : std_logic;
-- Clock period definitions
constant m_clk_period : time := 10 ns;
BEGIN
-- Instantiate the Unit Under Test (UUT)
uut: test_bench2 PORT MAP (
a => a,
reset => reset,
m_clk => m_clk,
d => d);
-- Clock process definitions
m_clk_process :process
begin
m_clk <= '0';
wait for m_clk_period/2;
m_clk <= '1';
wait for m_clk_period/2;
end process;
-- Stimulus process
stim_proc: process
begin
reset<='0';
wait for 25 ns;
reset<='1';
wait;
end process;
END;
```

Select behavioral simulation and the test_bench_simu at the hierarchy window and the open the modelSim simulator window. Just keep clicking the run button; the inputs will be applied as given in the test_bench code and the appropriate output will be generated.

12.9.3 Data/Text File Reading/Writing

In this section, we discuss data/text reading and writing from/to a text file. This type of reading and writing is not possible for synthesis since the FPGA cannot read/write from/to a computer file in real time. Thus, this technique is more popularly being used for pumping more data sets to test the performance of the design we built. Suppose our design has a large array of inputs. We cannot put them in the test-bench program because it will make the test-bench code very difficult to read in. In such cases, it is advisable to store and read them from a text file. We can even have an output file where we can store the results. For example, we have two files; the first is named 1.txt and kept as the input file. The values will be read from this file and simply copied to the second file, named 2.txt. There are many options in the text I/O features of VHDL that make it possible to open one or more data files, read lines from those files, and parse the lines to form individual data elements, such as elements in an array or record. To support the use of files, VHDL has the concept of a file data type and includes standard, built-in functions for opening, reading from, and writing to data type files.

The `textio` package, which is included in the standard library, expands on the built-in file type features by adding text parsing and formatting functions, functions and special file types for use with interactive "std_input" and "std_output" I/O operations and other extensions. The following section demonstrates how we can use the text I/O features of VHDL to read test data from an ASCII file using the standard text I/O features [161]. The vectors stored in the file are used to stimulate or drive test benches, and the output results are recorded to a file. The package file that needs to be included to make file IO work is `std.textio`. This allows the usage of all the keywords like `file_open`, `file_close`, `read`, `readline`, `write`, `writeline`, `flush` and `endfile`.

This library file allows reading and writing of types`bit`, `bit_vector_boolean`, `character`, `integer`, `real`, `string` and `time`. In order to compile the test-bench code correctly, all the files need to be in the same directory and compiled in the same directory. Rather than perform tedious analysis of simulation waveforms, it is often easier to examine the actual circuit data outputs that are recorded in a text file. The outline and sample VHDL test-bench code given below includes the code necessary for performing both input and output from/to VHDL text files. We need to first include the text IO libraries in the test bench as

```
use ieee.std_logic_textio.all;
use std.textio.all;
```

All the files that are to be used in the test-bench must be declared using the `file` statement. An example VHDL code below declares an input test file, "test1.txt", that should reside in the main project directory; otherwise, the full path name must be used in the quotes. The second file, "results1.txt", is the output results file that will be created by the test-bench. The file declaration code can be placed in the architecture declarations or in the main test-bench process declarations. The main test-bench process is often chosen since a number of variables are required for the actual IO operations.

```
file test_file: text open read_mode is "test1.txt";
file results: text open write_mode is "results1.txt";
```

Text IO occurs a line at a time and uses the `line` type; therefore, buffers of type `line` are declared for holding the data elements. For the "test1.txt" file, we declare the buffer in the main process as

```
Variable l_in: line;    -- your choice of name instead of l_in
```

Similarly, the output buffer for the results 1.txt file is declared as

```
Variable tx_out: line;    -- your choice of name instead of tx_out
```

Next, we will focus on inputting data from the text1.txt file. A line of the file is read using

```
Readline (test_file, l_in);
```

The buffer `l_in` now contains the first line of the file. Individual fields in the buffer must be accessed from the buffer using the VHDL read procedure defined in the text IO package. Each `read` procedure call is defined by two types of parameter sets distinguished by the omission/inclusion of the Boolean `good` parameter, which checks to see if the field has valid data or not. For example, a statement `read(l_in, Rec_Num, good => good_number);` contains the `good` parameter, where `good_number` is declared as a variable in the test-bench main process. If `Rec_Num` was defined as an integer variable in the process to receive the data, but a non-numeric base-10 ASCII value was contained in the field, `good` would return false. This provides some degree of validity checking. The use of the `good` parameter is optional, but recommended. If `good` is used, the next statement following the `read` statement should check the `good` return value. For example, the `if` statement `if not good_number then assert (false) report "bad record number in test file"; severity failure;` would terminate the test-bench after outputting the message on the simulator console. Continued execution of the `read` statement will produce all of the fields within the line.

In general, if `bit_vectors` and `integers` are separated by a space character in the line, the `read` statements will provide correct input for each field. This is not true for characters and string variables. All text input characters must be accounted for in the string input. After that, a new `read line` statement must be executed to get the next record in the file. Readline statements can be executed until an end-of-file condition is detected using `endfile(test_file) return`, where `test_file` is our input file declaration. In the test-bench, a `while` loop is often used as a primary loop within the main test bench process. Using this, the VHDL `while` statement `while not endfile(test_file) loop … end loop;` will test for the end-of-file at the top of the loop with all of the `read line` and `read` statements, assignments to circuit inputs and so on. Normally, the exit from the loop will begin the termination of the test-bench, unless additional input files are to be processed. The `endfile(test_file)` Boolean return can also be used in `if` statements to detect a normal or abnormal end-of-file condition. The `bit`, `bit_vector`, and `integer` type variables are very useful for receiving the input fields.

The `time` type may also be useful for some test-benches. However, we generally need to convert `bit_vectors` or `integers` to `std_logic_vectors` on input and `std_logic_vectors` to bit vectors going out. These conversions must be done using the VHDL libraries. `bit_vector to std_logic_vector` uses the `to_stdlogic_vector` function on input and the `to_bit_vector` function for converting `std_logic_vectors` for output as `bit_vectors`. If integers are used as input fields (often for convenience) and need to be converted to `std_logic_vectors` for input to the circuit, the function `conv_std_logic_vector(<INT_SIG>, <INTEGER_SIZE>)` is used, where `integer_size` is the size in bits of the returned `std_logic_vector`. Sometimes, the output file is handled well if `std_logic_vectors` are output as integers.

In this case, we have to use the `conv_integer` function. Other conversions for both input and output are also possible. The output file records are generated using the `write` function to build an output record or line. Once the line is populated, the `writeline` procedure is called to record the data in the file. The simulator should close the output file upon termination of the test-bench. For example, writing is `write(tx_out, clk_count);` ... `write(tx_out, instr_count);` and `teline(results, tx_out).` It is often nice to include descriptor fields in the output, which is done by using `write` statements of the form `write(tx_out, string'("clock count: "));`, which would normally precede the `write(tx_out, clk_count)` statement above to define the value in the output line record. Using descriptor strings in the results file makes it very easy to analyze.

DESIGN EXAMPLE 12.12

Design and develop a VHDL test-bench simulator to read text/data from a text file and write them to another text file.

Solution

```
library ieee;
use ieee.std_logic_1164.all;
--include this library for file handling in vhdl.
library std;
use std.textio.all;  --include package textio.vhd
-- uncomment the following library declaration if using
-- arithmetic functions with signed or unsigned values
--use ieee.numeric_std.all;
entity tb is
end tb;
architecture behavior of tb is
-- component declaration for the unit under test (uut)
component module_ut
port(
a : in  std_logic;
b : in  std_logic;
c : out  std_logic
);
end component;
--inputs
signal a : std_logic := '0';
signal b : std_logic := '0';
--outputs
signal c : std_logic;
--file read write related signal
--period of clock,bit for indicating end of file.
signal clock,endoffile : bit := '0';
--data read from the file.
--signal    dataread : string;
signal dataread : string(1 to 7);
--data to be saved into the output file.
--signal    datatosave : string;
signal datatosave : string(1 to 7);
--line number of the file read or written.
signal    linenumber : integer:=1;
begin
-- instantiate the unit under test (uut)
uut: module_ut port map (
a => a,
```

```
b => b,
c => c
);
clock <= not (clock) after 1 ns;     --clock with time period 2 ns
--read process
process
file   infile    : text is in  "1.txt";    --declare input file
variable  inline    : line; --line number declaration
--variable  dataread1    : string;
variable dataread1 : string(1 to 7);
begin
wait until clock = '1' and clock'event;
if (not endfile(infile)) then   --checking the "end of file" is not
reached.
readline(infile, inline);       --reading a line from the file.
--reading the data from the line and putting it in a real type variable.
read(inline, dataread1);
dataread <=dataread1;   --put the value available in variable in a
signal.
else
endoffile <='1';        --set signal to tell end of file read file is
reached.
end if;
end process;
--write process
process
file      outfile  : text is out "2.txt";  --declare output file
variable  outline  : line;   --line number declaration
begin
wait until clock = '0' and clock'event;
if(endoffile='0') then   --if the file end is not reached.
--write(linenumber,value(real type),justified(side),field(width),
digits(natural));
write(outline, dataread);--, right, 16, 12);
-- write line to external file.
writeline(outfile, outline);
linenumber <= linenumber + 1;
else
null;
end if;
end process;
end;
```

DESIGN EXAMPLE 12.13

Design and develop a digital system to interface a 2 × 16 LCD in 4-bit interface mode and display the character "Th" starting from first cell in line 1 with the cursor blinking on.

Solution

As given in this design example, the 2 × 16 LCD has to be interfaced in 4-bit operation mode. Thus, we can send only data with a 4-bit bus that is, D_7, D_6, D_5, D_4. Therefore, we need to first send 28H to inform the LCD internal processor that the interface mode of operation is 4-bit. Accordingly, we have to send the last 4 bits first, then only the first 4 bits, that is, $[b_7, b_6, b_5, b_4]$ first and then $[b_3, b_2, b_1, b_0]$ second, for every command and data value. Example VHDL code is given below.

```
library IEEE;
use IEEE.STD_LOGIC_1164.ALL;
```

```
use IEEE.STD_LOGIC_ARITH.ALL;
use IEEE.STD_LOGIC_UNSIGNED.ALL;
entity clkd is
Port (clk: in  STD_LOGIC;
c:out std_logic_vector(2 downto 0):="000";
d:out std_logic_vector(3 downto 0):="0000");
end clkd;
architecture Behavioral of clkd is
type state is(s0,s1,s2,s3,s4,s5,s6,s7,s8,s9,s10,s11,s12,
s13,s14,s15,s16,s17,s18,s19,s20,s21,s22,s23,s24,s25,s26,s27);
--signal no:std_logic_vector(7 downto 0):="00110000";
signal ps,ns:state;
signal clk_sig0:std_logic:='0';
begin
-----------------------------LF  clock  generater----------------
p1:process(clk)
variable cnt:integer:=0;
begin
if rising_edge(clk) then
if(cnt=2500000)then
clk_sig0<= not(clk_sig0);
cnt:=0;
else
cnt:= cnt + 1;
end if;
end if;
end process;

p2:process(clk_sig0)
begin
if rising_edge(clk_sig0) then
ps<=ns;
end if;
end process;
--------------------------------LCD Command and Data --------------------
p5:process(ps)
begin
case ps is
when s0=> d<="0010";c<="100"; ns<=s1;--28--h
when s1=> c<="000";ns<=s2;
when s2=> d<="1000";c<="100"; ns<=s3;--28--L
when s3=> c<="000";ns<=s4;
when s4=> d<="0000";c<="100"; ns<=s5;--0E--h
when s5=> c<="000";ns<=s6;
when s6=> d<="1100";c<="100"; ns<=s7;--0E--l
when s7=> c<="000";ns<=s8;
when s8=> d<="0000";c<="100"; ns<=s9;--01--h
when s9=> c<="000";ns<=s10;
when s10=> d<="0001";c<="100"; ns<=s11;--01--l
when s11=> c<="000";ns<=s12;
when s12=> d<="0000";c<="100"; ns<=s13;--06--h
when s13=> c<="000";ns<=s14;
when s14=> d<="0110";c<="100"; ns<=s15;--06--l
when s15=> c<="000";ns<=s16;
when s16=> d<="1100";c<="100"; ns<=s17;--80--h
when s17=> c<="000";ns<=s18;
when s18=> d<="0101";c<="100"; ns<=s19;--80--l
when s19=> c<="000";ns<=s20;
when s20=> d<="0101";c<="110"; ns<=s21;--T--h
when s21=> c<="010";ns<=s22;
when s22=> d<="0100";c<="110"; ns<=s23;--T--l
```

```
when s23=> c<="010";ns<=s24;
when s24=> d<="0110";c<="110"; ns<=s25;--h--h
when s25=> c<="010";ns<=s26;
when s26=> d<="1000";c<="110"; ns<=s27;--h--l
when s27=> c<="010";ns<=s16;
when others=> null;
end case;
end process;
end Behavioral;
```

DESIGN EXAMPLE 12.14

Design and develop a digital system to interface a 2 × 16 LCD in 4-bit interface mode and display the characters "ABCDEFGHIJKLMNO," and "123456789ABCDEF," in lines 1 and 2 respectively, with the cursor blinking on.

Solution

```
library ieee;
use ieee.std_logic_1164.all;
use ieee.numeric_std.all;
use ieee.std_logic_unsigned.all;
use ieee.std_logic_arith.all;
entity clkd is
port (m_clk: in  std_logic:='0';
c:out std_logic_vector(2 downto 0):="000";
d : out  std_logic_vector(0 to 3):="0000");
end clkd;
architecture behavioral of clkd is
-------------------------------------------------------
signal mess1: string (1 to 16):="ABCDEFGHIJKLMNO,";
signal mess2: string (1 to 16):="123456789ABCDEF,";
signal comm: std_logic_vector(0 to 39):=x"280E010680";
signal sta,nsta:integer range 0 to 15:=0;
signal temp: std_logic_vector(0 to 7):="00000000";
begin
--------------------------------------------------------
p0:process(m_clk)
variable cnt:integer:=0;
begin
if rising_edge(m_clk) then
if(cnt=2)then
sta<=nsta;
cnt:=0;
else
cnt:= cnt + 1;
end if;
end if;
end process;
--------------------------------------------------------
p1:process(sta)
variable i: integer:=0;
variable j: integer:=3;
begin
case sta is
when 0=> d<=comm(i to j); c<="100";nsta<=sta+1;
when 1=> if (j=39)then
c<="000";i:=1; j:=3; nsta<=sta+1;
else
c<="000";
```

```
i:=i+4; j:=j+4; nsta<=0;
end if;
when 2=> temp<=std_logic_vector(to_unsigned(character'pos(mess1(i)),8));
nsta<=sta+1;
when 3=> d<=temp(0 to 3); c<="110"; nsta<=sta+1;
when 4=> c<="010"; nsta<=sta+1;
when 5=> d<=temp(4 to 7); c<="110"; nsta<=sta+1;
when 6=> if (i=16)then
c<="010"; i:=1; nsta<=sta+1;
else
c<="010"; i:=i+1; nsta<=2;
end if;
when 7=> d<="1100"; c<="100"; nsta<=sta+1;
when 8=> c<="000"; nsta<=sta+1;
when 9=> d<="0000"; c<="100"; nsta<=sta+1;
when 10=> c<="000"; nsta<=sta+1;
when 11=> temp<=std_logic_vector(to_unsigned(character'pos(mess2(i)),8));
nsta<=sta+1;
when 12=> d<=temp(0 to 3); c<="110"; nsta<=sta+1;
when 13=> c<="010"; nsta<=sta+1;
when 14=> d<=temp(4 to 7); c<="110"; nsta<=sta+1;
when 15=> if (i=16)then
c<="010"; i:=32; j:=35; nsta<=0;
else
c<="010"; i:=i+1; nsta<=11;
end if;
when others=> nsta<=0;
end case;
end process;
-----------------------------------------------------------
end behavioral;
```

LABORATORY EXERCISES

1. Design and develop a digital system to realize the logarithm, exponential and square root functions using the SysGen tools and VHDL code.
2. Design and develop a digital system for reading/writing pixel values corresponding to an image of size N × N from/to a text file using the SysGen tools and VHDL code.
3. Design and develop a digital system to generate radar baseband signals using the SysGen tools and VHDL code with an appropriate DAC.
4. Design and develop a digital system to generate the frequency modulated continuous wave (FMCW) radar waveform using the SysGen tools.
5. Design and develop a Moore FSM using the SysGen tools to perform the following operations. Assume the Moore FSM gets two bits (x_0x_1) for every rising edge of the clock input and produces one output (y). The output y=0 if x_0x_1 are subsequently "01" and "11", y = 1 if x_0x_1 are subsequently "10" and "11", y = y′ if x_0x_1 are subsequently "10" and "01" and y = y otherwise.
6. Design and develop a digital system to convert the given polar data to rectangular form.
7. Design and develop an audio signal noise-cancellation system using the SysGen tools.

8. Design and develop an image/video noise-cancellation system using the SysGen tools.
9. Design and develop a digital system to obtain I and Q output from a radar IF signal at the receiver chain using the SysGen tools and VHDL code.
10. Design and develop a digital controller system to play a blackjack game using the SysGen tools and VHDL code.

13

Contemporary Design and Applications

13.1 Introduction and Preview

Incorporating digital designs in totality with a real-time system and validating its performance is important to check the reliability of the entire systems we build. Several real-time systems developed for various applications are given in this chapter. Basic introductions to and design methodologies for DPCM schemes, digital data encryption techniques, soft computing techniques (artificial neural networks and fuzzy logic controllers), bit error rate testers, optical up-/downlink controllers, channel coding, pick and place robot controls, and audio codec interfaces and controls are discussed. Discussing the above areas in detail is out of the scope of this book; thus, only necessary basic introductions are given at the beginning of all the sections. The readers are assumed to have basic knowledge of the above areas before reading this chapter. However, many books are available that are relevant to those fields where readers can find the necessary knowledge to understand the designs given in this chapter.

13.2 Differential Pulse Code Modulation System Design

Objective: DPCM is the process of converting an analog signal into a digital signal in which an analog signal is sampled and the difference between the actual sample value and its predicted value (predicted value is based on previous sample or samples) is quantized and then encoded, forming a digital value. DPCM code words represent differences between samples, unlike PCM, where code words represent a sample magnitude/value. The basic principle of DPCM coding is that most of the sources show significant correlation between successive samples; hence, encoding the redundancy of the sample values implies a lower bit rate. Realization of the basic concept (described above) is based on a technique in which we have to predict the current sample value based upon previous sample(s) and encode the difference between the actual value of the sample and the predicted value. Thus, the DPCM consists of a comparator, quantizer, encoder, adder, and delay element. Assume that x[n] is the input sample, e[n] is the comparator output, q[n] is the quantization noise, $e_q[n]$ is the quantizer output, $m_q[n]$ is the adder output, and $m_{q1}[n]$ is the predictor output. Then, the operation of a DPCM system can be given by [162]

$$e[n] = x[n] - m_{q1}[n], \tag{13.1}$$

$$e_q[n] = e[n] + q[n], \tag{13.2}$$

$$m_q[n] = m_{q1}[n] + e_q[n], \tag{13.3}$$

From Equations 13.2 and 13.3,

$$m_q[n] = m_{q1}[n] + e[n] + q[n]$$

Then, from Equation 13.1,

$$m_q[n] = m_{q1}[n] + x[n] - m_{q1}[n] + q[n]$$

$$m_q[n] = x[n] + q[n] \tag{13.4}$$

Equation 13.4 clearly means the entire DPCM system processes the input sample plus quantization noise. We develop a digital system to realize this operation.

Solution

```
library IEEE;
use IEEE.STD_LOGIC_1164.ALL;
use IEEE.STD_LOGIC_arith.ALL;
use IEEE.STD_LOGIC_unsigned.ALL;
entity DPCM_code is
port(clk:in std_logic;
data_out:std_logic_vector(8 downto 0));
end DPCM_code;
architecture Behavioral of DPCM_code is
type array_type is array(0 to 15) of std_logic_vector(7 downto 0);
signal x:array_type;
signal e,eq,y,xe:std_logic_vector(9 downto 0):=(others=>'0');
signal e1,y1,t1,t2:std_logic_vector(10 downto 0):=(others=>'0');
type state_type
is(rst_state,diff_state_1,diff_state_2,add_state_1,add_state_2,init_
state,final_state,equ_state,delay,wait1);
signal state:state_type:=rst_state;
signal count:integer:=0;
signal flag:std_logic_vector(1 downto 0):="00";
begin
process(clk)
begin
if(clk='1' and clk'event)then
case state is
when rst_state=>
flag<="00";
e<=(others=>'0');
x(0) <=x"11";
x(1) <=x"15";
x(2) <=x"33";
x(3) <=x"23";
x(4) <=x"40";
x(5) <=x"47";
x(6) <=x"31";
x(7) <=x"65";
x(8) <=x"44";
x(9) <=x"73";
```

```
x(10)<=x"19";
x(11)<=x"25";
x(12)<=x"44";
x(13)<=x"38";
x(14)<=x"55";
x(15)<=x"19";
count<=0;
xe<=(others=>'0');
eq<=(others=>'0');
state<=init_state;
when init_state=>
if(count=16)then
count<=0;
state<=final_state;
elsif(count=0)then
e<=("00" & x(0));
if(x(0)(0)='1')then
eq<=x(0)+"0000000001";
else
eq<="00" & x(0);
end if;
state<=wait1;
else
--count<=count+1;
state<=diff_state_1;
end if;
when wait1=>
y<=eq;
xe<=eq;
count<=1;
state<=init_state;
when diff_state_1=>
if(xe(9)='0')then
e1<=("000" & x(count))+('0' & ((xe xor "1111111111")+"0000000001"));
else
e1<=("000" & x(count))+('0' & xe);
end if;
t1<=("000" & x(count));
t2<=('0' & ((xe xor "1111111111")+"0000000001"));
state<=diff_state_2;
when diff_state_2=>
if(e1(10)='1')then
e<='0' & e1(8 downto 0);
else
e<='1' & ((e1(8 downto 0) xor "111111111")+"000000001");
end if;
state<=equ_state;
when equ_state=>
if(e(0)='1' and e(9)='0')then
eq(8 downto 0)<=e(8 downto 0)+"000000001";
eq(9)<='0';
elsif(e(0)='1' and e(9)='1')then
eq(8 downto 0)<=e(8 downto 0)+"000000001";
eq(9)<='1';
else
eq<=e;
end if;
state<=add_state_1;
when add_state_1=>
if(xe(9)='0' and eq(9)='0')then
y<='0' & (xe(8 downto 0)+eq(8 downto 0));
```

```
state<=delay;
elsif(xe(9)='1' and eq(9)='1')then
y<='1' & (xe(8 downto 0)+eq(8 downto 0));
state<=delay;
elsif(xe(9)='1' and eq(9)='0')then
y1<=('0' & eq)+('0' & ((('0' & xe(8 downto 0)) xor
"1111111111")+"0000000001"));
state<=add_state_2;
flag<="01";
else
y1<=('0' & xe)+('0' & ((('0' & eq(8 downto 0)) xor
"1111111111")+"0000000001"));
state<=add_state_2;
flag<="11";
end if;
when add_state_2=>
if (y1(10)='1')then
y<='0' & y1(8 downto 0);
else
y<='1' & ((y1(8 downto 0) xor "111111111")+"000000001");
end if;
state<=delay;
when delay=>
xe<=y;
count<=count+1;
state<=init_state;
when final_state=>
state<=final_state;
end case;
end if;
end process;
end Behavioral;
```

13.3 Data Encryption System

Objective: Data encryption translates data into another form/code so that only people with access to a secret/decryption key can read it. Encrypted data are commonly referred to as ciphertext, while unencrypted data are called plaintext. A standard encryption algorithm is used in the following design. Deriving the logical expression used in the following encryption design is left to the readers.

Solution

```
LIBRARY ieee;
USE ieee.std_logic_1164.all;
USE ieee.numeric_std.all;
use ieee.std_logic_unsigned.all;
use IEEE.STD_LOGIC_ARITH.ALL;
entity selva_project is
Port (m_clk,reset: in STD_LOGIC:='0';
EPC_data_disp_cont: in std_logic:='0';
EPC_disp_data: out std_logic_vector(0 to 15):="0000000000000000";
EPC:in std_logic_vector(1 downto 0):="00";
lcd_disp_data: out std_logic_vector(7 downto 0):="00000000";
lcd_disp_cont:out std_logic_vector(2 downto 0):="000");
```

```
end selva _project;
architecture Behavioral of selva _project is
------------------------------------------------------------------------------------------
constant Message0 : String(1 to 8) := "CPM+APM=";
constant Message1 : String(1 to 8) := "CPL+APL=";
signal st,nst: integer range 0 to 22:=0;
signal temp0,temp1,temp2,temp3,temp4,temp5,temp6,temp7,temp8,temp9:
std_logic_vector(0 to 3):="0000";
signal clk_sig,clk_sig0:std_logic:='0';
signal RTX,RMX,Apwdm,Apwdl,Kpwdm,Kpwdl,Rv,RTM,Rw,PAD1,Rvw,PAD2,CCPwdm,CC
Pwdl:std_logic_vector(0 to 15):="0000000000000000";
signal Apwd,Kpwd:std_logic_vector(0 to 31):="000000000000000000000000000
00000";
signal Message2 : String(1 to 8) := "3BCA6F54";
signal CCPwdm_disp0,CCPwdm_disp1,CCPwdm_disp2,CCPwdm_disp3,
CCPwdl_disp0,CCPwdl_disp1,CCPwdl_disp2,CCPwdl_disp3:std_logic_vector(7
downto 0):="00000000";
signal sta,nsta:integer range 0 to 76:=0;
signal d:std_logic_vector(7 downto 0):="00000000";
signal c:std_logic_vector(2 downto 0):="000";
begin
------------------------------------------------------------------------------------------
lcd_disp_cont<=c;
lcd_disp_data<=d;
------------------------------------------------------------------------------------------
p0:process(m_clk,reset)
variable cnt:integer:=1;
begin
if(reset='1') then
clk_sig<='0';
cnt:=1;
elsif rising_edge(m_clk) then
if(cnt=4)then
clk_sig<= not(clk_sig);
cnt:=1;
else
cnt:= cnt + 1;
end if;
end if;
end process;
------------------------------------------------------------------------------------------
p1:process(clk_sig)
begin
if rising_edge(clk_sig) then
st<=nst;
end if;
end process;
------------------------------------------------------------------------------------------
p2:process(st,EPC)
begin
case st is
when 0=>
case EPC is
when "00"=> RTX<=x"ABCD"; RMX<=x"A42D"; Apwd<=x"3BCA6F54";
Kpwd<=x"527F78DF";Message2 <="3BCA6F54";nst<=st+1;
when "01"=> RTX<=x"9745"; RMX<=x"957A"; Apwd<=x"A55A6FBD";
Kpwd<=x"256F932F";Message2 <="A55A6FBD";nst<=st+1;
when "10"=> RTX<=x"52CD"; RMX<=x"AF2D"; Apwd<=x"1747A6F5";
Kpwd<=x"256F7279";Message2 <="1747A6F5";nst<=st+1;
when "11"=> RTX<=x"7B4D"; RMX<=x"67C9"; Apwd<=x"7857A6FB";
Kpwd<=x"163F78BC";Message2 <="7857A6FB";nst<=st+1;
```

```
when others=> null;
end case;
when 1=>
Apwdl<=Apwd(0 to 15);Apwdm<=apwd(16 to 31);
Kpwdl<=Kpwd(0 to 15);Kpwdm<=Kpwd(16 to 31);
nst<=st+1;
when 2=>
Rv(0)<=Apwdl(to_integer(unsigned(RTX(0 to 3))));
Rv(1)<=Apwdl(to_integer(unsigned(RTX(4 to 7))));
Rv(2)<=Apwdl(to_integer(unsigned(RTX(8 to 11))));
Rv(3)<=Apwdl(to_integer(unsigned(RTX(12 to 15))));----------------
Rv(8)<=Apwdl(to_integer(unsigned(RMX(0 to 3))));
Rv(9)<=Apwdl(to_integer(unsigned(RMX(4 to 7))));
Rv(10)<=Apwdl(to_integer(unsigned(RMX(8 to 11))));
Rv(11)<=Apwdl(to_integer(unsigned(RMX(12 to 15))));
nst<=st+1;
when 3=>
temp0<=("000"& Apwdl(to_integer(unsigned(RTX(0 to 3)))));
temp1<=("000"& Apwdm(to_integer(unsigned(RTX(4 to 7)))));
temp2<=("000"& Apwdm(to_integer(unsigned(RTX(8 to 11)))));
temp3<=("000"& Apwdm(to_integer(unsigned(RTX(12 to 15)))));
temp4<=("000"& Apwdm(0));----------------------------------------
temp5<=("000"& Apwdl(to_integer(unsigned(RMX(0 to 3)))));
temp6<=("000"& Apwdm(to_integer(unsigned(RMX(4 to 7)))));
temp7<=("000"& Apwdm(to_integer(unsigned(RMX(8 to 11)))));
temp8<=("000"& Apwdm(to_integer(unsigned(RMX(12 to 15)))));
temp9<=("000"& Apwdm(0));
nst<=st+1;
when 4=>
Rv(4 to 7)<= (temp0 + temp1 + temp2 + temp3   + temp4);
Rv(12 to 15)<= (temp5 + temp6 + temp7 + temp8 + temp9);
nst<=st+1;
when 5=>
RTM<= (RTX xor RMX);
nst<=st+1;
when 6=>
Rw(0)<=Apwdl(to_integer(unsigned(RTX(0 to 3))));
Rw(1)<=Apwdl(to_integer(unsigned(RTX(4 to 7))));
Rw(2)<=Apwdl(to_integer(unsigned(RTX(8 to 11))));
Rw(3)<=Apwdl(to_integer(unsigned(RTX(12 to 15))));----------------
Rw(8)<=Apwdl(to_integer(unsigned(RTM(0 to 3))));
Rw(9)<=Apwdl(to_integer(unsigned(RTM(4 to 7))));
Rw(10)<=Apwdl(to_integer(unsigned(RTM(8 to 11))));
Rw(11)<=Apwdl(to_integer(unsigned(RTM(12 to 15))));
nst<=st+1;
when 7=>
temp0<=("000"& Apwdl(to_integer(unsigned(RTX(0 to 3)))));
temp1<=("000"& Apwdm(to_integer(unsigned(RTX(4 to 7)))));
temp2<=("000"& Apwdm(to_integer(unsigned(RTX(8 to 11)))));
temp3<=("000"& Apwdm(to_integer(unsigned(RTX(12 to 15)))));
temp4<=("000"& Apwdm(0));----------------------------------------
temp5<=("000"& Apwdl(to_integer(unsigned(RTM(0 to 3)))));
temp6<=("000"& Apwdm(to_integer(unsigned(RTM(4 to 7)))));
temp7<=("000"& Apwdm(to_integer(unsigned(RTM(8 to 11)))));
temp8<=("000"& Apwdm(to_integer(unsigned(RTM(12 to 15)))));
temp9<=("000"& Apwdm(0));
nst<=st+1;
when 8=>
Rw(4 to 7)<= (temp0 + temp1 + temp2 + temp3   + temp4);
Rw(12 to 15)<= (temp5 + temp6 + temp7 + temp8 + temp9);
nst<=st+1;
```

```
when 9=>
Rw(4 to 11)<=(Rw(4 to 7) * Rw(8 to 11));
nst<=st+1;
when 10=>
PAD1(0)<=Kpwdl(to_integer(unsigned(Rv(0 to 3))));
PAD1(1)<=Kpwdl(to_integer(unsigned(Rv(4 to 7))));
PAD1(2)<=Kpwdl(to_integer(unsigned(Rv(8 to 11))));
PAD1(3)<=Kpwdl(to_integer(unsigned(Rv(12 to 15))));---------------
PAD1(8)<=Kpwdl(to_integer(unsigned(Rw(0 to 3))));
PAD1(9)<=Kpwdl(to_integer(unsigned(Rw(4 to 7))));
PAD1(10)<=Kpwdl(to_integer(unsigned(Rw(8 to 11))));
PAD1(11)<=Kpwdl(to_integer(unsigned(Rw(12 to 15))));
nst<=st+1;
when 11=>
temp0<=("000"& Kpwdl(to_integer(unsigned(Rv(0 to 3)))));
temp1<=("000"& Kpwdm(to_integer(unsigned(Rv(4 to 7)))));
temp2<=("000"& Kpwdm(to_integer(unsigned(Rv(8 to 11)))));
temp3<=("000"& Kpwdm(to_integer(unsigned(Rv(12 to 15)))));
temp4<=("000"& Kpwdm(0));---------------------------------------
temp5<=("000"& Kpwdl(to_integer(unsigned(Rw(0 to 3)))));
temp6<=("000"& Kpwdm(to_integer(unsigned(Rw(4 to 7)))));
temp7<=("000"& Kpwdm(to_integer(unsigned(Rw(8 to 11)))));
temp8<=("000"& Kpwdm(to_integer(unsigned(Rw(12 to 15)))));
temp9<=("000"& Kpwdm(0));
nst<=st+1;
when 12=>
PAD1(4 to 7)<= (temp0 + temp1 + temp2 + temp3 + temp4);
PAD1(12 to 15)<= (temp5 + temp6 + temp7 + temp8      + temp9);
nst<=st+1;
when 13=>
PAD1(4 to 11)<=(PAD1(4 to 7) * PAD1(8 to 11));
nst<=st+1;
when 14=>
PAD1(4 to 15)<=(PAD1(4 to 11) * PAD1(12 to 15));
nst<=st+1;
when 15=>
Rvw<=(Rv xor Rw);
nst<=st+1;
when 16=>
PAD2(0)<=Kpwdl(to_integer(unsigned(Rv(0 to 3))));
PAD2(1)<=Kpwdl(to_integer(unsigned(Rv(4 to 7))));
PAD2(2)<=Kpwdl(to_integer(unsigned(Rv(8 to 11))));
PAD2(3)<=Kpwdl(to_integer(unsigned(Rv(12 to 15))));---------------
PAD2(8)<=Kpwdl(to_integer(unsigned(Rvw(0 to 3))));
PAD2(9)<=Kpwdl(to_integer(unsigned(Rvw(4 to 7))));
PAD2(10)<=Kpwdl(to_integer(unsigned(Rvw(8 to 11))));
PAD2(11)<=Kpwdl(to_integer(unsigned(Rvw(12 to 15))));
nst<=st+1;
when 17=>
temp0<=("000"& Kpwdl(to_integer(unsigned(Rv(0 to 3)))));
temp1<=("000"& Kpwdm(to_integer(unsigned(Rv(4 to 7)))));
temp2<=("000"& Kpwdm(to_integer(unsigned(Rv(8 to 11)))));
temp3<=("000"& Kpwdm(to_integer(unsigned(Rv(12 to 15)))));
temp4<=("000"& Kpwdm(0));---------------------------------------
temp5<=("000"& Kpwdl(to_integer(unsigned(Rvw(0 to 3)))));
temp6<=("000"& Kpwdm(to_integer(unsigned(Rvw(4 to 7)))));
temp7<=("000"& Kpwdm(to_integer(unsigned(Rvw(8 to 11)))));
temp8<=("000"& Kpwdm(to_integer(unsigned(Rvw(12 to 15)))));
temp9<=("000"& Kpwdm(0));
nst<=st+1;
when 18=>
```

```
PAD2(4 to 7)<= (temp0 + temp1 + temp2 + temp3 + temp4);
PAD2(12 to 15)<= (temp5 + temp6 + temp7 + temp8      + temp9);
nst<=st+1;
when 19=>
PAD2(4 to 11)<=(PAD1(4 to 7) * PAD1(8 to 11));
nst<=st+1;
when 20=>
PAD2(4 to 15)<=(PAD1(4 to 11) * PAD1(12 to 15));
nst<=st+1;
when 21=>
CCPwdm<=(Apwdm xor PAD1);
CCPwdl<=(Apwdl xor PAD2);
nst<=st+1;
when 22=>
if (CCPwdm(0 to 3)>= "1010") then CCPwdm_disp0<=("00110111" + ("0000"&
CCPwdm(0 to 3)));
else CCPwdm_disp0<= ("0011"& CCPwdm(0 to 3));
end if;
if (CCPwdm(4 to 7)>= "1010") then CCPwdm_disp1<=("00110111" + ("0000"&
CCPwdm(4 to 7)));
else CCPwdm_disp1<= ("0011"& CCPwdm(4 to 7));
end if;
if (CCPwdm(8 to 11)>= "1010") then CCPwdm_disp2<=("00110111" + ("0000"&
CCPwdm(8 to 11)));
else CCPwdm_disp2<= ("0011"& CCPwdm(8 to 11));
end if;
if (CCPwdm(12 to 15)>= "1010") then CCPwdm_disp3<=("00110111" +
("0000"& CCPwdm(12 to 15)));
else CCPwdm_disp3<= ("0011"& CCPwdm(12 to 15));
end if;
if (CCPwdl(0 to 3)>= "1010") then CCPwdl_disp0<=("00110111" + ("0000"&
CCPwdl(0 to 3)));
else CCPwdl_disp0<= ("0011"& CCPwdl(0 to 3));
end if;
if (CCPwdl(4 to 7)>= "1010") then CCPwdl_disp1<=("00110111" + ("0000"&
CCPwdl(4 to 7)));
else CCPwdl_disp1<= ("0011"& CCPwdl(4 to 7));
end if;
if (CCPwdl(8 to 11)>= "1010") then CCPwdl_disp2<=("00110111" + ("0000"&
CCPwdl(8 to 11)));
else CCPwdl_disp2<= ("0011"& CCPwdl(8 to 11));
end if;
if (CCPwdl(12 to 15)>= "1010") then CCPwdl_disp3<=("00110111" +
("0000"& CCPwdl(12 to 15)));
else CCPwdl_disp3<= ("0011"& CCPwdl(12 to 15));
end if;
nst<=0;
when others=> null;
end case;
end process;
--------------------------------------------------------------------------------------
p3:process(m_clk)
variable cnt: integer:=1;
begin
if rising_edge(m_clk) then
if(cnt=4000)then
clk_sig0<= not(clk_sig0);
cnt:=1;
else
cnt:= cnt + 1;
end if;
```

```
end if;
end process;
---------------------------------------------------------------------------------------
p4:process(clk_sig0)
begin
if rising_edge(clk_sig0) then
sta<=nsta;
end if;
end process;
---------------------------------------------------------------------------------------
p5:process(sta)
begin
case sta is
when 0=> d<="00111000";c<="100"; nsta<=sta+1;--38
when 1|3|5|7|9|43=> c<="000";nsta<=sta+1;
when 2=> d<="00001100";c<="100"; nsta<=sta+1;--0C
when 4=> d<="00000001";c<="100"; nsta<=sta+1;--01
when 6=> d<="00000110";c<="100"; nsta<=sta+1;--06
when 8=> d<="10000000";c<="100"; nsta<=sta+1;--80
when 10=> d<=std_logic_vector(to_unsigned(character'pos(Message0(1)),8));
c<="110"; nsta<=sta+1;--C
when 11|13|15|17|19|21|23|25|27|29|31|33|35|37|39|41|45|47|49|51|53|55|5
7|59|61|63|65|67|69|71|73|75=> c<="010";nsta<=sta+1;
when 12=> d<=std_logic_vector(to_unsigned(character'pos(Message0(2)),8));
c<="110"; nsta<=sta+1;--P
when 14=> d<=std_logic_vector(to_unsigned(character'pos(Message0(3)),8));
c<="110"; nsta<=sta+1;--M
when 16=> d<=std_logic_vector(to_unsigned(character'pos(Message0(4)),8));
c<="110"; nsta<=sta+1;--+
when 18=> d<=std_logic_vector(to_unsigned(character'pos(Message0(5)),8));
c<="110"; nsta<=sta+1;--A
when 20=> d<=std_logic_vector(to_unsigned(character'pos(Message0(6)),8));
c<="110"; nsta<=sta+1;--P
when 22=> d<=std_logic_vector(to_unsigned(character'pos(Message0(7)),8));
c<="110"; nsta<=sta+1;--M
when 24=> d<=std_logic_vector(to_unsigned(character'pos(Message0(8)),8));
c<="110"; nsta<=sta+1;--=
when 26=> d<=CCPwdm_disp0;c<="110"; nsta<=sta+1;--CCPwdm4
when 28=> d<=CCPwdm_disp1;c<="110"; nsta<=sta+1;--CCPwdm5
when 30=> d<=CCPwdm_disp2;c<="110"; nsta<=sta+1;--CCPwdm6
when 32=> d<=CCPwdm_disp3;c<="110"; nsta<=sta+1;--CCPwdm7
when 34=> d<=std_logic_vector(to_unsigned(character'pos(Message2(5)),8));
c<="110"; nsta<=sta+1;--APwdm4
when 36=> d<=std_logic_vector(to_unsigned(character'pos(Message2(6)),8));
c<="110"; nsta<=sta+1;--APwdm5
when 38=> d<=std_logic_vector(to_unsigned(character'pos(Message2(7)),8));
c<="110"; nsta<=sta+1;--APwdm6
when 40=> d<=std_logic_vector(to_unsigned(character'pos(Message2(8)),8));
c<="110"; nsta<=sta+1;--APwdm7
when 42=> d<="11000000";c<="100"; nsta<=sta+1;--C0
when 44=> d<=std_logic_vector(to_unsigned(character'pos(Message1(1)),8));
c<="110"; nsta<=sta+1;--C
when 46=> d<=std_logic_vector(to_unsigned(character'pos(Message1(2)),8));
c<="110"; nsta<=sta+1;--P
when 48=> d<=std_logic_vector(to_unsigned(character'pos(Message1(3)),8));
c<="110"; nsta<=sta+1;--L
when 50=> d<=std_logic_vector(to_unsigned(character'pos(Message1(4)),8));
c<="110"; nsta<=sta+1;--+
when 52=> d<=std_logic_vector(to_unsigned(character'pos(Message1(5)),8));
c<="110"; nsta<=sta+1;--A
```

```
when 54=> d<=std_logic_vector(to_unsigned(character'pos(Message1(6)),8));
c<="110"; nsta<=sta+1;--P
when 56=> d<=std_logic_vector(to_unsigned(character'pos(Message1(7)),8));
c<="110"; nsta<=sta+1;--L
when 58=> d<=std_logic_vector(to_unsigned(character'pos(Message1(8)),8));
c<="110"; nsta<=sta+1;--=
when 60=> d<=CCPwdl_disp0;c<="110"; nsta<=sta+1;--CCPwdL0
when 62=> d<=CCPwdl_disp1;c<="110"; nsta<=sta+1;--CCPwdL1
when 64=> d<=CCPwdl_disp2;c<="110"; nsta<=sta+1;--CCPwdL2
when 66=> d<=CCPwdl_disp3;c<="110"; nsta<=sta+1;--CCPwdL3
when 68=> d<=std_logic_vector(to_unsigned(character'pos(Message2(1)),8));
c<="110"; nsta<=sta+1;--APwdL0
when 70=> d<=std_logic_vector(to_unsigned(character'pos(Message2(2)),8));
c<="110"; nsta<=sta+1;--APwdL1
when 72=> d<=std_logic_vector(to_unsigned(character'pos(Message2(3)),8));
c<="110"; nsta<=sta+1;--APwdL2
when 74=> d<=std_logic_vector(to_unsigned(character'pos(Message2(4)),8));
c<="110"; nsta<=sta+1;--APwdL3
when 76=> nsta<=8;
when others => nsta<=8;
end case;
end process;
-----------------------------------------------------------------------------------------------------------
p6:process(EPC_data_disp_cont)
begin
case EPC_data_disp_cont is
when '0'=>EPC_disp_data<= RTX;
when '1'=>EPC_disp_data<=RMX;
when others => null;
end case;
end process;
-----------------------------------------------------------------------------------------------------------
end Behavioral;
```

13.4 Soft Computing Algorithms

Soft computing is a collection of algorithms that are employed for finding solutions for very complex problems. Soft computing yields low-cost and time-feasible solutions, unlike the more conventional methods which have not. As we know, how much space the data structures take and how much time an algorithm takes to run are very important factors since space and time are the two main parameters for measuring the efficiency of any algorithm. Soft computing algorithms take relatively smaller space and shorter computing time; hence, this technique is becoming more significant. The following are the primary types of algorithms coming under soft computing, (i) artificial neural network (ANN), (ii), fuzzy logic control (FLC) systems [24,39], (iii) evolutionary computation (EC), (iv) machine learning (ML), and (v) probabilistic reasoning (PR) and state vector machine (SVM). Soft computing is a popular technique for solving several higher degrees of nonlinear problems. Soft computing differs from conventional (hard) computing in the sense that it is tolerant of imprecision, uncertainty, partial truth and approximation. In effect, the role model for soft computing is the human mind. Hard computing, that is, conventional computing, requires a precise stated analytical model and often takes a lot of computation time. Many analytical models are valid only for ideal cases, and many real-world problems exist in a nonideal environment. Soft computing has wide applications, including handwriting recognition, automotive systems and manufacturing, image processing and data compression, efficient architecture design, decision-support systems, power systems, neuro-fuzzy systems, and fuzzy logic control. In this section, we discuss ANNs and FLC systems.

13.4.1 Artificial Neural Network

Objective: An ANN is a computational model based on the structure and functions of biological neural networks; thus, the ANN learns from experience. Information that flows through the network affects the structure of the ANN because a neural network changes or learns in a sense based on that input and output. ANNs are considered nonlinear statistical data modeling tools where the complex relationships between inputs and outputs are modeled or patterns are formed. An ANN has several advantages, but one of the most recognized is that it can actually learn from observing data sets. The ANN takes data samples rather than entire data sets to arrive at solutions, which saves both time and cost. Typical ANNs have three layers that are interconnected. The first layer consists of input neurons. Those neurons send data to the second layer, which in turn sends to the output neurons, that is, the third layer. The middle layer is called the hidden layer and the last layer is called the output layer [23,24,26,39,163–165]. ANNs, due to advances in biological research, promise an initial understanding of the natural thinking mechanism and to understand how brains store information patterns. Some of these patterns are very complicated but allow us to recognize individual faces from many different angles. This process of storing information as patterns, utilizing those patterns, and then solving problems, encompasses a new field in computing. An ANN cannot be programmed to perform a specific task because the network finds out how to solve the problem by itself, that is, its operation can be unpredictable. Now, we design a simple ANN for a data-mapping function. Store the following data in an Excel file columnwise, that is, with the values of x and y in two different columns. The value of x varies from 0 through 30 at an increment factor of 0.5, that is, x[n]=0:0.5:30. The corresponding values of y are y[n]= {0, 0.0099, 0.0395, 0.0882, 0.1558, 0.2419, 0.346, 0.4679, 0.6072, 0.7635, 0.9365, 1.1259, 1.3314, 1.5526, 1.7892, 2.0409, 2.3075, 2.5885, 2.8838, 3.1931, 3.516, 3.8523, 4.2018, 4.5642, 4.9392, 5.3265, 5.726, 6.1374, 6.5605, 6.9949, 7.4406, 7.8972, 8.3646, 8.8426, 9.3308, 9.8293, 10.3376, 10.8557, 11.3833, 11.9203, 12.4664, 13.0216, 13.5855, 14.1581, 14.7391, 15.3285, 15.926, 16.5314, 17.1446, 17.7656, 18.394, 19.0297, 19.6727, 20.3228, 20.9798, 21.6436, 22.314, 22.991, 23.6743, 24.3639, 25.0597}. Store the Excel file that contains the above data in C:\Program Files\MATLAB\R2014a\work\T_F_Book.xls as Sheet1, or store it where MATLAB® is installed on your computer. The following MATLAB code will read the data from the Excel file and design an ANN structure for mapping the given data set.

Solution

```
clear;
clc;
p=xlsread('C:\Program Files\MATLAB\R2014a\work\T_F_Book.xls','Sheet1');
t=p(:,[1]);
v=p(:,[2]);
Ex_data=[t v]
grid on;
figure(1);
subplot(221)
plot(t,v,'lineWidth',1.5);
set(gca,'fontsize',10);
title('Behavior of Unknown System');
```

```
xlabel('Pressure(Pa)'); ylabel('Arbitrary Output Voltage (v)');
%Define and plot training as well as checking data
data=[t v]; % 60 data
trndata=data(1:2:size(t),:); % 30 data
chkdata=data(2:2:size(t),:); %30 data
grid on;
subplot(222)
plot(trndata(:,1),trndata(:,2),'o',chkdata(:,1),chkdata(:,2),'x','lineWi
dth',1.5);
set(gca,'fontsize',10);
title('Training data and Testing data');
xlabel('Pressure(Pa)');
ylabel('Arbitrary Output Voltage (v)');
%initialize ANN
trainpoint=trndata(:,1)';
trainoutput=trndata(:,2)';
net=newff(minmax(trainpoint),[10,1],{'tansig','purelin'},'trainlm');
%simulate and plot the network output without training
y=sim(net,trainpoint);
grid on;
subplot(223)
plot(trainpoint,trainoutput,trainpoint,y,'o','lineWidth',1.5);
set(gca,'fontsize',10);
title('Original o/p & NN o/p without training');
xlabel('Pressure(Pa)');
ylabel('Arbitrary Output Voltage (v)');
grid on;
%Initialize parameter
net.trainparam.epochs=100;
net.trainparam.goal=0.0001;
net.trainparam.mc=0.3;
%Train the network and plot the error
net=train(net,trainpoint,trainoutput);
y=sim(net,trainpoint);
subplot(224)
plot(trainpoint,trainoutput,trainpoint,y,'o','lineWidth',1.5);
set(gca,'fontsize',10);
title('Original o/p & trained NN o/p ')
xlabel('Pressure(Pa)'); ylabel('Arbitrary Output Voltage (v)');
grid on;
%test the network and plot the error
checkpoint=chkdata(:,1)';
checkoutput=chkdata(:,2)';
w=sim(net,checkpoint);
e=w-checkoutput;
figure(2);
subplot(221)
bar(e);
set(gca,'fontsize',10);
title('Error profile ')
xlabel('Check data points');ylabel('Error');
grid on;
ylim([-0.1 0.15]);
xlim([0 30]);
```

The simulation results of this MATLAB code are given in Figures 13.1 and 13.2.

Designing the VHDL code for this ANN-based function approximation is left to the readers as a laboratory exercise.

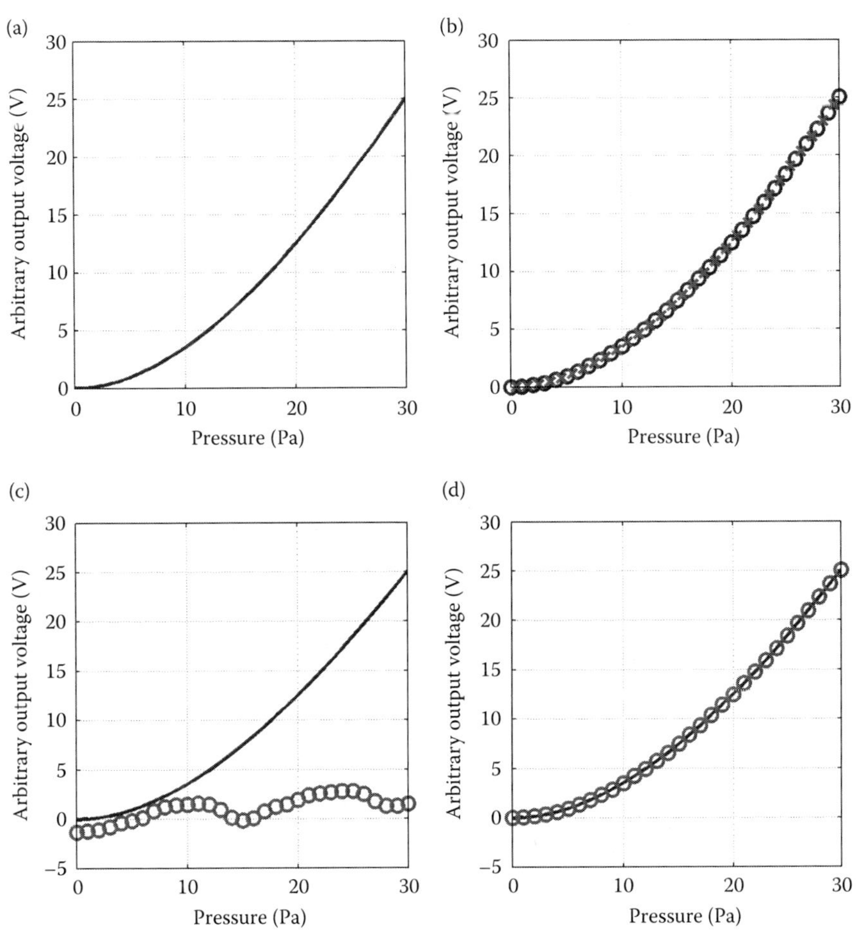

FIGURE 13.1
MATLAB simulation results of fuzzy logic controller. (a) Behavior of unknown system, (b) training data and testing data, (c) original o/p and NN o/p without training, and (d) original o/p and trained NN o/p.

13.4.2 Fuzzy Logic Controller

Objective: The main application of fuzzy logic in engineering is control systems. In recent years, the applications of fuzzy logic have increased significantly. These applications range from consumer products such as cameras, camcorders, washing machines, and microwave ovens to industrial process control, medical instrumentation, decision-support systems, and portfolio selection. In a wider sense, fuzzy logic is almost synonymous with the theory of fuzzy sets, which is a theory that relates to classes of objects with unsharp boundaries in which membership is a matter of degree [36,166,167]. The basic concept underlying FL is that of a linguistic variable, that is, a variable whose values are words rather than numbers, for example, very very high (VVH), very high (VH), low speed (LS), overcool (OC), overheat (OH), and so on. Fuzzy logic may be viewed as a methodology for computing with words rather than numbers. Although words are inherently less precise than numbers, their use is closer to human intuition. Furthermore, computing with words exploits the tolerance for imprecision and thereby lowers the cost of the solution. In fuzzy logic, this mechanism is provided by the calculus of fuzzy rules, which is a basis for the fuzzy

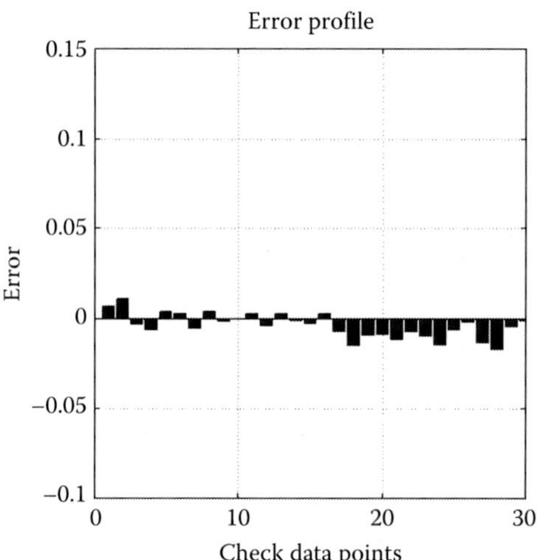

FIGURE 13.2
MATLAB simulation correlation errors.

dependency and command language. In the future, soft computing could play an increasingly important role in the conception and design of systems whose machine IQs (MIQs) are much higher than those of systems designed by conventional methods. In this section, we design a simple fuzzy logic controller as given below.

Solution

```
clc;
clear all;
x=[20:1:80];%Temperature
y=[10:1:90];%Cooling
%temperature membership function
VS=trimf(x,[20 20 35]);
S= trimf(x,[20 35 50]);
C= trimf(x,[35 50 65]);
F= trimf(x,[50 65 80]);
VF=trimf(x,[65 80 80]);
antecedent_mf=[VS;S;C;F;VF];
figure(1);
subplot(2,2,1)
plot(x,antecedent_mf,'lineWidth',1.5);
set(gca,'fontsize',10);
title('VS S C F VF ');
xlabel('Temperature');
ylabel('Membership');
grid on;
% %cooling
N= trimf(y,[10 10 30]);
S1= trimf(y,[10 30 50]);
HS= trimf(y,[30 50 70]);
VHS=trimf(y,[50 70 90]);
MS= trimf(y,[70 90 90]);
consequent_mf=[N;S1;HS;VHS;MS];
subplot(2,2,2)
grid on;
plot(y,consequent_mf,'lineWidth',1.5)
set(gca,'fontsize',10);
```

```
title('N S1 HS VHS MS');
grid on;
xlabel('Cooling');
ylabel('Membership');
temp=input('Enter the temperature value (10<= temp>=50):= ');
antecedent1=min(VS(find(x==temp)));
antecedent2=min(S(find(x==temp)));
antecedent3=min(C(find(x==temp)));
antecedent4=min(F(find(x==temp)));
antecedent5=min(VF(find(x==temp)));
antecedent=[antecedent1 antecedent2 antecedent3 antecedent4 antecedent5]
consequent1=(N.*antecedent5);
consequent2=(S1.*antecedent4);
consequent3=(HS.*antecedent3);
consequent4=(VHS.*antecedent2);
consequent5=(MS.*antecedent1);
consequent=[consequent1 consequent2 consequent3 consequent4 consequent5]
subplot(2,2,3)
plot(y, [consequent1;consequent2;consequent3;consequent4;consequent5],
'lineWidth',1.5);
set(gca,'fontsize',10);
grid on;
axis([10 90 0 1]);
title('consequent Fuzz set');
xlabel('Cooling');
ylabel('Membership');
output_mf=max([consequent1;consequent2;consequent3;consequent4;
consequent5]);
subplot(2,2,4)
plot(y,output_mf,'r','lineWidth',1.5);
set(gca,'fontsize',10);
grid on;
axis([10 90 0 1]);
xlabel('cooling');
ylabel('Membership');
output=sum(output_mf.*y)/sum(output_mf)
hold on;
stem(output,1,'lineWidth',1.5);
set(gca,'fontsize',10);
hold off;
```

The simulation results of the above fuzzy logic controller are given in Figure 13.3.

Designing the VHDL code for this fuzzy logic controller is left to the readers as a laboratory exercise.

13.5 Bit Error Rate Tester Design

Objective: The BER tester is an essential device to measure and quantify the performance and reliability of digital data links. Most commercial BER testers work based on the principle of model-based measurement of BER. That means the reference data pattern, normally a PRBS, is generated by the BER tester and will be input into one end of the channel and collected from the other end by the same device. Thus, the data comparison becomes simpler since the transmitted and received data patterns are available at the BER measurement

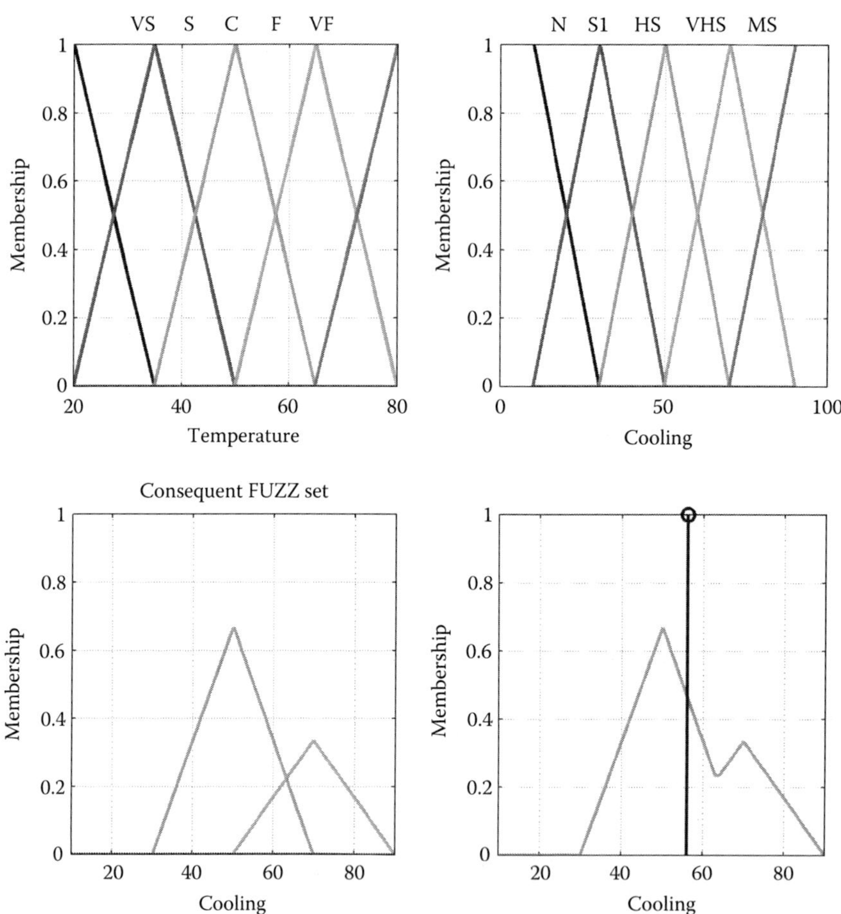

FIGURE 13.3
MATLAB simulation results of fuzzy logic controller.

device. In actual communication systems, for example, in a long-range FSO communication system [22,38,39,69], the PRBS reference data injection is at the transmitter and data collection is at the distant receiver. In this situation, we cannot compare the received sequence with the transmitted one. Therefore, we need a BER tester which has provisions to send the reference PRBS data at the transmitter, collect the data at the receiver and measure the BER at the receiver itself. This is possible only when we have perfect synchronization between the transmitting and receiving sequence. This synchronization can be achieved only by the pilot sequence. The transmitter and receiver have the same PRBS generators, and the transmitter first sends the pilot sequence and then the reference PRBS pattern. The receiver does not enable its PRBS generator until it detects a valid pilot sequence. Upon getting a valid pilot sequence, the PRBS generator of the receiver is enabled and starts comparing its bits with the received bits via an xor gate; thus, the decision of either correct or error bit is taken, which increments the error counter if an error is identified. Every predetermined period, the value of the error counter is sent to the computer or LCD and the counter is reset. Thus, we can design a simple BER tester just by keeping the same PRBS generator at the transistor and receiver with a proper pilot synchronization sequence. Design a digital system to realize the above operation.

Solution

```
library ieee;
use ieee.std_logic_1164.all;
use ieee.std_logic_arith.all;
use ieee.std_logic_unsigned.all;
entity bert_final is
port(clk,reset,si:in std_logic:='0';
x:inout std_logic:='0';
clkout:inout std_logic:='0';
txd : out std_logic:='0';
error1:out std_logic_vector(5 downto 0):="000000");
end bert_final;
architecture behavioral of bert_final is
signal state, nextstate : integer range 0 to 47;
signal clk_sig,baud_clk_sig : std_logic:='0';
signal po_reg, po_reg_temp: std_logic_vector(31 downto
0):="00000000000000000000000000000000";
signal mux_sig:std_logic:='0';
signal txd_temp:std_logic:='0';
signal sdata:std_logic_vector(10 downto 0):="00000000000";
signal error_temp:std_logic_vector(10 downto 0):="00000000000";
signal error:std_logic_vector(5 downto 0):="000000";
signal control_bit : std_logic :='0';
signal state1, nextstate1 : integer range 0 to 67;
begin
error1<=error;
error_temp <= ("1100"& error(5 downto 0) & '0');
---------------------------------------------------------------------
p1:process(clk)
variable cnt:integer:=0;
begin
if (rising_edge(clk)) and (reset='0') then
if(cnt=2000000) then
clk_sig <= not(clk_sig);
cnt:=0;
else
cnt:=cnt+1;
end if;
end if;
end process;
clkout <= clk_sig;
---------------------------------------
p2:process(clk)
variable cnt1:integer:=0;
begin
if rising_edge(clk) and (reset='0') then
if(cnt1=17)then
baud_clk_sig<= not(baud_clk_sig);
cnt1:=1;
else
cnt1:= cnt1 + 1;
end if;
end if;
end process;
----------------------------------
p3:process(clk_sig,reset)
begin
if reset ='1' then
state <= 0;
elsif clk_sig'event and clk_sig='1' then
```

```
state <= nextstate;
end if;
end process;
--------------------------------------
p4: process(state,si)
begin
case state is
when 0 =>
if si='0' then nextstate <= 1;
else nextstate <= 0;
end if;
-------------------------------------
when 1 =>
if si='1' then nextstate <= 2;
else nextstate <= 1;
end if;
--------------------------------------
when 2 =>
if si='1' then nextstate <= 3;
else nextstate <= 1;
end if;
--------------------------------------
when 3 =>
if si='0' then nextstate <= 4;
else nextstate <= 0;
end if;
---------------------------------------
when 4 =>
if si='0' then nextstate <= 5;
else nextstate <= 0;
end if;
---------------------------------------
when 5 =>
if si='1' then nextstate <= 6;
else nextstate <= 1;
end if;
--------------------------------------
when 6 =>
if si='1' then nextstate <= 7;
else nextstate <= 1;
end if;
--------------------------------------
when 7 =>
if si='0' then nextstate <= 8;
else nextstate <= 0;
end if;
--------------------------------------
when 8 =>
if si='0' then nextstate <= 9;
else nextstate <= 0;
end if;
-------------------------------------
when 9 =>
if si='1' then nextstate <= 10;
else nextstate <= 1;
end if;
-----------------------------------
when 10 =>
if si='1' then nextstate <= 11;
else nextstate <= 1;
end if;
```

```
-----------------------------------
when 11 =>
if si='0' then nextstate <= 12;
else nextstate <= 0;
end if;
-----------------------------------
when 12 =>
if si='0' then nextstate <= 13;
else nextstate <= 0;
end if;
-----------------------------------
when 13 =>
if si='1' then nextstate <= 14;
else nextstate <= 1;
end if;
---------------------------------------
when 14 =>
if si='1' then nextstate <= 15;
else nextstate <= 1;
end if;
--------------------------------------
when 15 =>
if si= '1' then nextstate <= 16;
else nextstate <= 1;
end if;
-------------------------------------
when
16|17|18|19|20|21|22|23|24|25|26|27|28|29|30|31|32|33|34|35|36|37|38|39|
40|41|42|43|44|45|46 =>
nextstate <= state+1;
-------------------------------------
when 47 =>
if si= '1' then nextstate <= 0;
else nextstate <= 1;
end if;
--------------------------------------
when others => nextstate <= 0;
end case;
end process;
x <= '1' when state > 15 else '0';
------------------------------------
p5:process(clk_sig,control_bit,si)
variable cnt2:integer:=0;
variable cnt3:integer:=0;
variable error_cnt:integer:=0;
begin
if rising_edge(clk_sig) then
if(cnt2=32) then
po_reg_temp <= po_reg;
cnt2:=0;
end if;
if(cnt3=33) then
cnt3:=0;
end if;
if (x='1') then
po_reg(31 downto 0) <= (po_reg(30
downto 0)& si);
cnt2:=cnt2+1;
end if;
if (control_bit='1') then
if (po_reg_temp(0)='1') then
```

```
error_cnt:=error_cnt+ 1;
cnt3:=cnt3+1;
else
error_cnt:=error_cnt;
end if;
po_reg_temp <= (po_reg_temp(0) &
po_reg_temp(31 downto 1));
if(cnt3=32) then
error<=conv_std_logic_vector(error_cnt,6);
end if;
else
error_cnt:=0;
end if;
end if;
end process;
----------------------------------------
p6:process(clk_sig,reset)
begin
if reset ='1' then
state1 <= 0;
elsif clk_sig'event and clk_sig='1' then
state1 <= nextstate1;
end if;
end process;
------------------------------------------
p7: process(state1,x)
begin
case state1 is
when 0 =>
if x='1' then
nextstate1 <= 1;
else nextstate1 <= 0;
end if;
-------------------------------------
when 1|2|3|4|5|6|7|8|9|10|11|12|13|14|15|16|17|18|19|20|21|22|23|24|25|26|
27|28|29|30|31|32|33|34|35|36|37|38|39|40|41|42|43|44|45|46|47|48|49|50|
51|52|53|54|55|56|57|58|59|60|61|62|63|64|65|66 =>
nextstate1 <= state1+1;
-----------------------------------------
when 67 =>
nextstate1 <= 0;
-------------------------------------
when others => nextstate1 <= 0;
end case;
end process;
control_bit <= '1' when state1 > 32 else '0';
-----------------------------------------------
p8:process(baud_clk_sig)
variable cnt2:integer:=0;
begin
if rising_edge(baud_clk_sig)then
if (cnt2>10) then
mux_sig<='1';
if (cnt2=115200) then
sdata<=error_temp;
cnt2:=1;
mux_sig<='0';
end if;
cnt2:=cnt2+1;
else
mux_sig<='0';
```

```
sdata<=('0'& sdata(10 downto 1));
cnt2:=cnt2+1;
end if;
case mux sig is
when '0'=>txd_temp<=sdata(0);
when '1'=>txd_temp<='1';
when others=> null;
end case;
end if;
end process;
txd<=txd_temp;
end behavioral;
```

13.6 Optical Up/Down Data Link

Objective: Design and develop a simple optical encryption and decryption system. Assume that there are two up/down data links that are connected through two separate FSO communication channels [39]. The performance of the FSO link is disturbed by a manmade atmospheric chamber. Two modulated optical beams are passed into the chamber at its front end and reflected back to the receiver by a reflector kept at another end of the chamber, which means the transmitter and receiver are located at a place so as to compare the transmitted data with the received data. Therefore, the downlink may not get the data that are transmitted due to the disturbance of the chamber. Develop a digital system to generate a PRBS which encrypts the original message (14 bits) that arrives from the external DIP switches. The encrypted data is transmitted over the FSO channel and received back as indicated above. The system has to compare both (Tx and Rx) data and calculate the BER, which has to be transmitted to the computer through the serial port once in a predetermined period. Two bits of selection inputs are used to choose one PRBS among three. Develop a digital system in the FPGA to perform this operation. The optical modulation followed in this work is ASK. A BER tester, discussed in the preceding section, is used to measure the BER.

Solution

```
LIBRARY ieee;
USE ieee.std_logic_1164.all;
USE ieee.numeric_std.all;
use ieee.std_logic_unsigned.all;
use IEEE.STD_LOGIC_ARITH.ALL;
--------------------------------------------------------
entity sheefa_pro is
port (m_clk:in std_logic:='0';
dss:in std_logic_vector(13 downto 0):="00000000000000";
onu_sel:std_logic_vector(1 downto 0):="00";
uss_tx,dss_tx: out std_logic:='0';
rx: in std_logic_vector(1 downto 0):="00";
uart_tx: out std_logic:='0');
end sheefa_pro;
architecture Behavioral of sheefa_pro is
signal clk_sig,prbs_onu1_feed,prbs_onu2_feed,prbs_onu3_feed,clk_sig0,
mux_sig,
txd_temp:std_logic:='0';
```

```
signal prbs_onu1:std_logic_vector(13 downto 0):="10110001100101";
signal prbs_onu2:std_logic_vector(13 downto 0):="11111110000000";
signal prbs_onu3:std_logic_vector(13 downto 0):="00000001111111";
signal sto_uss_onu1,sto_uss_onu2,sto_uss_onu3,rx_uss,dss_temp,
enc_dss,rx_enc_dss,decr_dss1,decr_dss2,decr_dss3,
error1,error2,error3:std_logic_vector(13 downto 0):="00000000000000";
signal h1,h2,er11,er12,er21,er22,er31,er32:std_logic_vector(7 downto 0):
="00000000";
signal sdata:std_logic_vector(10 downto 0):="11001100000";
begin
dss_tx<='1';
-----------------------------------------------------------------
P0:process(m_clk)
variable cnt1:integer:=0;
begin
if rising_edge(m_clk) then
if (cnt1=20000) then
clk_sig<=not clk_sig;
cnt1:=0;
else
cnt1:=cnt1+1;
end if;
end if;
end process;
-----------------------------------------------------------------
P1:process(clk_sig)
begin
if rising_edge(clk_sig) then
prbs_onu1_feed<=(prbs_onu1(12)xor prbs_onu1(8) xor prbs_onu1(6) xor
prbs_onu1(4)xor prbs_onu1(2));
prbs_onu2_feed<=(prbs_onu2(11)xor prbs_onu2(7) xor prbs_onu2(3) xor
prbs_onu2(1)xor prbs_onu2(5));
prbs_onu3_feed<=(prbs_onu3(3)xor prbs_onu3(12) xor prbs_onu3(1) xor
prbs_onu3(4)xor prbs_onu3(9));
end if;
if falling_edge(clk_sig) then
prbs_onu1<=(prbs_onu1_feed & prbs_onu1(13 downto 1));
prbs_onu2<=(prbs_onu2_feed & prbs_onu2(13 downto 1));
prbs_onu3<=(prbs_onu3_feed & prbs_onu3(13 downto 1));
end if;
end process;
--------------------------------------------------------------------------------------------------------------
p2:process(clk_sig,onu_sel,prbs_onu1,prbs_onu2,prbs_onu3,dss)
variable cnt2,cnt3:integer:=0;
begin
if rising_edge(clk_sig) then
if (cnt2=1) then
if (onu_sel="01") then
if (cnt3<14)then
sto_uss_onu1<=(sto_uss_onu1(12 downto 0) & prbs_onu1(0));
uss_tx<=prbs_onu1(0);
cnt3:=cnt3+1;
cnt2:=1;
else
cnt2:=cnt2+1;
cnt3:=0;
end if;
elsif (onu_sel="10") then
if (cnt3<14)then
sto_uss_onu2<=(sto_uss_onu2(12 downto 0) & prbs_onu2(0));
uss_tx<=prbs_onu2(0);
```

```
cnt3:=cnt3+1;
cnt2:=1;
else
cnt2:=cnt2+1;
cnt3:=0;
end if;
elsif (onu_sel="11") then
if (cnt3<14)then
sto_uss_onu3<=(sto_uss_onu3(12 downto 0) & prbs_onu3(0));
uss_tx<=prbs_onu3(0);
cnt3:=cnt3+1;
cnt2:=1;
else
cnt2:=cnt2+1;
cnt3:=0;
end if;
end if;
elsif (cnt2=2) then
case onu_sel is
when "01" =>rx_uss<=sto_uss_onu1;
when "10" =>rx_uss<=sto_uss_onu2;
when "11" =>rx_uss<=sto_uss_onu3;
when others=> null;
end case;
cnt2:=cnt2+1;
elsif (cnt2=3) then
dss_temp<=dss;
cnt2:=cnt2+1;
elsif (cnt2=4) then
enc_dss<=(dss_temp xor rx_uss);
cnt2:=cnt2+1;
elsif (cnt2=5) then
rx_enc_dss<=enc_dss;
cnt2:=cnt2+1;
elsif (cnt2=6) then
decr_dss1<=(enc_dss xor sto_uss_onu1);
decr_dss2<=(enc_dss xor sto_uss_onu2);
decr_dss3<=(enc_dss xor sto_uss_onu3);
cnt2:=cnt2+1;
elsif (cnt2=7) then
case onu_sel is
when "01"=> error1<= (decr_dss1 xor decr_dss1);
error2<= (decr_dss1 xor decr_dss2);
error3<= (decr_dss1 xor decr_dss3);
when "10"=> error1<= (decr_dss2 xor decr_dss1);
error2<= (decr_dss2 xor decr_dss2);
error3<= (decr_dss2 xor decr_dss3);
when "11"=> error1<= (decr_dss3 xor decr_dss1);
error2<= (decr_dss3 xor decr_dss2);
error3<= (decr_dss3 xor decr_dss3);
when others=> null;
end case;
cnt2:=cnt2+1;
elsif (cnt2=8) then
h1<=x"FF";h2<=x"FF";
er11<=error1(7 downto 0); er12<=("00"& error1(13 downto 8));
er21<=error2(7 downto 0); er22<=("00"& error2(13 downto 8));
er31<=error3(7 downto 0); er32<=("00"& error3(13 downto 8));
cnt2:=cnt2+1;
elsif (cnt2=9) then
cnt2:=0;
```

```
else
cnt2:=cnt2+1;
end if;
end if;
end process;
--------------------------------------------------------------------------
p3:process(m_clk,rx)
variable cnt4:integer:=0;
begin
if (rx="11") then
if rising_edge(m_clk) then
if(cnt4=208)then--208
clk_sig0<= not(clk_sig0);
cnt4:=1;
else
cnt4:= cnt4 + 1;
end if;
end if;
end if;
end process;
--------------------------------------------------------------------------
p4:process(clk_sig0)
variable cnt3,cnt4:integer:=1;
begin
if rising_edge(clk_sig0)then
if (cnt3>10) then
mux_sig<='1';
if (cnt3=11) then
sdata<=('0'& sdata(10 downto 1));
cnt3:=cnt3+1;
end if;
if (cnt3=1200) then
if (cnt4=1)then
sdata<="11000000000";
sdata(8 downto 1)<="11111111";--h1;
cnt3:=1;
mux_sig<='0';
cnt4:=cnt4+1;
elsif (cnt4=2)then
sdata<="11000000000";
sdata(8 downto 1)<="11111111";--h2;
cnt3:=1;
mux_sig<='0';
cnt4:=cnt4+1;
elsif (cnt4=3)then
sdata<="11000000000";
sdata(8 downto 1)<=er11;
cnt3:=1;
mux_sig<='0';
cnt4:=cnt4+1;
elsif (cnt4=4)then
sdata<="11000000000";
sdata(8 downto 1)<=er12;
cnt3:=1;
mux_sig<='0';
cnt4:=cnt4+1;
elsif (cnt4=5)then
sdata<="11000000000";
sdata(8 downto 1)<=er21;
cnt3:=1;
mux_sig<='0';
```

```
cnt4:=cnt4+1;
elsif (cnt4=6)then
sdata<="11000000000";
sdata(8 downto 1)<=er22;
cnt3:=1;
mux_sig<='0';
cnt4:=cnt4+1;
elsif (cnt4=7)then
sdata<="11000000000";
sdata(8 downto 1)<=er31;
cnt3:=1;
mux_sig<='0';
cnt4:=cnt4+1;
elsif (cnt4=8)then
sdata<="11000000000";
sdata(8 downto 1)<=er32;
cnt3:=1;
mux_sig<='0';
cnt4:=1;
end if;
else
cnt3:=cnt3+1;
end if;
else
mux_sig<='0';
sdata<=('0'& sdata(10 downto 1));
cnt3:=cnt3+1;
end if;
case mux_sig is
when '0'=>txd_temp<=sdata(0);
when '1'=>txd_temp<='1';
when others=> null;
end case;
end if;
end process;
uart_tx<=txd_temp;
end Behavioral;
```

The following MATLAB code reads the data arriving the serial port and displays the necessary values as written in the code.

MATLAB Code:

```
clc;
clear all;
fprintf ( ' ID.No: Date & Time ONU1 ERROR ONU2 ERROR ONU3 ERROR BER1
BER2 BER3 \n');
fprintf ( ' ==== ================= ========== ==========
================ === ==== ==== \n');
s = serial('com1');
s.BaudRate=9600;
fopen(s);
time = now;
stopTime = 'Mar-12 (Thu) 15:59:00';
count = 1;
ONU1=0;
ONU2=0;
ONU3=0;
BER1=0;
BER2=0;
BER3=0;
```

```
CL1=0;
CL2=0;
CL3=0;
i=0;
while ~isequal(datestr(now,'mmm-dd (ddd) HH:MM:SS'),stopTime)
date=datestr(now);
time(count) = datenum(clock);
a=fread(s,1,'uint8')';
if (a==255),
a=fread(s,1,'uint8')';
if (a==255),
e1=fread(s,2,'uint8')';
b=dec2bin(e1(1),8);
c=dec2bin(e1(2),8);
f=[c b];
ONU1=bin2dec(f);
BER1=(ONU1/16384)*100;
BER1_temp(count)=BER1;
e2=fread(s,2,'uint8')';
b=dec2bin(e2(1),8);
c=dec2bin(e2(2),8);
f=[c b];
ONU2=bin2dec(f);
BER2=(ONU2/16384)*100;
BER2_temp(count)=BER2;
e3=fread(s,2,'uint8')';
b=dec2bin(e3(1),8);
c=dec2bin(e3(2),8);
f=[c b];
ONU3=bin2dec(f);
BER3=(ONU3/16384)*100;
BER3_temp(count)=BER3;
end
end
fprintf('%6d %s %3E %3E %3E %3E %3E %3E\n',count,date,ONU1,ONU2,ONU3,BER1,
BER2,BER3);
count = count +1;
end
fclose(s);
CL1=(sum(BER1_temp)/length(BER1_temp));
CL1=(CL1/(14*length(BER1_temp)))
CL2=(sum(BER2_temp)/length(BER2_temp));
CL2=(CL2/(14*length(BER2_temp)))
CL3=(sum(BER3_temp)/length(BER3_temp));
CL3=(CL3/(14*length(BER3_temp)))
```

13.7 Channel Coding Techniques

In digital communication systems, there are two types of coding techniques, namely source coding and channel coding. Source coding is performed to convert the analog signal of the source into a digital signal, whereas channel coding is used to do some data manipulation at the transmitter so as to make the receiver detect and correct the error bits. There are several types of channel-coding techniques available and more are coming up every year. However, very popular and basic channel-coding techniques are (i) linear block codes and (ii) convolutional codes [168]. We discuss them in the following parts of this section.

13.7.1 Linear Block Code

Objective: A block code is a code in which k bits/symbols are the input and n bits/symbols are the output. We represent the code as an (n, k) code word. If we input k bits, then there are 2k distinct messages. Each message of n symbols associated with each input block is called a codeword. We could, in general, simply have a lookup table with k inputs and n outputs. However, as k gets large, this quickly becomes infeasible. The k-bits of message are processed with a generator matrix to get the (n – k)-bits of code bits. These code bits are fitted with that message, so now the total length of the coded message is n. This code word is transmitted through the channel after proper modulation. At the receiver, a Hamming matrix is designed using the same generator matrix used at the transmitter, and a syndrome calculator is also prepared. The received bits could be compared to the syndrome calculator, the error bit position would be detected and the error could be corrected [168,169]. Develop a digital system to realize this function.

Solution

```
library ieee;
use ieee.std_logic_1164.all;
entity hamc is
port(nrst,clk:in std_logic;
master reset and clock
load:in std_logic;
data:in std_logic_vector(3 downto 0);
cout:out std_logic;
cin:in std_logic;
decode:out std_logic_vector(3 downto 0));
end entity;
architecture mark0 of hamc is
type ram1 is array (0 to 3) of std_logic_vector(6 downto 0);
type ram2 is array (0 to 6) of std_logic_vector(2 downto 0);
type ram3 is array (0 to 6) of std_logic_vector(3 downto 0);
signal gen:ram1;
signal part:ram2;
signal gent:ram3;
signal c0,c1,c2,c3:std_logic_vector(6 downto 0);
signal code:std_logic_vector(6 downto 0);
signal treg:std_logic_vector(6 downto 0);
- receive signals --
signal s0,s1,s2,s3,s4,s5,s6:std_logic_vector(2 downto 0);
signal syn:std_logic_vector(2 downto 0);
signal crec:std_logic_vector(6 downto 0);
signal d0,d1,d2,d3,d4,d5,d6:std_logic_vector(3 downto 0);
begin
gen <= ("1101000","0110100","1110010","1010001");
c0 <= gen(0) when(data(3)='1') else (others=>'0');
c1 <= gen(1) when(data(2)='1') else (others=>'0');
c2 <= gen(2) when(data(1)='1') else (others=>'0');
c3 <= gen(3) when(data(0)='1') else (others=>'0');
code <= (((c0 xor c1) xor c2) xor c3);-- codeword
tx:process(clk,nrst,load)
begin
if(nrst='0') then
cout <= '0';
```

```
else
if(load='0') then
treg <= code;
else
if(falling_edge(clk)) then
cout <= treg(6);
treg <= treg(5 downto 0)&'0';
end if;
end if;
end if;
end process;
rx:process(clk,nrst,load)
begin
if(nrst='0') then
crec <= (others=>'0');
else
if(load='1') then
case(syn) is
error correction
when "000"=>
crec <= crec;
when "001"=>
crec(4) <= not crec(4);
when "010"=>
crec(5) <= not crec(5);
when "011"=>
crec(2) <= not crec(2);
when "100"=>
crec(6) <= not crec(6);
when "101"=>
crec(0) <= not crec(0);
when "110"=>
crec(3) <= not crec(3);
when "111"=>
crec(1) <= not crec(1);
when others=>
null;
end case;
else
if(falling_edge(clk)) then
crec <= crec(5 downto 0)&cin;
end if;
end if;
end if;
end process;
part <= ("100","010","001","110","011","111","101");
gent <= ("1011","1110","0111","1000","0100","0010","0001");
s0 <= part(0) when(crec(6)='1') else (others=>'0');
s1 <= part(1) when(crec(5)='1') else (others=>'0');
s2 <= part(2) when(crec(4)='1') else (others=>'0');
s3 <= part(3) when(crec(3)='1') else (others=>'0');
s4 <= part(4) when(crec(2)='1') else (others=>'0');
s5 <= part(5) when(crec(1)='1') else (others=>'0');
s6 <= part(6) when(crec(0)='1') else (others=>'0');
syn <= (((((s0 xor s1) xor s2) xor s3) xor s4) xor s5) xor s6;
d0 <= gent(0) when (crec(6)='1') else (others=>'0');
d1 <= gent(1) when (crec(5)='1') else (others=>'0');
d2 <= gent(2) when (crec(4)='1') else (others=>'0');
d3 <= gent(3) when (crec(3)='1') else (others=>'0');
d4 <= gent(4) when (crec(2)='1') else (others=>'0');
d5 <= gent(5) when (crec(1)='1') else (others=>'0');
```

```
d6 <= gent(6) when (crec(0)='1') else (others=>'0');
decode <= (((((d0 xor d1) xor d2) xor d3) xor d4) xor d5) xor d6;
end architecture;
```

13.7.2 Convolutional Code

Objective: A binary convolutional code is a set of infinite-length binary sequences which satisfy a certain set of conditions; for example, the sum of two code words is a code word. It is easiest to describe the set of sequences in terms of a convolutional encoder that produces these sequences. However for a given code, the encoder is not unique. Convolutional codes are generated using shift registers and xor additions. In general, the convolutional encoder is given by $g_1 = (1011)$, $g_2 = (1010)$ and $g_3 = (1110)$, where g_i represents the number of coded bits and 1 represents the connection between the respective position register and xor gate. For example, code bit 1, that is, g_1, is generated based on the values of register 1, register 3, and register 4, because of the connection (1011). In this way, n outputs can be generated. The code words will be transmitted only to the receiver. The Viterbi decoding algorithm is designed at the receiver based on the same convolutional structure used at the transmitter. The received code is compared or routed through the Viterbi algorithm, and the original data will be decoded based on the path having low-error metric weight. Design and develop a digital system to realize the above operation.

Solution

```
library IEEE;
use IEEE.STD_LOGIC_1164.ALL;
use IEEE.STD_LOGIC_ARITH.ALL;
use IEEE.STD_LOGIC_UNSIGNED.ALL;
entity kings_vitervi_circuit is
Port (paral_data_in : in STD_LOGIC_VECTOR (3 downto 0):="1001";
start_in : in STD_LOGIC:='0';
m_clk : in STD_LOGIC:='0';
code_byte_inout: inout STD_LOGIC_VECTOR (7 downto 0):="00000000";
decoded_data : out STD_LOGIC_VECTOR (3 downto 0):="0000");
end kings_vitervi_circuit;
architecture Behavioral of kings_vitervi_circuit is
signal clk_sig,ser_data_out,metric_error_clk_from_state :std_logic:='0';
signal lfsr_clk_sig,metric_error_clk,min_metric_error_calcula_clk:std_
logic:= '1';
signal slfsr:std_logic_vector(4 downto 0):="00000";
signal data_in:std_logic_vector(3 downto 0):="1001";
signal decoded_data_temp:std_logic_vector(3 downto 0):="0000";
signal code_out :STD_LOGIC_VECTOR (1 downto 0):="00";
signal state,nextstate:integer range -4 to 15:=-4;
signal metric_error0,metric_error1,metric_error2,metric_error3,metric_
error4,
metric_error5,metric_error6,metric_error7,metric_error8,metric_error9,
metric_error10,metric_error11,metric_error12,metric_error13,metric_
error14,metric_error15,metric_error_temp,
code_byte_temp,code_byte_temp_state: std_logic_vector(7 downto 0):
="00000000";
begin
p0:process(m_clk,start_in)
variable clk_deci_cnt:integer:=0;
begin
if (start_in='0') then
clk_sig<='0';
elsif rising_edge(m_clk) then
```

```
if(clk_deci_cnt=200) then
clk_sig<=not(clk_sig);
lfsr_clk_sig<=not(lfsr_clk_sig);
clk_deci_cnt:=0;
else
clk_deci_cnt:=clk_deci_cnt+1;
end if;
end if;
end process;
p1:process(clk_sig)
variable mux_sel_cnt:integer:=-1;
begin
if rising_edge(clk_sig) then
if (mux_sel_cnt=4) then
data_in<=paral_data_in;
mux_sel_cnt:=0;
else
mux_sel_cnt:=mux_sel_cnt+1;
end if;
case mux_sel_cnt is
when 0=> ser_data_out<=data_in(0);
when 1=> ser_data_out<=data_in(1);
when 2=> ser_data_out<=data_in(2);
when 3=> ser_data_out<=data_in(3);
when others=> null;
end case;
end if;
end process;
p2:process(lfsr_clk_sig)
variable slfsr_reset_cnt:integer:=0;
begin
if rising_edge(lfsr_clk_sig) then
if(slfsr_reset_cnt=4) then
slfsr<="00000";
slfsr_reset_cnt:=0;
else
slfsr<=(ser_data_out & slfsr(4 downto 1));
slfsr_reset_cnt:=slfsr_reset_cnt+1;
end if;
end if;
end process;
-------------------------------------------------------------------------
p3:process(slfsr)
begin
code_out(0)<=(slfsr(3) xor slfsr(0));
code_out(1)<=(slfsr(4) xor slfsr(1));
end process;
-------------------------------------------------------------------------
p4:process(lfsr_clk_sig)
variable code_byte_state_cnt:integer:=-1;
begin
if falling_edge(lfsr_clk_sig) then
if(code_byte_state_cnt=4) then
code_byte_state_cnt:=0;
else
code_byte_state_cnt:=code_byte_state_cnt+1;
end if;
case code_byte_state_cnt is
when 1=> code_byte_inout(0)<=code_out(0);
code_byte_inout(1)<=code_out(1);
when 2=> code_byte_inout(2)<=code_out(0);
```

```
code_byte_inout(3)<=code_out(1);
when 3=> code_byte_inout(4)<=code_out(0);
code_byte_inout(5)<=code_out(1);
when 4=> code_byte_inout(6)<=code_out(0);
code byte_inout(7)<=code_out(1);
when others=> null;
end case;
end if;
end process;
p5:process(lfsr_clk_sig)
variable metric_error_cnt:integer:=1;
begin
if rising_edge(lfsr_clk_sig) then
if(metric_error_cnt=3) then
metric_error_clk<=not (metric_error_clk);
metric_error_cnt:=metric_error_cnt+1;
elsif(metric_error_cnt=5) then
metric_error_clk<=not (metric_error_clk);
metric_error_cnt:=1;
else
metric_error_cnt:=metric_error_cnt+1;
end if;
end if;
end process;
p6:process(metric_error_clk)
begin
if rising_edge(metric_error_clk) then
code_byte_temp<=code_byte_inout;
end if;
end process;
p7:process(metric_error_clk_from_state)
begin
if rising_edge(metric_error_clk_from_state) then
code_byte_temp_state<=code_byte_temp;
end if;
end process;
p8:process(metric_error_clk_from_state)
begin
if falling_edge(metric_error_clk_from_state) then
metric_error0<=(code_byte_inout xor "00000000");
metric_error1<=(code_byte_inout xor "10000110");
metric_error2<=(code_byte_inout xor "00011000");
metric_error3<=(code_byte_inout xor "10011110");
metric_error4<=(code_byte_inout xor "01100000");
metric_error5<=(code_byte_inout xor "11100110");
metric_error6<=(code_byte_inout xor "01111000");
metric_error7<=(code_byte_inout xor "11111110");
metric_error8<=(code_byte_inout xor "10000000");
metric_error9<=(code_byte_inout xor "00000110");
metric_error10<=(code_byte_inout xor "10011000");
metric_error11<=(code_byte_inout xor "00011110");
metric_error12<=(code_byte_inout xor "11100000");
metric_error13<=(code_byte_inout xor "01100110");
metric_error14<=(code_byte_inout xor "11111000");
metric_error15<=(code_byte_inout xor "01111110");
end if;
end process;
p9:process(m_clk)
variable min_metric_error_calcula_cnt:integer:=1;
begin
if rising_edge(m_clk) then
```

```
if(min_metric_error_calcula_cnt=2) then
min_metric_error_calcula_clk<=not (min_metric_error_calcula_clk);
min_metric_error_calcula_cnt:=1;
else
min_metric_error_calcula_cnt:=min_metric_error_calcula_cnt+1;
end if;
end if;
end process;
p10:process(min_metric_error_calcula_clk)
begin
if rising_edge(min_metric_error_calcula_clk) then
state<=nextstate;
case state is
when -4=> metric_error_clk_from_state <='1'; nextstate<=state + 1;
when -3=> nextstate<=state + 1;
when -2=> metric_error_clk_from_state <='0';nextstate<=state + 1;
when -1=> nextstate<=state + 1;
when 0=> if (metric_error0 > metric_error1) then
metric_error_temp<=metric_error1;decoded_data_temp<="0001";
else
metric_error_temp<=metric_error0;decoded_data_temp<="0000";
end if;nextstate<=state + 1;
when 1=> if (metric_error_temp > metric_error2) then metric_error_
temp<=metric_error2;decoded_data_temp<="0010";
end if;nextstate<=state + 1;
when 2=> if (metric_error_temp > metric_error3) then
metric_error_temp<=metric_error3;decoded_data_temp<="0011";
end if;nextstate<=state + 1;
when 3=> if (metric_error_temp > metric_error4) then
metric_error_temp<=metric_error4;decoded_data_temp<="0100";
end if;nextstate<=state + 1;
when 4=> if (metric_error_temp > metric_error5) then
metric_error_temp<=metric_error5;decoded_data_temp<="0101";
end if;nextstate<=state + 1;
when 5=> if (metric_error_temp > metric_error6) then
metric_error_temp<=metric_error6;decoded_data_temp<="0110";
end if;nextstate<=state + 1;
when 6=> if (metric_error_temp > metric_error7) then
metric_error_temp<=metric_error7;decoded_data_temp<="0111";
end if;nextstate<=state + 1;
when 7=> if (metric_error_temp > metric_error8) then
metric_error_temp<=metric_error8;decoded_data_temp<="1000";
end if;nextstate<=state + 1;
when 8=> if (metric_error_temp > metric_error9) then
metric_error_temp<=metric_error9;decoded_data_temp<="1001";
end if;nextstate<=state + 1;
when 9=> if (metric_error_temp > metric_error10) then
metric_error_temp<=metric_error10;decoded_data_temp<="1010";
end if;nextstate<=state + 1;
when 10=> if (metric_error_temp > metric_error11) then
metric_error_temp<=metric_error11;decoded_data_temp<="1011";
end if;nextstate<=state + 1;
when 11=> if (metric_error_temp > metric_error12) then
metric_error_temp<=metric_error12;decoded_data_temp<="1100";
end if;nextstate<=state + 1;
when 12=> if (metric_error_temp > metric_error13) then
metric_error_temp<=metric_error13;decoded_data_temp<="1101";
end if;nextstate<=state + 1;
when 13=> if (metric_error_temp > metric_error14) then
metric_error_temp<=metric_error14;decoded_data_temp<="1110";
end if;nextstate<=state + 1;
```

```
when 14=> if (metric_error_temp > metric_error15) then
metric_error_temp<=metric_error15;decoded_data_temp<="1111";
end if;nextstate<=state + 1;
when 15=>decoded_data<=decoded_data_temp;nextstate<=-4;
end case;
end if;
end process;
end Behavioral;
```

13.8 Pick-and-Place Robot Controller

Objective: In the scientific world, robotics play a vital role in many applications, including defense, industrial control and manufacturing, packing and transportation, biomedical systems, security, unmanned vehicles, space research, and underground and submarine research. In this section, we discuss developing a controller to operate a pick-and-place robot. Assume that a robot is instructed to move forward on a programmed path. If it detects any object or obstacle, it has to pick it up and place it to either the right or left side, then continue its forward movement. The robot movement is controlled by two motors at each side, with a dummy wheel support at the front and one motor for controlling the pick-and-place operation, and the obstacle is detected using IR proximate sensors. Develop a control system to realize the above control action.

Solution

```
library ieee;
use ieee.std_logic_1164.all;
entity ppr is
port(clk,nrst:in std_logic;
det:in std_logic; -- detector input(1=>object detected, 0=>object not
detected)
m1,m2,m3:out std_logic_vector(1 downto 0));
end entity;
architecture mark0 of ppr is
signal pos:integer range 0 to 8;
signal cnt:integer range 0 to 75000000;
begin
o1:process(clk,nrst,det)
begin
if(nrst='0') then
m1 <= "00";
m2 <= "00";
m3 <= "00";
cnt <= 0;
pos <= 0;
else
if(falling_edge(clk)) then
case(pos) is
when 0=>
if(det='1') then
pos <= 1;
else
pos <= 0;
end if;
when 1=>
```

```
m1 <= "10";
m2 <= "00";
m3 <= "00";
cnt <= cnt + 1;
if(cnt=25000000) then
pos <= 2;
cnt <= 0;
else
pos <= 1;
end if;
when 2=>
m1 <= "00";
m2 <= "10";
m3 <= "00";
cnt <= cnt + 1;
if(cnt=25000000) then
pos <= 3;
cnt <= 0;
else
pos <= 2;
end if;
when 3=>
m1 <= "01";
m2 <= "00";
m3 <= "00";
cnt <= cnt + 1;
if(cnt=25000000) then
pos <= 4;
cnt <= 0;
else
pos <= 3;
end if;
when 4=>
m1 <= "00";
m2 <= "00";
m3 <= "10";
cnt <= cnt + 1;
if(cnt=75000000) then
pos <= 5;
cnt <= 0;
else
pos <= 4;
end if;
when 5=>
m1 <= "10";
m2 <= "00";
m3 <= "00";
cnt <= cnt + 1;
if(cnt=25000000) then
pos <= 6;
cnt <= 0;
else
pos <= 5;
end if;
when 6=>
m1 <= "00";
m2 <= "01";
m3 <= "00";
cnt <= cnt + 1;
if(cnt=25000000) then
pos <= 7;
```

```
cnt <= 0;
else
pos <= 6;
end if;
when 7=>
m1 <= "01";
m2 <= "00";
m3 <= "00";
cnt <= cnt + 1;
if(cnt=25000000) then
pos <= 8;
cnt <= 0;
else
pos <= 7;
end if;
when 8=>
m1 <= "00";
m2 <= "00";
m3 <= "01";
cnt <= cnt + 1;
if(cnt=75000000) then
pos <= 0;
cnt <= 0;
else
pos <= 8;
end if;
end case;
end if;
end if;
end process;
end architecture;
```

13.9 Audio Codec (AC97) Interfacing

Objective: The AC97 audio codec can record and play back high-quality stereophonic sound. The codec's ADCs and DACs operate at a 48-kHz sample rate with 18 bits of precision. The easiest way to generate alert beeps and other simple tones is to use the FPGA to generate a digital square wave on the "beep" input to the audio codec. As long as the reset pin of the codec is held low, the beep input is feed directly to the line-output jacks. The AC97 is a serial interface: data are transmitted to and from the codec one bit at a time. On every cycle of the AC97 bit clock, 1 bit of data is transferred from the AC97 controller to the codec over the SDATA_OUT line, and one bit of data is transferred from the codec to the FPGA over the SDATA_IN line. The constant streams of data passing between the codec and the FPGA are divided into frames. The bit clock is generated by the codec and runs at 12.288 MHz. There are 256 bits per frame; hence, 48,000 frames are sent per second. Each frame sent to the codec provides one 20-bit sample for each of the DACs in the codec, and each frame sent by the codec provides one 20-bit sample from each of the codec's ADCs. Frames are divided into 12 slots of 20 bits each, plus a 16-bit tag field, which serves as the frame header. The start of each frame is indicated by a rising edge of the SYNC signal. The SYNC signal goes high one clock cycle before the first bit of a frame and goes low at the same time as the last bit of the tag field is sent. Design and develop a digital system to perform the above interfacing and control the volume of the audio out signal [170,171].

Solution

Component 1

```
library IEEE;
use IEEE.STD_LOGIC_1164.ALL;
use work.Spar6_Parts.all;
entity Spar6_Talkthrough_TL is
Port (clk : in STD_LOGIC;
n_reset : in STD_LOGIC;
SDATA_IN : in STD_LOGIC;
BIT_CLK : in STD_LOGIC;
SOURCE : in STD_LOGIC_VECTOR(2 downto 0);
VOLUME : in STD_LOGIC_VECTOR(4 downto 0);
SYNC : out STD_LOGIC;
SDATA_OUT : out STD_LOGIC;
AC97_n_RESET : out STD_LOGIC);
end Spar6_Talkthrough_TL;
architecture arch of Spar6_Talkthrough_TL is
signal L_bus, R_bus, L_bus_out, R_bus_out :std_logic_vector
(17 downto 0);
signal cmd_addr :std_logic_vector(7 downto 0);
signal cmd_data :std_logic_vector(15 downto 0);
signal ready :std_logic;
signal latching_cmd :std_logic;
begin
ac97_cont0 : entity work.ac97(arch)
port map(n_reset =>n_reset, clk =>clk, ac97_sdata_out => SDATA_OUT,
ac97_sdata_in => SDATA_IN, latching_cmd =>latching_cmd,
ac97_sync => SYNC, ac97_bitclk => BIT_CLK,
ac97_n_reset => AC97_n_RESET, ac97_ready_sig => ready,
L_out =>L_bus, R_out =>R_bus,
L_in =>L_bus_out, R_in =>R_bus_out,
cmd_addr =>cmd_addr, cmd_data =>cmd_data);
ac97cmd_cont0 : entity work.ac97cmd(arch)
port map (clk =>clk, ac97_ready_sig => ready,
cmd_addr =>cmd_addr, cmd_data =>cmd_data,
volume => VOLUME, source => SOURCE,
latching_cmd =>latching_cmd);
process (clk, n_reset, L_bus_out, R_bus_out)
begin
if (clk'event and clk = '1') then
if n_reset = '0' then
L_bus<= (others => '0');
R_bus<= (others => '0');
elsif(ready = '1') then
L_bus<= L_bus_out;
R_bus<= R_bus_out;
end if;
end if;
end process;
end arch;
```

Main Code

```
library IEEE;
use IEEE.STD_LOGIC_1164.all;
use IEEE.numeric_std.all;
package Spar6_Parts is
component ac97
port (n_reset : in std_logic;
clk : in std_logic;
ac97_sdata_out : out std_logic;
```

```
ac97_sdata_in : in std_logic;
ac97_sync : out std_logic;
ac97_bitclk : in std_logic;
ac97_n_reset : out std_logic;
ac97_ready_sig : out std_logic;
L_out : in std_logic_vector(17 downto 0);
R_out : in std_logic_vector(17 downto 0);
L_in : out std_logic_vector(17 downto 0);
R_in : out std_logic_vector(17 downto 0);
latching_cmd : in std_logic;
cmd_addr : in std_logic_vector(7 downto 0);
cmd_data : in std_logic_vector(15 downto 0);
ac97cmd state machine);
end component;
component ac97cmd
port (clk          : in std_logic;
ready              : in std_logic;
cmd_addr           : out std_logic_vector(7 downto 0);
cmd_data           : out std_logic_vector(15 downto 0);
latching_cmd       : out std_logic;
volume             : in std_logic_vector(4 downto 0);
source             : in std_logic_vector(2 downto 0));
end component;
end Spar6_Parts;
library ieee;
use ieee.std_logic_1164.all;
use ieee.numeric_std.all;
entity ac97 is
port (n_reset : in std_logic;
clk : in std_logic;
-- ac97 interface signals
ac97_sdata_out : out std_logic;
ac97_sdata_in : in std_logic;
ac97_sync : out std_logic;
ac97_bitclk : in std_logic;
ac97_n_reset : out std_logic;
ac97_ready_sig : out std_logic;
L_out : in std_logic_vector(17 downto 0);
R_out : in std_logic_vector(17 downto 0);
L_in : out std_logic_vector(17 downto 0);
R_in : out std_logic_vector(17 downto 0);
latching_cmd : in std_logic;
cmd_addr : in std_logic_vector(7 downto 0);
ac97cmd state machine
cmd_data : in std_logic_vector(15 downto 0));
end ac97;
architecture arch of ac97 is
signal Q1, Q2            : std_logic;
signal bit_count       :std_logic_vector(7 downto 0);
signal rst_counter     :integer range 0 to 4097;
signal latch_cmd_addr   :std_logic_vector(19 downto 0);
signal latch_cmd_data   :std_logic_vector(19 downto 0);
signal latch_left_data :std_logic_vector(19 downto 0);
signal latch_right_data:std_logic_vector(19 downto 0);
signal left_data       :std_logic_vector(19 downto 0);
signal right_data      :std_logic_vector(19 downto 0);
signal left_in_data    :std_logic_vector(19 downto 0);
signal right_in_data   :std_logic_vector(19 downto 0);
begin
left_data<= L_out& "00";
right_data<= R_out& "00";
```

```
L_in<= left_in_data(19 downto 2);
R_in<= right_in_data(19 downto 2);
process (clk, n_reset)
begin
if (clk'event and clk = '1') then
if n_reset = '0' then
rst_counter<= 0;
ac97_n_reset <= '0';
elsifrst_counter = 3789 then
ac97_n_reset <= '1';
rst_counter<= 0;
else
rst_counter<= rst_counter + 1;
end if;
end if;
end process;
process (clk, n_reset, bit_count)
begin
if(clk'event and clk = '1') then
Q2 <= Q1;
if(bit_count = "00000000") then
Q1 <= '0';
Q2 <= '0';
elsif(bit_count>= "10000001") then
Q1 <= '1';
end if;
ac97_ready_sig <= Q1 and not Q2;
end if;
end process;
-----------------------------------------------------------------------------
process (n_reset, bit_count, ac97_bitclk)
begin
if(n_reset = '0') then
bit_count<= "00000000";
end if;
if (ac97_bitclk'event and ac97_bitclk = '1') then
if bit_count = "11111111" then
ac97_sync <= '1';
end if;
if bit_count = "00001111" then
ac97_sync <= '0';
end if;
-- At the end of each frame the user data is latched in
if bit_count = "11111111" then
latch_cmd_addr<= cmd_addr& "000000000000";
latch_cmd_data<= cmd_data& "0000";
latch_left_data<= left_data;
latch_right_data<= right_data;
end if;
if (bit_count>= "00000000") and (bit_count<= "00001111") then
case bit_count is
when "00000000" => ac97_sdata_out <= '1';
when "00000001" => ac97_sdata_out <= latching_cmd;
when "00000010" => ac97_sdata_out <= '1';
when "00000011" => ac97_sdata_out <= '1';
when "00000100" => ac97_sdata_out <= '1';
when others => ac97_sdata_out <= '0';
end case;
-- starting at slot 1 add 20 bit counts each time
elsif (bit_count>= "00010000") and (bit_count<= "00100011") then
if latching_cmd = '1' then
```

```
ac97_sdata_out <= latch_cmd_addr(35 - to_integer(unsigned(bit_count)));
else
ac97_sdata_out <= '0';
end if;
elsif (bit_count>= "00100100") and (bit_count<= "00110111") then
if latching_cmd = '1' then
ac97_sdata_out <= latch_cmd_data(55 - to_integer(unsigned(bit_count)));
else
ac97_sdata_out <= '0';
end if;
elsif ((bit_count>= "00111000") and (bit_count<= "01001011")) then
ac97_sdata_out <= latch_left_data(19);
latch_left_data<= latch_left_data(18 downto 0) &latch_left_data(19);
elsif ((bit_count>= "01001100") and (bit_count<= "01011111")) then
ac97_sdata_out <= latch_right_data(95
- to_integer(unsigned(bit_count)));
else
ac97_sdata_out <= '0';
end if;
bit_count<= std_logic_vector(unsigned(bit_count) + 1);
end if;
end process;
process (ac97_bitclk)
begin
if (ac97_bitclk'event and ac97_bitclk = '0') then
if (bit_count>= "00111001") and (bit_count<= "01001100") then
left_in_data<= left_in_data(18 downto 0) & ac97_sdata_in;
elsif (bit_count>= "01001101") and (bit_count<= "01100000") then
right_in_data<= right_in_data(18 downto 0) & ac97_sdata_in;
end if;
end if;
end process;
end arch;
--STATE MACHINE TO CONFIGURE THE AC97 --
library ieee;
use ieee.std_logic_1164.all;
use ieee.numeric_std.all;
-- one can add extra inputs and signals to control the input
to LM4550 (ac97)
-- register map below see volume and source for example
entity ac97cmd is
port (clk          : in std_logic;
ac97_ready_sig : in std_logic;
cmd_addr         : out std_logic_vector(7 downto 0);
cmd_data         : out std_logic_vector(15 downto 0);
latching_cmd : out std_logic;
volume  :in std_logic_vector(4 downto 0);
source :in std_logic_vector(2 downto 0));
end ac97cmd;
architecture arch of ac97cmd is
signal cmd                : std_logic_vector(23 downto 0);
signal atten       : std_logic_vector(4 downto 0);
ML4:0/MR4:0
type state_type is (S0, S1, S2, S3, S4, S5, S6, S7, S8, S9, S10, S11);
signal cur_state, next_state :state_type;
begin
-- parse command from data
cmd_addr<= cmd(23 downto 16);
cmd_data<= cmd(15 downto 0);
atten<= std_logic_vector(31 - unsigned(volume));
with cmd(23 downto 16) select
```

```
latching_cmd<=
'1' when X"02" | X"04" | X"06" | X"0A" | X"0C" | X"0E" | X"10" | X"12" |
X"14" | X"16" | X"18" | X"1A" | X"1C" | X"20" | X"22" | X"24" | X"26" |
X"28" | X"2A" | X"2C" | X"32" | X"5A" | X"74" | X"7A" | X"7C" | X"7E" |
X"80",
'0' when others;
process(clk, next_state, cur_state)
begin
if(clk'event and clk = '1') then
if ac97_ready_sig = '1' then
cur_state<= next_state;
end if;
end if;
end process;
------------------------------------------------------------------------------------------
process (next_state, cur_state, atten, source)
begin
case cur_state is
when S0 =>
cmd<= X"02_8000";
next_state<= S2;
when S1 =>
cmd<= X"04" & "000" &atten& "000" &atten;
next_state<= S4;
when S2 =>
cmd<= X"0A_0000";
next_state<= S11;
when S3 =>
cmd<= X"0E_8048";
next_state<= S10;
when S4 =>
cmd<= X"18_0808";
next_state<= S6;
when S5 =>
cmd<= X"1A" & "00000" & source & "00000" & source;
next_state<= S7;
when S6 =>
cmd<= X"1C_0F0F";
next_state<= S8;
when S7 =>
cmd<= X"20_8000";
next_state<= S0;
when S8 =>
cmd<= X"2C_BB80";
next_state<= S5;
when S9 =>
cmd<= X"32_BB80";
next_state<= S3;
when S10 =>
cmd<= X"80_0000";
next_state<= S9;
when S11 =>
cmd<= X"80_0000";
next_state<= S1;
end case;
end process;
end arch;
```

LABORATORY EXERCISES

1. Design and develop a digital system to realize the function of delta modulation (DM) and adaptive delta modulation (ADM).
2. Design and develop a digital system to realize the various line coding and decoding schemes with necessary external components.
3. Design and develop a digital system to realize various artificial neuron and fuzzy membership functions and build a simple ANN and FLC controller in the FPGA using VHDL code.
4. Design and develop a digital system to realize a cell averaged-constant false alarm rate (CA-CFAR) radar receiver. Assume that the inputs are arriving at the digital system from DIP switches.
5. Design and develop a digital system to interface with a camera OV7670, capture the video frames, and display them on the computer monitor through the VGA interface.
6. Design and develop a digital system to implement the control structure of a PID controller inside the FPGA.
7. Design and develop a digital system to implement a low-density parity check (LDPC)-based channel coding system.
8. Design and develop a digital system to generate a barcode and combined Barker codes.
9. Design and develop a digital system to control an ultrasonic-based target detection system. Use the computer monitor as a display device to show the range, speed, and coordinates of the detected target.
10. Design and develop a digital system to interface a gyroscope with the FPGA and store the measurement data appropriately in an Excel file.

Appendix A

A.1 Summary of Very-High-Speed Integrated Circuit Hardware Description Language Libraries

```
library STD;
library IEEE;
use IEEE.std_logic_1164.all;
use IEEE.std_logic_textio.all;
use IEEE.std_logic_arith.all;
use IEEE.numeric_bit.all;
use IEEE.numeric_std.all;
use IEEE.std_logic_signed.all;
use IEEE.std_logic_unsigned.all;
use IEEE.math_real.all;
use IEEE.math_complex.all;
use ieee.std_logic_textio.all;
use STD.textio.all;
library library_name;
use library_name.unit_name.all;
use workmy_pkg.s_inc;
```

A.2 Summary of Very-High-Speed Integrated Circuit Hardware Description Language Operators

`**`	Exponentiation, numeric `**` integer, result numeric
`abs`	Absolute value, `abs` numeric, result numeric
`not`	Complement, `not` logic or boolean, result same
`*`	Multiplication, numeric `*` numeric, result numeric
`/`	Division, numeric/numeric, result numeric
`mod`	Modulo, integer `mod` integer, result integer
`rem`	Remainder, integer `rem` integer, result integer
`+`	Unary plus, + numeric, result numeric
`-`	Unary minus, - numeric, result numeric
`+`	Addition, numeric + numeric, result numeric
`-`	Subtraction, numeric - numeric, result numeric
`&`	Concatenation, array or element & array or element, result array
`sll`	Shift left logical, logical array `sll` integer, result same
`srl`	Shift right logical, logical array `srl` integer, result same
`sla`	Shift left arithmetic, logical array `sla` integer, result same
	sra– Shift right arithmetic, logical array `sra` integer, result same
`rol`	Rotate left, logical array `rol` integer, result same
`ror`	Rotate right, logical array `ror` integer, result same
`=`	Test for equality, result is boolean

`/=`	Test for inequality, result is boolean
`<`	Test for less than, result is boolean
`<=`	Test for less than or equal, result is boolean
`>`	Test for greater than, result is boolean
`>=`	Test for greater than or equal, result is boolean
`and`	Logical `and`, logical array or boolean, result is same
`or`	Logical `or`, logical array or boolean, result is same
`nand`	Logical complement of `and`, logical array or boolean, result is same
`nor`	Logical complement of `or`, logical array or boolean, result is same
`xor`	Logical exclusive `or`, logical array or boolean, result is same
`xnor`	Logical complement of exclusive `or`, logical array or boolean, result is same

A.3 Summary of Very-High-Speed Integrated Circuit Hardware Description Language Commands

`Abs`	Operator, absolute value of right operand; no () needed
`Access`	Used to define an access type, pointer
`after`	Specifies a time after NOW
`alias`	Create another name for an existing identifier
`all`	Dereferences what precedes the .all
`and`	Operator, logical "`and`" of left and right operands
`architecture`	A secondary design unit
`array`	Used to define an array, vector, or matrix
`assert`	Used to have a program check on itself
`attribute`	Used to declare attribute functions
`begin`	Start of a begin–end pair
`block`	Start of a block structure
`body`	Designates a procedure body rather than declaration
`buffer`	A mode of a signal, holds a value
`bus`	A mode of a signal, can have multiple drivers
`case`	Part of a case statement
`component`	Starts the definition of a component
`configuration`	A primary design unit
`constant`	Declares an identifier to be read only
`disconnect`	Signal driver condition
`downto`	Middle of a range: 31 `downto` 0
`else`	Part of "if" statement, if cond then… else… end if;
`elsif`	Part of "if" statement, if cond then… elsif cond…
`end`	Part of many statements, may be followed by word and id
`entity`	A primary design unit
`exit`	Sequential statement, used in loops
`falling_edge (m_clk) then`	Used to trigger a process for the falling edge of the `m_clc` input
`file`	Used to declare a file type
`for`	Start of a for-type loop statement

`function`	Starts declaration and body of a function
`generate`	Makes copies, possibly using a parameter
`generic`	Introduces generic part of a declaration
`group`	Collection of types that can get an attribute
`guarded`	Causes a wait until a signal changes from False to True
`if`	Used in "if" statements
`impure`	An impure function is assumed to have side effects
`in`	Indicates a parameter is only input, not changed
`inertial`	Signal characteristic, holds a value
`inout`	Indicates a parameter is used and computed in and out
`is`	Used as a connective in various statements
`label`	Used in attribute statement as entity specification
`library`	Context clause, designates a simple library name
`linkage`	A mode for a port, used like buffer and inout
`literal`	Used in attribute statement as entity specification
`loop`	Sequential statement, loop… end loop;
`map`	Used to map actual parameters, as in port map
`mod`	Operator, left operand modulo right operand
`nand`	Operator, `"nand"` of left and right operands
`new`	Allocates memory and returns access pointer
`next`	Sequential statement, used in loops
`nor`	Operator, `"nor"` of left and right operands
`not`	Operator, complement of right operand
`null`	Sequential statement and a value
`of`	Used in type declarations: of Real;
`on`	Used as a connective in various statements
`open`	Initial file characteristic
`or`	Operator, logical `"or"` of left and right operands
`others`	Fill in missing, possibly all, data
`out`	Indicates a parameter is computed and output
`package`	A design unit, also package body
`port`	Interface definition, also port map
`postponed`	Make process wait for all nonpostponed processes to suspend
`procedure`	Typical programming procedure
`process`	Sequential or concurrent code to be executed
`pure`	A pure function may not have side effects
`range`	Used in type definitions, range 1 to 10;
`record`	Used to define a new record type
`register`	Signal parameter modifier
`rising_edge (m_clk) then`	Used to trigger a process for the rising edge of `m_clk` input
`reject`	Clause in delay mechanism, followed be a time
`rem`	Operator, remainder of left operand divided by right op
`report`	Statement and clause in assert statement, string output
`return`	Statement in procedure or function
`rol`	Operator, left operand rotated left by right operand
`ror`	Operator, left operand rotated right by right operand
`select`	Used in selected signal assignment statement
`severity`	Used in assertion and reporting, followed by a severity

`signal`	Declaration that an object is a signal
`shared`	Used to declare shared objects
`sla`	Operator, left operand shifted left arithmetic by right op
`sll`	Operator, left operand shifted left logical by right op
`sra`	Operator, left operand shifted right arithmetic by right
`srl`	Operator, left operand shifted right logical by right op
`subtype`	Declaration to restrict an existing type
`then`	Part of if condition then…
`to`	Middle of a range 1 to 10
`transport`	Signal characteristic
`type`	Declaration to create a new type
`unaffected`	Used in signal waveform
`units`	Used to define new types of units
`until`	Used in wait statement
`use`	Make a package available to this design unit
`variable`	Declaration that an object is a variable
`wait`	Sequential statement, also used in case statement
`when`	Used for choices in case and other statements
`while`	Kind of loop statement
`with`	Used in selected signal assignment statement
`xnor`	Operator, exclusive "`nor`" of left and right operands
`xor`	Operator, exclusive "`or`" of left and right operands

A.4 Summary of MATLAB Operators and Commands

+	Addition
–	Subtraction
^	Multiplication (scalar and array)
/	Division (right)
^	Power or exponentiation
:	Colon; creates vectors with equally spaced elements
;	Semicolon; suppresses display; ends row in array
,	Comma; separates array subscripts
...	Continuation of lines
%	Percent; denotes a comment; specifies output format
'	Single quote; creates string; specifies matrix transpose
=	Assignment operator
( )	Parentheses; encloses elements of arrays and input arguments
[ ]	Brackets; encloses matrix elements and output arguments
.^	Array multiplication
./	Array (right) division
.^	Array power
.\	Array (left) division
.'	Array (nonconjugated) transpose
<	Less than
≤	Less than or equal to

`>` Greater than
`≥` Greater than or equal to
`==` Equal to
`~=` Not equal to
`&` Logical or element-wise AND
`|` Logical or element–wise OR
`&&` Short-circuit AND
`||` Short-circuit OR
`Cd` Change current directory
`Clc` Clear the command window
`clear(all)` Removes all variables from the workspace
`clear x` Remove x from the workspace
`copyfile` Copy file or directory
`delete` Delete files
`dir` Display directory listing
`exist` Check if variables or functions are defined
`help` Display help for MATLAB functions
`lookfor` Search for specified word in all help entries
`mkdir` Make new directory
`movefile` Move file or directory
`pwd` Identify current directory
`rmdir` Remove directory
`type` Display contents of file
`what` List MATLAB files in current directory
`which` Locate functions and files
`who` Display variables currently in the workspace
`whos` Display information on variables in the workspace
`ans` Value of last variable (answer)
`eps` Floating–point relative accuracy
`i` Imaginary unit of a complex number
`Inf` Infinity (∞)
`Eps` Floating-point relative accuracy
`j` Imaginary unit of a complex number
`NaN` Not a number
`Pi` The number π (3.14159…)
`Disp` Display text or array
`Isempty` Determine if input is empty matrix
`isequal` Test arrays for equality
`length` Length of vector
`ndims` Number of dimensions
`numel` Number of elements
`size` Size of matrix
`cross` Vector cross product
`diag` Diagonal matrices and diagonals of matrix
`dot` Vector dot product
`end` Indicate last index of array
`find` Find indices of nonzero elements
`kron` Kronecker tensor product
`max` Maximum value of array

`min`	Minimum value of array
`prod`	Product of array elements
`reshape`	Reshape array
`sort`	Sort array elements
`sum`	Sum of array elements
`cond`	Condition number with respect to inversion
`det`	Determinant
`inv`	Matrix inverse
`linsolve`	Solve linear system of equations
`lu`	LU factorization
`norm`	Matrix or vector norm
`null`	Null space
`orth`	Orthogonalization
`rank`	Matrix rank
`rref`	Reduced row echelon form
`trace`	Sum of diagonal elements

A.5 Converting MATLAB Figure/Plot into Tiff Image

The following MATLAB code can be used to convert the MATLAB plot/figure into a required format (e.g., .tiff file) of required size and resolution. For example, Figures 10.21 and 11.5.

```
x=input('How many figures:= ');
if x==1,
width = 3.01; % Width in inches
height = 2.84; % Height in inches
elseif x==2,
width = 4.87; % Width in inches
height = 4.57; % Height in inches
elseif x==4,
width = 6.23; % Width in inches
height = 5.86; % Height in inches
end
set(gcf, 'PaperPosition', [0 0 width height]);
set(gcf,'color','C');
print('Scope', '-dtiff', '-r300');
hold on;
```

A.6 Converting MATLAB Simulink Model (.mdl file) Design into Tiff Image

The following MATLAB code can be used to convert the MATLAB Simulink/SysGen model into a required format (e.g., .tiff file) of required resolution. For example, Figures 10.32 and 12.6.

```
<Simulink Model File Name>;
Print('-s<Simulink Model File Name>', '-dtiff', '-r600', '<Simulink
Model File Name>.tiff');
```

For Example:

```
Scope; % to open the model
print('-sScope','-dtiff','-r600','Scope.tiff'); % to print as a tiff file
Fig_12_23; % to open the model
print('-sFig_12_23','-dtiff','-r600','Fig_12_23.tiff');
```

A.7 Converting SysGen Model (.mdl) File's Scope Plot into Tiff Image

After completing/simulating the Xilinx SysGen design (.mdl file), open the scope display. Click on the second icon, that is, the parameters icon. Then, a window will pop up which has three heads, namely general, history and style. For the purpose of converting the scope plots into a .tiff image, we need to work only with history. As shown there, store the value in the workspace of MATLAB. The following MATLAB code can be used to read the stored values of a MATLAB Simulink's/SysGen's Scope's plot from the Workspace and convert into a required format (e.g., .tiff file) of required resolution. For example, Figures 10.11 and 10.19.

```
Clc;
Clear all;
figure(1);
subplot(2,1,1);
plot(data1.signals(1).values);
set(gca,'fontsize',8);
xlabel('Time (mse)');
ylabel('Amplitude (V)');
subplot(2,1,2);
plot(data1.signals(2).values);
set(gca,'fontsize',8);
xlabel('Time (mse)');
ylabel('Amplitude (V)');
x=input('Do you want to get it in TIFF format? ');
if x==1,
width = 8; % Width in inches
height = 8; % Height in inches
set(gcf, 'PaperPosition', [0 0 width height]);
print('Scope', '-dtiff', '-r600');
end;
```

References

1. http://www.techulator.com/resources/13398-What-is-VLSI-Technology.aspx
2. http://www2.elo.utfsm.cl/~lsb/elo102/datos/microelectronics.pdf
3. https://www.allaboutcircuits.com/textbook/digital/chpt-3/cmos-gate-circuitry/
4. https://www.edgefx.in/understanding-cmos-fabrication-technology/
5. http://vlsibyjim.blogspot.in/2015/03/vlsi-design-flow.html
6. http://www.electronics-tutorials.ws/combination/comb_1.html
7. https://www.elprocus.com/tutorial-on-sequential-logic-circuits/
8. http://wdc65xx.com/asic-design-flow/
9. https://www.mepits.com/tutorial/169/VLSI/Application-Specific-Integrated-Circuit
10. https://www.doulos.com/knowhow/vhdl_designers_guide/a_brief_history_of_vhdl/
11. https://www.xilinx.com/products/technology/dsp.html
12. Jan Van der Spiegel, VHDL Tutorial, University of Pennsylvania, Philadelphia, USA, 2006.
13. http://www.seas.upenn.edu/~ese171/vhdl/vhdl_primer.html
14. http://www.euroelectronica.ro/9-dataflow-modeling-concurrent-statements/
15. https://www.nandland.com/vhdl/tips/tip-convert-numeric-std-logic-vector-to-integer.html
16. http://www.bitweenie.com/listings/vhdl-type-conversion/
17. http://www.daenotes.com/electronics/digital-electronics/flip-flops-types-applications-woking
18. http://www.circuitstoday.com/flip-flop-conversion
19. http://www.electronics-tutorials.ws/sequential/seq_5.html
20. http://www.electronics-tutorials.ws/counter/count_1.html
21. http://www.electronics-tutorials.ws/combination/comb_7.html
22. Muthiah D and Raj AAB, Implementation of high-speed LFSR design with parallel architectures, *International Conference on Computing, Communication and Applications (ICCCA), IEEE Explore*, pp. 1–6, 2012, India.
23. Tamilarasi S et al., ANFIS implementation for autonomous insulin injection systems, *International Journal of Advanced Research in Electronics and Communication*, 5(1), pp. 123–129, 2016.
24. Manoj M, et al., Survey on soft computing assisted controller driven insulin injection gadget, *International Conference on Emerging Trends in Engineering, Technology and Science (ICETETS-2016), Conference Proceedings*, pp. 535–538, 2016, India.
25. Meyer-Base, Digital Signal Processing with Field Programmable Gate Array, Springer, 2001, USA.
26. Anbuselvi T and Raj AAB, Modeling and hardware implementation of artificial neural network (ANN) controller for prediction of control variables, *Journal of Communication & Electronics*, 2(3), pp. 20–28, 2013.
27. Raj AAB, Design and implementation of low complexity digital circuits using CORDIC, *National Level Technical Conference on Communication Systems, Conference Proceedings*, pp. 1–4, India, 2011.
28. http://studymaterial4ec.blogspot.in/2012/05/combinational-logic-circuits-navigation.html
29. http://www.electronics-tutorials.ws/counter/bcd-counter-circuit.html
30. http://www.best-microcontroller-projects.com/real-time-clock-ic.html
31. Bazil Raj A et al., Low cost beam steering system for FSOC to SMF coupling, *International Conference on Advances in Engineering, Science and Management (ICAESM), IEEE Explore*, pp. 49–54, India, 2012.
32. Anthonisamy ABR et al., Performance analysis of free space optical communication in open-atmospheric turbulence conditions with beam wandering compensation control, *IET Communications*, 10(9), pp. 1096–1103, 2016.

33. Raj AAB and Selvi JAV, Lower-order adaptive beam steering system in terrestrial free space point-to-point laser communication using fine tracking sensor, *International Conferences on Signal Processing, Communication, Computing and Networking Technologies, IEEE Explore*, pp. 699–704, India, 2011.
34. http://www.electronics-tutorials.ws/blog/pulse-width-modulation.html
35. http://www.ece.mcmaster.ca/~shirani/2di4/chapter8p2.pdf
36. Raj AAB et al., Implementation of adaptive fuzzy logic controller in FPGA for steering tip-tilt mirror in free space optical communication, *International Conference on Computational Intelligence and Computing, Conference Proceedings*, pp. 1–5, India, 2011.
37. Raj A et al., Terrestrial free space line of sight optical communication (TFSLSOC) using adaptive control steering system with laser beam tracking, aligning and positioning (ATP), *International Conference on Wireless Communication and Sensor Computing (ICWCSC), IEEE Explore*, pp. 1–5, India, 2010.
38. Anbuselvi T and Raj AAB, Design and verification of pipelined parallel architecture implementation in FPGA for bit error rate tester, *International Journal of Research in Engineering and Technology*, 3(2), pp. 28–34, 2014.
39. Raj AAB, Free Space Optical Communication: System Design, Modeling, Characterization and Dealing with Turbulence, De Gruyter, Germany, 2016.
40. https://www.electrical4u.com/parallel-subtractor/
41. AAB Raj et al., Experimental analysis & mitigation of atmospheric turbulence effects in TFSLSoC using pipelined opto-electronic architecture, *International Journal of Information and Communication Technology*, 1(2), pp. 1–5, 2011.
42. Bazil AAB, Intensity feedback-based beam wandering mitigation in free-space optical communication using neural control technique, *EURASIP Journal on Wireless Communications and Networking*, (1), pp. 1–18, 2014.
43. http://www.eng.ucy.ac.cy/theocharides/Courses/ECE210/Mulitpliers_Dividers_supp4.pdf
44. https://www.allaboutcircuits.com/technical-articles/an-introduction-to-the-cordic-algorithm/
45. https://en.wikibooks.org/wiki/Digital_Circuits/CORDIC
46. https://www.allsyllabus.com/aj/note/ECE/DSP%20Algorithms%20and%20Architecture/unit2/Multiply%20and%20Accumulate%20Unit.php#.WjqXUVWWbIU
47. Raj AAB et al., A direct and neural controller performance study with beam wandering mitigation control in free space optical link, *Optical Memory and Neural Networks (Information Optics)*, 23(3), pp. 111–129, 2014.
48. Raj AAB et al., Design of cognitive decision making controller for autonomous online adaptive beam steering in free space optical communication system, *Wireless Personal Communications*, 84(1), pp. 765–799, 2015.
49. http://coep.vlab.co.in/?sub=28&brch=81&sim=604&cnt=1
50. http://sce2.umkc.edu/csee/hieberm/281_new/lectures/combinational/storage-components.html
51. http://www-inst.eecs.berkeley.edu/~cs150/sp13/agenda/lec/lec11-sram.pdf
52. https://www.engineersgarage.com/electronic-components/ir-infrared-led
53. https://www.edgefx.in/scrolling-led-display-circuit/
54. http://www.instructables.com/id/PIR-Controlled-Buzzer-Using-Arduino/
55. http://www.microcontrollerboard.com/lcd.html
56. https://exploreembedded.com/wiki/Interfacing_GLCD(128x64)_with_PIC16F877A
57. http://www.futureelectronics.com/en/switches/dip-switches.aspx
58. http://www.electronics-tutorials.ws/logic/logic_9.html
59. http://www.edgefxkits.com/blog/metal-detector-working-with-applications/
60. https://www.electrical4u.com/light-dependent-resistor-ldr-working-principle-of-ldr/
61. Raj AAB and Selvi AV, Formulation of atmospheric optical attenuation model in terms of weather data, *Journal of Optics*, 45(2), pp. 120–135, 2016.

62. Jennifer JS and Raj AAB, Formulation of empirical model for atmospheric turbulence strength (Cn2) prediction, *National Conference on Emerging Trends in Communication and Networking (ETCAN'15), Conference Proceedings*, pp. 1–4, 2015.
63. Jenifer JS et al., Certain investigations on atmospheric turbulence strength (Cn2) prediction models for FSO applications, *International Journal for Technological Research in Engineering*, 2(8), pp. 1579–1582, 2015.
64. Raj AAB, Real-time measurement of meteorological parameters for estimating low-altitude atmospheric turbulence strength (Cn2), *IET Science Measurement & Technology*, 8(6), pp. 459–469, 2014.
65. http://www.cs.princeton.edu/courses/archive/spr06/cos116/FSM_Tutorial.pdf
66. Raj AAB, Darusalam U, Performance improvement of terrestrial free-space optical communications by mitigating the focal-spot wandering, *Journal of Modern Optics*, 63(21), pp. 2339–2347. 2016.
67. https://www.tutorialspoint.com/automata_theory/moore_and_mealy_machines.htm
68. Anbuselvi T, Raj AAB, BER tester design in FPGA for quantitative analysis of FSO data link, *National Conference on Advanced Communications, Conference Proceedings*, pp. 28–34, India, 2014.
69. Raj AAB et al., Low cost BER measurement in wireless digital laser communication link with autonomous beam steering system, *3rd International Conference on Intelligent Information Systems and Management, Conference Proceedings*, pp. 146–151, India, 2012.
70. https://www.ece.umd.edu/class/enee245.F2016/Lab9.pdf
71. https://www.electricaltechnology.org/2014/10/traffic-light-control-electronic-project.html
72. http://www.newworldencyclopedia.org/entry/Escalator
73. https://funattic.com/dice-games/
74. Ames O. Hamblen et al., Rapid Prototyping of Digital Systems, Springer, 2010.
75. http://people.sabanciuniv.edu/erkays/el310/DesignwithASM_06a.pdf
76. http://www.dejazzer.com/ee478/lecture_notes/lec11_fsmd.pdf
77. http://ece.gmu.edu/coursewebpages/ECE/ECE545/F10/viewgraphs/ECE545_lecture12_Controller.pdf
78. Raj AAB et al., Comparison of different models for ground-level atmospheric turbulence strength (Cn2) prediction with a new model according to local weather data for FSO applications, *Applied Optics*, 54(4), pp. 802–815, 2015.
79. Raj AAB et al., Comparison of different models for ground-level atmospheric attenuation prediction with new models according to local weather data for FSO applications, *Journal of Optical Communications*, 36(2), pp. 181–196, 2015.
80. Raj AAB and Padmavathi S, Statistical analysis of accurate prediction of local atmospheric optical attenuation with a new model according to weather together with beam wandering compensation system: A season-wise experimental investigation, *Journal of Modern Optics*, 63(13), pp. 1286–1296, 2016.
81. http://www.circuitstoday.com/digital-to-analog-converters-da
82. http://what-when-how.com/8051-microcontroller/dac-interfacing/
83. https://www.hamamatsu.com/resources/pdf/ssd/si_pd_circuit_e.pdf
84. http://www.ti.com.cn/cn/lit/ds/symlink/lm35.pdf
85. https://www.teamwavelength.com/info/thermistors.php
86. http://elektron.pol.lublin.pl/elekp/ap_notes/NI_AN078_Strain_Gauge_Meas.pdf
87. http://www.electronics-tutorials.ws/opamp/op-amp-comparator.html
88. http://what-when-how.com/8051-microcontroller/sensor-interfacing-and-signal-conditioning/
89. Raj AAB, Experimental analysis & mitigation of atmospheric turbulence effects in TFSLSoC using pipelined opto-electronic architecture-overview, *National conference on Communication and Signal processing, Conference Proceedings*, pp. 1–8, India, 2013.
90. https://arcelect.com/rs232.htm
91. Raj AAB, Mitigation of beam fluctuation due to atmospheric turbulence and prediction of control quality using intelligent decision-making tools, *Applied Optics*, 53(17), pp. 3796–3806, 2014.

92. Joel M. Esposito, Tutorial: Serial Communication in Matlab, Esposito 2009, pp. 1–16.
93. Raj AAB and Lancelot JP, Seasonal investigation on prediction accuracy of atmospheric turbulence strength with a new model at Punalkulam, Tamil Nadu, *Journal of Optical Technology*, 83(1), pp. 55–68, 2016.
94. https://www.8051projects.net/wiki/ADC0808_Interfacing_Tutorial
95. http://what-when-how.com/8051-microcontroller/parallel-and-serial-adc/
96. https://www.maximintegrated.com/en/products/analog/data-converters/analog-to-digital-converters/MAX1112.html
97. Raj AAB, Mono-pulse tracking system for active free space optical communication, *Optik – International Journal for Light and Electron Optics*, pp. 7752–7761, 2016.
98. Raj AAB, Padmavathi S, Quality metrics and reliability analysis of laser communication system, *Defence Science Journal*, 66(2), pp. 175–185, 2016.
99. https://electrosome.com/ht12e-encoder-ic-remote-control-systems/
100. http://www.instructables.com/id/Make-a-RF-Transmitter-and-Receiver-With-HT12E-HT12/
101. https://www.maximintegrated.com/en/app-notes/index.mvp/id/4400
102. http://www.wu.ece.ufl.edu/books/EE/communications/CDMA/CDMA.htm
103. http://intra.itiltd-india.com/quality/TelecomBasics/tdma.pdf
104. http://www.ti.com/lit/ds/symlink/dac7728.pdf
105. http://what-when-how.com/8051-microcontroller/ds12887-rtc-interfacing/
106. http://www.ti.com/product/TCA9535/datasheet/detailed_description
107. http://courses.cs.tau.ac.il/embedded/docs/TelosB/SHT11.pdf
108. http://www.circuitstoday.com/interface-gsm-module-with-arduino
109. https://exploreembedded.com/wiki/B0.8051_Interfacing_GPS
110. https://eewiki.net/pages/viewpage.action?pageId=28278929
111. https://eewiki.net/pages/viewpage.action?pageId=15925278
112. http://www.electronics-tutorials.ws/blog/relay-switch-circuit.html
113. https://www.bc-robotics.com/tutorials/controlling-a-solenoid-valve-with-arduino/
114. http://www.electronics-tutorials.ws/blog/optocoupler.html
115. https://www.electrical4u.com/speed-control-of-dc-motor/
116. http://howtomechatronics.com/tutorials/arduino/arduino-dc-motor-control-tutorial-l298n-pwm-h-bridge/
117. https://www.jameco.com/jameco/workshop/howitworks/how-servo-motors-work.html
118. https://www.electricaltechnology.org/2016/05/bldc-brushless-dc-motor-construction-working-principle.html
119. https://www.tigoe.com/pcomp/code/circuits/motors/stepper-motors/
120. http://www.circuitstoday.com/liquid-level-indicator
121. https://circuitglobe.com/current-transformer-ct.html
122. http://lrf.fe.uni-lj.si/fkkt_ev/Literatura/Current_and_Voltage_Measurements.pdf
123. https://www.allaboutcircuits.com/textbook/semiconductors/chpt-7/silicon-controlled-rectifier-scr/
124. http://www.electronics-tutorials.ws/power/triac.html
125. http://www.electronics-tutorials.ws/power/diac.html
126. https://www.controleng.com/single-article/designing-real-time-process-controllers/735cb45fd6d2d53a4e14ca4802714e44.html
127. http://www-inst.eecs.berkeley.edu/~cs61c/sp06/handout/fixedpt.html
128. https://andybargh.com/fixed-and-floating-point-binary/
129. http://babbage.cs.qc.cuny.edu/IEEE-754.old/Decimal.html
130. https://en.wikibooks.org/wiki/Floating_Point/Fixed-Point_Numbers
131. https://www.cs.umd.edu/class/sum2003/cmsc311/Notes/Data/float.html
132. http://homepage.divms.uiowa.edu/~atkinson/m170.dir/overton.pdf
133. http://cs.boisestate.edu/~alark/cs354/lectures/ieee754.pdf
134. http://pages.cs.wisc.edu/~smoler/x86text/lect.notes/arith.flpt.html
135. http://kias.dyndns.org/comath/14.html

136. https://www.doc.ic.ac.uk/~eedwards/compsys/float/
137. http://mathworld.wolfram.com/Floating-PointArithmetic.html
138. http://china.xilinx.com/support/documentation/sw_manuals/xilinx14_7/sysgen_ref.pdf
139. http://www.xilinx.com/support.html
140. https://www.xilinx.com/support/answer-navigation/design-tools/vivado-complex-ip.html
141. https://documents.mx/download/link/lab-5-55844db234d84
142. https://www.xilinx.com/support/documentation/sw_manuals/xilinx2015_3/ug897-vivado-sysgen-user.pdf
143. https://www.xilinx.com/support/documentation/sw_manuals/xilinx11/sysgen_ref.pdf
144. https://www.xilinx.com/support/documentation/sw_manuals/xilinx13_2/ug819_live_emac_tutorial.pdf
145. https://www.xilinx.com/support/answers/65728.html
146. http://comparch.doc.ic.ac.uk/publications/files/tim05pcdt_abstract_2.txt
147. https://www.coursehero.com/file/4941729/Floatieee/
148. http://practicalcryptography.com/miscellaneous/machine-learning/intuitive-guide-discrete-fourier-transform/
149. https://link.springer.com/chapter/10.1007%2F978-3-540-72613-5_3
150. http://www.eas.uccs.edu/~mwickert/ece2610/lecture_notes/ece2610_chap8.pdf
151. https://www.ti89.com/cryptotut/mod_arithmetic.htm
152. https://web.stanford.edu/class/ee486/doc/chap2.pdf
153. http://www.studytonight.com/computer-architecture/booth-multiplication-algorithm
154. Jose A. Apolinário and Sergio L. Netto, Introduction to Adaptive Filters, Chapter-2, pp:23–49, Springer Science+Business Media, 2009.
155. http://en.dsplib.org/content/fft_introduction/fft_introduction.html
156. https://china.xilinx.com/support/documentation/sw_manuals/xilinx14_7/sysgen_ref.pdf
157. https://www.design-reuse.com/articles/10028/understanding-cascaded-integrator-comb-filters.html
158. http://www.analyzemath.com/polarcoordinates/polar_rectangular.html
159. http://www.lce.hut.fi/teaching/S-114.240/k2001/reports/Seppala_wavelet.pdf
160. https://www.xilinx.com/itp/xilinx10/isehelp/ise_c_simulation_test_bench.htm
161. http://vhdlguru.blogspot.in/2010_03_01_archive.html
162. http://einstein.informatik.uni-oldenburg.de/rechnernetze/dpcm.htm
163. Das KN et al., *Proceedings of Fourth International Conference on Soft Computing for Problem Solving*, 2015, Springer.
164. https://www.techopedia.com/definition/5967/artificial-neural-network-ann
165. Pasupathi T et al., FPGA implementation of adaptive integrated spiking neural network for efficient image recognition system, *ICTACT Journal on Image and Video Processing*, 4(4), pp. 848–852, 2014.
166. https://in.mathworks.com/help/fuzzy/what-is-fuzzy-logic.html
167. Raj AAB et al., Implementation of adaptive fuzzy logic controller in FPGA for steering tip-tilt mirror in free space optical communication, *National Conference on Optical Communications, Conference Proceedings*, pp. 1–4, India, 2014.
168. https://www.scribd.com/document/110989758/Notes-on-Coding-Theory-J-I-hall
169. http://ocw.usu.edu/Electrical_and_Computer_Engineering/Error_Control_Coding/lecture2.pdf
170. https://ocw.mit.edu/ans7870/6/6.111/s04/NEWKIT/audio.htm
171. https://eewiki.net/display/LOGIC/AC%2797+Codec+Hardware+Driver+Example

Index

A

Adaptive filter/equalizer design (DSP), 690–698
ADCs, *see* Analog to digital converters
Addressable Shift Register (ASR) block, 576–577, 713
Algorithmic state machine (ASM); *see also* Finite and algorithmic state machine approaches, system design with-based digital system design, 332–336
  charts, 328–332
ALUs, *see* Arithmetic logic units
Analog to digital (A/D) conversion, 340–341
Analog to digital converters (ADCs), 101, 396, 399–403, 616
Application-specific integrated circuits (ASICs), 17–19, 622
Arithmetic and logical programming, 135–203
  arithmetic and logical unit, 188–194
  arithmetic operations (adders and subtractors), 136–152
  arithmetic operations (dividers), 163–170
  arithmetic operations (multipliers), 152–163
  coordinate rotation digital computer algorithm, trigonometric computations using, 170–185
  double data rate SDRAM, 201
  double data rate two SDRAM, 201
  dynamic random access memory, 201
  extended data out DRAM, 201
  fast page mode DRAM, 201
  laboratory exercises, 203
  look up table, 139
  multiply-accumulation circuit, 185–188
  parallel and pipelined adders, 141–146
  parallel in serial out shift registers, 137
  random access memory design, 200–202
  read-only memory design and logic implementations, 194–200
  serial adder, 137–141
  static random access memory, 201
  subtractors, 146–152
  synchronous dynamic RAM, 201
Arithmetic logic units (ALUs), 12, 101, 189, 190
Arithmetic unit (AU), 188, 189, 191, 192
Artificial neural network (ANN), 741, 750, 751–752
ASICs, *see* Application-specific integrated circuits
ASM, *see* Algorithmic state machine
ASR block, *see* Addressable Shift Register block
AU, *see* Arithmetic unit
Audio codec (AC97) interfacing, 775–780
Automatic teller machines (ATMs), 478

B

Ball grid array (BGA), 13
Baud rate counter (BRC), 366, 372
BC, *see* Binary cell
BCD, *see* Binary coded decimal
BCH codes, *see* Bose-Chaudhuri-Hocquenghem codes
BGA, *see* Ball grid array
Bidirectional analog to digital converter interfacing, 399–403
Bidirectional port/switch design, 231–233
Bidirectional shift register (BSR), 78–79
Binary cell (BC), 135, 201, 202, 400
Binary coded decimal (BCD)
  conversion, 261
  converters, 277–286
  values, 99
Binary sequence recognizer, 286–293
Bit error rate (BER) tester design, 286–287, 741, 755–761
BLDC motor, *see* Brush less direct current motor
Block random access memory (BRAM), 600, 728
Booth multiplication algorithm and design, 684–689
Bose-Chaudhuri-Hocquenghem (BCH) codes, 580
BRAM, *see* Block random access memory
BRC, *see* Baud rate counter
Brushless direct current (BLDC) motor, 493, 508, 511–515, 540
BSR, *see* Bidirectional shift register
Buzzer control, 215–217, 295

C

Carry look-ahead adder (CLAA), 94, 142
Cascaded integrator-comb (CIC), 652
  Compiler, 576–577
  filters, 582, 701, 715–718
Central processing unit (CPU), 14, 19, 201

Channel coding techniques, 766–773
Chemical vapor deposition (CVD), 10
Chinese remainder theorem (CRT), 627, 665, 666
ChipScope Pro Analyzer, 622, 701, 728–729
CIC, *see* Cascaded integrator-comb
Ciphertext, 744
CISC, *see* Complex instruction set computer
CLAA, *see* Carry look-ahead adder
Clock, real-time, *see* Real-time clock and interface protocol programming
Clock Enable (CE) Probe, 578, 590
Combinational logic circuits, 8, 14–15, 70, 91–92, 97, 152
Complementary metal oxide semiconductor (CMOS), 4, 16
  fabrication and layout, 10–11
  technology and gate configuration, 7–9
Complex instruction set computer (CISC), 12
Computer numerical control (CNC) machine control, 498, 539
Concurrent code, 50–55
Contemporary design and applications, 741–781
  artificial neural network, 751–753, 754
  audio codec (AC97) interfacing, 775–780
  bit error rate tester design, 755–761
  channel coding techniques, 766–773
  ciphertext, 744
  convolutional code, 769–773
  data encryption system, 744–750
  differential pulse code modulation system design, 741–744
  fuzzy logic controller, 753–755
  laboratory exercises, 781
  linear block code, 767–769
  optical up/down data link, 761–766
  pick-and-place robot controller, 773–775
  soft computing algorithms, 750–755
Control device interfacing (real-world), 493–545
  brushless direct current motor control, 511–515
  capacitance-level sensors, 519
  crowbar circuit, 531
  Diac controller, 536–538
  direct current motor control, 504–508
  electromagnetic interference, 533
  Hall sensors, 514
  laboratory exercises, 544–545
  liquid/fuel level control, 518–522
  opto-isolator interfacing, 501–504
  phototransistor, 501
  positive-negative-positive transistors, 496
  power electronic device interfacing and control, 526–531
  power electronics bidirectional switch interfacing and control, 532–538
  radar level measurement systems, 519
  real-time process controller design, 538–544
  relay control, 494–497
  servo motor control, 508–511
  silicon controlled rectifiers, 511
  solenoid valve control, 497–501
  stepper motor control, 515–518
  trapezoidal permanent magnet motors, 511
  Triac control, 532–536
  voltage and current measurement, 522–526
Convolutional code, 767, 769–773
Coordinate rotation digital computer (CORDIC), 135
  algorithm, trigonometric computations using, 170–185
  design, 719–721
CPU, *see* Central processing unit
Crowbar circuit, 531
CRT, *see* Chinese remainder theorem
Current and voltage measurement, 522–526
CVD, *see* Chemical vapor deposition

D

DA-based computations, *see* Distributed arithmetic-based computations
DAFIR filter, *see* Distributive arithmetic finite impulse response filter
Data encryption system, 744–750
Data link path, 329
Data sequence detector (DSD) module, 372
Data shift registers, 77–84
Data storage register (DSR), 372
DC motor control, *see* Direct current motor control
DCT, *see* Discrete cosine transform
DDR SDRAM, *see* Double data rate SDRAM
DDR2 SDRAM, *see* Double data rate two SDRAM
DDS, *see* Direct digital synthesizer
Device-under-test (DUT), 349, 418, 728, 729
DFT, *see* Discrete Fourier transform
Dice game controller design, 314–321
Differential global positioning system (DGPS), 480
Differential pulse code modulation (DPCM) system design, 741–744
Digital to analog (D/A) conversion, 341–344, 426–434, 434–438
Digital clock design and interfacing, 115
Digital signal processing (DSP), 4, 21, 135, 549, 622

Digital signal processing (DSP) with field-programmable gate array, 627–699
  adaptive filter/equalizer design, 690–698
  Booth multiplication algorithm and design, 684–689
  Chinese remainder theorem, 627
  digital finite impulse response filter design, 641–644
  digital infinite impulse response filter design, 644–650
  discrete Fourier transform, 628–641
  distributed arithmetic-based computations, 672–683
  inverse Fourier transform, 630
  laboratory exercises, 698–699
  modulo adder design, 663–665
  multirate signal processing, 650–663
  residue number system, 665–672
Diode alternating current (Diac), 493, 532, 536–538
DIP, *see* Dual in-line package
Direct current (DC) motor control, 504–508
Direct digital synthesizer (DDS), 577
Direct sequence spread spectrum (DSSS), 415
Discrete cosine transform (DCT), 713, 724
Discrete Fourier transform (DFT), 578, 628–641, 701, 702, 703, 722
Discrete wavelet transform (DWT), image processing using, 722–727
Distributed arithmetic (DA)-based computations, 672–683
Distributive arithmetic finite impulse response (DAFIR) filter, 577
Dividers, 163–170, 171, 623
Double data rate (DDR) SDRAM, 201
Double data rate two (DDR2) SDRAM, 201
Double pole double throw (DPDT), 494, 495, 506
DRAM, *see* Dynamic random access memory
DSD module, *see* Data sequence detector module
DSP, *see* Digital signal processing
DSR, *see* Data storage register
DSSS, *see* Direct sequence spread spectrum
Dual in-line package (DIP), 13, 207, 228–231, 293
Dual-tone multifrequency (DTMF) decoder, 207, 235–238
DUT, *see* Device-under-test
Dynamic random access memory (DRAM), 201

E

EDA tool, *see* Electronics design automation tool
EDO DRAM, *see* Extended data out DRAM
EEPROM, *see* Electrically erasable programmable read-only memory
Electrically erasable programmable read-only memory (EEPROM), 196
Electromagnetic interference (EMI), 525, 533
Electronic model train controller design, 322–328
Electronics design automation (EDA) tool, 18
Electronics numerical integrator and computer (ENIAC), 5
EMIF, *see* External memory interface
Erasable programmable read-only memory (EPROM), 196
Escalator controller design, 309–314
Evolutionary computation (EC), 750
Extended data out (EDO) DRAM, 201
External memory interface (EMIF), 604

F

Fast page mode (FPM) DRAM, 201
FECD, *see* Forward Error Correction
Field effect transistors (FETs), 14
Finite and algorithmic state machine approaches, system design with, 261–337
  algorithmic state machine-based digital system design, 332–336
  algorithmic state machine charts, 328–332
  binary-to-binary-coded-decimal converter and its arithmetic, 282–286
  binary sequence recognizer, 286–293
  code classifier and binary to binary-coded decimal converters, 277–286
  data link path, 329
  dice game controller design, 314–321
  electronic model train controller design, 322–328
  escalator controller design, 309–314
  finite state machine design (models), 261–277
  finite state machine model conversion, 276–277
  input code classifier, 278–282
  laboratory exercises, 336–337
  Mealy finite state machine design, 271–275
  Moore finite state machine design, 269–271
  traffic light controller, 301–309
  vending machine controller, 293–301
Finite impulse response (FIR) filter design, 641–644
Flip-flops, 65–76

Floating-point arithmetic, 547, 548, 561–575; *see also* Very-high-speed integrated circuit hardware description language and Xilinx system generator tools, floating-point computations with
floating-point addition, 562–565
floating-point division, 573–575
floating-point multiplication, 568–573
floating-point subtraction, 565–568
Forward Error Correction (FEC), 578
FPM DRAM, *see* Fast page mode DRAM
Free space optical (FSO) communication systems, 403, 761
Full adder (FA), 15 16, 27, 91–96, 153
Fuzzy logic controller, 527–528, 741, 753–755

G

General packet radio service (GPRS), 469
Giant-scale integration (GSI), 3
Global navigation satellite system (GNSS), 480
Global positioning system (GPS) interface programming, 441, 478–480, 492
Global system for mobile (GSM) communications interface programming, 467–478
GPRS, *see* General packet radio service
Graphical liquid crystal display (GLCD), 223–228
GSI, *see* Giant-scale integration
GSM communications interface programming, *see* Global system for mobile communications interface programming

H

Half adder (HA), 15, 57, 91–96, 138, 146–153
High-current Darlington transistor array (HCDTA), 211
Horizontal dilution of precision (HDOP), 480

I

IC, *see* Integrated circuit
IDFT, *see* Inverse DFT
IEEE754 double-precision floating-point number system, 556–558
IEEE754 single-precision floating-point number system, 553–556
IFT, *see* Inverse Fourier transform
Image processing using discrete wavelet transform, 722–727
Infinite impulse response (IIR) filter design, 644–650, 710–712
Infrared (IR) sensors, 208, 238–242, 245, 340, 501, 502
Input/output bank programming and interfacing, 207–259
bidirectional port/switch design, 231–233
buzzer control, 215–217
dual inline package switch, 228–231
dual-tone multifrequency decoder, 235–238
general-purpose switch interfacing, 228–235
graphical liquid crystal display, 223–228
high-current Darlington transistor array, 211
infrared sensors, 238–242
interactive voice response, 235
laboratory exercises, 259
light-dependent resistor, 251–255
light-emitting diode displays, 208–210
liquid crystal display interfacing and programming, 217–228
matrix keypad interfacing, 233–235
metal detector, 248–251
multisegment display, 210–215
optical display interfacing, 208–215
optical sensor interfacing, 238–244
passive infrared sensor, 245–248
proximity sensor, 242–244
special sensor interfacing, 244–255
voltage divider circuit, 238
wind-speed sensor interfacing, 255–258
Integral square error (ISE) function, 576–577, 583, 586, 615, 620, 692
Integrated circuit (IC), 3, 185, 201–202
Interactive voice response (IVR), 235
Interfacing digital logic to the real world (sensors, analog to digital and digital to analog), 339–438
ADC0808/ADC0809 interfacing, 379–384
ADC0804 interfacing, 351–356
ADC0848 interfacing, 384–395
analog to digital conversion, 340–341
analog to digital converter MAX1112 interfacing and serial data fetching, 396–399
baud rate counter, 372
bidirectional analog to digital converter interfacing, 399–403
bipolar signal conditioning and data logging, 399–411
data sequence detector module, 372
data storage register, 372
device-under-test, 349
digital to analog conversion, 341–344

direct sequence spread spectrum, 415
high-voltage digital to analog conversion using DAC7728, 434–438
Kasami sequence generator, 422–423
laboratory exercises, 438
linear feedback shift registers, 415
low voltage digital to analog conversion using DAC0808, 426–434
magnitude comparator, 350–351
multichannel data logging, 378–399
optical power measurement, 344–346
optical Tx/Rx wireless control, 411–412
opto-electronic position detector interfacing, 403–411
parallel pseudorandom binary sequence generator, 418–421
programmable logic circuits, 424–425
pseudorandom binary sequence generator and time-division multiple access, 415–426
radio frequency transmitter/receiver wireless control, 412–415
receiver design, 371–378
remote control applications, encoder/decoder interfacing for, 411–415
sensor interfacing and measurement techniques, 344–356
serial communication (data reading/writing using MATLAB®), 358–365
serial pseudorandom binary sequence generator, 415–418
signal conditioning for sensor interfacing, 339–344
signal generator design and interfacing, 426–438
start bit sequence detector module, 372
strain measurement, 349
temperature measurement, 346–348
time division multiplexing, 424–426
transmitter design, 365–371
universal asynchronous receiver-transmitter design, 356–378
Inverse DFT (IDFT), 578, 628, 702–703
Inverse Fourier transform (IFT), 630
IR sensors, *see* Infrared sensors
ISE function, *see* Integral square error function
IVR, *see* Interactive voice response

J

JK flip-flop (JK-FF), 66, 67, 70, 72, 257
Joint photographic expert group (JPEG), 604
Joint text action group (JTAG), 620

K

Kasami sequence (Ks) generator, 422–423

L

Laboratory exercises
arithmetic and logical programming, 203
contemporary design and applications, 781
control device interfacing (real-world), 544–545
digital signal processing with field-programmable gate array, 698–699
finite and algorithmic state machine approaches, system design with, 336–337
input/output bank programming and interfacing, 259
interfacing digital logic to the real world (sensors, analog to digital and digital to analog), 438
real-time clock and interface protocol programming, 492
simple system design techniques, 134
SysGen-based system designs (advanced), 739–740
VHDL, digital circuit design with, 89
VHDL and Xilinx system generator tools, floating-point computations with, 624
VLSI (history and features), 21
Large-scale integration (LSI), 3, 6
LCD interfacing and programming, *see* Liquid crystal display interfacing and programming
LDR, *see* Light-dependent resistor
LEDs, *see* Light-emitting diodes
LFSRs, *see* Linear feedback shift registers
Light-dependent resistor (LDR), 244, 251–255
Light-emitting diodes (LEDs), 104, 208–210, 211, 238, 302
Linear block code, 767–769
Linear feedback shift registers (LFSRs), 415
Liquid crystal display (LCD) interfacing and programming, 217–228
Liquid/fuel level control, 518–522
Logical programming, *see* Arithmetic and logical programming
Logic unit (LU), 188
Look up table (LUT), 139, 161
LSI, *see* Large-scale integration

M

MAC, *see* Multiply-accumulation circuit
Machine learning (ML), 750

Magnitude comparator, 99, 350–351
MAS, *see* Micro-architectural specification
Master in slave out (MISO), 441
Master out slave in (MOSI), 441
MATLAB®
  code interfacing with SysGen tools, 610–614
  figure/plot, conversion of, 788
  MATLAB®-Xilinx SysGen, 701
  operators and commands, 786–788
  Simulink model design, conversion of, 788–789
Matrix keypad interfacing, 233–235
Mealy finite state machine design, 271–275
Medium-scale integration (MSI), 3
Metal detector, 244, 248–251
Micro-architectural specification (MAS), 12
Microelectronics, review of, 4–7
MISO, *see* Master in slave out
ML, *see* Machine learning
Modulo adder design, 663–665
Monopulse arithmetic circuit (MPAC), 403
Moore finite state machine design, 269–271
MOS field effect transistor (MOSFET), 7
MOSI, *see* Master out slave in
Most significant bit (MSB), 78
Motion pictures expert group (MPEG), 604
MPAC, *see* Monopulse arithmetic circuit
MSI, *see* Medium-scale integration
Multifrequency generator, 84–88
Multipliers, 152–163
Multiply-accumulation (MAC)
  circuit, 185–188
  finite impulse response filter, 712–714
Multirate signal processing, 650–663

N

Negative-positive-negative (NPN) relay switching circuit, 496

O

Optical display interfacing, 208–215
Optical power measurement, 344–346
Optical sensor interfacing, 238–244
  infrared sensors, 238–242
  proximity sensor, 242–244
  voltage divider circuit, 238
Optical up/down data link, 761–766
Opto-electronic position detector (OPD) interfacing, 403–411
Opto-isolator interfacing, 501–504

P

Parallel in parallel out (PIPO) shift register, 77, 78, 713
Parallel in serial out shift registers (PISO-SRs), 77, 78, 137, 366
Parallel and pipelined adders, 141–146
Parallel pseudorandom binary sequence generator, 418–421
Passive infrared (PIR) sensor, 245–248
Permanent magnet (PM), 504
Personal System/2 (PS/2) interface programming, 480–484
PGA, *see* Pin grid array
Phototransistor (PT), 238, 242, 501, 502
Pick-and-place robot controller, 773–775
Pin grid array (PGA), 13
PIPO shift register, *see* Parallel in parallel out shift register
PIR sensor, *see* Passive infrared sensor
PISO-SRs, *see* Parallel in serial out shift registers
PLCs, *see* Programmable logic circuits
PM, *see* Permanent magnet
Point-to-point wiring, 5
POS, *see* Products of sum
Positive-negative-positive (PNP) transistors, 496
PPS generator, *see* Pulse per second generator
Probabilistic reasoning (PR), 750
Products of sum (POS), 15
Programmable logic circuits (PLCs), 424
Programmable read-only memory (PROM), 196
Proportional, integral and derivative (PID), 527
Proximity sensor, 242–244
Pseudorandom binary sequence (PRBS)
    generator and time-division multiple access, 415–426
  direct sequence spread spectrum, 415
  Kasami sequence generator, 422–423
  linear feedback shift registers, 415
  parallel pseudorandom binary sequence generator, 418–421
  programmable logic circuits, 424
  serial pseudorandom binary sequence generator, 415–418
  time division multiplexing, 424–426
PT, *see* Phototransistor
Pulse per second (PPS) generator, 115
Pulse width modulation (PWM) signal generation, 91, 123–127

Q

Quad flat package (QFP), 13

R

Radio frequency (RF) transmitter/receiver wireless control, 412–415
Random access memory (RAM) design, 3, 200–202
  double data rate SDRAM, 201
  double data rate two SDRAM, 201
  dynamic random access memory, 201
  extended data out DRAM, 201
  fast page mode DRAM, 201
  static random access memory, 201
  synchronous dynamic RAM, 201
Read-only memory (ROM) design, implementations and, 194–200
Real-time clock (RTC) and interface protocol programming, 441–492
  general packet radio service, 469
  global positioning system interface programming, 478–480
  global system for mobile communications interface programming, 467–478
  inter-integrated circuit interface programming, 449–456
  laboratory exercises, 492
  Personal System/2 interface programming, 480–484
  real-time clock (DS12887) interface programming, 442–449
  serial clock, 449
  serial peripheral interface (SCP1000D) programming, 461–467
  two-wire interface (SHT11 sensor) programming, 456–461
  video graphics array interface programming, 484–492
Reduced instruction set computer (RISC), 12
Reed-Solomon (RS) decoder, 577, 579
Register transfer level (RTL), 13, 19
Relay control, 494–497
Remote control applications, encoder/decoder interfacing for, 411–415
  optical Tx/Rx wireless control, 411–412
  radio frequency transmitter/receiver wireless control, 412–415
Residue number system (RNS), 665–672
RISC, *see* Reduced instruction set computer
Robot controller, pick-and-place, 773–775
ROM design, *see* Read-only memory design, implementations and
RS decoder, *see* Reed-Solomon decoder
RTC and interface protocol programming, *see* Real-time clock and interface protocol programming
RTL, *see* Register transfer level

S

SBSD module, *see* Start bit sequence detector module
SCL, *see* Serial clock
SCRs, *see* Silicon controlled rectifiers
SDR, *see* Software defined radios
SD RAM, *see* Synchronous dynamic RAM
Sequential code, 55–65
Sequential logic circuits, 14, 15
Serial adder, 137–141
Serial clock (SCL), 449
Serial in parallel out (SIPO) shift register, 77, 78
Serial in serial out (SISO) shift register, 77
Serial pseudorandom binary sequence generator, 415–418
Servo motor control, 508–511
Short-term Fourier transform (STFT), 722
Silicon controlled rectifiers (SCRs), 493, 511, 528, 532
Single pole double throw (SPDT), 494
Single pole single throw (SPST), 494
SIPO shift register, *see* Serial in parallel out shift register
SISO shift register, *see* Serial in serial out shift register
Small-scale integration (SSI), 3
SoCs, *see* Systems on chips
Soft computing algorithms, 750–755
Software defined radios (SDR), 604
Solenoid valve control, 497–501
Solid state device (SSD), 3
SOP, *see* Sum of product
SPDT, *see* Single pole double throw
SPST, *see* Single pole single throw
SRAM, *see* Static random access memory
SR-flip-flop (SR-FF), 66, 70
SSD, *see* Solid state device
SSI, *see* Small-scale integration
Start bit sequence detector (SBSD) module, 372
Static random access memory (SRAM), 201
Stepper motor control, 515–518
STFT, *see* Short-term Fourier transform
Strain measurement, 349
Subtractors, 146–152

Sum of product (SOP), 15
Switch interfacing (general purpose), 228–235
  bidirectional port/switch design, 231–233
  dual inline package switch, 228–231
  matrix keypad interfacing, 233–235
Synchronous dynamic (SD) RAM, 201
SysGen-based system designs (advanced), 701–740
  cascaded integrator comb filter design, 715–718
  ChipScope Pro Analyzer, 728–729
  coordinate rotation digital computer design, 719–721
  data/text file reading/writing, 732–739
  fast Fourier transform computation, 701–704
  finite impulse response digital filter design, 704–710
  image processing using discrete wavelet transform, 722–727
  infinite impulse response digital filter design, 710 –712
  laboratory exercises, 739–740
  multiply accumulation finite impulse response filter, 712–714
  short-term Fourier transform, 722
  VHDL design debugging techniques, 727–739
  VHDL test-bench design, 729–732
SysGen tools, *see* Xilinx system generator (SysGen) tools
System design techniques (simple), 91–134
  analog to digital converters, 101
  binary coded decimal values, 99
  carry look-ahead adder, 94
  counter design and interfacing, 107–115
  digital clock design and interfacing, 115–123
  functions and procedures, 131–134
  half and full adder, 91–96
  half and full subtractor, 96–98
  laboratory exercises, 134
  magnitude comparator, 99
  packages and libraries, 127–131
  pulse width modulation signal generation, 123–127
  seven-segment display interfacing, 104–107
  signed magnitude comparator, 98–104
Systems on chips (SoCs), 21

T

TDM, *see* Time division multiplexing
Temperature measurement, 346–348
T flip-flop (T-FF), 66, 68
Tiff image
  converting MATLAB figure/plot into, 788
  converting MATLAB Simulink model design into, 788–789
  converting SysGen model scope plot into, 789
Time division multiplexing (TDM), 424–426
Traffic light controller, 301–309
Transistor to transistor level (TTL), 357
Trapezoidal permanent magnet motors, 511
Triac control, 532–536

U

Ultra-large-scale integration (ULSI), 3
Universal asynchronous receiver-transmitter (UART) design, 356–378
Universal serial bus (USB), 357

V

VDC, *see* Voltage divider circuit
Vending machine controller, 293–301
Very-high-density cable interconnect (VHDCI), 404
Very-high-speed integrated circuit hardware description language (VHDL)
  commands, 784–786
  design debugging techniques, 727–739
  libraries, 783
  operators, 783–784
  procedural interface (VHPI), 20
  test-bench design, 729–732
Very-high-speed integrated circuit hardware description language (VHDL), digital circuit design with, 23–89
  code design structures, 24–33
  concurrent code, 50–55
  data shift registers, 77–84
  data types and their conversions, 34–45
  flip-flops and their conversions, 65–76
  JK flip-flop (JK-FF), 67
  laboratory exercises, 89
  multifrequency generator, 84–88
  operators and attributes, 45–50
  sequential code, 55–65
  SR-flip-flop (SR-FF), 66
  T flip-flop, 68
Very-high-speed integrated circuit hardware description language (VHDL) and Xilinx system generator (SysGen) tools, floating-point computations with, 547–624

customized floating-point number system, 558–561
design flow for hardware co-simulation, 620–622
fixed and floating-point binary numbers, representation of, 549–561
fixed-point number system, 549–553
floating-point addition, 562–565
floating-point division, 573–575
floating-point multiplication, 568–573
floating-point subtraction, 565–568
fractional-point computation using SysGen tools, 594–603
IEEE754 double-precision floating-point number system, 556–558
IEEE754 single-precision floating-point number system, 553–556
laboratory exercises, 624
MATLAB® code interfacing with SysGen tools, 610–614
real-time verification and reconfigurable architecture design, 620–624
reconfigurable architecture design, 622–624
system design and implementation using SysGen tools, 583–594
use and interfacing methods of some blocksets, 577–583
VHDL code interfacing with SysGen tools, 614–619
Xilinx Simulink block sets, system engine model using, 603–610
Xilinx system generator (SysGen) tools, 575–594

Very-large-scale integration technology (VLSI) (history and features), 3–21
application-specific integrated circuits and their design flow, 17–19
CMOS fabrication and layout, 10–11
combinational logic circuits, 14–15
combinational and sequential circuit design, 14–15
complementary metal oxide semiconductor technology and gate configuration, 7–9
electronics design automation tool, 18
laboratory exercises, 21
microelectronics, review of, 4–7
MOS field effect transistor, 7
point-to-point wiring, 5
sequential logic circuits, 15
silicon chips, 6
subsystem design and layout, 15–16
VHDL requirements and features, 19–21
VLSI design flow, 12–13

VGA interface programming, *see* Video graphics array interface programming
VHDCI, *see* Very-high-density cable interconnect
VHDL, *see* Very-high-speed integrated circuit hardware description language
Video graphics array (VGA) interface programming, 484–492
VLSI, *see* Very-large-scale integration technology (history and features)
Voltage and current measurement, 522–526
Voltage divider circuit (VDC), 238

W

Wind-speed sensor interfacing, 255–258

X

Xilinx Simulink block sets, 603–610
Xilinx system generator (SysGen) tools, 575–594
AddSub block, 585
BitBasher, 577
BPSK AWGN Channel reference block, 581–582
Cascaded Integrator-Comb Compiler, 577–578
Channel Decimate by 2 MAC FIR Filter reference block, 581
Clock Enable Probe, 578
Constant block, 584
Convolution Encoder, 578
CORDIC SINCOS reference block, 582
CORDIC SQRT reference block, 582
Delay block, 580, 602
Down-Sample block, 578
Dual Port RAM block, 578
EDK Processor block, 579
Fast Fourier Transform 7.1 block, 578
FIFO block, 578
Gateway In blocks, 584
Gateway Out blocks, 585
Indeterminate Probe, 579
MATLAB Simulink block, 602
Mult block, 584–585
Network-Based Ethernet Co-Simulation block, 579
Opmode block, 579
Pause Simulation block, 579
PicoBlaze Microcontroller block, 579
Reed-Solomon decoder, 579–580
Register block, 580

Relational block, 580
Reset Generator block, 580
Resource Estimator block, 600
ROM block, 580
Scale block, 580
serial-to-parallel block, 598
Shared Memory block, 580–584
Shift block, 581
Slice block, 581
system design and implementation using, 583–594
System Generator block, 576–583
System Generator WaveScope block, 581
time division demultiplexer block, 598
time division multiplexer block, 598
Up-Sample block, 581
use and interfacing methods of some blocksets, 577–583